BIOLOGY
FOR TEXAS

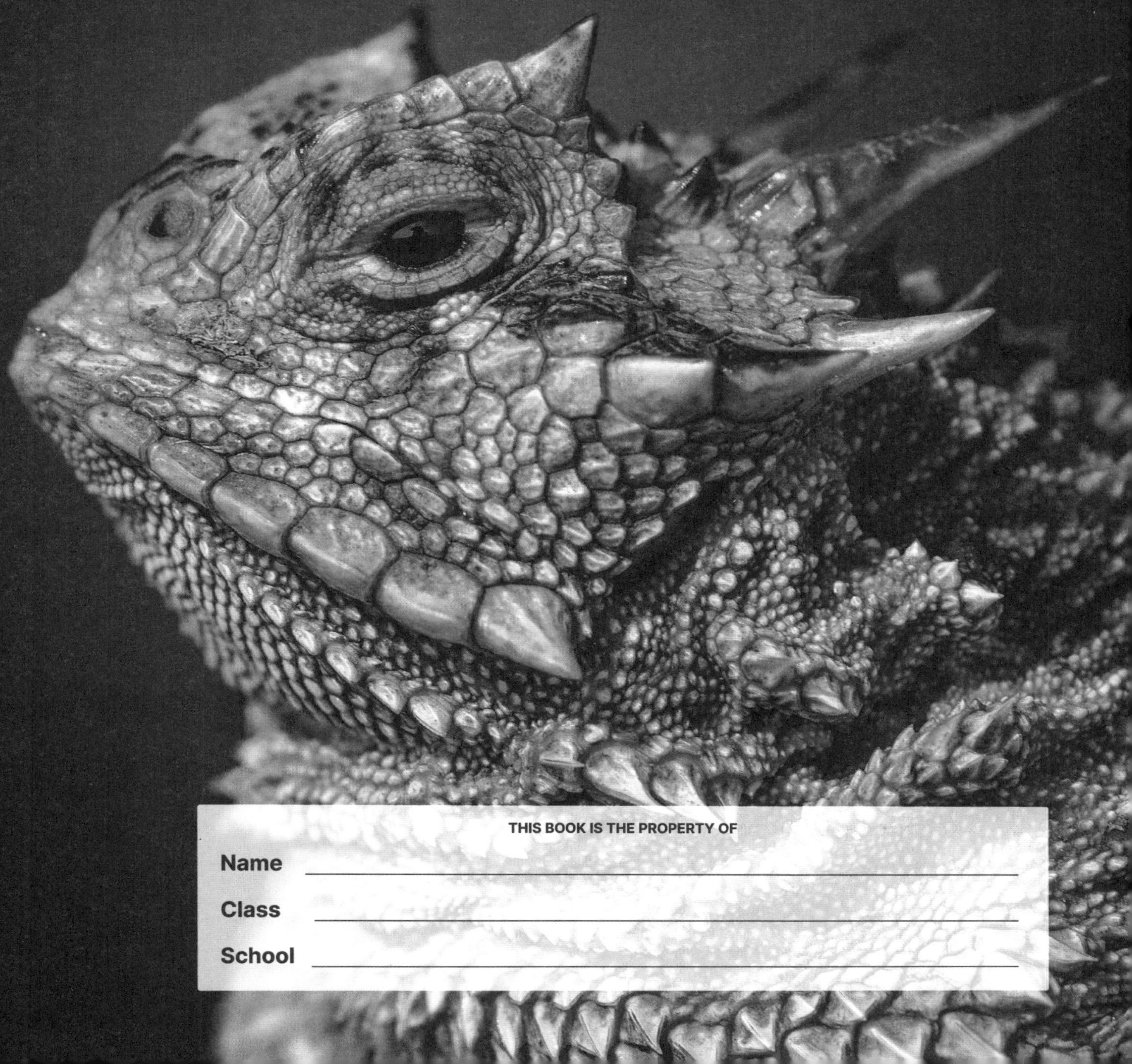

THIS BOOK IS THE PROPERTY OF

Name _____
Class _____
School _____

BIOLOGY
FOR TEXAS

About the Authors

Jillian Mellanby *Editor*
Jill began her science career with a degree in biochemistry and, after a short spell in research labs, became a science teacher both in the UK and then New Zealand. She spent many years managing the Royal Society of New Zealand's academic publishing programme of eight science journals which allowed her to hone her project management and editorial skills. She was also a part of the Expert Advice writing team at the Royal Society of New Zealand, producing science pieces for a public audience. She joined the BIOZONE team in late 2021, as editor.

Kent Pryor *Author*
Kent has a BSc from Massey University, majoring in zoology and ecology and taught secondary school biology and chemistry for 9 years before joining BIOZONE as an author in 2009.

Sarah Gaze *Author*
Sarah has 16 years experience as a Science and Chemistry teacher, recently completing MEd. (1st class hons) with a focus on curriculum, science, and climate change education. She has a background in educational resource development, academic writing, and art. Sarah joined the BIOZONE team, at the start of 2022.

Lissa Bainbridge-Smith *Author*
Lissa graduated with a Masters in Science (hons) from the University of Waikato. After graduation she worked in industry in a research and development capacity for eight years. Lissa joined BIOZONE in 2006 and is hands-on developing new curricula. Lissa has also taught science theory and practical skills to international and ESL students.

ISBN 978-1-99-101405-4

First Edition 2024

Copyright © 2024 Richard Allan
Published by **BIOZONE International Ltd**

First Printing
Printed by REPLIKA Press
using paper from renewable resources

Purchases of this book may be made direct from the publisher:

BIOZONE Corporation
USA and Canada
FREE phone: 1-855-246-4555
FREE fax: 1-855-935-3555
Email: sales@biozone.com
Web: www.BIOZONE.com

Cover photograph
Horned Lizard *Phrynosoma*

The Texas Horned Lizard, also known as the horny toad, is a reptile species that is native to the southwestern United States and northern Mexico. They are known for their distinctive appearance, featuring a flattened body, short tail, and numerous spines along their back and sides. They also have a unique defense mechanism: they can shoot blood out of their eyes to scare off predators. Unfortunately, due to habitat loss and other factors, Texas Horned Lizards are now considered a threatened species in some areas. Conservation efforts are ongoing to help protect and preserve these fascinating creatures.

PHOTO: https://stock.adobe.com
Photo ID: 321596050

Acknowledgements:
BIOZONE wishes to thank and acknowledge the whole team, including our designers, for their efforts and contributions to the production of this title. We also thank our reviewer, Tala'Shandria L. Allen, M.Ed.

All rights reserved. No part of this publication may be reproduced, stored in a retrieval system, or transmitted in any form or by any means, electrical, mechanical, photocopying, recording or otherwise, without the permission of BIOZONE International Ltd. This book may not be re-sold. The conditions of sale specifically prohibit the photocopying of exercises, worksheets, and diagrams from this book for any reason.

Contents

Using This Book vii
Using The Tab System ix
Using BIOZONE'S Resource Hub x

1 Cells and Cellular Processes

Learning Outcomes 1

- 1 **CONTENT ANCHOR**
  What is a Sponge? 3
- 2 Biomolecules in the Cell 4
- 3 Carbohydrates in Cells 6
- 4 Nucleic Acids in Cells 8
- 5 Proteins are Formed from Amino Acids 9
- 6 ● Investigating the Structure of Proteins 10
- 7 The Functions of Proteins in Cells 12
- 8 Lipids in Cells 16
- 9 The Development of Microscopes 17
- 10 Microscopes and Magnification 18
- 11 ● Studying Cells 20
- 12 Life Arises from Life 22
- 13 The Cell is the Unit of Life 24
- 14 Distinguishing Features of Prokaryotic Cells ... 26
- 15 Distinguishing Features of Eukaryotic Cells ... 27
- 16 Prokaryotic vs Eukaryotic cells 28
- 17 Comparing Cell and Virus Sizes 29
- 18 Why be Multicellular? 30
- 19 Eukaryotes have Complex Cells 31
- 20 Cell Membrane Structure 33
- 21 ● Diffusion in Cells - Passive Transport 37
- 22 Osmosis in Cells - Diffusion of Water 39
- 23 Active Transport in Cells 40
- 24 What is an Ion Pump? 41
- 25 Cytosis 42
- 26 Comparing Virus and Cell Structure 44
- 27 Viral Reproduction and Disease 46
- 28 How is Viral Disease Transmitted? 48
- 29 Epidemics and Pandemics 50
- 30 ● Modeling Viral Disease Outbreak and Spread ... 53
- 31 Viral Disease Case Study: Covid-19 55
- 32 What is a Sponge? Revisited 57
- 33 Summing Up 58

2 Cell Cycle

Learning Outcomes 61

- 34 **CONTENT ANCHOR**
  The Power to Rebuild 63
- 35 Growth and Repair of Cells 64
- 36 The Eukaryotic Cell Cycle 65
- 37 Mitosis and Cytokinesis 67
- 38 ● Modeling Mitosis 69
- 39 DNA Replication 70
- 40 Stages of DNA Replication 71
- 41 Evidence for Semi-conservative DNA Replication 72
- 42 Differentiation of Cells 75
- 43 Blood Cell Differentiation 76
- 44 Specialization in Plant Cells 77
- 45 Specialization in Animal Cells 78
- 46 Cells and the Environment 79
- 47 Role of Environment in Cell Development ... 80
- 48 Cell Cycle Disruptions and Cancer 82
- 49 The Power to Rebuild Revisited 85
- 50 Summing Up 86

3 Photosynthesis and Cellular Respiration

Learning Outcomes 89

- 51 **CONTENT ANCHOR**
  Mouse Trap 91
- 52 A Closer Look at Chloroplasts and Mitochondria 92
- 53 Energy in Cells 94
- 54 Introduction to Photosynthesis 96
- 55 Stages in Photosynthesis 98
- 56 ● Investigating Photosynthetic Rate 100
- 57 Energy Transfers Between Systems 101
- 58 Energy From Glucose 102
- 59 Aerobic Cellular Respiration 104
- 60 ● Measuring Respiration 106
- 61 ● Modeling Photosynthesis and Respiration ... 109
- 62 Reactions in Cells 113
- 63 What are Enzymes? 114
- 64 How Enzymes Work 115
- 65 ● Enzymes Have Optimal Conditions to Work ... 116
- 66 Design an Experiment to Test Catalase Activity 119
- 67 Mouse Trap Revisited 121
- 68 Summing Up 122

CODES: Activity is marked: ☐ to be done   ✓ when completed   ● Practical component

Contents

4 Animal and Plant Structure and Function

Learning Outcomes 125

- 69 **CONTENT ANCHOR**
  Complex Interactions 127
- 70 The Hierarchy of Life 128
- 71 Overview of Body Systems 129
- 72 The Body's Systems Work Together 131
- 73 Homeostasis 132
- 74 Negative Feedback Regulates the Body 133
- 75 Nervous Regulatory Systems 135
- 76 Hormonal Regulation 136
- 77 Nervous and Endocrine Interactions 137
- 78 Interactions Regulating the Blood 138
- 79 Interactions Regulating Respiratory Gases ... 140
- 80 ● Effect of Exercise on Heart Rate and Breathing 142
- 81 Interactions for Nutrient Absorption 145
- 82 Regulating Blood Glucose Levels 148
- 83 Interacting Systems: The Menstrual Cycle ... 150
- 84 Interacting Systems: Pregnancy and Birth ... 152
- 85 The Immune System 154
- 86 The Body's Defences: A Layered System ... 155
- 87 Blood Clotting and Defense 156
- 88 Interacting Systems: Responding to Infection .. 157
- 89 Plant Organ Systems 159
- 90 ● Interacting Systems in Plants 160
- 91 Xylem and Phloem 162
- 92 Stem and Root Structure 163
- 93 Transpiration 164
- 94 ● Investigating Transpiration 166
- 95 Uptake at the Roots 169
- 96 Translocation 171
- 97 Asexual Reproduction in Plants 172
- 98 ● Investigating Plant Propagation 173
- 99 Insect Pollinated Flowers 174
- 100 Wind Pollinated Flowers 175
- 101 Pollination and Fertilization 177
- 102 ● Seed Structure and Germination 178
- 103 Seed Dispersal 180
- 104 Responses in Plants 181
- 105 Tropisms and Growth Responses 182
- 106 Auxins, Gibberellins, and ABA 183
- 107 Plant Hormones as Signal Molecules 184
- 108 Investigating Phototropism 186
- 109 Giberellins and Stem Elongation 187
- 110 Investigating Gravitropism 188
- 111 Investigating Gravitropism in Seeds 189
- 112 Nastic Responses 190
- 113 Complex Interactions Revisited 192
- 114 Summing Up 193

5 DNA and Gene Expression

Learning Outcomes 196

- 115 **CONTENT ANCHOR**
  Real-Life Superpowers 198
- 116 DNA and Chromosomes 199
- 117 DNA and RNA 200
- 118 ● Modeling DNA Structure 201
- 119 Discovering DNA 206
- 120 The Origin of DNA 209
- 121 Introduction to Gene Expression 212
- 122 Transcription 213
- 123 mRNA Editing 214
- 124 The Genetic Code 215
- 125 Translation 216
- 126 ● Modeling Gene Expression 217
- 127 DNA Sequence and Traits 220
- 128 Mutations 221
- 129 Changes to DNA 223
- 130 Effect of Mutations 225
- 131 Molecular Technologies and DNA 227
- 132 Polymerase Chain Reaction 228
- 133 Gel Electrophoresis 229
- 134 Making Recombinant DNA 231
- 135 Gene Editing with CRISPR 233
- 136 Genetic Engineering for Insulin 234
- 137 Testing For Covid-19 236
- 138 Molecular Technologies and Research 237
- 139 Real-Life Superpowers Revisited 240
- 140 Summing Up 241

6 Patterns of Inheritance

Learning Outcomes 244

- 141 **CONTENT ANCHOR**
  Anyone for Chocolate? 246
- 142 What is a Trait? 247
- 143 Different Alleles for Different Traits 248
- 144 Sources of Variation 249
- 145 ● Examples of Genetic Variation 250
- 146 Sexual Reproduction Produces

CODES: Activity is marked: ▢ to be done   ✓ when completed   ● Practical component

Contents

		Genetic Variation	252
☐	147	Meiosis	253
☐	148	Meiosis and Variation	254
☐	149 ●	Modeling Meiosis	256
☐	150	Linked Genes and Variability	258
☐	151	Mendelian Genetics	259
☐	152	Monohybrid Crosses	262
☐	153	Probability	264
☐	154	Practicing Monohybrid Crosses	265
☐	155	Dihybrid Crosses	266
☐	156	Non Mendelian Genetics	269
☐	157	Incomplete Dominance	270
☐	158	Codominance	272
☐	159	Sex Linkage	274
☐	160	Testing the Outcomes of Genetic Crosses	276
☐	161	Anyone for Chocolate? Revisited	278
☐	162	Summing Up	279

7 Common Ancestry

		Learning Outcomes	283
☐	163	**CONTENT ANCHOR**	
		Dinosaur or Bird?	285
☐	164	Evolution and Common Ancestry	286
☐	165	Fossil Formation	287
☐	166	The Fossil Record	288
☐	167	Interpreting the Fossil Record	289
☐	168	Transitional Fossils	290
☐	169	Anatomical Homology	292
☐	170	Biogeography and Common Ancestry	294
☐	171	DNA Evidence for Common Ancestry	296
☐	172	Protein Evidence for Common Ancestry	297
☐	173	Developmental Homology	298
☐	174	Changes in the Fossil Record	299
☐	175	Punctuated Equilibrium and Gradualism	301
☐	176	Developing the Theory of Evolution	304
☐	177	Dinosaur or Bird? Revisited	306
☐	178	Summing Up	307

8 Evolution and Natural Selection

		Learning Outcomes	310
☐	179	**CONTENT ANCHOR**	
		How Does an Elephant Lose its Tusks?	312
☐	180	How Does Natural Selection Work?	313
☐	181 ●	Modeling Natural Selection with M&M's®	315
☐	182	The Role of Variation in Populations	316
☐	183	Natural Selection in Galápagos Finches	317
☐	184	Selection Pressure in Populations	319
☐	185	Directional Selection in Moth Populations	321
☐	186	Measuring Gene Pool Change	322
☐	187	Natural Selection in Rock Pocket Mice	323
☐	188 ●	Modeling Natural Selection in Rock Pocket Mice	325
☐	189	What is a Species?	328
☐	190	How Species Form	329
☐	191	Patterns of Evolution	331
☐	192	Evolutionary Mechanisms in Gene Pools	332
☐	193	Gene Flow	333
☐	194	Genetic Drift	336
☐	195	The Founder Effect	337
☐	196	Genetic Bottlenecks	339
☐	197	Mutations and the Gene Pool	340
☐	198	Genetic Recombination and the Gene Pool	341
☐	199	How Does an Elephant Lose its Tusks? Revisited	342
☐	200	Summing Up	343

9 Ecological Interactions

		Learning Outcomes	346
☐	201	**CONTENT ANCHOR**	
		A Mammoth Task	348
☐	202	Components of an Ecosystem	349
☐	203	Habitat and Tolerance Range	350
☐	204	The Ecological Niche	351
☐	205	Ecosystem Dynamics	352
☐	206	The Resilient Ecosystem	352
☐	207	A Case Study in Ecosystem Resilience	355
☐	208	Species Interactions	356
☐	209	Predator-Prey Relationships	358
☐	210	Predation and Destabilized Ecosystems	360
☐	211 ●	Investigating Predator-Prey Stability	361
☐	212	Competition for Resources	363
☐	213	Intraspecific Competition	364
☐	214	Interspecific Competition	366
☐	215	The Impact of Competing Invasive Species	367

CODES: Activity is marked: ☐ to be done   ✓ when completed   ● Practical component

Contents

☐	216	Parasitism: One-sided Benefits	368
☐	217	Commensalism: Free for the Taking	370
☐	218 ●	Mutualism: A Beneficial Dependance	371
☐	219	Eat or be Eaten	373
☐	220	Photoautotrophs and Heterotrophs	374
☐	221	Trophic Levels	377
☐	222	Matter Cycles Through an Ecosystem	378
☐	223	Disruption of Matter Cycles	379
☐	224	Energy Flows Through an Ecosystem	380
☐	225	Ecological Pyramids	382
☐	226 ●	Investigating Ecological Pyramids	384
☐	227	Disruption to Biomass and Energy Flow	385
☐	228	Nutrient Cycles	386
☐	229	The Carbon Cycle	388
☐	230 ●	Modeling the Carbon Cycle	389
☐	231	Disruptions to the Carbon Cycle	390
☐	232 ●	Ocean Acidification	392
☐	233	The Nitrogen Cycle	394
☐	234	Disruptions to the Nitrogen Cycle	396
☐	235	Ecosystem Changes can be Permanent	398
☐	236	Biodiversity in an Ecosystem	400
☐	237	Keystone Species and Ecosystem Stability	402
☐	238 ●	Human Activity and Biodiversity	404
☐	239 ●	Human Impacts on Marine Biodiversity	408
☐	240	Deforestation and Species Survival	412
☐	241	Can't See the Wood for the Trees	414
☐	242	The Effects of Damming on Biodiversity	416
☐	243	Humans Depend on Biodiversity	418
☐	244	A Mammoth Task Revisited	419
☐	245	Summing Up	420

10 Science Practices

		Learning Outcomes	423
☐	246	How Do We Do Science?	424
☐	247	Systems and System Models	425
☐	248	Hypotheses, Laws, and Theories	426
☐	249	Observations and Assumptions	427
☐	250	Accuracy and Precision	428
☐	251	Working With Numbers	429
☐	252	Tallies, Percentages, and Rates	430
☐	253	Fractions and Ratios	431
☐	254	Dealing With Large Numbers	432
☐	255	Apparatus and Measurement	433
☐	256	Types of Data	434
☐	257	Variables and Controls	435
☐	258	Recording Results	436
☐	259	Constructing Tables	437
☐	260	Which Graph to Use?	438
☐	261	Drawing Line Graphs	439
☐	262	Drawing Scatter Graphs	440
☐	263	Correlation or Causation?	441
☐	264	Drawing Bar Graphs	442
☐	265	Drawing Histograms	443
☐	266	Drawing Pie Graphs	444
☐	267	Mean, Median, and Mode	445
☐	268	What is Standard Deviation?	447
☐	269	Detecting Bias in Samples	448
☐	270	Biological Drawings	449
☐	271	Practicing Biological Drawings	451
☐	272	Safety and Ethics in Investigations	452
☐	273	A Qualitative Practical Task	455
☐	274	Analyzing Experimental Data	456

Glossary	459
Equipment List	466
Image Credits	468
Index	469

CODES: Activity is marked: ☐ to be done   ☑ when completed   ● Practical component

Using This Book

Activities make up most of this book. These are presented as integrated instructional sequences over multiple pages, allowing you to build a deeper understanding of phenomena as you progress through each chapter. Each chapter begins with the **Content Anchor** of the chapter. This is something you would have seen or experienced, but may not be able to explain fully. The content anchor is revisited at the end of the chapter. The intervening activities in a chapter are designed to let you explore some aspect of the content anchor, contributing to your deeper understanding.

Structure of a chapter

Chapter introduction
Identifies the activities relating to the learning outcomes. Relevant TEKS and ELPS are identified.

Content anchor
The first activity is an anchor for the chapter. It introduces a phenomenon that you come to understand by working through the activities in the chapter.

Content anchor revisited
Once you have completed the activities in the chapter should be able to explain various aspects of the content anchor more fully.

Summing up
Find out what you know about the ideas, connections, and skills you have explored in the chapter.

Activity pages

Chapter Introductions

Learning objectives
These provide guidance for the chapter content and help to focus on important areas of study.

The activity in the book related to these questions or statements.

ELPS
Identify the ELPS covered in the chapter and provide guidance for enhanced language learning.

The chapter number is identified for easy navigation.

TEKS Scientific and engineering practices covered.

TEKS Science concepts covered.

Mark the check boxes to indicate the outcomes you should complete. Check them off when you have finished.

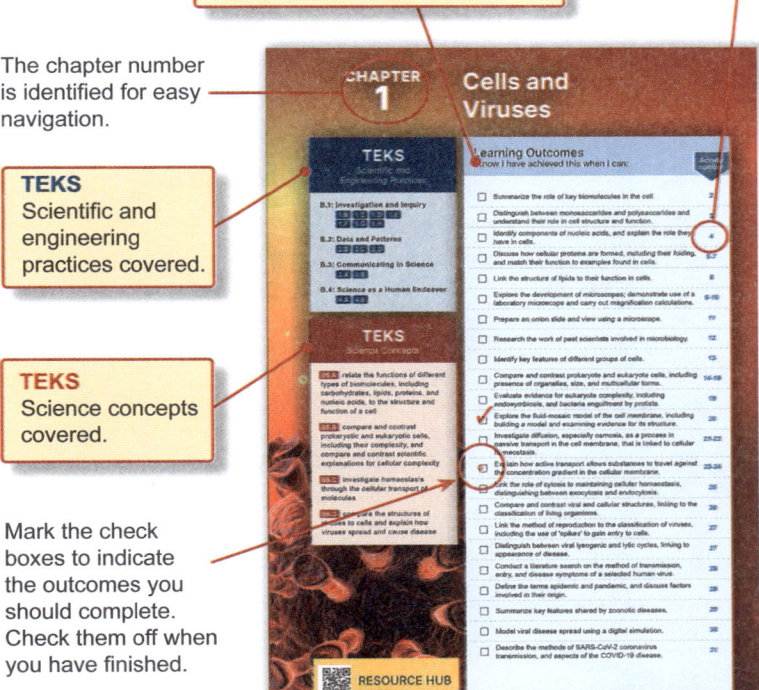
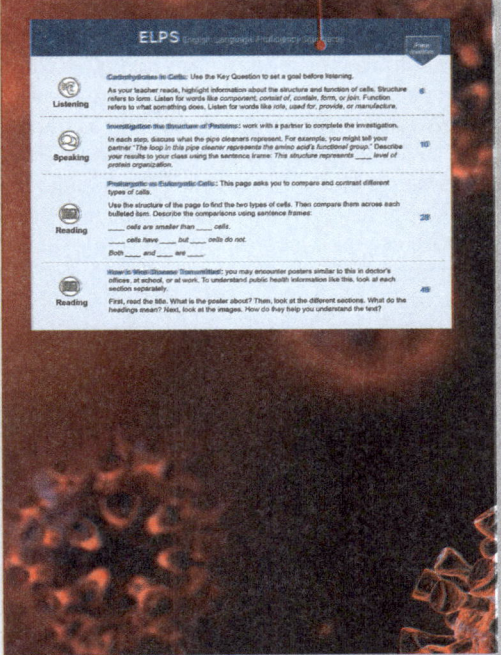

©2024 **BIOZONE** International
Photocopying Prohibited

Glossary Terms

Building communication skills and scientific literacy is an important feature of any science course. By speaking with, listening to your peers and teachers, and writing answers, you naturally practice and develop communication skills. To help develop **scientific literacy** we have included a **glossary** at the back of this worktext (pages 459-465). The glossary provides a definition in English and also in Spanish. Refer to the glossary to help you understand the meaning of a key term. It is easy to see which key terms are in the glossary, the terms have been **bolded in blue** within an activity (see below). Note: Key terms are only bolded the first time they appear within an activity.

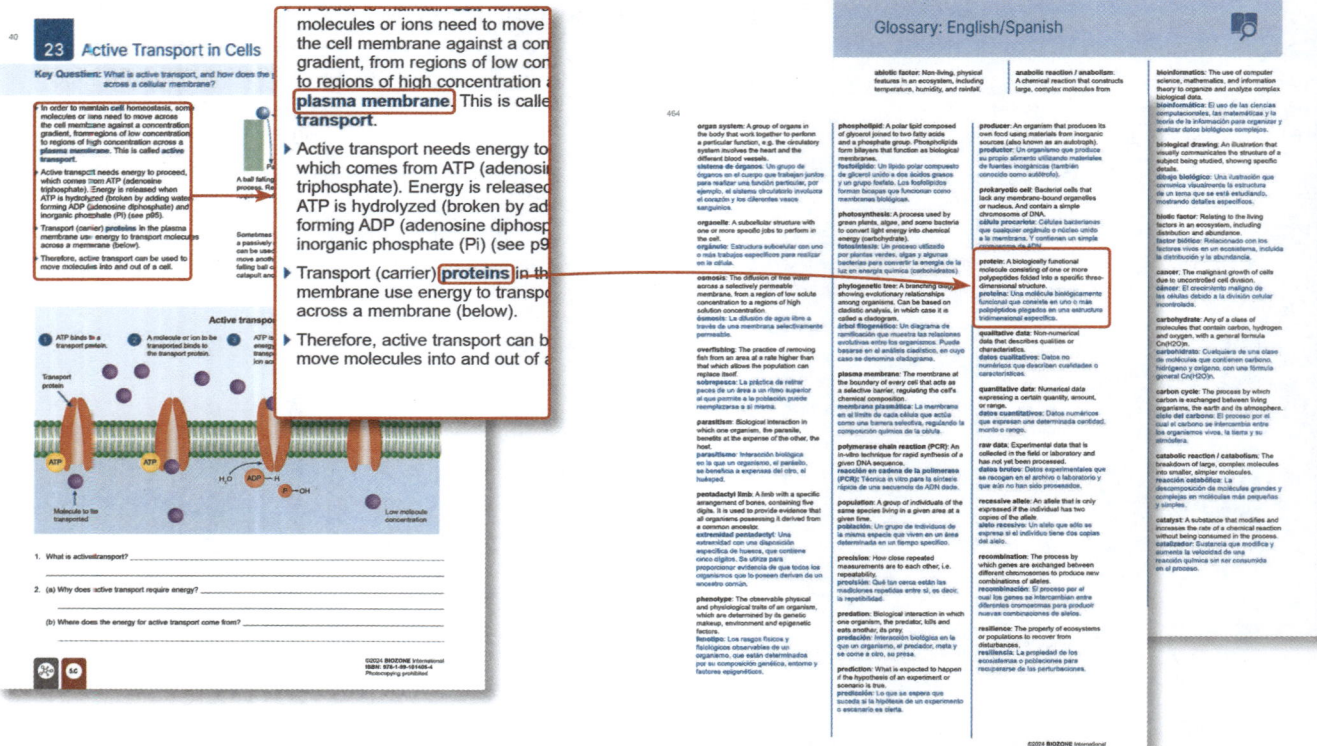

Practical Investigations

An important part of science involves carrying out investigations and carefully observing and recording what occurs during them. Throughout the book you will notice green investigation panels (like the one shown right). Each investigation has been designed using simple equipment found around the home or in most high school laboratories. The investigations provide opportunities for you to investigate phenomena for yourself. The investigations have different purposes depending on where they occur within the chapter. Some provide stimulus material or ask questions to encourage you to think about a particular phenomenon before you study it in detail. Others build on work you have already carried out and provide a more complex scenario for you to explain. Equipment lists are provided as an appendix at the back of the book. The investigations will help you develop:

- Skills in observation
- Skills in critical analysis and problem solving
- Skills in mathematics and numeracy
- Skills in collecting and analyzing data and maintaining accurate records
- Skills in working independently and collaboratively as part of a group
- Skills in communicating and contributing to group discussions

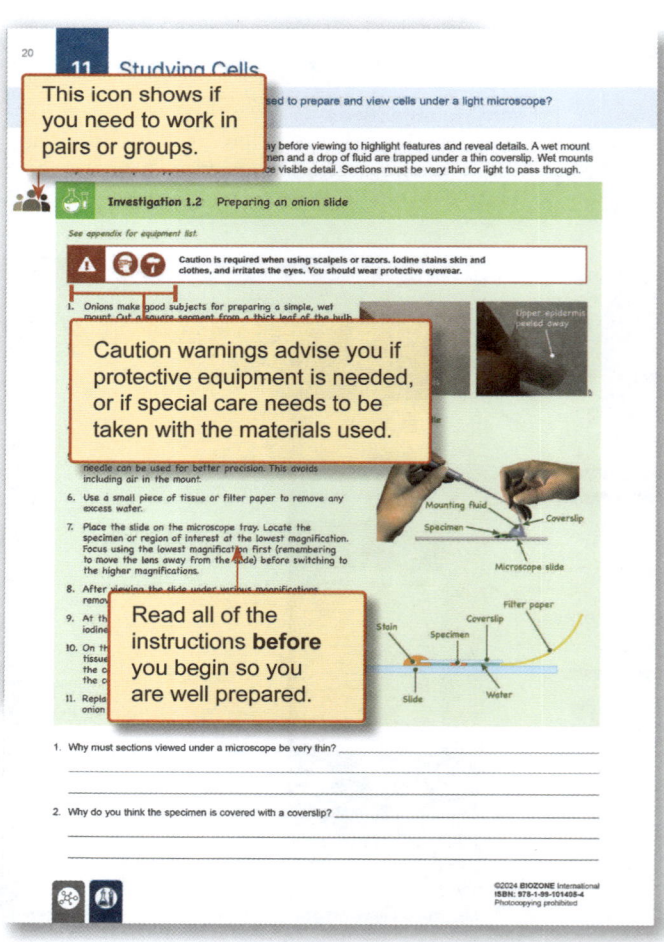

©2024 **BIOZONE** International
Photocopying Prohibited

Understanding the Tab System

The tab system helps you identify the TEKS science concepts and scientific and engineering practices within each activity, and whether there are supporting features in the **BIOZONE RESOURCE HUB**.

The gray hub tab indicates that the activity has online support via the **BIOZONE RESOURCE HUB**. This may include videos, animations, articles, 3D models, and computer models.

The **blue TEKS** tabs use picture codes to identify the scientific and engineering practices relevant to the activity. You will use scientific and engineering practices in the course of completing the activities.

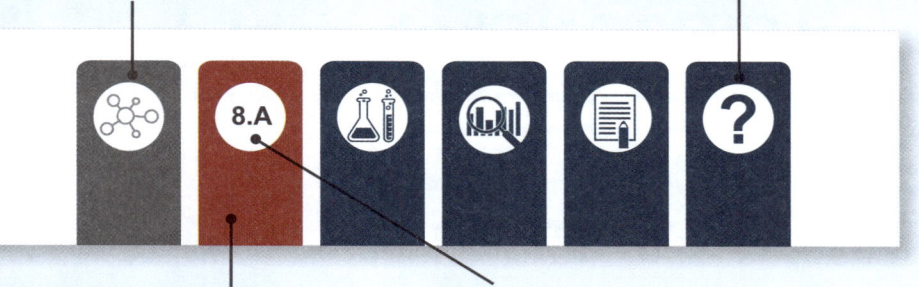

The **red TEKS** tabs indicate the Science Concepts that are covered in the activity. These are detailed in the introduction to each chapter, and linked to appropriate activities.

The TEKS code refers specifically to the TEK covered in the activity.

TEKS
Scientific and Engineering Practices

B.1: Investigating
The student asks questions, identifies problems, and plans and safely conducts classroom, laboratory, and field investigations to answer questions, explain phenomena, or design solutions using appropriate tools and models.

B.2: Patterns
The student analyzes and interprets data to derive meaning, identify features and patterns, and discover relationships or correlations to develop evidence-based arguments or evaluate designs.

B.3: Communicating
The student develops evidence-based explanations and communicates findings, conclusions, and proposed solutions.

B.4: Discovery
The student knows the contributions of scientists and recognizes the importance of scientific research and innovation on society.

ELPS
English Language Proficiency Standards

1. Learning Strategies: developing awareness of the learning process through a variety of strategies.

2. Listening: gaining increased comprehension of spoken language presented in a range of mediums.

3. Speaking: using both formal and informal speech to craft fluent and accurate communication with appropriate vocabulary.

4. Reading: utilizing a variety of text types with different purposes to gain increasing comprehension.

5. Writing: effectively address specific purposes to a range of audience types using accurate written communication.

Margin Icons

A group symbol indicates where students can work together.

A computer symbol indicates a computer is needed to complete the task (e.g. using a spreadsheet).

Scan the QR codes to go directly to an interactive 3D model.

Using BIOZONE's Resource Hub

BIOZONE's Resource Hub provides links to online content supporting the activities in the book. From this page, you can also check for any errata or clarifications to the book since printing.

The external websites are mostly narrowly focused animations and video clips directly relevant to that part of the activity identified by the hub icon. They provide great support to help your understanding.

www.BIOZONEhub.com

Then enter the code in the text field: **TXB1-4054**

Or scan this QR code

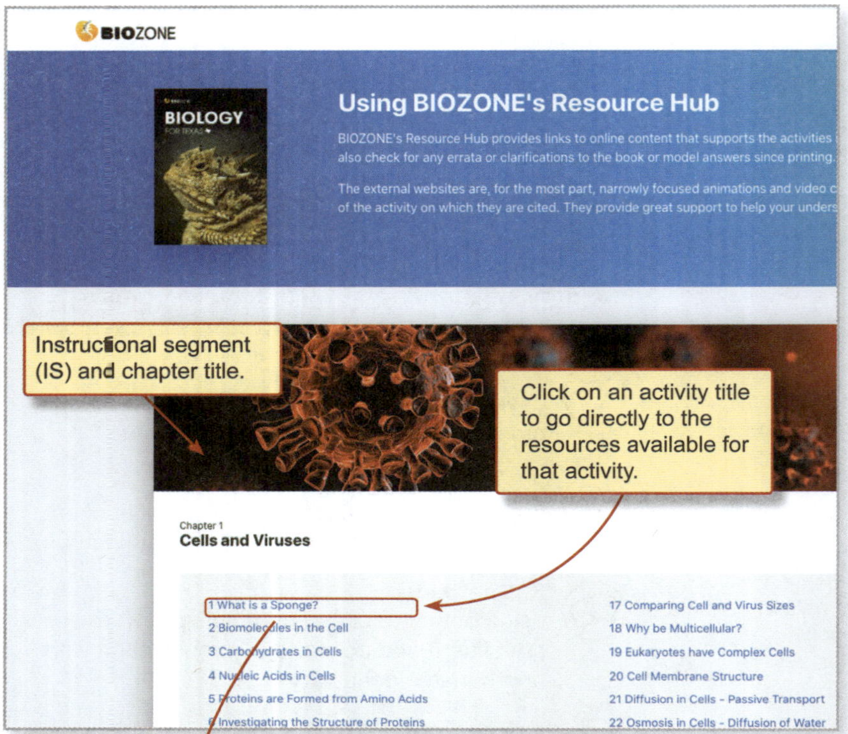

The Resource Hub icons

- Simulation
- Weblink
- PDF
- 3D Model
- Spreadsheet
- Video

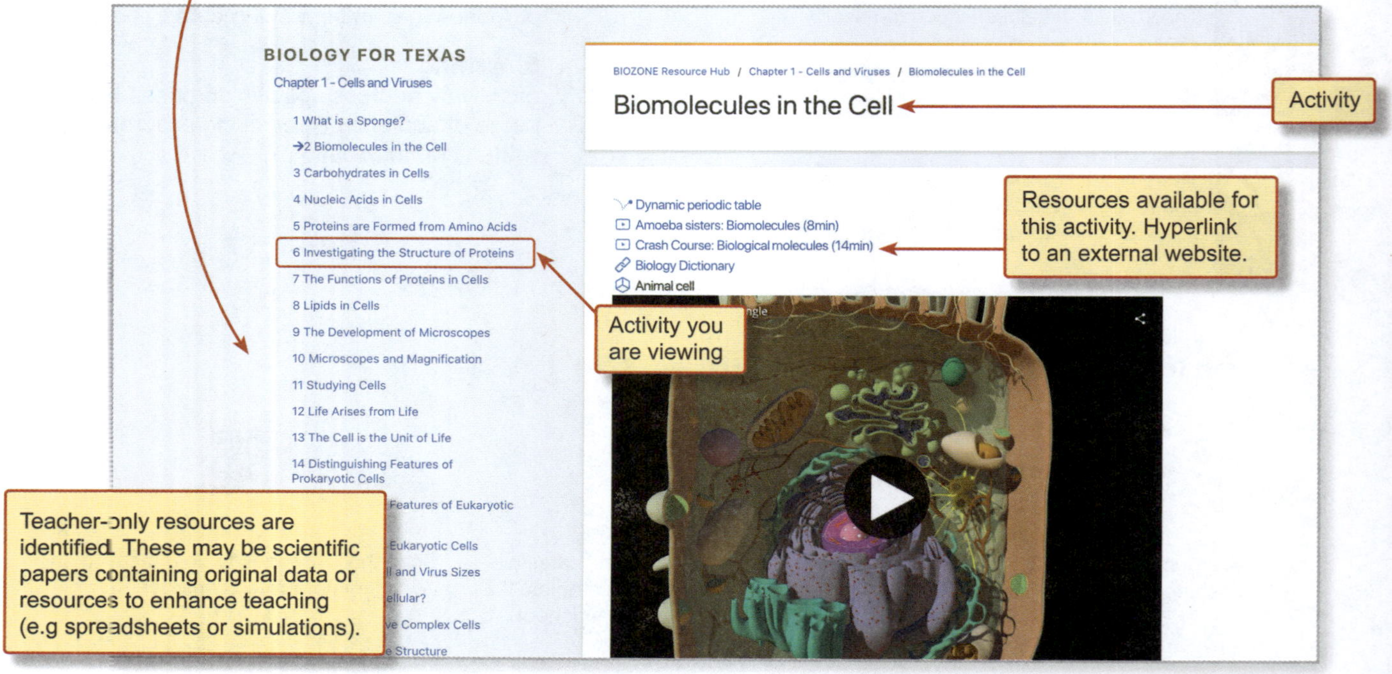

Instructional segment (IS) and chapter title.

Click on an activity title to go directly to the resources available for that activity.

Activity

Resources available for this activity. Hyperlink to an external website.

Activity you are viewing

Teacher-only resources are identified. These may be scientific papers containing original data or resources to enhance teaching (e.g. spreadsheets or simulations).

©2024 BIOZONE International
Photocopying Prohibited

CHAPTER 1
Cells and Viruses

TEKS
Scientific and Engineering Practices

B.1: Investigation and Inquiry
1.B 1.C 1.D 1.E
1.F 1.G 1.H

B.2: Data and Patterns
2.B 2.C 2.D

B.3: Communicating in Science
3.A 3.B

B.4: Science as a Human Endeavor
4.A 4.B

TEKS
Science Concepts

B5.A relate the functions of different types of biomolecules, including carbohydrates, lipids, proteins, and nucleic acids, to the structure and function of a cell

B5.B compare and contrast prokaryotic and eukaryotic cells, including their complexity, and compare and contrast scientific explanations for cellular complexity

B5.C investigate homeostasis through the cellular transport of molecules

B5.D compare the structures of viruses to cells and explain how viruses spread and cause disease

Learning Outcomes
I know I have achieved this when I can:

	Learning Outcome	Activity number
☐	Summarize the role of key biomolecules in the cell.	2
☐	Distinguish between monosaccharides and polysaccharides and understand their role in cell structure and function.	3
☐	Identify components of nucleic acids, and explain the role they have in cells.	4
☐	Discuss how cellular proteins are formed, including their folding, and match their function to examples found in cells.	5-7
☐	Link the structure of lipids to their function in cells.	8
☐	Explore the development of microscopes; demonstrate use of a laboratory microscope and carry out magnification calculations.	9-10
☐	Prepare an onion slide and view using a microscope.	11
☐	Research the work of past scientists involved in microbiology.	12
☐	Identify key features of different groups of cells.	13
☐	Compare and contrast prokaryote and eukaryote cells, including presence of organelles, size, and multicellular forms.	14-18
☐	Evaluate evidence for eukaryote complexity, including endosymbiosis, and bacteria engulfment by protists.	19
☐	Explore the fluid-mosaic model of the cell membrane, including building a model and examining evidence for its structure.	20
☐	Investigate diffusion, especially osmosis, as a process in passive transport in the cell membrane, that is linked to cellular homeostasis.	21-22
☐	Explain how active transport allows substances to travel against the concentration gradient in the cellular membrane.	23-24
☐	Link the role of cytosis to maintaining cellular homeostasis, distinguishing between exocytosis and endocytosis.	25
☐	Compare and contrast viral and cellular structures, linking to the classification of living organisms.	26
☐	Link the method of reproduction to the classification of viruses, including the use of 'spikes' to gain entry to cells.	27
☐	Distinguish between viral lysogenic and lytic cycles, linking to appearance of disease.	27
☐	Conduct a literature search on the method of transmission, entry, and disease symptoms of a selected human virus.	28
☐	Define the terms epidemic and pandemic, and discuss factors involved in their origin.	29
☐	Summarize key features shared by zoonotic diseases.	29
☐	Model viral disease spread using a digital simulation.	30
☐	Describe the methods of SARS-CoV-2 coronavirus transmission, and aspects of the COVID-19 disease.	31

RESOURCE HUB
bit.ly/3K1Vzkh

ELPS English Language Proficiency Standards

		Page number
Listening	**Carbohydrates in Cells:** Use the Key Question to set a goal before listening. As your teacher reads, highlight information about the structure and function of cells. Structure refers to form. Listen for words like *component*, *consist of*, *contain*, *form*, or *join*. Function refers to what something does. Listen for words like *role*, *used for*, *provide*, or *manufacture*.	6
Speaking	**Investigation the Structure of Proteins:** Work with a partner to complete the investigation. In each step, discuss what the pipe cleaners represent. For example, you might tell your partner "*The loop in this pipe cleaner represents the amino acid's functional group.*" Describe your results to your class using the sentence frame: *This structure represents ____ level of protein organization.*	10
Reading	**Prokaryotic vs Eukaryotic Cells:** This page asks you to compare and contrast different types of cells. Use the structure of the page to find the two types of cells. Then compare them across each bulleted item. Describe the comparisons using sentence frames: ____ *cells are smaller than* ____ *cells.* ____ *cells have* ____ *but* ____ *cells do not.* *Both* ____ *and* ____ *are* ____.	28
Reading	**How is Viral Disease Transmitted:** You may encounter posters similar to this in doctor's offices, at school, or at work. To understand public health information like this, look at each section separately. First, read the title. What is the poster about? Then, look at the different sections. What do the headings mean? Next, look at the images. How do they help you understand the text?	49

1 What is a Sponge?

Content Anchor: How can we tell what type of organism a sponge is by using a microscope?

▸ Although most sponges for washing are synthetic and made from types of plastic, some of you may use a real sponge (right) to wash with. They soak up lots of water and contain natural antibacterial and antifungal properties.

▸ These sponges are the dead internal remains of once-living organisms and share at least 70% of our genetic material, the information that makes us human!

1. What key structural feature does the sponge share with all living organisms, and what importance does this have to being alive?

   _____
   _____
   _____
   _____
   _____
   _____

2. What features differ between animals, plants, fungi, and bacteria that would tell us which of these a sponge belongs to?

   _____
   _____
   _____

3. Imagine you were given a very small, thin piece of sponge to look at under a powerful microscope. Discuss with your classmates what you might see that would help tell you what type of organism it was. In the box below, draw and label, in pencil, any structures of the magnified sponge piece that you think you might see and say why you think the sponge is animal, plant, bacteria, or other (we will return to review your drawing at the end of the chapter):

2 Biomolecules in the Cell

Key Question: What biomolecules are present in the cells of organisms, and what role do they play in the function and structure of the cell?

- Water is the main component of **cells** and organisms, providing an aqueous environment in which metabolic reactions, the metabolism, can occur.
- Most other substances in cells are compounds of carbon, hydrogen, oxygen, and nitrogen. Carbon can combine with many other elements to form a large number of carbon-based (or organic) molecules.
- The organic molecules that make up living things, called biomolecules, can be grouped into four broad classes: **carbohydrates**, **lipids**, **proteins**, and nucleic acids. In addition, a small number of inorganic ions, such as Mg^{2+} are also components of larger molecules.

The biomolecules in cells

Centrioles
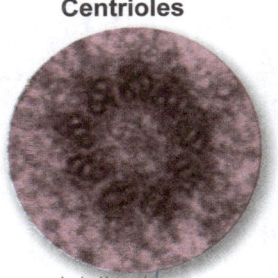
Louisa Hayward

Proteins have an enormous number of structural and functional roles in cells, e.g. as enzymes, as structural materials such as collagen, in transport, and movement, e.g. cytoskeleton and centrioles.

Components: C, H, O, N, S, P

Chloroplasts in plant cells
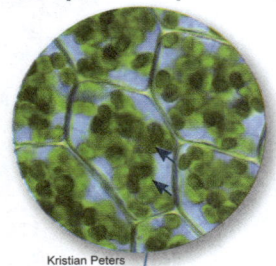
Kristian Peters

Inorganic ions: Dissolved ions participate in metabolic reactions and are components of larger, organic molecules, e.g. Mg^{2+} is a component of the green chlorophyll pigment in the chloroplasts of green plants.

Plant epidermis
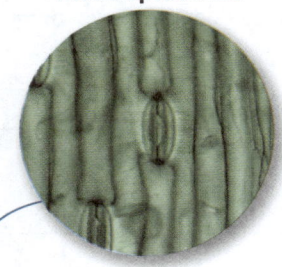

Water is a major component of cells. Many substances dissolve in it and metabolic reactions occur in it. In plant cells, fluid pressure against the cell wall provides turgor, which supports the cell. **Components: H, O**

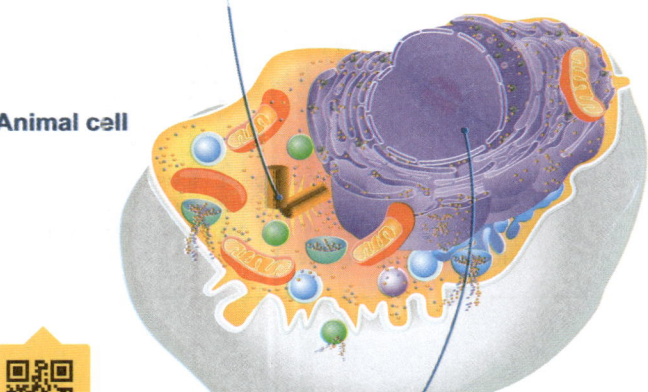

Animal cell | Plant cell

Chromosome

Nucleotides and nucleic acids: Nucleic acids encode information for the construction and functioning of an organism (DNA and RNA). ATP, a nucleotide derivative, is the energy carrier of the cell.

Components: C, H, O, N, P

Plant cell wall

Carbohydrates form the structural components of cells, e.g. cellulose **cell walls** (arrowed). They are important in providing usable energy as **glucose**, in energy storage and they are involved in cellular recognition.

Components: C, H, O

Chloroplast membranes
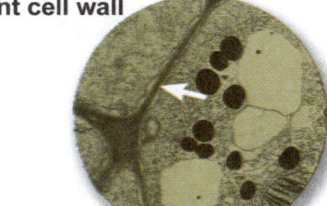
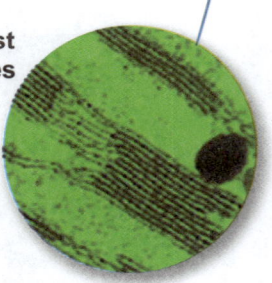

Simple lipids provide a concentrated source of energy. **Phospholipids** (a complex lipid) are a major component of cellular membranes, including the membranes of **organelles**, such as chloroplasts and mitochondria.

Components: C, H, O (lipids)

C, H, O, P, N (phospholipids)

The elements of life

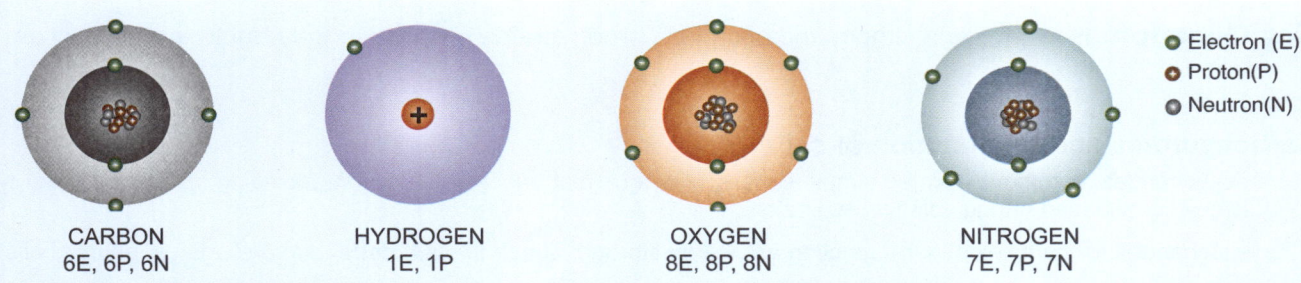

▸ Carbon is abundant. It has four valence (outer shell) electrons that are available to form up to four covalent (shared electron) bonds with other atoms. Complex biological molecules consist of carbon atoms bonded with other elements, especially oxygen and hydrogen, but also nitrogen, phosphorus, and sulfur.

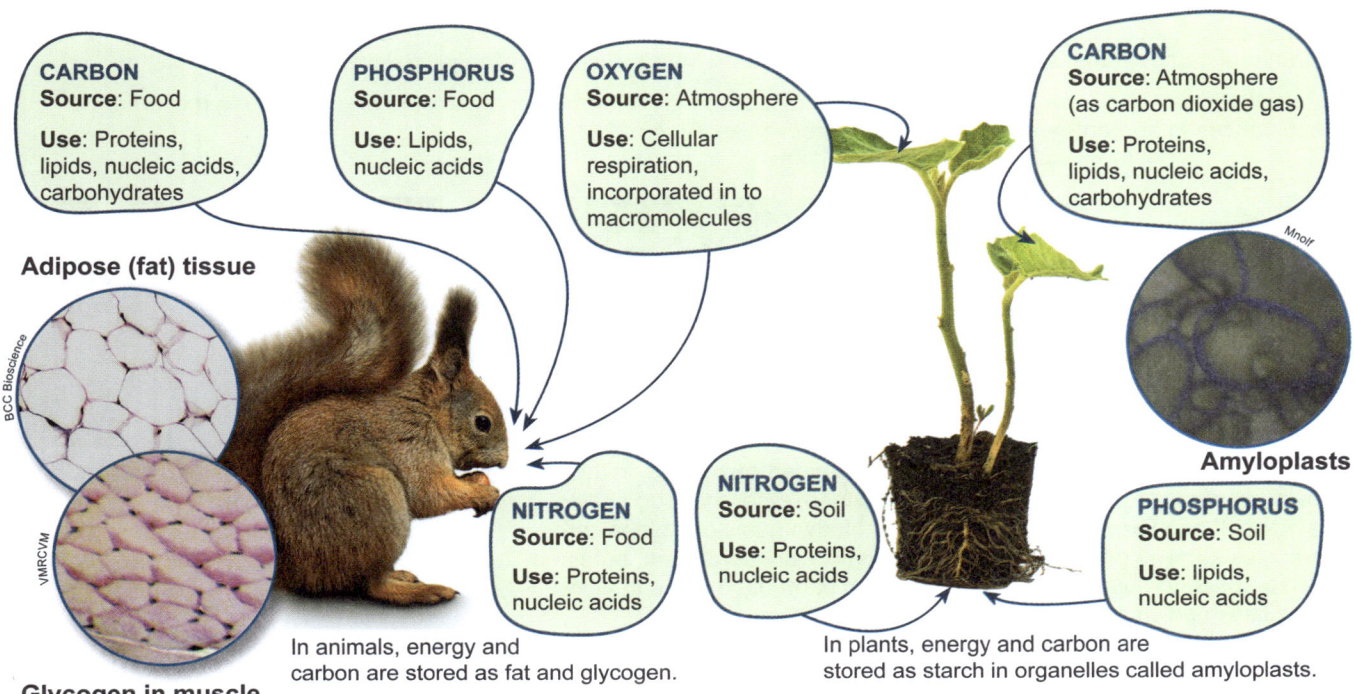

1. Summarize the role of each of the following cell components:

    (a) Carbohydrates: _____

    (b) Lipids: _____

    (c) Proteins: _____

    (d) Nucleic acids: _____

    (e) Inorganic ions: _____

    (f) Water: _____

2. Explain why carbon is so important for building the molecular components of an organism: _____
_____
_____
_____

3. State the main source of carbon, phosphorus, and nitrogen for animals: _____

4. (a) State the main source of carbon for plants: _____

    (b) State the main source of phosphorus and nitrogen for plants: _____

3 Carbohydrates in Cells

Key Question: How are both simple and complex carbohydrates involved in the structure and function of cells?

Carbohydrates and the function of cells
- Monosaccharides, commonly termed simple sugars, play a central role in cell function, providing a source of energy that is quickly released during cellular respiration.
- Polysaccharides, or complex sugars, function as energy storage: starch in plant **cells**, and glycogen in animal cells.
- Plants, and some single-celled organisms, are called producers. They manufacture simple **carbohydrates** in the form of **glucose** during photosynthesis. All other organisms, called heterotrophs, acquire carbohydrates by eating.
- Simple carbohydrates are the building blocks that join together to form complex carbohydrate macromolecules.

Carbohydrates and the structure of cells
- Some complex carbohydrates have multiple strong bonds between molecules. This allows them to form rigid cellular structural components, such as plant **cell walls**, made from cellulose, or fungal cell walls made from chitin.
- Carbohydrates also form an important component of the cell membrane, controlling and facilitating movement of substances in and out, and allowing cells to recognize each other.

Glucose is a versatile molecule. It provides energy to power cellular reactions. It can form energy storage molecules such as glycogen, or it can be used to build structural molecules in polymer form.

Simple carbohydrates, in the form of glucose, enter the food chain when photosynthesis combines carbon, oxygen, and hydrogen, obtained from carbon dioxide and water.

Fructose, often called fruit sugar, is a simple monosaccharide. It is often made from sugar cane (above). Both fructose and glucose can be directly absorbed into the bloodstream to travel to cells.

Monosaccharides
- Monosaccharides (mono = one) are single-sugar biomolecules and include glucose and fructose.
- They are used as a primary energy source for fueling cell metabolism.
- Monosaccharides can be classified by the number of carbon atoms they contain, and join together to form carbohydrate macromolecules, such as starch and glycogen.

Glucose isomers
Molecules such as glucose can have many different isomers.

Isomers are compounds with the same chemical formula (same types and numbers of atoms) but different arrangements of atoms. The different arrangement of the atoms means that each isomer has different properties.

1. Describe the two major functions of monosaccharides in the cell:

   (a) _____

   (b) _____

2. A large amount of energy is released during the breakdown of the glucose molecule into smaller water and carbon dioxide molecules. How is this relevant to the role that glucose plays in the cell?

   _____

   _____

3. Glucose is highly soluble (easily dissolved) so it can easily be carried in the blood. Discuss how this makes it suitable as a biomolecule that is used to supply cells with an energy source:

   _____

   _____

Polysaccharides: their structure and function in a cell

▸ Polysaccharides (many sugars) are macromolecules consisting of straight or branched chains of many monosaccharides. They can consist of one or more types of monosaccharides. The most common polysaccharides are cellulose, starch, and glycogen. Polysaccharides can also be referred to as a polymer (many monomers).

▸ The macromolecules below contain only glucose but their properties and roles are very different. These differences are a function of the glucose isomer involved and the types of linkages joining the monomers. Thus, different polysaccharides can be a source of readily available glucose or a structural material that resists digestion.

Cellulose

▸ Cellulose is a structural material found in the cell walls of plants. It is made up of many glucose molecules linked together to form a straight chain polymer. Cellulose is very strong, so is an ideal structural material. Its structure makes it difficult for herbivores to digest. Plant eating animals often rely on bacteria in their guts to help break down the cellulose in plant cell walls. This helps them extract the nutrients that can be obtained from it.

Cotton fibers contain more than 90% cellulose fiber.

The structure of polysaccharides (also called complex carbohydrates) can be compared using molecular visualization software.

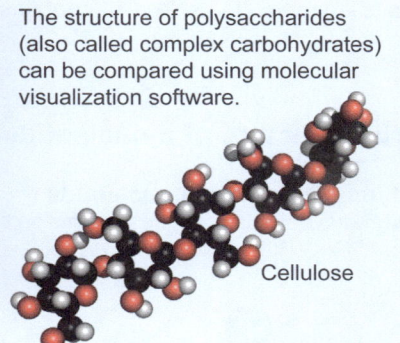

Cellulose

Starch

▸ Starch is also a polymer of glucose. It acts as an energy storage molecule in plants and is found concentrated in insoluble starch granules within specialized **organelles** in plant cells (see photo, right). Starch is composed of a mixture of two polymers of glucose: amylose (unbranched) and amylopectin (branched). Starch can easily be hydrolyzed (broken apart) by enzymes into smaller soluble sugars, such as glucose, when required.

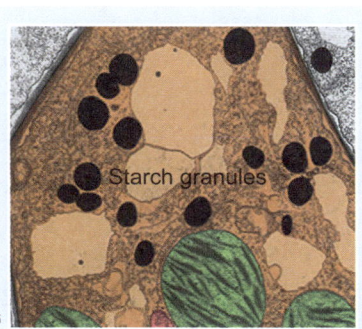

Black starch granules in a plant cell (false color TEM).

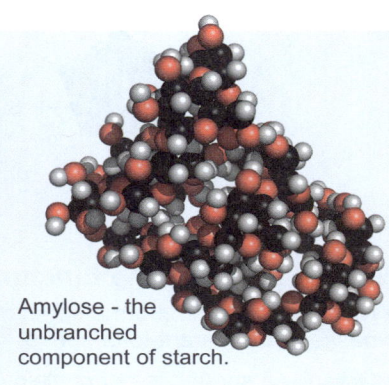

Amylose - the unbranched component of starch.

Glycogen

▸ Glycogen is a branched polysaccharide. It is more highly branched and more water-soluble than starch. Glycogen is also a carbohydrate storage compound in animal tissues, and is found mainly in liver and muscle cells (photo, right). It is readily hydrolyzed (broken apart) by enzymes to form glucose, making it an ideal energy storage molecule for active animals.

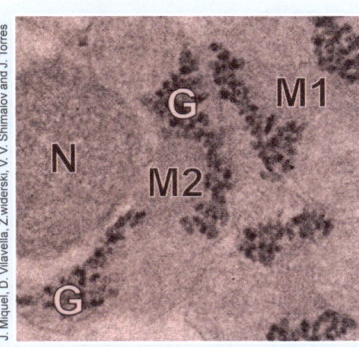

Glycogen (G) in the spermatozoa of a flatworm. M1, M2=mitochondria, N=**nucleus**.

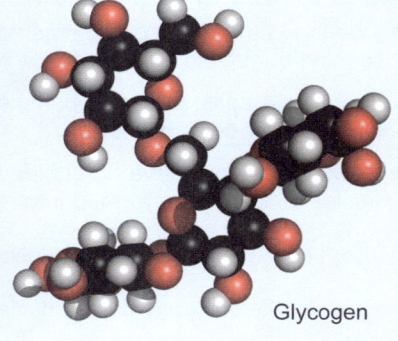

Glycogen

4. Link the properties of cellulose to the role it plays in plant cell structure:

___
___
___
___
___

4 Nucleic Acids in Cells

Key Question: What are nucleic acids, and how do they contribute to the structure and function of cells?

- **Nucleotides** are the building blocks of nucleic acids (DNA and RNA). Nucleotides have three components forming their structure: a nitrogen-containing base, a five carbon sugar, and a phosphate group.
- Five different types of nitrogen bases are found in nucleotides. DNA contains adenine (A), guanine (G), cytosine (C), and thymine (T). RNA also contains adenine, guanine, and cytosine, but uracil (U) is present instead of thymine.

The structure of a nucleotide

Symbolic form of a nucleotide
(showing positions of the 5 C atoms on the sugar)

Sugars

Nucleotides contain one of two different types of sugar. **Deoxyribose** sugar is only found in DNA. **Ribose** sugar is found in RNA.

Deoxyribose sugar (found in DNA)

Ribose sugar (found in RNA)

Nucleic acid and the function of cells

- Nucleic acids are information-carrying biomolecules enabling **genes**, 'instructions' for body form and function, to be passed from parent to offspring.
- Organized sequences of nucleic acid, in the form of DNA or RNA, allow **proteins** to be formed in the **cell** (see activity 5).
- All living organisms on earth contain the same nucleic acid biomolecules, indicating a common ancestry.

Nucleic acid and the structure of cells

- Two long strands of 'complementary' nucleic acids can join together, using hydrogen bonding, to form DNA (deoxyribonucleic acid). A single strand forms most types of RNA (ribonucleic acid).
- **Eukaryote** cells, the more complex cell types found in plants, animals, fungi, and single celled protists, contain the DNA in their cell's **nucleus**.
- **Prokaryote** cells have no nucleus, so the nucleic acids are located in the gel-like cytoplasm instead. Bacterial nucleic acid is often formed into rings, called plasmids.

1. (a) List the components of a nucleotide in DNA: _____

   (b) List the components of a nucleotide in RNA: _____

2. List the nucleotide bases present:

   (a) In DNA: _____

   (b) In RNA: _____

3. Name the sugar present:

   (a) In DNA: _____   (b) In RNA: _____

4. Explain the role that the nucleic acid biomolecule plays in cell structure and function: _____

_____

_____

©2024 **BIOZONE** International
ISBN: 978-1-99-101405-4
Photocopying prohibited

5 Proteins are Formed from Amino Acids

Key Question: How is protein formed in the cell, and how is the final form of the protein linked to its function?

Proteins are made of amino acids

▸ The sequence of nucleotides in DNA determines the sequence of amino acids that make up a particular **protein**.
▸ Proteins are large molecules made up of many smaller units called **amino acids**. The amino acids are joined together by peptide bonds (between the amine and carboxyl groups). Therefore, a chain of amino acids is called a polypeptide. The sequence of amino acids in a protein is determined by the order of **nucleotides** in DNA or RNA.
▸ All amino acids have a common structure (below left) with an amine group (blue), a carboxyl group (red), a hydrogen atom, and a functional or 'R' group (orange). Each type of amino acid has a different functional R group (side chain). Each functional R group has a different chemical property.

The general structure of an amino acid

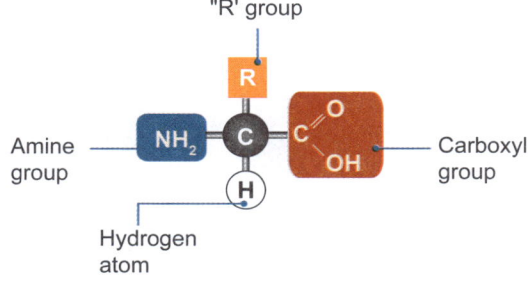

A polypeptide chain

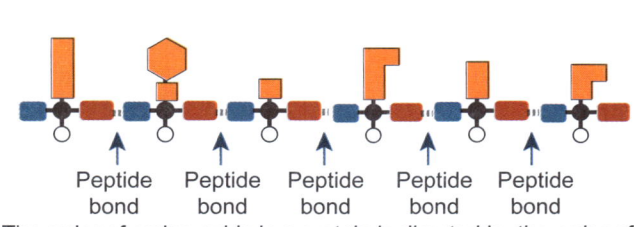

The order of amino acids in a protein is directed by the order of nucleotides in DNA or RNA.

The shape of a protein determines its role

▸ The sequence of amino acids determines how a protein will fold up, i.e. the shape it will form.
▸ The shape of a protein determines its role. Proteins generally fall into two groups, globular and fibrous (below).
▸ The shape of a protein is very important to its function. If the structure of a protein is denatured **(destroyed)** it can no longer carry out its function.

Globular proteins
Globular proteins are round and water soluble. Their functions include:
- Catalytic, e.g. enzymes.
- Regulation, e.g. hormones.
- Transport, e.g. hemoglobin.
- Protective, e.g. antibodies.

Insulin: a globular protein

Fibrous proteins
Fibrous proteins are long and strong. Their functions include:
- Support and structure, e.g. connective tissue.
- Contractile, e.g. myosin, actin.

Collagen: a fibrous protein

1. (a) Name the four components of an amino acid: _____

   (b) What makes each type of amino acid unique? _____

2. Why are proteins important in organisms? _____

3. (a) Why is the shape of a protein important? _____

   (b) What happens to a protein if it loses its shape? _____

6 Investigating the Structure of Proteins

Key Question: How does modeling help us understand the structure of a protein?

Proteins fold up into a functional structure

- The **amino acid** sequence of a **protein** is only the first step in making a functional protein. A protein must fold into a functional structure in order to carry out its biological role. This is where the 'R' groups become important in how the protein folds.

- The amino acid chain will fold up into a specific shape depending on the interactions between the different 'R' groups (below). These interactions include hydrogen bonds, disulfide (S – S) bonds, and hydrophobic (water-hating) and hydrophilic (water-loving) interactions.

- First, the amino acid chain folds into coils (helices) and sheets to create a secondary structure. These shapes are created and maintained by hydrogen bonds between CO and NH groups.

- More distant parts of the folded chain can then interact to create a highly organized, tertiary structure. Disulfide bridges are important in maintaining the folded tertiary structure.

- Some functional proteins, such as hemoglobin (right) consist of two or more polypeptide chains. When multiple polypeptides come together, the protein has a quaternary structure. The functional structure of hemoglobin also includes four iron atoms. These enable it to bind oxygen.

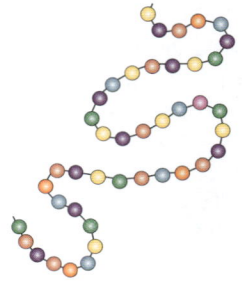
Primary structure
Amino acid chain

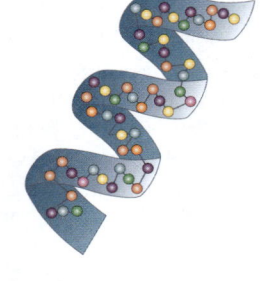
Secondary structure
Coiled helix

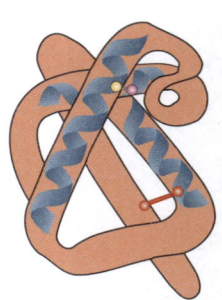
Tertiary structure
Folded helices

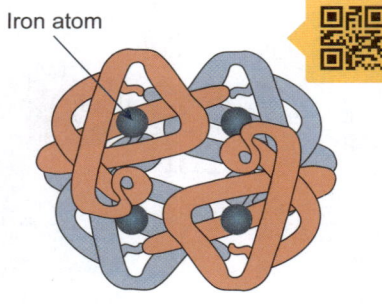
Iron atom
Quaternary structure
Multi-unit protein

Investigation 1.1   Modeling protein structure

See appendix for equipment list.

Work in pairs for this activity.

1. You will need pipe cleaners with four colors. We have used 2 white, 2 pink, 2 purple, and 4 blue but you can swap out for the colors you have. Each color represents a different amino acid.

2. Twist a loop in the center of each pipe cleaner (figure 1). The twist represents the amino acid's functional group.

3. Join the amino acids together (figure 2) by twisting their arms together in the following sequence:
   1) white 2) pink 3) blue 4) purple 5) blue 6) pink 7) blue 8) white 9) blue 10) purple.

What level of protein organization does the structure in figure 2 represent? _____

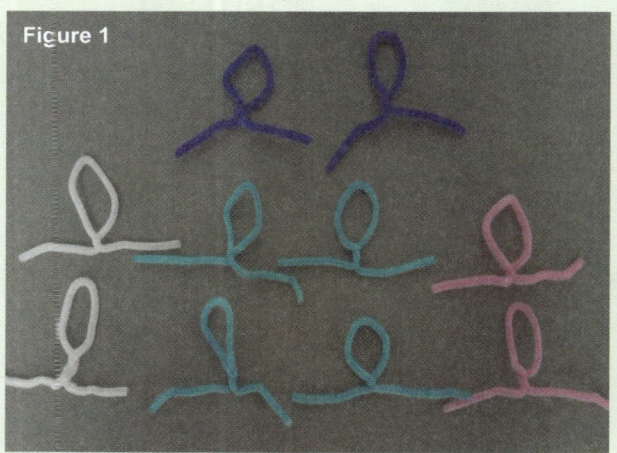
Figure 1

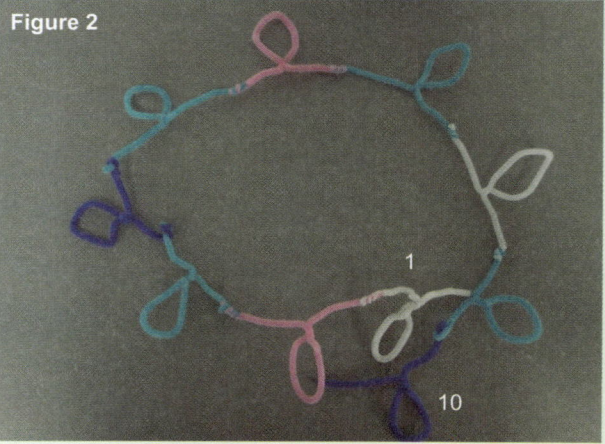
Figure 2

4. Attach sticky tape to the loops of the purple pipe cleaners and to one arm of each of the blue pipe cleaners. These represent places where hydrogen bonding can occur.

5. Join the sticky tape together between amino acids 3 and 5 and also between amino acids 7 and 9 (figure 3).

   Describe what happens to the shape of the model when you do this: _____
   _____
   _____

   What level of protein structure does this represent? _____

6. Attach binder clips or paper clips to the loops of the two pink amino acids and then use the clips to join the two pink amino acids together. The clips represent a disulfide bond.

7. Join the sticky tape together on the two purple amino acids (figure 4). Your protein has now formed its fully functional structure.

   What level of protein structure does this represent? _____

Figure 3 H = hydrogen bond

Figure 4

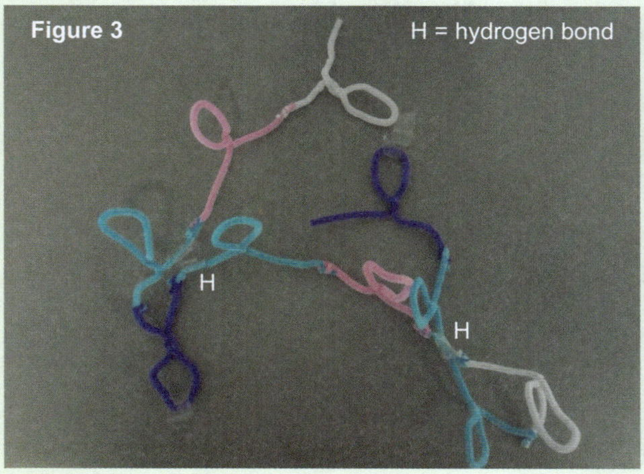

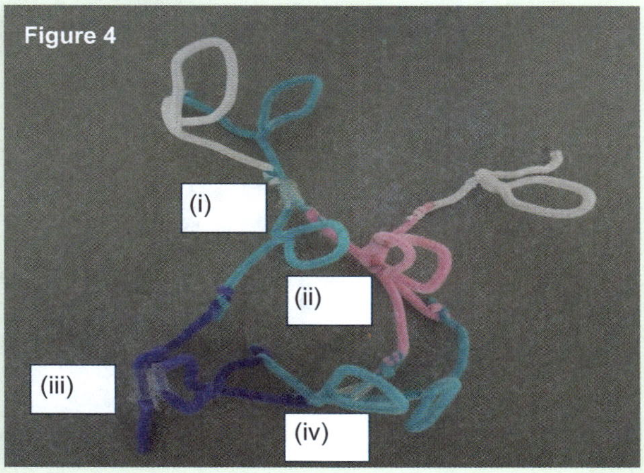

8. (a) Label figure 4 to show the location of all of the hydrogen bonds (H) and the disulfide bond (S).
   (b) Based on the properties of your model and its components, which of these bonds is likely the strongest?
   _____

9. Break the hydrogen bonds between amino acids 3 and 5 and also between 7 and 9 in your molecule.

   (a) What happens to the shape of the protein? _____
   _____

   (b) What process does breaking these bonds represent? _____
   _____

   (c) What effect will this process have on the protein's ability to carry out its job? _____
   _____
   _____
   _____

1. How could you adapt your model to demonstrate quaternary structure? _____
_____
_____
_____

7 The Functions of Proteins in Cells

Key Question: What kinds of proteins are found in the cells and what are their numerous roles?

▸ Recall that, in eukaryotic **cells**, most of the genetic information (DNA) is found in the **nucleus**. DNA provides instructions that code for the formation of proteins. **Proteins** carry out most of a cell's work. A cell produces many different types of proteins and each carries out a specific task in the cell.

▸ Proteins are involved in the structure, function, and regulation of the body's cells, tissues, and organs. Without functioning proteins, a cell cannot carry out its specialized role and the organism may die.

The nucleus is the control center of a cell

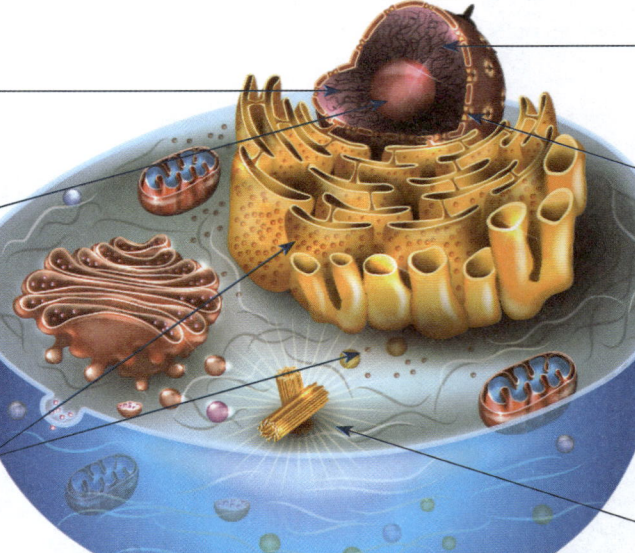

A generalized animal cell

The DNA within the nucleus carries the instructions for the cell's structure and function. This involves producing proteins.

Ribosomes are made in the nucleolus (a dense region within the nucleus).

Proteins are made outside the nucleus by ribosomes. These may be free in the cytoplasm or associated with the rough endoplasmic reticulum (rER).

Genes (sections of DNA) code for specific proteins. A cell can control the type of protein it makes by only expressing the genes for the proteins it needs.

The double-layered **nuclear membrane** has pores to allow materials to move between the nucleus and the cytoplasm.

Microtubules made of protein form the cell's internal skeleton. This includes the centrioles and spindle fibers involved in cell division.

▸ While a generalized cell produces a range of proteins, some body cells are highly specialized to produce large amounts of a specific protein. This specialization defines their functional role. Three examples are pictured below.

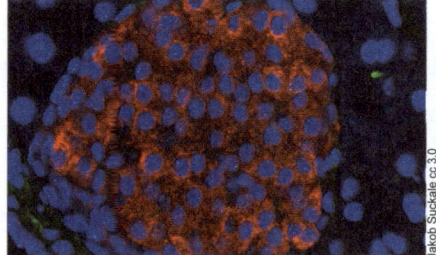
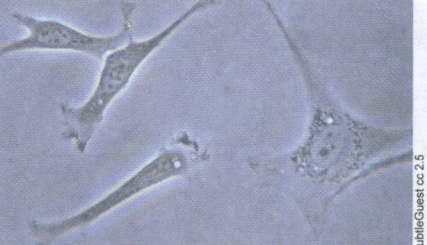
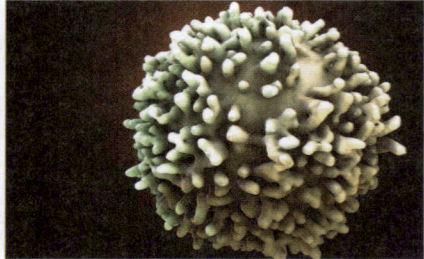

Cells within specialized regions of the pancreas produce and release a protein hormone called insulin. Insulin (red in photo) helps regulate blood **glucose**.

Fibroblasts are specialized cells that continuously produce and secrete the materials that form connective tissue. This includes a protein called collagen.

B lymphocytes (B cells) are white blood cells that are specialized to produce and secrete proteins called antibodies. These protect the body against diseases.

1. Suggest what might happen to a protein's functionality if it was incorrectly encoded by the DNA. Explain your answer:

   _____
   _____
   _____

2. The following page shows six models of proteins in action, six protein functions, six protein examples, and six photographs. These are not in any matched order. Cut out the 24 boxes and paste or tape them into the grid on the next page so that each pictogram is matched with its correct function, example, and illustrative photograph.

MODELS	FUNCTION	EXAMPLE	PHOTO EXAMPLE
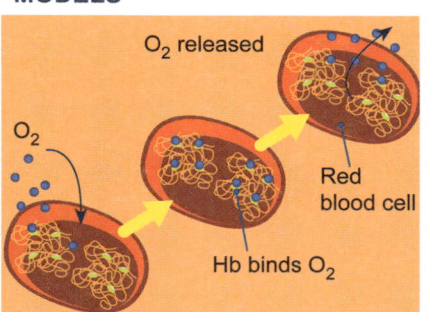	**Internal defense** Antibodies (also called immunoglobulins) are "Y" shaped proteins that protect the body by identifying and killing disease-causing organisms such as bacteria and **viruses**.	**Immunoglobulin A** IgA is found in the gut and airways. It destroys disease-causing organisms growing in these areas.	
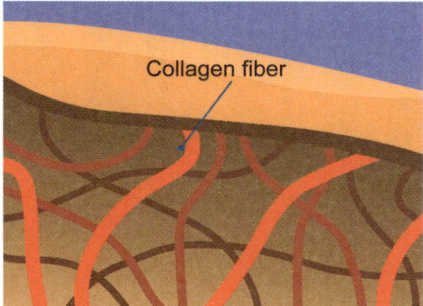	**Catalytic** Thousands of different chemical reactions take place in an organism every minute. Each chemical reaction is catalyzed by enzymes. The word-ending "ase" indicates an enzyme.	**Actin & myosin** Two proteins that work together to bring about contraction (movement) in all the muscles of the body, including those that work without your awareness.	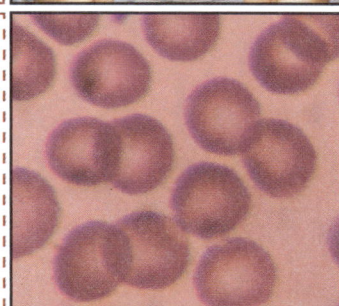
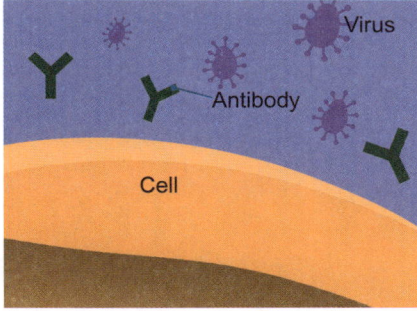	**Regulation** Regulatory proteins such as hormones act as signal molecules to control biological processes and coordinate responses in cells, tissues, and organs.	**Hemoglobin** A protein found in red blood cells. It binds oxygen and carries it through the blood, delivering it to cells.	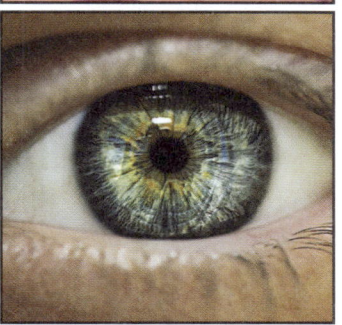
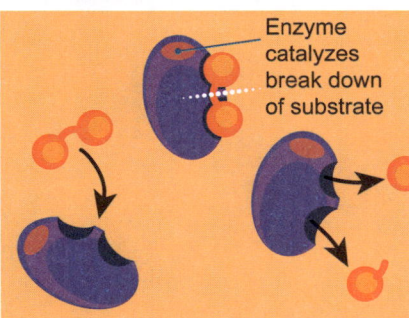	**Movement** Contractile proteins are involved in movement of muscles and form the internal, supporting structures of cells.	**Collagen** Found in the skin and connective tissues, including bones, tendons, and ligaments. It is the most abundant protein in the body.	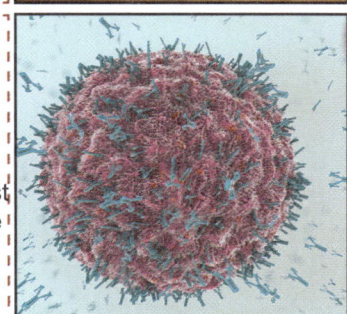
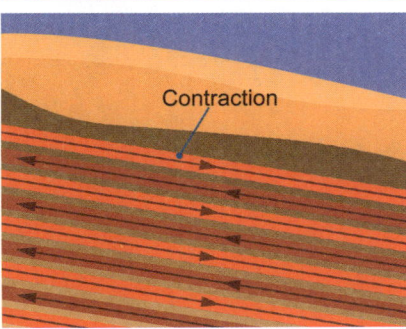	**Transport** Proteins can carry substances around the body or across membranes. In the blood, they transport and store oxygen. In cell membranes, they help molecules move into and out of cells.	**Estrogen** A hormone that is critical for reproduction in females. Estrogen levels increase during pregnancy to maintain a healthy pregnancy.	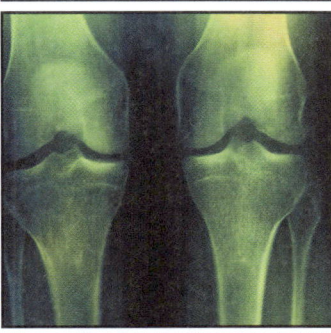
	Structural Structural proteins provide physical support or protection. They are strong, fibrous (thread like) and stringy.	**Amylase** An enzyme that breaks down starch into sugars in the first stage of digestion.	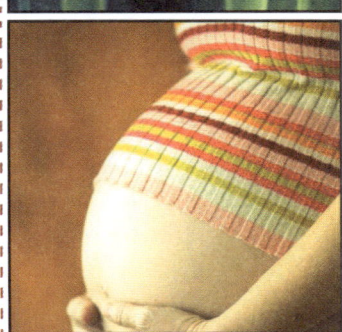

This page is deliberately left blank

PICTOGRAM	FUNCTION	EXAMPLE	PHOTO EXAMPLE

8 Lipids in Cells

Key Question: What features characterize lipid molecules, how are they formed, and what are their biological roles in cells?

- **Lipids** are organic compounds that are mostly nonpolar (have no overall charge) and hydrophobic (repel water), so they do not dissolve in water.
- Simple lipids (fats and waxes) are distinct from complex lipids, such as the fatty acids in **phospholipids** which form the **cell** membranes and fat-soluble cell components like steroids. Steroids are a type of hormone made by the body that is involved in the control of growth, reproduction, metabolism of energy in the cells, and homeostasis.
- Most fatty acids consist of an even number of carbon atoms, with hydrogen bound along the length of the chain.
- All organisms, except archaea, have fatty acids as a main component of their cellular membranes.

The biological roles of lipids

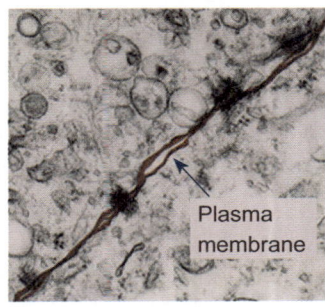   

Phospholipids form the basic structure of cellular membranes that surround each cell. They control the movement of substances in and out.

Waxes and oils secreted on to surfaces provide waterproofing in plants and animals. They prevent water loss from cells on the surface.

Fat provides large amounts of energy and metabolic water. In some desert dwelling animals (e.g. kangaroo rat) this provides all the water they need.

As well as storing energy, fat stores provide insulation. This reduces heat losses to the environment, to maintain thermal homeostasis.

1. (a) Lipids contain more energy per gram than glucose, yet their energy release is slower. Glucose is a simple molecule that can travel quickly to cells across the cellular membrane, while lipids can be quite large and need to be broken down before providing energy. Why are lipids more suitable as a cellular energy storage biomolecule?

    _____
    _____
    _____

    (b) Link the hydrophobic property of lipids to their suitability as part of a cell membrane structure: _____
    _____
    _____

2. Triglycerides (shown to the right) are broken at the ester bond by both acids and alkaline solutions to form three fatty acids. Why do fatty acids have the potential to provide a large amount of energy to cells (think about the molecular structure of fatty acids):

    _____
    _____
    _____
    _____
    _____
    _____

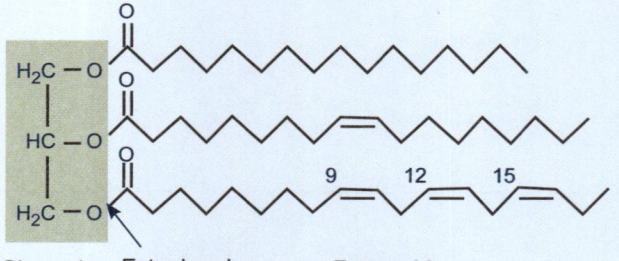

Triglycerides (triacylglycerols)

Glycerol — Ester bond — Fatty acids

Triglyceride: an example of a neutral fat

Neutral fats and oils are the most abundant lipids in living things. They make up the fats and oils found in plants and animals. Neutral fats and oils consist of a glycerol attached to one (mono-), two (di-) or three (tri-) fatty acids by **ester bonds**.

9 The Development of Microscopes

Key Question: When did humans discover cells, and what role did the microscope play?

Van Leeuwenhoek and the 'animalcules'

- Scientists have always been interested in trying to visualize smaller and smaller objects.
- In 1665, the English scientist Robert Hooke, built a compound microscope that allowed him to see the cellular structure of a bottle cork, originally the wood from the cork oak tree. He was the first to name the small repeating units he saw as '**cells**', as they resembled the cramped rooms of monks living together in a monastery.
- Ten years later, Dutch scientist Anton van Leeuwenhoek was able to view and describe the cellular structure of living organisms, such as single-celled protists. He developed a single lens microscope (right) that was quite different from Hooke's compound microscope. The design enabled him to look through liquids.
- In 1678, van Leeuwenhoek used his microscope to see bacteria for the first time. He termed all these tiny creatures 'animalcules' and was the only scientist at the time who was able to see bacterial cells and unicellular organisms at such high magnification.
- The work of Hooke and van Leeuwenhoek inspired other scientists to try and understand what these tiny structures actually were, and what implications there were for all living organisms.

Anton van Leeuwenhoek (1632-1723) and a version of his microscope (insert). It had a magnification of 270 x.

Examples of microscopic protists

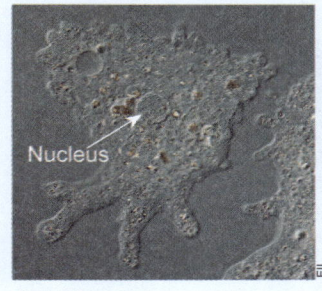
Amoeba

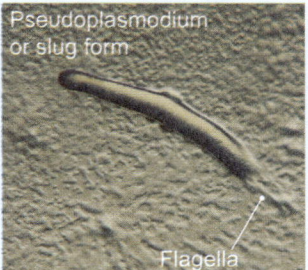
Paramecium

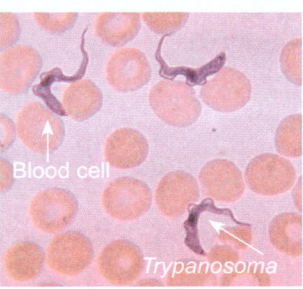
Dictyostelium slug (slime mold)   *Trypanosoma* in blood

1. What was the reason that scientists did not know about protists or cells prior to the discoveries of Hooke and van Leeuwenhoek?

   _____

   _____

   _____

2. Van Leeuwenhoek made over 500 different versions of his microscope and observed protists, blood cells, sperm, nematodes (parasitic worms), and rotifers (microscopic animals) over many years. What characteristics of a good scientist did van Leeuwenhoek demonstrate?

   _____

   _____

   _____

   _____

10 Microscopes and Magnification

Key Question: What are the important features of a light microscope, and how do you calculate the magnification of the image they produce?

- The light (or optical) microscope (LM) is an important tool in biology. Light microscopy involves illuminating a sample and passing the light that is transmitted or reflected through lenses to give a magnified view of the sample.

- High power, compound light microscopes use visible light and a combination of lenses to magnify objects up to several hundred times. Bright field microscopy (below) is the simplest, and involves illuminating the specimen from below and viewing from above. Specimens must be thin and mostly transparent so that light can pass through. No detail will be seen in specimens that are thick or opaque.

- The wavelength of light limits the resolution of light microscopes to around 0.2 µm (µm = micrometer, 1 millionth of a meter). Objects closer than this will not be distinguished as separate. Electron microscopes provide higher resolutions as they use a shorter wavelength electron beam, rather than light.

Structure of a typical compound light microscope

Word list: *In-built light source, arm, coarse focus knob, fine focus knob, condenser, mechanical stage, eyepiece lens, objective lens*

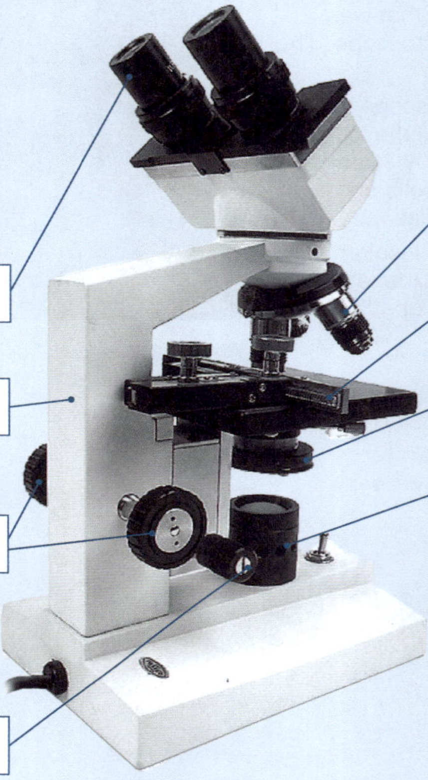

(a) ____
(b) ____
(c) ____
(d) ____
(e) ____
(f) ____
(g) ____
(h) ____

A stoma (pore) in the leaf epidermis is visible with a light microscope

What is magnification?
Magnification refers to the number of times larger an object appears compared to its actual size.

Magnification is calculated as follows:

Objective lens power **X** Eyepiece lens power

What is resolution?
Resolution is the ability to distinguish between close together but separate objects. Examples of high and low resolution for separating two objects viewed under the same magnification are shown on the right. Factors such as dirty lenses can reduce resolution.

High resolution

Low resolution

1. Label the image above of the compound light microscope (a) to (h). Use words from the list supplied in image.

2. Determine the magnification of a microscope using:

   (a) 15 X eyepiece and 40 X objective lens: ____

   (b) 10 X eyepiece and 60 X objective lens: ____

How do we calculate linear magnification?

▸ Magnification is how much larger an object appears compared to its actual size. It can be calculated from the ratio of image height to object height.

▸ If the ratio is greater than one, the image is enlarged. If it is less than one, it is reduced. In calculating magnification, all measurements should be in the same units.

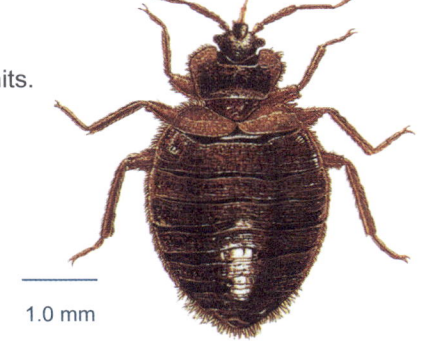

1.0 mm

Worked example

1. Measure the body length of the bed bug image (right). Your measurement should be 40 mm (not including the body hairs and antennae).

2. Measure the length of the scale line marked 1.0 mm. You will find it is 10 mm long. The magnification of the scale line can be calculated using equation 1.

   The magnification of the scale line is **10** (10 mm ÷ 1 mm)
   The magnification of the image will also be x10 because the scale and image are magnified to the same degree.

3. Calculate the actual size of the bed bug using equation 2. The actual size of the bed bug is **4 mm** (40 mm ÷ 10)

Microscopy equations

1. Magnification = $\dfrac{\text{measured size of object}}{\text{actual size of object}}$

2. Actual object size = $\dfrac{\text{size of the image}}{\text{magnification}}$

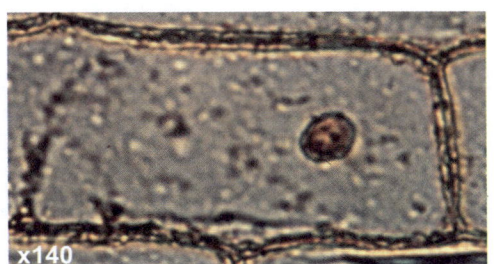

x140

3. The bright field microscopy image (left) shows an onion epidermal cell. Its measured length is 52,000 µm (52 mm). The image has been magnified x140. Calculate the actual size of the cell:

   _____
   _____
   _____

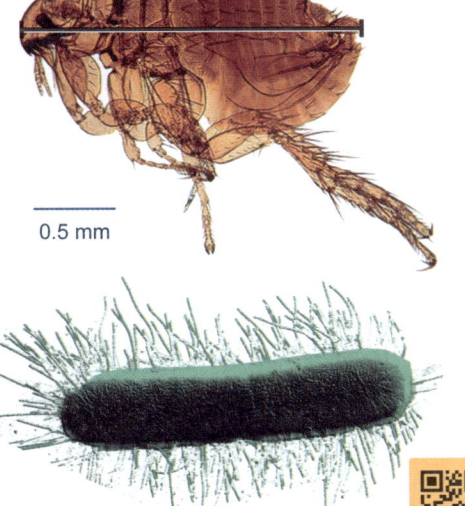

0.5 mm

4. The image of the flea (left) has been captured using light microscopy.

   (a) Calculate the magnification using the scale line on the image:

   _____
   _____
   _____

   (b) The body length of the flea is indicated by a line. Measure along the line and calculate the actual length of the flea:

   _____
   _____

5. The image size of the *E.coli* cell (left) is 43 mm, and its actual size is 2 µm. Using this information, calculate the magnification of the image:

   _____
   _____

6. Explain why a higher magnification is not particularly useful if the resolution is poor: _____
   _____
   _____

7. When focusing a specimen, it is necessary to focus on the lowest magnification first, before switching to higher magnifications. Why do you think this is important?
   _____
   _____

11 Studying Cells

Key Question: What techniques are used to prepare and view cells under a light microscope?

▸ Specimens are usually prepared in some way before viewing to highlight features and reveal details. A wet mount is a temporary preparation in which a specimen and a drop of fluid are trapped under a thin coverslip. Wet mounts improve a sample's appearance and enhance visible detail. Sections must be very thin for light to pass through.

Investigation 1.2   Preparing an onion slide

See appendix for equipment list.

> ⚠ Caution is required when using scalpels or razors. Iodine stains skin and clothes, and irritates the eyes. You should wear protective eyewear.

1. Onions make good subjects for preparing a simple, wet mount. Cut a square segment from a thick leaf of the bulb using a razor or scalpel.
2. Bend the segment towards the upper epidermis (upper cell layer) until the lower epidermis and inner leaf tissue (the parenchyma) snaps, leaving the upper epidermis attached.
3. Carefully peel off the parenchyma from one side of the snapped leaf and then the other, leaving a peel of just the upper epidermis.
4. Place the peel in the center of a clean glass microscope slide and cover it with a drop of water.
5. Carefully lower a coverslip over the peel. A mounted needle can be used for better precision. This avoids including air in the mount.
6. Use a small piece of tissue or filter paper to remove any excess water.
7. Place the slide on the microscope tray. Locate the specimen or region of interest at the lowest magnification. Focus using the lowest magnification first (remembering to move the lens away from the slide) before switching to the higher magnifications.
8. After viewing the slide under various magnifications, remove the slide and place it on the bench.
9. At the edge of the coverslip, place a small drop of iodine stain.
10. On the opposite side of the coverslip use a piece of tissue or filter paper to draw the water out from under the coverslip. The iodine will be drawn under the coverslip.
11. Replace the slide on the microscope, view the stained onion peel, and scientifically draw the cells you observe.

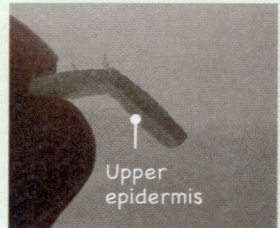
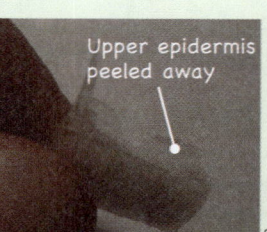
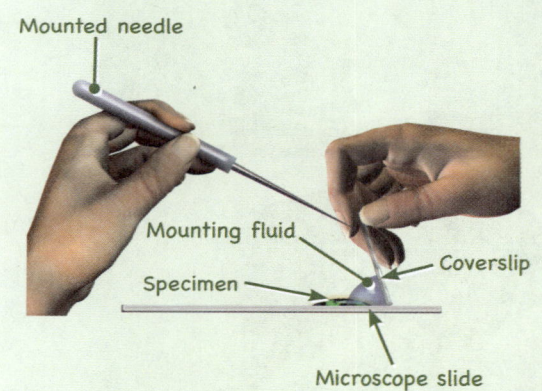
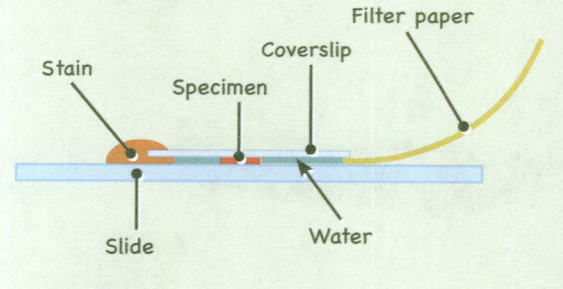

1. Why must sections viewed under a microscope be very thin? _____
   _____
   _____

2. Why do you think the specimen is covered with a coverslip? _____
   _____
   _____

Stains and their uses

- Staining material for viewing under a microscope can make it easier to distinguish particular **cell** structures.
- Stains and dyes can be used to highlight specific components or structures. Stains contain chemicals that interact with molecules in the cell. Some stains bind to a particular molecule, making it easier to see where those molecules are. Others cause a change in a target molecule, which changes the color, making them more visible.
- Most stains are non-viable, meaning they are used on dead specimens. Harmless, viable stains can be applied to living material.

Some commonly used stains

Stain	Final color	Used for
Iodine solution	Blue-black	Starch
Crystal violet	Purple	Gram staining
Aniline sulfate	Yellow	Lignin
Methylene blue	Blue	Nuclei
Hematoxylin and eosin (H&E)	H=dark blue/violet E=red/pink	H=Nuclei E=**Proteins**

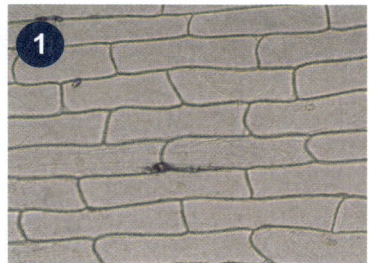

The light micrographs 1 and 2 (above) show how the use of a stain can enhance certain structures. The left image (1) is unstained and only the **cell wall** is easily visible. Adding iodine (2) makes the cell wall and nuclei stand out.

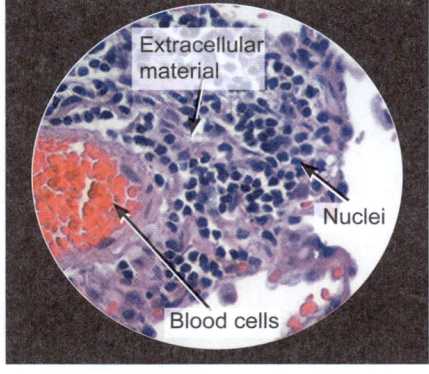
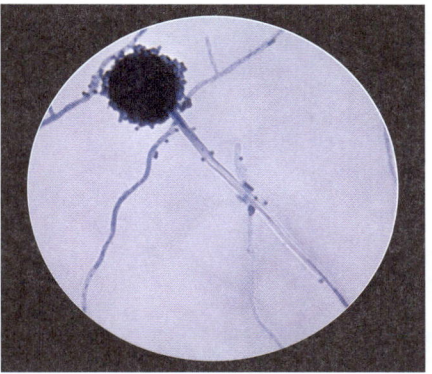
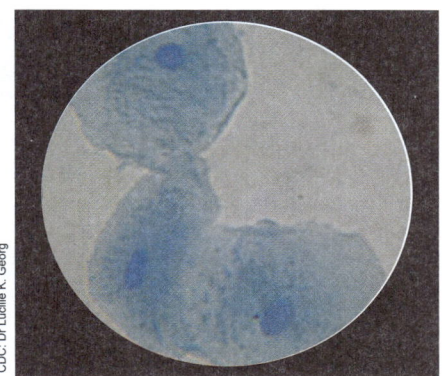

H&E stain is a common stain for animal tissues. Nuclei stain dark blue, whereas proteins, extracellular material, and red blood cells stain pink or red.

Viable stains do not immediately harm living cells. Trypan blue is a vital stain that stains dead cells blue but is excluded by live cells. It is also used to study fungi.

Methylene blue is a common temporary stain for animal cells, such as these cheek cells. It stains DNA, making the nuclei more visible.

3. Suggest a stain that could be used to show up nuclei in cells being viewed under a microscope?
   _____
   _____

4. Describe the difference the iodine stain made when viewing the onion cells under the microscope, compared to when they were viewed without the stain:
   _____
   _____
   _____

5. What is the main purpose of using a stain? _____
   _____

6. What is the difference between a viable and non-viable stain? _____
   _____
   _____

7. Identify a stain that would be appropriate for distinguishing each of the following:

   (a) Live vs dead cells: _____ (c) Lignin in a plant root section: _____

   (b) Red blood cells in a tissue preparation: _____ (d) Nuclei in cheek cells: _____

12 Life Arises from Life

Key Question: How have the experiments of Louis Pasteur and Robert Koch contributed to our understanding of microbiology, the study of microscopic life?

▶ For a long time people thought that new life could spontaneously evolve from non-living matter (spontaneous generation) and that clouds of poisonous gas caused disease (Miasma Theory).

▶ The work of two scientists, Frenchman, Louis Pasteur and German, Robert Koch, methodically disproved both of these theories. Their work significantly contributed to our knowledge of how all **cells** originate from other living cells.

Louis Pasteur

Robert Koch

Pasteur's swan neck flask experiment

▶ Prior to the observation of cells dividing, by German scientist Rudolf Virchow, scientists did not have a clear understanding of how new organisms arose. For example, when a piece of meat was left on a bench, maggots appeared a day or two later. Under spontaneous generation theory, the maggots had spontaneously generated from the components of the meat. Louis Pasteur disproved this theory when he carried out his very simple swan neck flask experiments (below).

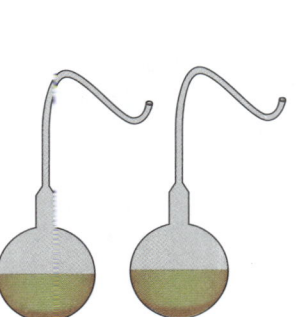

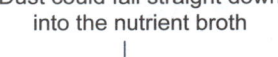
Dust could fall straight down into the nutrient broth

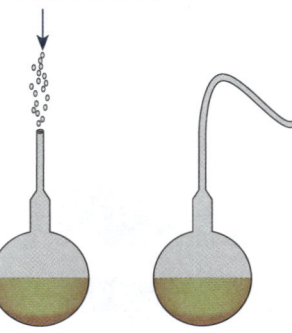

Microbes growing in broth.

Microbes trapped in the neck of the flask could not reach the broth.

1 Louis Pasteur filled two swan necked flasks with a nutrient broth and then boiled them to kill any microbes present.

2 He broke off the neck of one of the flasks, allowing air and dust (on which microbes are carried) to fall straight down onto the broth. The neck of the other flask prevented dust falling onto the broth.

3 The broth in the broken flask eventually turned dark, indicating microbial growth. The broth in the unbroken flask remained unchanged. Pasteur concluded that spontaneous generation could not have occurred or both flasks would have turned dark.

1. A student set up an experiment to replicate Pasteur's experiment. Nutrient broth was added to two test tubes and the tubes were sealed. Both tubes were heated for several minutes over a Bunsen burner. After cooling, tube 1 was uncovered, and tube 2 was left covered. The students observations are described in the table (right).

Test tube	Observation of the broth at day 1	Observation of the broth at day 10
1 uncovered	Clear	Cloudy
2 covered	Clear	Cloudy

(a) Did the student collect qualitative or quantitative data? _____

(b) Explain your reason for your answer in 2(a): _____

2. Suggest a reason why the student's results were different from Pasteur's results: _____

The work of Robert Koch

▶ Miasma theory proposed that noxious clouds of gas (also called bad air or night air) contained diseases such as cholera or the black plague, and that people became sick by breathing in the gas. The theory was popular and commonly believed for many centuries until Robert Koch's work in the 1800s demonstrated that microorganisms, such as bacteria, caused disease. This finding became known as 'germ theory'.

▶ Robert Koch's work with the anthrax bacterium allowed him to propose four basic criteria, known as Koch's postulates, for demonstrating that a disease is caused by a particular organism. These are described below.

Koch's postulates

Pathogenic microorganisms are isolated from a dead animal.

The microorganisms are injected into a healthy animal.

The disease is reproduced in the second animal. Microorganisms are isolated.

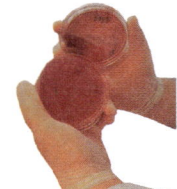
Isolated pathogenic microorganisms are identical to original pathogens.

Koch isolated bacteria from a diseased animal, then injected them into a healthy animal, causing it to exhibit identical symptoms to the first. This demonstrated that a specific infectious disease (e.g. anthrax) was caused by a specific microorganism (*Bacillus anthracis*). Koch used the procedure to identify the bacteria that caused anthrax and tuberculosis.

Koch's findings are summarized as Koch's postulates:

1. The same pathogen must be present in every case of the disease.
2. The pathogen must be isolated from the diseased host and grown in pure culture.
3. The pathogen from the pure culture must then cause the disease when it is inoculated into a healthy, susceptible animal.
4. The pathogen must be isolated from the inoculated animal and shown to be the original organism.

Postulate: to suggest a theory or idea as a basic principle from which a further idea is developed.

3. Briefly outline how Louis Pasteur's discovery that heating kills microbes can be used in food manufacturing, such as milk pasteurization, to make food safer to consume:

_____

4. Carry out your own research and critique how the work of Louis Pasteur and Robert Koch has contributed to our understanding of microbiology and infectious disease:

_____

13 The Cell is the Unit of Life

Key Question: How do we classify cell types, and what do they all need for survival?

- All living organisms are composed of **cells**. Cells are broadly classified as **prokaryotic** or **eukaryotic**.
- All cells have a cell membrane, cytoplasm, and contain genetic material, either in a **nucleus** or within the cytoplasm itself.
- **Viruses** are not considered living as they do not consist of a cell, and therefore cannot carry out all of the life processes. Instead, they require other living cells to perform these processes for them. This includes reproduction.

All cells show the functions of life

Cells use food (e.g. **glucose**) to maintain a stable internal environment, grow, reproduce, and produce wastes. The sum total of all the chemical reactions that sustain life is called metabolism.

- Movement
- Respiration
- Sensitivity
- Growth
- Reproduction
- Excretion
- Nutrition

Living things → **Cells**

Prokaryotic (bacterial) cells

- Autotrophic or heterotrophic
- Single celled
- Lack a membrane-bound nucleus and membrane-bound **organelles**
- Cells 0.5-10 µm
- DNA is a single, circular chromosome. There may be small accessory chromosomes called plasmids.
- **Cell walls** containing peptidoglycan, a polysaccharide.

Viruses are non-cellular

- Typical size range: 20-300 nm.
- Contain no cytoplasm or organelles.
- No chromosome, just RNA or DNA strands.
- Enclosed in a **protein** coat.
- Depend on the cellular machinery of the host cell for metabolism and reproduction.

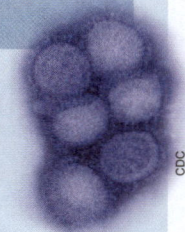

Influenzavirus

Eukaryotic cells
- Cells 30-150 µm • Membrane-bound nucleus and membrane-bound organelles • Linear chromosomes

Plant cells
- Exist as part of a multicellular organism with specialization of cells into many types.
- Autotrophic (make their own food): photosynthetic cells with chloroplasts.
- Cell walls of cellulose.

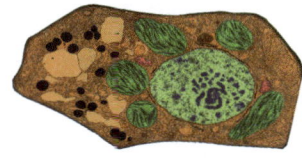

Generalized plant cell

Animal cells
- Exist as part of a multicellular organism with specialization of cells into many types.
- Heterotrophic (obtain nutrition from external sources.
- Lack cell walls.

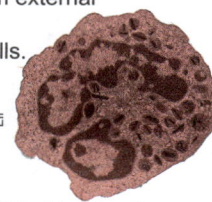

White blood cell

Protist cells
- Mainly single-celled or exist as cell colonies.
- Some make food by photosynthesis (autotrophic).
- Some are heterotrophic.

Amoeba cell

Fungal cells
- Rarely exist as discrete cells, except for some unicellular forms, e.g. yeasts.
- Plant-like, but lack chlorophyll.
- Rigid cell walls containing chitin.
- Heterotrophic.

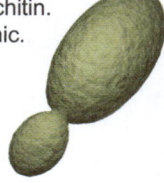

Yeast cell

1. What are the characteristic features of a prokaryotic cell? _____

2. What are the characteristic features of a eukaryotic cell? _____

3. Why are viruses considered to be non-cellular (non-living)? _____

What cells need for survival

▸ All cells have specific requirements for survival. These include obtaining nutrients and removing wastes. Cells require a range of molecules and ions to build and maintain cell structures and energy to power the reactions for building and maintenance. They also need to remove wastes created from these reactions.

Cells need energy
Cells have evolved to use two basic forms of energy: light or chemical energy.

▸ Organisms called producers use light energy from the Sun to power the chemical reactions that build organic molecules. These organic molecules can then be used to power other reactions in the cell or build macromolecules.

▸ Organisms called consumers use chemical energy to power cellular reactions. In plants and animals, glucose is used in the process of cellular respiration to produce ATP which powers most cellular reactions.

Cells require resources
Cells require molecules and ions to build macromolecules and help carry out cellular reactions.

▸ Carbon dioxide is needed by plants to build organic molecules during photosynthesis.

▸ Oxygen is needed by plants and animals as an electron acceptor at the end of cellular respiration (see ch3).

▸ In plants, nitrates provide nitrogen, which is incorporated into **amino acid** molecules. Animals use these (by eating plants or plant eaters) to obtain building blocks for their proteins.

▸ Various metal ions are also needed. Some are needed in relatively large amounts, e.g. Na^+ for nerve cell function in animals, while others are needed only in very small amounts.

Cells need to remove wastes
Cells need to remove wastes generated during cellular reactions. What is regarded as a waste depends on the type of cell.

▸ Oxygen is a waste product of photosynthesis, but is required for cellular respiration. Other waste products include nitrogen wastes such as urea, ammonia, and uric acid (from metabolic processes).

▸ Most cellular reactions generate heat, which must be managed so that an organism does not overheat. In animals, metabolic heat is removed from cells by the blood and transferred to places where it can radiate into the environment, e.g. the skin.

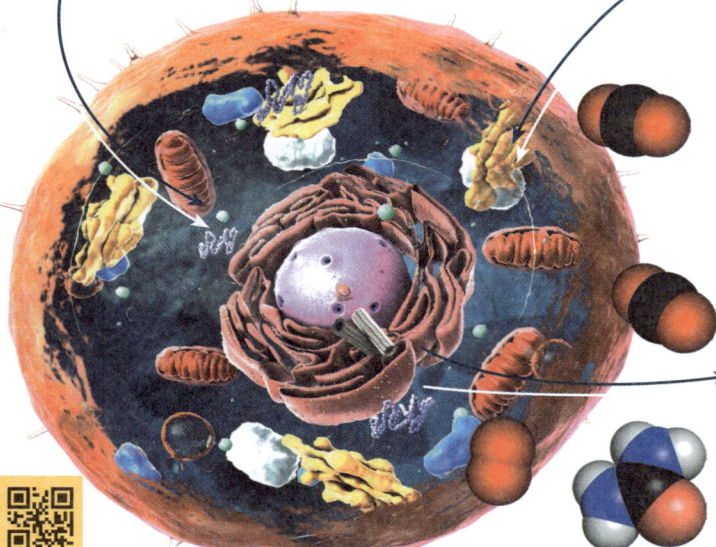

Cellular environments
The exact conditions needed by a cell depends on many factors, including whether the organism is unicellular or multicellular, and what environment it has evolved to survive in.

Thermophiles in hot thermal pools.

Some unicellular organisms (called thermophiles) can survive in temperatures as high as 122°C. Their enzymes can not function at the lower temperatures outside hot environments.

Halophiles require environments with high salt concentrations (up to five times as concentrated as the sea). These cells are specially adapted to retain water. If placed in fresh water, they quickly swell and burst.

Cells in multicellular organisms require the homeostatic environment provided by the organism. This provides the cells with nutrients, waste removal, and a relatively constant temperature.

4. Why do cells need energy? _____

_____

5. Why must cells be able to remove wastes? _____

_____

6. Describe an example of where waste products of one cellular process can be used as a resource for another:

_____

_____

14 Distinguishing Features of Prokaryotic Cells

Key Question: What are the distinguishing features of prokaryotic cells?

▶ Bacterial (**prokaryotic**) **cells** are relatively small, and do not contain a distinct **nucleus** and membrane-bound cellular **organelles**. The **cell wall** is an important feature. It is a complex, multi-layered structure.

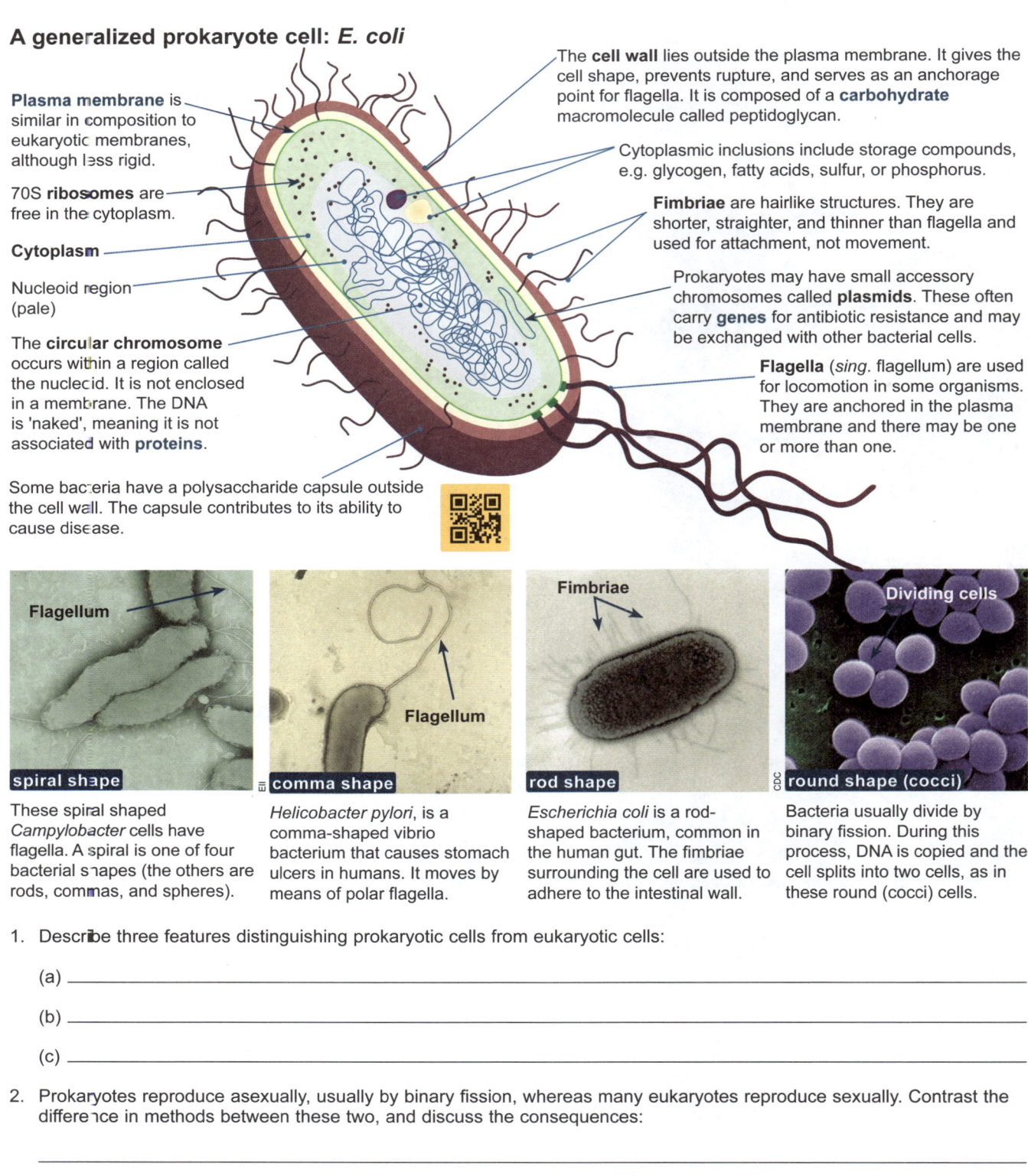

A generalized prokaryote cell: *E. coli*

Plasma membrane is similar in composition to eukaryotic membranes, although less rigid.

70S **ribosomes** are free in the cytoplasm.

Cytoplasm

Nucleoid region (pale)

The **circular chromosome** occurs within a region called the nucleoid. It is not enclosed in a membrane. The DNA is 'naked', meaning it is not associated with **proteins**.

Some bacteria have a polysaccharide capsule outside the cell wall. The capsule contributes to its ability to cause disease.

The **cell wall** lies outside the plasma membrane. It gives the cell shape, prevents rupture, and serves as an anchorage point for flagella. It is composed of a **carbohydrate** macromolecule called peptidoglycan.

Cytoplasmic inclusions include storage compounds, e.g. glycogen, fatty acids, sulfur, or phosphorus.

Fimbriae are hairlike structures. They are shorter, straighter, and thinner than flagella and used for attachment, not movement.

Prokaryotes may have small accessory chromosomes called **plasmids**. These often carry **genes** for antibiotic resistance and may be exchanged with other bacterial cells.

Flagella (*sing.* flagellum) are used for locomotion in some organisms. They are anchored in the plasma membrane and there may be one or more than one.

spiral shape — These spiral shaped *Campylobacter* cells have flagella. A spiral is one of four bacterial shapes (the others are rods, commas, and spheres).

comma shape — *Helicobacter pylori*, is a comma-shaped vibrio bacterium that causes stomach ulcers in humans. It moves by means of polar flagella.

rod shape — *Escherichia coli* is a rod-shaped bacterium, common in the human gut. The fimbriae surrounding the cell are used to adhere to the intestinal wall.

round shape (cocci) — Bacteria usually divide by binary fission. During this process, DNA is copied and the cell splits into two cells, as in these round (cocci) cells.

1. Describe three features distinguishing prokaryotic cells from eukaryotic cells:

    (a) _____

    (b) _____

    (c) _____

2. Prokaryotes reproduce asexually, usually by binary fission, whereas many eukaryotes reproduce sexually. Contrast the difference in methods between these two, and discuss the consequences:

    _____
    _____
    _____
    _____

15 Distinguishing Features of Eukaryotic Cells

Key Question: What are the distinguishing features of eukaryotic cells?

▸ Plants, fungi, protists, and animals are all **eukaryotes**. Their **cells** are more complex than those of **prokaryotes**, and contain a **nucleus** and membrane-bound **organelles**, such as mitochondria. Autotrophs (organisms that produce **glucose** using photosynthesis), also have chloroplasts.

A generalized eukaryote cell: animal

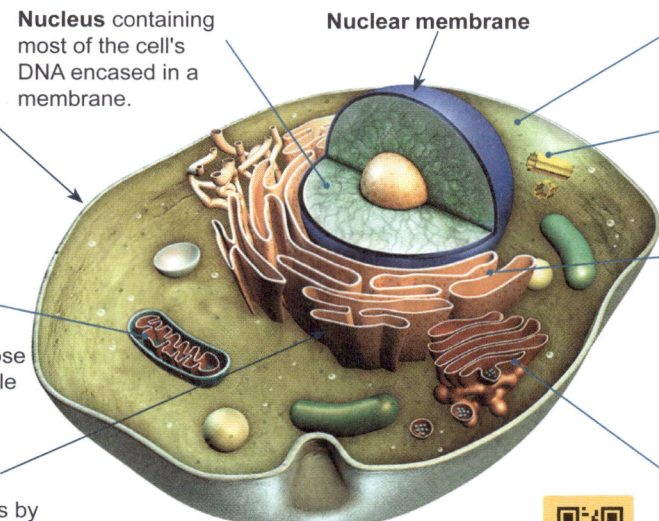

Plasma membrane (located inside the cell wall in plants) controls the movement of materials into and out of the cell.

Nucleus containing most of the cell's DNA encased in a membrane.

Nuclear membrane

Cytoplasm: a watery solution containing dissolved materials, enzymes, and the cell organelles.

Centrioles organize the structural microtubule endoskeleton and assist in cell division.

Endoplasmic reticulum (ER): A network of tubes and flattened sacs continuous with the nuclear membrane. There are two types of ER. Rough ER has ribosomes attached. Smooth ER has no ribosomes (so it appears smooth).

Mitochondria are the cell's energy producers. They use the chemical energy in glucose to make ATP (the cell's usable energy).

Ribosomes (80S) are small structures that make proteins by joining amino acids. They are often found on the surface of the endoplasmic reticulum

The **Golgi apparatus** is a structure made up of membranous sacs. It stores, modifies, and packages proteins.

Note: A **cell wall** is present in fungi, plants, and some protists. Most plants, fungi, and protist cells do not have centrioles, or they are modified for different functions. Plant cells, and autotroph protists also have chloroplasts. Plant cells often have large vacuoles.

Plant cells

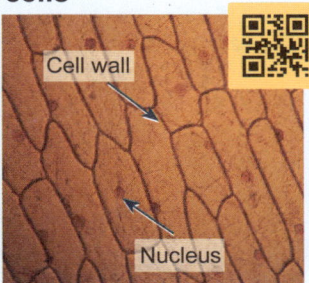

Palisade mesophyll cells — Cell wall made of cellulose, Chloroplasts

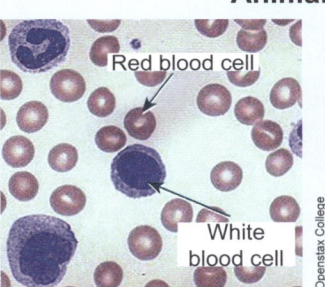

Epidermal (leaf) cell — Cell wall, Nucleus

Animal cells

Blood cells — Red blood cell, White blood cell

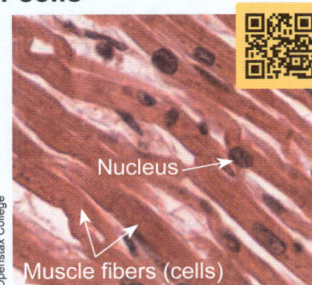

Muscle cells — Nucleus, Muscle fibers (cells)

1. Describe three features common to all eukaryotic cells (animal, plant, fungi, and protists):

   (a) _____

   (b) _____

   (c) _____

2. Prokaryotes are much less complex in cellular structure than eukaryotes. What does this tell you about the common ancestry of both groups of organisms (this will be explored in more detail in activity 19)?

   _____

   _____

   _____

   _____

16 Prokaryotic vs Eukaryotic Cells

Key Question: How can we compare and contrast prokaryotic and eukaryotic cells?

Prokaryotic cells

- **Prokaryotic cells** are small (~0.5-10 μm) single cells.
- They lack any membrane-bound **organelles**.
- They are relatively basic cells and have very little cellular organization. Their DNA, ribosomes, and enzymes are free floating within the cell cytoplasm.
- The ribosomes (70S) are smaller than eukaryotic ribosomes.
- Photosynthetic bacteria have enzymes and light capturing membranes.
- Single circular chromosome of naked DNA.

Eukaryotic cells

- Eukaryotic cells are large (30-150 μm). They may exist as single cells or as part of a multicellular organism.
- **Eukaryotes** have membrane-bound organelles, e.g. mitochondria and (photosynthetic organisms) chloroplasts.
- Eukaryotic cells are complex and have a membrane-bound **nucleus**.
- Ribosomes (80S) are larger than in prokaryotes, except those in mitochondria and chloroplasts, which are 70S.
- Photosynthesis occurs only in chloroplast organelles.
- Multiple linear chromosomes consisting of DNA and associated **proteins**.

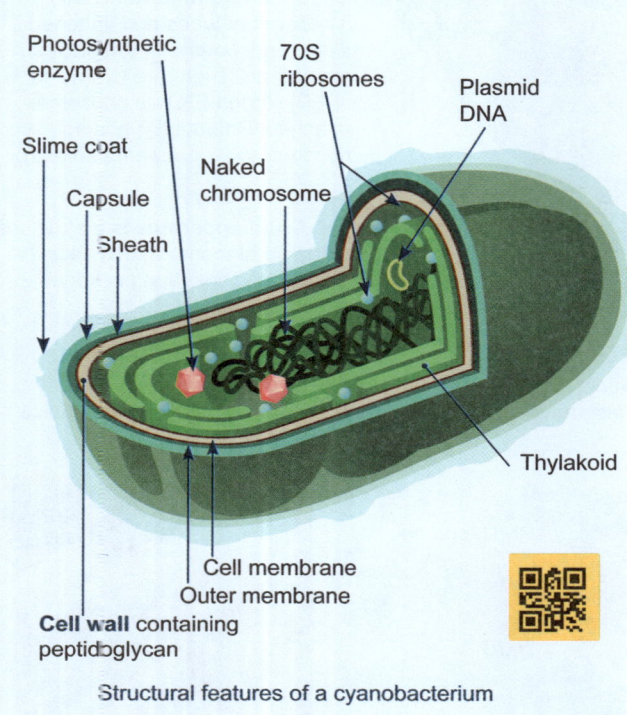

Structural features of a cyanobacterium

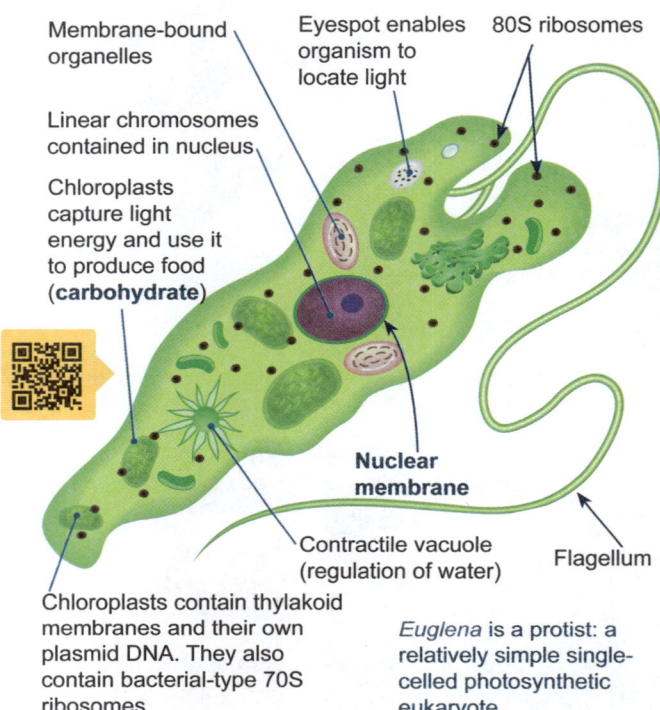

Chloroplasts contain thylakoid membranes and their own plasmid DNA. They also contain bacterial-type 70S ribosomes.

Euglena is a protist: a relatively simple single-celled photosynthetic eukaryote.

1. The cells of prokaryotes and eukaryotes are diverse. Identify the features that distinguish them: _____
   _____
   _____

2. What is interesting about the ribosomes of prokaryotes and the ribosomes found in eukaryote chloroplasts?
   _____

3. Cyanobacterial cells (above left) and *Euglena* (above right) are both photosynthetic organisms. Use information from the models to explain how the cellular structures enable them to capture light energy:
   _____
   _____
   _____

17 Comparing Cell and Virus Sizes

Key Question: How do different types of cells vary in size, and how do they compare to the size of a typical virus?

▸ Different types of **cells** are different sizes. **Eukaryotic cells** are much larger than **prokaryotic cells**, but even they vary widely in size. Cells also have different shapes. Many have no fixed shape, but others have shapes approximating spheres, e.g. *Streptococcus*, cylinders, e.g. *E. coli*, or rectangular prisms, e.g. plant cells. **Viruses** are much smaller than the smallest prokaryote cell.

Typical sizes of cells and viruses

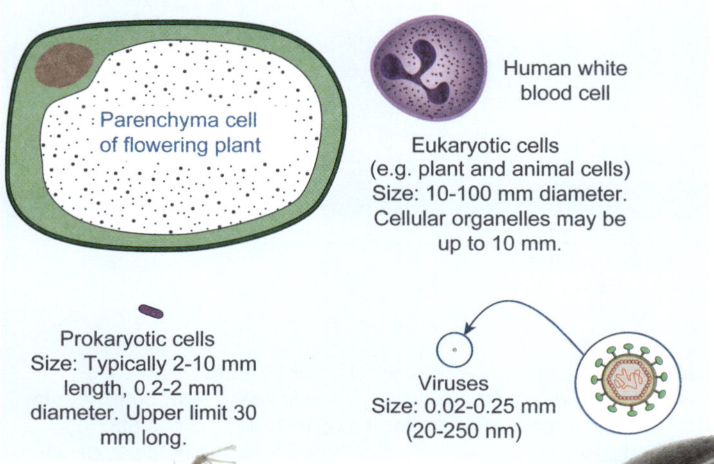

Parenchyma cell of flowering plant

Human white blood cell

Eukaryotic cells (e.g. plant and animal cells) Size: 10-100 mm diameter. Cellular organelles may be up to 10 mm.

Unit of length (international system)		
Unit	**Meters**	**Equivalent**
1 meter (m)	1 m	= 1000 millimeters
1 millimeter (mm)	10^{-3} m	= 1000 micrometers
1 micrometer (µm)	10^{-6} m	= 1000 nanometers
1 nanometer (nm)	10^{-9} m	= 1000 picometers

Prokaryotic cells Size: Typically 2-10 mm length, 0.2-2 mm diameter. Upper limit 30 mm long.

Viruses Size: 0.02-0.25 mm (20-250 nm)

Micrometers are sometimes referred to as microns. Smaller structures are usually measured in nanometers (nm) e.g. molecules (1 nm) and **plasma membrane** thickness (10 nm).

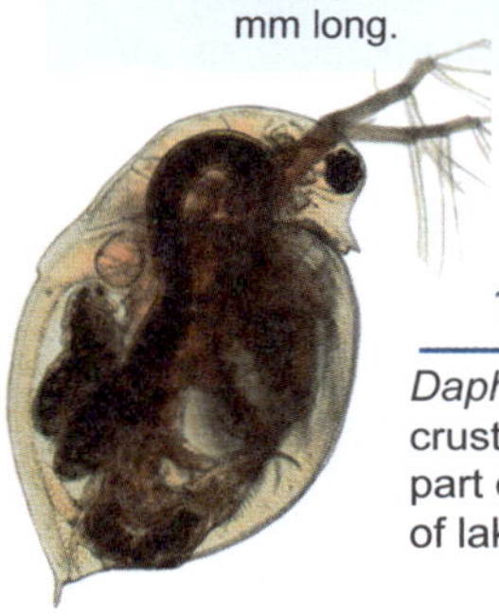

Daphnia is a small crustacean found as part of the zooplankton of lakes and ponds. — 1.0 mm

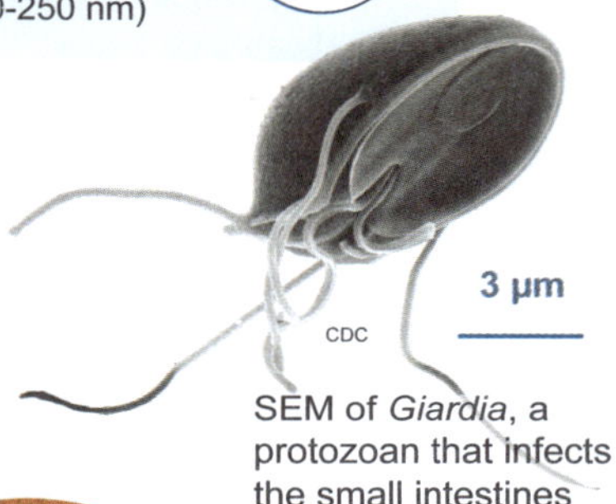

SEM of *Giardia*, a protozoan that infects the small intestines of many vertebrate groups. — 3 µm

Paramecium is a protozoan commonly found in ponds. — 50 µm

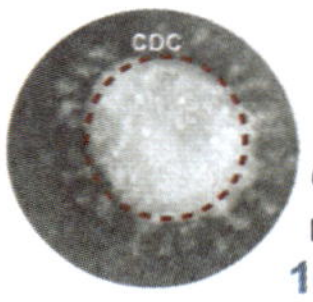

Coronavirus is the virus responsible for COVID-19 — 10 nm

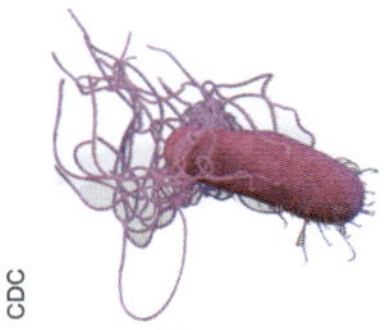

Salmonella is a bacterium found in many environments and causes food poisoning in humans. — 10 µm

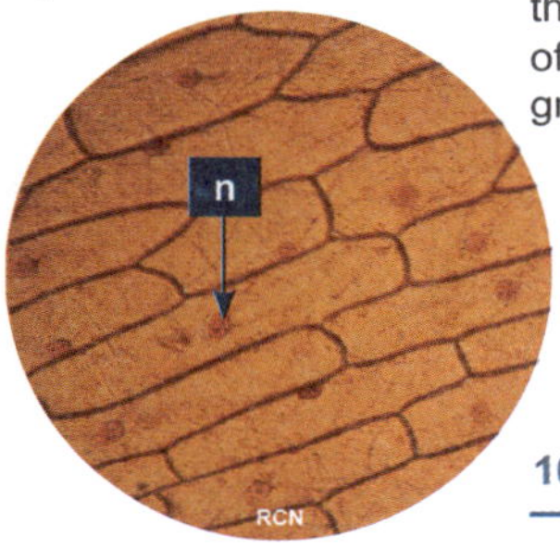

Onion epidermal cells: the **nucleus** (n) is just visible. — 100 µm

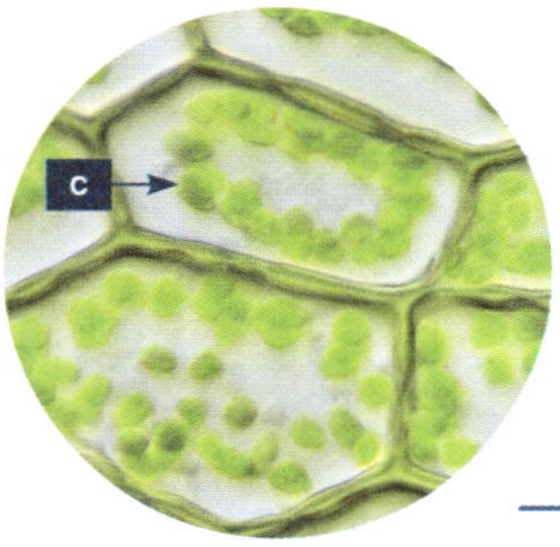

Mosses are low growing primitive plants. In these cells, the chloroplasts (c) can be seen, mostly around the cell edges. — 50 µm

1. Using the measurement scales provided on each of the photographs above, determine the longest dimension (length or diameter) of the cell/animal/organelle indicated in µm and mm. Do not include cilia or flagella. Attach your working:

    (a) *Daphnia*: _____ µm _____ mm   (e) Chloroplast: _____ µm _____ mm

    (b) *Giardia*: _____ µm _____ mm   (f) *Paramecium*: _____ µm _____ mm

    (c) Nucleus: _____ µm _____ mm   (g) *Salmonella*: _____ µm _____ mm

    (d) Moss leaf cell: _____ µm _____ mm   (h) *Coronavirus*: _____ µm _____ mm

2. Mark and label the examples above on the log scale below according to their size:

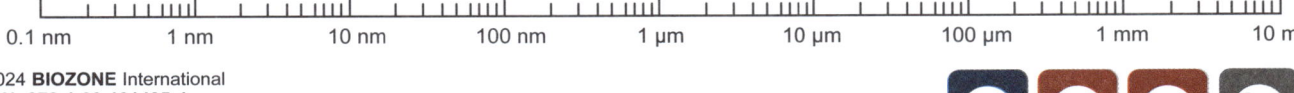

0.1 nm   1 nm   10 nm   100 nm   1 µm   10 µm   100 µm   1 mm   10 mm

©2024 **BIOZONE** International
ISBN: 978-1-99-101405-4
Photocopying prohibited

18 Why be Multicellular?

Key Question: What are some advantages of multicellular form for both prokaryotes and eukaryotes?

▶ Although many single celled organisms exist, being larger and multicellular (with many **cells**) has advantages, as the repeated evolution of this form shows. Analyze the information below and use it to explain the advantages of being multicellular.

Prokaryotic cells can work together

▶ Most **prokaryotic cells** 'behave' as unicellular individuals. However, many form colonies, which may act in a basic multicellular way. Some prokaryotes are able to form groups and some cells specialize to perform specific tasks.

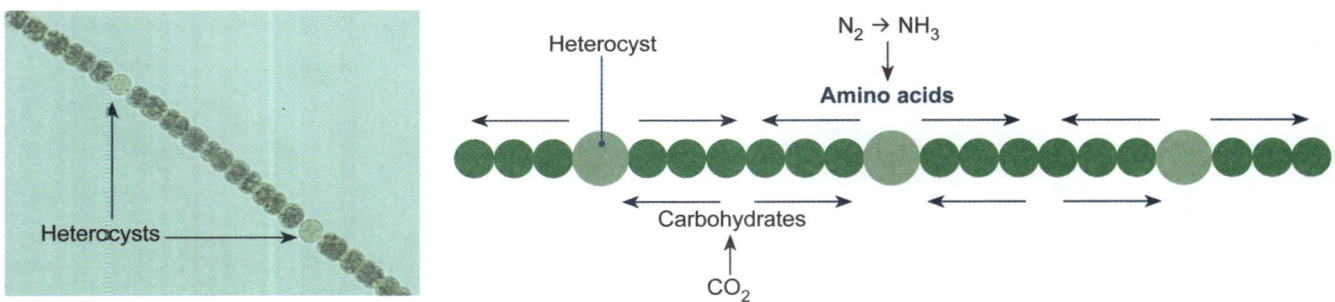

Cyanobacteria (e.g. *Anabaena* shown above) are a group of bacteria that can photosynthesize. The bacteria often form long filaments of individual cells joined together. Under low-nitrogen conditions, some of these cells will specialize to form **heterocysts**. These cells are able to fix nitrogen from molecular nitrogen (N_2). They show quite different **gene expression** from neighboring, unspecialized cells, in that they produce the enzyme **nitrogenase** and cannot photosynthesize. What's more, the heterocysts share the nitrogen they fix with neighboring cells and receive other nutrients from them, indicating basic cooperation.

Unicellular eukaryotes can behave as a single organism

▶ Although most eukaryotes are multicellular, some are permanently unicellular, and some spend some of their life cycle as single independent cells, but in certain conditions come together and behave as a single entity.

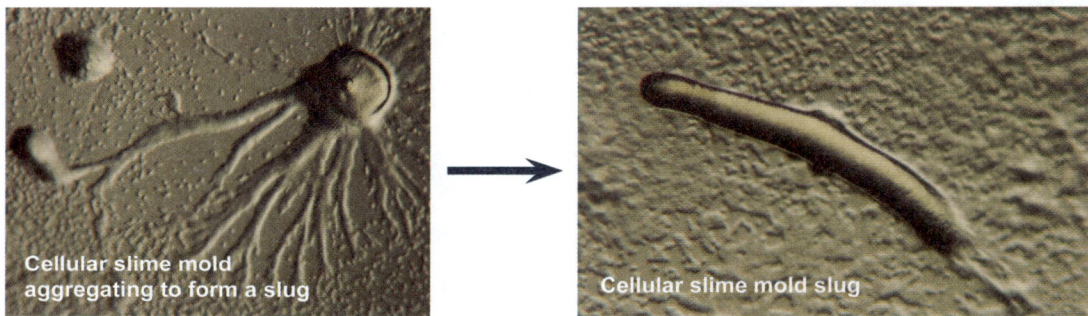

Cellular slime molds spend much of their life cycle as individual, amoeboid-like cells, often living in the soil and feeding on bacteria. When food becomes scarce, the single cells group together to form a slug, which is capable of movement. The slug may move some distance before finding a suitable area to form a fruiting body. The fruiting body is held a few millimeters off the ground by a stalk. The cells in the stalk die, whereas the cells in the fruiting body form spores, which will be dispersed to new areas.

1. What are some similarities in how prokaryotic and eukaryotic unicellular organisms gain an **advantage** by joining together to form multicellular organisms?

   _____
   _____
   _____

2. What are some similarities in the **process** of prokaryotic and eukaryotic unicellular organisms joining together to form multicellular organisms?

   _____
   _____
   _____

19 Eukaryotes have Complex Cells

Key Question: What was the origin of complexity in eukaryote cells and what is the evidence to support the claims?

▸ Endosymbiosis theory is used to explain how the complexity of **eukaryote cells** came about by the engulfment of prokaryote cells in early common ancestors (from endo: internal, symbiosis: advantageous relationship).
▸ It is hypothesized that eukaryotic cells evolved from pre-eukaryotic (**prokaryote**) cells, similar to archaea, that ingested other free-floating bacteria. They formed a symbiotic relationship with the cells they engulfed.
▸ As the engulfed bacteria provided useful functions for the host cells, such as respiration and photosynthesis, they were 'retained'. In return, the engulfed bacteria were supplied with energy and a safe, regulated habitat.
▸ The two **organelles** that evolved in eukaryotic cells as a result of bacterial endosymbiosis were mitochondria, for aerobic respiration, and chloroplasts, for photosynthesis, also in aerobic conditions. Primitive eukaryotes probably acquired mitochondria by engulfing purple bacteria. Similarly, chloroplasts may have been acquired by engulfing photosynthetic cyanobacteria. Other organelles may have formed from infolding of the **plasma membrane**.
▸ An alternative theory for the evolution of eukaryotic cells is the "inside-out theory," in which protrusions from the eukaryotic ancestor cell wrapped around bacteria and fused creating the cell's membrane structures.

Evolution of eukaryotic cells

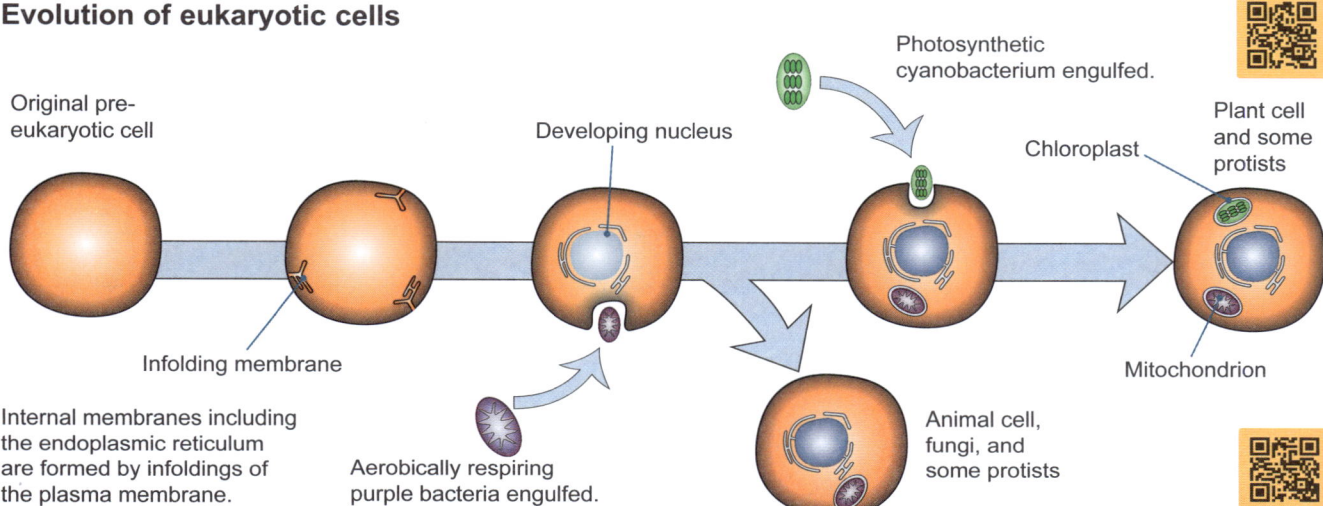

Evidence for the endosymbiosis theory:
Evidence for the bacterial origin of chloroplasts and mitochondria by endosymbiosis includes:

- Mitochondria and, in particular, chloroplasts, have a similar morphology (structure) to bacteria.
- Mitochondria and chloroplasts divide by binary fission, splitting in half to form new organelles, just like bacteria. Thus new mitochondria and chloroplasts arise from preexisting ones; they are not manufactured by the cell.
- Both mitochondria and chloroplasts have a chemically distinct inner membrane. The outer membrane is similar to the plasma membrane (as if a vesicle formed around the engulfed cell), but the inner membrane is similar to the bacterial membrane.
- Bacterial DNA is a single circular molecule. Mitochondria and chloroplasts also have their own single circular DNA. Like bacterial DNA, this DNA has no intervening, non-**protein**-coding regions or associated proteins. Also, the organelle DNA mutates at a different rate to the nuclear DNA.
- Mitochondria and chloroplasts contain ribosomes that are more similar in size (70S) to bacterial ribosomes than ribosomes in the cytoplasm.
- Antibiotics that inhibit protein synthesis in bacteria also inhibit protein synthesis in mitochondria and chloroplasts.
- Analysis of chloroplast DNA has shown that they are related to cyanobacteria.

1. Which endosymbiosis occurred first in the evolution of eukaryotic cells? Explain your reasoning: _____
   _____
   _____

2. How would **symbiosis** benefit both the engulfed bacterium and the eukaryotic cell? _____
   _____
   _____

Engulfment of bacteria by protists

Paramecium bursaria

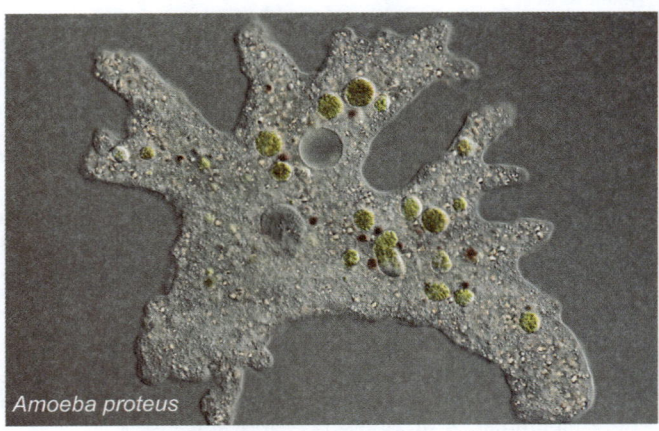
Amoeba proteus

Paramecium bursaria (above) is a single celled protozoan ('animal-like' protist). It engulfs cells of *Zoochlorella*, a photosynthetic green alga. It houses the algae and carries them to lighted areas in a pond where they can photosynthesize. In return, the protozoan uses the food made by the algae.

In the 1970s, microbiologist Kwang Jeon studied the infection of *Amoeba proteus* by *Legionella*-like bacteria. He found that most infected amoebae died. The few that survived, and their descendants, functioned well with the bacteria still thriving inside. When antibiotics were used to kill the bacteria inside the amoebae, the amoebae also died, having become dependent on a bacterial protein. A symbiosis had evolved.

3. Outline four pieces of evidence for endosymbiosis as a hypothesis for eukaryote evolution:

   (a) _____

   (b) _____

   (c) _____

   (d) _____

4. How do the examples of *Paramecium bursaria* and *Amoeba proteus* support the endosymbiotic origin of chloroplasts and other organelles found in eukaryotes?

   _____

5. One alternative hypothesis for the mechanism of how the organelles came to be inside the eukaryote cells is called the 'inside-out' hypothesis, where the cell membrane was extended and wrapped around the engulfed bacteria, so forming the nuclear and endoplasmic reticulum membranes at the same time.

   NEED HELP? See Activity 248

   (a) What is the difference between a theory, hypothesis, and a law in science?

   _____

   (b) Why is it considered 'good science practice' to continue developing and testing theories when the current theory seems to already explain the phenomenon well?

   _____

20 Cell Membrane Structure

Key Question: What are the key components of plasma membranes and how do they enable cellular homeostasis?

▶ The **plasma membrane** encloses the contents of a **cell**. It is a key structure in regulating cellular homeostasis: the process of maintaining a steady state of conditions inside the cell. The membrane does this by enabling and controlling movement of substances in and out of the cell.

▶ Recall **lipid** structure from activity 8. The fluid-mosaic model of membrane structure (below) describes a **phospholipid** bilayer with **proteins** of different types moving freely within it.

▶ The double layer of lipids is quite fluid. It is a dynamic structure and is actively involved in cellular activities.

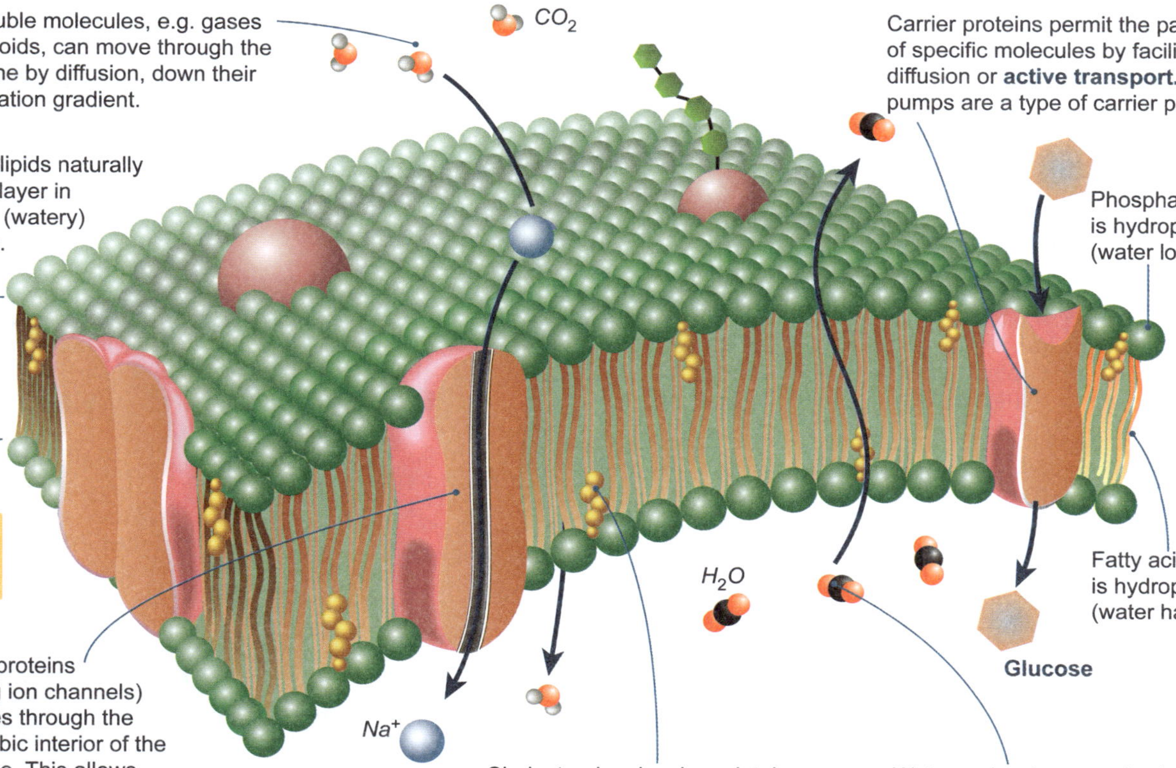

Lipid soluble molecules, e.g. gases and steroids, can move through the membrane by diffusion, down their concentration gradient.

CO_2

Carrier proteins permit the passage of specific molecules by facilitated diffusion or **active transport**. Ion pumps are a type of carrier protein.

Phospholipids naturally form a bilayer in aqueous (watery) solutions.

Phosphate head is hydrophilic (water loving).

H_2O

Fatty acid tail is hydrophobic (water hating).

Glucose

Channel proteins (including ion channels) form pores through the hydrophobic interior of the membrane. This allows water soluble molecules to pass by facilitated diffusion.

Na^+

Cholesterol molecule maintains membrane integrity, preventing it becoming too fluid or too firm.

Water molecules pass freely between the phospholipid molecules by osmosis.

1. List the important components of the plasma membrane: _____

2. Identify which kind of molecule on the diagram:

   (a) Can move through the plasma membrane by diffusion: _____

   (b) Forms a channel through the membrane: _____

3. List the types of proteins pictured in the diagram: _____

4. (a) On the diagram (right) label the hydrophobic and hydrophilic ends of the phospholipid.

   (b) Which end is attracted to water? _____

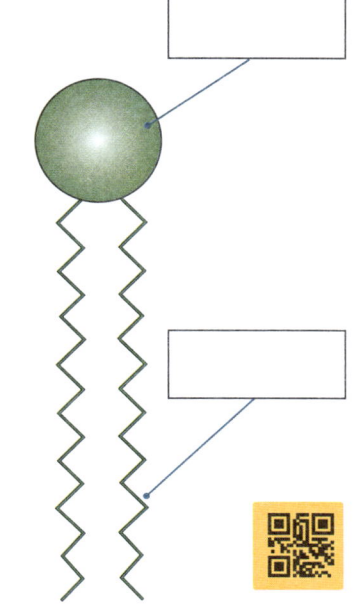

©2024 **BIOZONE** International
ISBN: 978-1-99-101405-4
Photocopying prohibited

Freeze-fracture method of viewing the plasma membrane

▶ The freeze-fracture technique for preparing and viewing cell membranes has provided evidence to support the fluid mosaic model of the plasma membrane. Cellular membranes play many extremely important roles in cells and understanding their structure is essential to understand cellular function. In fact, understanding the structure and function of membrane proteins is essential to understand cellular transport processes, and cell recognition and signaling. Cellular membranes are far too small to be seen clearly using light microscopy, and any detail is impossible to resolve. Since early last century, scientists have known that membranes were composed of a lipid bilayer with associated proteins. But how did they find out just how these molecules were organized?

▶ The answers were provided with electron microscopy, and one technique in particular: freeze fracture. As the name implies, freeze fracture is the freezing of a cell and then fracturing it so the inner surface of the membrane can be seen using electron microscopy. Membranes are composed of two layers of phospholipids held together by weak intermolecular bonds. These split apart during fracture.

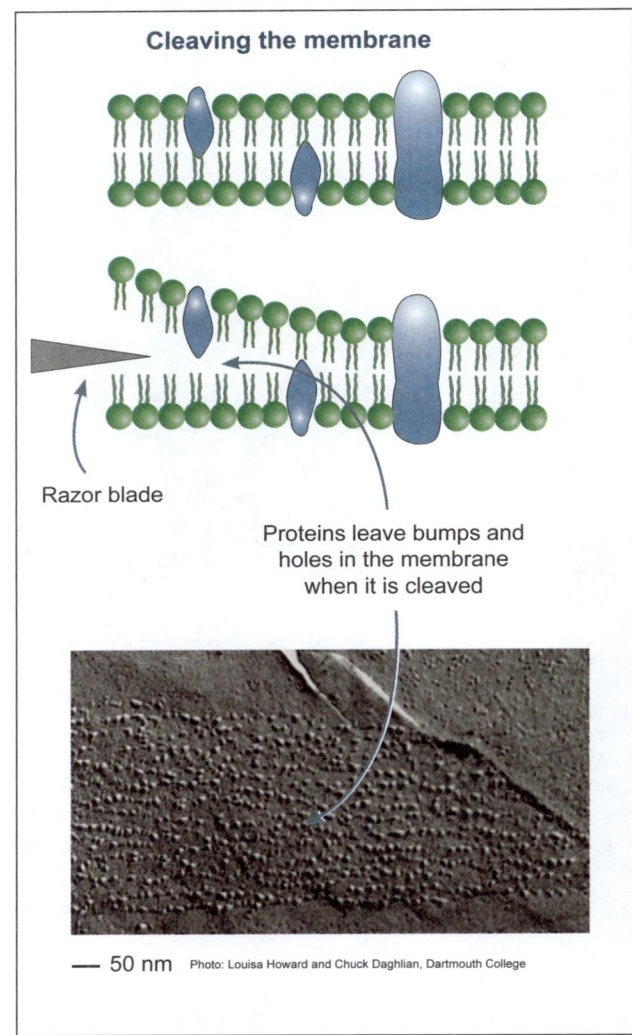

Cleaving the membrane

Razor blade

Proteins leave bumps and holes in the membrane when it is cleaved

— 50 nm   Photo: Louisa Howard and Chuck Daghlian, Dartmouth College

The procedure involves several steps:

- Cells are immersed in chemicals that alter the strength of the internal and external parts of the plasma membrane and immobilize any mobile macromolecules.

- The cells are passed through a series of glycerol solutions of increasing concentration. This protects the cells from bursting when they are frozen.

- The cells are mounted on gold supports and frozen using liquid propane.

- The cells are fractured in a helium-vented vacuum at -150° C. A razor blade cooled to -170° C acts as both a cold trap for water and the fracturing instrument.

- The surface of the fractured cells may be evaporated a little to produce some relief on the surface (known as etching) so that a three-dimensional effect occurs.

- For viewing under an electron microscope (EM), a replica of the cells is made by coating them with gold or platinum to ~3 nm thick. A layer of carbon around 30 nm thick is used to provide contrast and stability for the replica.

- The samples are then raised to room temperature and placed into distilled water or digestive enzymes, which separates the replica from the sample. The replica is then rinsed in distilled water before it is ready for viewing.

- The freeze fracture technique provided the necessary supporting evidence for the current fluid mosaic model of membrane structure. When cleaved, proteins in the membrane left impressions that showed they were embedded into the membrane and not a continuous layer on the outside as earlier models proposed.

5. Write a simplified method for the freeze fracture technique: _____

_____

_____

_____

6. What evidence do freeze fracture studies provide for the fluid mosaic model of membrane structure? _____

_____

_____

_____

Modeling the plasma membrane

▶ Plasma membranes are often shown as two-dimensional structures (as shown on the previous page). Even when drawn to represent a three-dimensional structure, the nature of the plasma membrane may not be obvious. In this part of the activity, you will build a simple, three-dimensional plasma membrane model.

1. Cut out the plasma membrane along the red lines. Cut out the solid black circles. Fold along the black lines. Use clear tape to stick the sides together to produce a 3D, slightly curved box.

2. Cut out the three proteins along the red lines. Fold along the black lines and use clear tape to produce three cylinders.

3. Cut out both **carbohydrate** chains. Fold over the black squares. Stick one to the black square on the end of the glycoprotein. Stick the other to the black square on the plasma membrane surface to produce a glycolipid.

4. Slide the two transmembrane proteins into the channels created by cutting out the circles from the plasma membrane.

5. Slide the peripheral protein about halfway into the final hole. This completes your plasma membrane model.

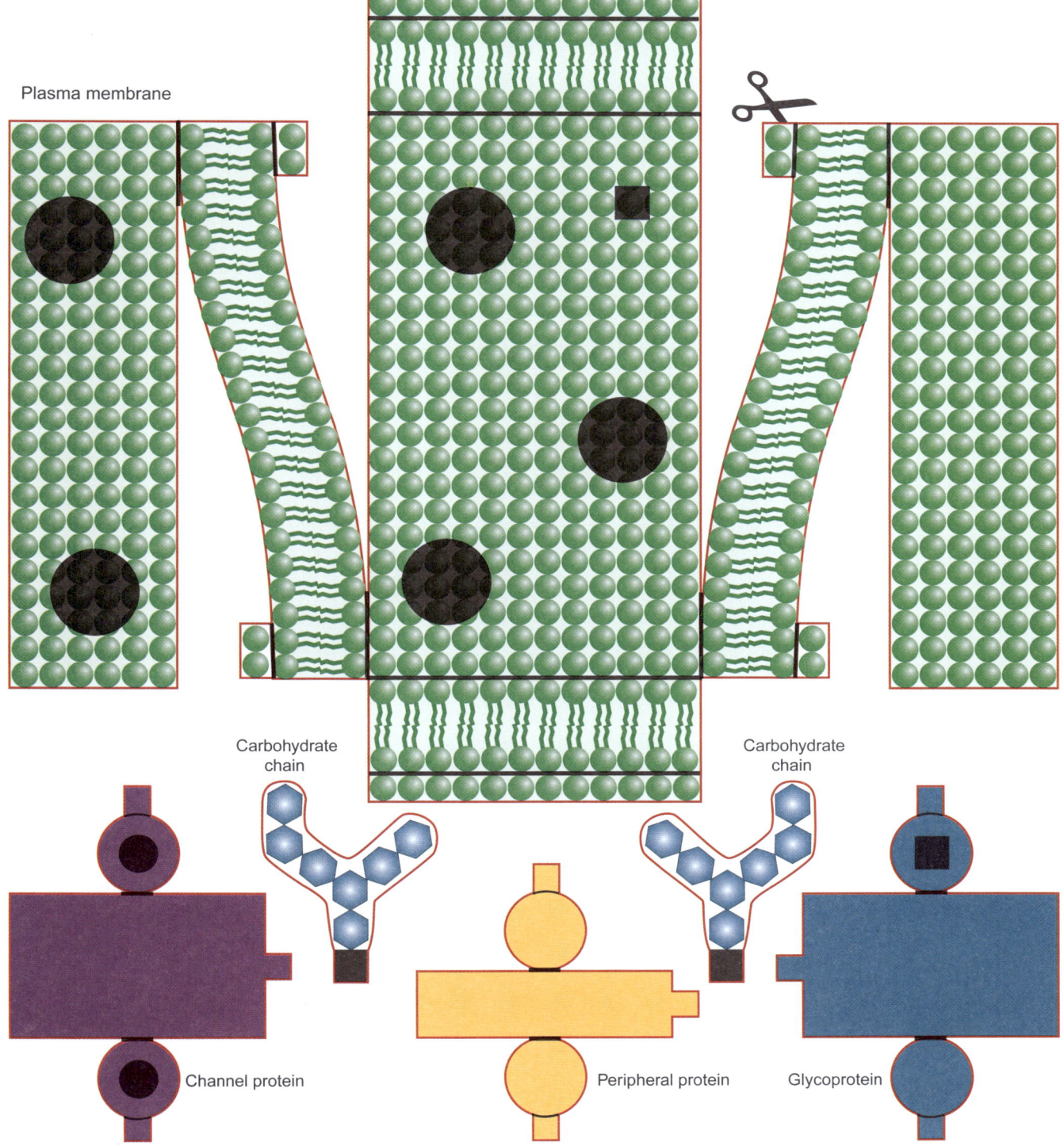

This page is deliberately left blank

21 Diffusion in Cells - Passive Transport

Key Question: What is diffusion, and what factors affect the rate of diffusion of a particle from one point to another across a cell membrane?

What is diffusion?

▸ Diffusion is the movement of particles from regions of high concentration to regions of low concentration. Diffusion is a passive process, meaning it needs no input of energy to occur. During diffusion, molecules move around randomly, becoming evenly dispersed.

▸ Most diffusion in biological systems occurs across **cell** membranes. Simple diffusion occurs directly across a membrane, whereas facilitated diffusion involves helper **proteins**. Neither requires the cell to expend energy.

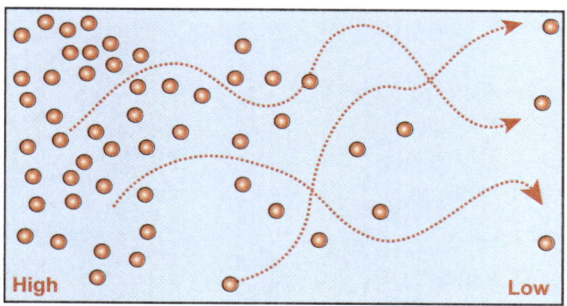

Concentration gradient

If molecules can move freely, they move from high to low concentration (down a concentration gradient) until evenly dispersed. Net movement then stops.

Factors affecting the rate of diffusion

Concentration gradient	Diffusion rate is higher when there is a greater concentration difference between two regions.
The distance moved	Diffusion occurs at a greater rate over shorter distances than over larger distances.
The surface area involved	The larger the area across which diffusion occurs, the greater the rate of diffusion.
Barriers to diffusion	Rate of diffusion is slower across thick barriers than across thin barriers.
Temperature	Rate of diffusion increases with temperature.

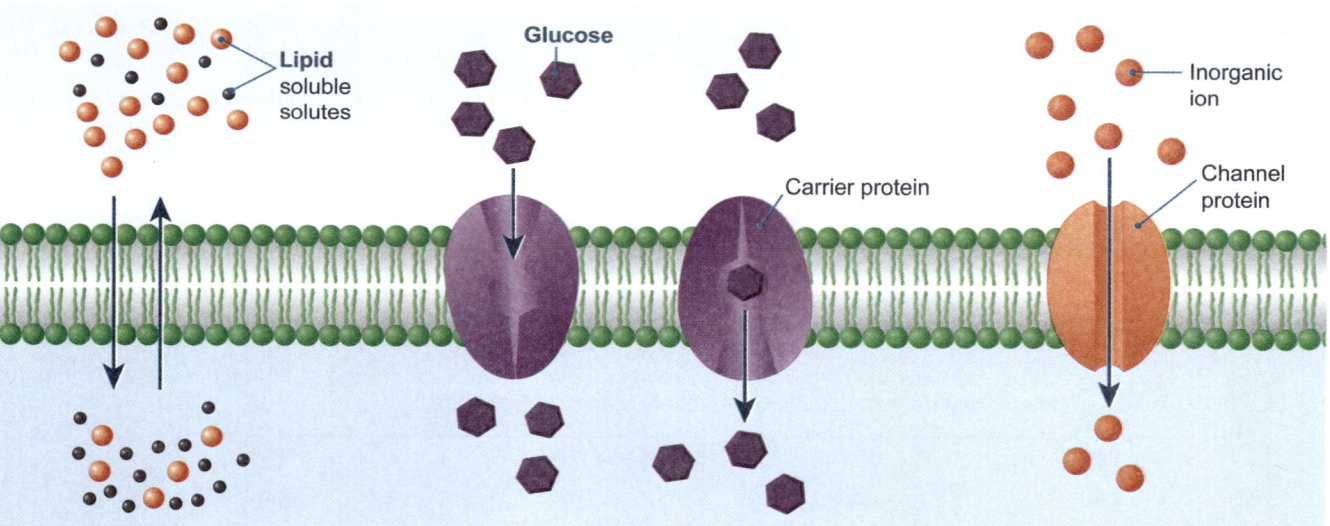

Simple diffusion
Molecules move directly through the membrane without assistance and without any energy expenditure. **Homeostasis example**: O_2 diffuses into the blood and CO_2 diffuses out.

Facilitated diffusion by carriers
Carrier proteins allow large lipid-insoluble molecules that cannot cross the membrane by simple diffusion to be transported into the cell. **Homeostasis example**: the transport of glucose into red blood cells.

Facilitated diffusion by channels
Channel proteins (hydrophilic pores) in the membrane allow inorganic ions to pass through the membrane. **Homeostasis example**: K^+ ions leaving nerve cells to restore membrane resting potential.

1. What is diffusion? _____

_____

2. (a) How is facilitated diffusion different from simple diffusion? _____

_____

   (b) How is it the same? _____

Investigating diffusion

▶ Diffusion through a selectively permeable cell membrane can be modeled using dialysis tubing. The pores of the dialysis tubing determine the size of the molecules that can pass through. In the experiment described below, you will investigate how glucose will diffuse down its concentration gradient from a high glucose concentration to a low glucose concentration and demonstrate, via the model, the selective permeability of the **plasma membrane**.

Investigation 1.3 Simple diffusion across a membrane

See appendix for equipment list.

1. Add 200 mL of distilled water to a clean 200 mL beaker. Remove a 1 mL sample and place in a clean test tube. Use a glucose dipstick to test for the presence and concentration of glucose in the 1 mL sample. If glucose is present, the indicator window will change color. The color change can be compared against a reference to determine the concentration of glucose present.

2. Now add a few drops of Lugol's indicator to test for the presence of starch. Lugol's indicator contains iodine, and turns blue/black in the presence of starch.

3. Obtain a short section of dialysis tubing, approximately 10 cm long. Use thread or nylon line to tie off one end (or tie a knot in the tubing if long enough).

4. You may need to rinse the tubing under water to make it pliable enough to open.

5. Fill the dialysis tubing with 5 mL each of a 1% starch solution and a 10% glucose solution.

6. Remove a 1 mL sample and place in a clean test tube. Tie off the top of the dialysis tubing, rinse well with distilled water, then place in the beaker of distilled water.

7. Test for the presence and concentration of glucose and then starch in the sample from the dialysis tubing as in steps 1 and 2.

8. Leave the dialysis tubing in the distilled water for 30 minutes.

9. Remove 1 mL of water from the beaker and place in a clean test tube. Use a glucose dipstick to test for the presence and concentration of glucose. Test for the presence of the starch using Lugol's indicator.

10. Remove a 1 mL sample from the dialysis tubing and place in a clean test tube. Use a glucose dipstick to test for the presence and concentration of glucose. Test for the presence of the starch using Lugol's indicator.

3. What is the aim of the experiment? _____

4. What part of a cell does the dialysis tubing represent? _____

5. In the spaces provided (below) draw the distribution of starch and glucose at the start and at the end of the experiment. Use the colored symbols shown under the table to represent starch and glucose:

Dialysis tubing start	Beaker start		Dialysis tubing end	Beaker end

● Starch
● Glucose

6. Explain your results: _____

7. In summary, how does this investigation demonstrate passive transport across a membrane as a mechanism to maintain cellular homeostasis? _____

22 Osmosis in Cells - Diffusion of Water

Key Question: How does osmosis transport water across a semi-permeable membrane to ensure water balance homeostasis in the cell?

Osmosis

- Osmosis is the diffusion of water molecules from regions of lower solute concentration (higher free water concentration) to regions of higher solute concentration (lower free water concentration) across a partially permeable **cell** membrane.
- Homeostatic water balance in the cell is crucial to its function. Osmosis is a means of passively transporting water in and out of cells.
- A partially permeable membrane lets some, but not all, molecules pass through. The **plasma membrane** of a cell is an example of a partially permeable membrane.
- Although osmosis is a passive process, cells can actively move solute to 'induce' osmotic flow of water - such as in the root hairs of plant cells.

Osmotic potential

The presence of solutes (dissolved substances) in a solution increases the tendency of water to move into that solution. This tendency is called the osmotic potential or osmotic pressure. The greater a solution's concentration, i.e. the more total dissolved solutes it contains, the greater the osmotic potential.

Dialysis tubing ready for use

Demonstrating osmosis

Osmosis can be demonstrated using the simple experiment described below.

A **glucose** solution (high solute concentration) is placed into dialysis tubing, and the tubing is placed into a beaker of water (low solute concentration). The difference in concentration of glucose (solute) between the two solutions creates an osmotic gradient. Water moves by osmosis into the glucose solution and the volume of the glucose solution inside the dialysis tubing increases.

The dialysis tubing acts as a partially permeable membrane, allowing water to pass freely, while keeping the glucose inside the dialysis tubing.

Remember that:
A solvent dissolves other substances.
A solute is the substance dissolved by a solvent to form a solution.

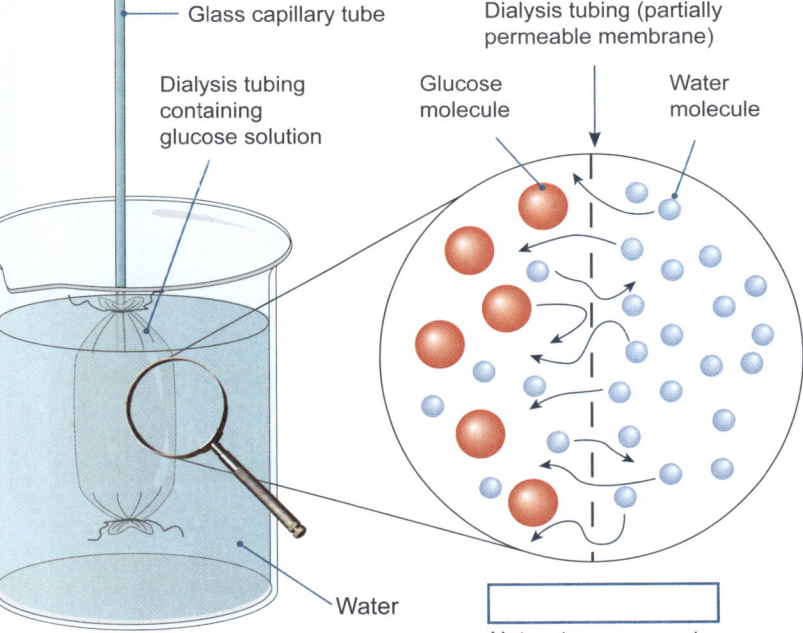

1. What is osmosis? _____
   _____
   _____

2. (a) In the blue box on the model above, draw an arrow to show the direction of net water movement.

   (b) Why did water move in this direction? _____
   _____
   _____
   _____

23 Active Transport in Cells

Key Question: What is active transport, and how does the process transport molecules and ions across a cellular membrane?

- In order to maintain **cell** homeostasis, some molecules or ions need to move across the cell membrane against a concentration gradient, from regions of low concentration to regions of high concentration across a **plasma membrane**. This is called **active transport**.

- Active transport needs energy to proceed, which comes from ATP (adenosine triphosphate). Energy is released when ATP is hydrolyzed (broken by adding water) forming ADP (adenosine diphosphate) and inorganic phosphate (Pi) (see p95).

- Transport (carrier) **proteins** in the plasma membrane use energy to transport molecules across a membrane (below).

- Therefore, active transport can be used to move molecules into and out of a cell.

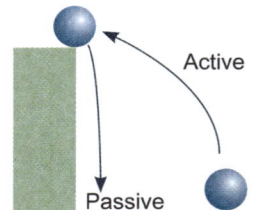

A ball falling is a passive process. Replacing the ball requires active energy input.

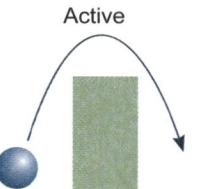

Energy is needed to move an object over a physical barrier.

Sometimes the energy of a passively moving object can be used to actively move another, e.g. a falling ball can be used to catapult another.

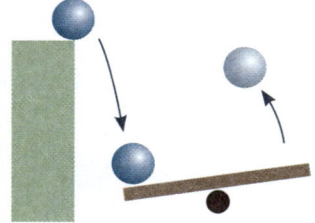

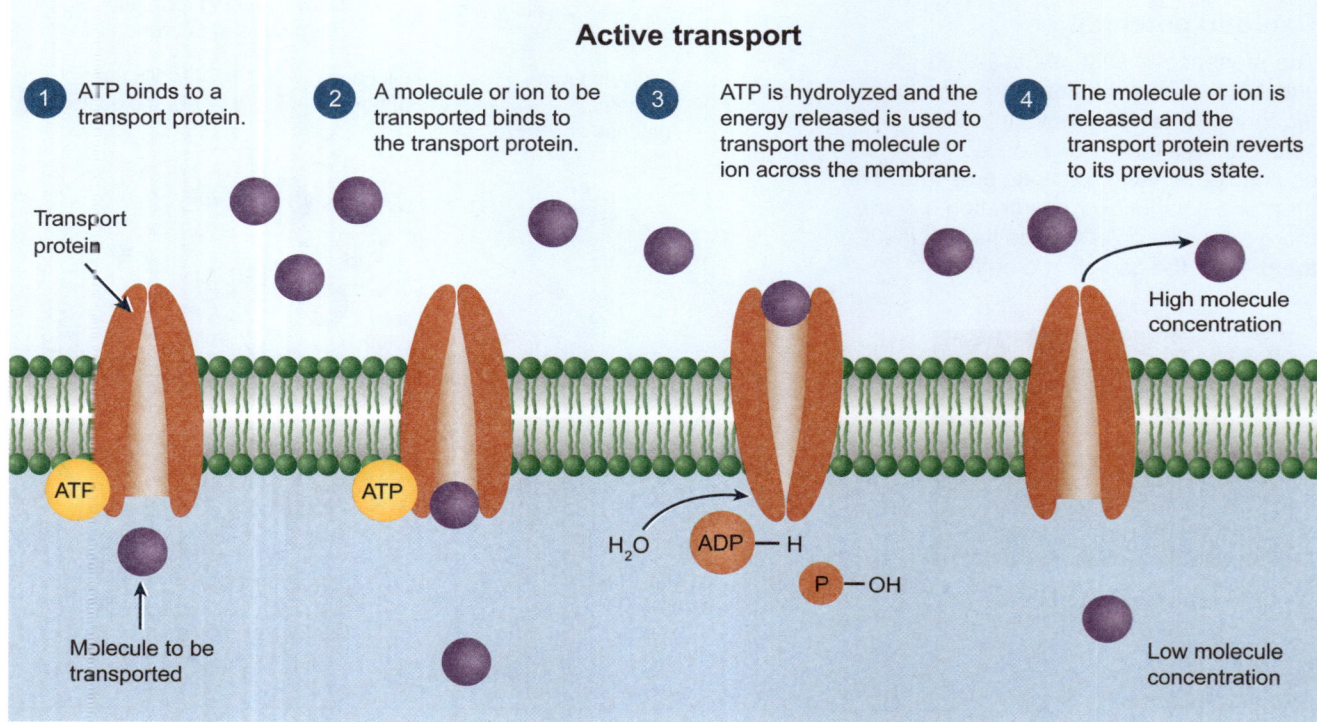

1. What is active transport? _____

2. (a) Why does active transport require energy? _____

    (b) Where does the energy for active transport come from? _____

24 What is an Ion Pump?

Key Question: How do ion pumps transport ions and molecules across cellular membranes?

- **Proteins** play an important role in the movement of molecules into and out of **cells**. Recall, when molecules are moved against their concentration gradient, energy is needed.
- Some **lipid**-soluble molecules can readily cross the cell membrane. Water soluble or charged molecules cross the membrane by facilitated diffusion or by **active transport** involving special membrane proteins called ion pumps.
- Ion pumps directly, or indirectly, use energy to transport ions across the membrane against a concentration gradient.
- The sodium-potassium pump (below, left) is found in almost all animal cells and is also common in plant cells. The concentration gradient created by ion pumps is often coupled to the transport of other molecules, such as **glucose** or sucrose, across the membrane (below right).

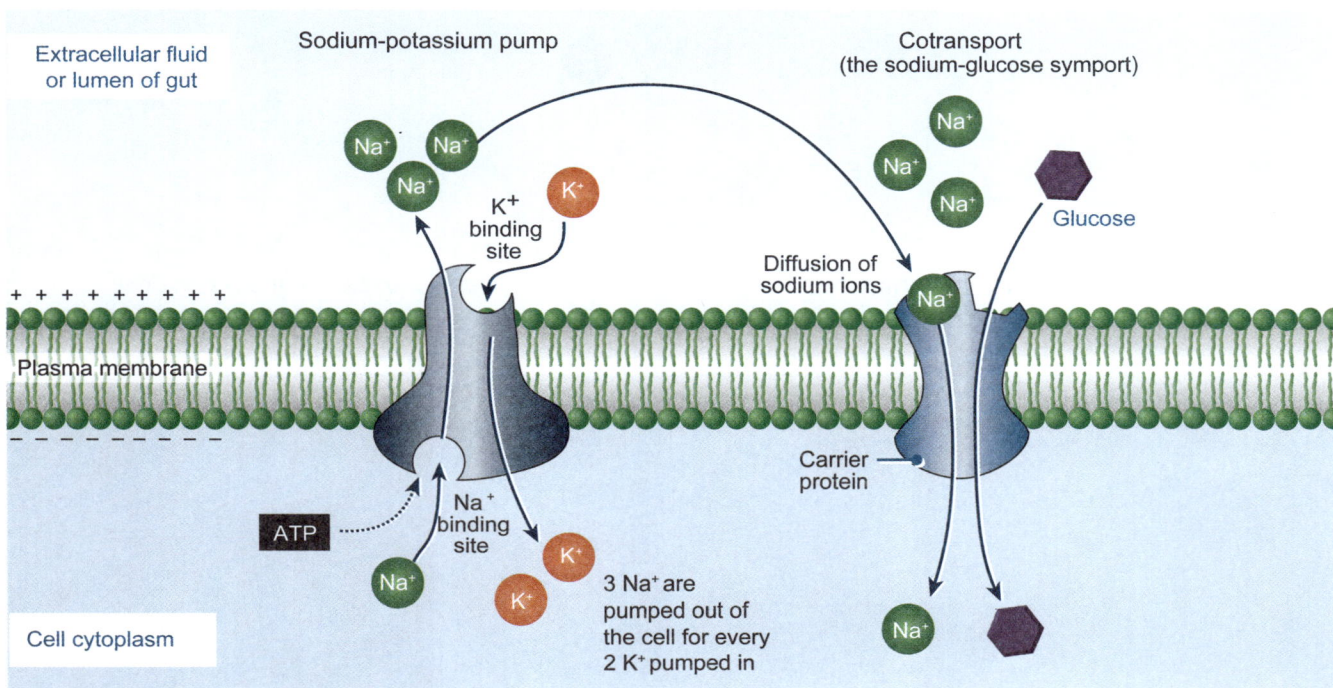

Sodium-potassium (Na+/K+) pump
The Na$^+$/K$^+$ pump is a protein in the membrane that uses energy in the form of ATP to exchange sodium ions (Na$^+$) for potassium ions (K$^+$) across the membrane. The unequal balance of Na$^+$ and K$^+$ across the membrane creates large concentration gradients that can be used to drive transport of other substances, e.g. cotransport of glucose. The Na$^+$/K$^+$ pump also helps to maintain the right balance of ions and so helps regulate the cell's water balance.

Cotransport (coupled transport)
A specific carrier protein controls the entry of glucose into the intestinal epithelial cells from the gut where digestion is taking place. The energy for this is provided indirectly by a gradient in sodium ions. The carrier "couples" the return of Na$^+$ down its concentration gradient to the transport of glucose into the cell. The process is therefore called cotransport. A low intracellular concentration of Na$^+$ (and therefore the concentration gradient for transport) is maintained by a sodium-potassium pump.

1. What is an ion pump? _____
   _____
   _____

2. (a) Explain what is meant by cotransport: _____
   _____

   (b) How is cotransport used to move glucose into the intestinal epithelial cells? _____
   _____
   _____

25 Cytosis

Key Question: How does the folding of the plasma membrane enable the cell to bring in or export material?

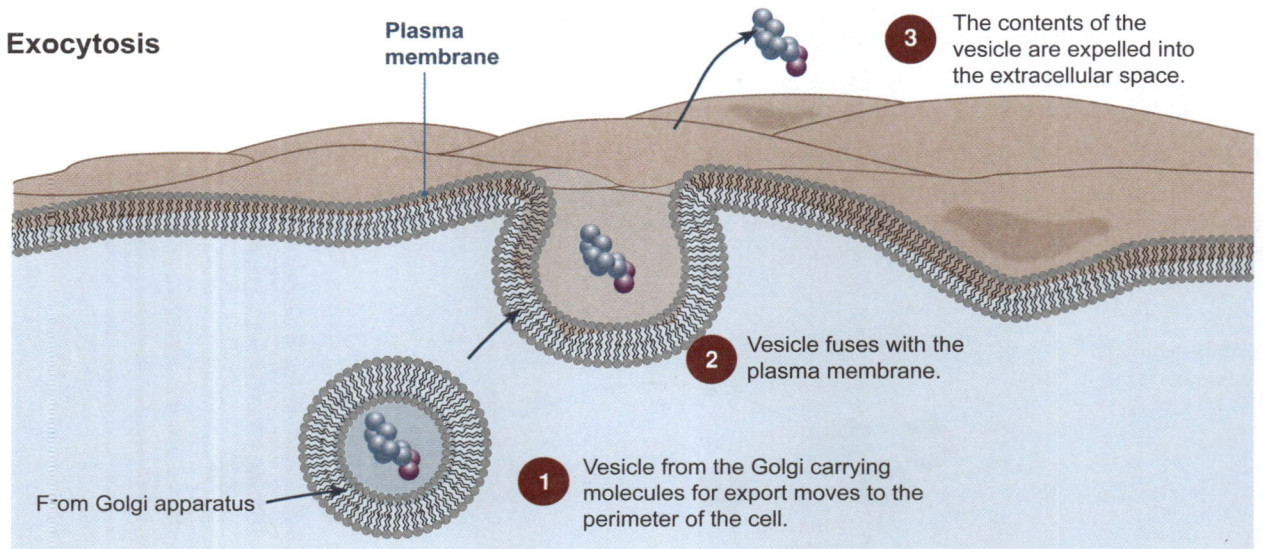

Exocytosis

1. Vesicle from the Golgi carrying molecules for export moves to the perimeter of the cell.
2. Vesicle fuses with the plasma membrane.
3. The contents of the vesicle are expelled into the extracellular space.

From Golgi apparatus

▶ **Exocytosis** (above) is an **active transport** process in which a secretory vesicle fuses with the plasma **cell membrane** and expels its contents into the extracellular space. In multicellular organisms, various types of cells e.g. endocrine cells and nerve cells, are specialized to manufacture products, such as **proteins**, and then export them from the cell to elsewhere in the body or outside it.

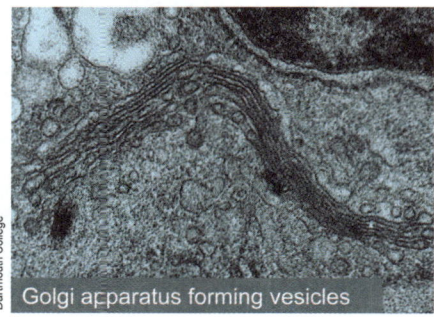

Golgi apparatus forming vesicles

The transport of Golgi vesicles to the edge of the cell and their expulsion from the cell occurs through the activity of the cytoskeleton. This requires energy (ATP).

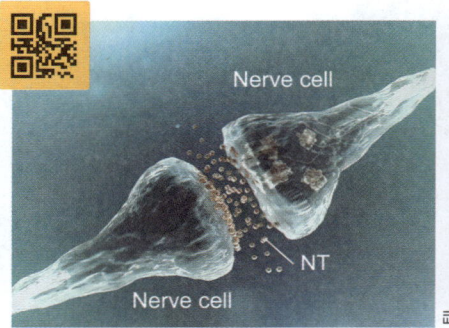

Exocytosis is important in the transport of neurotransmitters (NT) into the junction (synapse) between nerve cells to transmit nervous signals, as shown in this illustration.

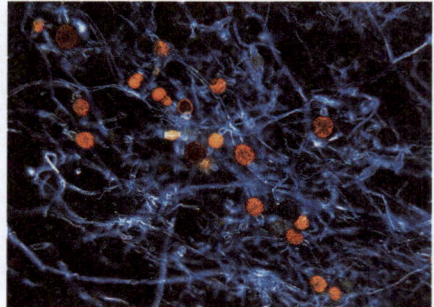

Fungi and bacteria use exocytosis to secrete digestive enzymes, which break down substances extracellularly so that nutrients can be absorbed (by **endocytosis**).

1. (a) What is the purpose of exocytosis? _____

   (b) How does it occur? _____

2. Describe two examples of the role of exocytosis in cells to maintain homeostasis:

   (a) _____

   (b) _____

Endocytosis

▸ Endocytosis is a type of active transport in which the **plasma membrane** folds around a substance to transport it across the membrane into the cell. The ability of cells to do this is a function of the fluid nature of the plasma membrane. Some **viruses** enter cells by this process, 'tricking' the cell into identifying them as harmless.

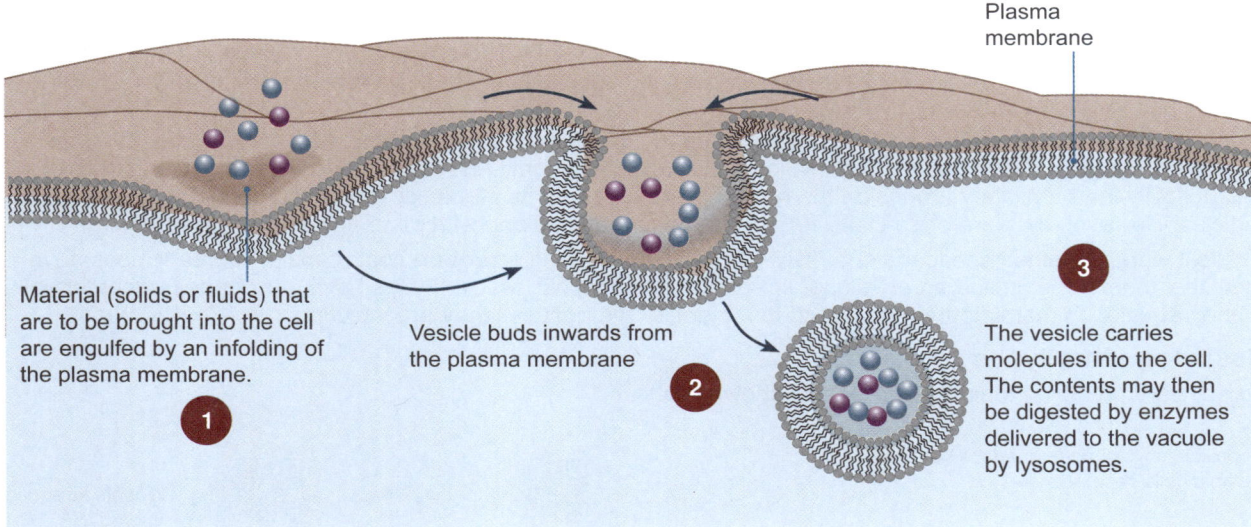

1. Material (solids or fluids) that are to be brought into the cell are engulfed by an infolding of the plasma membrane.

2. Vesicle buds inwards from the plasma membrane

3. The vesicle carries molecules into the cell. The contents may then be digested by enzymes delivered to the vacuole by lysosomes.

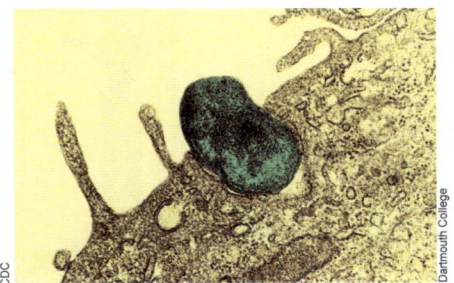

Phagocytosis (or 'cell-eating') involves the cell engulfing solid material to form large phagosomes or vacuoles (e.g. food vacuoles). It may be non-specific or receptor-mediated. **Examples**: Feeding in *Amoeba*, phagocytosis of foreign material and cell debris by neutrophils and macrophages.

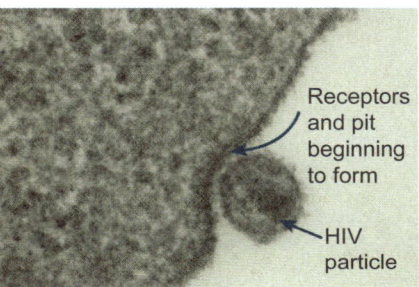

Receptor mediated endocytosis is triggered when certain metabolites, hormones, or viral particles bind to specific receptor proteins on the membrane so that the material can be engulfed. **Examples**: The uptake of lipoproteins by mammalian cells and endocytosis of viruses (above).

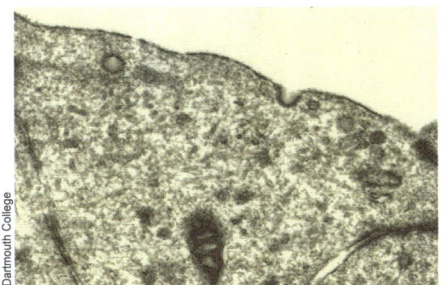

Pinocytosis (or 'cell-drinking') involves the non-specific uptake of liquids or fine suspensions into the cell to form small pinocytic vesicles. Pinocytosis is used primarily for absorbing extracellular fluid. **Examples**: Uptake in many protozoa, some cells of the liver, and some plant cells.

3. What is the purpose of endocytosis? _____

4. Is endocytosis active or passive transport? _____

5. Describe the following types of endocytosis:

   (a) Phagocytosis: _____

   (b) Receptor mediated endocytosis: _____

   (c) Pinocytosis: _____

6. Explain how the plasma membrane can form a vesicle: _____

26 Comparing Virus and Cell Structure

Key Question: How does the structure of viruses compare to the structure of cells?

- A **virus** is an extremely small, infectious, and highly specialized intracellular parasite. Viruses are disease-causing agents (pathogens) that replicate (reproduce themselves) only inside the living **cells** of other organisms. They are not considered living themselves.
- Viruses are acellular, meaning they are not made up of cells like the **prokaryotes** or **eukaryotes**, so they do not conform to the existing criteria upon which a five or six kingdom classification system is based. Viruses are metabolically inert until they are inside the host cell and hijacking its metabolic machinery to make new viral particles. However, they are often classified as microorganisms, along with other tiny living organisms.
- A typical virus contains genetic material (DNA or RNA) encased in a **protein** coat (capsid). Some viruses have an additional membrane, called an envelope, surrounding the capsid. Many viruses have glycoprotein receptor spikes on their envelopes that help them to attach to surface of the host cell they are infecting.

Classifying virus types

- Viruses vary greatly in their appearance as shown below.

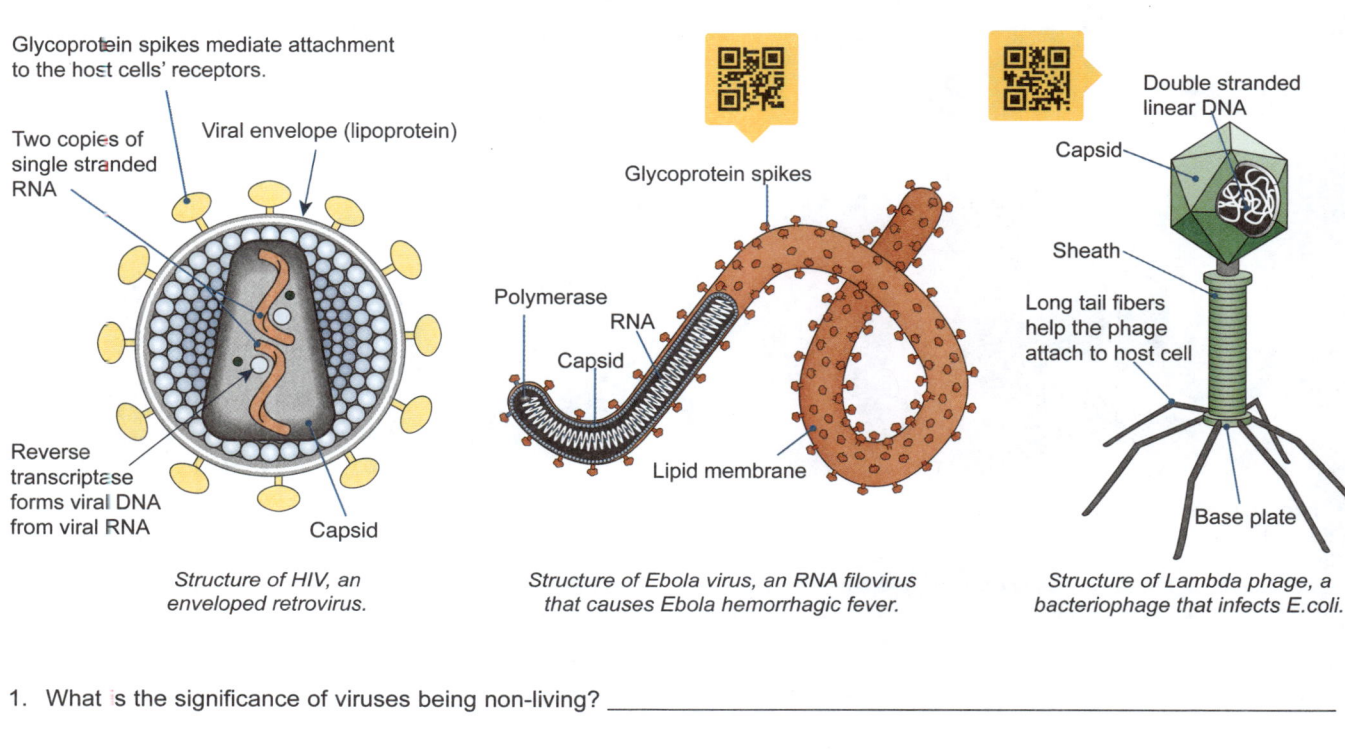

Structure of HIV, an enveloped retrovirus.

Structure of Ebola virus, an RNA filovirus that causes Ebola hemorrhagic fever.

Structure of Lambda phage, a bacteriophage that infects E.coli.

1. What is the significance of viruses being non-living? _____
   _____
   _____

2. How do many viruses attach themselves to the host cell that they are infecting? _____
   _____
   _____

3. Describe the basic structure of a generalized virus, identifying the structures the three virus examples above have in common with each other: _____
   _____
   _____

4. Describe the purpose of the following:

   (a) Glycoprotein spikes: _____

   (b) A bacteriophage's tail fibers: _____

   (c) Protein capsid: _____

The founders of virology

▸ Virology is the study of viruses. Prior to the 1890s no one knew of the existence of viruses.
▸ Dimitri Ivanovsky (1864-1920), a Russian botanist, was particularly interested in what was causing disease to tobacco plants. He used a very fine mesh to filter out bacteria from an infectious solution but discovered the particle causing the disease was small enough to travel through it.
▸ Following on from these findings, Dutch biologist Martinus Beijerinck (1851-1931) repeated the experiments on the tobacco plants. He discovered the 'infecting solution' could also infect other plants. Beijerinck identified the first recorded virus, TMV, tobacco mosaic virus, and also coined the term 'virus'.
▸ It wasn't until after electron microscopy was developed that microscopes could visualize viruses. In 1939, scientists viewed a virus, the TMV, for the first time.

Martinus Beijerinck left and Dmitri Ivanovsky below, inset.

Comparing virus and typical cell structure

▸ Although both viruses and single-celled organisms are grouped together as microorganisms, they are quite different in structure.
▸ Many key structures present in cells, that are required to perform life functions, are missing in viruses.

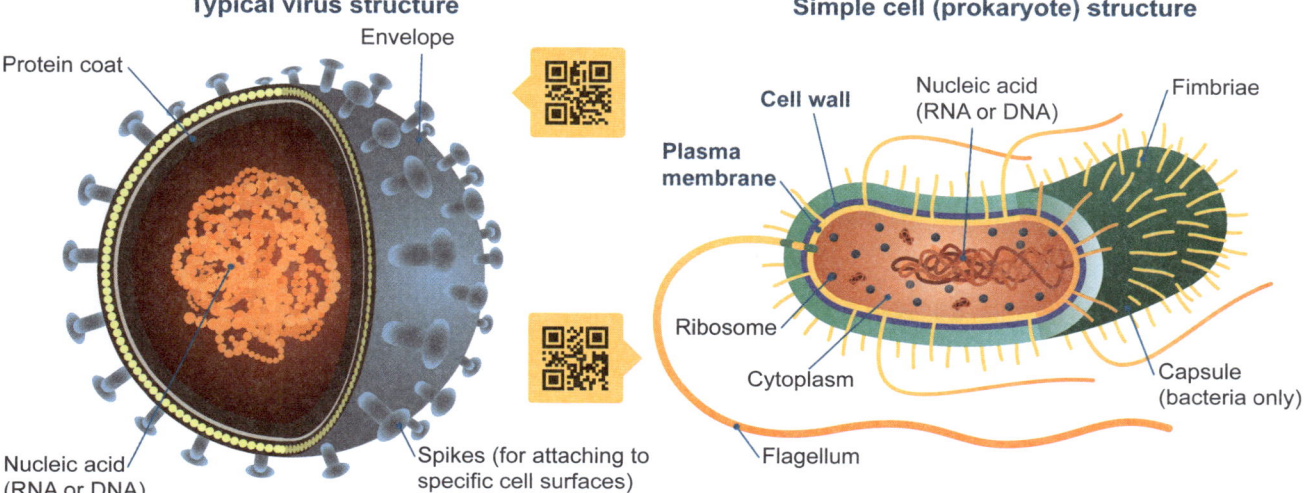

Size: Typically 0.02-0.25 mm length

Size: Typically 2-10 mm length

5. Use information from the labeled models above and previously in the chapter, of a typical virus and typical cell structure, to answer the questions below:

   (a) What structures are **present** in viruses and all cells? _____

   _____

   (b) What structural features are **absent** in a virus, but present in all cells? _____

   _____

   _____

6. Select two structural features from Q5 (b) and discuss how their absence in viruses is linked to the lack of life processes:

   _____

   _____

   _____

   _____

27 Viral Reproduction and Disease

Key Question: How do viruses enter cells, reproduce and cause disease?

- **Viruses** cause disease when they, or their genetic material, enter a **cell** and replicate. This process often kills the cell in the process, and sometimes the organism. Disease is a series of symptoms and immune system responses.
- Viruses will only cause disease in specific hosts, whose cells they can 'unlock' with adapted spikes or other structures. Viruses have no effect on organisms that are not hosts, i.e. tobacco mosaic virus causes humans no harm. Families of viruses can often cause similar disease symptoms, such as the coronavirus causing COVID-19 and the common cold.

Viral entry to cell host

- The specific mechanism and steps of viral entry depends upon the type of virus. For example, an enveloped RNA or DNA based virus (below), or a bacteriophage (opposite page), as well as the host it infects.
- Three processes (attachment, penetration, and uncoating, shown below) are shared by both DNA- and RNA-containing animal viruses. Generally, DNA viruses replicate their DNA in the **nucleus** of the host cell using viral enzymes, and synthesize their capsid and other **proteins** in the cytoplasm using the host cell's enzymes.

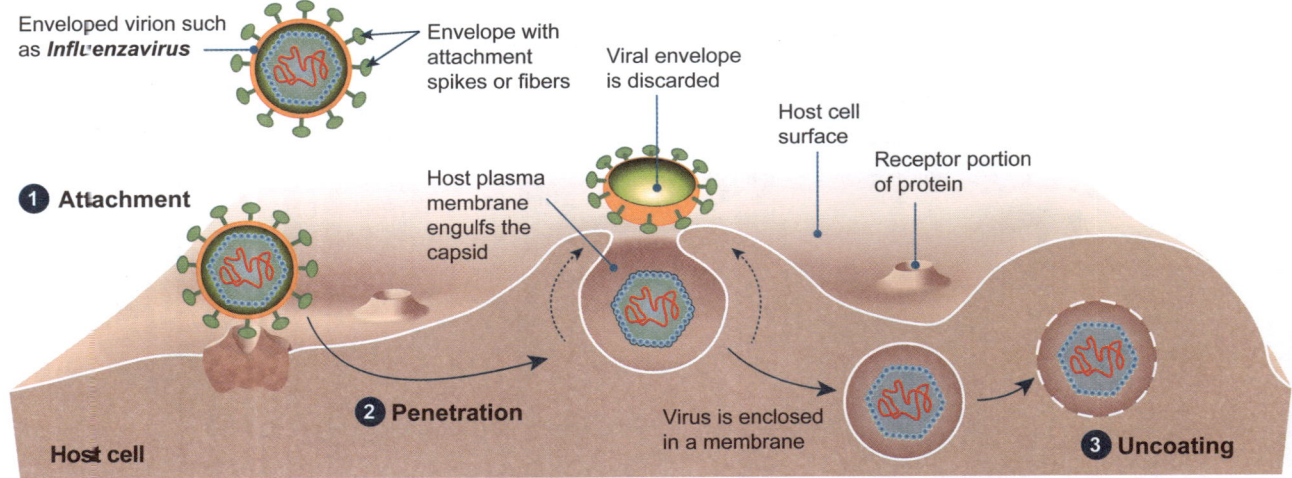

A variety of viral reproduction strategies

- Generally, DNA viruses replicate their DNA in the nucleus of the host cell using viral enzymes, and synthesize their capsid and other proteins in the cytoplasm using the host cell's enzymes.
- Viruses that infect animals are more complex and varied in structure than the viruses that infect bacteria (bacteriophage). Likewise, animal host cells are more diverse in structure and metabolism than bacterial cells.

Animal viruses
- Animal viruses exhibit a number of different mechanisms for replicating, i.e. entering a host cell and producing and releasing new virions. Enveloped viruses bud out from the host cell, whereas those without an envelope are released by rupture of the cell membrane.

Bacterial virus (bacteriophage)
- When infecting bacterial cells, bacteriophages produce an enzyme called endolysin using the cell's own protein synthesis mechanism in the ribosomes.
- This manufactured enzyme then targets particular areas in the cell membrane to rupture, or lyse, from the inside, and release the newly made virions.
- Endolysins have future potential as anti-viral drugs, as an alternative or supplement to antibiotics, as they are bacteria specific, and can be used on human pathogens.

Viral exit from cell host

Budding

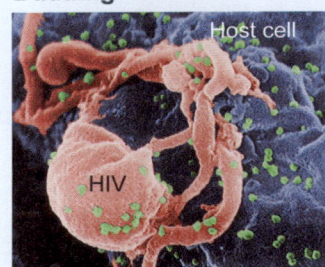

After replication, new viral particles (virions) leave the host cell to infect more cells. In animals, enveloped viruses bud from the host cell, e.g. HIV (shown left).

Rupturing

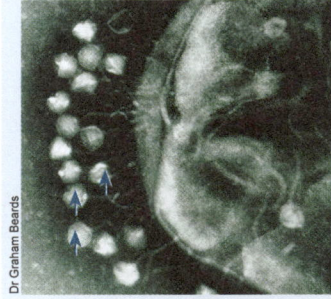

Bacteriophages (arrowed) are viruses that infect bacteria. They use tail fibers to attach to the host cell and a contractile region below the capsid to inject their DNA into the cell. To move from cell to cell these viruses rupture the cell membrane of the host.

Lysogenic and lytic cycles

▸ Viruses have two main reproductive processes, the lysogenic and lytic cycles. Some viruses use both, while other viruses only use the lytic cycle:
- In the lysogenic cycle, the viral genetic material is incorporated into the host's. It is then replicated when the cell is replicated - no virus particles are made. Disease is often in a 'dormant' state, with no symptoms seen. Because the lysogenic cycle allows a phage to reproduce without killing its host, the host could appear not to be infected.
- In the lytic cycle, the viral genetic material produces virus particles, that combine to form new viruses which then spread to other cells, causing damage to them. The bacteriophage shows both lysogenic and lytic cycles below.

Life cycle of λ–phage, a lysogenic bacteriophage

1 The phage attaches itself to a specific host cell, and inserts its nucleic acid and some enzymes into the bacterium.

2 A **prophage** forms when the viral nucleic acid integrates into the bacterial DNA.

3 **Lysogenic cycle**
The prophage's nucleic acid is fully integrated and replicates when the bacterial cells replicate. The bacteria may develop new properties as a result.

Lytic cycle
Prophages can be induced (by mutagens) to change from the lysogenic cycle to the lytic cycle. The lytic cycle results in the death of the host and the production of more phages.

4 Viral components are produced and assembled by the host cell.

5 The host cell bursts and new phages emerge to infect new cells.

1. Describe the purpose of the glycoprotein spikes found on some enveloped viruses: _____

2. (a) Explain the significance of endocytosis to the entry of an enveloped virus into an animal cell: _____

   (b) State where an enveloped virus replicates its viral DNA: _____

   (c) State where an enveloped virus synthesizes its proteins: _____

3. What is the significance of viral reproduction to the ability of a virus to cause disease? _____

4. Why might an organism appear to NOT have a disease if the virus it was infected by was in the lysogenic cycle? _____

5. Why is the viral lytic cycle so destructive to organisms? _____

28 How is Viral Disease Transmitted?

Key Question: How do viruses transmit disease from host to host?

▸ As you will recall, all **viruses** are pathogens (disease-causing agents) and cause infectious disease. An infectious disease is a disease which can be spread between individuals, known as transmission. Infectious disease from viruses can spread rapidly within and between hosts and regions, given the right conditions.

▸ The human body, like that of other large animals, is constantly exposed to a wide range of potential viral pathogens.

▸ Transmission and spread of a viral pathogen depends on its rate of growth, the density of the host population, the mobility of the host population, and the mode of transmission (see right).

Transmission and spread

Cough and fever are common symptoms of viral infection

Most pathogens, once inside the body, multiply rapidly, producing symptoms and making the host infectious within a few days. Others take longer to present symptoms. The infectious period can last from a few days to weeks, but in some cases the host may be infectious for long periods of time.

Human cities can contain millions of people, often living very closely together. In these overcrowded conditions infectious diseases can spread rapidly, especially if sanitation or personal hygiene is poor, or if seasonal weather produces conditions favorable to the spread of the pathogen. High speed transport (e.g. air travel) can help spread a pathogen around a region very quickly.

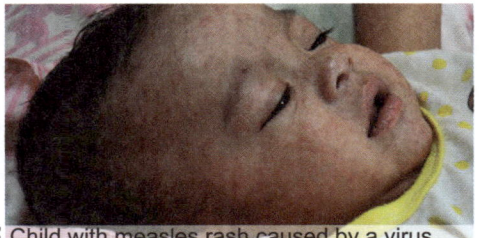
Child with measles rash caused by a virus

The type of transmission (direct, indirect, or vector) affects how quickly a pathogen can spread and also how easy (or difficult) it is to control its spread. Spread is also dictated by how infectious the pathogen is. Highly infectious pathogens spread much more rapidly than others. For example, measles is much more infectious than flu.

Portals of viral entry in humans

Respiratory tract
The mouth and nose are major entry points for pathogens, particularly airborne viruses, which are inhaled from the expelled mucus of infected people.

Virus examples: Covid-19, common cold, RSV, influenza, measles, rubella, and chickenpox.

Ebolavirus causes ebola disease

Gastrointestinal tract
Food and water are often contaminated with microorganisms, but most of these are destroyed in the stomach.

Virus examples: stomach flu (gastroenteritis), mumps, hepatitis A, poliomyelitis, and norovirus.

Flavivirus sp. causes zika virus

Breaking the skin surface
The skin provides an effective barrier to most pathogens, but cuts and abrasions allow pathogens to penetrate.

Virus examples: hepatitis B, rabies, zika, HIV, and Ross River virus.

Urinogenital openings
Urinogenital openings provide entry points for the pathogens responsible for sexually transmitted infections (STIs) and other opportunistic infections.

Virus examples: Human papillomavirus (HPV), and HIV.

Influenzavirus

HIV

Modes of viral transmission in humans

Direct contact	Indirect contact	Vector transmission
Direct person to person contact can occur when an infectious person touches or exchanges fluids with another.	Indirect transmission occurs when an infected person infects another without having direct contact with that person.	Vectors carry a pathogen from one person to another, or from an animal to a human. Different vectors carry different pathogens.

Person to person transmission (e.g. touching, kissing).

Contact with droplets produced when coughing or sneezing.

Fecal-oral transmission, e.g. someone not washing hands properly after using the toilet.

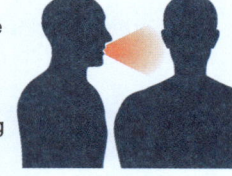

In some cases the pathogen can become airborne. Transmission can occur by breathing or coughing.

Transmission by contact with a contaminated object touched by an infected person.

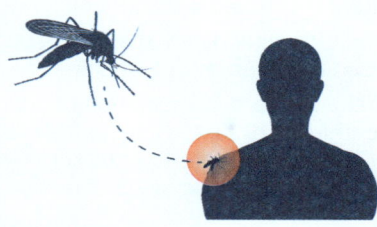

Vector-borne pathogens are commonly carried by insects but sometimes other animals (e.g. ticks). They not only infect a person but can transfer the pathogen from person to person.

A well know example is the Zika virus transmitted by the *Aedes* mosquito.

1. Why can disease spread quickly in congested human cities? _____
   _____
   _____

2. Why would transmission by direct touch be slower than transmission by coughing or sneezing? _____
   _____
   _____
   _____

3. Research a virus, other than Covid-19, that affects humans. Record your findings in the table below:

Name of virus	How is the disease transmitted to humans?	How does the virus enter the body?
What are the symptoms of the viral disease?		

29 Epidemics and Pandemics

Key Question: How does transmission of viral disease become an epidemic or pandemic?

- The ability of a viral disease to spread and reach epidemic status depends on how easily the pathogen can spread, how resistant the population is to it, and how effective control measures are to contain it.
- An epidemic is an increase in the number of cases of a disease above what is normally expected in a population. Often, the increase occurs suddenly.
- When the viral disease spreads uncontrollably over multiple countries or continents at once it is termed a pandemic. Covid-19 disease was declared a pandemic by the World Health Organization on the 11th March, 2020.
- Viral disease outbreaks have occurred throughout history. The 'Spanish flu', an H1N1 variant that spread across the world from 1918-1919 as soldiers returned home from the Great War, infected up to 500 million people. At least 50 million deaths were attributed to the disease. Today, medical and technological advances, disease modeling, and improved infrastructure and health services can be used to help predict and control disease transmission.

Predicting future disease is often a case of identifying diseases in animals that could cross over to humans. Many viral infectious diseases have an animal origin, including various strains of influenza (e.g. avian flu H5N1 and swine flu H1N1). Identifying these pathogens in animals, especially livestock and poultry living in close proximity to humans, can help prepare for possible outbreaks.

Population density is important to how quickly an infectious disease can spread. Cities generally have very high population densities. Disease spreads most quickly in areas with poor living conditions, poor sanitation, and low levels of immunity. For example, the 1918 Spanish flu initially spread quickly due to the cramped conditions of military camps and hospital wards in Europe.

How quickly an infectious disease spreads also depends on the population's mobility. Covid-19 spread around the world quickly due to fast and extensive international travel, mainly by air. Predicting the movements of groups of people can help predict where and when diseases will occur.

What causes a viral epidemic?

- An epidemic of infectious viral disease occurs when a **virus** and its susceptible host(s) are present in sufficient numbers to allow the virus to move between, and infect, individuals reasonably quickly.
  Viral epidemics can be caused by several factors:
- The virus has increased in numbers.
- The virus has become more infectious. This may be due to a genetic change (mutation).
- The virus has been introduced into an area or into a population where it has not previously been.
- The host has become more susceptible to the virus, e.g. through overcrowding, presence of other diseases, reduced immunity, low vaccination rates, poor sanitation.

The spread of disease also depends on the mode of transmission. Is it spread by airborne particles or by touch? The most feared scenario is an airborne pathogen that is highly contagious, has a long infectious period, and is ultimately deadly.

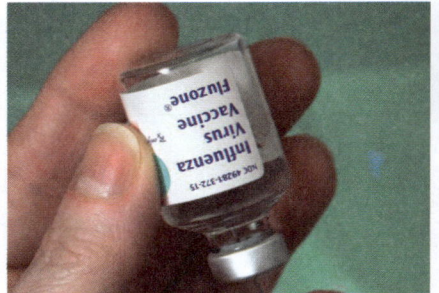

Constant changes occur in viral genetic material due to mutation. This means that, even with prior infection and vaccination, waves of new virus variants in diseases such as flu, colds, and Covid-19, continue to spread, over and over.

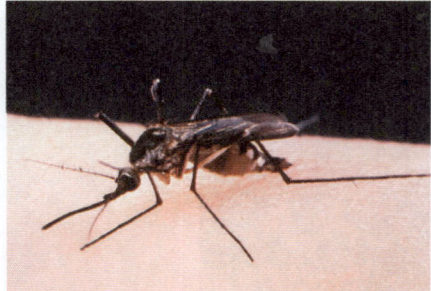

Increasing global warming is likely to increase instances of new disease outbreaks. For example, zika may spread further north and south from the tropics as the climate becomes more favorable for its *Aedes sp.* mosquito vectors.

Predicting viral disease spread

▸ Understanding how a **virus** is transmitted and how quickly it can spread can help authorities plan and prepare for an increase in case numbers.

▸ In ancient times, the quickest way to reach a destination was to travel on land, usually by walking. Distance predicted how quickly a disease would spread. A disease often spread in a circular pattern from a central point or followed common trade routes.

▸ Today, it is not always distance that dictates the spread of a disease, but time. More efficient travel (cars, buses, trains, and planes) allow people to travel further, more quickly. If they are infected, the disease is also traveling at a more rapid pace.

▸ Global air travel routes often make it faster and easier to reach a destination by flying from a small hub to a larger one, then on to another small hub, rather than flying directly between two small hubs.

▸ The spread of a disease can be modeled based on its origin, where the major travel hubs are located, and the time taken to get between those hubs.

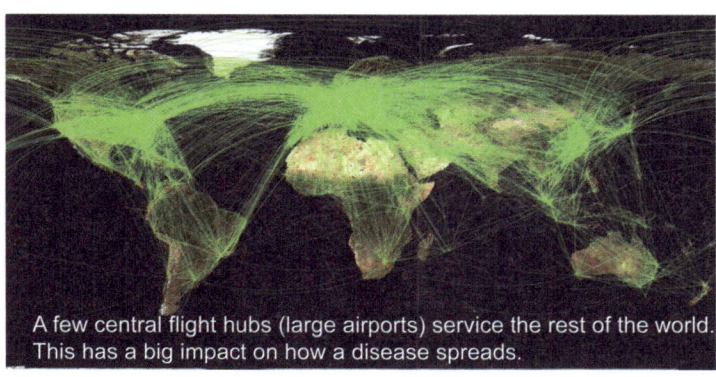

A few central flight hubs (large airports) service the rest of the world. This has a big impact on how a disease spreads.

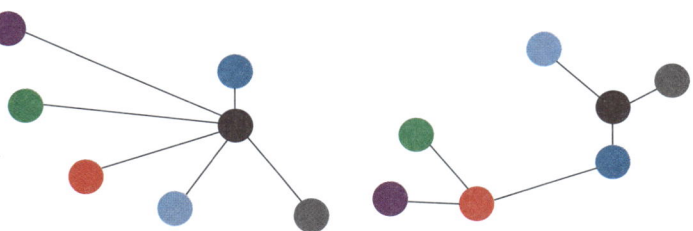

Based on distance, a travel map may look like this. Based on flight times, a travel map may look like this.

1. Explain how each of the following are important in predicting where the next epidemic may originate:

   (a) Virus in livestock related to human diseases: _____

   (b) Population density: _____

   (c) Global travel networks: _____

   (d) Potential novel or new virus: _____

   (e) Climate change: _____

2. In the boxes below, draw a diagram to show how a disease might spread, based solely on distance from the source, and then based on modern travel times from the source, as shown above:

   Travel distance

   Travel time

Zoonotic viral diseases and transmission

▸ Some viruses that infect other species can also be spread and infect humans. The diseases they cause in humans are called zoonotic diseases, and can be particularly dangerous, as humans have no natural immunity to them.

▸ The SARS-CoV-2 virus that has led to Covid-19 disease in humans is zoonotic. Recent research has suggested that the raccoon dog (*Nyctereutes procyonoides*) could have been the natural host. The transmission host is thought to be a cat-like civet, also found in the same location. Finally, humans became the spillover host in late 2019, and person-to-person transmission began. The virus has since mutated, many times, into different variants that have become more contagious than the original virus.

▸ Other examples of viral zoonotic diseases are shown below.

Zika virus is primarily transmitted to people when they are bitten by an infected *Aedes* mosquito. Symptoms are usually mild but infection during pregnancy carries a high risk of miscarriage, pregnancy complications, and fetal abnormalities. Zika virus can also be transmitted from the mother to fetus, through blood transfusions and organ donations, and through sexual contact.

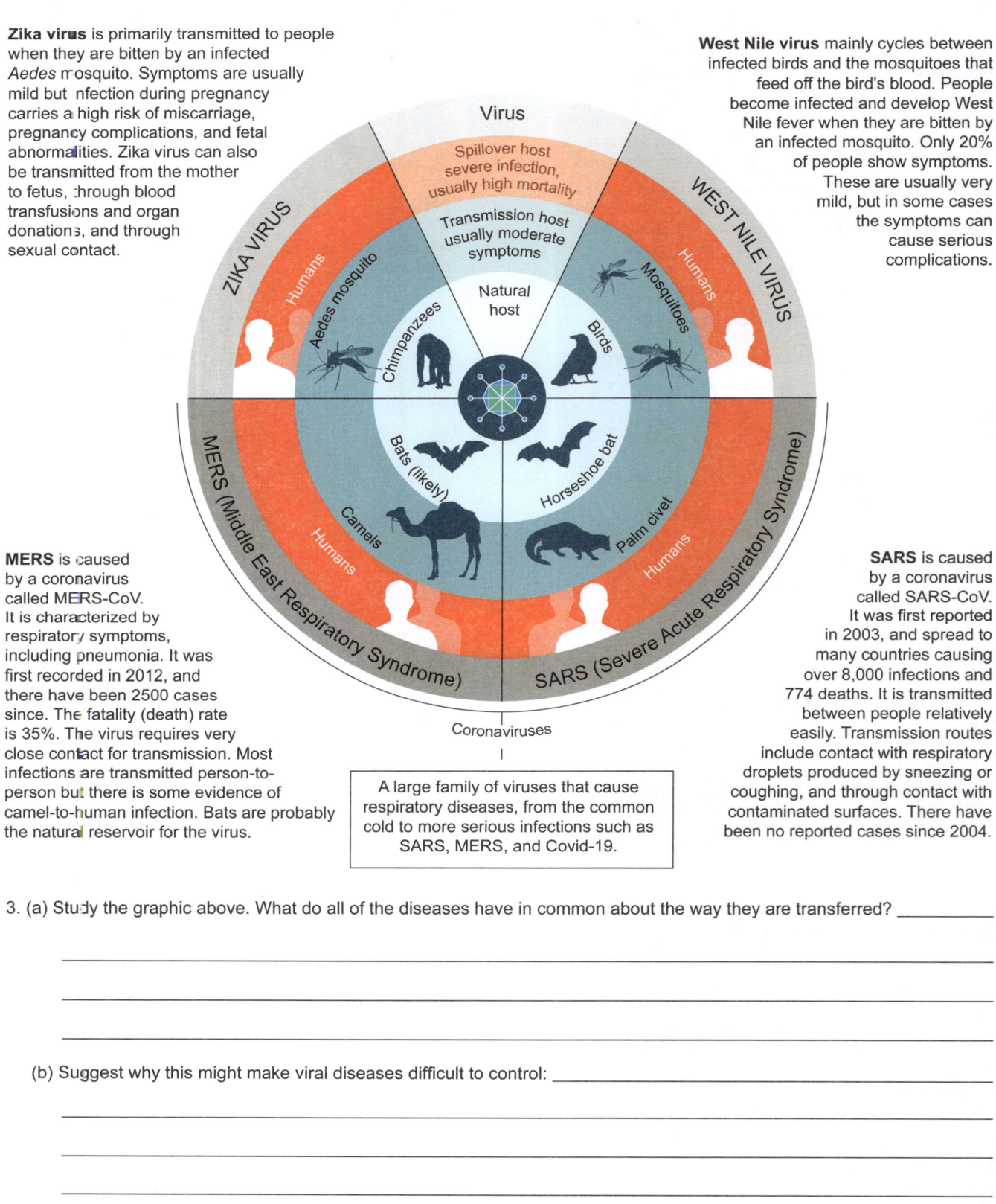

West Nile virus mainly cycles between infected birds and the mosquitoes that feed off the bird's blood. People become infected and develop West Nile fever when they are bitten by an infected mosquito. Only 20% of people show symptoms. These are usually very mild, but in some cases the symptoms can cause serious complications.

MERS is caused by a coronavirus called MERS-CoV. It is characterized by respiratory symptoms, including pneumonia. It was first recorded in 2012, and there have been 2500 cases since. The fatality (death) rate is 35%. The virus requires very close contact for transmission. Most infections are transmitted person-to-person but there is some evidence of camel-to-human infection. Bats are probably the natural reservoir for the virus.

SARS is caused by a coronavirus called SARS-CoV. It was first reported in 2003, and spread to many countries causing over 8,000 infections and 774 deaths. It is transmitted between people relatively easily. Transmission routes include contact with respiratory droplets produced by sneezing or coughing, and through contact with contaminated surfaces. There have been no reported cases since 2004.

Coronaviruses: A large family of viruses that cause respiratory diseases, from the common cold to more serious infections such as SARS, MERS, and Covid-19.

3. (a) Study the graphic above. What do all of the diseases have in common about the way they are transferred? _____

_____

_____

_____

(b) Suggest why this might make viral diseases difficult to control: _____

_____

_____

_____

30 Modeling Viral Disease Outbreak and Spread

Key Question: How can computer modeling be used to demonstrate viral disease outbreak and spread?

Modeling a disease

▶ A spreadsheet can be used to model the spread of viral disease. Numerous online models can also be used.
▶ In the most simple model (right), whenever an infected person meets another, a new infection occurs. The number of interactions at each infection cycle affects the spread of the disease.
▶ Using a spreadsheet, you will first model an infected person meeting (and infecting) two other people. In this model, once the infected person has infected two people they are no longer infectious.

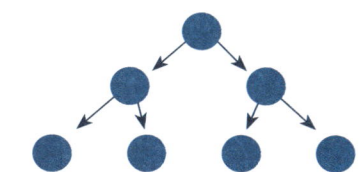

A simple infection model. One person infects two, who infect two more...

Investigation 1.4 Modeling disease outbreak and spread

See appendix for equipment list.

1. Working in pairs, enter the following into a spreadsheet:

	A	B
1	New infections	Total Infections
2	1	=SUM($A$2:A2)
3	=A2*2	=SUM($A$2:A3)
4	=A3*2	=SUM($A$2:A4)
5	=A4*2	=SUM($A$2:A5)
6	=A5*2	=SUM($A$2:A6)
7	=A6*2	=SUM($A$2:A7)
8	=A7*2	=SUM($A$2:A8)
9	=A8*2	=SUM($A$2:A9)

One infection cycle. Copy this down to row 12 (10 cycles of interactions).

1. How many new infections are there per infection cycle after 10 infection cycles? _____

2. How many infected people are there in total after 10 infection cycles? _____

2. Now set the interactions per infected person to 3 (A2*3) and reset the model.

3. How many new infections are there per cycle of infection after 10 infection cycles? _____

4. How many infected people are there after 10 cycles of infection? _____

3. We can now extend the model by adding in a little randomness. The number of people interacting with each infected person may not always be the same. In our extended model, we shall randomize the number of people interacting to between 1 and 4.

	A	B	C
1	New infections	People interacted with per person	Total infected people
2	1	=RANDBETWEEN(1,4)	=SUM($A$2:A2)
3	=A2*B2	=RANDBETWEEN(1,4)	=SUM($A$2:A3)
4	=A3*B3	=RANDBETWEEN(1,4)	=SUM($A$2:A4)
5	=A4*B4	=RANDBETWEEN(1,4)	=SUM($A$2:A5)
6	=A5*B5	=RANDBETWEEN(1,4)	=SUM($A$2:A6)
7	=A6*B6	=RANDBETWEEN(1,4)	=SUM($A$2:A7)
8	=A7*B7	=RANDBETWEEN(1,4)	=SUM($A$2:A8)
9	=A8*B8	=RANDBETWEEN(1,4)	=SUM($A$2:A9)
10	=A9*B9	=RANDBETWEEN(1,4)	=SUM($A$2:A10)

Add new infections to total from previous row (cycle) — Generates a random number between 1 and 4 — Calculates the total number of people infected

5. Run the model five times by recalculating the spreadsheet using the **recalculate** or **calculate now** option (depending on your spreadsheet). On average, how many people in total have been infected after ten cycles?

_____

_____

©2024 **BIOZONE** International
ISBN: 978-1-99-101405-4
Photocopying prohibited

4. Not all interactions will result in an infection. The pathogen may not be highly infectious or the correct mode of transmission may not have occurred (for example, a person with a cold may have been careful where and how they coughed).

5. First we need to decide the probability of each interacting person being infected. For this model, we will say there is a 50% chance that any interacting person will be infected. We shall first produce a random number between 0 and 1 (see * below). We can now use this block of infected (1) or not infected (0) cells in our model. Once the formula is set up, you can recalculate the spreadsheet to obtain different infection scenarios.

	A	B	C	D
1	New infections	People interacted with per person	Infected people	Total infected people
2	1	=RANDBETWEEN(1,4)	=IF(B2=1,$B$15,IF(B2=2,$B$15+$B$16,IF(B2=3,$B$15+$B$16+$B$17,IF(B2=4,$B$15+$B$16+$B$17+$B$18))))	=SUM($A$2:A2)
3	=A2*C2	=RANDBETWEEN(1,4)	=IF(B3=1,$B$15,IF(B3=2,$B$15+$B$16,IF(B3=3,$B$15+$B$16+$B$17,IF(B3=4,$B$15+$B$16+$B$17+$B$18))))	=SUM($A$2:A3)
4	=A3*C3	=RANDBETWEEN(1,4)	=IF(B4=1,$B$15,IF(B4=2,$B$15+$B$16,IF(B4=3,$B$15+$B$16+$B$17,IF(B4=4,$B$15+$B$16+$B$17+$B$18))))	=SUM($A$2:A4)
5	=A4*C4	=RANDBETWEEN(1,4)	=IF(B5=1,$B$15,IF(B5=2,$B$15+$B$16,IF(B5=3,$B$15+$B$16+$B$17,IF(B5=4,$B$15+$B$16+$B$17+$B$18))))	=SUM($A$2:A5)
6	=A5*C5	=RANDBETWEEN(1,4)	=IF(B6=1,$B$15,IF(B6=2,$B$15+$B$16,IF(B6=3,$B$15+$B$16+$B$17,IF(B6=4,$B$15+$B$16+$B$17+$B$18))))	=SUM($A$2:A6)
7	=A6*C6	=RANDBETWEEN(1,4)	=IF(B7=1,$B$15,IF(B7=2,$B$15+$B$16,IF(B7=3,$B$15+$B$16+$B$17,IF(B7=4,$B$15+$B$16+$B$17+$B$18))))	=SUM($A$2:A7)
8	=A7*C7	=RANDBETWEEN(1,4)	=IF(B8=1,$B$15,IF(B8=2,$B$15+$B$16,IF(B8=3,$B$15+$B$16+$B$17,IF(B8=4,$B$15+$B$16+$B$17+$B$18))))	=SUM($A$2:A8)
9	=A8*C8	=RANDBETWEEN(1,4)	=IF(B9=1,$B$15,IF(B9=2,$B$15+$B$16,IF(B9=3,$B$15+$B$16+$B$17,IF(B9=4,$B$15+$B$16+$B$17+$B$18))))	=SUM($A$2:A9)
10	=A9*C9	=RANDBETWEEN(1,4)	=IF(B10=1,$B$15,IF(B10=2,$B$15+$B$16,IF(B10=3,$B$15+$B$16+$B$17,IF(B10=4,$B$15+$B$16+$B$17+$B$18))))	=SUM($A$2:A10)
11	=A10*C10	=RANDBETWEEN(1,4)	=IF(B11=1,$B$15,IF(B11=2,$B$15+$B$16,IF(B11=3,$B$15+$B$16+$B$17,IF(B11=4,$B$15+$B$16+$B$17+$B$18))))	=SUM($A$2:A11)
12	=A11*C11	=RANDBETWEEN(1,4)	=IF(B12=1,$B$15,IF(B12=2,$B$15+$B$16,IF(B12=3,$B$15+$B$16+$B$17,IF(B12=4,$B$15+$B$16+$B$17+$B$18))))	=SUM($A$2:A12)
13				
14				
15	=RAND()	=IF(A15>0.5,1,0)		
16	=RAND()	=IF(A16>0.5,1,0)		
17	=RAND()	=IF(A17>0.5,1,0)		
18	=RAND()	=IF(A18>0.5,1,0)		

* Produces a 50% probability of infection

The "IF" statement incorporates the number of interactions and the probability of infection into the model

6. Run the model five times by recalculating the spreadsheet as before. On average, how many people in total have been infected after ten cycles now?

_____

7. The third model above is much more realistic than the first, but still lacks many factors that would affect the model outcome. List at least three factors that could be added to the model to make it even more realistic:

_____
_____
_____
_____

Modeling with S, I, and R

▶ A more advanced predictive mathematical model than your spreadsheet called SIR can be used to show the transmission of infectious diseases. In this model there are three compartments: **S** (the number of susceptible individuals), **I** (the number of infected individuals), and **R** the number removed (those who have been removed through recovery or death).

▶ The data in the table (below right) is a theoretical example. It assumes a closed system, e.g. a single state with no travel, no prior immunity (everyone is susceptible), no vaccine, and no physical distancing or other precautionary measures in place.

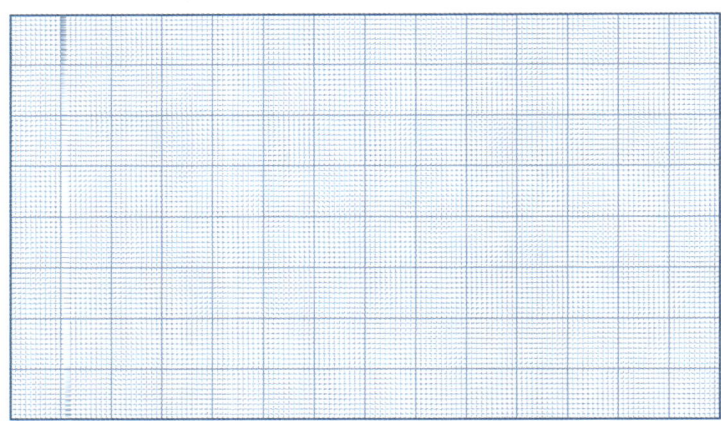

Week	S	I	R
0	7,000,000	2	0
1	6,999,986	15	1
2	6,999,881	113	8
3	6,999,090	847	65
4	6,993,162	6352	488
5	6,948,741	47,597	3664
6	6,618,002	354,538	27,462
7	4,271,669	2,523,602	204,731
8	0	5,533,470	1,466,532
9	0	2,766,735	4,233,267
10	0	1,383,368	5,616,634
11	0	691,684	6,308,318
12	0	345,842	6,654,160
13	0	172,921	6,827,081
14	0	86,460	6,913,542

8. Plot the tabulated SIR data (right) on the grid provided. Plot all three data sets on one axis and add a key.

31 Viral Disease Case Study: Covid-19

Key Question: How is SARS-CoV-2 coronavirus transmitted, and how does it cause disease in humans?

▶ Reports of a novel viral respiratory disease in Wuhan, China were reported on the 31st December 2019. Early in January 2020, a new coronavirus, SARS-CoV-2, was identified as the cause of the infections. By 11th March, 2020, it had spread quickly across the world and was declared a pandemic by the World Health Organization.

What is Covid-19?

▶ Covid-19 is the human disease caused by infection with the SARS-CoV-2 **virus** (right).
▶ The virus affects the respiratory, and other, systems.
▶ Most infected people recover without specialized care, but some develop severe breathing problems and may require high level hospital care. The elderly, and people with underlying medical problems, are most at risk of becoming very sick.
▶ The virus is transmitted person-to-person, and is highly contagious.
▶ Vaccines have now been developed and have helped reduce the severity of symptoms in many people.
▶ Numerous mutations, or variants, of the virus have emerged since 2019.

A representation of the SARS-CoV-2 virus

Proteins
Viral envelope (mostly **lipid**)
Glycoprotein spikes

Transmission of SARS-CoV-2

▶ The SARS-CoV-2 virus, which causes COVID-19, can be spread in the tiny liquid particles given off when an infected person coughs, sneezes, speaks, or breathes.
▶ Transmission can be by direct or indirect contact. Droplets may land on surfaces that are then touched by others. The virus needs to be breathed in, or make contact with the mouth or nose. A person may touch a contaminated surface with their hands then touch their nose or use their hands to eat, and so become infected by the virus.
▶ The modes of transmission of SARS-CoV-2 make wearing masks and washing hands before eating important factors in slowing the spread of the virus.

1. (a) Name the virus that causes Covid-19: _____

   (b) How is the virus spread? _____
   _____
   _____
   _____

2. Which disease is SARS-CoV-2 most similar to, in terms of its transmission? _____
   _____

3. Why is coughing into your elbow more effective at stopping the spread of a virus than coughing into your hands?
   _____
   _____

4. Where do vector-borne pathogens like SARS-CoV-2 originate? _____
   _____

5. Why did the virus suddenly start infecting humans when it had already been present in the environment? _____
   _____

Covid-19 disease can lead to Acute Respiratory Distress Syndrome (ARDS)

▶ In serious cases, SARS-CoV-2, the coronavirus that causes Covid-19, causes the development of pneumonia and Acute Respiratory Distress Syndrome (ARDS).

▶ Viral pneumonia, a common consequence of severe Covid-19, impacts breathing. The lungs fill with fluid and become inflamed. Surviving patients usually face many weeks of lingering symptoms, such as coughing and shortness of breath.

▶ ARDS is not a single respiratory disease, but a set of related syndromes. Developing Covid-19 induced ARDS is commonly fatal. Life-threatening damage to both the alveoli and surrounding capillaries normally requires the patient to be placed on mechanical ventilation while the lungs heal.

▶ The SARS-CoV-2 protein binds to ACE receptors on the epithelial calls of the alveoli, setting off a 'cytokinetic storm'. This is an exaggerated immune response, and creates 'leaky' capillaries surrounding the alveoli. Cytokines, released by immune **cells** in the blood, cause an inflammatory response. Alveoli swell, and fluid from the capillaries leak in to fill the alveoli sacs, preventing oxygen from entering the capillaries. Covid-19 induced ARDS resulted in a mortality rate of 39% of those hospitalized at the peak of the pandemic. This mainly impacted older patients and those with other underlying health conditions.

Covid-19 induced ARDS and the effect on the respiratory system

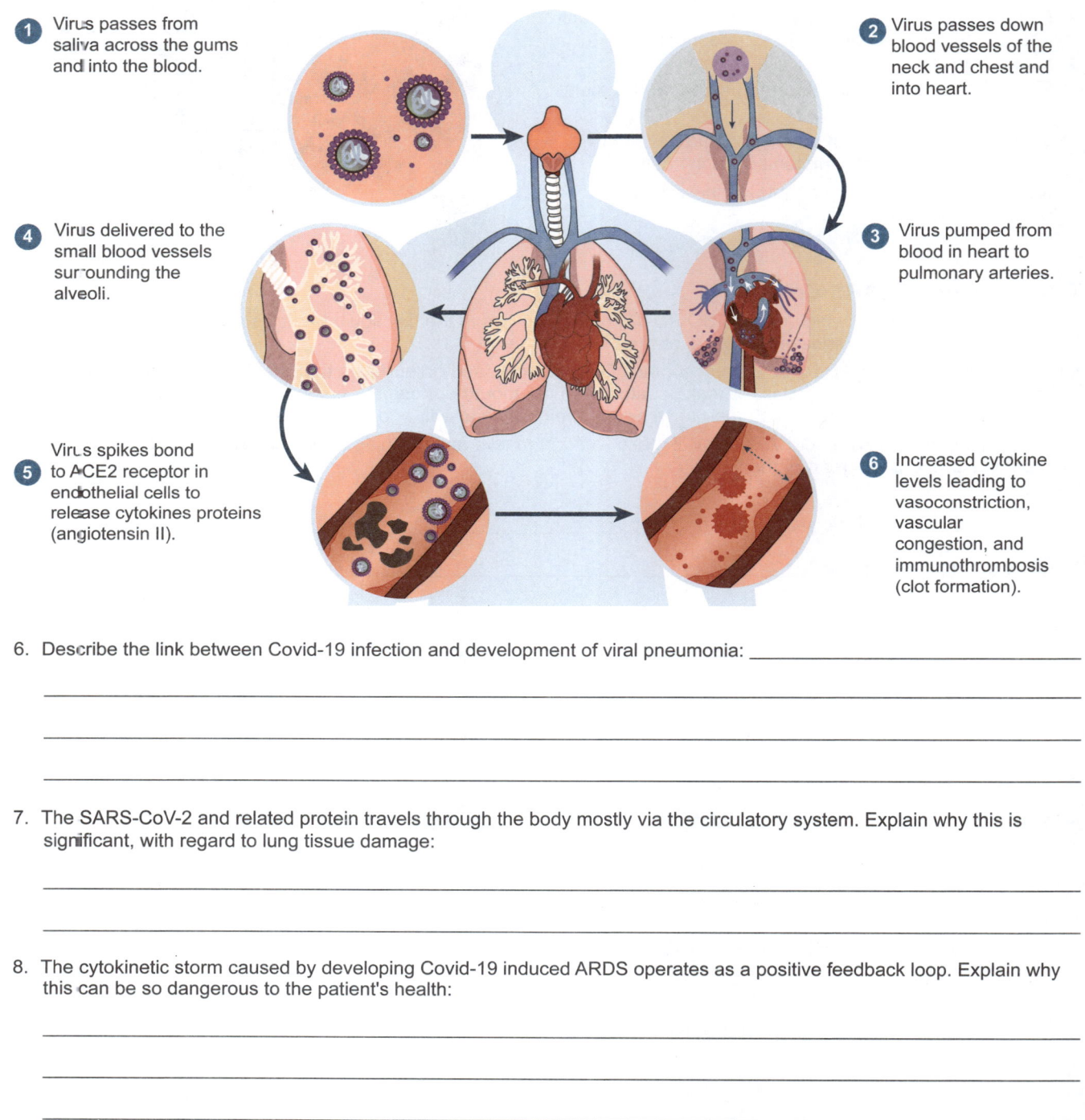

1. Virus passes from saliva across the gums and into the blood.
2. Virus passes down blood vessels of the neck and chest and into heart.
3. Virus pumped from blood in heart to pulmonary arteries.
4. Virus delivered to the small blood vessels surrounding the alveoli.
5. Virus spikes bond to ACE2 receptor in endothelial cells to release cytokines proteins (angiotensin II).
6. Increased cytokine levels leading to vasoconstriction, vascular congestion, and immunothrombosis (clot formation).

6. Describe the link between Covid-19 infection and development of viral pneumonia: _____
_____
_____
_____

7. The SARS-CoV-2 and related protein travels through the body mostly via the circulatory system. Explain why this is significant, with regard to lung tissue damage:
_____
_____

8. The cytokinetic storm caused by developing Covid-19 induced ARDS operates as a positive feedback loop. Explain why this can be so dangerous to the patient's health:
_____
_____
_____

32 What is a Sponge? Revisited

Content Anchor Revisited: How can we tell what type of organism a sponge is by using a microscope?

Several sponges filter feeding in their marine ecosystem

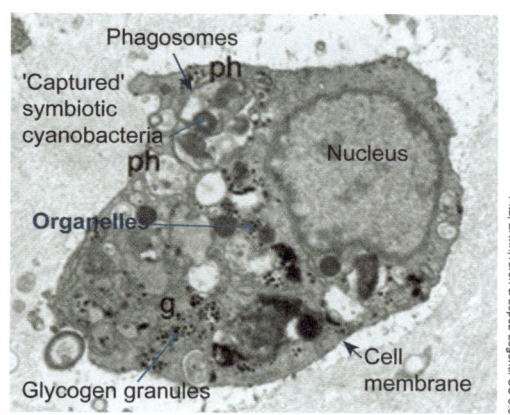
Microscopic image of Mediterranean sponge *Ircinia sp.* **cell** showing captured cyanobacteria.

Sponges: plant, animal, fungi, or something else?

▶ At first glance a sponge looks like some type of ocean plant or algae (above left). There are no visible sense organs or any of the other common structures that would identify it as an animal.

▶ However, if a sample of sponge was placed under the microscope (above right), then very obvious features would soon indicate its true biological classification.

1. Beside each microscopic sponge structure or structures, explain in detail why each observation is evidence of a sponge being classified as an animal and why it excludes it from being something else:

    (a) A visible cell, one of many differentiated cells in a large multicellular structure: _____

    _____

    _____

    (b) A nucleus: _____

    _____

    _____

    (c) A cell membrane, but no cell wall: _____

    _____

    _____

    (d) Has mitochondria, but no chloroplasts (except those in captured cyanobacteria): _____

    _____

    _____

2. The sponge is a filter feeder, pumping seawater through its multicellular body to capture small plankton. It also has a **symbiotic** relationship with photosynthetic bacteria, as seen in the phagosomes in the cell image. The sponge cell image above has glycogen granules, a stored form of glucose. Discuss how both the photosynthetic prokaryote bacteria and eukaryote sponge obtain their energy, related to structures in their cells:

    _____

    _____

    _____

    _____

3. Using what you have learned in this chapter, return to activity 1 and redraw what you might see in a typical sponge cell.

33 Summing Up

Read each question carefully. Place a cross in the box beside the best answer to the question from the four answer choices provided.

1. Which of the following elements is **not found** in carbohydrates:
   - (a) Carbon
   - (b) Nitrogen
   - (c) Oxygen
   - (d) Hydrogen

2. Which of the following elements is **not found** in nucleic acids:
   - (a) Oxygen
   - (b) Nitrogen
   - (c) Sulfur
   - (d) Phosphorus

3. Which of the following form the structure of biological membranes:
   - (a) Neutral fats
   - (b) Phospholipids
   - (c) Fibrous proteins
   - (d) Carbohydrates

4. The cell component **not found** in a prokaryotic cell is:
   - (a) Plasma membrane
   - (b) Chromosome
   - (c) Ribosomes
   - (d) Nucleus

5. Which of the following relationships between cell structures and their respective functions is correct?
   - (a) Cell wall: support, protection
   - (b) Chloroplasts: chief sites of cellular respiration
   - (c) Chromosomes: cytoskeleton of the nucleus
   - (d) Ribosomes: secretion

6. From the diagram below, what is likely to **occur next**?

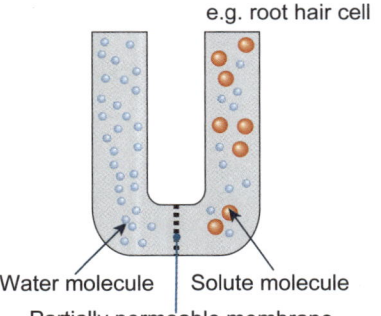

   - (a) Nothing
   - (b) Water moves from right to left
   - (c) Solute spreads out from right to left
   - (d) Water moves from left to right

7. What is the **most correct** term for the process that is likely to occur in Q6?
   - (a) Passive transport: Osmosis
   - (b) Active transport: Osmosis
   - (c) Facilitated diffusion
   - (d) Coupled transport (cotransport)

8. Scientists originally thought that some diseases were due to toxins from bacteria rather than viruses. The link between viruses and disease was not confirmed by scientists until after 1931. Why was this?
   - (a) Viruses are not that common
   - (b) No-one was looking for viruses
   - (c) Microscopes with high enough magnification had not been developed to see viruses
   - (d) Only some viruses cause disease

9. Viruses cause disease to hosts when they undergo a particular 'life' process near or inside a cell. What is that process called?
   - (a) Reproduction
   - (b) Movement
   - (c) Respiration
   - (d) Excretion

10. What **can** the graph below tell us about the pattern of spread of the H1N1 virus causing swine flu in Australia in 2009?

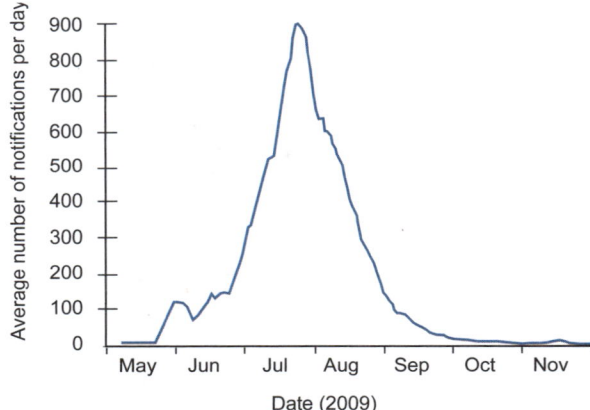

   - (a) H1N1 disease was declared a pandemic
   - (b) H1N1 disease was eliminated due to vaccination
   - (c) Swine flu was likely caused by a zoonotic virus
   - (d) Australia had an outbreak of swine flu in 2009

11. Zoonotic viruses are those that pass through an animal host through to humans, usually through an intermediate transition host. The disease outbreak is normally rapid in human populations, and can cause severe health impacts. What is the **most likely** reason?
   - (a) The viruses from other animals are dangerous
   - (b) Humans are likely to have no immunity to them
   - (c) Zoonotic diseases are more contagious
   - (d) Zoonotic diseases occur in crowded populations

12. (a) List the four main macromolecule components of living organisms: _____

_____

(b) List the elements that all these macromolecules share: _____

13. (a) What is a polysaccharide? _____

_____

(b) Give some examples of polysaccharides involved in cellular structure or function: _____

_____

14. Suggest why polysaccharides are such a good source of energy for cells: _____

_____

_____

15. (a) What are the main features of a prokaryotic cell? _____

_____

_____

(b) What are the main features of a eukaryotic cell? _____

_____

_____

16. Use the information from the model below to answer the following questions:

| Original pre-eukaryotic cell | Oxygen-using bacterium engulfed | Photosynthetic cyanobacterium engulfed | Plant cell Chloroplast |

↓ Animals   Mitochondrion

(a) What cellular evidence in eukaryotes can be observed to support the endosymbiosis theory? _____

_____

_____

_____

(b) Archaea and bacteria, both prokaryotes, were present on earth before plants and animals. What cellular evidence supports this statement:

_____

_____

_____

17. Discuss the key differences in structure between a typical virus and a typical prokaryote cell. How do these relate to viruses being considered as non-living?

_____

_____

_____

_____

18. Describe the difference between a pandemic and an epidemic: _____
_____
_____

19. The diagram (right) shows a simplified model of disease transmission of a virus.

    (a) Pathogen A spreads very easily between people and its most likely entry route is through a person's nose and mouth. What is pathogen A's most likely mode of transmission?
    _____
    _____
    _____

    (b) Researchers suspect that a new pathogen (pathogen B) infecting humans may have been transferred from a particular species of bats to humans, which rapidly resulted in a pandemic. How did the pandemic occur when not every country has the species of bat present?
    _____
    _____
    _____

    (c) What would the researchers need to prove in order to confirm their suspicion about the transmission of pathogen B?
    _____
    _____
    _____

20. Select a human disease caused by a virus, other than Covid-19. Research to find information about the virus structure, how it causes disease in humans, and how it has spread through human populations. Summarize your findings with text and diagrams below. Attach a reference list of where your information was found, including at least 5 sources.

CHAPTER 2: Cell Cycle

TEKS
Scientific and Engineering Practices

B.1: Investigation and Inquiry
1.A 1.C 1.F 1.G

B.2: Data and Patterns
2.B 2.C

B.3: Communicating in Science
3.A 3.B

B.4: Science as a Human Endeavor
4.A 4.B

TEKS
Science Concepts

B6.A explain the importance of the cell cycle to the growth of organisms, including an overview of the stages of the cell cycle and deoxyribonucleic acid (DNA) replication models

B6.B explain the process of cell specialization through cell differentiation, including the role of environmental factors

B6.C relate disruptions of the cell cycle to how they lead to the development of diseases such as cancer

Learning Outcomes
I know I have achieved this when I can:

	Learning Outcome	Activity number
☐	Explain the three functions of mitotic cell division.	35
☐	Identify the main phases, and their sub-divisions, in the eukaryote cell cycle.	36
☐	Describe the events that occur in each phase of the cell cycle.	36
☐	Explain how the cell cycle is regulated by the checkpoints.	36
☐	Name, order, and summarize the key steps of mitosis.	37
☐	Model the process of mitosis.	38
☐	Identify and describe the key steps of DNA replication, including the role of enzymes.	39-40
☐	Model Meselson and Stahl's semi-conservative DNA replication experiments.	41
☐	Use evidence to support the semi-conservative model.	41
☐	Describe the properties of stem cells, and their role in differentiation of cells.	42
☐	Analyze the differentiation of blood cells.	43
☐	Link the function of different plant and animal cells to their specialization.	44-45
☐	Explain the advantage of cell specialization to an organism.	45
☐	Investigate the effect of environmental factors, such as temperature, on cell differentiation, and subsequent phenotype changes in selected case studies.	46
☐	Plot and analyze data to make an evidence-based statement on the link between environmental influences and changes in gene expression.	47
☐	Identify genes responsible for cell cycle checkpoint regulation.	48
☐	Describe how disruptions in regulatory genes can lead to development of cancer.	48
☐	Explain the link between decreased apoptosis rate and the development of cancer.	48
☐	Describe how cancer causing viruses affect the cell cycle.	48

RESOURCE HUB

bit.ly/3IUemgO

ELPS English Language Proficiency Standards

		Page number
Learning	**The Power to Rebuild.** When you read questions, look for signal words to help you analyze your task. For example, question 2 uses the word *compare*. This word tells you to look for similarities and differences. Question 5 uses the words *discuss* and *record*, so you know you will talk about something and then write down your conclusions.	**63**
 **Listening**	**The Power to Rebuild.** Listen actively to your partner's comments. When you do not understand, ask for the help you need. You can use the sentence frames: *I don't understand _____. What do you mean by _____?* You can also ask your partner to speak more slowly, repeat their words, or point to a word on the page.	**63**
Speaking	**Modeling Mitosis.** Work in a group to complete the investigation. The key question and instructions mention stages in a process. Think about other words you use to describe processes such as *before*, *during*, and *after*. As you build and discuss your model, use process words in sentence frames such as: Before mitosis begins _____. During prophase _____. After telophase _____.	**69**
Learning	**Cell Cycle Disruptions and Cancer.** Find the two terms in the bulleted list. These words are the key vocabulary you must understand for this section. With a partner, find the place on the page where the words are defined. Restate the definitions in your own words. Ask your partner to check your definitions. If you need more help, ask your teacher to explain the terms.	**82**
Learning	**Summing Up.** Discuss the questions with your teacher and classmates. Ask them to help you remember content from the chapter. For example, your classmates can go back to models you created during your investigation to help you describe the stages of mitosis. Your teacher can help point to pages and images in the book that address each question.	**86-88**

34 The Power to Rebuild

Content Anchor: What are the axolotl's superpowers of regeneration that allow it to regrow amputated limbs and even parts of its brain?

- Humans and most other vertebrates are able to repair damaged tissue when injured, but that's about the limit of their regenerative powers. Many invertebrates, such as octopuses and crabs, can grow new limbs if they are damaged or amputated. However, the axolotl has advanced regenerative powers. Young axolotls often attack and bite each other, inflicting severe wounds. Their amazing powers of healing allow them to recover, without scarring, from intense injury.

- If a limb is damaged, or amputated it grows back with no scarring - over and over again, if necessary.

- Axolotls can also grow a new tail, a new jaw, replace a damaged spinal cord and even some parts of their brain. Organs can be transplanted from one axolotl to another without any rejection issues and the axolotl is more resistant to cancer than mammals.

1. Think about last time you scraped your knees or accidentally cut yourself. How did the wound heal - from the center outwards or from the edges to the center?

2. What is special about the axolotl's ability to heal itself, compared to humans?

3. Do you know of an organ in the human body that has the power to regenerate if a large piece of it has to be removed due to damage or disease?

4. What stages do cells go through, in order to divide and replicate correctly?

5. Working in pairs, discuss how a cell "knows" what it is to develop into: skin, bone, brain tissue, blood, etc and record the main points of your discussion below:

35 Growth and Repair of Cells

Key Question: How do cells reproduce and repair themselves, and how do organisms grow?

Mitotic **cell division** has three purposes:

▸ Growth: multicellular organisms grow from a single fertilized cell into a mature organism. Depending on the organism, the mature form may consist of several thousand to several trillion cells. These cells that form the building blocks of the body are called somatic cells.

▸ Repair: damaged and old cells are replaced with new cells.

▸ Asexual reproduction: some unicellular eukaryotes, such as yeasts and some multicellular organisms, e.g. *Hydra*, reproduce asexually by mitotic division.

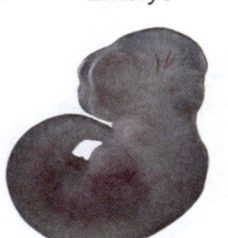

Fertilized egg cell   *Embryo*   *Adult*

Asexual reproduction
Some simple eukaryotic organisms reproduce asexually by **mitosis**. Yeasts, such as baker's yeast, can reproduce by budding. The parent cell buds to form a daughter cell (right). The daughter cell continues to grow, and eventually separates from the parent cell.

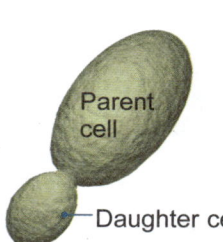

Parent cell — Daughter cell

Growth
Multicellular organisms develop from a single fertilized egg cell and grow by increasing cell numbers. Cells complete a **cell cycle** in which the cell copies its DNA and then divides to produce two identical cells. During the period of growth, the production of new cells is faster than the death of old ones. Organisms, such as the 12 day old mouse embryo (above, middle), grow by increasing their total cell number and the cells become specialized as part of development. Cell growth is highly regulated and once the mouse reaches its adult size (above, right), physical growth stops and the number of cell deaths equals the number of new cells produced.

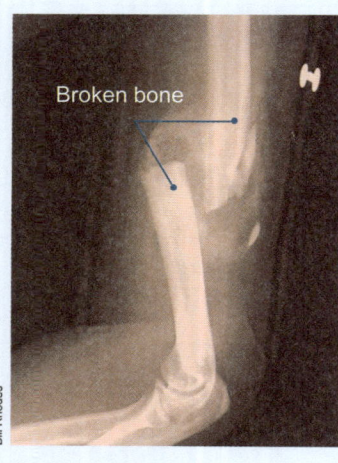
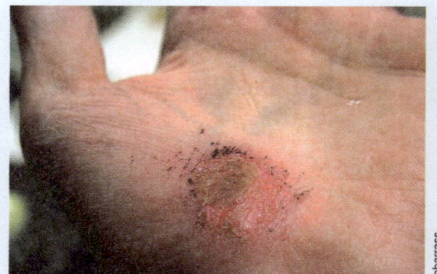

Broken bone   Damaged limbs

Repair
Mitosis is vital in the repair and replacement of damaged cells. When you break a bone or graze your skin, new cells are generated to repair the damage. Some organisms, like the sea star (above right), are able to generate new limbs if they are broken off.

1. Use some examples you may have encountered to explain the role of mitosis in:

   (a) Growth of an organism: _____

   (b) Replacement of damaged cells: _____

   (c) Asexual reproduction: _____

36 The Eukaryotic Cell Cycle

Key Question: What are the phases of the eukaryotic cell cycle, and what specific cellular events occur in each phase?

▸ The life cycle of a eukaryotic cell is called the **cell cycle**. The cell cycle can be divided into two broad phases: interphase and M phase. Specific activities occur in each phase.

Interphase

▸ Cells spend most of their time in **interphase**, which is divided into three stages:
- The first gap phase (G_1)
- The S-phase (S)
- The second gap phase (G_2)

▸ During interphase, the cell increases in size, carries out its normal activities, and replicates its DNA in preparation for **cell division**. Interphase is not a stage in **mitosis**.

Mitosis and cytokinesis (M-phase)

▸ Mitosis and **cytokinesis** occur during M-phase. During mitosis, the cell nucleus containing the replicated DNA divides in two equal parts. (see activity 39). Cytokinesis occurs at the end of M-phase. During cytokinesis, the cell cytoplasm divides and two new daughter cells are produced.

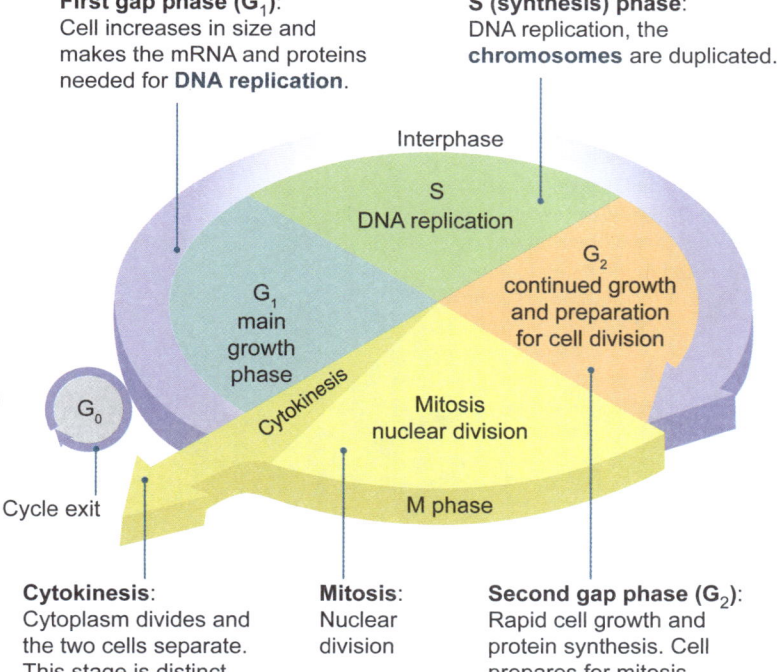

First gap phase (G_1): Cell increases in size and makes the mRNA and proteins needed for **DNA replication**.

S (synthesis) phase: DNA replication, the **chromosomes** are duplicated.

Cytokinesis: Cytoplasm divides and the two cells separate. This stage is distinct from mitosis.

Mitosis: Nuclear division

Second gap phase (G_2): Rapid cell growth and protein synthesis. Cell prepares for mitosis.

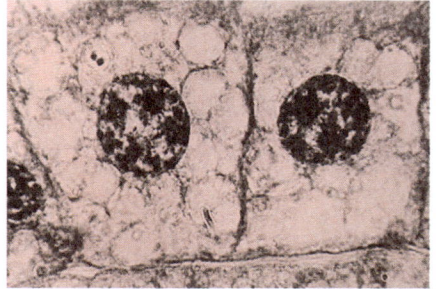

During interphase, the cell grows and acquires the materials needed to undergo mitosis. It also prepares the nuclear material for separation by replicating it.

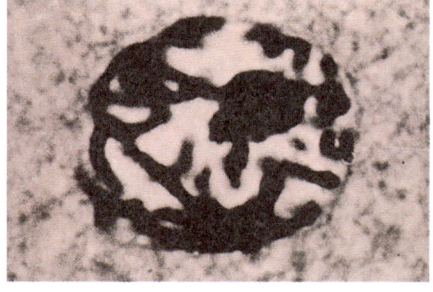

During interphase, the nuclear material is unwound. As mitosis approaches, the nuclear material begins to reorganize in readiness for nuclear division.

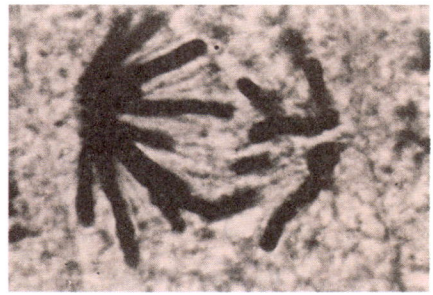

During mitosis, the chromosomes are separated. Mitosis is a highly organized process and the cell must pass "checkpoints" before it proceeds to the next phase.

1. Briefly outline what occurs during the following phases of the cell cycle:

    (a) Interphase: _____

    (b) Mitosis: _____

    (c) Cytokinesis: _____

Regulating the cell cycle

▶ Cell checkpoints give cells a way to ensure that all cellular processes have been completed correctly before entering the next phase. A collection of genes control the checkpoints. At each checkpoint, a set of conditions, including absence of flaws in DNA replication and chromosome separation, determines whether or not the cell will continue into the next phase. Cancer can result when the pathways regulating the checkpoints fail, due to mutation in the controlling genes.

▶ Non-dividing cells enter a resting phase (G_0), where they may remain for a few days or up to several years. Under specific conditions, they may re-enter the cell cycle.

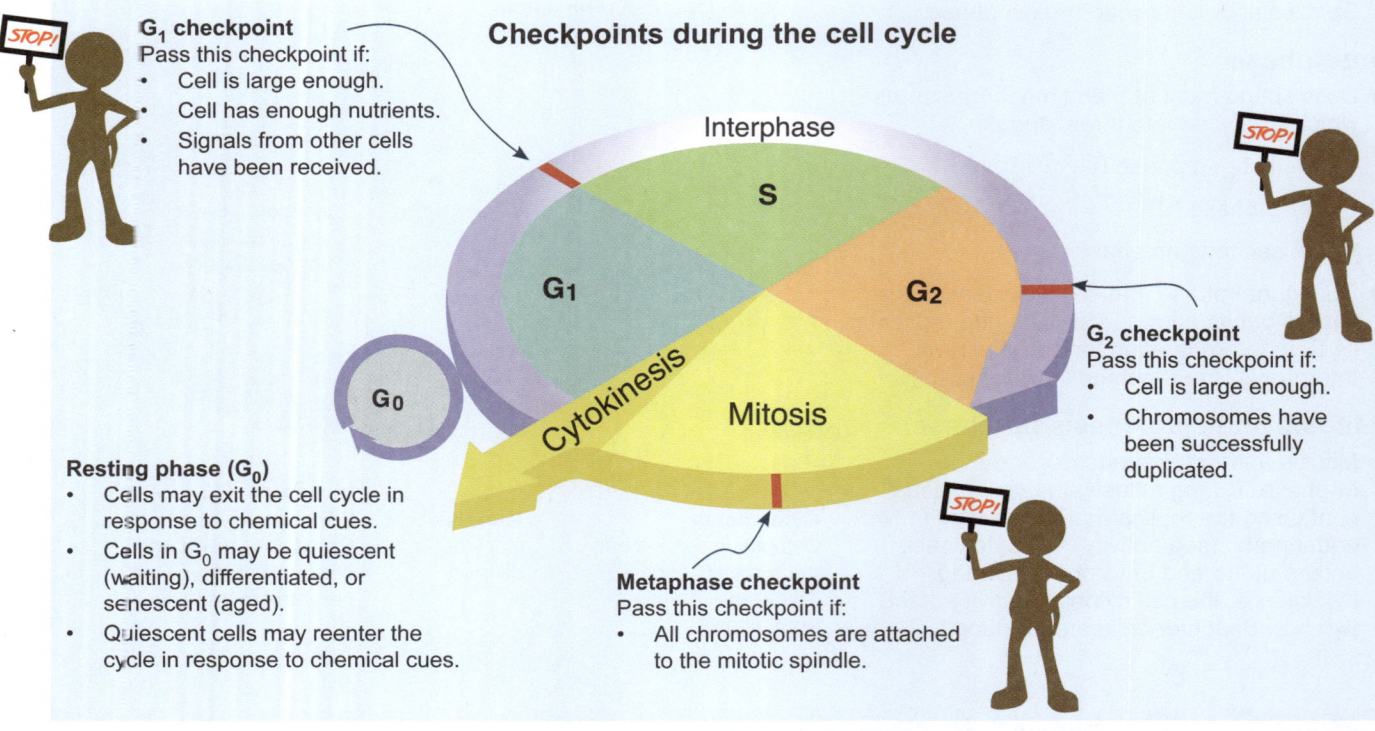

Checkpoints during the cell cycle

G_1 checkpoint
Pass this checkpoint if:
- Cell is large enough.
- Cell has enough nutrients.
- Signals from other cells have been received.

G_2 checkpoint
Pass this checkpoint if:
- Cell is large enough.
- Chromosomes have been successfully duplicated.

Metaphase checkpoint
Pass this checkpoint if:
- All chromosomes are attached to the mitotic spindle.

Resting phase (G_0)
- Cells may exit the cell cycle in response to chemical cues.
- Cells in G_0 may be quiescent (waiting), differentiated, or senescent (aged).
- Quiescent cells may reenter the cycle in response to chemical cues.

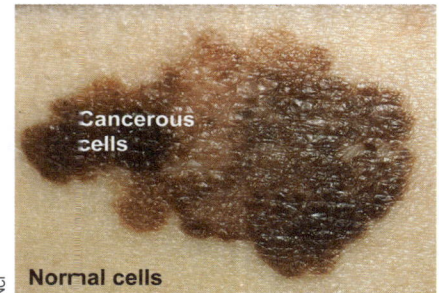

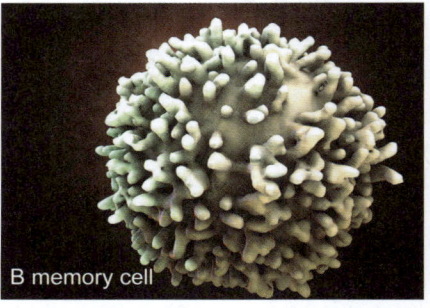

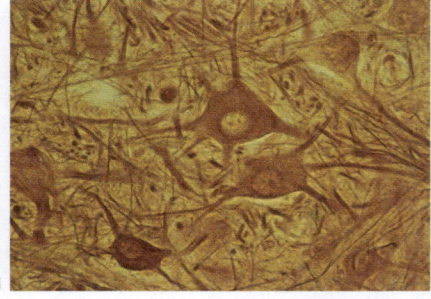

Skin cancer (melanoma). The cancer cells grow more rapidly than the normal skin cells because normal cell regulation checkpoints are ignored. This is why the cancerous cells sit higher than the normal cells and can rapidly spread (a process called metastasis).

Most lymphocytes in human blood are in the resting G_0 phase and remain there unless they are stimulated by specific antigens to re-enter the cell cycle via G_1. G_0 phase cells are not completely dormant, continuing to carry out essential cell functions in reduced form.

Many fully differentiated (specialized) cells, e.g. neurons (above), exit the cell cycle permanently and stay in G_0. These cells continue their functional role in the body, but do not continue to divide by mitosis. Senescent cells have accumulated mutations, lose function, and die.

2. Explain the importance of cell cycle checkpoints: _____

_____

_____

3. What is the link between cell cycle checkpoint disruptions and the development of cancer? _____

_____

_____

_____

_____

37 Mitosis and Cytokinesis

Key Question: What happens in the different stages of mitosis leading up to the formation of two daughter cells, and is it different for plant and animal cells?

The cell cycle and stages of mitosis

▶ **Mitosis** is continuous, but is divided into stages for easier reference (1-6 below). Enzymes are critical at key stages. The example below illustrates the **cell cycle** in an animal cell.

▶ In animal cells, centrioles (located in the centrosome) form the spindle. During **cytokinesis** (division of the cytoplasm) a constriction forms, dividing the cell in two. Cytokinesis is part of M-phase, but it is distinct from mitosis.

▶ Plant cells lack centrioles, and the spindle is organized by structures associated with the plasma membrane. In plant cells, cytokinesis involves formation of a cell plate in the middle of the cell. This will form a new cell wall.

The animal cell cycle and stages of mitosis

Interphase

Nucleus — Nuclear membrane
Centrosome containing centrioles (forms spindle)

Interphase refers to events between mitosis. The cell replicates its DNA and prepares for mitosis. The centrosome (containing two centrioles) also divides.

DNA condenses into distinct **chromosomes.** The centrosomes migrate to the poles.

1 Early prophase

Centrosomes move to opposite poles

Cytokinesis

Division of the cytoplasm. When cytokinesis is complete, there are two separate daughter cells.

Chromosomes appear as two **chromatids** held together at the centromere. The nuclear membrane breaks down. The centrioles begin to form the spindle (made up of microtubules and proteins).

2 Late prophase

Homologous pair of replicated chromosomes

6 Telophase

Two new nuclei form. A furrow forms across the midline of the parent cell, pinching it in two.

Some spindle fibers attach to and organize the chromosomes on the equator of the cell. Some spindle fibers span the cell.

3 Metaphase

Spindle

5 Late anaphase

Other spindle fibers lengthen, pushing the poles apart and causing the cell to elongate.

Spindle fibers attached to chromatids shorten, pulling the chromatids apart. Spindle shortening is catalyzed by enzymes.

4 Anaphase

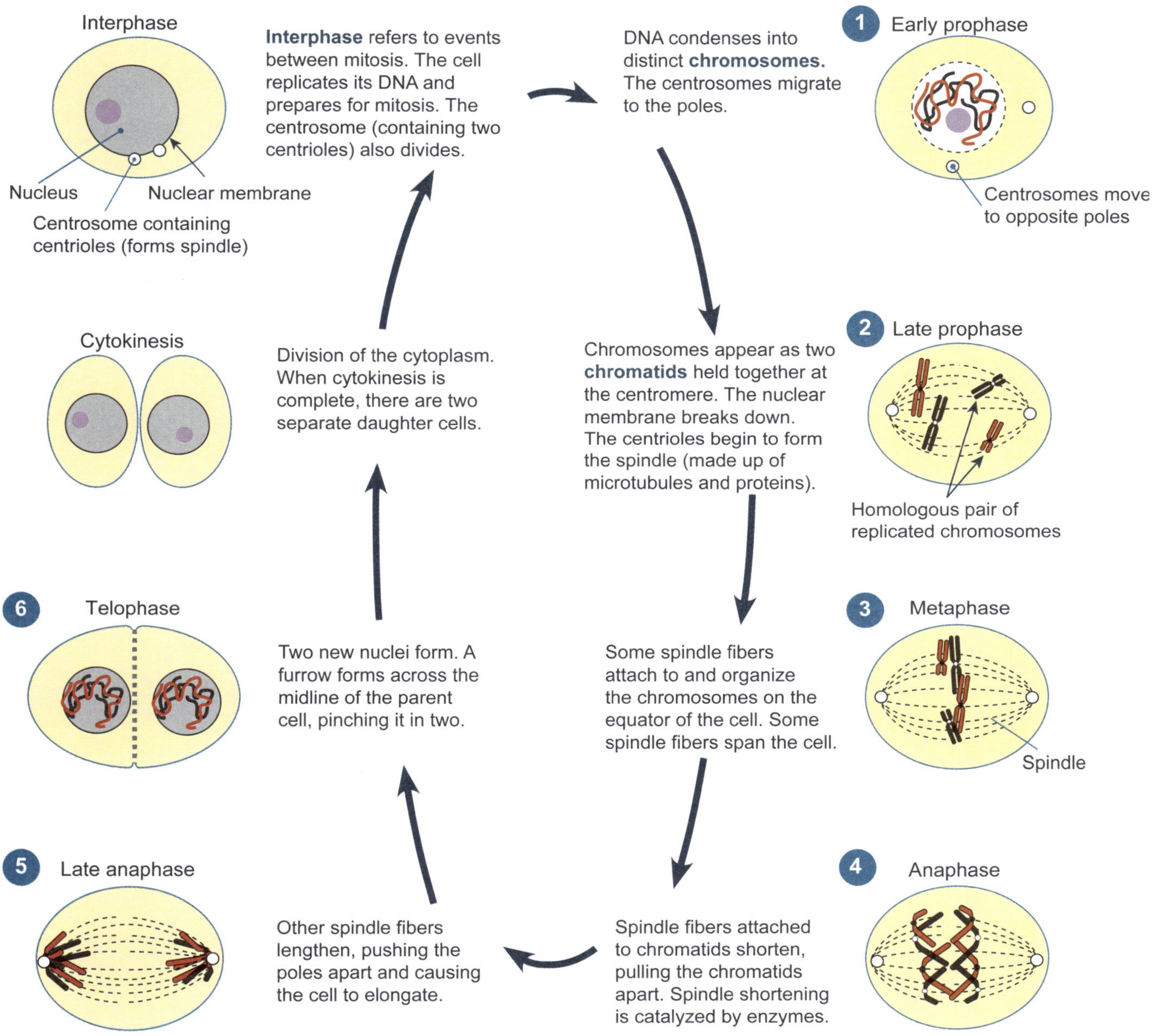

1. What must occur before mitosis takes place? _____

_____

_____

Cytokinesis

▸ In plant cells (below right), cytokinesis (division of the cytoplasm) involves construction of a cell plate (a precursor of the new cell wall) in the middle of the cell. The cell wall materials are delivered by vesicles derived from the Golgi. The vesicles join together to become the plasma membranes of the new cell surfaces. Animal cell cytokinesis (below left) begins shortly after the sister chromatids have separated in anaphase of mitosis. A ring of microtubules assembles in the middle of the cell, next to the plasma membrane, constricting it to form a cleavage furrow. In an energy-using process, the cleavage furrow moves inwards, forming a region of separation where the two cells will separate.

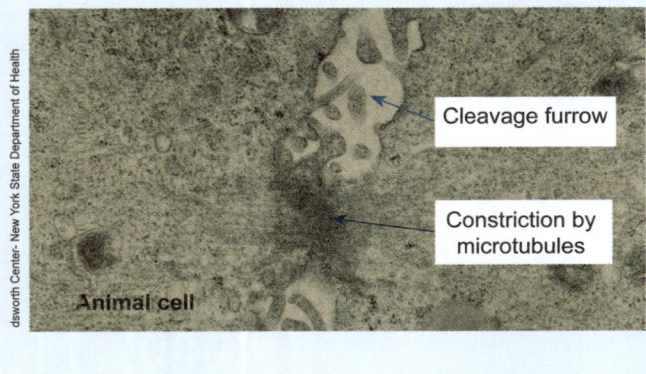
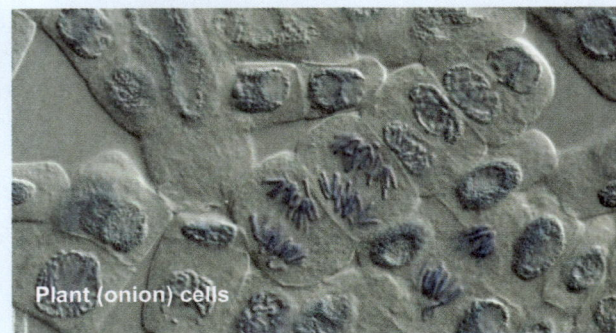

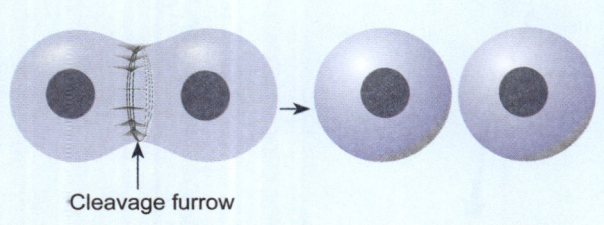
Cytokinesis in an animal cell

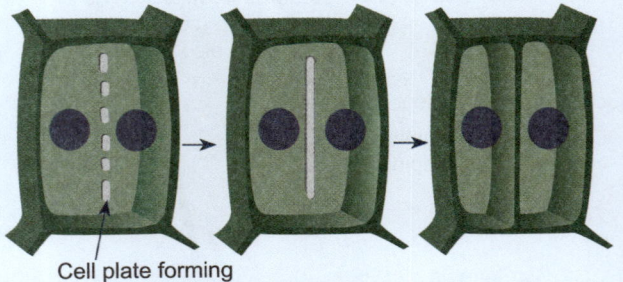
Cytokinesis in a plant cell

2. Summarize what happens in each of the following phases:

   (a) Prophase: _____

   (b) Metaphase: _____

   (c) Anaphase: _____

   (d) Telophase: _____

3. (a) What is the purpose of cytokinesis? _____

   (b) Summarize the differences between cytokinesis in an animal cell and a plant cell: _____

38 Modeling Mitosis

Key Question: How can I model the stages of mitosis to help to visualize and understand the process?

Investigation 2.1 Modeling mitosis

See appendix for equipment list.

1. Use the information on the previous pages to model **mitosis** in an animal cell using pipe cleaners and string. Work in pairs and use four chromosomes for simplicity (2N = 4). Photograph or film each stage.

2. Take care using and moving around with scissors. Do not wrap string around any body parts, especially the neck region.

3. Photo 1 (below) can be used as a starting point for your model. It represents a cell in interphase before mitosis begins. The circular structures are the replicated centrosomes.

4. Before you start, identify the structures A-C in photo

   A: _____ B: _____ C: _____

5. Remember to label your photos as you place them on the page.

Photo 1

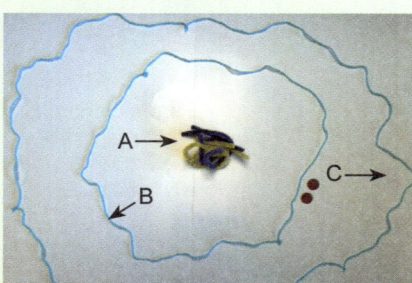

39 DNA Replication

Key Question: How does a cell make a copy of its DNA before mitosis occurs?

- Before a cell can divide by **mitosis**, its DNA must be copied (replicated). **DNA replication** ensures that the two daughter cells receive identical genetic information.
- In eukaryotes, DNA exists in the cell nucleus and is organized into structures called **chromosomes**.
- DNA replication takes place in the time between **cell divisions**. After the DNA has replicated, each chromosome is made up of two **chromatids** which are joined at the centromere.
- The process of DNA replication is known as **semi-conservative**, meaning that each chromatid contains half the original (parent) DNA and half the new (daughter) DNA. The two chromatids will become separated during cell division to form two separate chromosomes.

DNA replication duplicates chromosomes

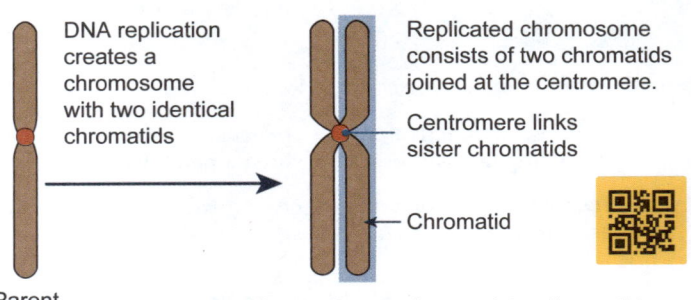

DNA replication creates a chromosome with two identical chromatids

Replicated chromosome consists of two chromatids joined at the centromere.

Centromere links sister chromatids

Chromatid

Parent chromosome

The centromere keeps sister chromatids together in an organized way until they are separated before nuclear division.

1. What is the purpose of DNA replication? _____
   _____
   _____
   _____
   _____

DNA replication is semi-conservative

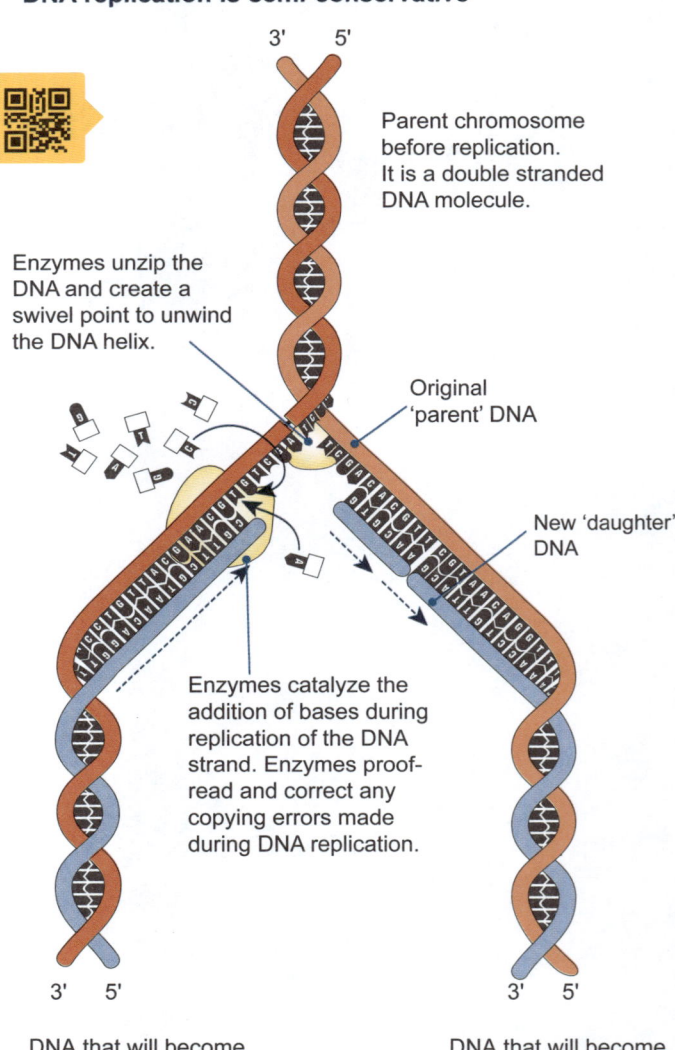

Parent chromosome before replication. It is a double stranded DNA molecule.

Enzymes unzip the DNA and create a swivel point to unwind the DNA helix.

2. What would happen if DNA was not replicated prior to cell division?
   _____
   _____
   _____
   _____

Original 'parent' DNA

New 'daughter' DNA

3. (a) What does a replicated chromosome look like?
   _____
   _____

   (b) What is the purpose of the centromere?
   _____
   _____
   _____

Enzymes catalyze the addition of bases during replication of the DNA strand. Enzymes proof-read and correct any copying errors made during DNA replication.

4. Explain why identical copies of DNA are important for the growth process to occur correctly?
   _____
   _____
   _____
   _____
   _____

DNA that will become one chromatid

DNA that will become the other chromatid

DNA replication is called semi-conservative. This is because each resulting DNA molecule is made up of one parent strand and one daughter strand of DNA.

40 Stages of DNA Replication

Key Question: How does DNA unwind for replication and how are enzymes involved?

▸ The individual units that make up the DNA molecule are called nucleotides (see activity 4). During **DNA replication**, new nucleotides are added at a region called the replication fork. This replication fork moves along the **chromosome** as replication progresses.

▸ Nucleotides are added through complementary base-pairing. The base pairing rule ensures that nucleotide A is always paired with nucleotide T and nucleotide C is always paired with nucleotide G.

▸ The DNA strands can only be replicated in one direction, so one strand has to be copied in short segments which are joined together later.

▸ This whole process occurs simultaneously for each chromosome of a cell and the entire process is tightly controlled by enzymes.

Stages in DNA replication

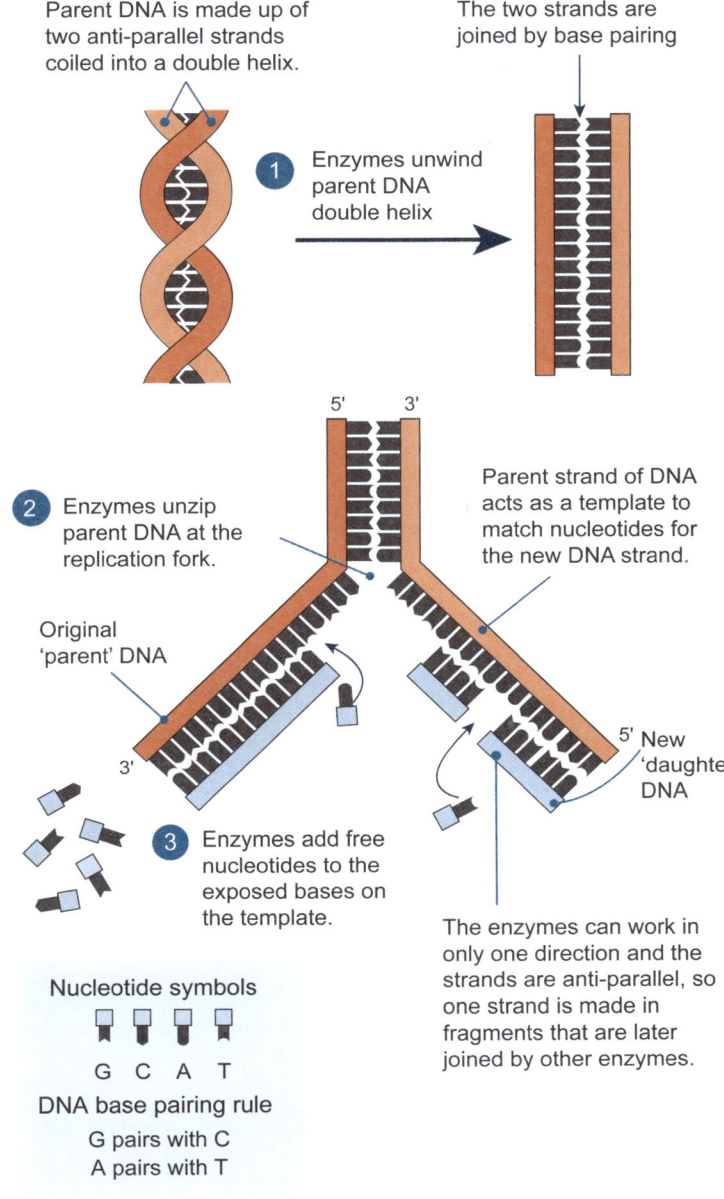

1. Enzymes unwind parent DNA double helix
2. Enzymes unzip parent DNA at the replication fork.
3. Enzymes add free nucleotides to the exposed bases on the template.

Parent DNA is made up of two anti-parallel strands coiled into a double helix.

The two strands are joined by base pairing

Parent strand of DNA acts as a template to match nucleotides for the new DNA strand.

The enzymes can work in only one direction and the strands are anti-parallel, so one strand is made in fragments that are later joined by other enzymes.

Nucleotide symbols
G C A T

DNA base pairing rule
G pairs with C
A pairs with T

4. Two new double-stranded DNA molecules

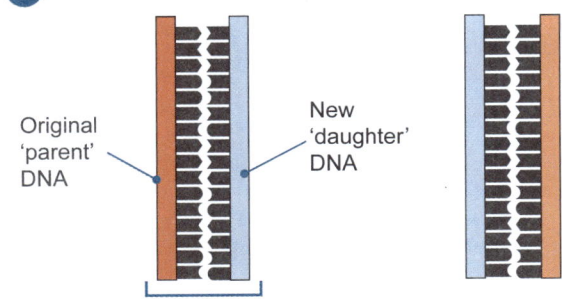

Enzymes are involved at every step of DNA replication. They unzip the parent DNA, add the free nucleotides to the 3' end of each single strand, join DNA fragments, and check and correct the new DNA strands.

1. How are the new strands of DNA lengthened?

___

2. What rule ensures that the two new DNA strands are identical to the original strand?

___

3. Why does one strand of DNA need to be copied in segments?

___

4. Describe three activities carried out by enzymes during DNA replication:

(a) ___

(b) ___

(c) ___

41 Evidence for Semi-conservative DNA Replication

Key Question: How do we know that DNA replication is semi-conservative?

Initially, three models were proposed to explain the process of **DNA replication**:

1 – a **semi-conservative** model in which each DNA strand served as a template, forming a new DNA molecule that was half old and half new DNA;

2 – a conservative model which suggested that the original DNA served as a complete template so that the resulting DNA was completely new; and

3 – a dispersive model which suggested that the two new DNA molecules had part new and part old DNA interspersed throughout them.

In 1958, two scientists, Matthew Meselson and Franklin Stahl, carried out an experiment that proved the semi-conservative model to be the correct one.

- They grew bacteria in a solution containing a heavy nitrogen isotope (^{15}N) until their DNA contained only ^{15}N.
- The bacteria were then placed into a growth solution containing the nitrogen isotope (^{14}N), which is lighter than ^{15}N.
- After a set number of generation times, the DNA was extracted and centrifuged in a solution that provides a density gradient.
- Heavy DNA (containing only ^{15}N) sinks to the bottom, light DNA (containing only ^{14}N) rises to the top, and intermediate DNA (one light and one heavy strand) settles in the middle, as shown below:

Meselson and Stahl's experiment

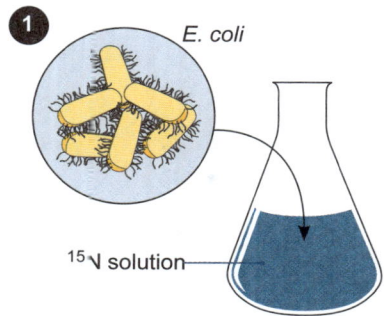

E. coli were grown in a nutrient solution containing ^{15}N. After 14 generations all the bacterial DNA contained ^{15}N. A sample was removed. This was **generation 0**.

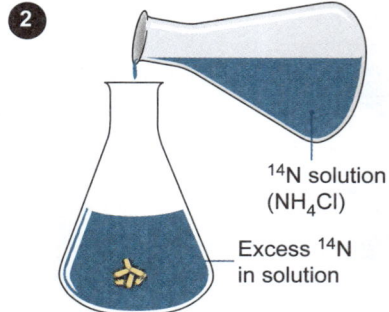

Generation 0 was added to a solution with excess ^{14}N (as NH_4Cl). During replication, new DNA would incorporate ^{14}N and be 'lighter' than the original DNA (which had only ^{15}N).

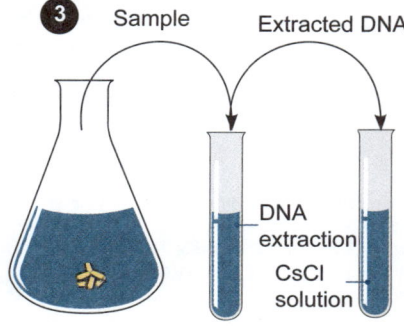

Each generation (~ 20 minutes), a sample was taken and treated to release the DNA. The DNA was placed in a CsCl solution which provided a density gradient for separating the DNA.

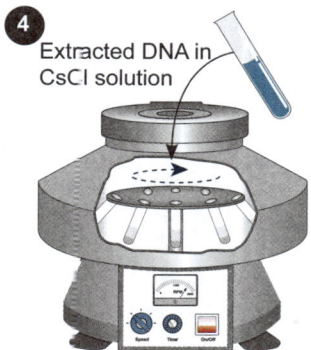

Samples were spun in a high speed ultracentrifuge at 140,000 g for 20 hours. Heavier ^{15}N DNA moved closer to the bottom of the test tube than light ^{14}N DNA or ^{14}N, ^{15}N intermediate DNA.

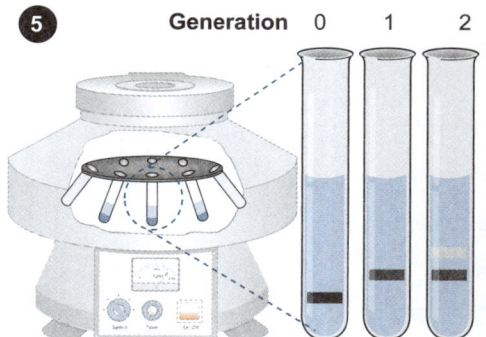

All the DNA in the generation 0 sample moved to the bottom of the test tube. All the DNA in the generation 1 sample moved to an intermediate position. At generation 2, half the DNA was at the intermediate position and half was near the top of the test tube. In subsequent generations, more DNA was near the top and less was in the intermediate position.

Models for DNA replication

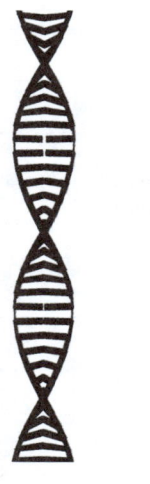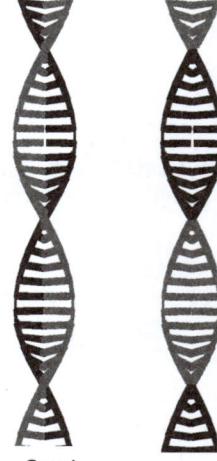

Conservative   Semi-conservative   Dispersive

1. In this activity, you will model the semi-conservative model of DNA replication. 1(a) and (dი(i)) are done for you:

   (a) The DNA model below represents the DNA of the bacteria after growing in the solution containing the heavy nitrogen (Generation 0). The relative mass of the DNA can be modeled by adding together the nitrogen masses.

   (b) In the space below (center) (Generation 1), split the DNA from Generation 0 along its center, then write in the complementary base pairs to form two DNA strands (i and ii). This represents the DNA after it has been grown in the solution containing ^{14}N for one generation.

   (c) In the space below (right) (Generation 2), split the two DNA chains along their centers, then write the complementary base pairs to form four new DNA strands (i, ii, iii, and iv). This represents the DNA after it has been growing in the solution containing the ^{14}N for two generations.

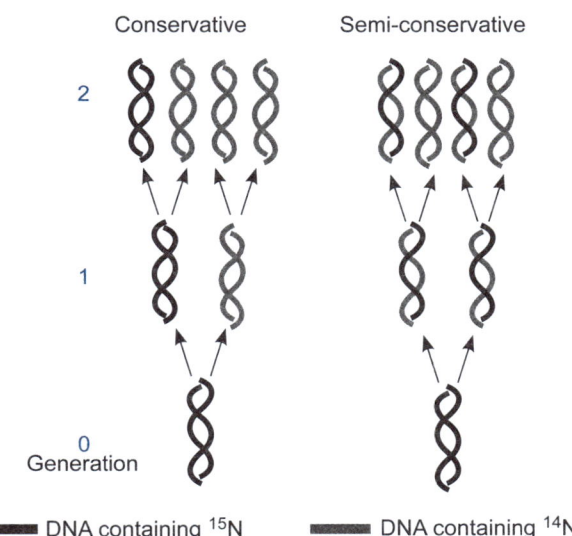

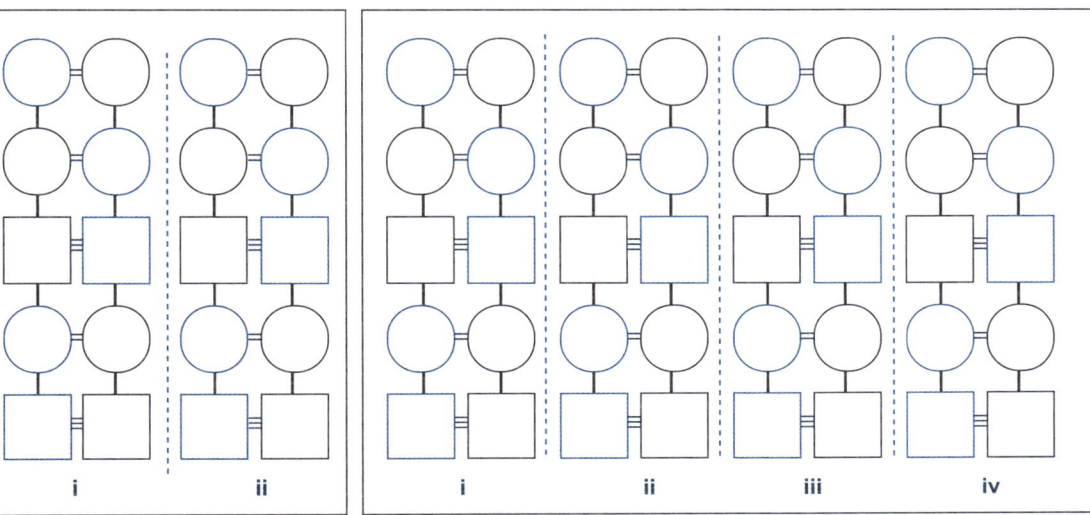

(d) Mass of nitrogen in DNA strand, Generation 0:   i: **150**

Mass of nitrogen in DNA strands, Generation 1:   i: _____   ii: _____

Mass of nitrogen in DNA strands, Generation 2:   i: _____   ii: _____

iii: _____   iv: _____

(e) On the test tubes (right), mark a bar representing the mass of each of the DNA strands from (a), (b), and (c).

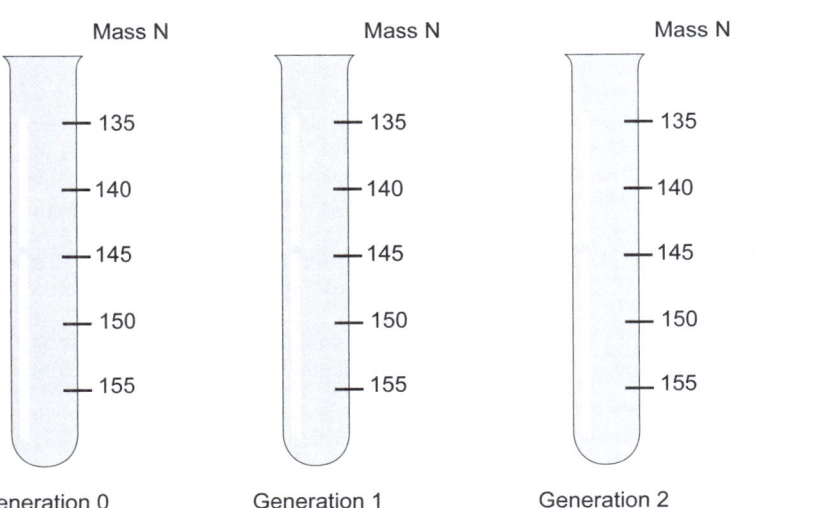

2. In this part of the activity, you will model the conservative model of DNA replication:

   (a) The DNA model below represents the DNA of the bacteria after it has grown in the solution containing the heavy nitrogen (Generation 0). The relative mass of the DNA can be modeled by adding together the nitrogen masses.

   (b) In the box below (center), recreate the original DNA strand. Beside it, create a matching strand with bases using the light nitrogen. This represents the DNA after it has been replicated in the solution with the ^{14}N for one generation.

   (c) In the space below (right), recreate the original DNA strand and the first generation strand. Beside these, create matching strands with bases using the ^{14}N. This represents the DNA after it has been growing in the solution containing the ^{14}N for two generations.

Generation 0

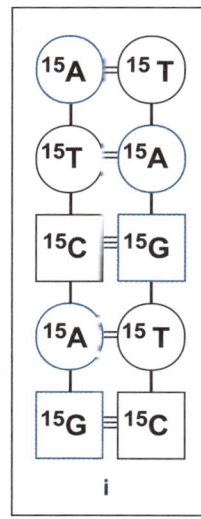

i

Generation 1

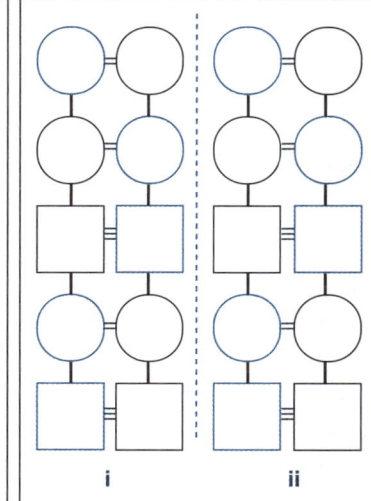

i   ii

Generation 2

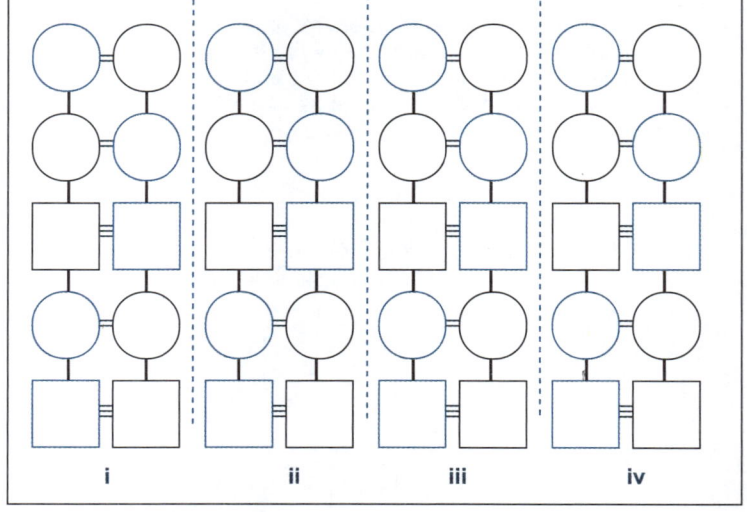

i   ii   iii   iv

(d) Mass of nitrogen in DNA strand, Generation 0:   i: _____

    Mass of nitrogen in DNA strands, Generation 1:   i: _____   ii: _____

    Mass of nitrogen in DNA strands, Generation 2:   i: _____   ii: _____

                                                                           iii: _____   iv: _____

(e) On the appropriate test tubes (right), mark a bar representing the mass of each of the DNA strands from Generation 0, Generation 1, and Generation 2

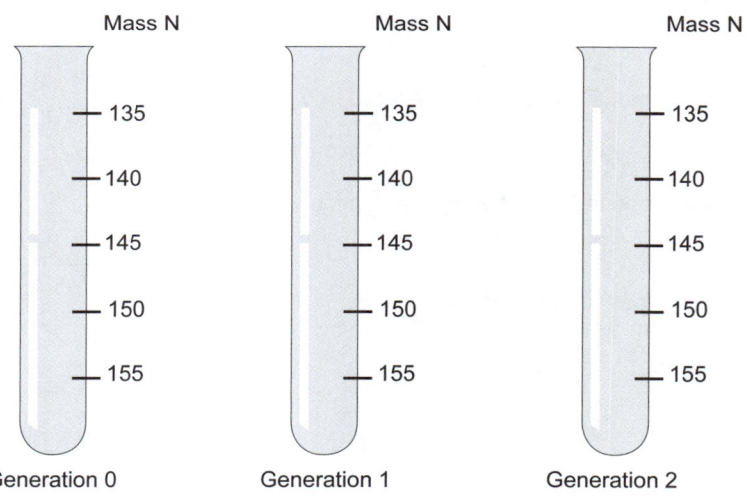

3. In their experiment, Meselson and Stahl obtained the following results: Generation 0 = 100% 'heavy DNA'; **Generation 1**= 100% 'intermediate DNA'; Generation 2 = 50% 'intermediate DNA', 50% 'light DNA'.

   From the results of your two modeling exercises, use the evidence to make a claim (conservatively or semi-conservatively) about the model of DNA replication which matches the result of Meselson and Stahl:

   _____
   _____
   _____

42 Differentiation of Cells

Key Question: How do many different cell types arise during development of the embryo?

▸ When a cell divides by **mitosis**, it produces genetically identical cells. However, a multicellular organism is made up of many different types of cells, each specialized to carry out a particular role. How can it be that all of an organism's cells have the same genetic material, but the cells have a wide variety of shapes and functions? The answer is through cellular differentiation (transformation) of unspecialized cells called **stem cells**.

▸ Although each cell has the same genetic material (genes), different genes are turned on (activated) or off in different patterns during development in particular types of cells. The differences in gene activation controls what type of cell forms (below). Once the developmental pathway of a cell is determined, it cannot alter its path and change into another cell type.

How stem cells give rise to different cell types

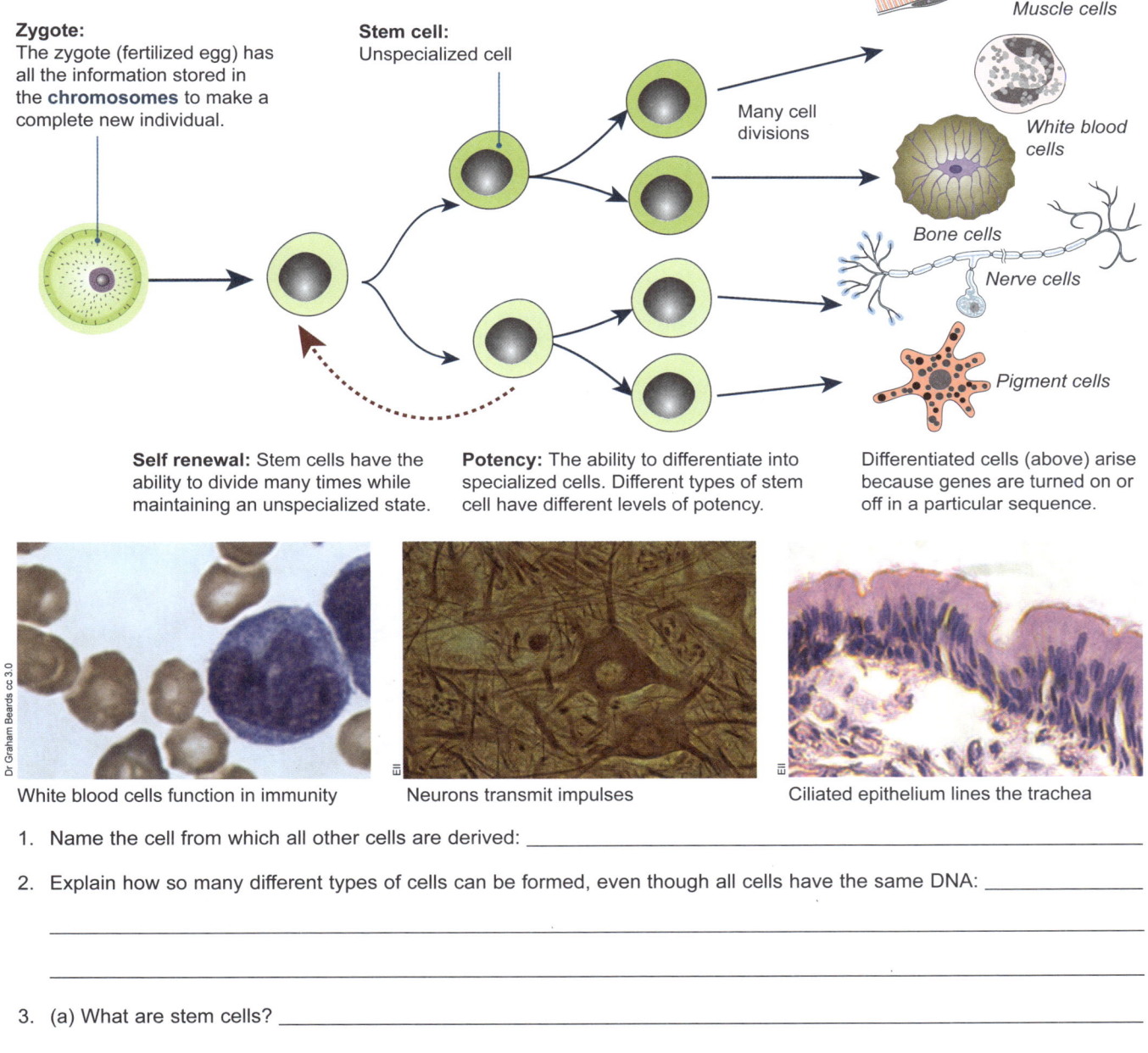

Zygote: The zygote (fertilized egg) has all the information stored in the **chromosomes** to make a complete new individual.

Stem cell: Unspecialized cell

Many cell divisions

Muscle cells
White blood cells
Bone cells
Nerve cells
Pigment cells

Self renewal: Stem cells have the ability to divide many times while maintaining an unspecialized state.

Potency: The ability to differentiate into specialized cells. Different types of stem cell have different levels of potency.

Differentiated cells (above) arise because genes are turned on or off in a particular sequence.

White blood cells function in immunity | Neurons transmit impulses | Ciliated epithelium lines the trachea

1. Name the cell from which all other cells are derived: _____

2. Explain how so many different types of cells can be formed, even though all cells have the same DNA: _____

3. (a) What are stem cells? _____

   (b) What are the two defining properties of stem cells? _____

      i _____

      ii _____

43 Blood Cell Differentiation

Key Question: How do stem cells, which are undifferentiated, develop into many different blood cell types?

Totipotent stem cells
These stem cells can differentiate into all the cells in an organism. Example: in humans, the zygote and its first few divisions. The tissue at the root and shoot tips of plants is also totipotent.

Pluripotent stem cells
These stem cells can give rise to any cells of the body, except extra-embryonic cells, e.g. placenta and chorion. Example: embryonic stem cells.

Multipotent stem cells
These adult stem cells can give rise to a limited number of cell types, related to their tissue of origin. Example: bone marrow stem cells (below), skin stem cells, bone stem cells, umbilical cord blood.

The zygote and its first few divisions into the morula (~16 cell stage) — 4 cell divisions → 3 cell divisions → The inner cell mass of the blastocyst (~128 cells)

Function of blood cells

Immunity
White blood cells (leukocytes) are part of the immune system. They defend the body against infectious disease and foreign materials.

Lymphocytes
Specialized white blood cells involved in the specific immune response.
- T lymphocyte (Matures in thymus)
- Natural killer lymphocyte
- B lymphocyte

Granulocytes
Specialized white blood cells which destroy foreign material, e.g. bacterial cells, by phagocytosis.
- Neutrophil
- Basophil
- Eosinophil

Monocytes and macrophages

Precursor cell → Stem cell → Precursor cell

Gas exchange
Transports oxygen around the body.
- Red blood cells

Blood clotting
Form clots to prevent excessive bleeding.
- Platelets

1. (a) What type of stems cells are blood cells produced from? _____

   (b) What are the features of the stem cells you described in (a): _____
   _____

2. Describe the functional roles of blood cells: _____
   _____
   _____
   _____

44 Specialization in Plant Cells

Key Question: How does cell specialization allow plant cells to carry out specialist functions?

Cell specialization

- A specialized cell is a cell with the specific features needed to perform a particular function in the organism.
- **Cell specialization** occurs during development when specific genes are switched on or off.
- Multicellular organisms have many types of specialized cells. These work together to carry out the essential functions of life.
- The size and shape of a cell allows it to perform its function. The number and type of organelles in a cell is also related to the cell's role in the organism.
- In plants, **mitosis** takes place in meristematic cells, located at the shoot and root tips.

Cells in the leaves of plants are often green because they contain a pigment called chlorophyll, which is needed for photosynthesis.

Specialized cells in vascular tissue are needed to transport water and sugar around the plant.

Some cells are strengthened to provide support for the plant, allowing it to keep its form and structure.

Plants have root-hair cells to obtain water and nutrients (mineral ions) from the soil.

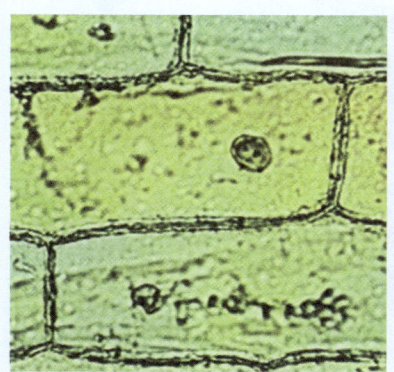

Many plant cells have a regular shape because of their semi-rigid cell wall. Specialized cells form different types of tissues. Simple tissues, like this onion epidermis, have only one cell type.

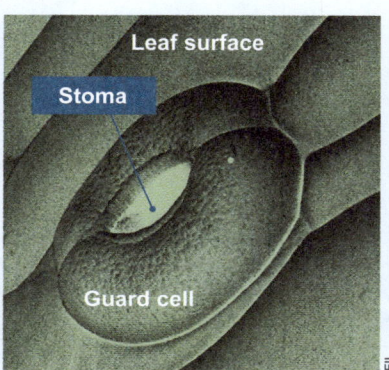

Specialized guard cells surround the stomata (pores) on plant leaves. The guard cells control the opening and closing of stomata and prevent too much water being lost from the plant.

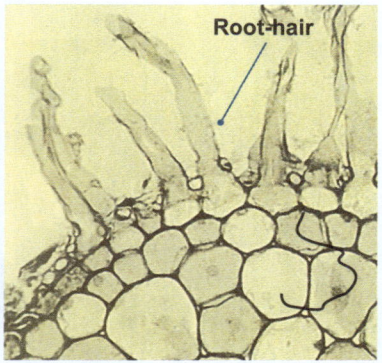

A plant root-hair is a tube-like outgrowth of a plant root cell. Their long, thin shape greatly increases their surface area. This allows the plant to absorb water and minerals efficiently.

1. What is a specialized cell? _____

   _____

2. (a) Name the specialized cell that helps to prevent water loss in plants: _____

   (b) How does this cell prevent water being lost? _____

   _____

   _____

3. How do specialized root hairs help plants to absorb more water and minerals from the soil? _____

   _____

   _____

45 Specialization in Animal Cells

Key Question: How does cell modification allow animal cells to carry out specialist functions?

Specialization in animal cells

- There are over 200 different types of cell in the human body.
- Animal cells lack a cell wall, so they can take on many different shapes. Therefore, there are many more types of animal cells than there are plant cells.
- Specialized cells often have modifications or exaggerations to a normal cell feature to help them do their job. For example, nerve cells have long, thin extensions to carry nerve impulses over long distances in the body.
- **Cell specialization** improves efficiency because each cell type is highly specialized to perform a particular task.
- In animal cells, **mitosis** takes place in somatic (body) cells.

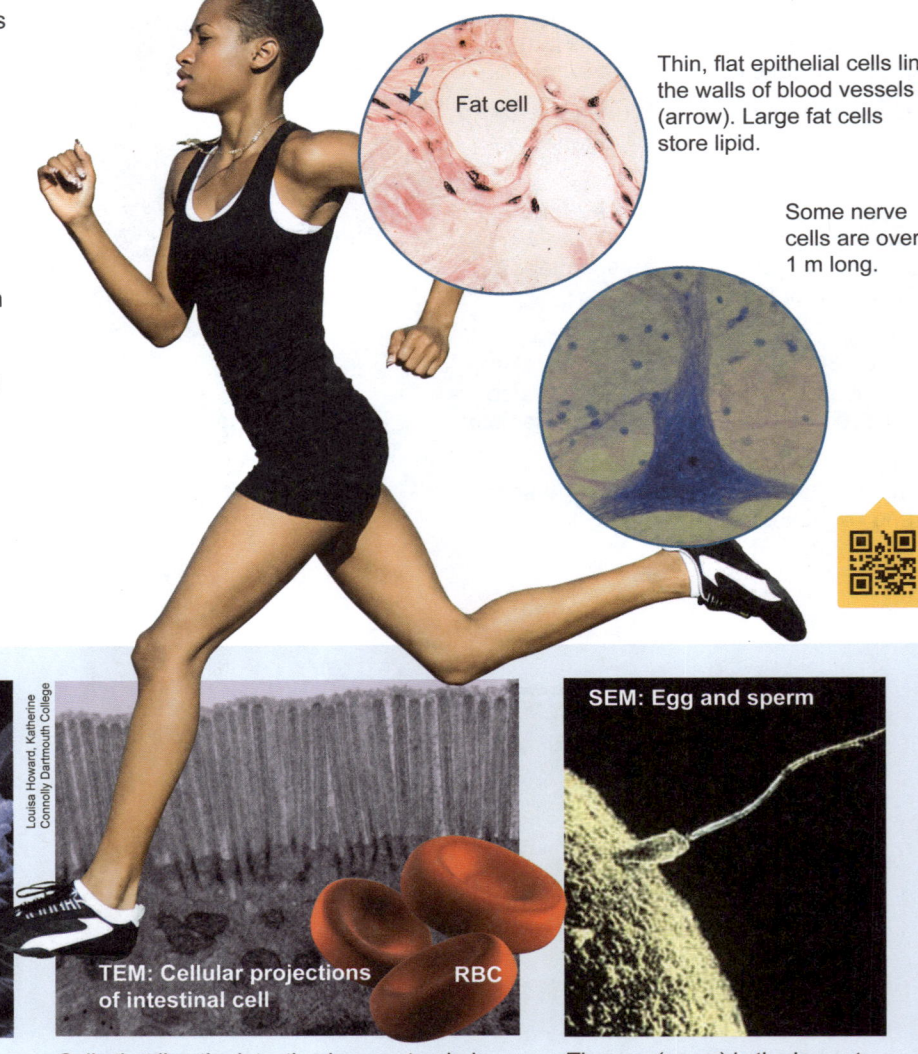

Thin, flat epithelial cells line the walls of blood vessels (arrow). Large fat cells store lipid.

Some nerve cells are over 1 m long.

SEM: White blood cell

Some animal cells can move or change shape. A sperm cell must be able to swim so that it can fertilize an egg. A white blood cell changes its shape to engulf and destroy foreign materials, e.g. bacteria.

TEM: Cellular projections of intestinal cell — RBC

Cells that line the intestine have extended cell membranes. This increases their surface area so that more nutrients (food) can be absorbed. Red blood cells (RBCs) have no nucleus so they have more room inside to carry oxygen around the body.

SEM: Egg and sperm

The egg (ovum) is the largest human cell. It is about 0.1 mm in diameter and can be seen with the naked eye. The smallest human cells are sperm cells and red blood cells.

1. What is the advantage of cell specialization in a multicellular organism? _____

_____

_____

2. For each of the following specialized animal cells, name a feature that helps it carry out its function:

   (a) White blood cell: _____

   (b) Sperm cell: _____

   (c) Nerve cell: _____

   (d) Red blood cell: _____

46 Cells and the Environment

Key Question: How can the environment affect an organism's phenotype?

▶ Cell development and differentiation is encoded by genes. Cellular differentiation is a product both of the genes themselves, and their internal and external environment, and the variations in the way those genes are controlled. This is know as **epigenetics**.

▶ The phenotype of an organism is its observable, physical characteristics, or traits. Environmental factors can modify the cellular differentiation that is encoded by genes and hence the phenotype, without changing the genotype. Environmental factors that affect cellular differentiation, leading to phenotypic changes in plants and animals include nutrients or diet, temperature, altitude or latitude, and the presence of other organisms. This can occur both during development and later in life.

The effect of temperature

▶ The sex of some animals is determined by the incubation temperature during their embryonic development. Examples include turtles, crocodiles, and the American alligator. In some species, high incubation temperatures produce males and low temperatures produce females. In other species, the opposite is true. Temperature regulated sex determination may provide an advantage by preventing inbreeding, since all siblings will tend to be of the same sex.

▶ Color-pointing is a result of a temperature sensitive mutation to one of the melanin-producing enzymes. The dark pigment is only produced in the cooler areas of the body (face, ears, feet, and tail), while the rest of the body is a pale color, or white. Color-pointing is seen in some breeds of cats and rabbits, e.g. Siamese cats and Himalayan rabbits.

The effect of other organisms

▶ The presence of other individuals of the same species may control sex determination for some animals. Some fish species, including Sandager's wrasse (right), show this characteristic. The fish live in groups consisting of a single male with attendant females and juveniles. In the presence of a male, all juvenile fish of this species grow into females. When the male dies, the dominant female will undergo physiological changes to become a male. The male and female look very different.

▶ Some organisms respond to the presence of other, potentially harmful, organisms by changing their body shape. Invertebrates, such as some *Daphnia* species (right), grow a helmet when invertebrate predators are present. The helmet makes *Daphnia* more difficult to attack and handle. Such changes are usually in response to chemicals produced by the predator (or competitor) and are common in plants as well as animals.

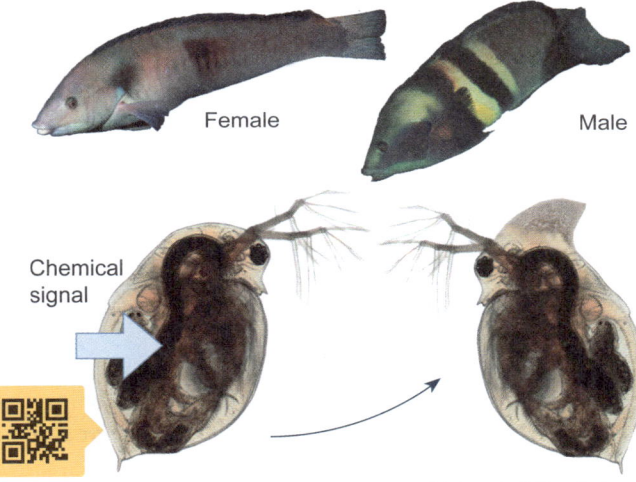

1. (a) Give two examples of how temperature affects a phenotypic characteristic in an organism: _____

   _____

   (b) Why are the darker patches of fur in color-pointed cats and rabbits found only on the face, paws, and tail? _____

   _____

   _____

2. How is helmet development in *Daphnia* an adaptive response to the environment? _____

   _____

   _____

47 Role of the Environment in Cell Development

Key Question: How can the environment or experiences of an individual affect the development of following generations?

▸ Studies of heredity, i.e. how characteristics are passed on, have found that the environment or lifestyle of an ancestor can have an effect on future generations. Certain environmental factors or diets can affect the packaging of the DNA (rather than the DNA itself), determining which genes are switched on or off. This determines the cellular development of the individual. These effects can be passed on to offspring, and even to future generations. It is thought that these inherited effects may provide a rapid way to adapt to particular environmental situations.

The destruction of New York's Twin Towers on September 11, 2001, traumatized thousands of people. In those thousands were 1700 pregnant women. Some of them suffered severe, post-traumatic stress disorder (PTSD), others did not. Studies on the mothers who developed PTSD found very low levels of the stress-related hormone cortisol in their saliva. Low levels of cortisol can be caused by very high stress, as the body uses it faster than it can be produced. The children of these mothers also had much lower levels of cortisol than those whose mothers had not suffered PTSD, particularly those who had been in the third trimester of pregnancy. This indicates a developmental response to severe stress.

In the winter of 1944-45, towards the end of the Second World War, widespread starvation affected people in the western Netherlands, including pregnant women. After the war in Europe ended in May 1945, diets quickly returned to normal. However, children whose mothers were in the first trimester of pregnancy during the famine period showed long term health effects. These children grew up to have higher than average rates of obesity later in life, as well as higher incidence rates of cardiovascular disease. In the first trimester of fetal development, DNA methylation, a process that changes a gene's activity but not its structure, had occurred on genes responsible for energy metabolism. Malnutrition in mothers caused **epigenetic** changes to their offspring.

Our ancestors' environment can have a long lasting effect

▸ How much do genes contribute to phenotype? What about environment? This question is often called nature versus nurture.

▸ A 2004 study of the grooming of rat pups by their mothers provides some insight. In this study, the quality of care by a pup's mother affected how the pup behaved when it reached adulthood.

▸ Rat pups that were groomed more often by their mother were better at coping with stress than pups that received less grooming. The effect was caused by changes in the expression of a hormone receptor gene with a role in the response to stress.

▸ DNA analysis found differences in the way the DNA was chemically tagged to regulate gene expression. Rats that received a lot of grooming had higher expression of the gene for the hormone receptor. The opposite was true for rats who received little grooming.

1. (a) Describe how grooming by mother rats on their pups affect the pups in the long term: _____

_____

   (b) How was this achieved? _____

_____

_____

Studying the effect of environment in generations of rats

The effect of the environment and diet of mothers on later generations exposed to a breast cancer trigger (a cancer-causing chemical) was investigated in rats fed a diet high in estrogen. The length of time taken for breast cancer to develop in later generations after the trigger for breast cancer was given was recorded and compared. The data are presented below.

F_1 = daughters, F_2 = granddaughters, F_3 = great granddaughters.

Weeks since trigger	Cumulative percentage rats with breast cancer (rat mothers on high estrogen diet (HED))					
	F_1%		F_2%		F_3%	
	Mothers on high estrogen diet	Control	Mothers on high estrogen diet	Control	Mothers on high estrogen diet	Control
6	5	0	10	0	0	0
8	10	0	10	0	15	10
10	30	15	15	20	30	20
12	38	19	30	30	40	20
14	50	22	30	40	50	20
16	50	22	30	40	50	30
18	60	35	40	40	75	40
20	60	42	50	50	80	45
22	80	55	50	50	80	60

Data source: S. De Assis: Nature Communications 3 (Article 1053) (2012)

2. Plot the tabulated data above on to the grids below. Include a key in the box provided:

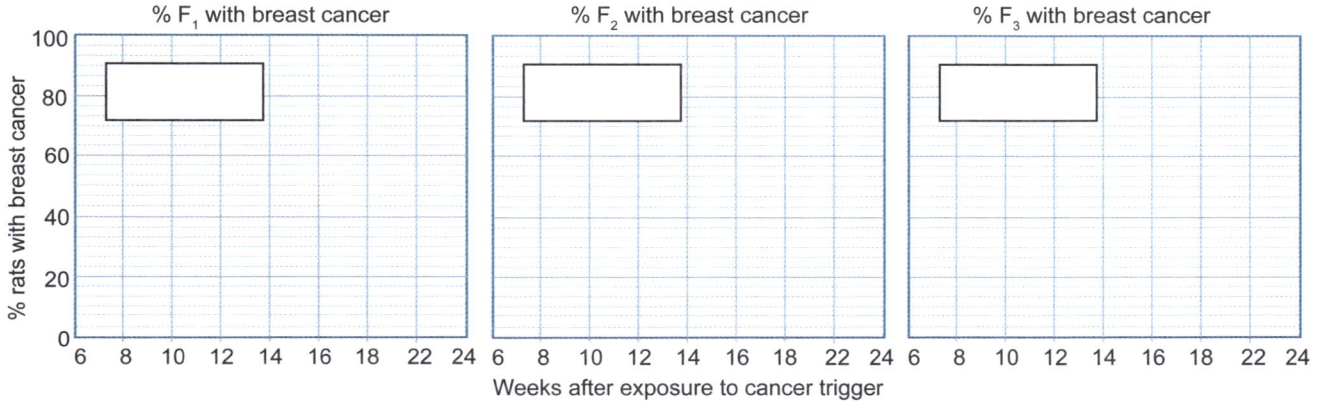

3. (a) Which generations are affected by the mother eating a high estrogen diet? Explain: _____

    (b) What do these experiments show with respect to diet and generational effects? _____

4. What evidence is there that epigenetics can have long term to permanent effects on gene expression?

48 Cell Cycle Disruptions and Cancer

Key Question: What happens when cell cycle checkpoints fail?

Formation of cancerous cells

▶ The formation of cancer cells results from changes in the genes controlling normal cell growth and division. The resulting cells become immortal and no longer carry out their functional role.

▶ Two types of gene are normally involved in controlling the **cell cycle**:
- Proto-oncogenes
- Tumor-suppressor genes

Cancer: cells out of control
Cancerous transformation results from changes in the genes controlling normal cell growth and division. The resulting cells are no longer destroyed at the normal end of their life span and malfunction.

Proto-oncogenes and tumor-suppressor genes
▶ Proto-oncogenes start **cell division** and are essential for normal cell development.
▶ Tumor-suppressor genes switch off cell division.
▶ In their normal form, these types of genes work together, enabling the body to repair defective cells and replace dead ones. Mutations in these genes can disrupt this regulation.
▶ Proto-oncogenes, through mutation, can give rise to oncogenes, which cause uncontrolled cell division. Mutations to tumor-suppressor genes initiate most human cancers. The best studied tumor-suppressor gene is p53, which codes for a protein that halts the cell cycle so that DNA can be repaired before division. The p53 gene acts at the G1-S checkpoint and initiates DNA repair or apoptosis.

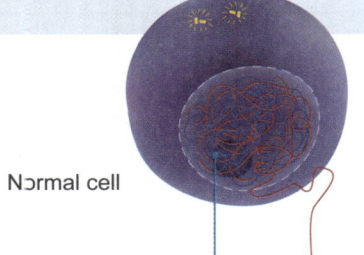

Normal cell

If the damage is too serious to repair, the p53 gene activates other genes to cause the cell to enter apoptosis (programmed cell death).

Tumor-suppressor genes
When damage occurs, the tumor suppressor gene p53 commands other genes to bring cell division to a halt. If repairs are made, then the p53 gene allows the cell cycle to continue.

DNA molecule

Damaged DNA

Proto-oncogenes
These genes that turn on cell division. A mutated form, or oncogene, leads to unregulated cell division. A mutation to one or two controlling genes might cause a benign (non-malignant) tumor. A large number of mutations can cause loss of control, causing a cell to become cancerous (left).

Cancerous cell showing the membrane protrusions that are important in cancer cell adhesion and migration.

1. How do cancerous cells differ from normal cells? _____

2. Describe the involvement of regulatory genes in control of the cell cycle: _____

Proteins help regulate the cell cycle

▶ Proteins produced by genes help to regulate the cell cycle. Cancerous cells can develop when the proteins are damaged due to gene mutation.

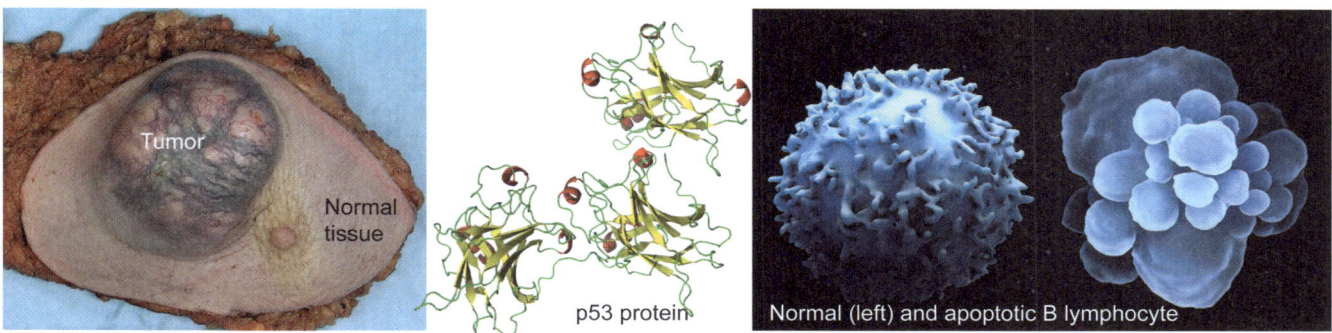

p53 protein

Normal (left) and apoptotic B lymphocyte

The protein product of the gene BRCA1 is involved in repairing damaged DNA and BRCA1 deficiency is associated with abnormalities in cell cycle checkpoints. Mutations to this gene and BRCA2 are found in about 10% of all breast cancers and 15% of ovarian cancers.

One of the most important proteins in regulating the cell cycle is the protein produced by the gene p53. The p53 tumor-suppressor protein helps to regulate the cell cycle, including apoptosis (programmed cell death) and genomic stability. Mutations to the p53 gene are found in about 50% of cancers. Apoptosis is a highly controlled process that involves cell shrinkage, and DNA fragmentation.

3. (a) Explain how the normal controls over the cell cycle can be lost: _____
   _____
   _____

   (b) How can these failures result in cancer? _____
   _____
   _____

4. Why are cancer cells so harmful to the human body? _____
   _____
   _____
   _____

Features of cancer cells

▶ The diagram below shows a single lung cell that has become cancerous. It no longer carries out the role of a lung cell, and instead takes on a parasitic lifestyle, taking nutrients from the body and contributing nothing in return. The rate of cell division of the cancerous cell is greater than in normal cells in the same tissue because there is no resting phase between divisions.

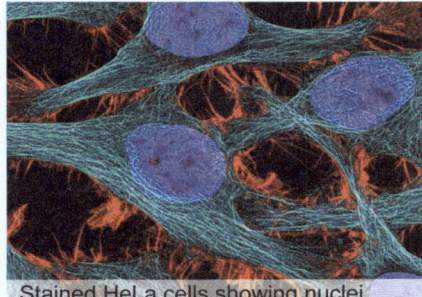

Stained HeLa cells showing nuclei (blue), actin filaments (red), and microtubules (cyan).

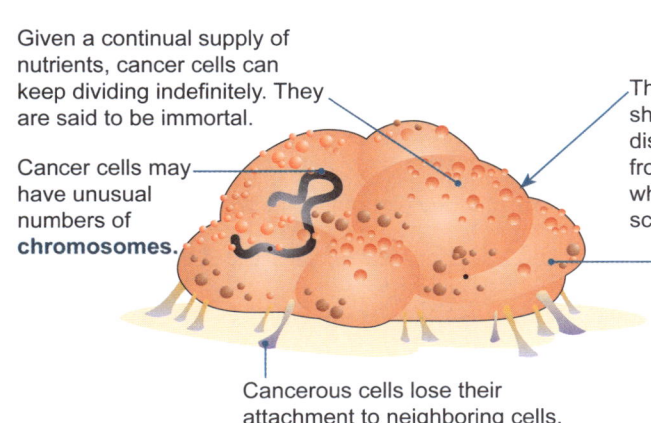

Given a continual supply of nutrients, cancer cells can keep dividing indefinitely. They are said to be immortal.

Cancer cells may have unusual numbers of **chromosomes**.

The bloated, lumpy shape is readily distinguishable from a healthy cell, which has a flat, scaly appearance.

Metabolism is disrupted and the cell ceases to function constructively.

Cancerous cells lose their attachment to neighboring cells.

HeLa is an immortal cell line widely used in scientific research. The line is derived from the cervical cancer cells of Henrietta Lacks (HeLa cells) after being taken from her and cultured without her consent just before her death from cervical cancer. This cell line was the first to be cultured successfully for scientific research, and is robust and very aggressive in its growth.

Reduction in rates of apoptosis can cause cancer

▶ Apoptosis, or cell death, is a process that removes damaged or abnormal cells before they can multiply. However, when apoptosis malfunctions, it can cause a number of diseases, including cancer.

▶ When cell cycle checkpoints fail, the normal rate of apoptosis falls. This allows a damaged cell to divide without regulation and it can lead to the formation of tumors.

▶ A number of factors can disrupt the cell cycle and cause a cell to become cancerous. These include defective genes, some viruses, and a number of chemical and environmental factors, called carcinogens.

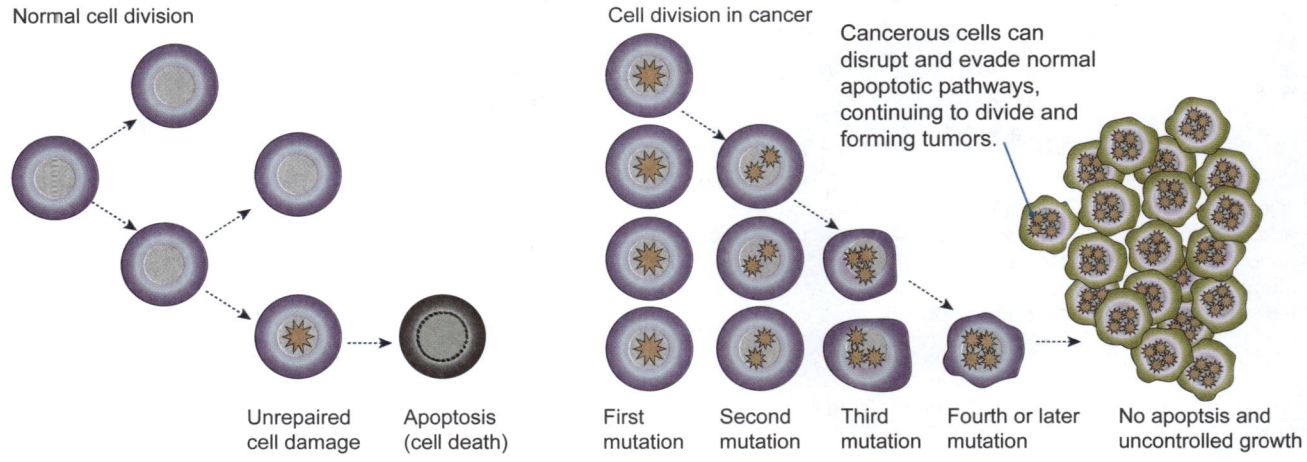

Tumor suppressor genes, e.g. the p53 gene, normally halt cell division of DNA damaged cells until the damage is repaired. If the damage cannot be repaired, apoptosis, the process of controlled cell death, is triggered.

Cancerous cells may inhibit the expression of the p53 gene. Around 50% of all human tumors contain p53 gene mutations.

Viruses can inhibit apoptosis

Some viruses, such as human papillomavirus (HPV) (near right), invade human cells and inhibit apoptosis. HPV is a known carcinogen. Nearly all cases of cervical cancer can be attributed to HPV infection. HPV integrates its DNA into the host's DNA promoting genomic instability and producing a protein that binds to and inactivates the p53 protein. Inactivation of p53 promotes unregulated cell division, cell growth, and cell survival.

Several studies have suggested a link between cytomegalovirus (CMV, image far right) and a type of brain tumor. Other herpes viruses are also suspected carcinogens. The ability of the virus to inhibit apoptosis is one mechanism involved in a cell turning cancerous.

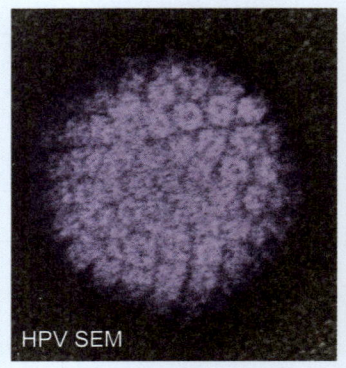

HPV SEM

False color SEM of human papillomavirus, a causative agent in cervical cancer.

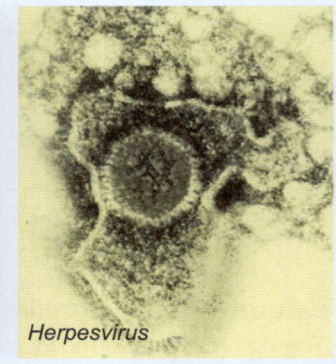

Herpesvirus

The human herpesviruses, invade human cells and can inhibit apoptosis.

5. How can a decreased rate of apoptosis lead to cancer? _____

_____

_____

_____

6. A number of Herpesviruses are recognized cancer-forming agents. How might they exert this effect? _____

_____

_____

_____

49 The Power to Rebuild Revisited

Content Anchor Revisted: What are the axolotl's superpowers of regeneration that allow it to regrow amputated limbs and even parts of its brain?

▶ At the beginning of this chapter you were asked to consider how a skin wound healed and think about the ability of the axolotl to heal from injuries, compared to how humans do this.

▶ Now that you have learned more about **cell division** and cell differentiation you should have a better understanding of how the axolotl is able to regrow complex tissues.

1. Briefly describe the stages that skin cells go through, in order to divide and replicate correctly, to heal after an injury:

   _____
   _____
   _____
   _____
   _____

2. How do cells surrounding an axolotl's injured limb 'know' to develop into skin, bone, brain tissue, blood etc?

   _____
   _____
   _____

3. When an axolotl loses its limb, initial healing begins with blood clotting and then cells from nearby tissues move to the wound site. These cells undergo a change so that they have the ability to become any other tissue. What do we call unspecialized cells that have the ability to differentiate into any cell type?

   _____

4. Use the information that you have learned to describe the stage of mitosis shown in each of the photographs below. Underneath your identification, briefly state the reason for your choice:

   (a) _____   (b) _____   (c) _____   (d) _____

5. Where are the different locations of mitosis in animal cells and plant cells? _____
   _____
   _____
   _____
   _____

50 Summing Up

Read each question carefully. Place a cross in the box beside the best answer to the question from the four answer choices provided.

Questions 1-2 relate to the image below.

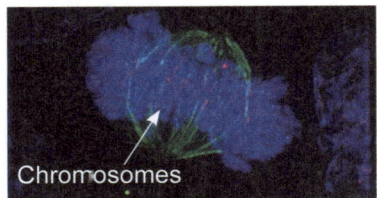
Chromosomes

1. The phase of mitosis pictured is.
   - (a) Prophase
   - (b) Telophase
   - (c) Metaphase
   - (d) Anaphase

2. The green fluorescent structures are:
   - (e) Centromeres
   - (f) Spindle fibers
   - (g) Kinetochores
   - (h) An artifact of slide preparation

3. Cells that no longer divide (such as certain specialized cells in the brain) stay in which phase of the cell cycle?
   - (a) S phase
   - (b) G_0 phase
   - (c) G_1 phase
   - (d) Interphase

4. Which statement about daughter cells following mitosis and cytokinesis is correct:
   - (a) They are genetically different from each other and from the parent cell.
   - (b) They are genetically identical to each other and to the parent cell.
   - (c) They are genetically identical to each other but different from the parent cell.
   - (d) Only one of the daughter cells is identical to the parent cell.

5. DNA replicates prior to mitosis. Which model best explains the DNA replication process:
   - (a) Conservative
   - (b) Dispersive
   - (c) Semi-conservative
   - (d) Complementary

6. Totipotent stem cells are important in cell and body development because:
   - (a) They can differentiate into any type of body cell.
   - (b) They can divide by mitosis more rapidly than any other type of cell.
   - (c) They are needed to form brain cells.
   - (d) They form the placenta and umbilical cord in the growing fetus.

7. Errors in which of the following genes are involved in cancer?
   - (a) BRCA1
   - (b) p53
   - (c) (a) and (b)
   - (d) None of the above

8. The photograph below depicts a Siamese cat and kittens. The kittens are born creamy white, and do not develop dark markings until a few weeks after birth:

What is the reason for this phenomenon?
   - (a) The cells making the darker markings take longer to form.
   - (b) The environment (cooler temperatures) activates enzymes in differentiated cells in the extremities.
   - (c) The enzyme making darker markings takes time to build up in the cells.
   - (d) The cooler environmental temperatures damage the cells at extremities, making them darker.

9. The purpose of cell cycle checkpoints is to:
   - (a) Regulate the timing of mitosis stages.
   - (b) Control DNA replication so that cells can divide correctly.
   - (c) Ensure that all cellular processes have been completed correctly before entering the next phase.
   - (d) Remove mutated cells so that they cannot divide.

10. Apoptosis is a process controlled by genes, and occurs during the cell cycle. What is the key purpose of apoptosis?
    - (a) Remove damaged or abnormal cells before they can multiply.
    - (b) Target and kill cancer cells in the body.
    - (c) Act as a type of cell cycle checkpoint.
    - (d) Cause stem cells to differentiate into cell types required in the body.

11. If the DNA content of a non-dividing cell is X, what would be its content at the start of interphase, during telophase, and after cytokinesis respectively:
    - (a) 2X, X, X
    - (b) X, 2X, X
    - (c) X, X, 2X
    - (d) 2X, 2X, X

12. Briefly outline the events in mitosis: _____
_____
_____

13. Where does mitosis take place in:

   (a) Animals: _____

   (b) Plants: _____

14. A cell with 10 chromosomes undergoes mitosis.

   (a) How many daughter cells are created? _____

   (b) How many chromosomes does each daughter cell have? _____

15. Briefly explain how multicellular organisms can develop from a single cell: _____
_____
_____

16. What two things must occur for a new cell to be produced? _____
_____
_____

17. (a) Explain how the semi-conservative model of DNA replication ensures identical daughter cells forming in mitosis:
_____
_____

   (b) Why is it important for two copies of identical DNA to be made before mitosis? _____
_____
_____

18. (a) Label the cell cycle (right) with the following labels: G_1, G_2, M, S, cytokinesis.

   (b) Briefly describe what happens in each of the following phases:

   G_1: _____
   _____
   _____

   G_2: _____
   _____
   _____

   M phase: _____
   _____
   _____
   _____

INTERPHASE

M PHASE

19. Every cell in your body has the same genetic information. How can there be so many different types of cell in your body? The answer is that, during different cell divisions, only certain parts of the genetic information is used (much like reading only some books in the library), thus producing the many different cell types.

    On the following page, a hypothetical cell has 11 genes in its DNA. Each gene initiates certain processes in the cell as shown in the **table**. Use these "genes" to fill in the boxes on the next page showing the cell as it progresses through two divisions. Note: if a gene is switched on, the cell follows the instruction for that gene.

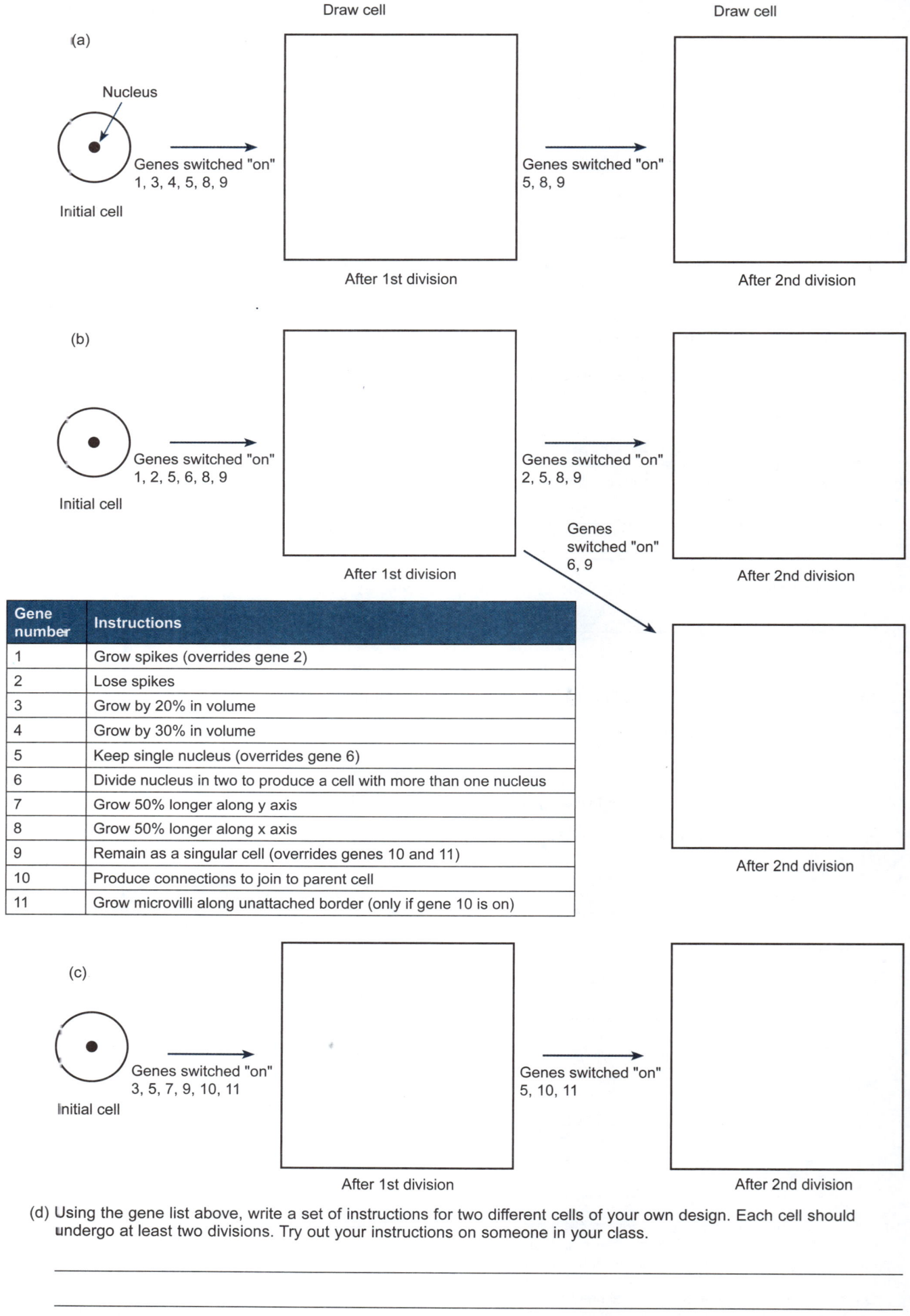

(d) Using the gene list above, write a set of instructions for two different cells of your own design. Each cell should undergo at least two divisions. Try out your instructions on someone in your class.

_____

_____

_____

Gene number	Instructions
1	Grow spikes (overrides gene 2)
2	Lose spikes
3	Grow by 20% in volume
4	Grow by 30% in volume
5	Keep single nucleus (overrides gene 6)
6	Divide nucleus in two to produce a cell with more than one nucleus
7	Grow 50% longer along y axis
8	Grow 50% longer along x axis
9	Remain as a singular cell (overrides genes 10 and 11)
10	Produce connections to join to parent cell
11	Grow microvilli along unattached border (only if gene 10 is on)

CHAPTER 3
Photosynthesis and Cellular Respiration

TEKS
Scientific and Engineering Practices

B.1: Investigation and Inquiry
1.A 1.B 1.C 1.D
1.E 1.F 1.G

B.2: Data and Patterns
2.B 2.C 2.D

B.3: Communicating in Science
3.A 3.B

B.4: Science as a Human Endeavor

TEKS
Science Concepts

B11.A explain how matter is conserved and energy is transferred during photosynthesis and cellular respiration using models, including the chemical equations for these processes

B11.B Investigate and explain the role of enzymes in facilitating cellular processes

Learning Outcomes
I know I have achieved this when I can:

Outcome	Activity number
☐ Explain the role of mitochondria and chloroplasts in cells.	52
☐ Describe the structure and function of ATP in cells.	53
☐ Draw a model showing the chemical transformation between ATP and ADP molecules.	53
☐ Write a simple word equation for photosynthesis.	54
☐ Identify the energy forms and the transformations occurring during photosynthesis.	54
☐ Describe the location, the component functions, and the steps that occur in the light dependent and light independent phases of photosynthesis.	55
☐ Investigate the photosynthetic rate in *Cabomba* plants when altering light levels.	56
☐ Construct a schematic diagram representing energy flow and conservation of matter during photosynthesis and respiration.	57
☐ Write a simple word equation for respiration.	58
☐ Describe the general location of respiration and the transportation of the reactants and products.	58
☐ Compare and contrast aerobic and anaerobic respiration, including fermentation.	58
☐ Detail the steps of respiration, including reactants/products, and the specific location where they each occur.	59
☐ Discuss the transfer of energy and pathways from storage as glucose to the cells of body tissue.	59
☐ Calculate the energy efficiency of respiration.	59
☐ Investigate respiration rates in germinated and non-germinated seeds.	60
☐ Model photosynthesis and respiration reactions.	61
☐ Compare anabolic and catabolic reactions.	62
☐ Explain the importance of enzymes to human metabolism.	63
☐ Describe the induced fit model of enzyme activity.	63
☐ Link enzyme activity to lowering activation energy of reactions.	64
☐ Explain how concentration and temperature affect enzyme reaction rates.	65
☐ Investigate the effect of temperature on amylase enzyme reaction rates.	65
☐ Plan an investigation to test how the number of days of mung bean germination affect the catalase enzyme reaction rate.	66
☐ Evaluate a planned investigation on enzyme reaction rate.	66

RESOURCE HUB
bit.ly/3ZmNAoe

ELPS English Language Proficiency Standards

		Page number
Learning	**Investigating Photosynthetic Rate.** Notice the words in this investigation. You know the words *place* and *position* in *Place the tube in a rack and position a lamp*... but have you heard them used as verbs? In this context, *place* means "put" and *position* means "put in a certain way." Look at *run, record,* and *mean*. What do they mean in this context?	100
Reading	**Energy from Glucose.** Before reading the page, look carefully at the images including the direction of arrows. What do you already know about air that goes in and out of an animal's mouth? Read the key question. How do humans carry out gas exchange? Make a prediction about the content of this page based on what you already know.	102
Speaking	**Measuring Respiration.** Take turns with your classmates as you complete the investigation. Have each group member describe their actions as they complete each task. Listen for the key words, expressions, and scientific terminology your classmates use. When it is your turn, try to use words and expressions in the same way they do.	106
Reading	**What are Enzymes?** Ask your classmates or teacher to point to each section of the model as they describe how enzymes work. Then read through the text describing each image. Notice that passive sentence structures are common in scientific text. Ask your teacher to rephrase the sentences using active voice. For example, "Two substrate molecules move into the active site…"	114
Learning	**Enzymes have Optimal Conditions to Work.** Before attempting to answer the questions, work with your group to clarify just what the questions are asking. Where possible, rewrite the question. For example, you can simplify question 5 to say, "What is the purpose of this investigation?"	118

51 Mouse Trap

Content Anchor: Under what conditions can an animal survive in a sealed system?

Mouse in a jar

▸ Around 1772, Joseph Priestley carried out a series of experiments. He wanted to see if there was a relationship between the survival of plants and animals in a closed system. One of his experiments is shown below.

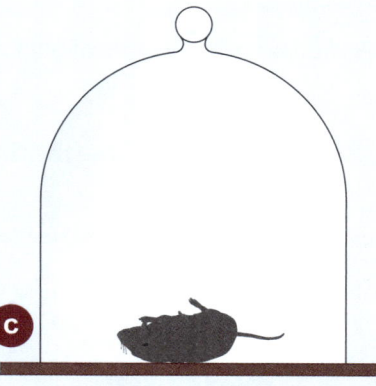

a. Priestley placed a mouse in a sealed, empty bell jar. The mouse quickly collapsed and died.

b. Priestley placed a mint plant in a sealed bell jar and left it for several days before adding a mouse. After several minutes the mouse was still alive.

c. Priestley removed the mouse from jar (b) and placed it into a sealed, empty bell jar. The mouse died very quickly.

1. (a) Can you explain why the mouse died in jars (a) and (c), but not in jar (b)? _____

_____

_____

_____

_____

(b) What metabolic or chemical processes might explain the results that Joseph Priestley obtained? _____

_____

_____

_____

_____

2. Draw a very simple diagram to show what is happening in jar (b):

3. In another experiment, Joseph Priestley left a plant covered with a bell jar for many days. He then placed a candle with a glowing wick into the jar. The wick ignited and began to burn. What was present to allow the wick to ignite?

_____

52 A Closer look at Chloroplasts and Mitochondria

Key Question: Why are chloroplasts and mitochondria important and what do they look like?

- **Chloroplasts** and **mitochondria** are membranous **organelles** in which specialized biochemical reactions occur.
- Chloroplasts are the organelles responsible for **photosynthesis** and mitochondria are involved in the production of **ATP** during **cellular respiration**.
- Recall that chloroplasts are found only in plant cells and some protists, whereas mitochondria are found in all eukaryotic cells.
- The structure and features of both organelles are described below and on the following page.

The structure of a mitochondrion

Mitochondria are enclosed by a double membrane envelope (inner and outer membrane). The inner membrane is highly folded. This increases its surface area.

Some steps of cellular respiration (electron transport chain) and ATP synthesis occur on the inner membrane. An **enzyme** called ATP synthase is involved in catalyzing ATP production.

The inward foldings are called cristae.

Like chloroplasts, mitochondria have their own circular DNA.

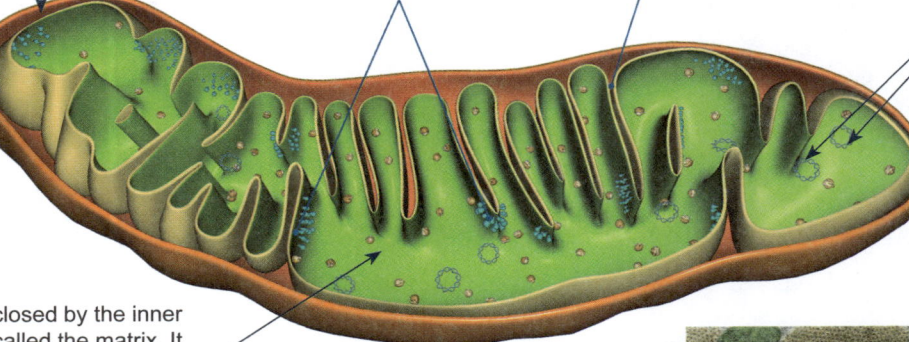

Mitochondria (singular mitochondrion) vary in diameter from 0.75 - 3.0 µm. They can be quite long compared to their diameter.

The space enclosed by the inner membrane is called the matrix. It is where the **Krebs cycle** (part of cellular respiration) occurs.

Cells that require a lot of ATP for cellular processes have a lot of mitochondria. Sperm cells, heart cells, liver cells, and muscle cells all have high numbers of mitochondria. The false color TEM (right) shows muscle cells (yellow) and many mitochondria (green).

1. What is the function of mitochondria? _____

_____

2. Red blood cells contain no mitochondria, whereas a single heart cell can have up to 5000 mitochondria. What does this information suggest about the relative metabolic activity of each cell type?

_____

_____

_____

3. (a) Identify the enzyme mentioned in the mitochondrion diagram above: _____

   (b) What is the role of this enzyme? _____

_____

©2024 **BIOZONE** International
ISBN: 978-1-99-101405-4
Photocopying prohibited

The structure of a chloroplast

The internal structure of chloroplasts is characterized by a system of membranous structures called thylakoids arranged into stacks called grana.

Liquid stroma contains the enzymes for the light independent phase (see activity 55) as well as the chloroplast's DNA.

Stroma lamellae connect the grana. They make up 20% of the thylakoid membrane.

Lipid droplet

A double membrane envelope (inner and outer membrane) encloses the chloroplast.

Grana are stacks of thylakoids.

Thylakoid membranes are the site of light absorption and provide a large surface area for binding chlorophyll (below). They are organized so they do not to shade each other.

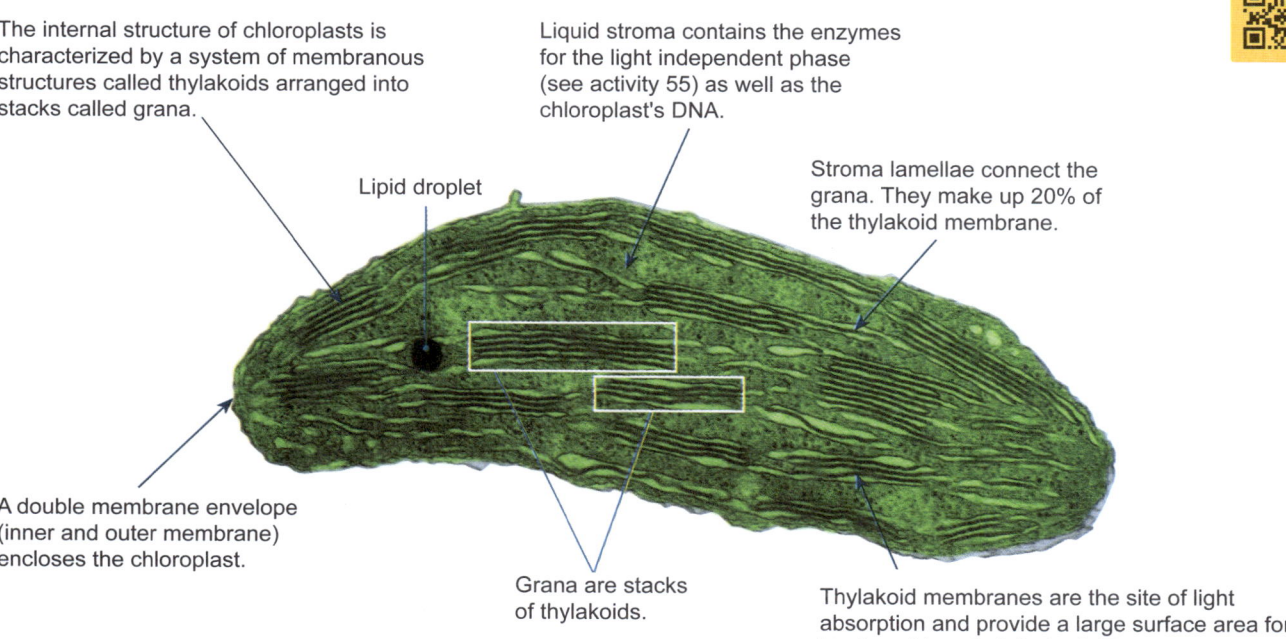

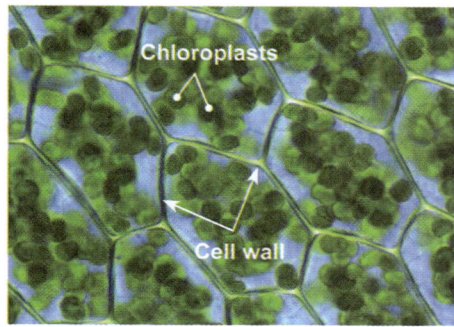

Chloroplasts are the organelles responsible for photosynthesis. A mesophyll leaf cell contains between 50-100 chloroplasts. Chloroplasts contain light capturing pigments, e.g. chlorophyll, that are used for photosynthesis. Chloroplasts are generally aligned so that their broad surface runs parallel to the cell wall to maximize the surface area available for light absorption. The image (left) shows chloroplasts in leaf cells. They appear green because they absorb blue and red light and reflect green light.

4. Use the diagram of the chloroplast above to help you label this transmission electron microscope image of a chloroplast:

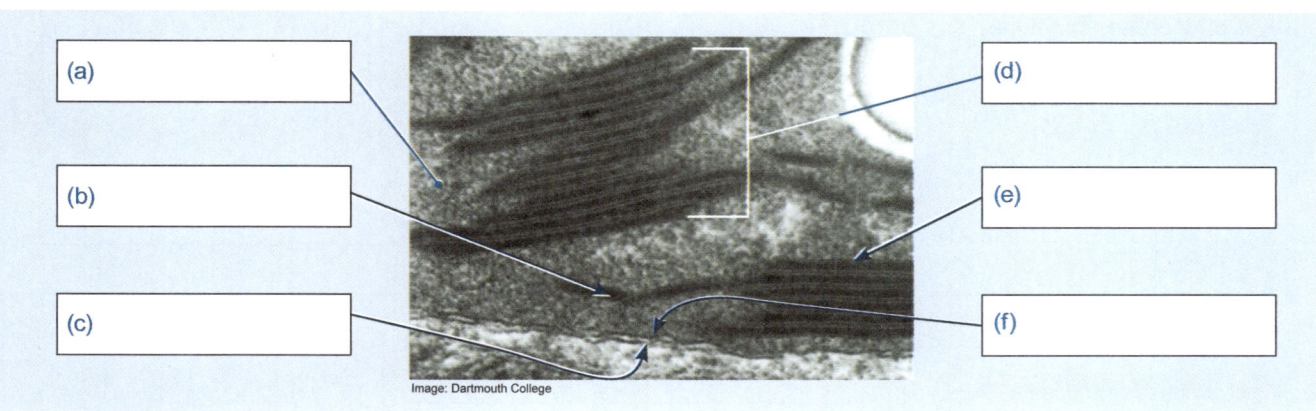

5. What does chlorophyll do? _____

6. What features of chloroplasts help maximize the amount of light that can be absorbed? _____

53 Energy in Cells

Key Question: How does ATP provide the energy needed to perform essential life functions?

Energy for metabolism

▶ All organisms require energy to be able to perform the metabolic processes required for them to function and reproduce. Some examples are provided below.

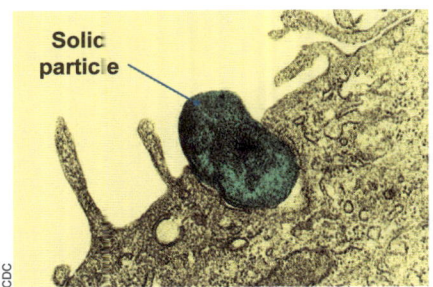

Energy is needed to actively transport molecules across the cellular membrane. The above image shows phagocytosis, where a large particle is being engulfed.

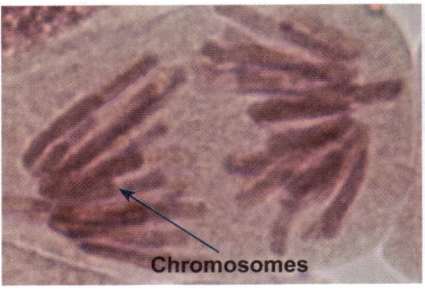

Cell division (mitosis) (above), requires energy to proceed. ATP provides energy for mitotic spindle formation and chromosome separation.

The maintenance of body temperature requires energy. The body requires energy for both heating and cooling, as it either shivers or produces sweat.

Adenosine triphosphate (ATP)

▶ Energy to carry out metabolic processes is provided by **adenosine triphosphate (ATP)**.

▶ ATP is considered to be a universal energy carrier, transporting chemical energy within the cell for use in metabolic processes. It is produced during **cellular respiration**: a set of metabolic reactions which ultimately convert biochemical energy from "food" into ATP.

▶ The ATP molecule (right) is a nucleotide derivative. It has three components:
- A purine base (adenine)
- A pentose sugar (ribose)
- Three phosphate groups.

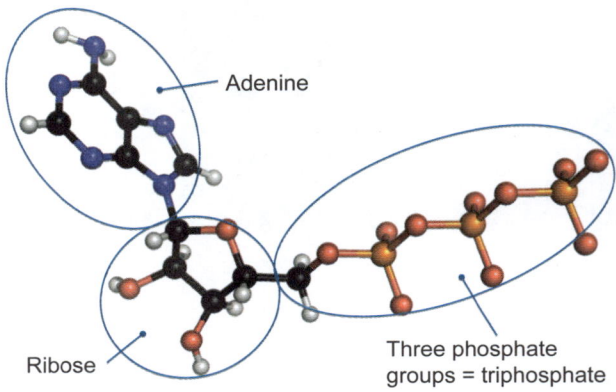

Note: Adenine + ribose = adenosine

1. What is the biological role of ATP? _____

_____

2. What process produces ATP in a cell? _____

3. Identify the three distinct components of the space filling model of ATP, labeled (a)-(c) below right:

   (a) _____

   (b) _____

   (c) _____

4. Which two of the components you labeled in question 3 make up adenosine?

_____

How does ATP provide energy?

▸ ATP acts as a store of energy within the cell. The bonds between the phosphate groups contain electrons in a high energy state. These electrons store a large amount of energy that is released during a chemical reaction. The removal of one phosphate group from ATP results in the formation of adenosine diphosphate (ADP), shown right.

▸ The bonds between the phosphate groups of ATP are unstable and very little energy is needed to break them.

▸ The energy in the ATP molecule is transferred to a target molecule, e.g. a protein, by a hydrolysis reaction. Water is split as part of the reaction to remove the phosphate at the end of the ATP. This produces ADP and an inorganic phosphate molecule (Pi). These are recycled to reform ATP during cellular respiration.

▸ When the Pi molecule combines with a target molecule, energy is released. Most of the energy (about 60%) is lost as heat (this helps keep you warm). The rest of the energy is transferred to the target molecule, allowing it to do work, e.g. joining with another molecule (right).

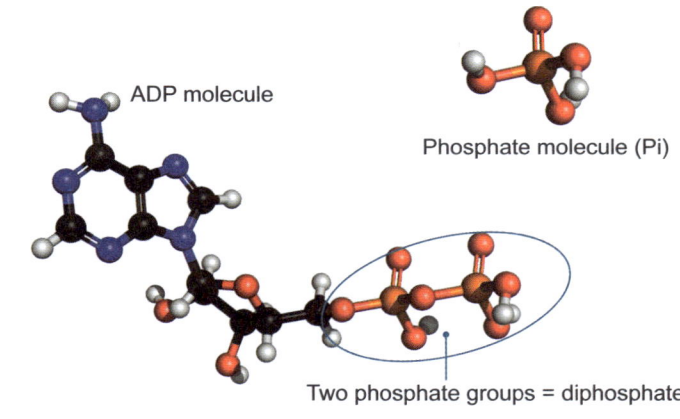

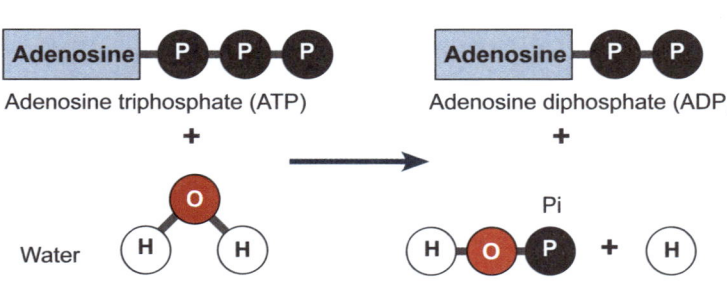

Note! The phosphate bonds in ATP are often referred to as being high energy bonds. This can be misleading. The bonds contain *electrons* in a high energy state (making the bonds themselves relatively weak). A small amount of energy is required to break the bonds, but when the intermediaries recombine and form new chemical bonds, a large amount of energy is released. The final product is less reactive than the original reactants.

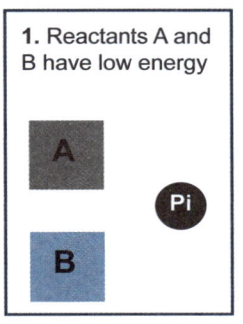

1. Reactants A and B have low energy

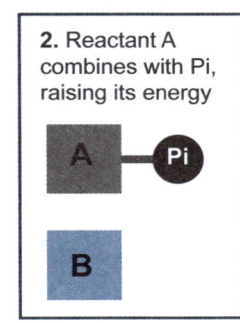

2. Reactant A combines with Pi, raising its energy

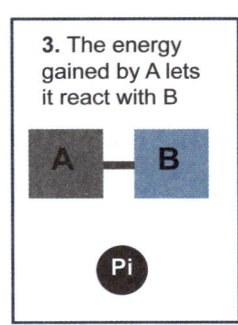

3. The energy gained by A lets it react with B

5. Where is the energy stored in ATP? _____
_____

6. (a) What products are formed during hydrolysis of ATP? _____

   (b) What happens to the ADP molecule once it has been utilized in a reaction? _____
   _____

   (c) Why it is important that ATP is regenerated?_____
   _____
   _____

7. In the box on the right, draw a simple diagram to show the cyclic relationship between ATP and ADP. Add the numbers of phosphate groups in both ATP and ADP:

54 Introduction to Photosynthesis

Key Question: How does photosynthesis convert sunlight, carbon dioxide, and water into glucose and oxygen?

▶ Plants, algae, and some bacteria are photoautotrophs. They use pigments called **chlorophylls** to capture light energy which is used in a process called **photosynthesis.**

▶ During photosynthesis, carbon dioxide and water are converted into **glucose** and oxygen. The reaction requires energy obtained from sunlight, which is transformed into chemical energy within the bonds of the glucose molecule. This chemical energy fuels life's essential processes.

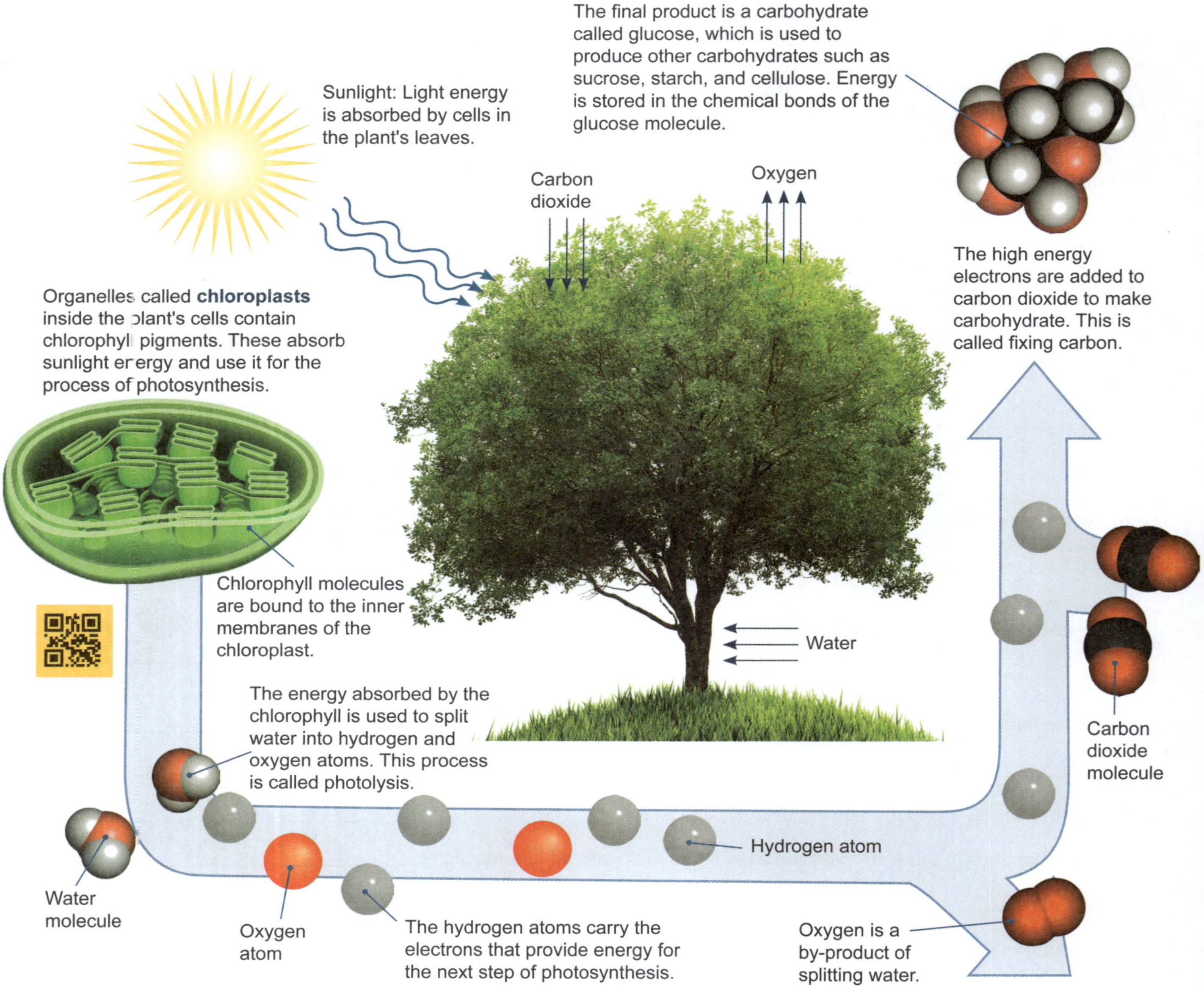

Sunlight: Light energy is absorbed by cells in the plant's leaves.

The final product is a carbohydrate called glucose, which is used to produce other carbohydrates such as sucrose, starch, and cellulose. Energy is stored in the chemical bonds of the glucose molecule.

Carbon dioxide

Oxygen

Organelles called **chloroplasts** inside the plant's cells contain chlorophyll pigments. These absorb sunlight energy and use it for the process of photosynthesis.

The high energy electrons are added to carbon dioxide to make carbohydrate. This is called fixing carbon.

Chlorophyll molecules are bound to the inner membranes of the chloroplast.

Water

The energy absorbed by the chlorophyll is used to split water into hydrogen and oxygen atoms. This process is called photolysis.

Carbon dioxide molecule

Hydrogen atom

Water molecule

Oxygen atom

The hydrogen atoms carry the electrons that provide energy for the next step of photosynthesis.

Oxygen is a by-product of splitting water.

Photosynthesis is commonly displayed as either of the two equations on the right. Both are correct, but the bottom equation cancels the extra water molecules.

$$6CO_2 + 12H_2O \xrightarrow{Light} C_6H_{12}O_6 + 6O_2 + 6H_2O$$

$$6CO_2 + 6H_2O \xrightarrow{Light} C_6H_{12}O_6 + 6O_2$$

---

1. Write the word equation for photosynthesis: _____

   _____

2. Where does the oxygen released during photosynthesis come from? _____

Requirements for photosynthesis

Plants need only a few raw materials to make their own food:

- *Light energy from the Sun.*
- *Chlorophyll absorbs light energy.*
- *CO_2 gas is used to make carbohydrate.*
- *Water is split to provide the electrons for the fixation of carbon as carbohydrate.*

Photosynthesis is not a single process but two complex processes (the **light dependent** and **light independent** reactions) each with multiple steps.

Production of carbohydrate occurs in the fluid stroma of the chloroplast. Commonly called carbon "fixation", it **does not require sunlight** (light independent phase).

Energy capture occurs in the inner membranes (thylakoids) of the chloroplast and **requires sunlight** (light dependent phase).

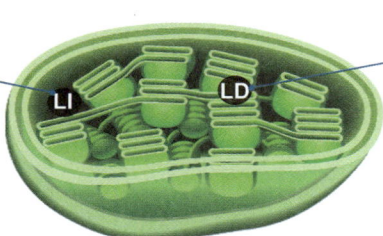

The photosynthesis of marine algae, such as these diatoms, supplies a substantial portion of the world's oxygen. The oceans also act as sinks for absorbing large amounts of CO_2.

Macroalgae, like this giant kelp, are important marine producers. Algae living near the ocean surface have access to light for photosynthesis.

On land, vascular plants are the main producers of food. Plants at different levels in a forest receive different intensities and quality of light.

3. (a) What form of energy is used to drive photosynthesis? _____

   (b) What is the name of the molecule that captures this energy? _____

   (c) Where in the plant cell does photosynthesis take place? _____

   (d) What form of energy is your answer to (a) converted into? _____

   (e) What happens in each of the two phases of photosynthesis? _____

4. (a) Primary production is the production of carbon compounds from carbon dioxide, generally by photosynthesis. Study the graph (right) showing primary production in the oceans. Describe what the graph is showing:

   _____

   (b) Explain the shape of the curves described in (a): _____

   (c) About 90% of all marine life lives in the photic zone (the depth to which light penetrates). Suggest why this is so:

   _____

Ocean primary production

Primary production ($mgC/m^3/day$)

2011
2012
2013

55 Stages in Photosynthesis

Key Question: What are the two main reactions in photosynthesis?

- **Photosynthesis** has two phases, the **light dependent phase** and the **light independent phase**.
- In the reactions of the light dependent phase, light energy is converted to chemical energy in the form of **ATP** and NADFH. This phase occurs in the thylakoid membranes of the chloroplasts.
- In the reactions of the light independent phase, chemical energy is used to make carbohydrate. This occurs in the stroma of **chloroplasts**.
- An overview of the two stages of photosynthesis is shown in the diagram of a chloroplast below.

Light dependent phase (LDP):
Location: Thylakoid membranes of the grana.
Process: In the first phase of photosynthesis, chlorophyll captures light energy, which is used to split water, producing O_2 gas (expelled as a waste product), electrons and H^+ ions, which are transferred to the molecule NADPH. ATP is also produced.

Light independent phase (LIP):
Location: Stroma.
Process: The second phase of photosynthesis uses the NADPH and ATP produced in the LDP to drive a series of enzyme-controlled reactions (the Calvin cycle) that fix carbon dioxide to produce triose phosphate. This phase does not need light to proceed.

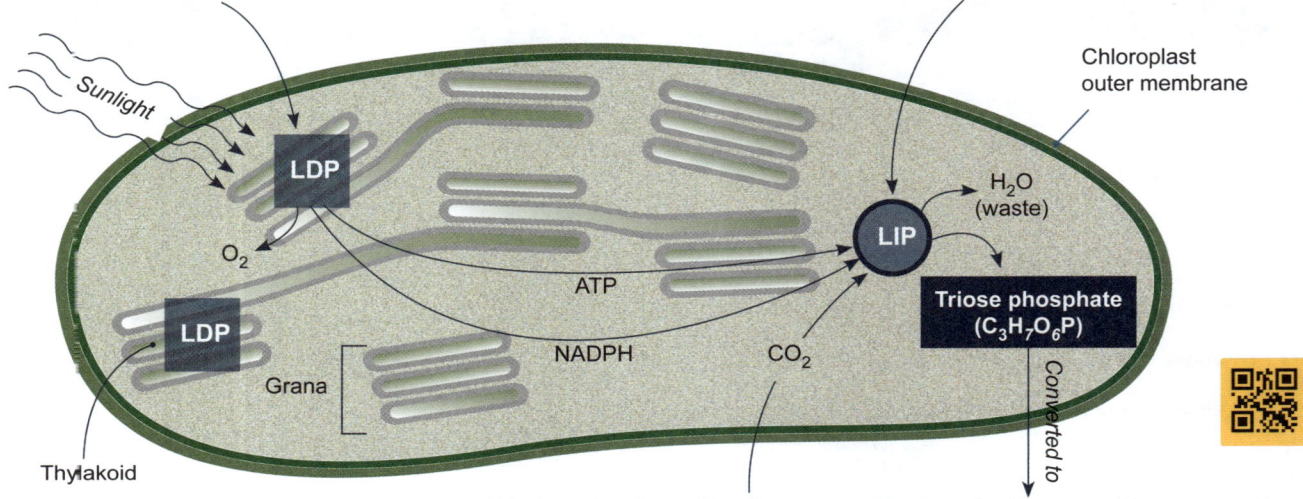

CO_2 from the air provides the raw materials for glucose production.

Monosaccharides, e.g. glucose, and other carbohydrates, lipids, and amino acids.

Enzymes are required for photosynthesis to proceed

Enzymes facilitate cellular processes, including photosynthesis. RuBisCo (right) is the central enzyme in the LIP of photosynthesis (carbon fixation) catalyzing the first step in the Calvin cycle. However, it is remarkably inefficient, processing just three reactions a second. To compensate, RuBisCo makes up almost half the protein content of chloroplasts.

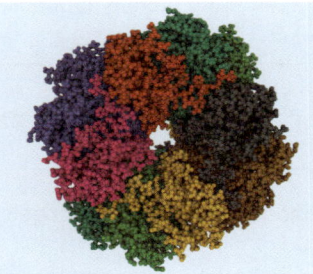

1. (a) Where does the light dependent phase of photosynthesis occur? _____

   (b) Where does the light independent phase of photosynthesis occur? _____

2. How are the light dependent and light independent phases linked? _____
   _____
   _____

3. What is the role of the enzyme RuBisCo? _____
   _____

4. Use the information on the previous page to fill in the diagram below, including the raw material (inputs), products (outputs), and processes.

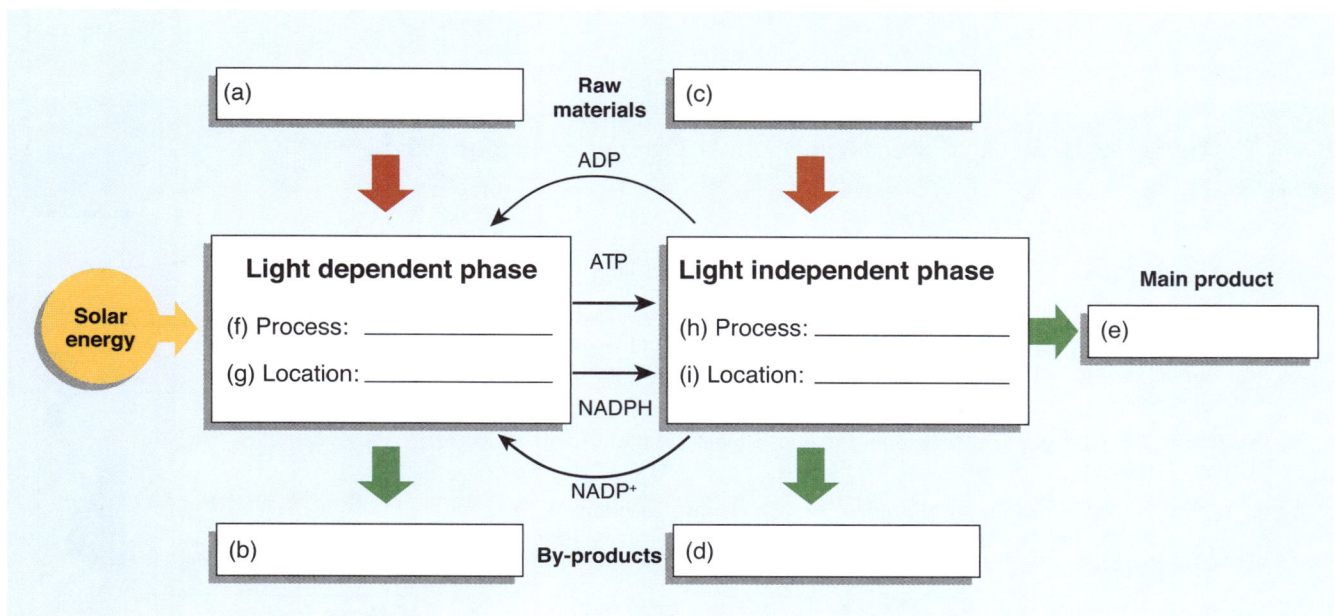

5. In two experiments, radioactively-labeled oxygen (shown in red in the equations below) was used to follow oxygen through the photosynthetic process. The results of the experiment are shown below:

Experiment A: $6CO_2 + 12H_2\textcolor{red}{O} + \text{sunlight energy} \rightarrow C_6H_{12}O_6 + 6\textcolor{red}{O}_2 + 6H_2O$

Experiment B: $6C\textcolor{red}{O}_2 + 12H_2O + \text{sunlight energy} \rightarrow C_6H_{12}\textcolor{red}{O}6 + 6O_2 + 6H_2O$

From these results, what would you conclude about the source of the oxygen in:

(a) The carbohydrate produced? _____

(b) The oxygen released? _____

6. Describe what happens during:

(a) The light dependent phase of photosynthesis: _____

(b) The light independent phase of photosynthesis: _____

7. What is the function of each of the following in photosynthesis:

(a) ATP: _____

(b) NADPH: _____

(c) Light: _____

(d) Chlorophyll: _____

(e) Water: _____

56 Investigating Photosynthetic Rate

Key Question: How does light intensity affect photosynthesis rate?

Investigation 3.1 — Measuring bubble production in *Cabomba*

See appendix for equipment list.

1. Fill a boiling tube 2/3 full with a 20°C solution of 1% sodium hydrogen carbonate ($NaHCO_3$).
2. Cut ~ 7 cm long piece of *Cabomba* stem (cut underwater). Place the *Cabomba* into the boiling tube (cut end up). Carefully push the *Cabomba* down.
3. Place the boiling tube in a rack and position a lamp so that it will shine on the tube when switched on.
4. To test the set-up, switch on the lamp for one minute to check that bubbles emerge freely from the stem. If they don't, you may have to recut the stem to open it.
5. When you have checked your set-up, switch off the lamp and, **after 5 minutes**, use a stopwatch to record the number of bubbles emerging from the stem in one minute. Repeat.
6. Use a ruler to mark out distances 0, 5, 10, 15, 20, and 25 cm from the boiling tube.
7. Starting at 25 cm, move the lamp to each of the distances in turn and use a stopwatch to record the number of bubbles emerging from the stem in one minute. Run two tests at each distance and allow 5 minutes after moving to a new distance before recording (this allows for acclimation).
8. Record your results in the table (right). Calculate the mean rate of gas production for each distance (and lamp OFF).
9. After you have finished recording, remove the stopper from the tube and test the gas with a glowing splint. What happens?

NEED HELP? See Activity 267

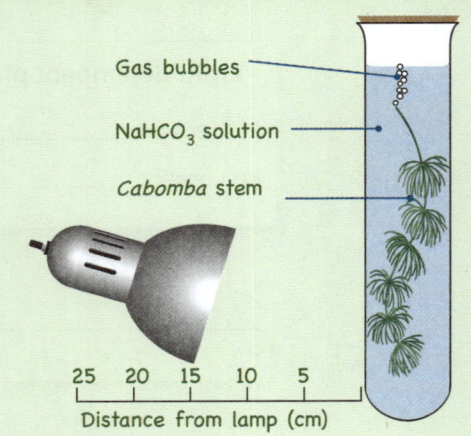

Distance (cm)	Bubbles per minute		
	Test 1	Test 2	Mean
OFF			
25			
20			
15			
10			
5			
0			

_____

1. Use your calculated means to draw a graph of gas production vs light intensity (distance).

2. What did your splint test tell you about the gas produced by the *Cabomba* plant?

   NEED HELP? See Activity 261

   _____

3. From this experiment what can you say about photosynthesis, light, and the gas produced?

   _____

   _____

   _____

4. How could you improve the design of this investigation?

   _____

   _____

   _____

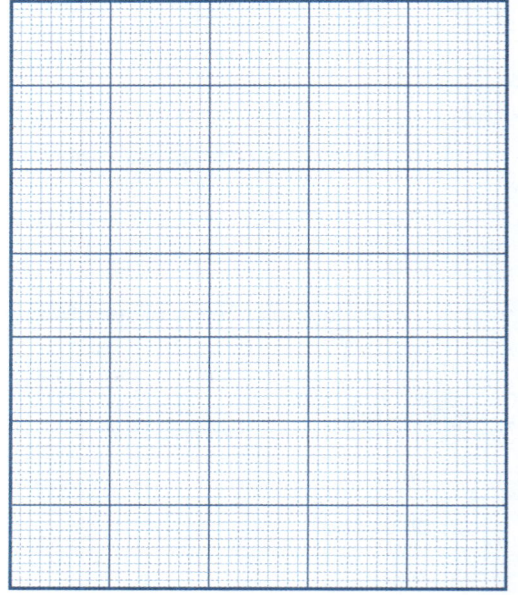

57 Energy Transfers Between Systems

Key Question: How is the stored energy in glucose used to power the chemical reactions which occur in living organisms?

▸ During **photosynthesis**, light energy is converted into chemical energy in the form of **glucose**. Glucose is used by plants and animals to provide the energy for **cellular respiration**.

▸ During cellular respiration, **ATP** is formed through a series of chemical reactions. ATP provides the energy needed to drive life's essential processes.

▸ Heterotrophs (organisms that cannot make their own food) obtain their glucose by eating plants or other organisms.

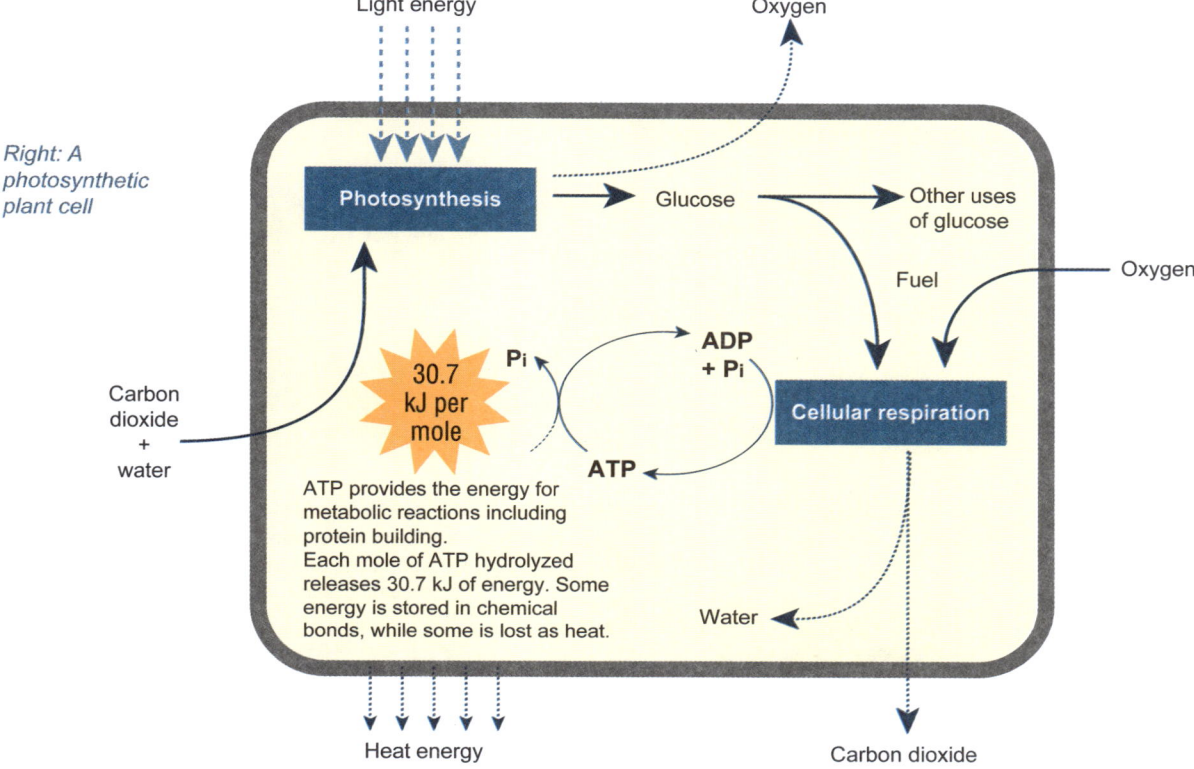

Right: A photosynthetic plant cell

1. Use the diagram above to explain where heterotrophs join the energy transfer system: _____
   _____
   _____

2. Complete the schematic diagram (below) of the transfer of energy and production of macromolecules using the following word list: *water, ADP, protein, carbon dioxide, amino acid, glucose, ATP*.

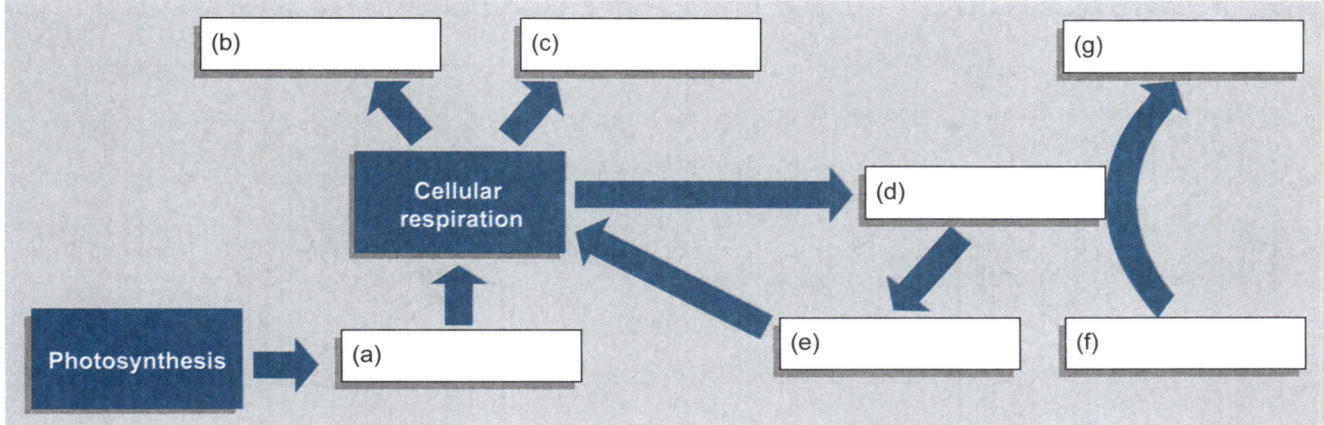

58 Energy From Glucose

Key Question: How is energy released from glucose during the process of cellular respiration?

▶ Energy is released in cells by the breakdown of sugars and other substances in **cellular respiration**. During **aerobic respiration**, oxygen is consumed and carbon dioxide is released. These gases need to be exchanged with the environment by **diffusion**. Diffusion gradients are maintained by transport of gases away from the gas exchange surface, e.g. the cell surface. In anaerobic pathways, **ATP** is generated but oxygen is not used.

The overall equation for cellular respiration is:

Glucose + Oxygen → Carbon dioxide + Water + Energy

$C_6H_{12}O_6 + 6O_2 \longrightarrow 6CO_2 + 6H_2O + Energy$

To carry out aerobic respiration, the body needs oxygen and must remove carbon dioxide. Gas exchange surfaces provide a way for these respiratory gases to enter and leave the body by diffusion. Some organisms, such as this frog, can use the body surface for gas exchange, but many have specialized structures, e.g. lungs, gills, or stomata.

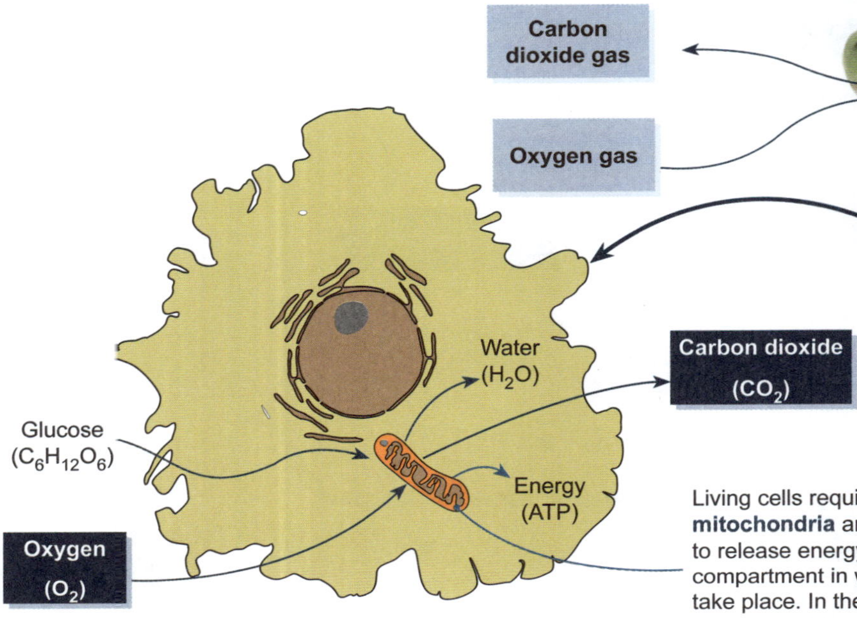

Every living cell of an organism's body respires. Aerobic cellular respiration creates a constant demand for oxygen and a need to eliminate carbon dioxide gas.

Living cells require energy for the activities of life. In eukaryotes, **mitochondria** are the main site where **glucose** is broken down to release energy. The mitochondrion provides a membranous compartment in which the reactions of cellular respiration can take place. In the process, oxygen is used to make water and carbon dioxide is released as a waste product.

1. (a) Write the word equation for aerobic cellular respiration: _____

   (b) Write the chemical equation for aerobic cellular respiration: _____

2. What is the main site for cellular respiration in a eukaryotic cell? _____

3. Distinguish between cellular respiration and gas exchange: _____

4. (a) What gases are involved in cellular respiration? _____

   (b) Why must these gases be continuously supplied to or transported away from cells? _____

   (c) By which transport process do these gases move through the gas exchange surface? _____

5. What is the main difference between aerobic and anaerobic pathways for ATP generation? _____

©2024 **BIOZONE** International
ISBN: 978-1-99-101405-4
Photocopying prohibited

Aerobic and anaerobic pathways for ATP production

A Aerobic respiration

Aerobic respiration produces the energy (as ATP) needed for metabolism. The rate of aerobic respiration is limited by the amount of oxygen available. In animals and plants, most of the time oxygen supply is sufficient to maintain aerobic metabolism. Aerobic respiration produces a high yield of ATP per molecule of glucose (path A below).

B Lactic acid fermentation

During maximum physical activity, when oxygen is limited, **anaerobic metabolism** provides ATP for working muscle. In mammalian muscle, metabolism of a respiratory intermediate produces lactate, which provides fuel for working muscle and produces a low yield of ATP. This process is called lactic acid **fermentation** (path B below).

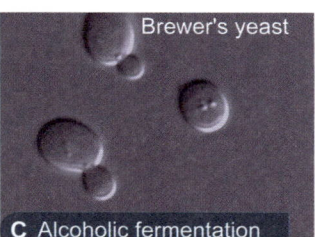

C Alcoholic fermentation

The process of brewing utilizes the anaerobic metabolism of yeasts. Brewers' yeasts preferentially use anaerobic metabolism in the presence of excess sugars. This process, called alcoholic fermentation, produces ethanol and CO_2 from the respiratory intermediate pyruvate. It is carried out in vats that prevent entry of O_2 (path C below).

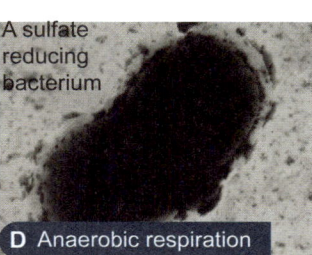

D Anaerobic respiration

Many bacteria and archaea are anaerobic, using molecules other than oxygen, e.g. nitrate or sulfate, as a terminal electron acceptor of their electron transport chain. These electron acceptors are not as efficient as oxygen (less energy is released per oxidized molecule) so the energy (ATP) yield from anaerobic respiration is generally quite low (path D below).

In most energy-yielding pathways the initial source of chemical energy is glucose. The first step, **glycolysis**, is an almost universal pathway. The paths differ in what happens after glucose has been converted to pyruvate.

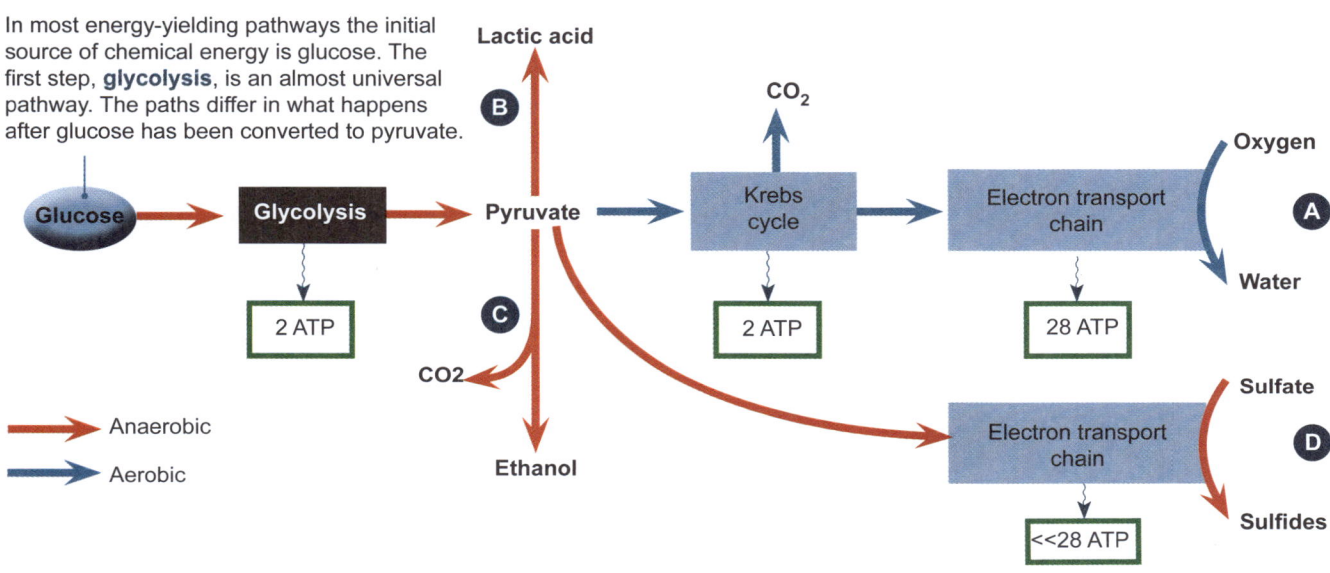

6. Compare anaerobic pathways in eukaryotes, e.g. yeasts, and anaerobic respiration in anaerobic microbes: _____

_____

_____

_____

7. Explain why aerobic respiration is energetically more efficient than fermentation and anaerobic respiration: _____

_____

_____

8. (a) How many ATP molecules are produced from one glucose molecule during aerobic respiration? _____

   (b) How many ATP molecules are produced during lactic acid or alcoholic fermentation? _____

   (c) Calculate the efficiency of fermentation compared to aerobic respiration: _____

   _____

9. When brewing alcohol, why is it important to prevent entry of oxygen to the fermentation vats? _____

_____

_____

59 Aerobic Cellular Respiration

Key Question: What are the stages of aerobic cellular respiration?

- **Cellular respiration** is the process of extracting the energy stored in the chemical bonds in **glucose** and storing it in **ATP** molecules. The process includes many chemical reactions, some of which produce ATP molecules and some that prepare molecules for further chemical reactions.

- Cellular respiration can be divided into four major steps, each with its own set of chemical reactions. The four steps are: **glycolysis**, the **link reaction**, the **Krebs cycle**, and the electron transport chain (ETC). Every step, except the link reaction, produces ATP.

- Glycolysis occurs in the cytoplasm, the other steps take place within the **mitochondrion**.

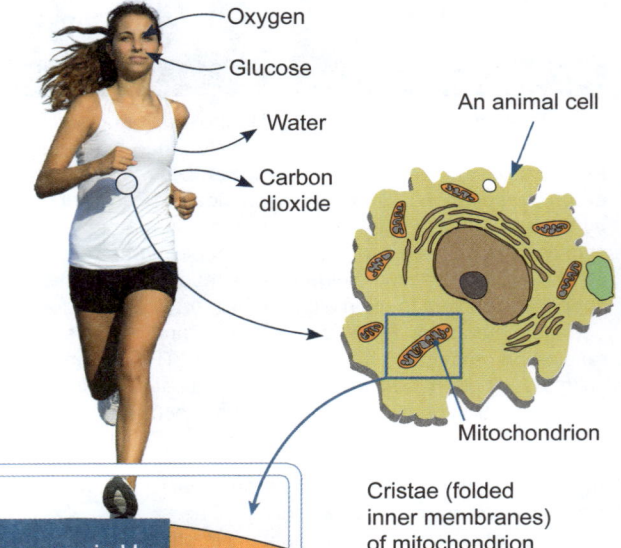

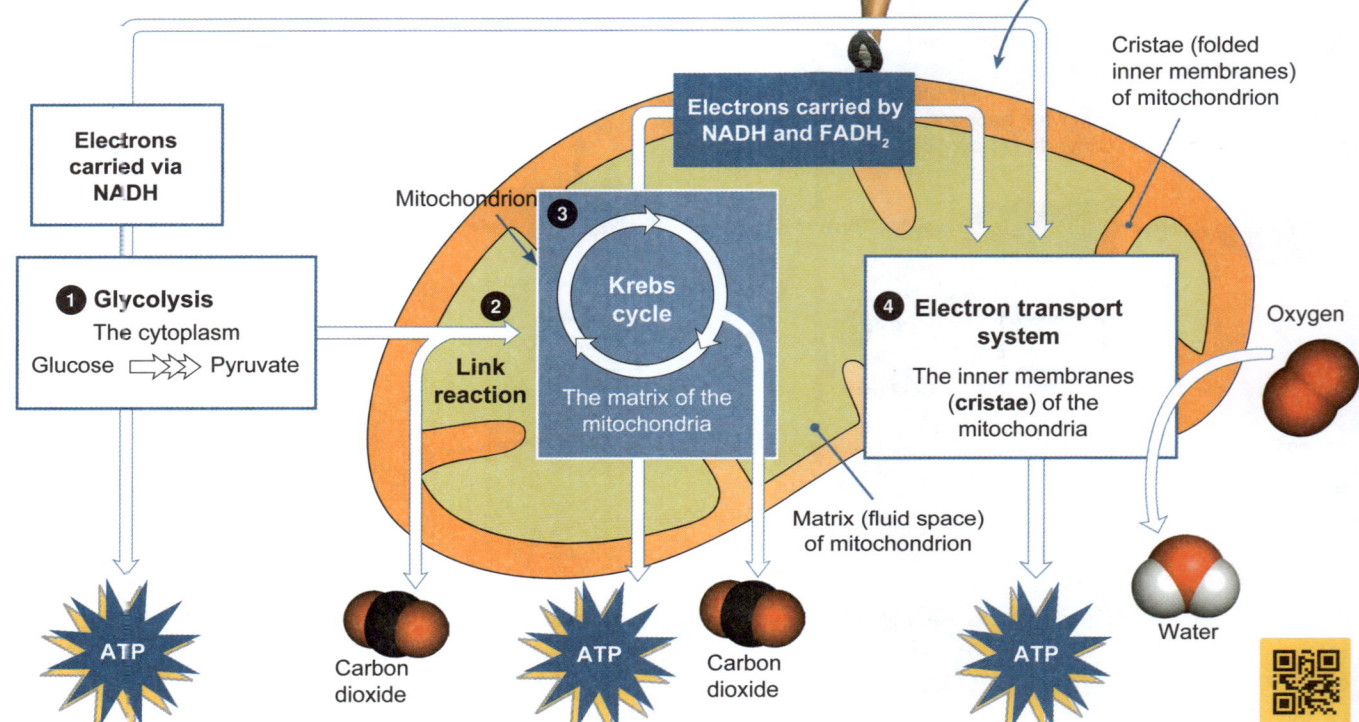

The general equation for cellular respiration

$$C_6H_{12}O_6 + 6O_2 \longrightarrow 6CO_2 + 6H_2O + \text{energy}$$

1. Which of the four steps in cellular respiration yield ATP? _____

2. (a) What are the main reactants (inputs) for cellular respiration? _____

   (b) What are the main products (outputs) for cellular respiration? _____

3. Which of the four steps of cellular respiration occur in the mitochondria? _____

©2024 **BIOZONE** International
ISBN: 978-1-99-101405-4
Photocopying prohibited

How does cellular respiration provide energy?

▸ A molecule's energy is contained in the electrons within its chemical bonds. During a chemical reaction, energy, e.g. heat, can break the bonds of the reactants.

▸ When the reactants form products, the new bonds within the product will contain electrons with less energy, making the bonds more stable. The difference in energy is usually lost as heat. However, some of the energy can be captured to do work.

▸ Glucose contains 16 kJ of energy per gram (2870 kJ/mol). The step-wise breakdown of glucose through a series of chemical reactions yields ATP. In total, 32 ATP molecules can be produced from 1 glucose molecule.

A model for ATP production and energy transfer from glucose

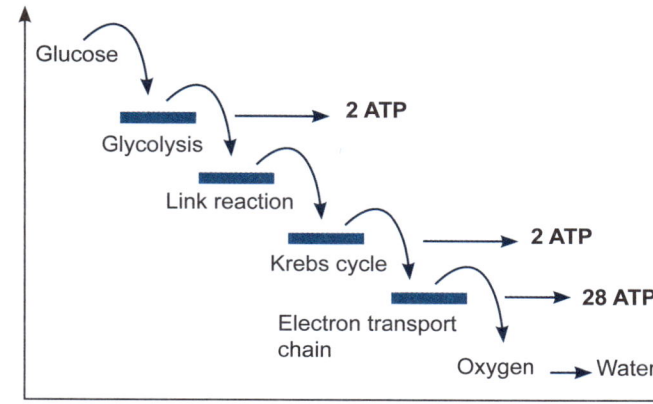

A model for ATP use in the muscles

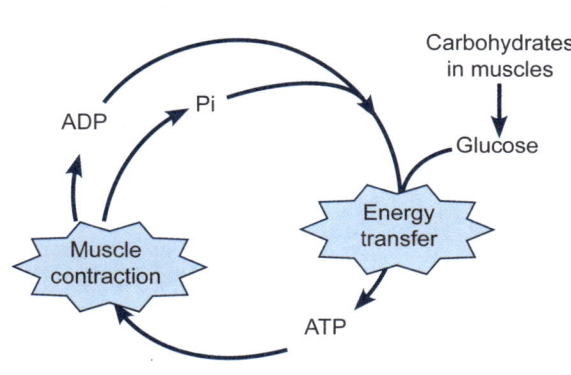

4. Explain how the energy in glucose is converted to useful energy in the body. Use the example of muscle contraction to help illustrate your ideas:

   _____
   _____
   _____
   _____
   _____

5. (a) One mole of glucose contains 2870 kJ of energy. The hydrolysis of one mole of ATP releases 30.7 kJ of energy. Calculate the percentage of energy that is transformed to useful energy in the body. Show your working.

   _____
   _____
   _____
   _____

   (b) Use your calculations above to explain why shivering keeps you warm and extreme muscular exertion causes you to get hot:

   _____
   _____
   _____
   _____
   _____

60 Measuring Respiration

Key Question: How can a respirometer be used to measure the rate of cellular respiration in germinating seeds?

▸ A respirometer can be used to measure the amount of oxygen consumed by an organism during **cellular respiration** and so can be used to measure respiration rate.

▸ A simple respirometer is shown in the diagram below. The carbon dioxide produced during respiration is absorbed by the potassium hydroxide. As the oxygen is used up, the colored bubble in the glass tube moves. Measuring the movement of the bubble, e.g. with a ruler or taped graph paper, allows us to estimate the change in volume of gas and therefore the rate of cellular respiration.

Investigation 3.2 — Measuring respiration in germinating seeds

See appendix for equipment list.

⚠ Caution is required when handling potassium hydroxide as it is caustic and can cause chemical burns. You should wear protective eyewear and gloves.

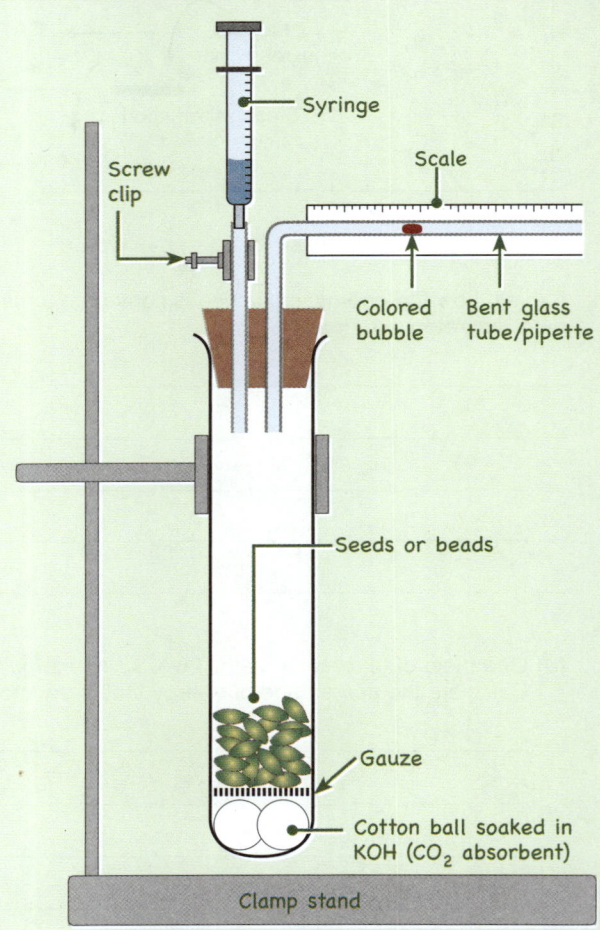

1. Work in groups of four to set up three respirometers using the set-up shown (right) as a guide.

2. Collect three boiling tubes and place two cotton balls in the bottom of each. Label the tubes A, B, and C.

3. Use a dropper to add 15% potassium hydroxide (KOH) solution on the cotton balls until they are saturated (there should be no liquid in the boiling tube). Add the same amount of KOH to the cotton balls in each boiling tube.

4. Place gauze on top of the cotton balls in each tube. This prevents the KOH coming into contact with the seeds and killing them.

5. Quarter fill tube A with germinated bean seeds. These seeds will be wet because they have been germinated under wet paper towels for four days.

6. Quarter fill tube B with ungerminated (dry) seeds.

7. Quarter fill tube C with glass beads.

8. Place a two-hole stopper firmly in each boiling tube. In one hole, insert a bent glass tube or bent pipette. In the second hole, insert a tube that can be clamped shut using a screw clip.

9. Use a dropper or fine pipette to place a drop of colored liquid into the bent tube/pipette of each set up. Attach a syringe to the clamped tube. Open the screw clip and use the syringe to draw the colored bubble into the middle of the bent tube/pipette.

10. Place all three tubes in a water bath at 25°C. Secure them with a clamp stand or in racks.

11. Leave the apparatus to acclimatize for 10 minutes.

12. At the end of the acclimation period, close the screw clip on the boiling tubes. Mark the position of the bubble with a marker pen. This is your time zero position. Start the timer.

13. Use a ruler or the pipette's scale (if there is one) to measure the distance the colored bubble moved at 5, 10, 15, 20, and 25 minutes.

14. Record your results on the table at the top of the next page.

> Respirometers of this sort are very sensitive to poor procedure because the volumes involved are so small.
>
> Be very careful with your set-up and when taking readings. Have one person responsible taking the measurements of the bubble movement.

Time (minutes)	Distance bubble moved (mm)		
	Germinated seeds	Ungerminated seeds	Glass beads
0			
5			
10			
15			
20			
25			

1. What is the purpose of the test tube with the beads? _____

_____

_____

2. (a) Calculate the corrected distance the bubble moved in tubes A and B by subtracting the distance moved in tube C from each value. Record these values in the table below.

   (b) Use the corrected distance the bubble moved to calculate the rate of respiration. Record this in the table below:

Time (minutes)	Corrected distance bubble moved (mm)		Rate (mm/min)	
	Germinated seeds	Ungerminated seeds	Germinated seeds	Ungerminated seeds
0				
5				
10				
15				
20				
25				

NEED HELP? See Activity 252

(c) Plot the rate of respiration on the grid (right). Include appropriate titles and axis labels:

NEED HELP? See Activity 261

(d) What does your plot show? _____

_____

_____

_____

_____

_____

3. Why does the bubble in the capillary tube move? _____

4. What conclusion can you make about cellular respiration in germinated and ungerminated seeds? _____

5. How would you have to modify the experiment if you were measuring respiration in a plant instead of seeds? _____

6. Explain the purpose of:

   (a) KOH: _____

   (b) The acclimation period: _____

   (c) The ungerminated seeds: _____

Time (minutes)	Distance bubble moved (mm)	Rate (mm/min)
0	0	
5	25	
10	65	
15	95	
20	130	
25	160	

7. A student decided to repeat the respirometer experiment but used maggots instead of seeds. Their results are shown on the table (above, right).

   (a) Calculate the rates and record them in the table (right):

   (b) Graph the rates on the grid (below, right):

   (c) Describe the results: _____

61 Modeling Photosynthesis and Respiration

Key Question: How can models be used to demonstrate photosynthesis and cellular respiration?

Investigation 3.3 Modeling photosynthesis and cellular respiration

See appendix for equipment list.

1. Work by yourself for this task. If you have beads or molecular models you could use these instead of the shapes on the next page.

2. Cut out the atoms and shapes on the following page. They are color coded as follows:

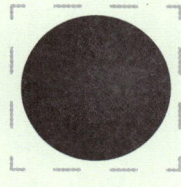
Carbon

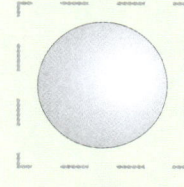
Hydrogen

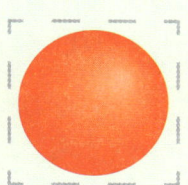
Oxygen

3. Use the cutouts to model photosynthesis and cellular respiration by following the steps below.

1. (a) Write the chemical equation for photosynthesis here: _____

   (b) State the starting reactants in photosynthesis: _____

   (c) State the total number of atoms of each type needed to make the starting reactants:

   Carbon: _____   Hydrogen: _____   Oxygen: _____

   (d) Use the atoms you have cut out to make the starting reactants in photosynthesis.

   (e) State the end products of photosynthesis: _____

   (f) State the total number atoms of each type needed to make the end products of photosynthesis:

   Carbon: _____   Hydrogen: _____   Oxygen: _____

   (g) Use the atoms you have cut out to make the end products of photosynthesis.

   (h) What do you notice about the number of C, H, and O atoms on each side of the photosynthesis equation? _____
   _____

   (i) Name the energy source for this process and add it to the model you have made: _____

2. (a) Write the chemical equation for cellular respiration: _____

   (b) State the starting reactants in cellular respiration: _____

   (c) State the total number of atoms of each type needed to make the starting reactants:

   Carbon: _____   Hydrogen: _____   Oxygen: _____

   (d) Use the atoms you have cut out to make the starting reactants in cellular respiration.

   (e) State the end products of cellular respiration: _____

   (f) State the total number of atoms of each type needed to make the end products of cellular respiration:

   Carbon: _____   Hydrogen: _____   Oxygen: _____

   (g) Use the atoms you have cut out to make the end products of cellular respiration.

   (h) Name the end products of cellular respiration that are utilized in photosynthesis: _____

3. Photosynthesis is the process by which carbon dioxide is fixed to form glucose. In the space below, draw a diagram to show the process of photosynthesis. Include reactants and products and the location of the reactions involved:

4. Cellular respiration is a continuous, integrated process. A simple diagram of the process in a eukaryote is shown below.

   (a) In the diagram below, fill in the rectangles with the process, and the ovals with the substance used or produced. Use the following word list (some words can be used more than once): *pyruvate, glycolysis, glucose, oxygen (O_2), link reaction, electron transport chain (ETC), Krebs cycle, ATP, carbon dioxide (CO_2), water (H_2O)*

   (b) Add in a pathway to show fermentation. Write the two possible products of this pathway in eukaryotes.

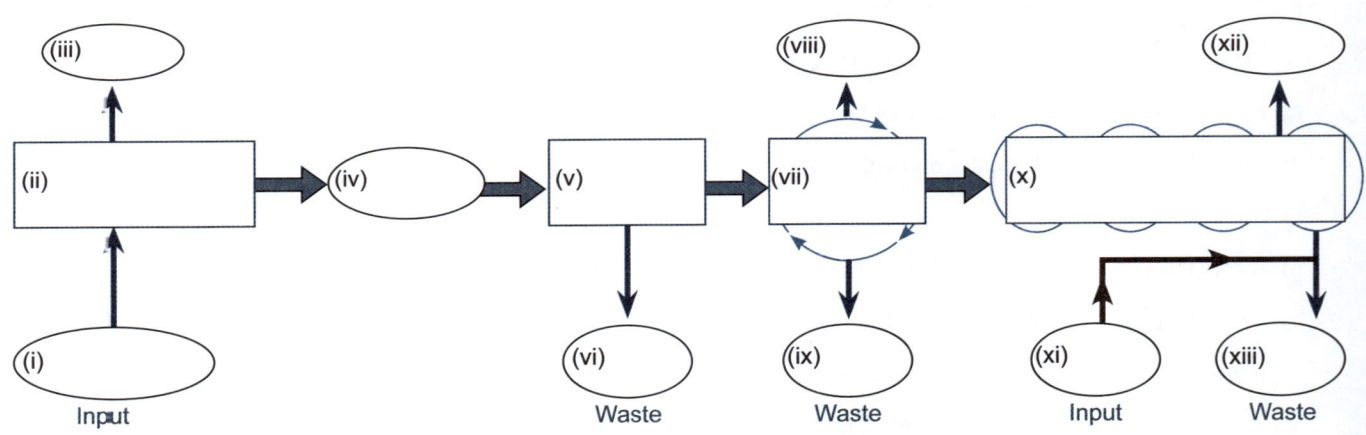

   (c) Use the completed diagram to explain the difference in ATP yield between aerobic and anaerobic pathways:

   _____
   _____
   _____
   _____
   _____
   _____

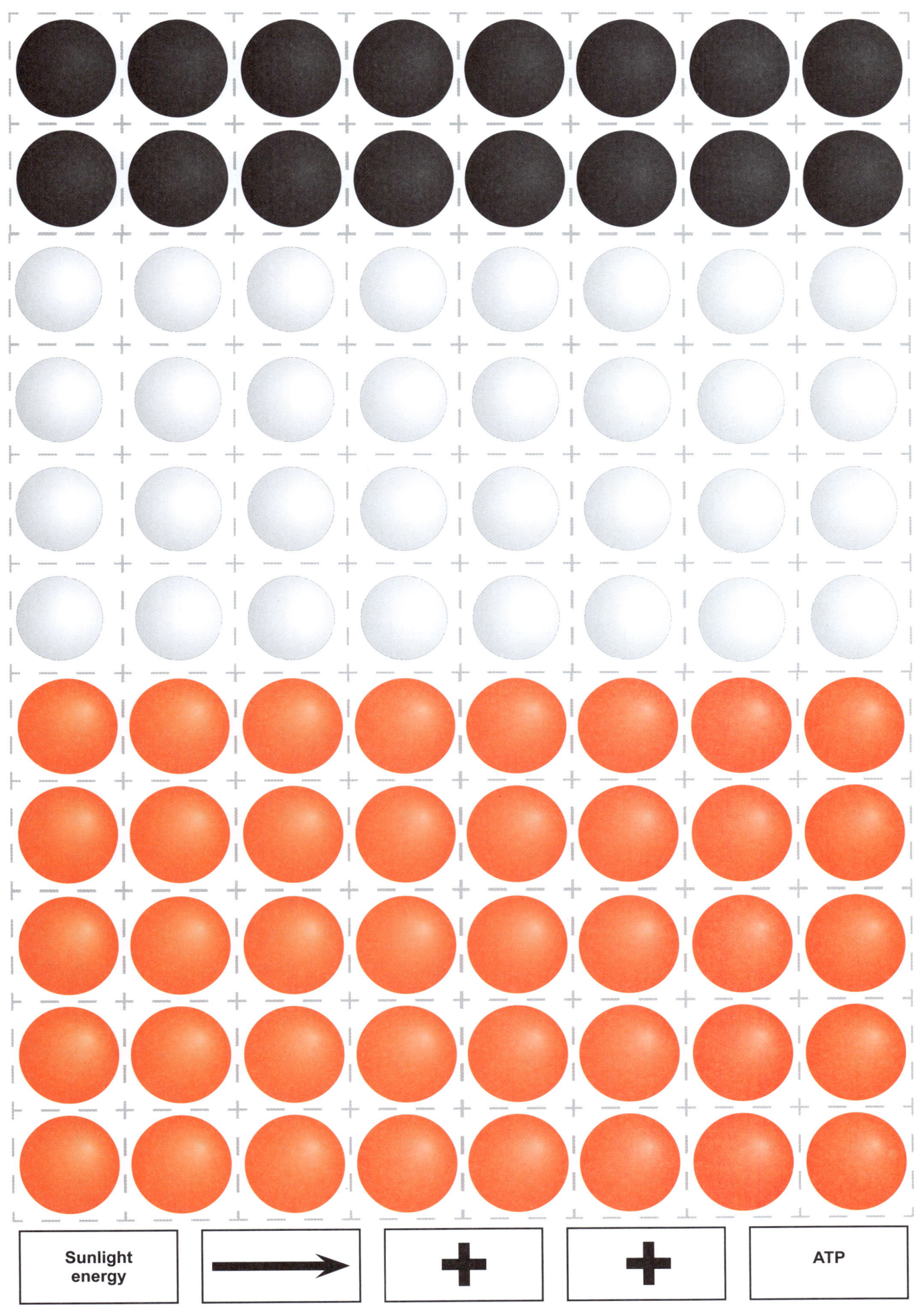

This page is left blank deliberately

62 Reactions in Cells

Key Question: How do anabolic and catabolic reactions build or break down molecules in the body?

▸ **Metabolism** refers to all of the chemical reactions and cellular processes carried out within a living organism to maintain life. All cellular processes are controlled by **enzymes**. Without enzymes, cellular processes would happen very slowly, too slowly to keep us alive! There are thousands of different types of enzymes. Each enzyme catalyzes only one type of chemical reaction.

▸ There are two categories of metabolic reactions: anabolic and catabolic.

Anabolic reactions

▸ **Anabolic reactions** are reactions that result in the production (synthesis) of a more complex molecule from smaller components or smaller molecules. During anabolic reactions, simple molecules are joined to form a larger, more complex molecule.

▸ Anabolic reactions need a net input of energy to proceed. They are called endergonic reactions.

Catabolic reactions

▸ **Catabolic reactions** are reactions that break down large molecules into smaller components.

▸ Catabolic reactions involve a net release of energy. They are called exergonic reactions.

▸ Catabolic reactions are the opposite of anabolic reactions.

▸ The energy released from catabolic reactions can be used to drive other metabolic processes.

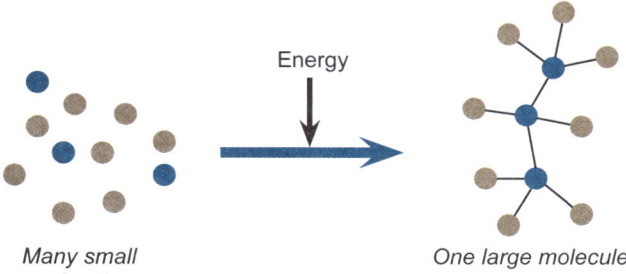
Many small molecules → One large molecule (Energy)

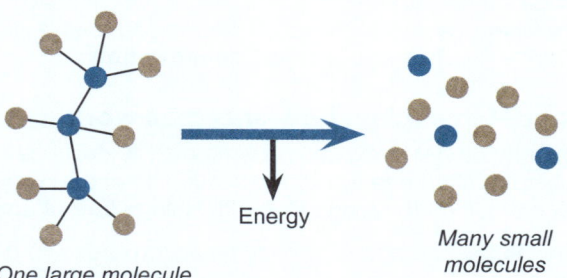
One large molecule → Many small molecules (Energy)

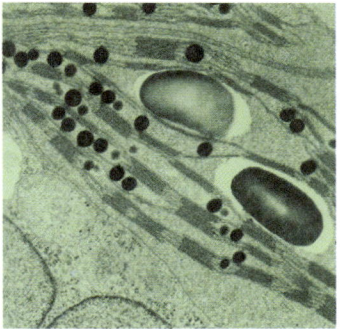
Plants carry out **photosynthesis** in **organelles** called **chloroplasts** (left). Photosynthesis is an anabolic process because it converts carbon dioxide and water into **glucose**. Energy from the sun is required to drive photosynthesis.

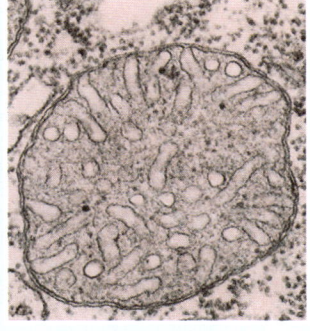
Cellular respiration is an example of a catabolic reaction. Glucose is broken down in a series of reactions to release carbon dioxide, water, and **ATP** (energy). The energy is used to fuel other activities in the cell. Some stages of cellular respiration take place in the **mitochondria** (left).

1. What is an anabolic reaction? _____

2. (a) What is a catabolic reaction? _____

   (b) Why are catabolic reactions considered to be the opposite to anabolic reactions? _____

3. Identify the following reactions as either catabolic or anabolic:

   (a) Protein synthesis: _____   (c) Digestion: _____

   (b) ATP conversion to ADP: _____   (d) DNA synthesis: _____

63 What are Enzymes?

Key Question: What are enzymes and what role do they play in biological reactions?

What are enzymes?

- **Enzymes** are proteins. Their shape is determined by their specific tertiary (3-dimensional) structure.
- Enzymes control all the metabolic reactions and cellular processes that take place in a cell.
- Each enzyme controls a very specific metabolic reaction, or series of related metabolic reactions.
- Enzymes may break down a single substrate molecule (**catabolism**), or join two or more substrate molecules together (**anabolism**).
- Enzymes are called biological **catalysts**, they speed up biochemical reactions. The enzyme itself is not used up in the reaction, and carries out the same reaction many times.
- Extremes of temperature or pH can change the shape of the enzyme and alter the enzyme's active site where catalysis occurs. This can lead to loss of enzyme function and is called **denaturation**.

The molecule(s) an enzyme acts on is called the substrate.

An enzyme acts on a specific substrate or group of related substrates.

The **active site** is the region of an enzyme where the reaction takes place. The substrate molecule(s) interact with the active site and undergo a chemical reaction.

The properties of the active site are specific to an enzyme and are a function of its tertiary structure. The active site accounts for an enzyme's specificity for its substrate (the substrate has a shape and charge to interact correctly with the active site).

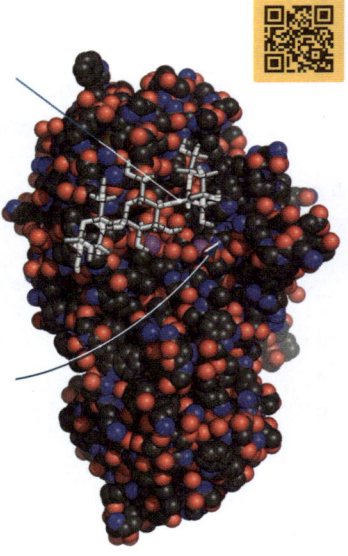

This enzyme is called amylase. Its substrate is starch.

A model to explain how enzymes work

- An early model to explain enzyme activity described the enzyme and its substrate as a lock and key, where the substrate fitted neatly into the active site of the enzyme. Evidence showed this model to be flawed and it has since been modified to recognize the flexible nature of enzymes. This revised model is called the induced fit model.

- The induced fit model for enzyme action is shown below. Rather than being inflexible, as proposed by the lock and key model, the shape of the enzyme changes when the substrate fits into the active site. The substrate becomes bound to the enzyme by weak chemical bonds. This weakens bonds within the substrate itself, allowing the reaction to proceed more readily.

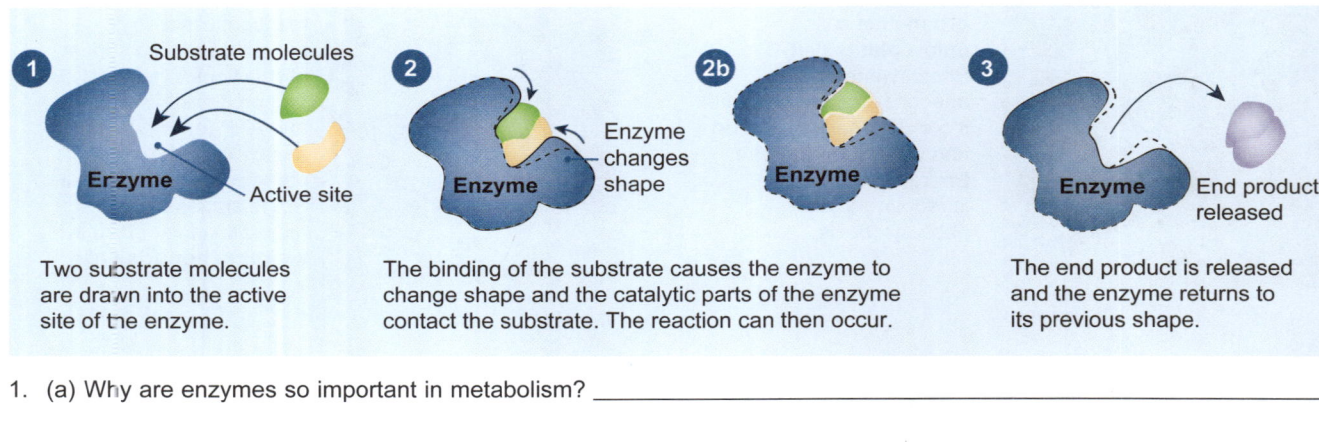

1. Two substrate molecules are drawn into the active site of the enzyme.
2. The binding of the substrate causes the enzyme to change shape and the catalytic parts of the enzyme contact the substrate. The reaction can then occur.
3. The end product is released and the enzyme returns to its previous shape.

1. (a) Why are enzymes so important in metabolism? _____

   (b) Why does an organism need so many different enzymes? _____

2. Describe the induced fit model of enzyme activity: _____

64 How Enzymes Work

Key Question: How do enzymes work and how can their activity be stopped?

Enzymes lower the activation energy for biochemical reactions

▶ The presence of an **enzyme** simply makes it easier for a reaction to take place. All **catalysts** speed up reactions by influencing the stability of bonds in the reactants. They may also provide an alternative reaction pathway, thus lowering the activation energy (E_a) needed for a reaction to take place (below). These reactions are accompanied by energy changes.

▶ When the reactants have lower energy than the product, the reaction requires energy input and is endergonic. When the reactants have higher energy than the product, energy is released in product formation and the reaction is exergonic.

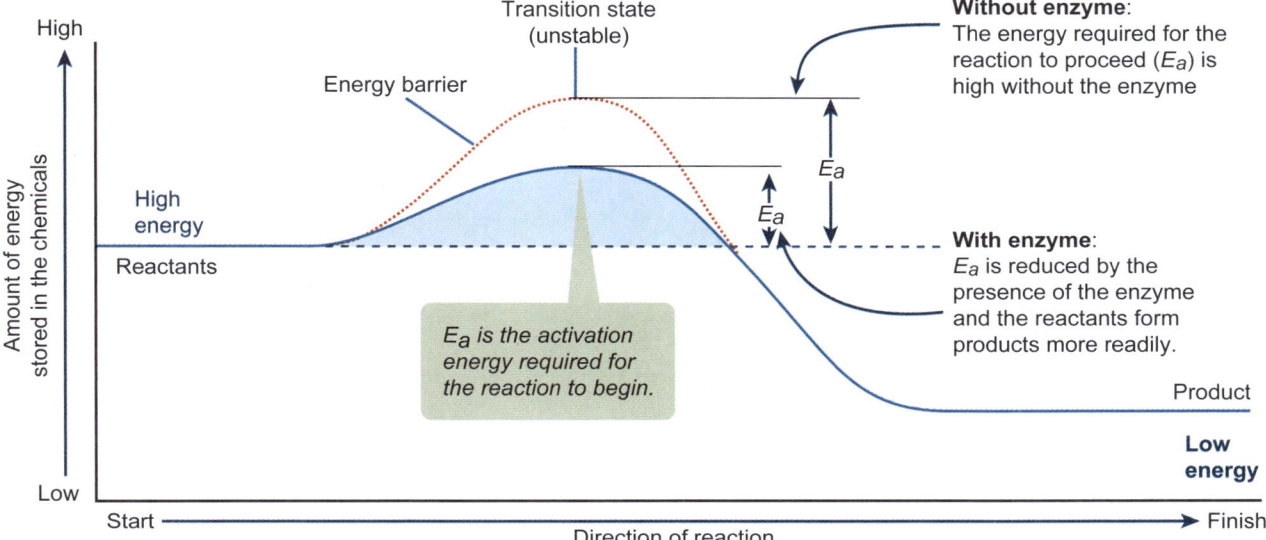

Enzyme activity can be stopped

▶ Several factors can prevent an enzyme from being able to catalyze a reaction. The first is **denaturation** and the second is enzyme inhibition.

▶ All enzymes have optimal conditions in which they work best, e.g. a specific pH or temperature range and outside of the optimal conditions, enzyme activity slows down. If an enzyme is exposed to extreme conditions, it can lose its tertiary structure so that it can no longer bind substrate and carry out its job. This is called denaturation (below). Extremes in pH and temperature outside of the normal range are common causes of denaturation.

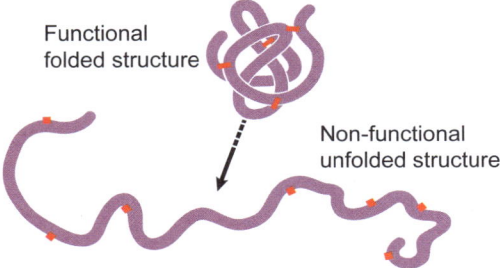

▶ Enzyme activity can be stopped, temporarily or permanently, by chemicals called enzyme inhibitors. Some inhibitors bind permanently and prevent the enzyme ever regaining activity again. Other inhibitors may bind temporarily and, once removed, the enzyme regains function.

1. How do enzymes lower the activation energy for a reaction?

   _____
   _____
   _____
   _____
   _____
   _____

2. The enzyme amylase breaks down starch (a glucose polymer) into smaller glucose units. Would this cellular process proceed if amylase was denatured? Explain your answer:

   _____
   _____
   _____
   _____
   _____
   _____
   _____

65 Enzymes Have Optimal Conditions to Work

Key Question: What conditions are optimal for enzymes, and what happens to their structure and function outside of these conditions?

- **Enzymes** usually have a set of conditions, e.g. pH and temperature, where their activity is greatest. This is called their **optimum**. At low temperatures, the activity of most enzymes is very slow, or does not proceed at all. This is because the enzyme and substrate molecules are moving slowly and do not collide as often. When temperature is higher, the enzyme and substrate molecules move more quickly and collide more often, so the enzyme-catalyzed reaction takes place more frequently.

- Enzyme activity increases with increasing temperature, but falls off after the optimum temperature is exceeded and the enzyme is **denatured** (change their shape and become inactive). Extremes in pH can also cause denaturation. Once an enzyme is denatured, it can no longer catalyze a cellular process and, if enough enzyme molecules are affected, the process stops.

- Within their normal operating conditions, enzyme reaction rates are influenced by enzyme and substrate concentration in a predictable way. In the graphs below, the rate of reaction or degree of enzyme activity is plotted against each of four factors that affect enzyme function.

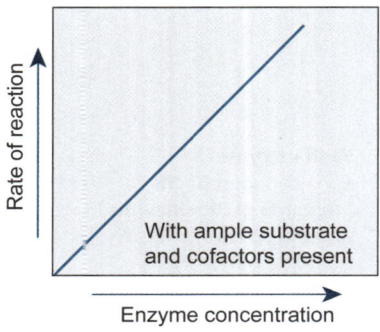

1. **Enzyme concentration**
   (a) Describe the change in the rate of reaction when the enzyme concentration is increased (assuming there is plenty of the substrate present):
   _____
   _____

   (b) Suggest how a cell may vary the rate of an enzyme controlled reaction:
   _____
   _____
   _____

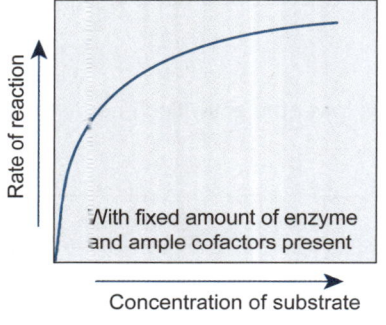

2. **Substrate concentration**
   (a) Describe the change in the rate of reaction when the substrate concentration is increased (assuming a fixed amount of enzyme):
   _____
   _____

   (b) Explain why the rate changes the way it does: _____
   _____
   _____
   _____

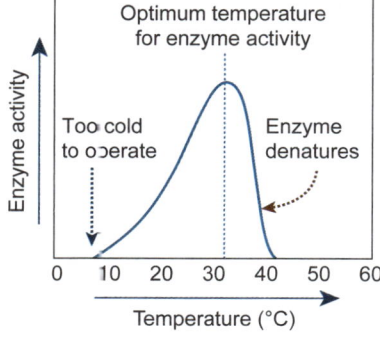

3. **Temperature**
   Higher temperatures speed up all reactions, but few enzymes can tolerate temperatures higher than 50–60°C. The rate at which enzymes are denatured increases with higher temperatures.

   (a) Describe what is meant by an optimum temperature for enzyme activity:
   _____
   _____

   (b) Explain why most enzymes perform poorly at low temperatures:
   _____
   _____
   _____

Investigating enzyme function

▸ As we have already seen, enzymes have an ideal "operating temperature". This is called the optimum temperature and it is where the enzyme's catalytic activity is at its highest.

▸ In the experiment below, you will investigate how temperature affects the activity of an enzyme called salivary amylase. Salivary amylase is a digestive enzyme found in saliva. It breaks down the starch in food into smaller sugar molecules, which are eventually converted into **glucose**. Amylase works best at pH 7.0 (the pH of the mouth). The reaction is described below:

$$\text{Starch (starting material/substrate)} \xrightarrow{\text{Amylase}} \text{Sugars (product)}$$

▸ The activity of amylase can be tracked using the iodine test. Iodine solution (usually potassium iodide) shows if starch is present. Iodine solution is a yellow/brown color.

▸ If the iodine indicator stays yellow/brown after the addition of a sample it means there is no starch present in the sample (image right).

▸ If the sample turns blue/black, it means that starch is present (image far right).

Investigation 3.4 Effect of temperature on enzyme activity

See appendix for equipment list.

⚠ Wear protective eyewear and gloves when handling chemicals.

Work in pairs or small groups for this activity.

1. Each pair or group should set up a spotting plate (photo, right) by adding a single drop of 0.1 M iodine solution (I_2KI) into each well.

2. Add 2 mL of 1% amylase solution and 1 mL of a buffer solution at pH 7.0 into a clean test tube.

3. Place the test tube in a water bath set at 10°C. Let the tube sit for 5 minutes until it is at the correct temperature.

4. Add 1 mL of 1% starch solution into the test tube. Mix well and immediately start a timer.

5. After one minute, use a plastic pipette to take one or two drops from the test tube.

6. Add a drop of the sample into the first well of the reaction plate. Record in the table below if starch is present (Y) or not present (N).

7. Continue to take samples every minute until no starch is detected (the color stays brown after the sample is added).

8. Repeat the procedure at temperatures of 20, 30, 40, 50, and 60°C, using a clean spotting plate each time.

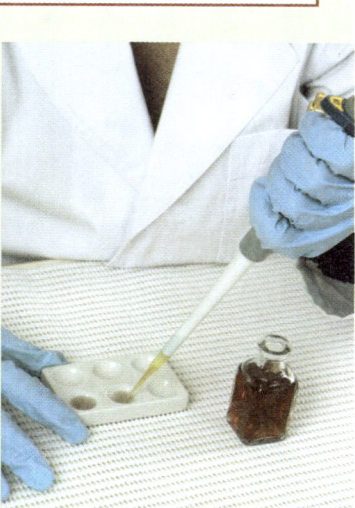

Temperature (°C)	\multicolumn{15}{c}{Time (minutes)}
	1	2	3	4	5	6	7	8	9	10	11	12	13	14	15
10															
20															
30															
40															
50															
60															

4. Use enzyme vocabulary, e.g. substrate, product, to describe how the iodine test is used to follow the progress of the reaction of amylase on starch:

___

5. Write a suitable aim for this investigation: ___

6. No controls have been included in this investigation. What would a suitable control have been? ___

7. (a) Was there any temperature(s) where denaturation occurred (there was no enzyme activity)? If so, identify it here: ___

   (b) How did you know denaturation had occurred? ___

8. On the grid, plot the time taken for all the starch to be digested against temperature. *If enzyme denaturation occurred, do not plot that result.*

   NEED HELP? See Activity 261

9. Describe how temperature affects the activity of the enzyme amylase: ___

10. Based on your experiment, what is the optimum temperature for amylase? ___

11. Predict amylase activity below 10°C and give a reason for your prediction: ___

66 Design an Experiment to Test Catalase Activity

Key Question: What factors must be taken into consideration when planning an investigation?

▸ In this activity you will plan an experimental investigation to ask and answer questions about the activity of catalase **enzyme** in germinating seeds of differing germination ages.

▸ The aim of this experiment is "*To investigate and compare the effect of germination age on the level of catalase activity in mung beans*". Use the aim, the background information provided, your own knowledge about enzymes, and the equipment list provided, to plan your own comparative investigation. Use the questions in this activity to guide you.

▸ Remember to formulate your hypothesis and then plan a scientifically robust experiment to test it. Safety issues must be taken into consideration as you design your experiment. Think carefully about how you will collect and analyze your data so that your results are meaningful and allow you to make valid conclusions about your findings.

NEED HELP? See Activity 246

Background information

▸ Germinating seeds are metabolically very active. The metabolism produces reactive oxygen species, including hydrogen peroxide (H_2O_2). H_2O_2 helps germination by breaking seed dormancy, but it is also toxic.

▸ To counter the toxic effects of H_2O_2 and prevent cellular damage, germinating seeds produce catalase, an enzyme that breaks down H_2O_2 to water and oxygen.

▸ The equation is given below:

$$2H_2O_2 \longrightarrow 2H_2O + O_2$$

▸ A class was divided into six groups, with each group testing the seedlings of each age. Each group's set of results (for 0.5, 2, 4, 6, and 10 days) represents one trial. The students decided to use a reaction time of 30 seconds.

Equipment list

- Rubber stopper with tubes
- Conical flask
- Small tube
- Syringe with 20 cm³ 20 vol H_2O_2
- Graduated measuring cylinder with water
- Glass dish
- Large tube
- Timer
- Germinating mung bean seeds at 0.5, 2, 4, 6, and 10 days germination

1. Use the aim and background information to form a hypothesis for your experiment: _____

_____

_____

_____

2. In the space below summarize your method as step by step instructions. You can draw your experimental set up on a separate piece of paper if you wish:

_____

_____

_____

_____

_____

_____

_____

_____

3. Identify any safety issues you need to take into consideration when planning this investigation: _____
_____
_____
_____
_____

4. (a) What data will you collect? _____
_____
_____

   (b) How often will you collect it? _____
_____

   (c) What data transformation(s) and statistical test(s) would you use to analyze the data? _____
_____
_____

5. In the space below, draw a table to record the data for all six groups. Include spaces to record the results of any data transformations and calculations performed:

6. (a) What sort of graph would you use to plot the results of this experiment? _____

   (b) Explain your choice: _____
_____

7. Describe any potential sources of errors in the apparatus or the procedure: _____
_____
_____
_____
_____

8. Describe two things that might affect the validity of findings in this experimental design: _____
_____
_____
_____

67 Mouse Trap Revisited

Content Anchor Revisited: Under what conditions can an animal survive in a sealed system?

Man in a box

- In 2012, researchers carried out a larger version of Joseph Priestley's famous mouse in a jar experiment.
- 274 plants were placed in a sealed container with oxygen-depleted air (12.4% oxygen). A healthy 47 year-old man entered the container for 48 hours. Gas levels were monitored throughout the experiment. The container was kept in constant light.
- The experiment was run to completion with no harm to the person within the box.
- Ethics approval was obtained beforehand and medical staff were on hand during the experiment.

1. Use your understanding from the information in this chapter to identify:

   (a) The two gases primarily being monitored in the experiment: _____

   (b) The two metabolic processes involved in this experiment: _____

2. (a) The graph on the right shows the change in oxygen concentration over the course of the experiment. Describe the trend in oxygen levels over time:

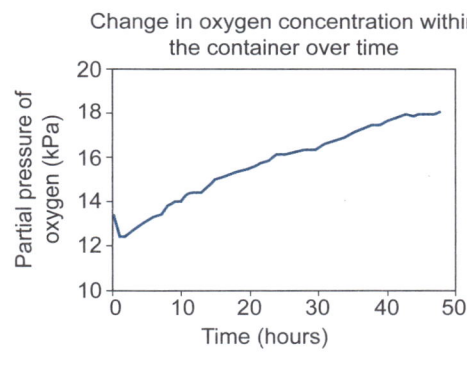

Change in oxygen concentration within the container over time

   (b) Explain why this change occurred (your answer should make reference to the gases and metabolic pathways involved):

3. Revisit the model you produced in activity 51. Refine it and add more detail to explain the relationship between cellular respiration and photosynthesis:

68 Summing Up

Read each question carefully. Place a cross in the box beside the best answer to the question from the four answer choices provided.

1. Which statement best describes the function of ATP?
   - (a) ATP is a structural component of plant cell walls
   - (b) ATP carries the genetic information of organisms
   - (c) ATP provides the energy for chemical reactions to occur
   - (d) ATP is a biological catalyst

2. Select the option which correctly identifies the organelle below AND the cellular process which takes place in it:

   - (a) Chloroplast, photosynthesis
   - (b) Chloroplast, cellular respiration
   - (c) Mitochondrion, photosynthesis
   - (d) Mitochondrion, cellular respiration

3. Enzymes speed up reactions by:
   - (a) Reducing the activation energy needed
   - (b) Increasing the activation energy needed
   - (c) Adding energy to the reaction
   - (d) Taking part in the reaction and forming part of the product(s)

4. The type of energy transformation occurring during photosynthesis is:
   - (a) Light to heat
   - (b) Light to chemical
   - (c) Chemical to heat
   - (d) None of the above

5. Which group of macromolecules do enzymes belong to?
   - (a) Lipids
   - (b) Proteins
   - (c) Carbohydrates
   - (d) Nucleotides

6. A solution of amylase was heated to 70°C for 10 minutes. When the treated amylase was added to a solution of starch, the iodine test showed no sugars had been produced. This is because:
   - (a) The enzyme has been denatured
   - (b) An enzyme inhibitor is preventing the enzyme from working
   - (c) There is no substrate present
   - (d) Amylase does not catalyze the reaction which converts starch in to sugars

7. The diagram below is showing:

   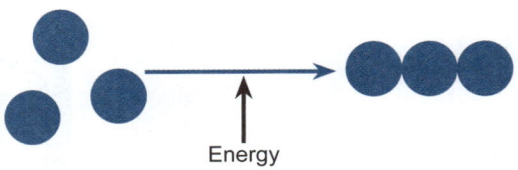

   - (a) A catabolic reaction, such as cellular respiration
   - (b) A catabolic reaction, such as photosynthesis
   - (c) An anabolic reaction, such as cellular respiration
   - (d) An anabolic reaction, such as photosynthesis

8. The model below is of a glucose molecule. During cellular respiration glucose is converted into:

   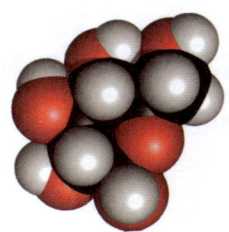

   - (a) Starch and carbon dioxide
   - (b) Starch and water
   - (c) Water and oxygen
   - (d) Water and carbon dioxide

9. Enzymes can change shape when exposed to extremes of temperature or pH. What are the most likely results if the shape of an enzyme changes?
   - (a) The enzyme will no longer be able to bind its substrate
   - (b) Enzyme activity will speed up
   - (c) The enzyme will bind a new substrate
   - (d) Different products will be produced during the reaction

10. Which molecules are both the products of cellular respiration and the raw materials for photosynthesis?
    - (a) Carbon dioxide, ATP, and oxygen
    - (b) Carbon dioxide and water
    - (c) Glucose and oxygen
    - (d) Glucose, ATP, and oxygen

11. Which answer correctly describes the equation below, and the organelle in which it takes place?

    $$H_2O + CO_2 \longrightarrow C_6H_{12}O_6 + O_2$$

    - (a) Photosynthesis, chloroplast
    - (b) Photosynthesis, mitochondrion
    - (c) Cellular respiration, chloroplast
    - (d) Cellular respiration, mitochondrion

Question 12 and 13 relate the to the process of photosynthesis, shown below:

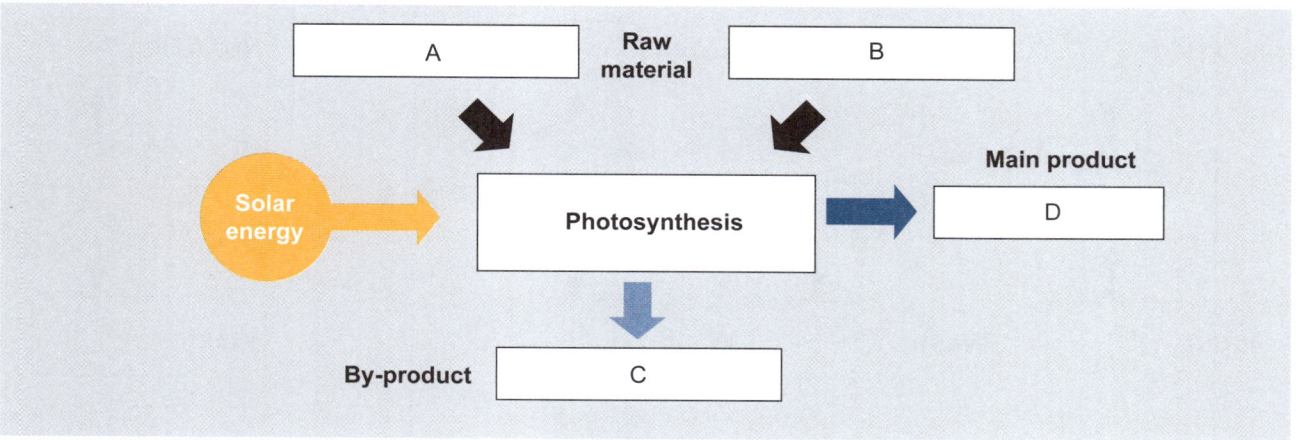

12. Raw materials A and B are:
 - (a) Oxygen, carbon dioxide
 - (b) Carbon dioxide, water
 - (c) Water, oxygen
 - (d) Water, glucose

13. Main product D and by-product C are
 - (a) D: Oxygen C: Glucose
 - (b) D: Carbon dioxide C: Oxygen
 - (c) D: Glucose C: Oxygen
 - (d) D: Water C: Glucose

14. Students investigated the effect of different light wavelengths (color) on the rate of photosynthesis. They used a leaf disk assay in which the rate of photosynthesis is measured indirectly by timing how long it takes for prepared leaf disks (right) to float to the surface when placed in an illuminated beaker of sodium hydrogen carbonate. The results are tabulated below:

(a) Why do you think photosynthesizing leaf disks would float?

_____

_____

_____

Preparing the leaf disks by applying pressure to a sodium hydrogen carbonate solution.

(b) Which light color was the most effective at driving photosynthesis? Explain:

_____

_____

_____

Light color	Time taken for 10 discs to float (s)
Blue	162
Red	558
Green	998
White	694

(c) Which light color was the least effective at driving photosynthesis?

_____

15. Outline the differences between photosynthesis and cellular respiration, including reference to the raw materials used and the waste products produced:

_____

_____

_____

_____

_____

_____

Refer to the diagram of cellular respiration below for questions 16-17

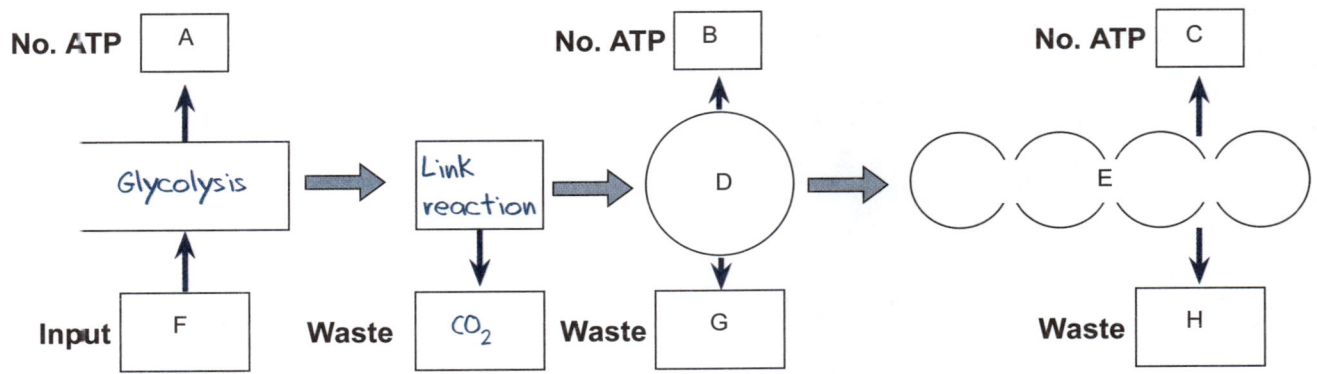

16. The total number of ATP molecules produced in steps A, B, and C is:
    - (a) Approximately 20
    - (b) Approximately 28
    - (c) Approximately 32
    - (d) Approximately 38

17. The waste products G and H are
    - (a) G: Oxygen H: Carbon dioxide
    - (b) G: Carbon dioxide H: Oxygen
    - (c) G: Water H: Carbon dioxide
    - (d) G: Carbon dioxide H: Water

18. Study the enzymatic word equation below and answer the following questions:

$$\text{Sucrose + Water} \xrightarrow{\textit{Sucrase}} \text{Glucose + Fructose}$$

(a) Identify the substrate: _____

(b) Identify the products: _____

(c) Identify the enzyme: _____

19. Identify the following statements as true of false (circle the correct answer)

    (a) Enzymes are biological catalysts. They lower the activation energy of a reaction.   True / False

    (b) Enzyme inhibitors allow enzymes to work faster.   True / False

    (c) The induced fit model states that the enzyme changes shape when a substrate fits into the active site.   True / False

20. The diagram below outlines the three main steps when an enzyme catalyzes a reaction. The steps are NOT in order.

    (a) In the boxes below, write the numbers 1 - 3 to indicate the correct order of sequence:

    (b) Write a brief description under each image to describe what is happening:

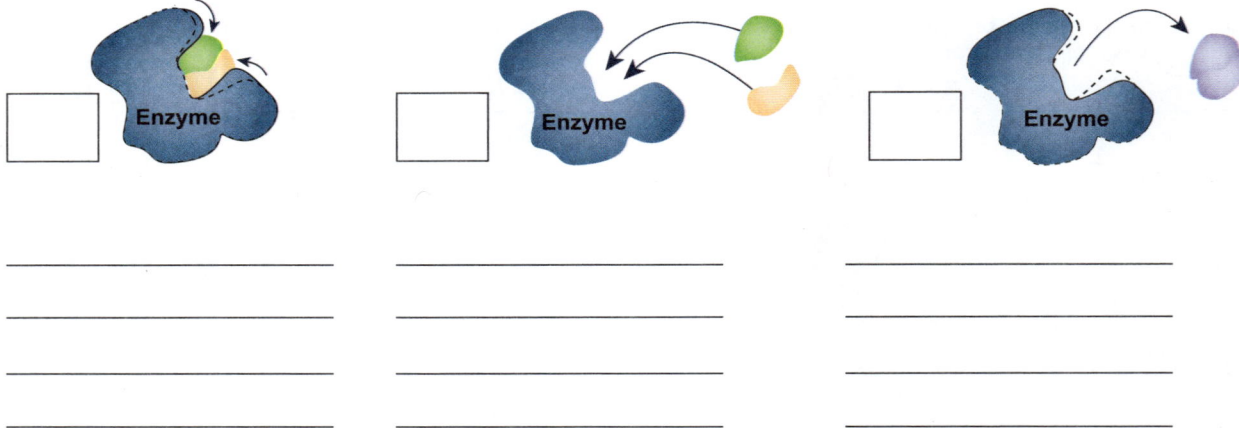

CHAPTER 4
Animal and Plant Structure and Function

TEKS
Scientific and Engineering Practices

B.1: Investigation and Inquiry
1.A 1.B 1.D 1.E 1.F 1.G

B.2: Data and Patterns
2.B 2.C 2.D

B.3: Communicating in Science
3.A

B.4: Science as a Human Endeavor
4.A 4.B

TEKS
Science Concepts

B12.A analyze the interactions that occur among systems that perform the functions of regulation, nutrient absorption, reproduction, and defense from injury or illness in animals

B12.B explain how the interactions that occur among systems that perform functions of transport, reproduction, and response in plants are facilitated by their structures

Learning Outcomes
I know I have achieved this when I can:

	Learning Outcome	Activity number
☐	Identify 11 human body organ systems and their functions.	71
☐	Define homeostasis and its key components, and explain the importance of homeostasis in the body.	72-73
☐	Discuss how feedback mechanisms control homeostasis, including blood glucose levels.	74, 82
☐	Compare and contrast how the nervous and hormonal regulatory systems function when interacting together to regulate the body.	75-77
☐	Explain the roles of the circulatory and urinary system when interacting together to remove wastes.	78
☐	Describe how the circulatory and respiratory system interact to regulate respiratory gases.	79
☐	Plan an investigation to test the effect of exercise on the circulatory and respiratory systems.	80
☐	Describe how the circulatory and digestive system interact together to provide the body's cells with nutrients.	81
☐	Summarize the roles of hormones in controlling the menstrual and ovarian cycles, and the interactions involved.	83
☐	Analyze how the circulatory and reproductive systems interact together to maintain pregnancy and induce birth.	84
☐	Outline how the immune, endocrine, and circulatory systems interact to defend the body from bleeding and infection.	85-88
☐	Draw a slide specimen of root and stem cells, involved in maintaining water balance in a plant.	90
☐	Describe structural features of xylem and phloem vascular tissue, found in roots and stems.	91-92
☐	Investigate how vascular tissue and stomata interact together in the process of transpiration to maintain water balance.	93-94
☐	Discuss how osmosis and translocation allow the movement of water, minerals, and sugars through the plant.	95-96
☐	Explore different methods of asexual reproduction in plants.	97
☐	Investigate the propagation of plants in different mediums.	98
☐	Analyze and describe the different mechanisms of plant pollination and fertilization, and the structures involved.	99-101
☐	Investigate factors affecting germination in seeds.	102
☐	Describe different examples of seed dispersal methods.	103
☐	Explain how plants respond to environmental change, and the hormones involved.	104-106
☐	Investigate the effect of auxin and gibberellins hormones on plant growth.	107, 109
☐	Investigate the effect of environmental stimuli, such as light and gravity, on phototropism and gravitropism responses.	108-111
☐	Explore how some plants use reversible nastic responses to respond to non-specific stimuli, such as touch.	112

RESOURCE HUB
bit.ly/3L13n7Z

ELPS English Language Proficiency Standards

		Page number
Learning	**Overview of Body Systems**. Work with a partner. Use the images on the page to recall the different organ systems in the human body. Write them into the table. Discuss how each system works and any illnesses or injuries you or your partner have experienced with these systems. Think about how these systems interact with other systems in your body, and begin your answer to question 2.	130
Listening	**Effect of Exercise on Heart Rate and Breathing**. Listen as your classmates read the instructions for Investigation 3.3. Use the images to help you understand by picturing the actions in your head. When instructions are complex, ask your classmates or teacher to clarify. Demonstrate your understanding by completing the investigation with your group.	142
Writing	**Plant Organ Systems**. As you read about plant organ systems, look for words that are new to you. Use images and information on the page or your book's glossary to create a list of words and meanings. Then, use these words as you write to explain the functions of a plant's root and shoot systems. Can these words be used outside of scientific studies?	159
Reading	**Insect Pollinated Flowers**. Before answering the first question, look carefully at each part of the pollinated flower. Circle the name of each structure and underline the function it plays in pollination. Then, highlight all parts with a female function in red and all parts with a male function in yellow. Which structures have neither a male nor female function?	174

69 Complex Interactions

Content Anchor: How do organisms maintain their internal environment and what interactions between organ systems occur to assist this?

A series of systems

▸ Every object that carries out a task is made up of smaller, simple objects or systems. For example, a car is made of several mechanical and electronic systems that work together to produce motion. These can be controlled in comfort from the interior compartment. Systems in a car include the engine, transmission, fuel or battery system, cooling (for both the car and you), entertainment, and various other control systems.

▸ In modern cars, most of these systems are controlled, or at least monitored, by the car's computers, which are able to process information and respond accordingly. For example, collision mitigation systems in a car will detect an impending collision and apply the brakes.

▸ Animals are a far more complex version of this kind of machine. They have **organ systems** that work together to enable an organism to continuously respond to the changing environment.

▸ Plants also have organ systems, although they are less complex than animals'.

1. Name the organ systems in an animal: _____
   _____
   _____

2. Name the organ systems in a plant: _____
   _____
   _____

3. (a) Pick two of the organ systems you identified in an animal and explain how they interact with each other:
   _____
   _____
   _____

   (b) Pick two of the organ systems you identified in a plant and explain how they interact with each other:
   _____
   _____
   _____

4. Organ systems in plants are less complex and less numerous than those of animals. Why might this be?
   _____
   _____

5. The photo shows a Venus flytrap that has responded to a person touching it. Can you identify the organ systems involved in the person touching the plant, and the responses involved in the Venus flytrap catching the person's finger?

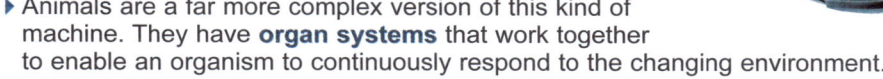

   _____
   _____
   _____

70 The Hierarchy of Life

Key Question: How are the cells of organisms organized so that they work together in a coordinated way?

- All multicellular organisms are organized in a hierarchy of structural levels, where each level builds on the one below it. It is traditional to start with the simplest components (parts) and build from there. Higher levels of organization are more complex than lower levels.
- Hierarchical organization enables specialization so that individual components perform a specific function or set of related functions. Specialization enables organisms to function more efficiently.
- The diagram below explains this hierarchical organization for a human.

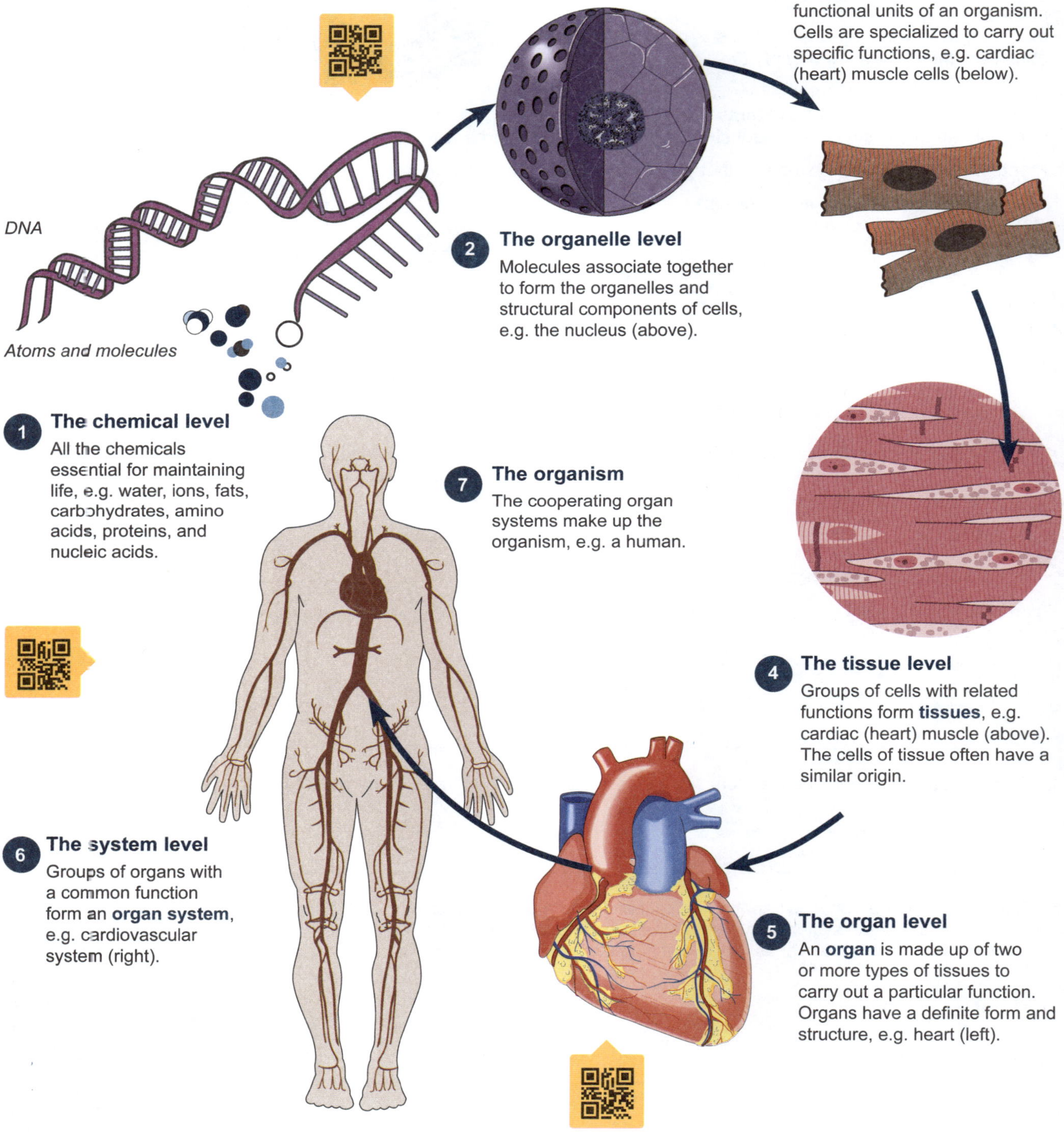

3 The cellular level
Cells are the basic structural and functional units of an organism. Cells are specialized to carry out specific functions, e.g. cardiac (heart) muscle cells (below).

DNA

Atoms and molecules

2 The organelle level
Molecules associate together to form the organelles and structural components of cells, e.g. the nucleus (above).

1 The chemical level
All the chemicals essential for maintaining life, e.g. water, ions, fats, carbohydrates, amino acids, proteins, and nucleic acids.

7 The organism
The cooperating organ systems make up the organism, e.g. a human.

4 The tissue level
Groups of cells with related functions form **tissues**, e.g. cardiac (heart) muscle (above). The cells of tissue often have a similar origin.

6 The system level
Groups of organs with a common function form an **organ system**, e.g. cardiovascular system (right).

5 The organ level
An **organ** is made up of two or more types of tissues to carry out a particular function. Organs have a definite form and structure, e.g. heart (left).

©2024 **BIOZONE** International
ISBN: 978-1-99-101405-4
Photocopying prohibited

71 Overview of Body Systems

Key Question: What are the organ systems of the body, and what are their main components?

Organs and organ systems

▸ There are 11 **organ systems** in the human body. These are the skeletal, muscular, integumentary (skin), nervous, endocrine, cardiovascular, respiratory, lymphatic, digestive, urinary, and reproductive systems. Each has evolved to carry out a specific set of tasks, but each also interacts with the other organ systems of the body.

▸ Each organ system has a number of different **organs** that carry out specific functions within it. Organs may carry out more than one function in more than one system, e.g. the pancreas, in both endocrine and digestive systems.

1. Identify each organ below, the organ system(s) it belongs to, and its function(s). You may need to research the images:

(a) i.   Organ: _____

   ii. Organ system: _____

   iii. Organ function: _____
   _____
   _____
   _____

(b) i.   Organ: _____

   ii. Organ system: _____

   iii. Organ function: _____
   _____
   _____
   _____

(c) i.   Organ: _____

   ii. Organ system: _____

   iii. Organ function: _____
   _____
   _____
   _____

(d) i.   Organ: _____

   ii. Organ system: _____

   iii. Organ function: _____
   _____
   _____
   _____

(e) i.   Organ: _____

   ii. Organ system: _____

   iii. Organ function: _____
   _____
   _____
   _____

2. The 11 organ systems in the human body each have a number of components with specific functions. Match up the organ system with the image (1-11) below and the function given in the table. Use the descriptions for help.

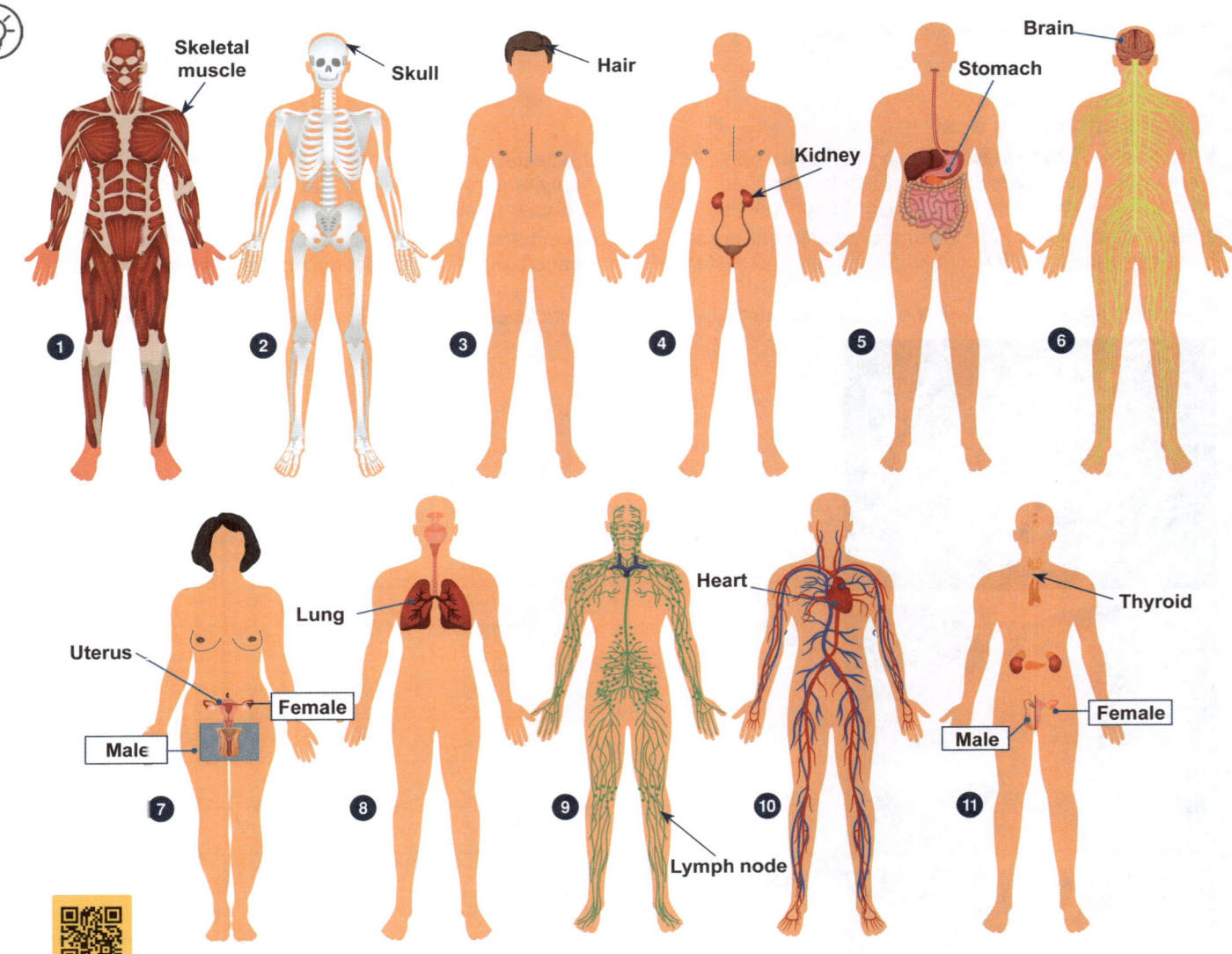

Organ system number	Organ system name
1	
2	
3	
4	
5	
6	
7	
8	
9	
10	
11	

Organ system number	Organ system function
	Support and protection of tissues and organs, movement (with muscular system), blood cells production.
	Excretion of nitrogenous wastes and other metabolic waste products. Maintains fluid and electrolyte balance.
	Physical and chemical digestion and absorption of ingested food to provide the body's fuel.
	Movement of body (limbs, locomotion), and its component parts, e.g. gut.
	Circulates tissue fluid, internal defense against pathogens.
	Production of gametes and offspring.
	Delivers O_2 and nutrients to **tissues** and organs and removes CO_2 and other waste products.
	Physical and chemical protection of tissues, thermoregulation, synthesis of vitamin D precursor.
	Regulates all visceral (organ) and motor functions of the body.
	Produces hormones that activate and regulate homeostatic functions, growth, and development
	Interface for gas exchange with the internal environment obtaining O_2 and expelling CO_2.

72 The Body's Systems Work Together

Key Question: How do the body's organ systems work together to maintain the body?

▸ The body's **organ systems** work together to maintain the constant internal environment necessary for the functioning of the body's **cells**. This is known as **homeostasis**. It includes processing information from the external environment (via the nervous system), responding to the environment, e.g. via the muscular system, and maintenance of the body, e.g. via the immune or digestive systems.

▸ The simplified example below illustrates how organ systems interact with each other to maintain homeostasis.

Once food has been digested (broken down) in the digestive system, it is absorbed (taken up) into the blood of the circulatory system.

The **digestive system** (gut and associated digestive glands) is responsible for the breakdown and absorption of food. Ultimately, it provides the energy and nutrients required by all the body's systems.

Unabsorbed digestive matter is expelled from the digestive system as feces.

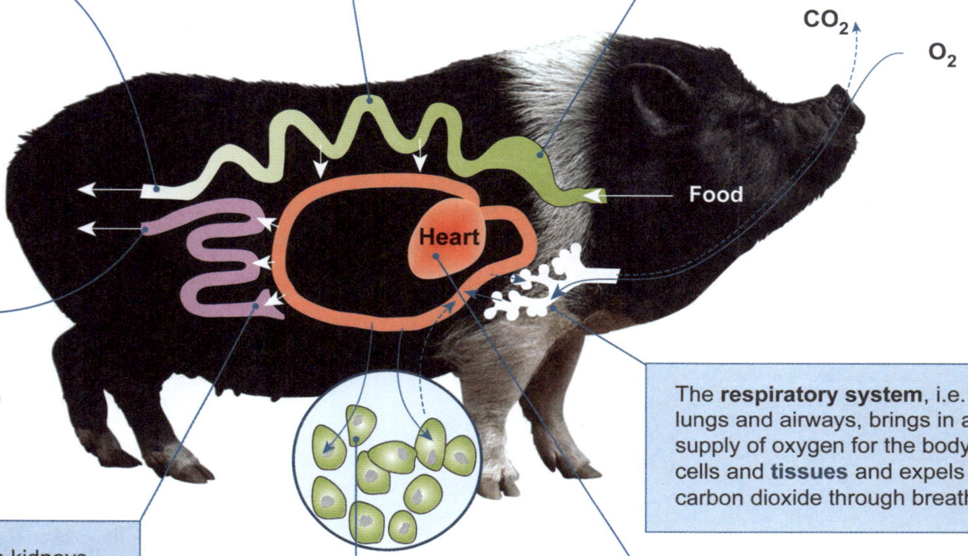

Urine is produced by the kidneys of the urinary system. It contains the waste products of metabolism, particularly nitrogen-containing wastes and excess ions.

The **respiratory system**, i.e. the lungs and airways, brings in a supply of oxygen for the body's cells and **tissues** and expels waste carbon dioxide through breathing.

The **urinary system** (the kidneys and associated ducts) has several roles including disposing of nitrogen-containing and other waste products from the body, regulating ion balance, and controlling the volume and pressure of the blood.

Once nutrients have been absorbed, they are carried in the blood and delivered to cells all around the body (solid arrows). Wastes (dashed arrow) move from the cells back into the blood and are transported and removed.

The **circulatory system** (the heart, blood vessels and blood) distributes respiratory gases, nutrients, and other substances, e.g. **hormones**, to the cells and tissues of the body.

1. Why is it important that body systems maintain homeostasis? _____
_____
_____
_____
_____

2. Using an example, briefly explain why maintaining homeostasis often involves more than one body system:
_____
_____
_____
_____

73 Homeostasis

Key Question: How do organisms maintain a constant internal environment despite changes in their external environment?

Maintaining a constant internal state

▸ **Homeostasis** literally means "constant state". Organisms maintain homeostasis, i.e. a relatively constant internal environment, even when the external environment is changing. This requires energy.

▸ For example, when you exercise, your body must keep your temperature constant, at about 37.0 °C, despite the increased heat generated by activity. Similarly, you must regulate blood glucose (sugar) levels and blood pH, water and electrolyte balance, and blood pressure. Your body's **organ systems** carry out these tasks.

▸ To maintain homeostasis, the body's receptors must detect changes in the environment, process this sensory information, and respond to it appropriately. The response provides new feedback to the receptor. These three components are illustrated below.

The analogy of a temperature setting on a heat pump is a good way to explain how homeostasis is maintained. A heat pump has sensors (a receptor) to monitor room temperature. It also has a control center to receive and process the data from the sensors. Depending on the data it receives, the control center activates the effector (heating/cooling unit), switching it either on or off.

When the room is too cold, the heating unit switches on, and the cooling unit is off. When it is too hot, the heating unit switches off and the cooling unit is switched on. This system maintains a constant temperature, similar to homeostasis in the body.

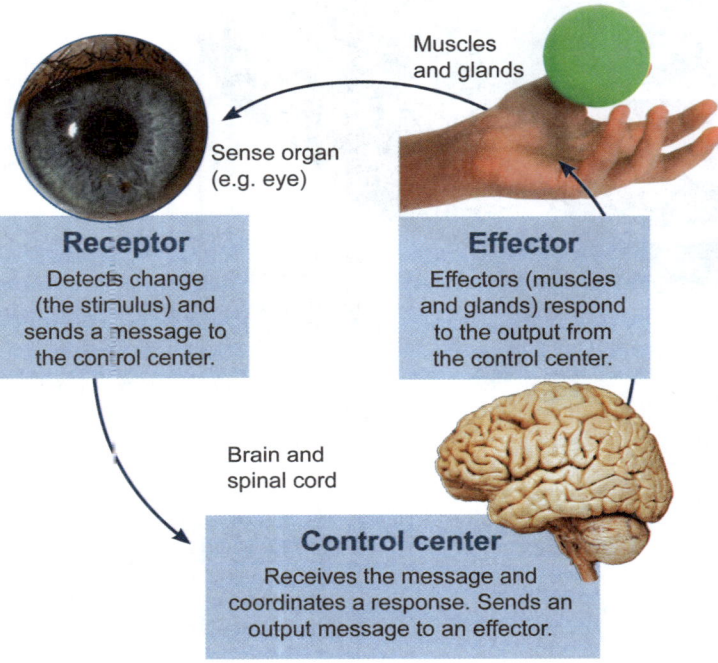

Receptor — Detects change (the stimulus) and sends a message to the control center.

Effector — Effectors (muscles and glands) respond to the output from the control center.

Control center — Receives the message and coordinates a response. Sends an output message to an effector.

The analogy of riding a mountain bike can be used to demonstrate that many homeostatic systems have multiple mechanisms to maintain a steady state. To stay upright on the bike you must use your body weight, arms, pedals, brakes, and steering.

1. What is homeostasis? _____

2. What are the roles of the following components in maintaining homeostasis:

   (a) Receptor: _____

   (b) Control center: _____

   (c) Effector: _____

74 Negative Feedback Regulates the Body

Key Question: How do negative feedback mechanisms detect changes in the body's internal environment away from normal and return it to a steady state?

Negative feedback is a control system which maintains the body's internal environment at a steady state. Negative feedback has a stabilizing effect and discourages variations from a set point. It works by returning internal conditions back to a steady state when variations are detected (right). Most body systems achieve **homeostasis** through negative feedback. Body temperature, blood glucose levels, and blood pressure are all controlled by negative feedback mechanisms.

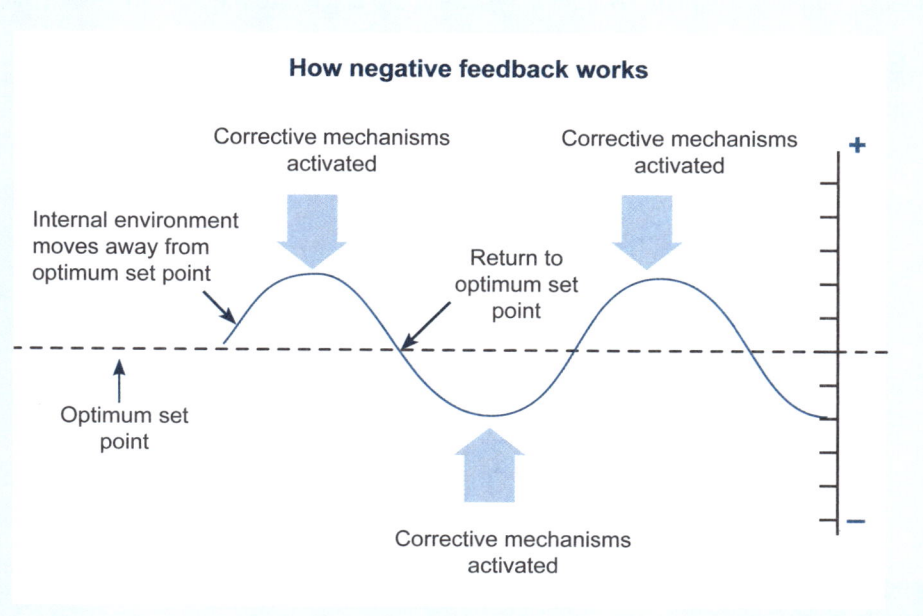

Negative feedback in blood pH

Regulation of ventilation rate (breathing) helps to maintain blood pH between 7.35 and 7.45. Low blood pH (high H^+) stimulates increased breathing rate, which reduces H^+ via exhalation. This reduces sensory input to the brain and breathing returns to normal.

Negative feedback in stomach emptying

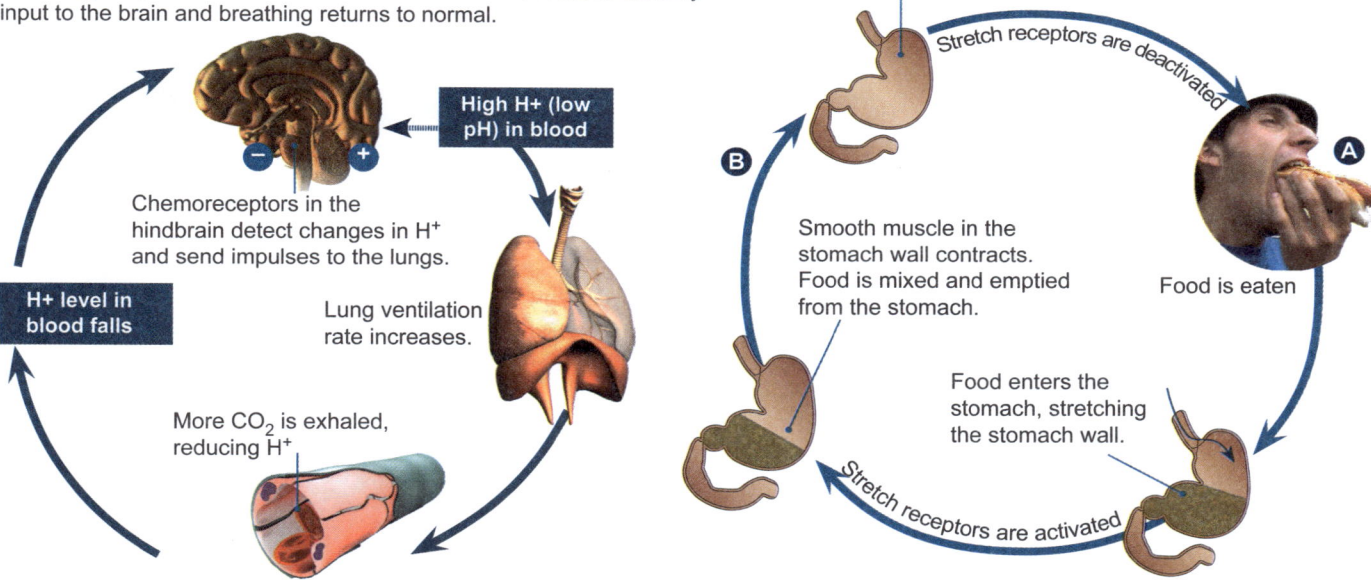

1. How do negative feedback mechanisms maintain homeostasis in a variable environment? _____
   _____
   _____

2. On the diagram of stomach emptying:

   (a) State the stimulus at A: _____ State the response at B: _____

   (b) Name the effector in this system: _____

   (c) What is the steady state for this example? _____

©2024 **BIOZONE** International
ISBN: 978-1-99-101405-4
Photocopying prohibited

How is body temperature regulated?

▶ In humans, the temperature regulation center is a region of the brain called the hypothalamus. It has thermoreceptors that monitor core body temperature and has a 'set-point' temperature of 36.7°C.

▶ The hypothalamus acts like a thermostat. It registers changes in the core body temperature and also receives information about temperature changes from thermoreceptors in the skin. It then coordinates nervous and hormonal responses to counteract the changes and restore normal body temperature, as shown in the feedback diagram below. When normal temperature is restored, the corrective mechanisms are switched off.

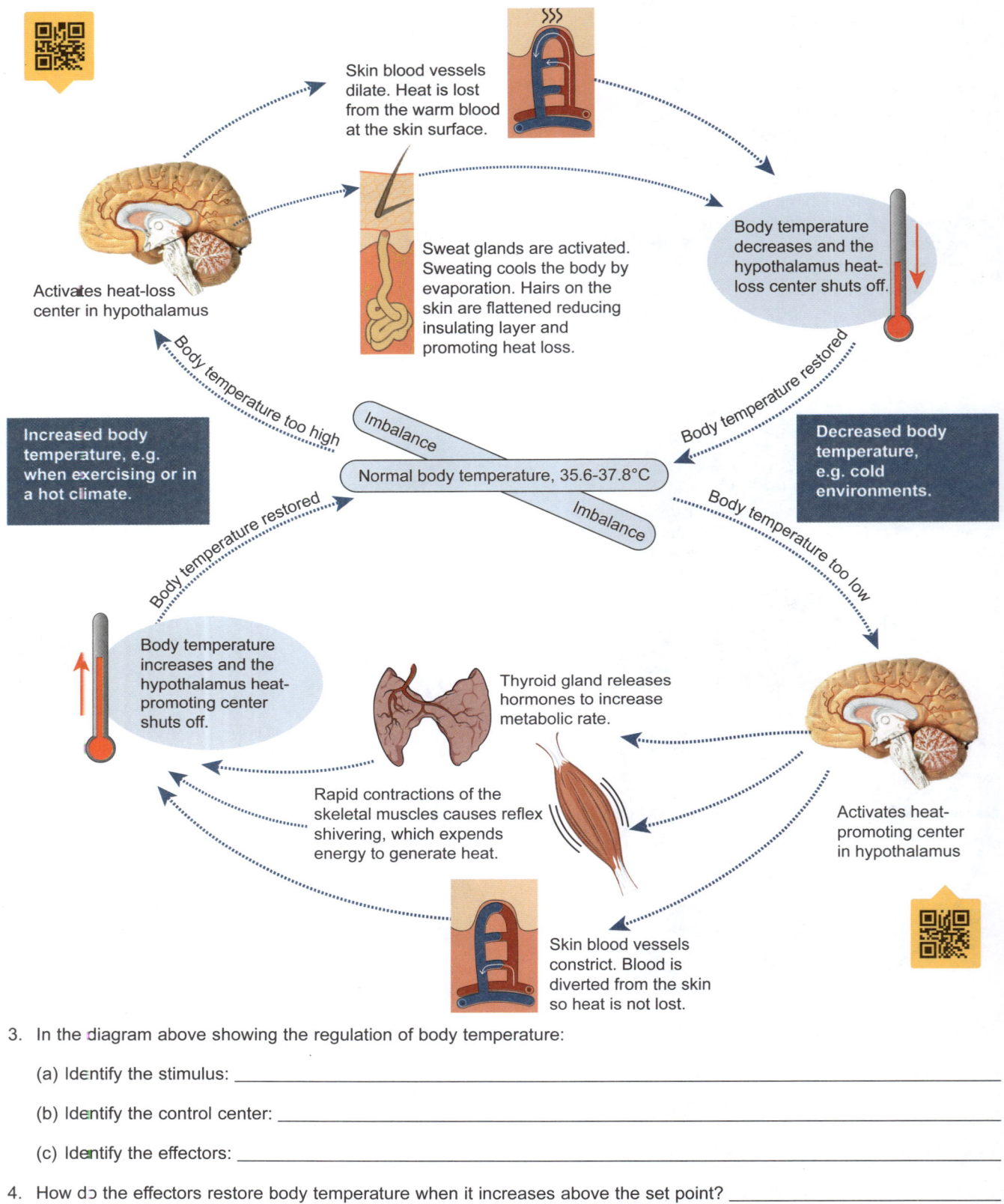

3. In the diagram above showing the regulation of body temperature:

   (a) Identify the stimulus: _____

   (b) Identify the control center: _____

   (c) Identify the effectors: _____

4. How do the effectors restore body temperature when it increases above the set point? _____

75 Nervous Regulatory Systems

Key Question: How does the nervous system regulate functions of the body?

Nervous regulation

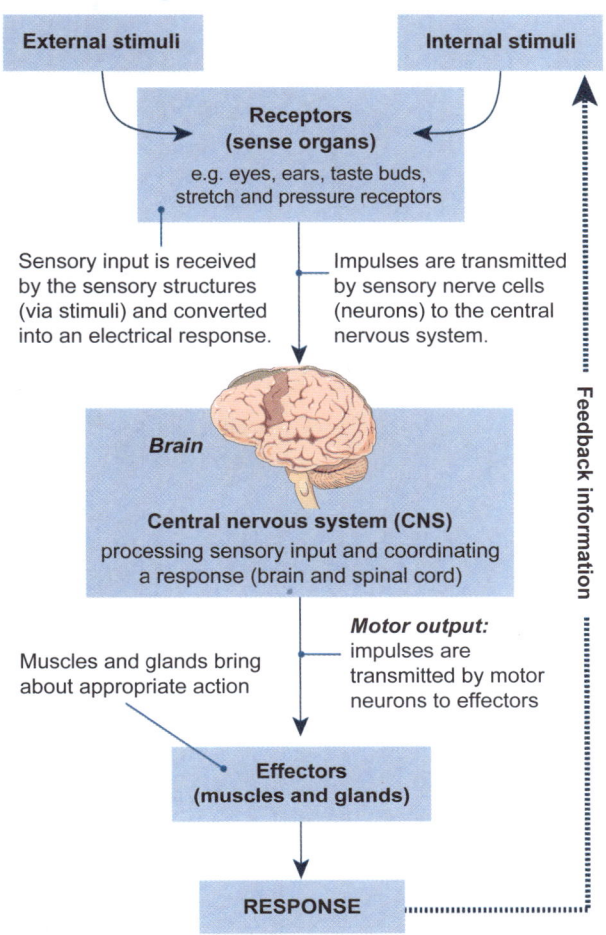

- In humans, the nervous and endocrine (hormonal) systems work together to maintain **homeostasis** in a fluctuating external environment.
- The nervous system is a signaling network, with branches carrying information directly to and from specific target **tissues**. Impulses can be transmitted over considerable distances, and the response is very precise and rapid.
- For example, think about a time a nuisance fly landed on you, or dust blew in your eyes. The reaction time for you to notice the disturbance and react by swatting at the fly, or turning your head and closing your eyes, is usually less than a second.

Comparison of nervous and hormonal control		
	Nervous control	Hormonal control
Communication	Impulses across synapses	Hormones in the blood
Speed	Very rapid (within a few milliseconds)	Relatively slow (over minutes, hours, or longer)
Duration	Short term and reversible	Longer lasting effects
Target pathway	Specific (through nerves) to specific cells	Hormones broadcast to target cells everywhere
Action	Causes glands to secrete or muscles to contract	Causes changes in metabolic activity

1. Use the information above to describe the steps or processes that must occur for the person in the photo to catch the approaching throwing disk:

   _____
   _____
   _____
   _____
   _____
   _____
   _____
   _____

2. Explain why hormonal regulation would not be appropriate for the signaling required to catch the throwing disk:

   _____
   _____
   _____

76 Hormonal Regulation

Key Question: How do the hormones regulate functions of the body?

▸ **Hormones** are produced by endocrine cells and secreted into the bloodstream where they are distributed throughout the body. Although hormones are sent throughout the body, they affect only specific target cells. These target cells have receptors on the plasma membrane which recognize and bind the hormone (see inset, below). The binding of hormone and receptor triggers the response in the target cell. Cells are unresponsive to a hormone if they do not have the appropriate receptors.

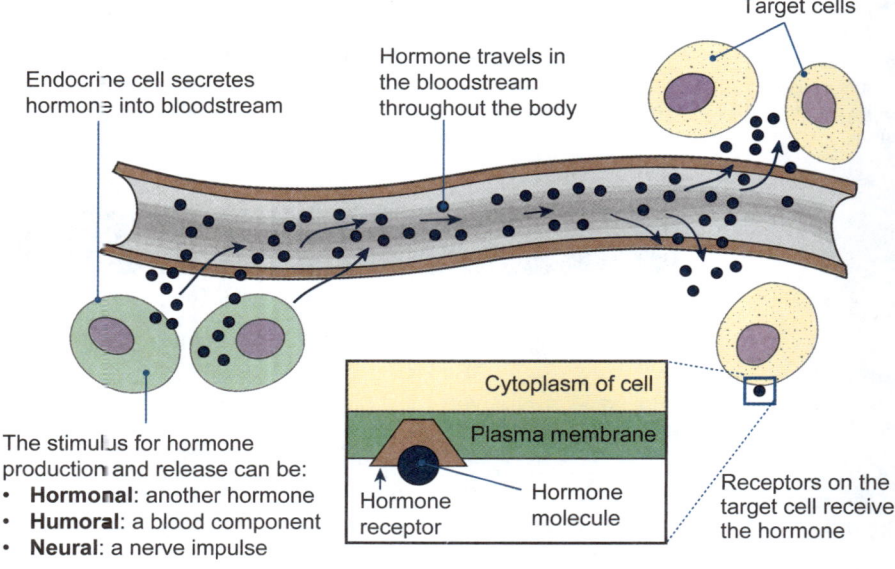

The stimulus for hormone production and release can be:
- **Hormonal:** another hormone
- **Humoral:** a blood component
- **Neural:** a nerve impulse

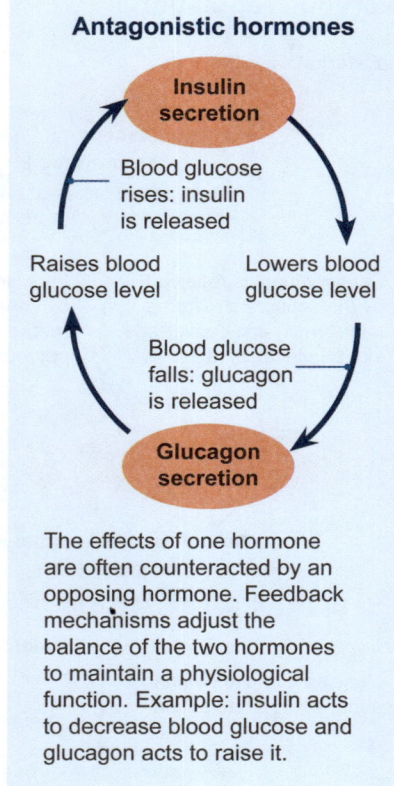

Antagonistic hormones

The effects of one hormone are often counteracted by an opposing hormone. Feedback mechanisms adjust the balance of the two hormones to maintain a physiological function. Example: insulin acts to decrease blood glucose and glucagon acts to raise it.

1. (a) What are antagonistic hormones? Describe an example of how two such hormones operate: _____
_____
_____
_____
_____

   (b) Describe the role of feedback mechanisms in adjusting hormone levels (explain using an example if this is helpful):
_____
_____
_____

2. How can a hormone influence only the target cells even though all cells may receive the hormone? _____
_____
_____
_____

3. Explain why hormonal control differs from nervous system control with respect to the following:

   (a) The speed of hormonal responses is slower: _____
_____

   (b) Hormonal responses are generally longer lasting: _____
_____

77 Nervous and Endocrine Interactions

Key Question: In what ways are the nervous and endocrine systems both similar and different, and how do they work together to maintain homeostasis?

▶ In mammals, the nervous system and endocrine (hormonal) systems act both independently and together to maintain **homeostasis**. The two systems are quite different in their modes of action, in the responses they elicit, and in their duration of action.

▶ The nervous system stimulates rapid, short-lived responses through electrical signals transmitted directly between adjacent **cells**. The endocrine system produces a slower, more long-lasting response through chemicals called **hormones**, carried in the blood. These hormones control many life processes.

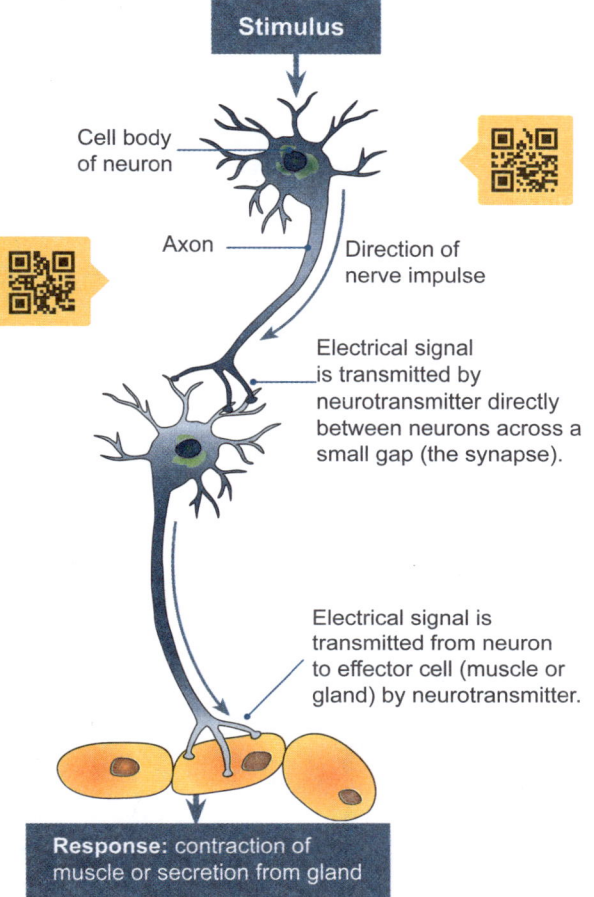

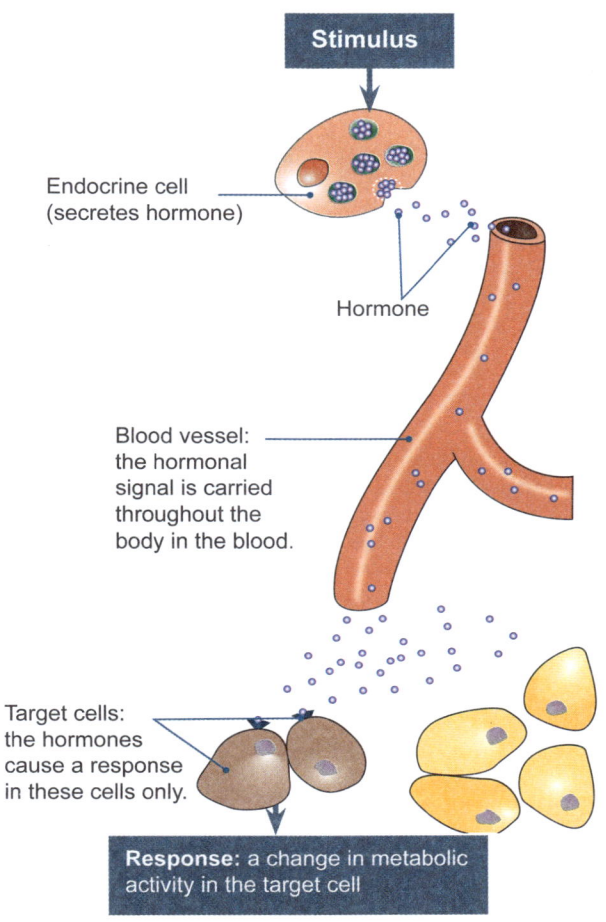

The nervous system transmits electrical impulses directly between cells through electrical junctions or via chemicals called neurotransmitters, which can diffuse across the gap (synapse) between cells. The response of a cell to nervous stimulation is rapid (milliseconds), short lived, and localized. Neural communication is important in rapid responses to stimuli, e.g. reflexes such as pain withdrawal, and may be involved in triggering the release of hormones.

Hormones secreted from endocrine cells are carried in the blood throughout the body, where they interact only with target cells carrying the correct receptor to bring about a response. The speed of hormonal signaling is relatively slow and it exerts its effects over minutes, hours, or days. Hormonal responses are important in regulating physiological processes such as growth and reproduction, as well as levels of glucose and electrolytes in the blood.

1. Contrast the mode of action of the nervous and endocrine systems: _____

_____

_____

2. Using examples, explain how the endocrine and nervous systems are involved in maintaining homeostasis:

_____

_____

_____

_____

78 Interactions Regulating the Blood

Key Question: How do the circulatory and urinary systems interact to remove wastes from the body's tissues and help maintain blood volume and pressure?

Circulatory system

Function

Delivers oxygen (O_2) and nutrients to all **cells** and **tissues**. Removes carbon dioxide (CO_2) and other waste products of metabolism.

Components

- Heart
- Blood vessels:
  - Arteries
  - Veins
  - Capillaries
- Blood

Interaction between systems

In mammals, the urinary system and cardiovascular system interact to remove metabolic wastes from the body.

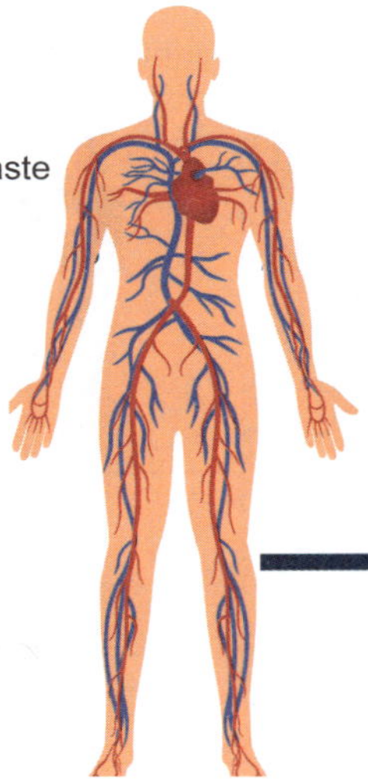

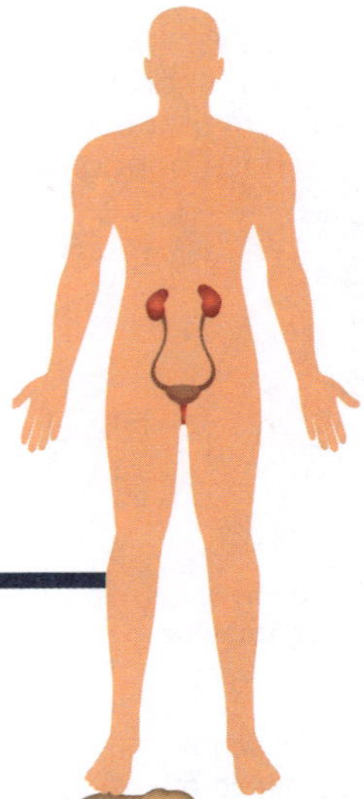

Urinary system

Function

Filters blood, retaining useful molecules and removing harmful ones. Regulates blood volume and ion content.

Components

- Kidneys
- Ureters
- Bladder
- Urethra

Metabolic wastes generated by the cells move into the blood plasma and are transported to the kidneys. Blood is forced at high pressure through the capillaries of the glomerulus (filter unit of the kidney), producing a filtrate. Both useful and harmful molecules are contained in the filtrate, although large proteins and blood cells are excluded.

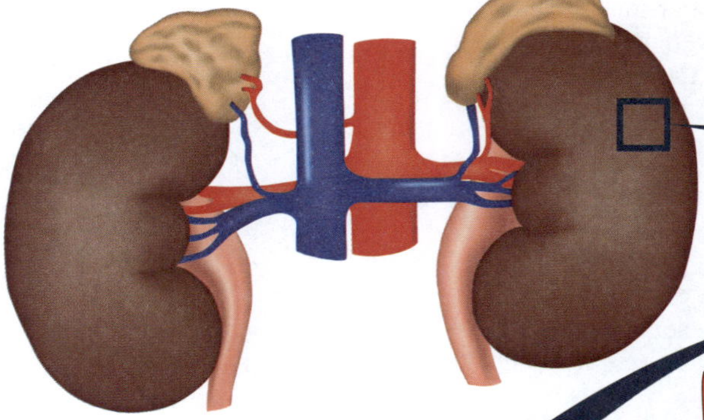

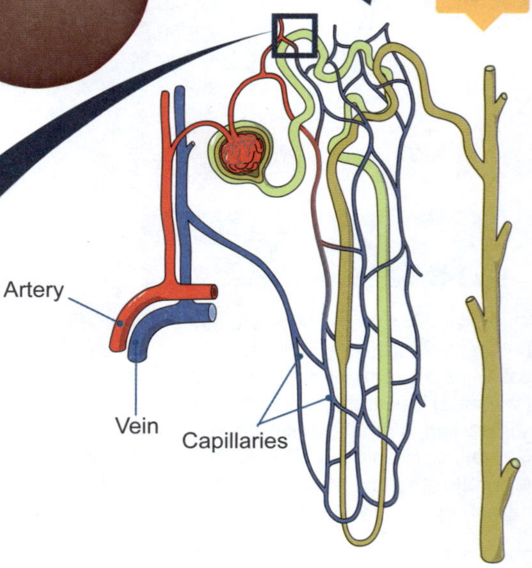

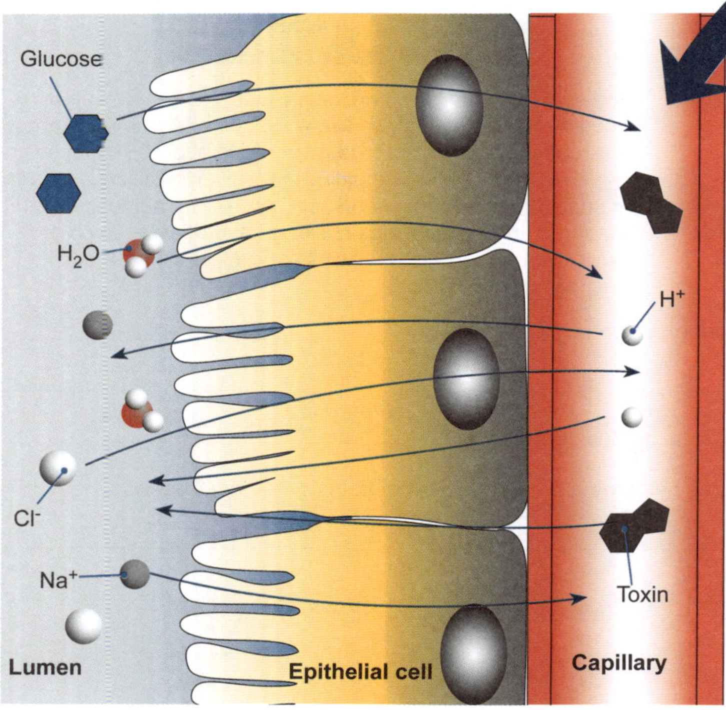

The tubules through which the filtrate travels are surrounded by capillaries (above). As the filtrate moves along the tubules, epithelial cells reabsorb useful molecules or ions, e.g. glucose and sodium ions, back into the blood in the capillaries (left). Unwanted ions (e.g. H^+) and toxins are also secreted into the filtrate by active transport and eliminated in the urine.

Sodium chloride crystals

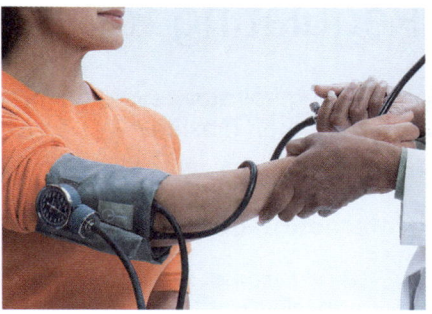

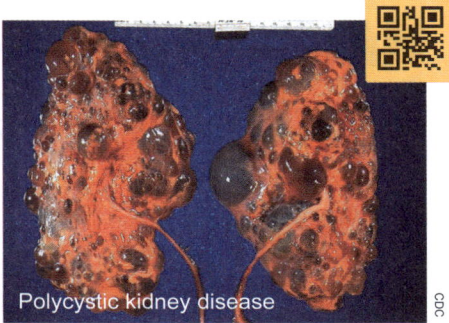
Polycystic kidney disease

The kidneys regulate blood volume by regulating the ion content of the blood. Retention of ions help retain water (via osmosis). Various **hormones** carried in the blood stimulate the kidneys to increase water retention.

The regulation of blood volume is important for the body. Optimal blood volume enables the blood to flow at the correct rate through the capillaries, helps maintain the correct concentrations of electrolytes, and maintains the pressure needed for glomerular filtration.

Chronic kidney disease makes it difficult for the body to regulate blood volume, and this can promote hypertension. Hypertension can lead to atherosclerosis of the arteries and, in turn, heart disease, heart attack, or peripheral arterial disease.

1. (a) Explain why regulating salt content of the blood, in turn regulates blood volume: _____

   (b) How would the active secretion of hydrogen ions into the filtrate (urine) help to regulate blood pH? _____

2. (a) Explain why kidney disease can make regulating blood volume and pressure difficult: _____

   (b) Explain why kidney disease can also lead to heart disease and heart attacks: _____

3. Use the graph on the right to answer the following questions:

   (a) What is the effect of reducing glomerular filtration rate on the risk of cardiovascular disease? _____

   (b) How many times greater is the risk of cardiovascular disease at a GFR of <15 compared to a GFR of >60? _____

   (c) What might cause reduced GFR? _____

   Risk of cardiovascular disease vs glomerular filtration rate

Estimated GFR (mL/min 1.73 m²)	Rate of cardiovascular event per 100 person years
>60	2.11
45-59	3.65
30-44	11.29
15-29	21.8
<15	36.6

4. In your own words, describe how the circulatory system and urinary system work together to remove wastes and regulate the volume, pressure, and composition of the blood: _____

©2024 **BIOZONE** International
ISBN: 978-1-99-101405-4
Photocopying prohibited

79 Interactions Regulating Respiratory Gases

Key Question: How do the circulatory and respiratory systems interact to provide the body's tissues with oxygen and remove carbon dioxide?

Circulatory system

Function
Delivers oxygen (O_2) and nutrients to all **cells** and **tissues**. Removes carbon dioxide (CO_2) and other waste products of metabolism. CO_2 is transported to the lungs.

Components
▶ Heart
▶ Blood vessels:
   • Arteries
   • Veins
   • Capillaries
▶ Blood

Interaction between systems

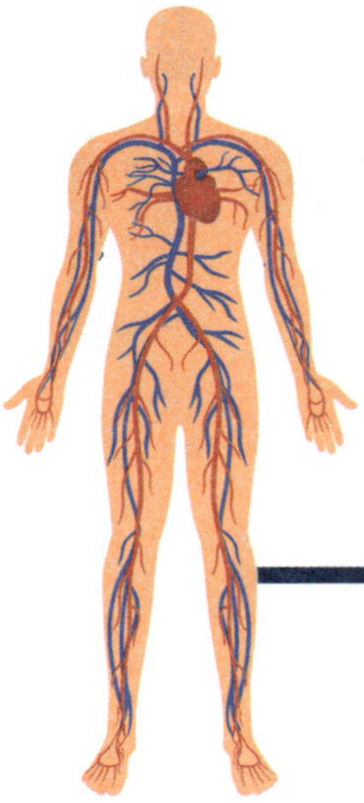

In vertebrates, the respiratory system and cardiovascular system interact to supply oxygen and remove carbon dioxide from the body.

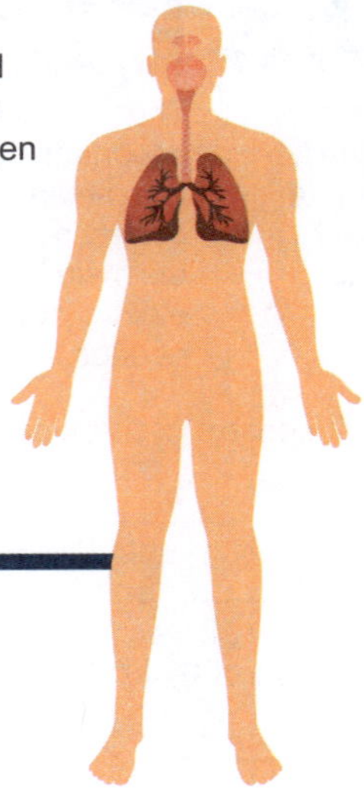

Respiratory system

Function
Provides surface for gas exchange. Moves fresh air into and stale air out of the body.

Components
▶ Airways:
   • Pharynx
   • Larynx
   • Trachea
▶ Lungs:
   • Bronchi
   • Bronchioles
   • Alveoli
▶ Diaphragm

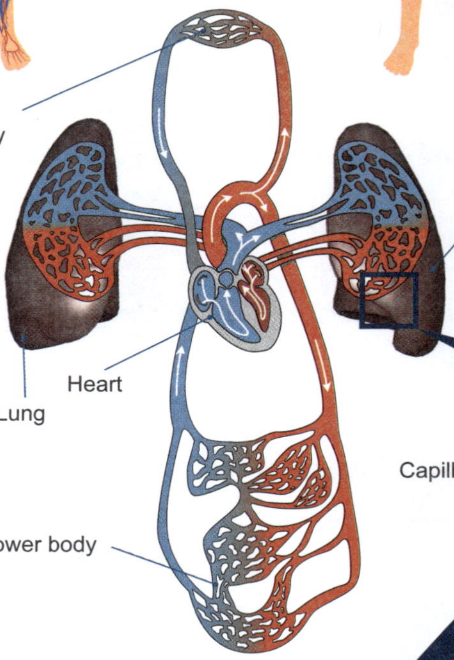

Head and upper body

Lung

Heart

Lung

Lower body

Bronchiole

Capillaries

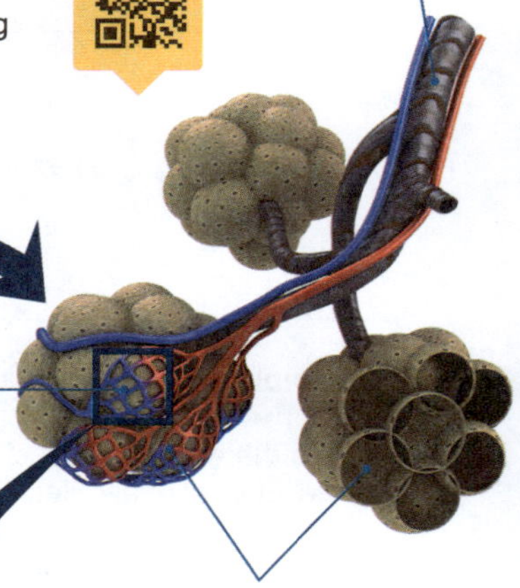

The airways of the lungs end at the alveoli (singular, alveolus). These are microscopic air sacs that enable gas exchange.

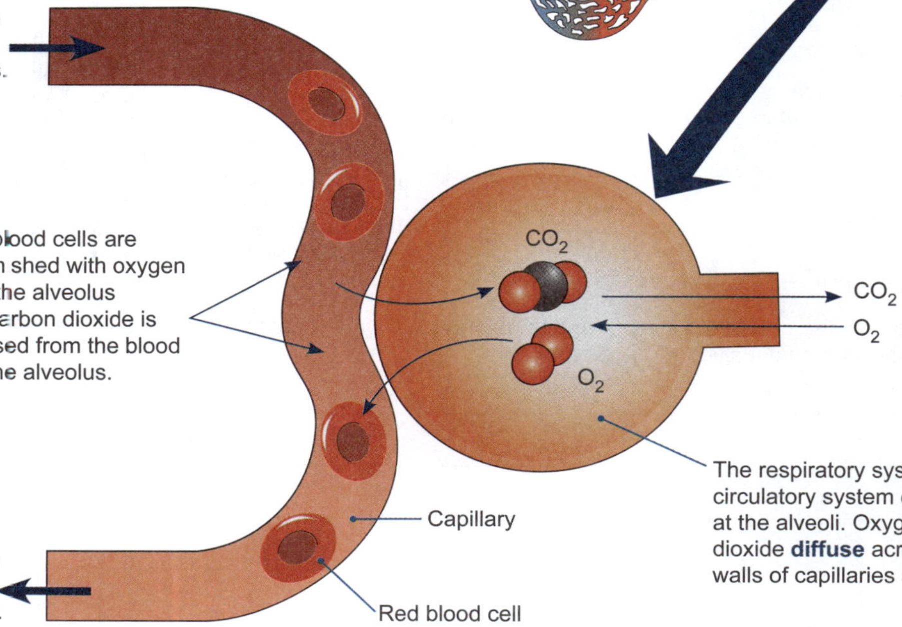

Oxygen (O_2) from inhaled air moves from the lungs into the circulatory system and is transported to the heart by red blood cells. The heart pumps the blood to the body where O_2 is released and carbon dioxide (CO_2) is picked up. The blood returns to the heart and is pumped to the lungs where CO_2 is released into the lungs to be breathed out.

From the heart to the lungs.

Red blood cells are replenished with oxygen from the alveolus and carbon dioxide is released from the blood into the alveolus.

From the lungs to the heart.

Capillary

Red blood cell

The carbon dioxide released from the blood exits the body during exhalation. Inhalation brings in fresh air, containing oxygen.

The respiratory system and the circulatory system come together at the alveoli. Oxygen and carbon dioxide **diffuse** across the thin walls of capillaries and alveoli.

©2024 BIOZONE International
ISBN: 978-1-99-101405-4
Photocopying prohibited

Responding to exercise

▸ During exercise, your body needs more oxygen to meet the extra demands placed on the muscles, heart, and lungs. At the same time, more carbon dioxide must be expelled. To meet these increased demands, blood flow must increase. This is achieved by increasing the rate of heart beat. As the heart beats faster, blood is circulated around the body more quickly, and exchanges between the blood and tissues increase.

▸ The arteries and veins must be able to cope with the extra pressure of higher blood flow and must expand (dilate) to accommodate the higher blood volume. If they didn't, they could rupture (break). During exercise, the muscular, circulatory and nervous systems interact to maintain the body's systems in spite of increased demands (right).

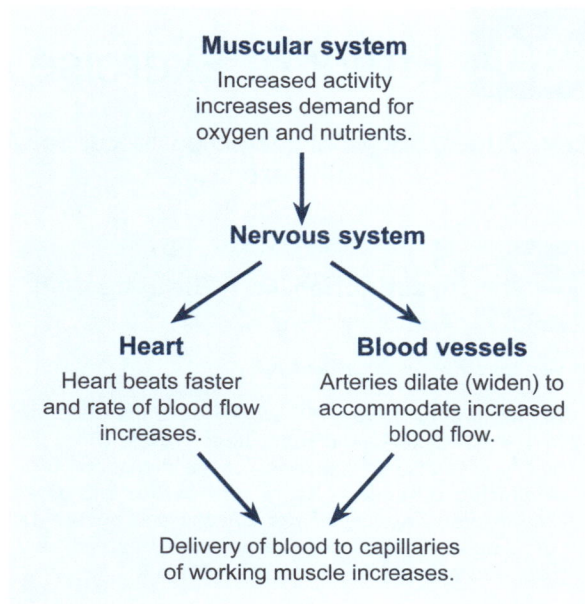

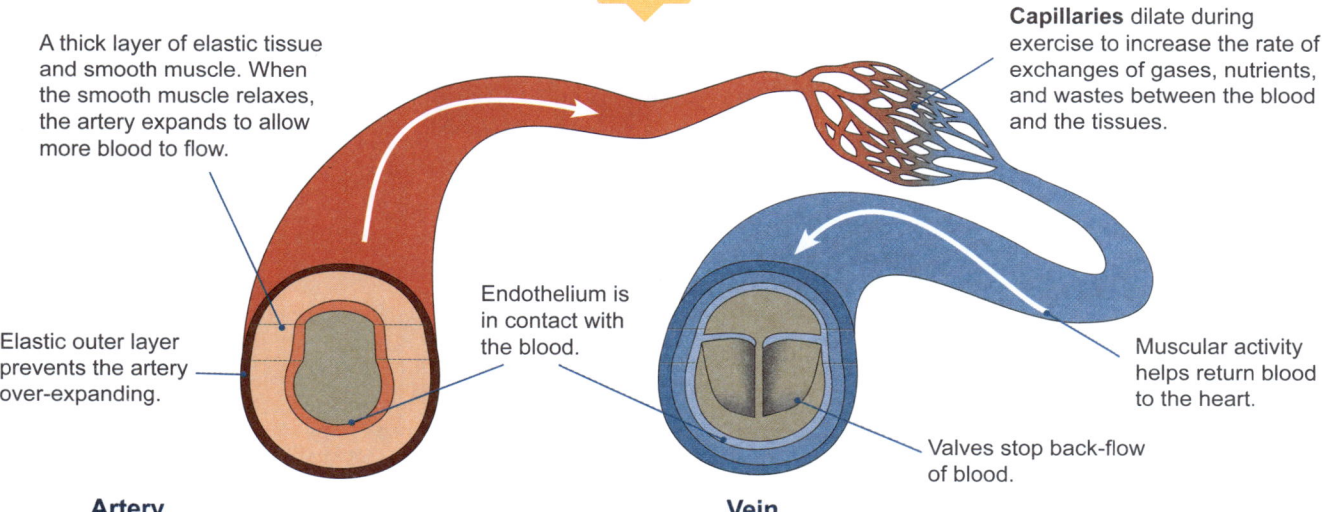

Artery
The strong, stretchy structure of arteries enables them to respond to increases in blood flow and pressure as more blood is pumped from the heart.

Vein
Veins return blood to the heart. They are less muscular than arteries, but valves and the activity of skeletal muscles, especially during exercise, help venous return.

1. In your own words, describe how the circulatory system and respiratory system work together to provide the body with oxygen and remove carbon dioxide:

___

2. (a) What happens to blood flow during exercise? ___

   (b) How do body systems interact to accommodate the extra blood flow needed when a person exercises?

___

80 Effect of Exercise on Heart Rate and Breathing

Key Question: What interactions occur between the circulatory, respiratory, and muscular systems during exercise?

Investigation 4.1   Investigating the effect of exercise on heart rate.

See appendix for equipment list.

In this practical, you will work in groups of three to see how exercise affects heart rates. The body's response to exercise can be measured by monitoring changes in heart rate before and after a controlled physical effort. Choose one person to carry out the exercise, one person to record heart rate, and one person to monitor timing.

Heart rate (beats per minute) is obtained by measuring the pulse (right) for 15 seconds and multiplying by four.

CAUTION: The person exercising should have no known pre-existing heart or respiratory conditions.

Gently press your index and middle fingers, not your thumb, against the carotid artery in the neck (just under the jaw) or the radial artery (on the wrist just under the thumb) until you feel a pulse.

Measuring the carotid pulse

Measuring the radial pulse

1. **Resting measurements:** The person carrying out the exercise should sit down on a chair for 5 minutes and try not to move. After 5 minutes of sitting, measure their heart rate. Record the resting data in the table (below).

2. **Exercising measurements:** Choose an exercise to perform. Some examples include: step ups onto a chair, skipping rope, jumping jacks, or running in place.

3. Begin the exercise, and take measurements after 1, 2, 3, and 4 minutes. The person exercising should stop just long enough for the measurements to be taken. Record the results in the table.

4. **Post-exercise measurements:** After the exercise period has finished, the exerciser should sit down on a chair. Take pulse measurements 1 and 5 minutes after finishing the exercise. Record the results in the table, below.

Activity	Resting before exercise	At 1 minute during exercise	At 2 minutes during exercise	At 3 minutes during exercise	At 4 minutes during exercise	1 minute after exercise	5 minutes after exercise
Pulse Rate							

1. (a) Graph your results on a piece of paper or use a spreadsheet. Use the vertical axis to plot heart rate. When you have finished answering the questions below, attach it to this page.

   (b) Analyze your graph and describe what happened to heart rate during exercise: _____
   _____

2. (a) Describe what happened to heart rate after exercise: _____
   _____
   _____

   (b) Why did this change occur? What were the interactions that occurred between the cardiovascular and muscular systems?
   _____
   _____
   _____
   _____

Investigation 4.2 Investigating the effect of exercise on breathing rate.

See appendix for equipment list.

You could work in both small groups and individually to see how exercise affects breathing rates.

1. Plan your investigation using the guidance below.
2. Conduct your investigation following your planned method and collect data.
3. Complete question (7) after the data collection, data processing, and investigation conclusion have been completed.

3. (a) How is breathing linked to exercise? _____

   _____

   _____

   (b) Write a hypothesis for your investigation: _____

   _____

4. Identifying evidence: Develop an investigation plan and consider the following:

   (a) What factor (stimulus) will be changed by the experimenter? What is the independent variable? _____

   _____

   (b) What factor (response) will be measured? What is the dependent variable? _____

   _____

   _____

   (c) How will this data contribute to investigating the phenomenon in question 3. (b)? _____

   _____

   _____

5. Planning for the investigation: Construct an investigation plan. Complete the chart to develop the structure:

Steps to consider	Your investigation
(a) What will you measure? i.e type, length of exercise, units that will be used?	
(b) How will the response of the living system (breathing) be measured, i.e. units, length, timing?	
(c) How are the internal conditions (homeostasis of oxygen demand) linked to what is being measured?	

6. Use the following space to construct a draft method. Develop a final investigation procedure (on separate paper), including a data table in which to record your results, incorporating the information provided previously. Ensure control of other variables. You may wish to include a labeled diagram.

7. Attach your written procedure and data table to this page. Carry out your investigation by following your procedure and collect your data; use your findings to answer the following questions:

   (a) Explain how your conclusion was able to answer "How does exercise affect breathing rates?" _____

   (b) How accurate and precise was your data and what steps did you take to ensure accuracy? _____

   (c) How generalizable was your data and how did you ensure reliability? _____

   (d) What limitations did your group experience in collecting the data? _____

   (e) How could you refine the design of the investigation to increase accuracy and generalizability of your data?

81 Interactions for Nutrient Absorption

Key Question: How do the circulatory and digestive systems interact to provide the body's tissues with the nutrients that they require?

Circulatory system

Function
Delivers oxygen (O_2) and nutrients to all **cells** and **tissues**. Removes carbon dioxide (CO_2) and other waste products of metabolism.

Components
- Heart
- Blood vessels:
  - Arteries
  - Veins
  - Capillaries
- Blood

Interaction between systems

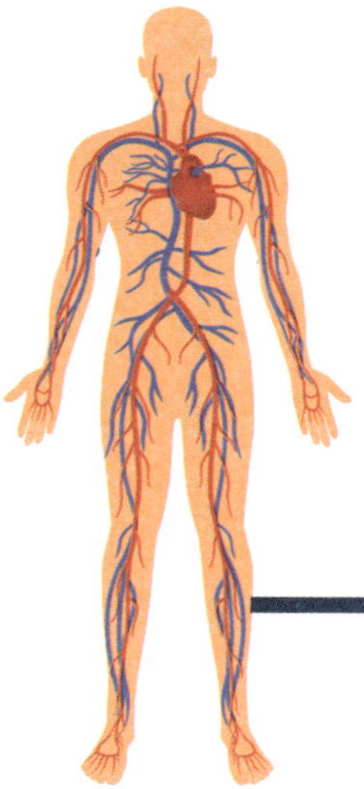

In mammals, the **digestive system** and **cardiovascular system** interact to supply nutrients to the body.

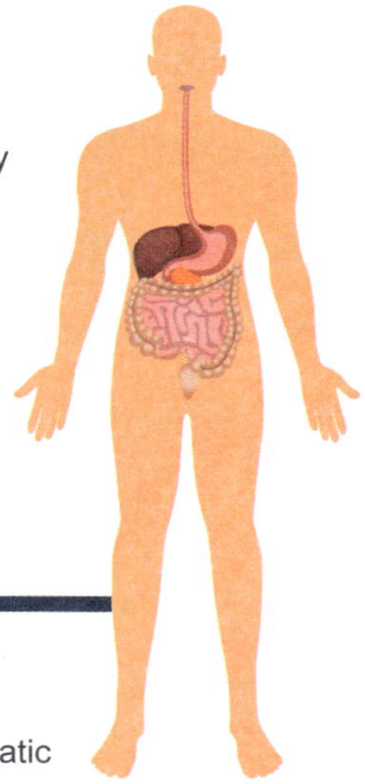

Digestive system

Function
Digest food and absorb useful molecules from it, and eliminate undigested material.

Components
- Mouth and pharynx
- Esophagus
- Stomach
- Liver and gall bladder (accessory organs)
- Pancreas (accessory organ)
- Small intestine
- Large intestine

Food is digested in the stomach and small intestine from where it is absorbed and passed to the circulatory system. The capillaries around the stomach and intestines collect nutrients and then drain into the hepatic portal vein. This vein carries the blood directly to the liver. The liver then processes this nutrient rich blood, e.g. it stores glucose as glycogen. The hepatic vein then transports nutrients from the liver to supply the other tissues of the body.

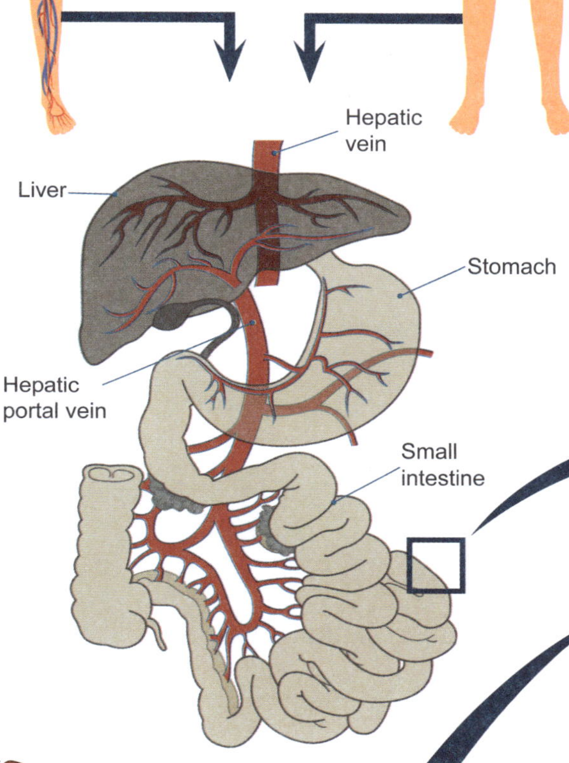

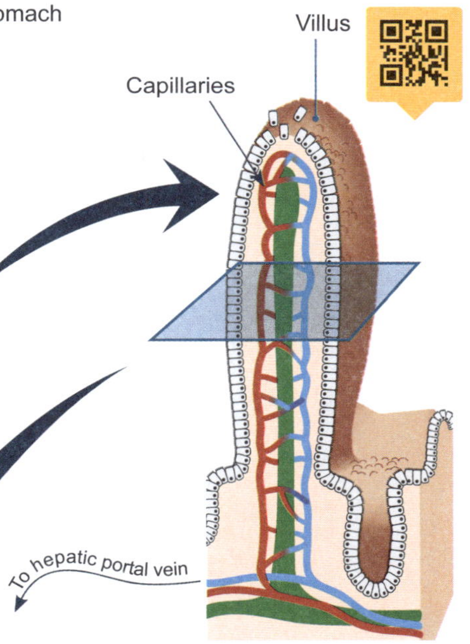

Villi project into the lumen of the small intestine and absorb nutrients. Villi contain capillary networks which receive the nutrients and transport them to the hepatic portal system.

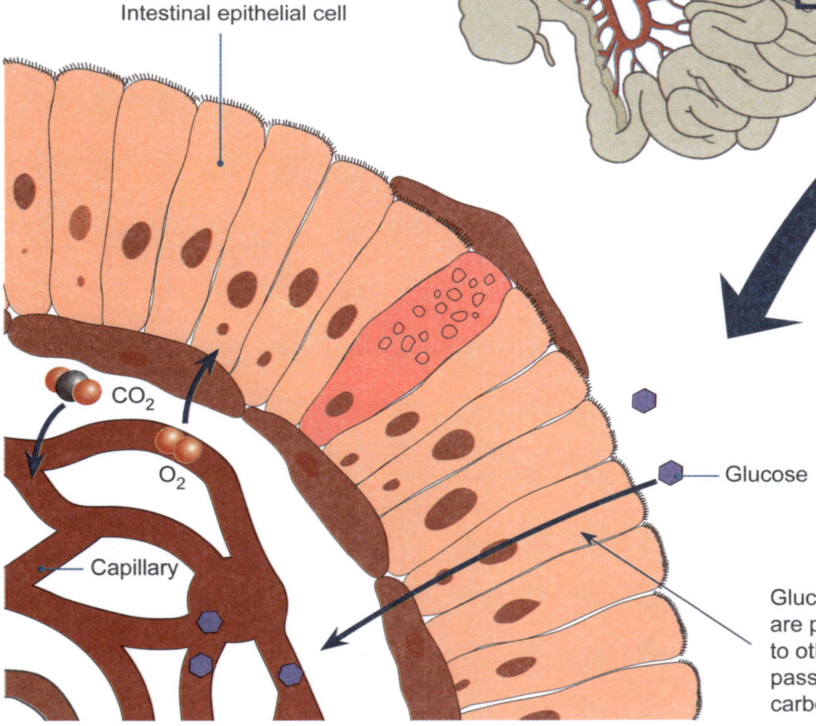

Glucose and other nutrient molecules are passed to the blood and transported to other parts of the body. Oxygen passes to the intestinal cells, while carbon dioxide passes into the blood.

 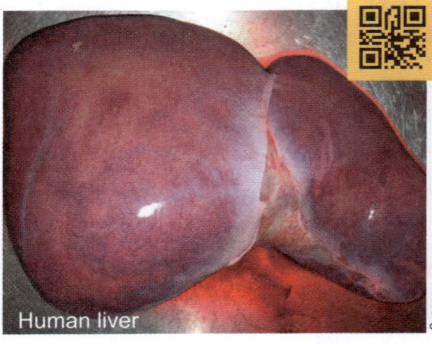 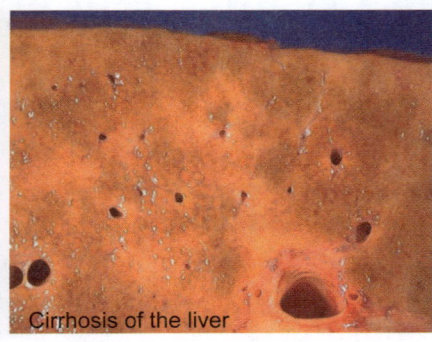

Human liver | Cirrhosis of the liver

Blood flow to the digestive tract increases steadily after a meal, reaching a maximum after about 30 minutes. It remains elevated for about 2.5 hours. During exercise, blood flow to the digestive tract is reduced as it is redirected to the muscles.

Nutrients, e.g. minerals, sugars, and amino acids, are transported in the blood plasma to the liver. The liver receives the nutrient-rich, deoxygenated blood from the digestive system via the hepatic portal vein, and oxygen rich blood from the hepatic artery.

Scarring of the liver tissue, or cirrhosis, can result in high blood pressure. The scarred tissue obstructs blood flow in the liver. This causes pressure to build up in upstream blood vessels, resulting in swelling and possible hemorrhage.

1. How are nutrients transported in the blood? _____

2. Explain how liver cirrhosis affects the circulatory system: _____

3. (a) At which two points in the body do the digestive and circulatory systems directly interact? _____

   (b) Explain what is happening at these points: _____

4. (a) What happens to blood flow to the digestive tract after a meal? _____

   (b) Explain why it is often recommended that a person should exercise within 2.5 hours of eating, or eat within half an hour after exercising, to gain most benefit from the exercise (in terms of muscle development): _____

5. In your own words, describe how the circulatory and digestive systems work together to provide the body with nutrients: _____

Nervous and endocrine interactions with the digestive system

▶ The endocrine and nervous systems are both involved in the regulation of digestion.
▶ Most digestive juices are only secreted when there is food in the gut and both nervous and hormonal mechanisms are involved in coordinating and regulating this activity.
▶ The digestive system is under unconscious control. Several **hormones**, which are released into the bloodstream in response to nervous or chemical stimuli, influence the activity of the gut and associated organs.

Feeding center:
The feeding center in the hypothalamus, found in the brain, continuously monitors metabolites in the blood and stimulates hunger when these metabolites reach low levels. After a meal, a neighboring part of the brain suppresses the activity of the feeding center for a period of time.

Pancreatic secretions and bile:
Cholecystokinin (CCK) stimulates secretion of enzyme-rich fluid from the pancreas and release of bile from the gall bladder. Secretin stimulates the pancreas to increase its secretion of alkaline fluid and the production of bile from the liver cells.

Vagus nerve

Gastrin

CCK and secretin

Intestinal secretion of hormones:
The entry of chyme (especially fat and gastric acid) into the small intestine stimulates the intestinal mucosa to secrete the hormones cholecystokinin (CCK) and secretin.

Salivation:
Entirely under nervous control. Some saliva is secreted continuously. Food in the mouth stimulates the salivary glands to increase their secretions.

The stomach and pancreas are stimulated by the nervous system, increasing or decreasing secretions. These are reflexes in response to the sight, smell, or taste of food.

Gastric secretion:
Physical expansion and the presence of food in the stomach cause release of the hormone called gastrin. Gastrin in the blood increases gastric secretion.

Summary of hormones acting in the gut

Hormone	Organ	Effect
Secretin	Pancreas	Increases secretion of alkaline fluid
Secretin	Liver	Increases bile production
CCK	Pancreas	Increases enzyme secretion
CCK	Liver	Stimulates release of bile
Gastrin	Stomach	Increases stomach motility and secretion

6. Describe the role of each of the following stimuli in the control of digestion, identifying both the response and its effect:

   (a) Presence of food in the mouth: _____

   (b) Presence of fat and acid in the small intestine: _____

   (c) Stretching of the stomach by the presence of food: _____

7. Describe the role of nerves and hormones in controlling digestion: _____

82 Regulating Blood Glucose Levels

Key Question: How is a constant blood glucose level maintained in the body?

The importance of blood glucose

▸ Glucose is the body's main energy source. It is chemically broken down during cellular respiration to generate ATP, which is used to power metabolism. Glucose is the main sugar circulating in blood, so it is often called blood sugar. Blood glucose levels are regulated by **negative feedback** involving two **hormones**, insulin and glucagon.

▸ Blood glucose levels are tightly controlled because **cells** must receive an adequate and regular supply of fuel. Prolonged high or low blood glucose causes serious physiological problems and even death. Normal activities, such as eating and exercise, alter blood glucose levels. However, the body's control mechanisms regulate levels so that fluctuations are minimized and generally occur within a physiologically acceptable range. For humans, this is 60-110 mg/dL (milligrams per decilitre), indicated by the shaded area in the graph below.

▸ British chemist Dorothy Hodgkin was awarded a Nobel Prize in 1964 for her determining the structure of insulin by x-ray crystallography. Her work allowed researchers to better understand and manufacture this life-saving protein.

Insulin enables cells to take up glucose

After a meal is eaten, food is broken down by the digestive system, and the components of the food, including glucose, are absorbed into the bloodstream and transported around the body.

The rise in blood glucose after a meal stimulates the release of the hormone insulin from the pancreas. Insulin stimulates cells to take up glucose, the fuel they need to carry out their functions.

When the cells take up glucose, the amount in the blood is reduced and blood glucose level returns to normal. Between meals, the liver can release glucose from stored glycogen to keep blood glucose stable.

What happens if your body does not produce insulin?

▸ In some people, the insulin-producing cells of the pancreas are damaged and the body cannot produce insulin. This life threatening disorder, which commonly affects children and teenagers, is called type 1 **diabetes mellitus**.

▸ The cells in people with type 1 diabetes cannot take up glucose from the blood. Glucose remains in the blood and blood glucose levels are elevated. The kidneys try to rid the body of the apparently "excess" glucose so sufferers produce large volumes of "sweet" urine, as glucose is excreted in the urine. They feel tired and weak and are constantly hungry and thirsty. Fats are metabolized for fuel.

▸ The only current treatment is regular injection with human insulin, together with careful dietary management to control blood glucose levels.

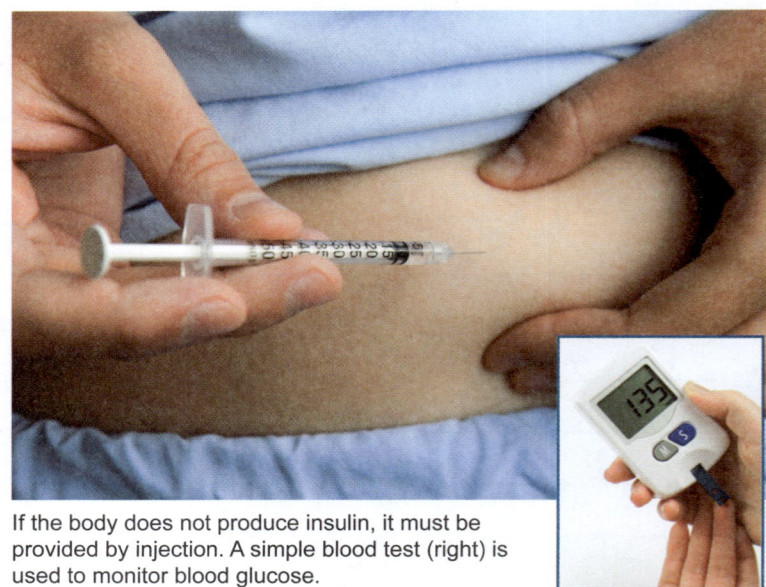

If the body does not produce insulin, it must be provided by injection. A simple blood test (right) is used to monitor blood glucose.

Controlling blood glucose levels

Blood glucose (BG) is controlled by two hormones produced by special endocrine cells in the pancreas. The hormones work antagonistically (oppose each other) and levels are tightly controlled by negative feedback.

▸ Insulin lowers blood glucose by promoting glucose uptake by cells and glycogen storage in the liver.
▸ Glucagon increases blood glucose by promoting release of glucose from the breakdown of glycogen in the liver.
▸ When normal blood glucose levels are restored, negative feedback stops hormone secretion.

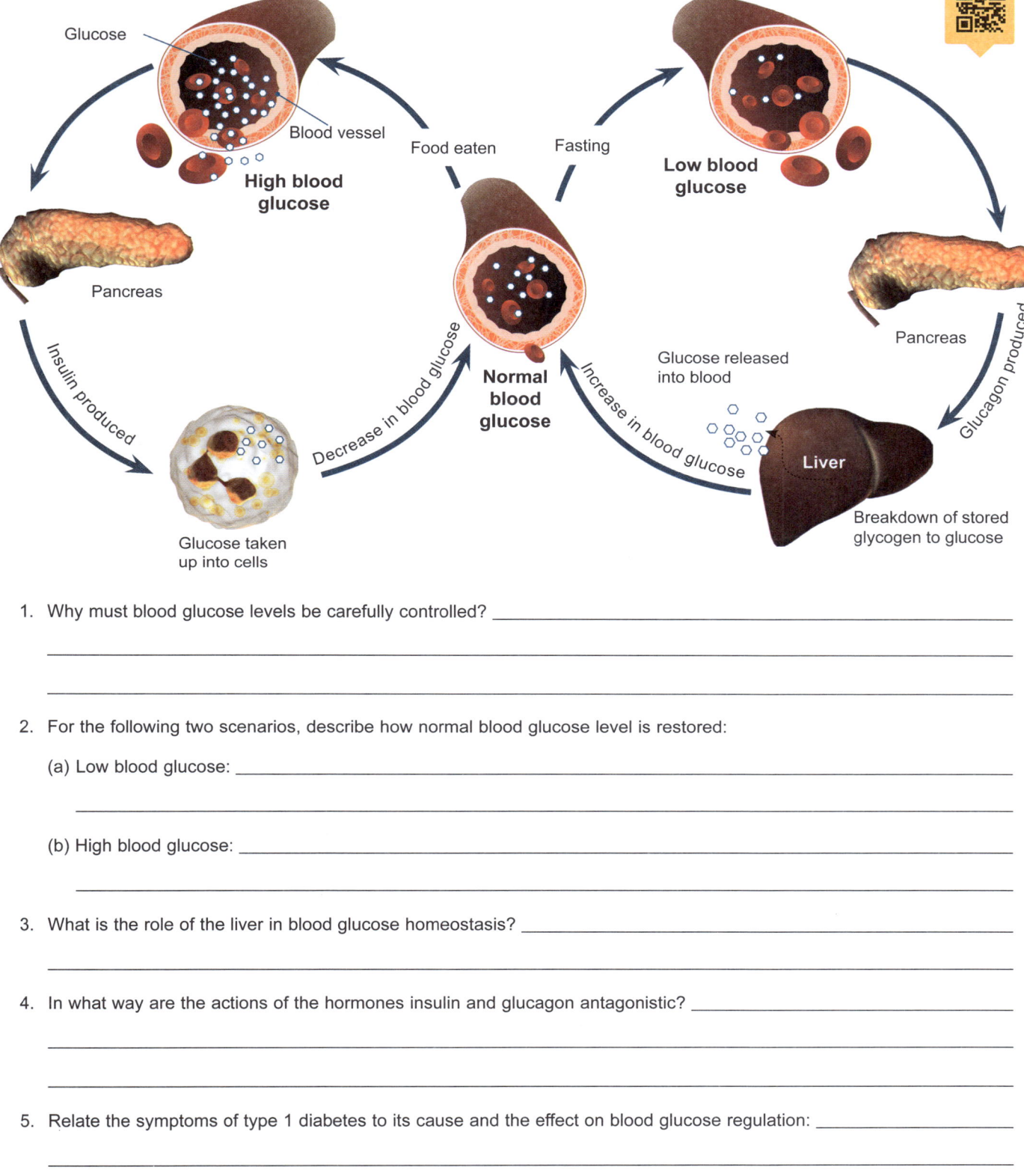

1. Why must blood glucose levels be carefully controlled? _____

2. For the following two scenarios, describe how normal blood glucose level is restored:

    (a) Low blood glucose: _____

    (b) High blood glucose: _____

3. What is the role of the liver in blood glucose homeostasis? _____

4. In what way are the actions of the hormones insulin and glucagon antagonistic? _____

5. Relate the symptoms of type 1 diabetes to its cause and the effect on blood glucose regulation: _____

83 Interacting Systems: The Menstrual Cycle

Key Question: What is the interaction between the endocrine system and the female reproductive system?

- The menstrual cycle lasts for about 28 days and involves development and release of the egg cells into fallopian tubes, development of the lining of the uterus, and the shedding of the lining if fertilization does not occur.

- There are four major **hormones** involved in the menstrual cycle. **Luteinising hormone** (LH) and **follicle stimulating hormone** (FSH) are released from the pituitary gland, whereas **estrogen** and **progesterone** are released from the ovaries. These hormones rise and fall in different ways over the length of the cycle.

Hormone levels

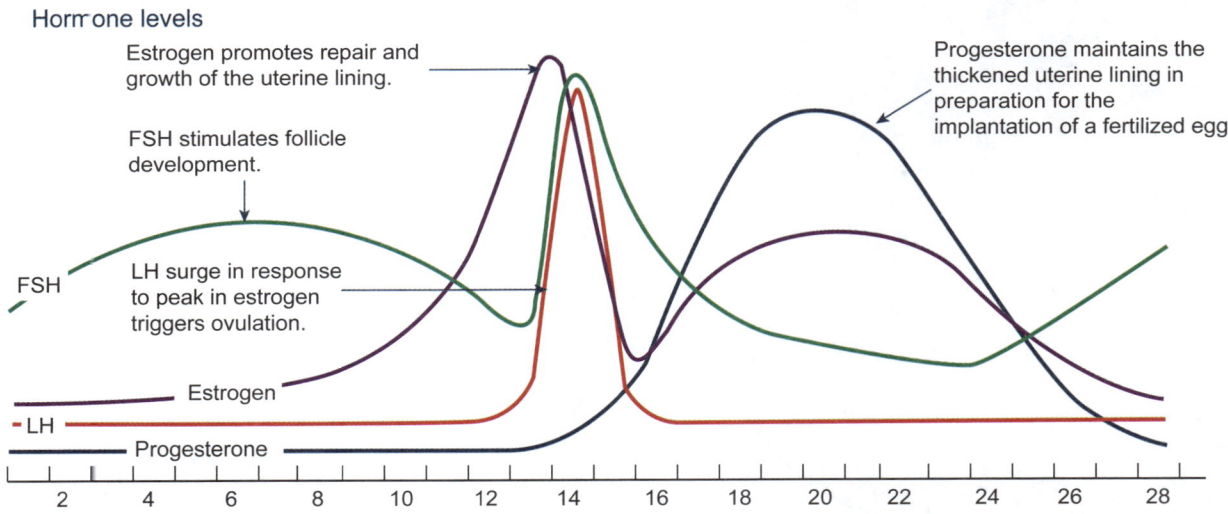

- One of the follicles begins developing in response to FSH and becomes the Graafian follicle, which releases estrogen. Estrogen levels peak, stimulating a surge in LH and triggering ovulation.
- After ovulation, the Graafian follicle develops into the corpus luteum, which secretes large amounts of progesterone (and smaller amounts of estrogen).

Ovarian cycle

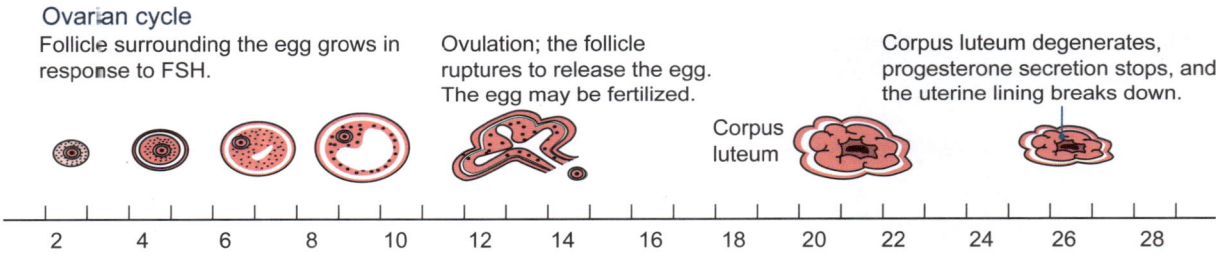

- The Graafian follicle continues to grow. At around day 14, it ruptures to release the egg (ovulation). LH causes the ruptured follicle to develop into a corpus luteum. The corpus luteum secretes progesterone which promotes full development of the uterine lining, maintains the embryo in the first 12 weeks of pregnancy, and inhibits the development of more follicles.

Menstrual cycle

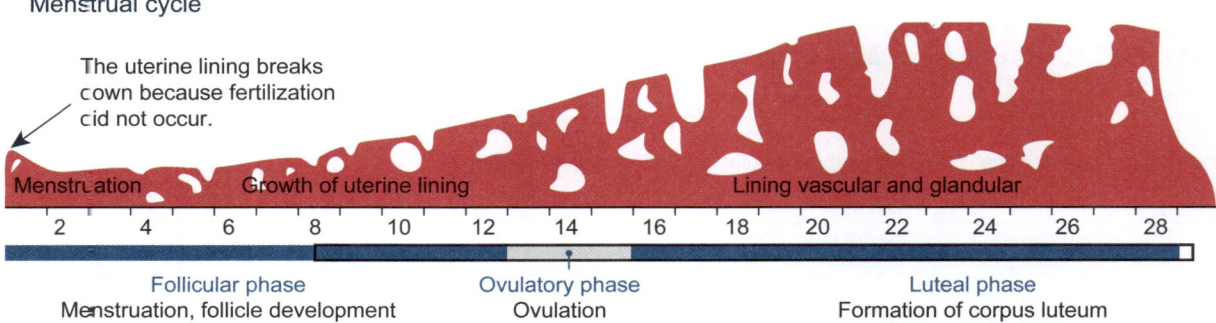

- If fertilization does not occur, the corpus luteum breaks down. Progesterone secretion declines, causing the uterine lining to be shed (menstruation). If fertilization occurs, high progesterone levels maintain the thickened uterine lining. The placenta develops and nourishes the embryo completely by 12 weeks.

Control of the menstrual cycle

▶ The female menstrual cycle is regulated by several reproductive hormones. The main control centers for this regulation are the hypothalamus and the anterior pituitary gland. The hypothalamus secretes gonadotropin releasing hormone (GnRH), which is transported in blood vessels to the anterior pituitary. Here, it induces the release of two hormones: follicle stimulating hormone (FSH) and luteinizing hormone (LH).

▶ These two hormones bring about the cyclical changes in the ovaries and uterus. Regulation of blood hormone levels during the menstrual cycle is achieved through **negative feedback** mechanisms.

▶ The exception to this is the mid cycle surge in LH, which is induced by the rapid increase in estrogen secreted by the developing follicle which surrounds the developing egg **cell** in the ovary.

- In the **first half of the cycle**, FSH stimulates follicle development in the ovary. The developing follicle secretes estrogen which acts on the uterus and, in the anterior pituitary, inhibits FSH secretion.
- In the **second half of the cycle**, LH induces ovulation and development of the corpus luteum. The corpus luteum secretes progesterone which acts on the uterus and also inhibits further secretion of LH and FSH.

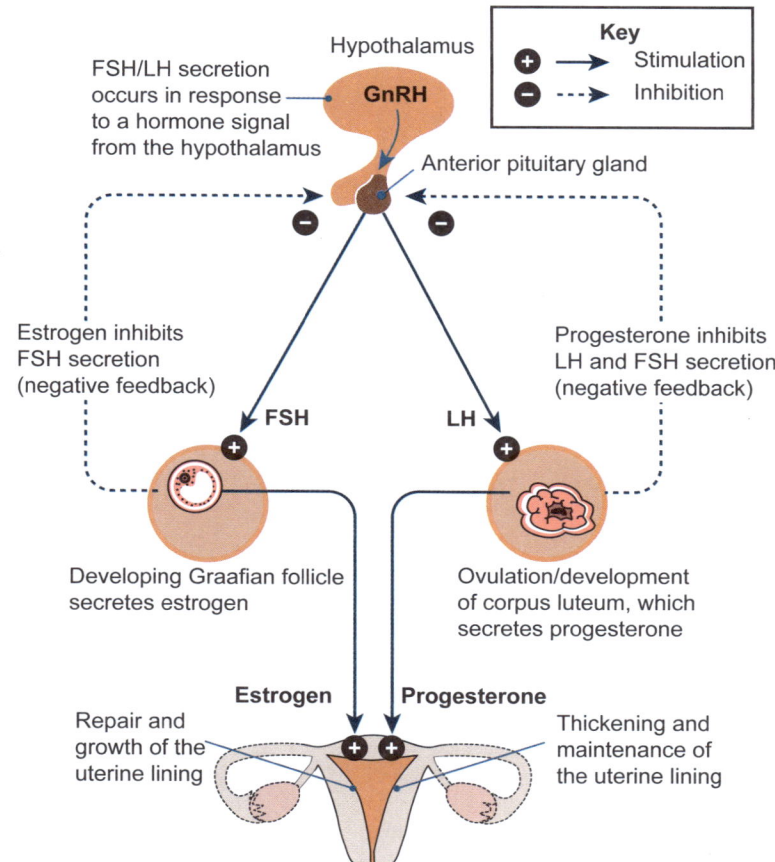

1. Summarize the role that the hormones below have in the control of the menstrual cycle and the site they secrete from:

   (a) GnRH: _____

   (b) FSH: _____

   (c) LH: _____

   (d) Estrogens: _____
   _____

   (e) Progesterone: _____
   _____

2. Describe the interactions of systems that control the menstrual cycle in the first and second halves of the cycle:

   (a) First half: _____
   _____
   _____
   _____
   _____

   (b) Second half: _____
   _____
   _____
   _____
   _____

84 Interacting Systems: Pregnancy and Birth

Key Question: What interactions occur between systems during fetal development and birth?

Circulatory system

Function
Delivers oxygen (O_2) and nutrients to all **cells** and **tissues**. Removes carbon dioxide (CO_2) and other waste products of metabolism.

Components
- Heart
- Blood vessels:
  - Arteries
  - Veins
  - Capillaries
- Blood

Interaction between systems

In (female) mammals, the **reproductive system** and **cardiovascular system** interact to help the fetus develop.

Reproductive system

Function
The female reproductive system houses the fetus and provides a link between the fetus and the mother's circulatory system.

Components
- Vagina
- Uterus
- Fallopian tubes
- Ovaries

The human fetus depends entirely on it's mother for nutrients, oxygen, and the elimination of wastes. The placenta is the specialized **organ** that performs this role. It enables exchange between fetal and maternal tissues and allows a prolonged period of fetal growth and development within the uterus.

The placenta also has an endocrine role, producing progesterone and estrogen to maintain the pregnancy.

Umbilical cord

Above: Fetus (near full term), showing placental attachment and position in the uterus.

Cervix

The maternal and fetal blood vessels are in such close proximity that oxygen and nutrients can diffuse from the maternal blood into the capillaries of the villi. From the villi, the nutrients circulate in the umbilical vein, returning to the fetal heart. Carbon dioxide and other wastes leave the fetus through the umbilical arteries, pass into the capillaries of the villi, and diffuse into the maternal blood. The fetal and maternal blood do not mix. The exchanges occur via **diffusion** through capillaries.

Below: Photograph of a human placenta, just after delivery.

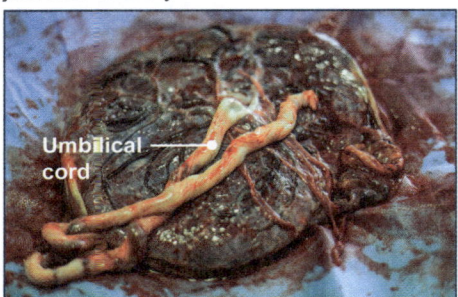

Umbilical cord

The placenta is a disk-like organ, about the size of a dinner plate and weighing about 1kg. It develops when finger-like projections (villi) from the fetal membranes grow into the lining of the uterus. The villi contain the capillaries connecting the fetal arteries and vein. They continue invading the maternal tissue until they are bathed in the maternal blood sinuses.

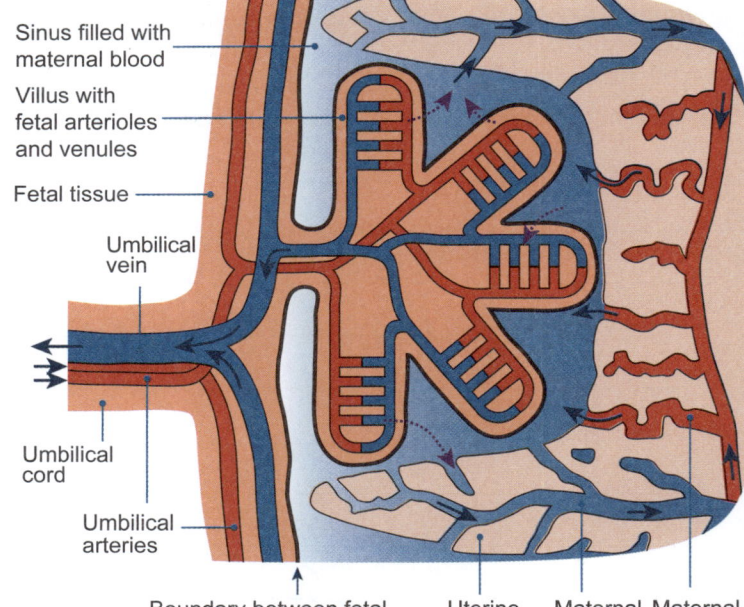

Sinus filled with maternal blood
Villus with fetal arterioles and venules
Fetal tissue
Umbilical vein
Umbilical cord
Umbilical arteries
Boundary between fetal and maternal tissues
Uterine lining
Maternal venule
Maternal arteriole

→ Blood flow
⋯▶ Exchange of wastes and nutrients via diffusion

©2024 **BIOZONE** International
ISBN: 978-1-99-101405-4
Photocopying prohibited

Hormonal regulation of birth

▶ A human pregnancy lasts about 38 weeks after fertilization. It ends in labor, the birth of the baby, and expulsion of the placenta. During pregnancy, progesterone maintains the placenta and inhibits contraction of the uterus. At the end of a pregnancy, increasing estrogen induces labor.

▶ The **hormone** prostaglandin, an aging placenta, and the state of the fetus itself also play a role in birth. At the same time, a hormone called oxytocin stimulates the contractions of the uterus that will expel the baby.

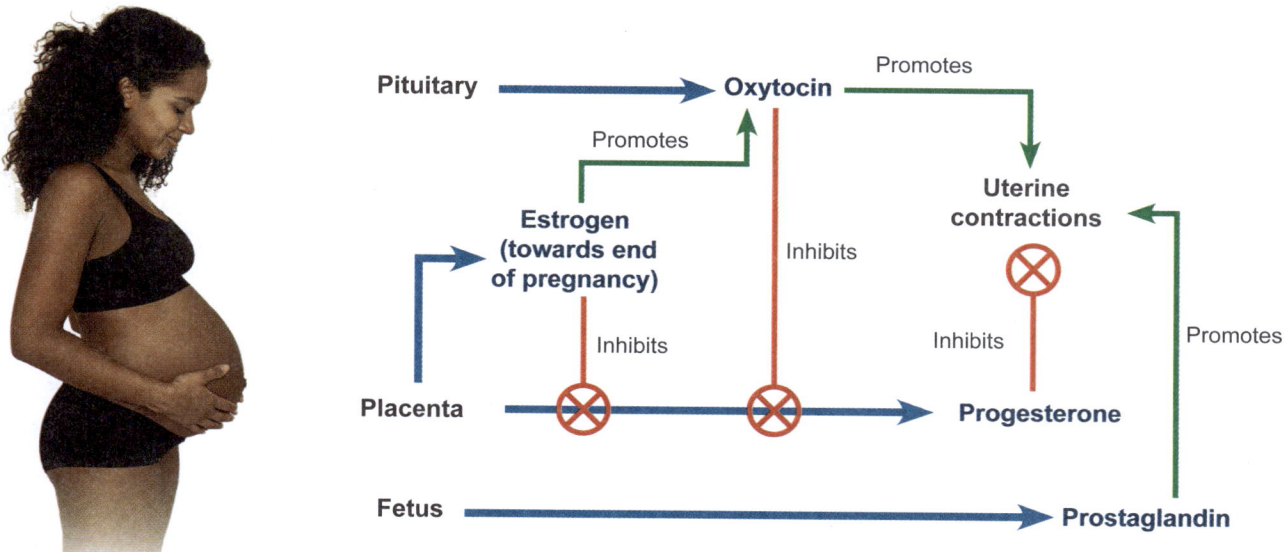

▶ During the nine months of pregnancy, the placenta produces progesterone which inhibits uterine contractions. At the end of the pregnancy, the breakdown of the placenta results in a fall in the level of progesterone, which enables uterine contractions.

▶ The fetus also begins to produce prostaglandin, which further promotes uterine contractions. Uterine contractions produce a positive feedback loop that ends with the birth of the baby.

1. The umbilical cord contains the fetal arteries and vein. Describe the status of the blood in each type of fetal vessel as either: *Oxygenated and containing nutrients* or *Deoxygenated and containing nitrogenous wastes*:

    (a) Fetal arteries: _____

    (b) Fetal vein: _____

2. Describe the structure of the human placenta and explain how it interacts with the mother's circulatory system:

    _____

3. Analyze the hormonal interactions that control pregnancy and birth: _____

4. Why does an increase in estrogen trigger birth? _____

85 The Immune System

Key Question: What is the structure of the immune system?

- The immune system is made up from many diverse parts of the body, including parts of the blood, the skin, the spleen, and lymphatic vessels.
- Its function is to protect the body from pathogens, such as bacteria or viruses. It has a range of physical, chemical, and biological defenses that provide resistance against these pathogens.
- The immune system can be split into the **innate** (always ready/nonspecific) defense system and the **adaptive** (specific) defense system (right). These can be further split into cellular or humoral (fluid based) systems.

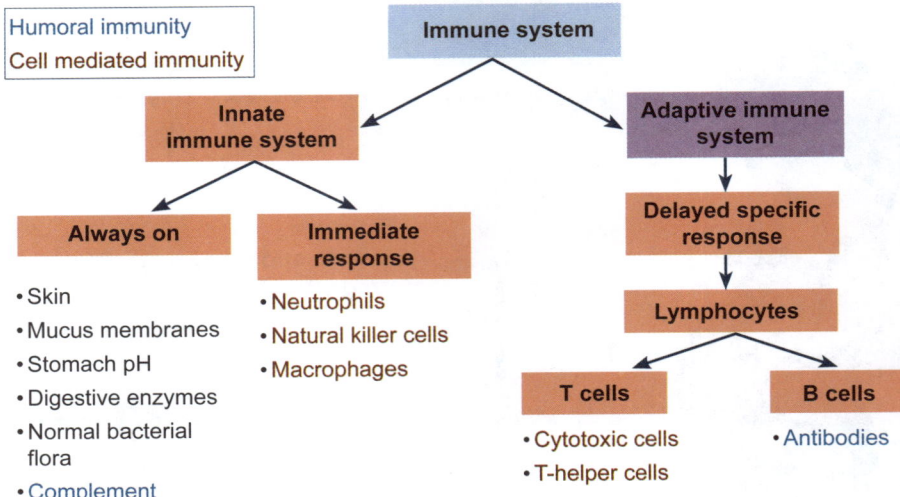

Humoral immunity
Cell mediated immunity

Immune system
- Innate immune system
  - Always on
    - Skin
    - Mucus membranes
    - Stomach pH
    - Digestive enzymes
    - Normal bacterial flora
    - Complement system
  - Immediate response
    - Neutrophils
    - Natural killer cells
    - Macrophages
- Adaptive immune system
  - Delayed specific response
  - Lymphocytes
    - T cells
      - Cytotoxic cells
      - T-helper cells
    - B cells
      - Antibodies

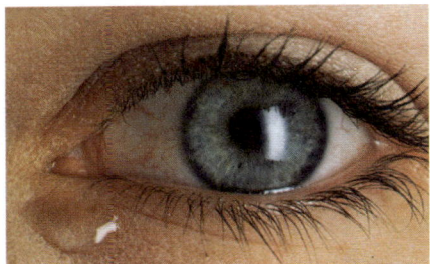

Some parts of the immune system are always on because they also carry out functions for other organs systems. Digestive enzymes and stomach acids destroy pathogens, but also help in digestion of food. Tears contain proteins that defend against pathogens, but also lubricate the eyes.

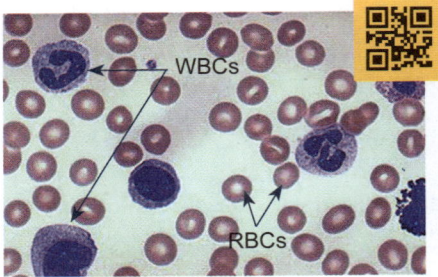

White blood cells (leukocytes) are part of the cell mediated immunity and are produce by the bone marrow. White blood cells can be divided in lymphoid and myeloid cells. Lymphoid cells include T-cells, B-cells and natural killer cells. Myeloid cells include monocytes and neutrophils, which are phagocytic.

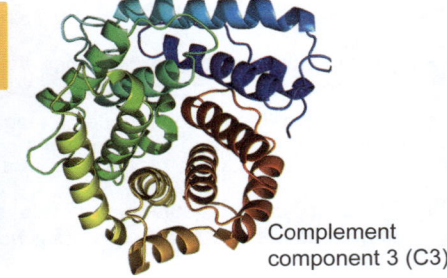

Complement component 3 (C3)

The complement system comprises a number of different proteins. The proteins circulate as inactive precursors until they are activated. Complement proteins have three main roles: phagocytosis, attracting macrophages and neutrophils to the infection site, and rupturing the membranes of foreign cells.

1. Describe the cellular and humoral systems: _____

2. Explain how "always on" parts of the immune system help defend against pathogens: _____

3. What are the differences between the innate and adaptive immune responses? _____

86 The Body's Defenses: A Layered Structure

Key Question: How does a layered defense system provide resistance against disease?

Layers of defense

▶ The human body has a range of physical, chemical, and biological defenses that provide resistance against pathogens. The first line of defense consists of external barriers to prevent pathogens entering the body. If this fails, a second line of defense targets any foreign bodies that enter. Lastly, the specific immune response provides a targeted third line of defense against the pathogen.

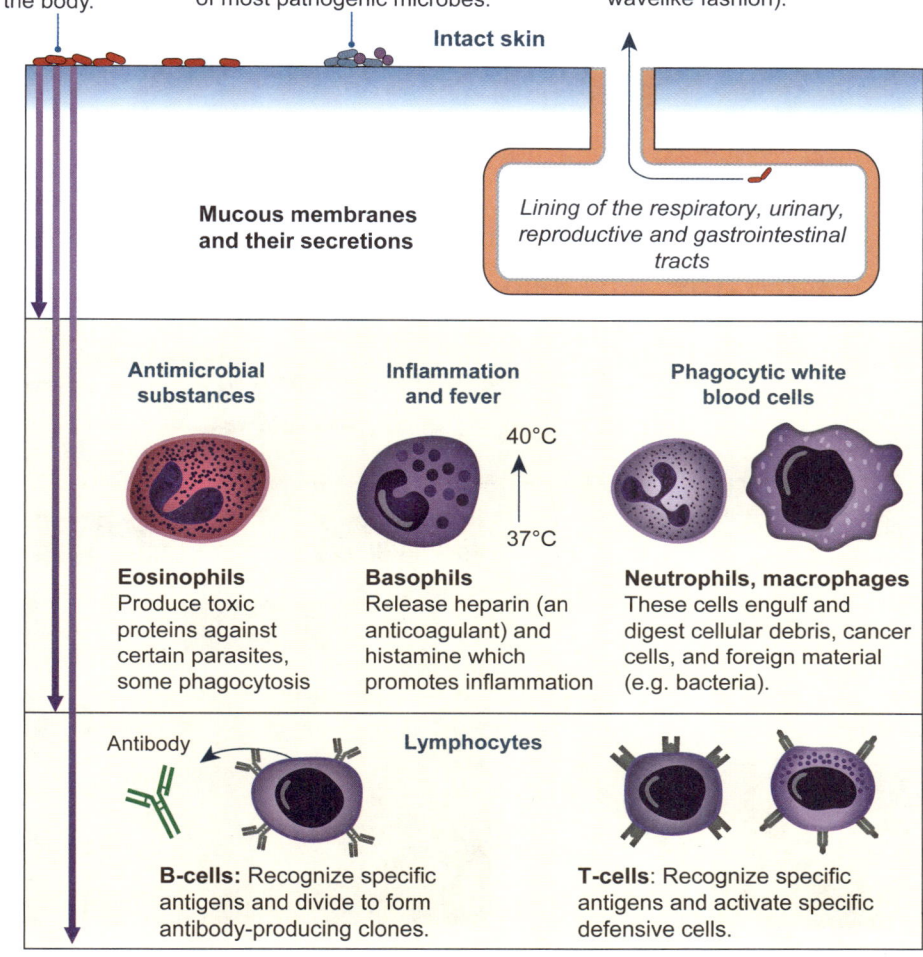

1st line of defense
The skin provides a physical barrier to the entry of pathogens. Its low pH is unfavorable to the growth of many bacteria and its chemical secretions, e.g. sebum, antimicrobial peptides, inhibit growth of bacteria and fungi. Tears, mucus, and saliva also help to wash bacteria away.

2nd line of defense
A range of defense mechanisms operate inside the body to inhibit or destroy pathogens. These responses are nonspecific. White blood **cells** are involved in most of these responses. The 2nd line of defense includes the complement system, whereby blood plasma proteins work together to bind pathogens and induce inflammation to help fight infection.

3rd line of defense
Once the pathogen has been identified by the immune system, lymphocytes (specialized white blood cells) produce specific responses to the pathogen, including the production of defensive proteins called **antibodies**, which are specific against a particular **antigen**.

1. Describe how the immune system has a layered structure: _____

_____

_____

_____

_____

2. How does having a layered defense help protect an organism from a pathogen? _____

_____

_____

_____

87 Blood Clotting and Defense

Key Question: How does blood clotting occur and how does it help repair injury and prevent further infection?

- Blood has a role in the body's defense against infection. Tearing or puncturing of a blood vessel initiates blood clotting, which quickly seals off the tear, preventing blood loss and the invasion of bacteria into the site.
- Clot formation is triggered by the release of clotting factors from the damaged **cells** at the site of the damage. A hardened clot forms a scab, which acts to prevent further blood loss and also as a mechanical barrier to pathogen entry.

1. Injury to the lining of a blood vessel exposes collagen fibers to the blood. Platelets stick to the collagen fibers.

3. As the platelets clump together, more chemicals are released, accelerating the clot formation (positive feedback).

When tissue is wounded, the blood quickly coagulates to prevent further blood loss and maintain the integrity of the circulatory system. For external wounds, clotting also prevents the entry of pathogens. Blood clotting involves a cascade of reactions involving many clotting factors in the blood.

4. A fibrin clot reinforces the seal. The clot traps blood cells and the positive feedback loop ends. The clot eventually dries to form a scab.

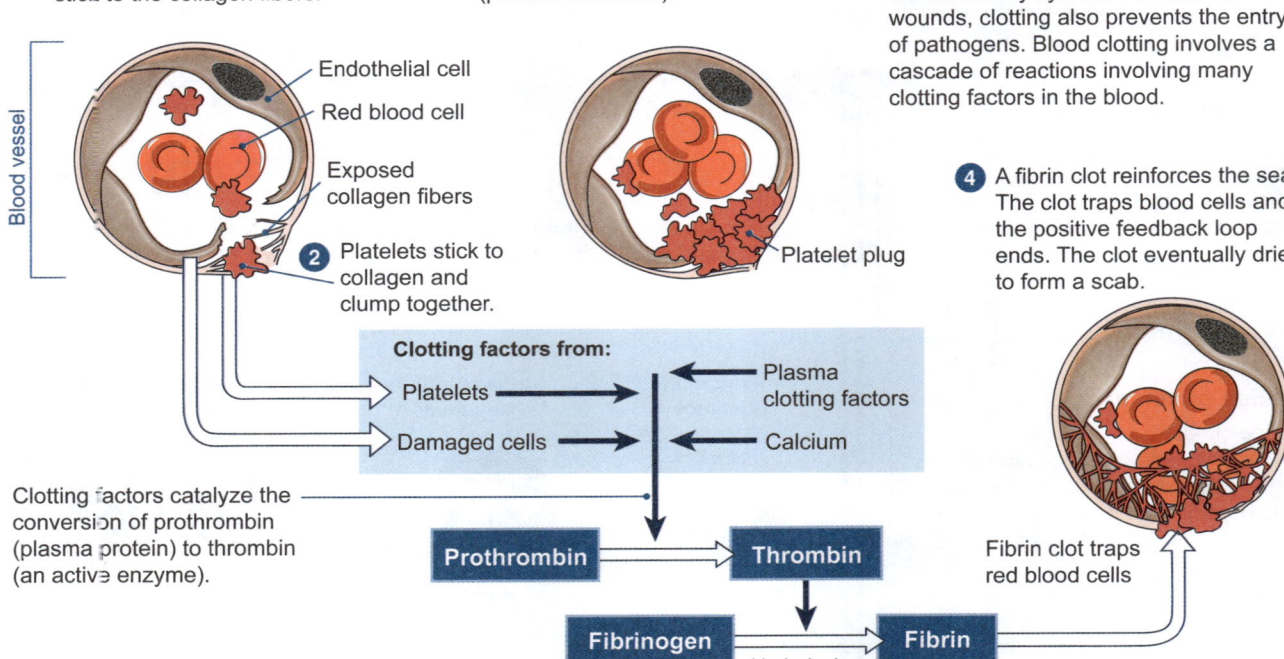

Clotting factors catalyze the conversion of prothrombin (plasma protein) to thrombin (an active enzyme).

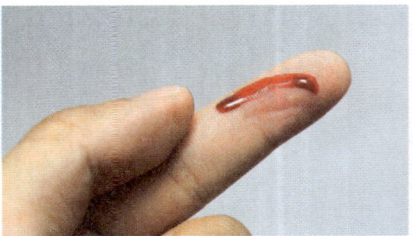

Bleeding helps clean a wound. Blood is under pressure (even in veins) and this helps force any foreign objects out of the wound and prevents further entry.

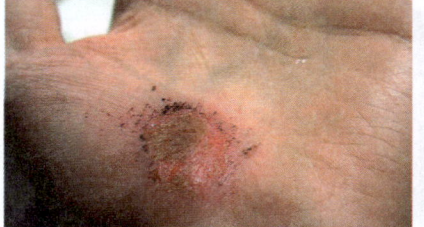

Blood clots harden and form scabs that act as a temporary barrier to prevent further infection while the skin and blood vessels underneath are repaired.

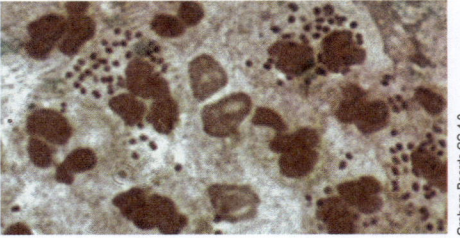

Blood clotting helps with the immune response by trapping bacteria and foreign objects, reducing their spread and helping their destruction by white blood cells.

1. What role does bleeding and blood clotting have in internal defense? _____

2. Give a brief description of the interactions that occur to produce a blood clot: _____

88 Interacting Systems: Responding to Infection

Key Question: What organ systems interact to prevent and respond to infection?

▸ Damage to the body's **tissues**, e.g. by sharp objects, heat, or microbial infection, triggers a defensive response called inflammation. It is usually characterized by four symptoms: pain, redness, heat, and swelling.
▸ The circulatory system will form blood clots if required, and transports cells of the immune system to the site of damage to respond to any infection (below).

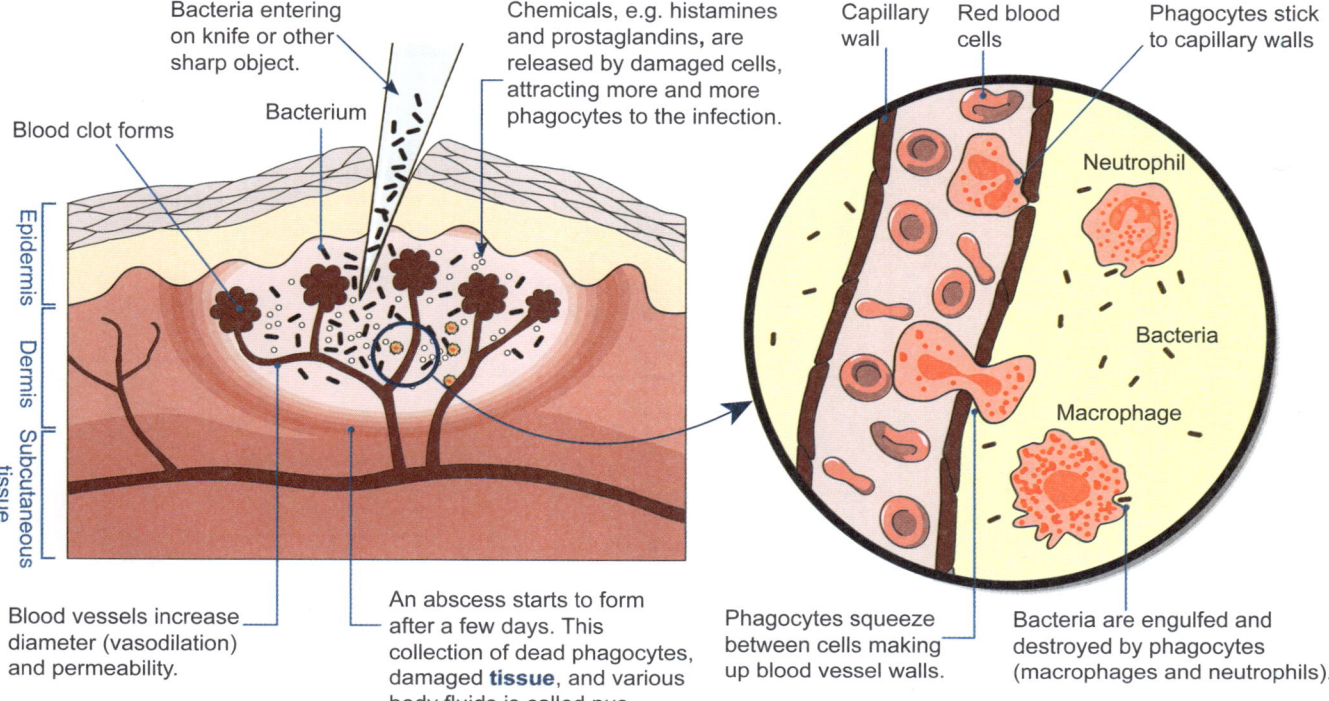

Stages in the inflammatory response

Increased diameter and permeability of blood vessels	**Phagocyte migration and phagocytosis**	**Tissue repair**
Blood vessels increase their diameter and permeability in the area of damage. This increases blood flow to the area and allows defensive substances to leak into tissue spaces.	Within one hour of injury, phagocytes appear on the scene. They squeeze between cells of blood vessel walls to reach the damaged area, where they destroy invading microbes.	Functioning cells or supporting connective cells create new tissue to replace dead or damaged cells. Some tissue regenerates easily, (skin) while others do not at all (cardiac muscle).

1. Outline the stages of inflammation and identify any interactions between systems that occur: _____
   _____
   _____
   _____
   _____
   _____

2. Identify two features of phagocytes important in the response to microbial invasion: _____
   _____
   _____

Fever

▶ Fever is defined as an increase in body temperature above the normal range (36.2 - 37.2°C). To a point, fever is beneficial, because it assists a number of the defense processes. The release of the protein interleukin-1 helps to reset the thermostat of the body to a higher level and increases production of T cells.

▶ High body temperature also enhances the effect of interferon (an antiviral protein) and may inhibit the growth of some bacteria and viruses. Because high temperatures speed up the body's metabolic reactions, it may promote more rapid tissue repair. Fever also increases heart rate so that white blood cells are delivered to sites of infection more rapidly.

1. Pathogen or toxin
The most frequent cause of fever is infection from bacteria (and their toxins) and viruses. A macrophage ingesting one of these will start the fever-causing process.

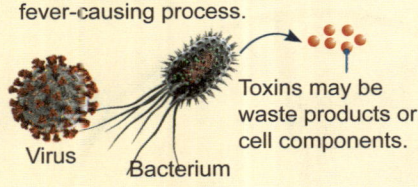

Toxins may be waste products or cell components.
Virus
Bacterium

2. Macrophages respond
Macrophage ingests bacterium, destroying it and releasing endotoxins. The endotoxins induce the macrophage to produce the protein interleukin-1.

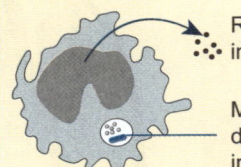

Release of interleukin-1

Macrophage digests bacterium in vacuole.

3. Thermostat is reset
Interleukin-1 circulates to the brain and induces the hypothalamus to produce more prostaglandins, which resets the 'thermostat' to a higher temperature, producing fever.

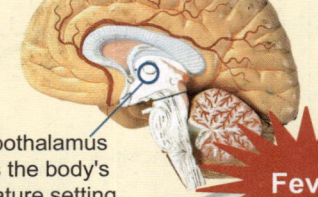

The hypothalamus controls the body's temperature setting.

Fever
38.5°C

6. Crisis phase
Body temperature is maintained at the higher setting until the interleukin-1 has been eliminated. As the infection subsides, the thermostat is reset to 37°C. Heat losing mechanisms, such as sweating and vasodilation cause the person to feel warm. This crisis phase of the fever indicates that body temperature is falling.

5. Chill phase
Although body temperature is elevated above normal, the skin remains cold, and shivering increases. This condition, called a chill, is a definite sign that body temperature is rising. When the body reaches the setting of the thermostat, the chill disappears.

We are all familiar with the feeling of chill associated with a fever and the sweating that precedes the fever's resolution. Fevers of less than 40°C do not need treatment but high, prolonged fevers require prompt attention, because death usually results if the body temperature rises above 45.5°C.

4. Fever onset
The body responds to a higher temperature set point by constricting blood vessels, increasing metabolic rate, and shivering. These responses raise body temperature beyond the normal range of 36.2 – 37.2 °C.

3. Which systems are involved in fever? Analyze and explain their interactions: _____

_____
_____
_____
_____
_____

4. The immune system is able to remember different pathogens via the production of memory B-cells. What is the advantage of this?

_____

89 Plant Organ Systems

Key Question: What are the different parts of the plant organ system?

▸ Plants have fewer **organ systems** than animals because they are simpler and have lower energy demands. The two primary plant organ systems are the shoot system and the root system.

Shoot system
The above-ground parts of the plant: including organs such as leaves, buds, stems, and the flowers and fruit (or cones) if present. All parts of the shoot system produce **hormones**.

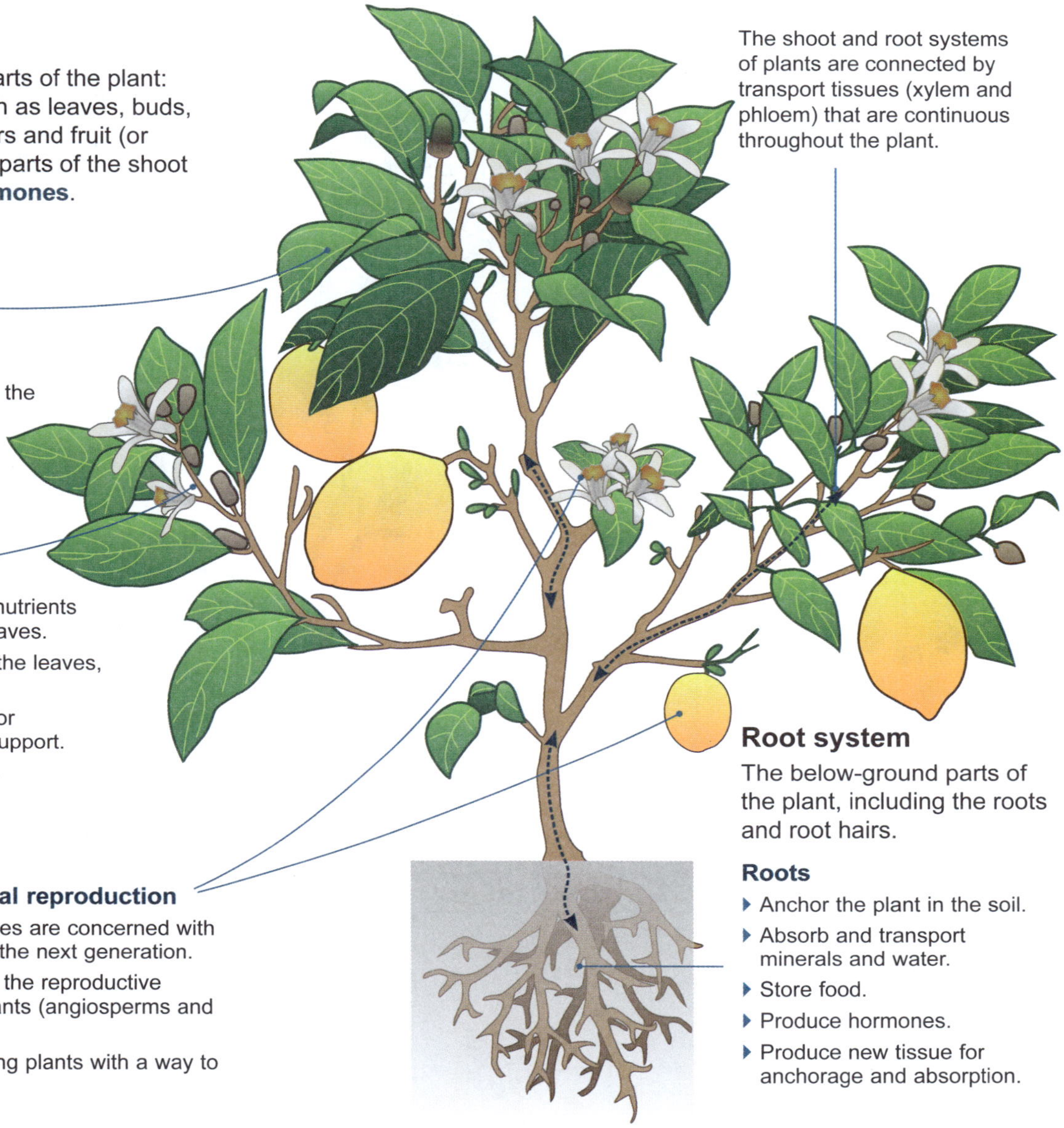

The shoot and root systems of plants are connected by transport tissues (xylem and phloem) that are continuous throughout the plant.

Leaves
▸ Manufacture food via photosynthesis.
▸ Exchange gases with the environment.
▸ Store food and water.

Stems
▸ Transport water and nutrients between roots and leaves.
▸ Support and hold up the leaves, flowers and fruit.
▸ Produce new tissue for photosynthesis and support.
▸ Store food and water.

Root system
The below-ground parts of the plant, including the roots and root hairs.

Roots
▸ Anchor the plant in the soil.
▸ Absorb and transport minerals and water.
▸ Store food.
▸ Produce hormones.
▸ Produce new tissue for anchorage and absorption.

Structures for sexual reproduction
▸ Reproductive structures are concerned with passing on **genes** to the next generation.
▸ Flowers or cones are the reproductive structures of seed plants (angiosperms and gymnosperms).
▸ Fruits provide flowering plants with a way to disperse the seeds.

1. Describe how each of the following systems provides for the essential functions of life for the plant:

   (a) Root system: _____

   _____

   (b) Shoot system: _____

   _____

2. Place the following list of plant functions into the correct boxes below: *Photosynthesis, transport, absorption, anchorage, storage, sexual reproduction, hormone production, growth*

Functions shared by the root and shoot system	Functions unique to the shoot system	Functions unique to the root system

90 Interacting Systems in Plants

Key Question: How do the shoot and root systems of plants interact to balance water uptake and loss, so that the plant can maintain the essential functions of life?

Loss of water vapor (H_2O) is a consequence of gas exchange. The plant can reduce water loss by closing the stomata, but this also stops photosynthesis because the carbon dioxide (CO_2) cannot enter the leaf. If the plant cannot replace the water it loses it will wilt and die.

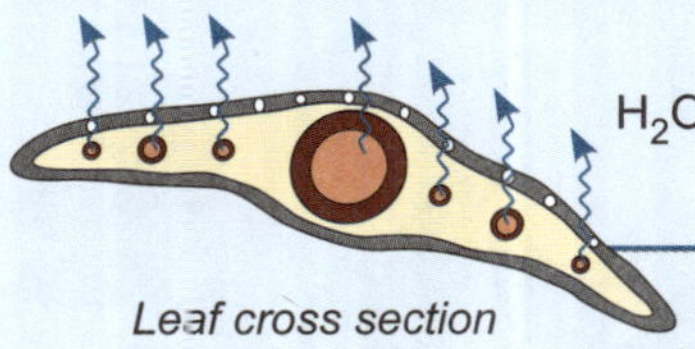

Leaf cross section

The evaporation of water from the leaves (**transpiration**) draws water up from the roots.

Oxygen (and water vapor) are produced by cellular respiration.

Carbon dioxide is needed for photosynthesis.

The plant exchanges gases with the environment by **diffusion** through pores in the leaf called stomata.

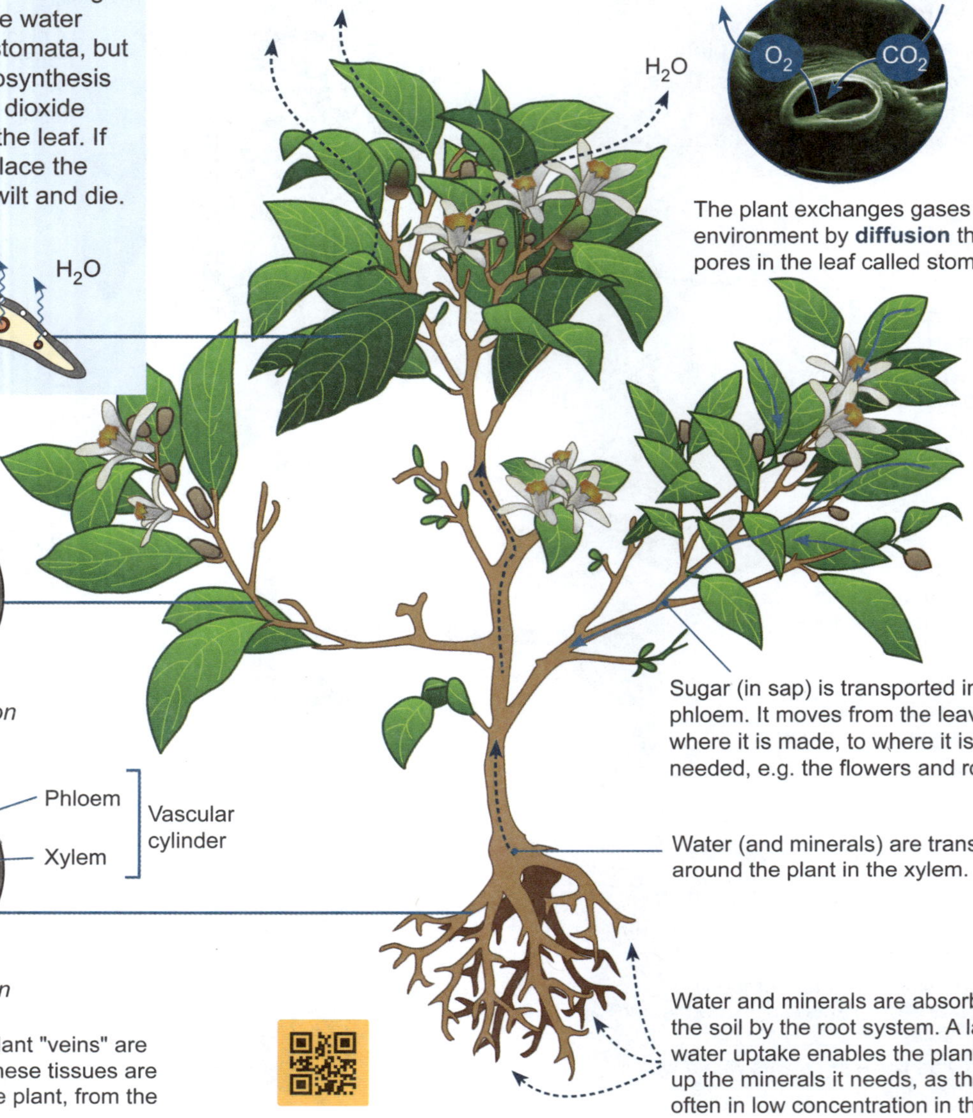

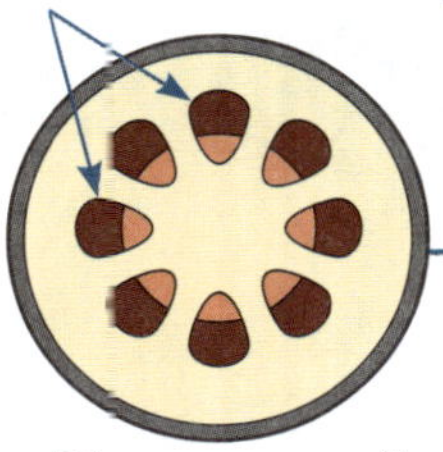

Stem cross section — Vascular bundles

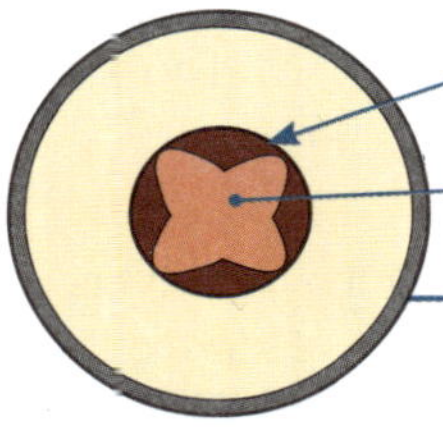

Root cross section — Phloem, Xylem, Vascular cylinder

Sugar (in sap) is transported in the phloem. It moves from the leaves, where it is made, to where it is needed, e.g. the flowers and roots.

Water (and minerals) are transported around the plant in the xylem.

Water and minerals are absorbed from the soil by the root system. A large water uptake enables the plant to take up the minerals it needs, as these are often in low concentration in the soil.

The vascular tissues or plant "veins" are the phloem and xylem. These tissues are continuous throughout the plant, from the roots to the shoots.

1. In what way are the shoot and root system connected? _____
   _____

2. (a) How do gases enter the shoot system? _____

   (b) How does gas exchange affect a plant's water balance? _____
   _____

3. How do the root and shoot systems work together to maintain water balance? _____
   _____
   _____
   _____

- The angiosperms (flowering plants) are commonly divided into two groups, the monocots (plants that produce seeds with one embryonic leaf), and the dicots (plants that produce seeds with two embryonic leaves).
- The arrangement of the vascular tissue is quite different between these two groups. In the stem this difference can be clearly seen. The vascular bundles in dicots are arranged around the periphery of the stem, while in monocots they appear more scattered. These arrangements can best be seen by making slides of herbaceous plants for viewing under a microscope.

cherry tree: dicot

Tulip: monocot

Investigation 4.3 Investigating vascular tissue

See appendix for equipment list.

1. Your teacher may provide you with prepared slides or you may need to make them yourself. Refer to investigation 1.2 if you need to make your own slides. You will also be provided with two unknown prepared slides. One will be a monocot and one will be a dicot. They will be slides showing vascular tissue in either stems or roots.

2. You will need to prepare slides with transverse sections (cut across the stem or root) of stems and roots or use the slides provided by your teacher. Useful dicot plants for this are buttercups and sunflowers and useful monocot plants are corn or maize.

3. Go to the **BIOZONE Resource Hub** and look at the 4 'Interacting Systems in Plants' images:
   Angiosperm morphology: monocotyledonous roots,
   Angiosperm morphology: monocotyledonous stems,
   Angiosperm morphology: herbaceous dicotyledonous roots,
   Angiosperm morphology: herbaceous dicotyledonous stems.
   These links have many high quality images. They will help you identify the vascular tissue in your slides.

   bit.ly/3L13n7Z

4. Place a slide on the microscope stage and focus on low power, and then on high power. In the space below record whether the specimen is a dicot or monocot, stem or root, and write a brief description of or draw the arrangement of the vascular tissue. Use extra paper if needed. Include a scale for your diagrams.

5. Repeat for all the plant roots and stems available.

6. Identify the unknown slides provided by your teacher. For each slide state whether it is a monocot or dicot, a root or stem slide and the reasons for your decision.

   Slide 1: _____

   Slide 2: _____

91 Xylem and Phloem

Key Question: What is the structure of the vascular tissue in plants?

▸ Xylem and phloem are the main supporting tissue in plants. Both are complex tissues composed of a number of cell types. Xylem tissue is largely composed of large vessels, which have thickened and strengthened walls and conduct water. These are made of cells called vessel elements and tracheids (diagram below). Xylem also contains packing cells (parenchyma) and fibers, which support the tissue. Xylem is dead when mature.

▸ Phloem tissue is composed of packing cells and supporting fiber cells, and two special cell types: sieve tubes and companion cells. Phloem transports dissolved sugars, and is alive when mature.

Xylem tissue

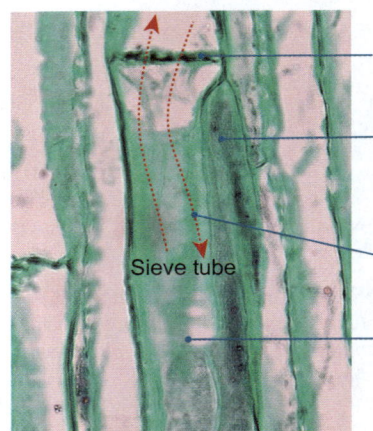

- Xylem vessels form continuous tubes throughout the plant.
- Spiral thickening around the walls of the vessels give extra strength allowing the vessels to remain rigid and upright.
- Mature xylem vessels are dead and the cytoplasm has gone.

Conducting cells of xylem

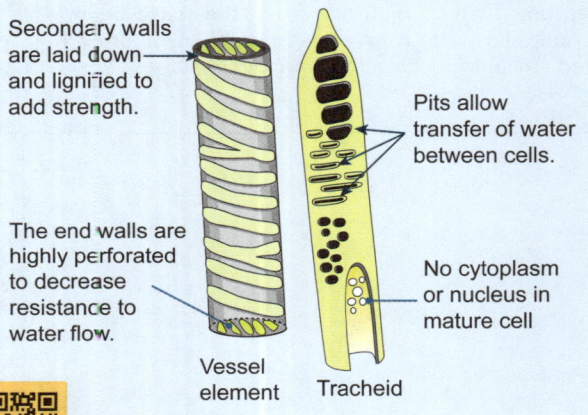

- Secondary walls are laid down and lignified to add strength.
- The end walls are highly perforated to decrease resistance to water flow.
- Vessel element
- Tracheid
- Pits allow transfer of water between cells.
- No cytoplasm or nucleus in mature cell

Phloem tissue

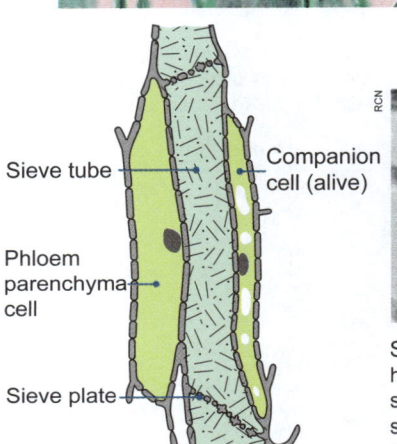

- Sieve tube end plate
- Companion cell: a cell adjacent to the sieve tube cell, responsible for keeping the phloem cell alive.
- Sugar solution flows in both directions
- Sieve tubes lose most of their organelles but the tissue is alive when mature.
- Sieve tube

Phloem cells

- Sieve tube
- Phloem parenchyma cell
- Sieve plate
- Companion cell (alive)

Cross-section through sieve tube end plate

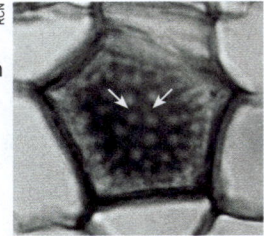

Sieve tube end plate: Small holes (arrows) perforate the sieve tube cells allowing the sugar solution to pass through.

---

1. Is xylem dead or alive when mature? _____

2. Is phloem dead or alive when mature? _____

3. What is the purpose of the holes in the sieve plate at the ends of each sieve tube cell? _____

4. (a) Name the cell type in the phloem that actually conducts the sugar solution: _____

   (b) What is the purpose of the companion cell in phloem tissue? _____

5. Describe the structural and functional differences between xylem and phloem: _____

92 Stem and Root Structure

Key Question: What is the structure of the vascular tissue in the stems and roots of plants?

- The structure of the vascular **tissue** in dicotyledons (dicots) has a very regular arrangement, with xylem and phloem found close together. In the stem, the vascular tissue is distributed in a regular fashion near the outer edge of the stem. In the roots, the vascular tissue is found near the center of the root.

Dicot stem structure

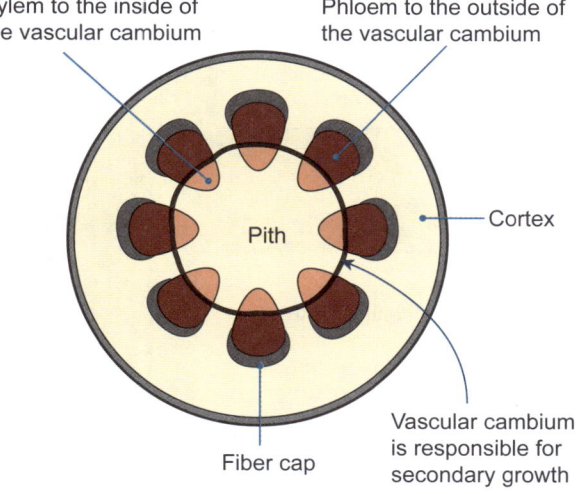

In dicots, the vascular bundles (xylem and phloem) are arranged in an orderly fashion around the stem. Each vascular bundle contains xylem (to the inside) and phloem (to the outside). Between the phloem and the xylem is the vascular cambium. This is a layer of **cells** that divide to produce the thickening of the stem.

Dicot root structure

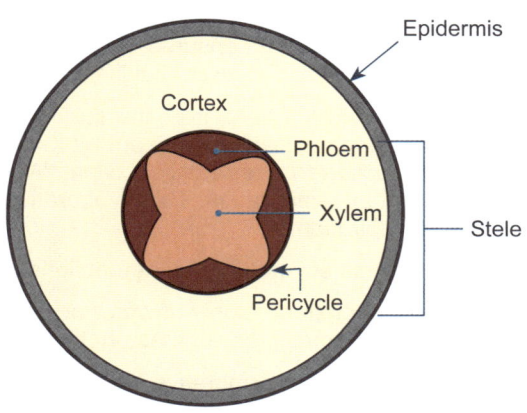

In a dicot root, the vascular tissue, (xylem and phloem) forms a central cylinder through the root called the stele. The large cortex is made up of parenchyma (packing) cells, which store starch and other substances. Air spaces between the cells are essential for aeration of the root tissue, which is non-photosynthetic.

1. In the stem micrograph below Identify the structures labeled A, B, and C:

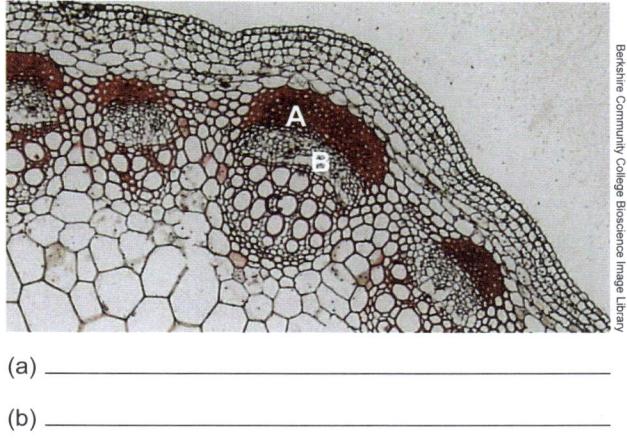

(a) _____

(b) _____

(c) _____

2. In the root micrograph below Identify the structures labeled A and B:

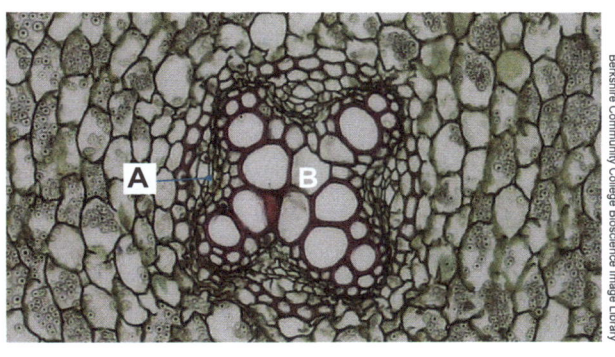

(a) _____

(b) _____

3. Describe the differences in the structure of the vascular tissue in stems and roots: _____

_____

_____

_____

4. What is the role of the vascular cambium? _____

_____

93 Transpiration

Key Question: How does the process of transpiration help maintain water homeostasis in plants?

Maintaining water balance

- Like animals, plants need water for life processes. Water gives **cells** turgor (rigidity from pressure of liquid), transports dissolved substances, and is a medium in which metabolic reactions can take place. Maintaining water balance is an important **homeostatic** function in plants. Evaporative water loss from stomata drives a **transpiration** stream that ensures plants have a constant supply of water to support essential life processes.

- Vascular plants obtain water from the soil. Water enters the plant via the roots, and is transported throughout the plant by a specialized **tissue** called xylem. Water is lost from the plant by evaporation. This evaporative water loss is called transpiration.

- Transpiration has several important functions:
  - Provides a constant supply of water needed for essential life processes such as photosynthesis.
  - Cools the plant by evaporative water loss.
  - Helps the plant take up minerals from the soil.

- However, if too much water is lost by transpiration, a plant will become dehydrated and may die.

The role of stomata

- Water loss occurs mainly through stomata (pores in the leaf). The rate of water loss can be regulated by specialized guard cells either side of the stoma, which open or close the pore.
  - Stomata open: transpiration rate increases.
  - Stomata closed: transpiration rates decrease.

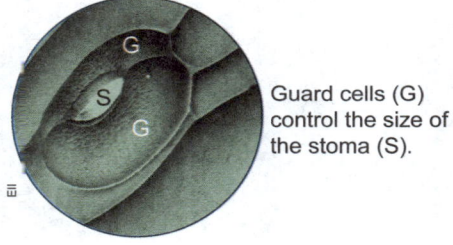

Guard cells (G) control the size of the stoma (S).

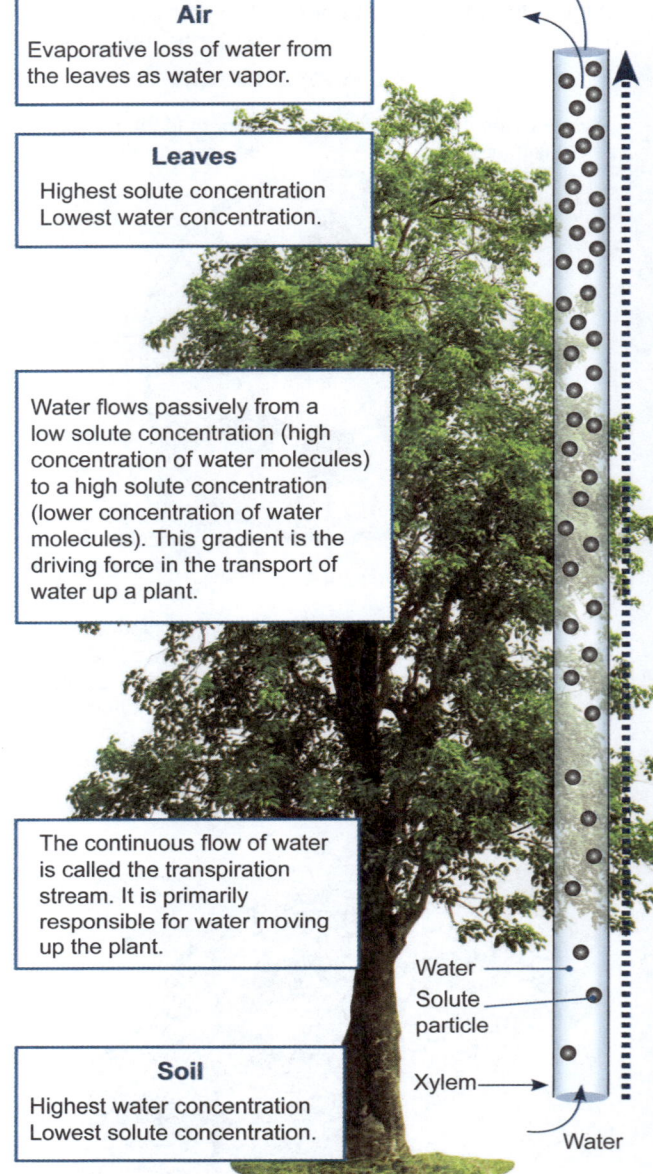

Air
Evaporative loss of water from the leaves as water vapor.

Leaves
Highest solute concentration
Lowest water concentration.

Water flows passively from a low solute concentration (high concentration of water molecules) to a high solute concentration (lower concentration of water molecules). This gradient is the driving force in the transport of water up a plant.

The continuous flow of water is called the transpiration stream. It is primarily responsible for water moving up the plant.

Soil
Highest water concentration
Lowest solute concentration.

Water
Solute particle
Xylem
Water

1. (a) What is transpiration? _____

    _____

   (b) How does transpiration provide water for essential life processes in plants? _____

    _____

    _____

    _____

2. How does the plant regulate the amount of water lost from the leaves? _____

    _____

    _____

Processes involved in moving water through the xylem

1. Transpiration pull
Water is lost from the air spaces by evaporation through stomata and is replaced by water from other cells. The constant loss of water to the air (and production of sugars) creates a solute concentration in the leaves that is higher than elsewhere in the plant. Water is pulled through the plant along a gradient of increasing solute concentration.

2. Cohesion-tension
The transpiration pull is assisted by a property of water called **cohesion**. Water molecules cling together as they are pulled through the plant. They also adhere to the walls of the xylem (**adhesion**). This creates one unbroken column of water through the plant. The upward pull on the cohesive sap creates a tension that helps water uptake and movement up the plant.

3. Root pressure
Water entering the vascular tissue from the soil creates a root pressure; a weak 'push' effect for the water's upward movement through the plant. Root pressure can force water droplets out of the leaves of some small plants under certain conditions, but generally it plays a minor part in the upwards movement of the water.

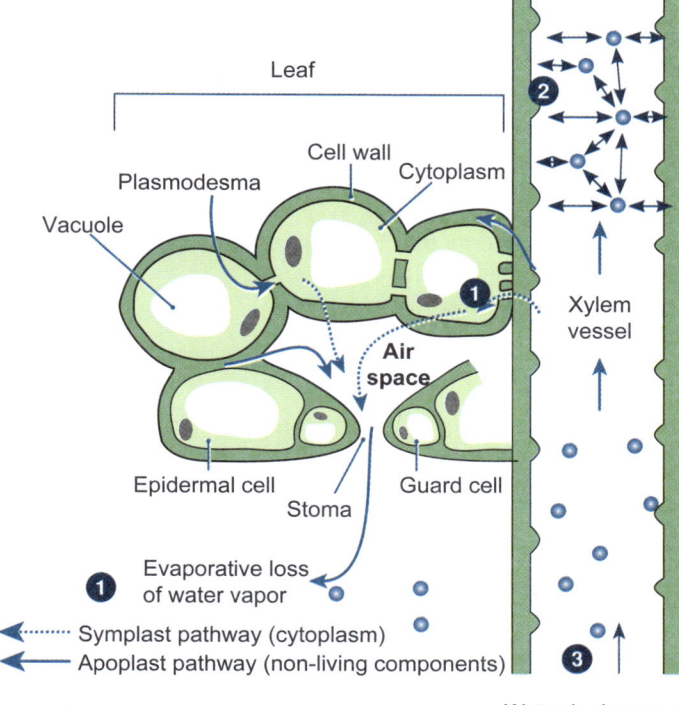

3. (a) What would happen if too much water was lost from the leaves? _____

(b) When might this happen? _____

4. Describe the three processes that assist the transport of water from the roots of the plant upward:

(a) Transpiration pull: _____

(b) Cohesion-tension: _____

(c) Root pressure: _____

5. Explain how the structure of the xylem and leaf helps the plant perform the function of moving water up the plant stem:

94 Investigating Transpiration

Key Question: What effects do physical factors in the environment, such as humidity, temperature, light level, and air movement, have on transpiration rate in plants?

The potometer

▶ A potometer is a simple instrument for investigating **transpiration** rate (water loss per unit time). The equipment is simple to use and easy to obtain. A basic potometer, such as the one shown right, can easily be moved around so that transpiration rate can be measured under different environmental conditions.

▶ Some questions a potometer can help answer are:
- What is the effect of humidity or vapor pressure (high or low) on transpiration?
- What is the effect of temperature (high or low) on transpiration?
- What is the effect of air movement (still or windy) on transpiration?
- What is the effect of light level (high or low) on transpiration?

▶ It is also possible to compare the transpiration rates of plants with different adaptations e.g. comparing transpiration rates in plants with rolled leaves to rates in plants with broad leaves. If possible, experiments like these should be conducted simultaneously using replicate equipment. If conducted sequentially, care should be taken to keep the environmental conditions the same for all plants used.

A potometer attached to a data logger

Investigation 4.4   Investigating plant transpiration

1. Four different conditions that influence transpiration will be tested: room conditions (ambient), wind, bright light, and high humidity.

2. Before starting, your teacher will decide if your group is to test one of these conditions (and which one) and pool class data for all four.

3. Set up the potometer and plant as in the diagram. It is best if the plant leaves used are large and few (4-6 leaves) rather than small and many. Alternatively, the plant can be placed in a 250 mL conical flask with 200 mL of water and a thin layer of cooking oil floated on top. This is weighed before the experiment and then every 3 minutes (or as the experiment requires). The difference in mass in grams is equal to the volume of water transpired in mL. After setting up the potometer, let the apparatus equilibrate for 10 minutes, and then record the position of the air bubble in the pipette (or the mass of the equipment for the alternative method). This is time 0 and position 0.

4. The plant can now be exposed to one of the four conditions. Record results in Table 1.

5. For the <u>ambient environment</u> the equipment can be placed on the bench away from bright light or wind. Record the net movement of the bubble every 3 minutes for 30 minutes.

6. For the <u>high wind environment</u> the equipment can be placed on the bench in front of a fan set on a moderate speed (away from bright light). Record the net movement of the bubble every 3 minutes for 30 minutes.

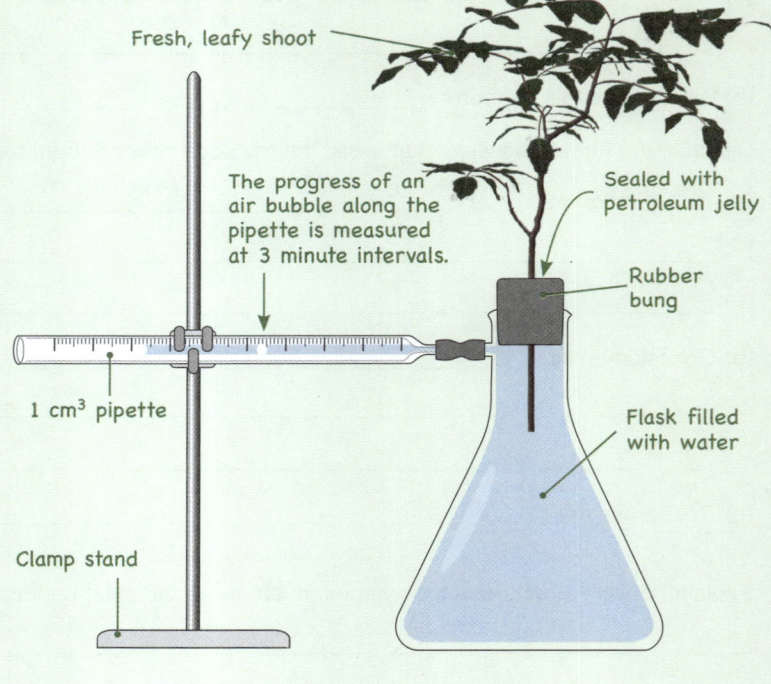

7. For the bright <u>light environment,</u> the equipment can be placed on the bench in front of a bright light (about 40 cm away). Record the net movement of the bubble every 3 minutes for 30 minutes.

8. For the <u>high humidity environment</u> the equipment can be placed on a bench away from bright light, in a sealed plastic bag with 2-3 sprays of water from a spray bottle. Record the net movement of the bubble every 3 minutes for 30 minutes.

9. It is important that for fair comparison of transpiration the area of leaf used in each environment (or by different groups) should be calculated and the volume of water lost per square centimeter compared (mL/cm^2).

10. Leaf area can be measured by tracing the leaves onto graph paper and counting the squares, or by tracing or photocopying the leaves onto a paper of a known mass per area, then cutting out the shapes and weighing them. **For both methods, multiply by 2 for both leaf surfaces**.

11. Once the area of the leaf is calculated, the transpiration (water lost) in mL/cm^2 can be calculated for each time recording and recorded in Table 2.

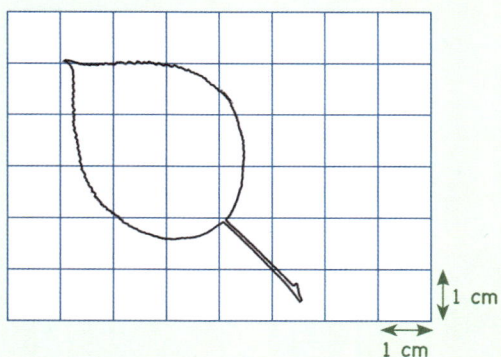

1. What question is your group trying to answer by doing this experiment? _____

Table 1. Potometer readings (in mL water loss)

Treatment \ Time (min)	0	3	6	9	12	15	18	21	24	27	30
Ambient											
Wind											
High humidity											
Bright light											

Table 2. Potometer readings in mL per cm^2

Treatment \ Time (min)	0	3	6	9	12	15	18	21	24	27	30
Ambient											
Wind											
High humidity											
Bright light											

2. Measure the area of the leaves you used: _____

3. Why is comparing water loss per square cm over time more important than just comparing the water loss over time?

4. Plot the data in Table 2 on the grid provided:

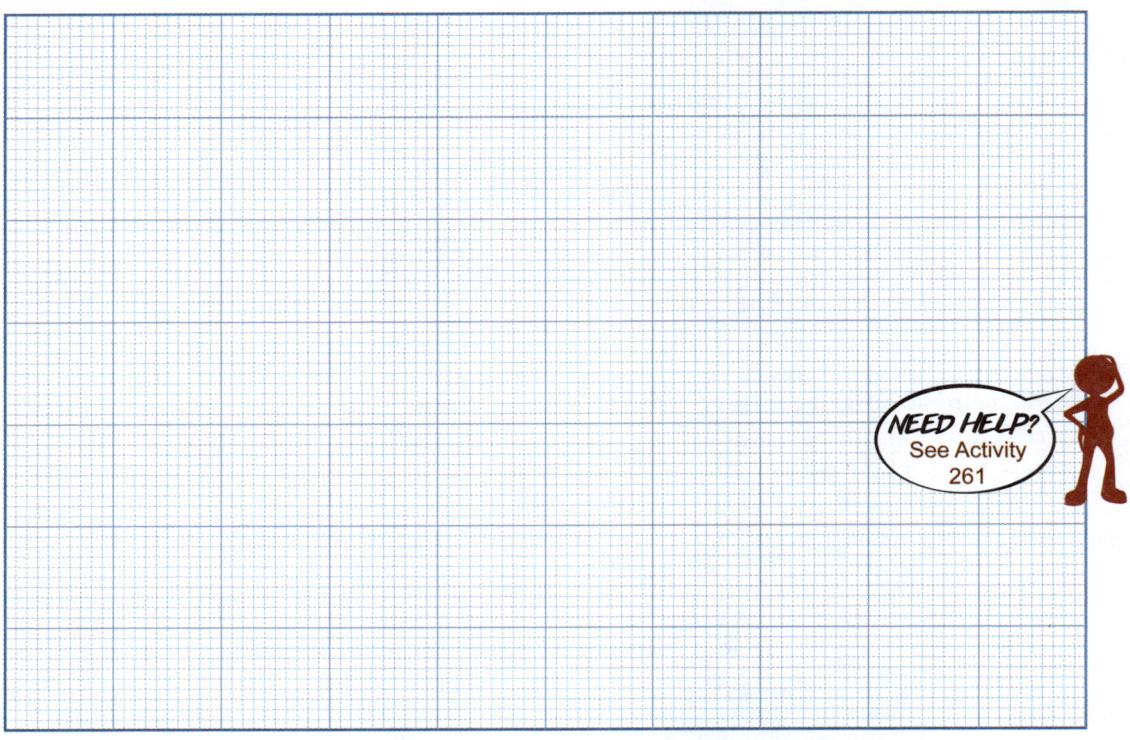

5. Identify the independent variable: _____

6. (a) Identify the control: _____

   (b) Explain the purpose of including an experimental control: _____

   _____

   _____

   _____

7. (a) Which factors increased water loss? _____

   (b) How does each environmental factor influence water loss? _____

   _____

   _____

   _____

   _____

   _____

8. From your results, predict how each of the following conditions might influence transpiration:

   (a) Low humidity, e.g. dry desert: _____

   (b) Low light levels, e.g. overcast day: _____

   (c) Hot dry winds: _____

9. How might different types of plants affect the results? _____

   _____

   _____

95 Uptake at the Roots

Key Question: How are water and minerals transported through the roots to the xylem of a plant?

- Plants constantly take up water, minerals, and ions to compensate for the continuous loss of water from the leaves by **transpiration** and to provide the materials they need for the manufacture of food. The uptake of water and minerals is mostly restricted to the younger, most recently formed **cells** of the roots and the root hairs.
- Water moves into the roots because the solute concentration is higher in the root **tissue** than in the soil. Water and ions are taken up by **diffusion** (**osmosis** in the case of water) and active transport.
- Some water moves through the plant tissues via cytoplasmic connections between cells (the plasmodesmata), but most passes through the free spaces outside the plasma membranes of the cells.
- Root hairs (right) provide a large surface area for absorption. They lack the waxy cuticle found on leaves so there is no barrier to water uptake.

Water and mineral uptake by roots

Water uptake
The uptake of water by roots occurs by osmosis. Water takes three paths through the root tissue:
- Most water travels through all the extra-cellular spaces outside the plasma membranes of cells, e.g. through the cell walls.
- A smaller amount moves through the cytoplasm of cells.
- A very small amount travels through the plant vacuoles.

Mineral uptake
- Many minerals are dissolved in water and absorbed passively by diffusion.
- Minerals in very low concentration in the soil must be taken up by active transport by the root cells. This requires energy expenditure by the plant.

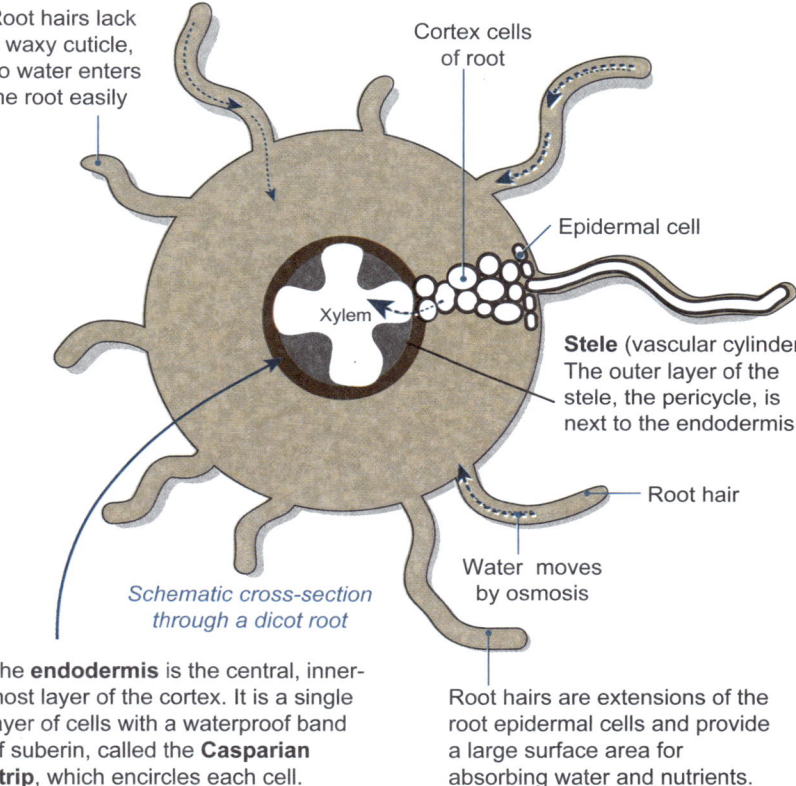

Schematic cross-section through a dicot root

The **endodermis** is the central, innermost layer of the cortex. It is a single layer of cells with a waterproof band of suberin, called the **Casparian strip**, which encircles each cell.

Root hairs are extensions of the root epidermal cells and provide a large surface area for absorbing water and nutrients.

1. (a) Why must plants constantly take up water from the soil? _____

    _____

    _____

   (b) Describe a benefit of a large water uptake: _____

    _____

    _____

    _____

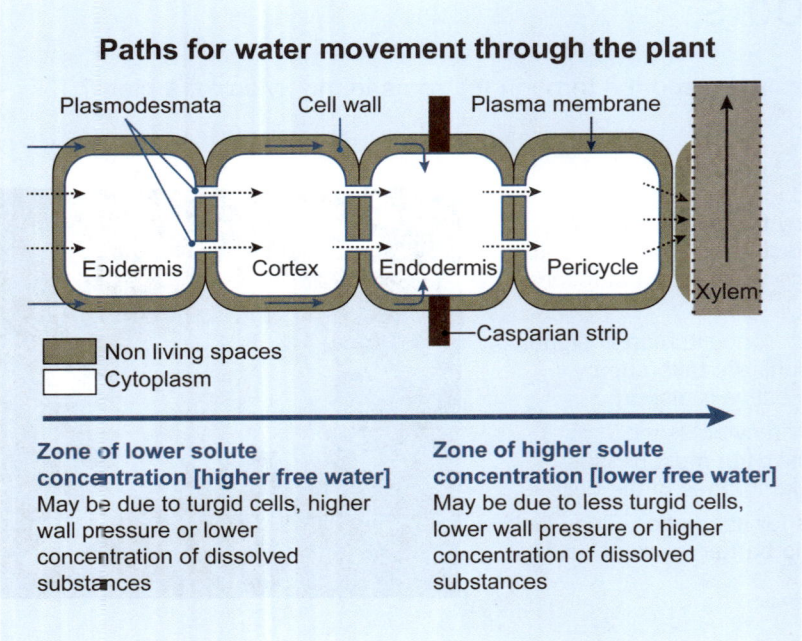

- The uptake of water through the roots occurs by osmosis, i.e. the diffusion of water from a lower solute concentration to a higher solute concentration. Most water travels through the non-living spaces, i.e. the spaces within the cellulose cell walls, the water-filled spaces of dead cells, and the hollow tubes of xylem vessels. A smaller amount moves through the living cytoplasm of cells. A very small amount travels through the plant vacuoles.

- Some dissolved mineral ions enter the root passively with water. Minerals that are in very low concentration in the soil are taken up by active transport. At the waterproof Casparian strip, water and dissolved minerals must pass into the cytoplasm, so the flow of materials into the stele can be regulated.

2. (a) What is the purpose of the root hairs? _____

    (b) What adaptations do root hairs show for their role? _____

3. (a) What is the root stele? _____

    (b) What is the waterproof barrier that surrounds the stele of the root? _____

4. What are plasmodemata and what is their role in the plant cell? _____

5. Describe how the structures of the root facilitate the movement of water and minerals from the soil to the xylem of the plant: _____

96 Translocation

Key Question: What model best explains the movement of the products of photosynthesis through the phloem of a plant?

Phloem transport

▸ In vascular plants, the products of photosynthesis move as phloem sap. Apart from water, phloem sap contains mainly sucrose (up to 30%). It may also contain minerals, **hormones**, and amino acids in transit around the plant. Movement of sap in the phloem is from a source (an **organ** where sugar is made or mobilized) to a sink (an organ where sugar is stored or used). The sap moves through the phloem sieve-tube members.

▸ Phloem sap moves from source to sink at rates as great as 100 m/h. The most acceptable model for phloem movement is the mass flow hypothesis (also know as the pressure flow hypothesis). Phloem sap moves by bulk flow, which creates a pressure (hence the term "pressure-flow"). The key elements in this model are outlined below and right. For simplicity, the cells that lie between the source (and sink) cells and the phloem sieve-tube have been omitted.

❶ Loading sugar into the phloem increases the solute concentration inside the sieve-tube cells. This causes the sieve-tubes to take up water by osmosis.

❷ The water uptake creates a hydrostatic pressure that forces the sap to move along the tube, just as pressure pushes water through a hose.

❸ The pressure gradient in the sieve tube is reinforced by the active unloading of sugar and consequent loss of water by osmosis at the sink, e.g. root cell.

❹ Xylem recycles the water from sink to source.

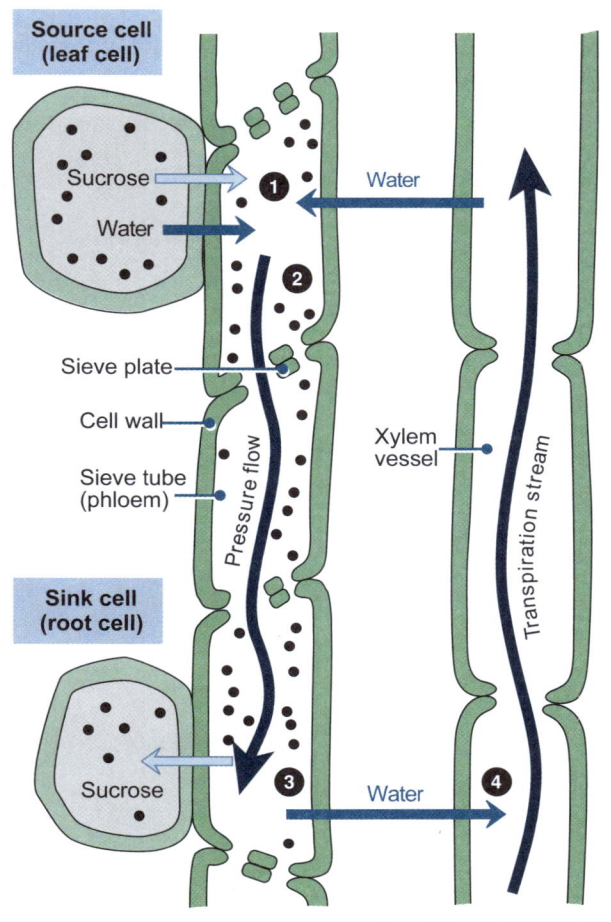

Source: Modified after Campbell Biology 1993

1. (a) From what you know about osmosis, explain why water follows the sugar as it moves through the phloem: _____

    (b) What is meant by 'source to sink' flow in phloem transport? _____

2. Why does a plant need to move food around, particularly from the leaves to other regions? _____

3. Describe the interactions between the phloem and xylem in the transport of plant sap: _____

97 Asexual Reproduction in Plants

Key Question: What structures do plants use to carry out asexual reproduction?

▸ Many flowering plants reproduce asexually by vegetative propagation. This is the process by which new plants arise from vegetative **tissues** of the parent plant. Vegetative propagation allows plants to spread rapidly in favorable conditions, avoiding the high energy cost of producing flowers, pollen, seeds, or fruits.

A bulb is really just a typical shoot compressed into a shortened form. Fleshy storage leaves are attached to a stem plate and form concentric circles around the growing tip. New roots form from the lower part of the stem. An onion is a bulb.

1. What is meant by vegetative propagation?
   _____
   _____
   _____
   _____

Tubers are the swollen part of an underground stem or root, usually modified for storing food. The potato (right) is a stem tuber, as indicated by the presence of terminal and lateral buds.

2. Describe the structures plants used for asexual reproduction:
   _____
   _____
   _____

In rhizomes, food is stored in the horizontal, underground stem. Rhizomes tend to be thick, fleshy or woody, and bear nodes with scale or foliage leaves and buds. Growth occurs at the buds on the ends of the rhizome or nearby nodes. Ginger, turmeric, irises, and lily-of-the-valley are rhizomes.

_____
_____
_____
_____
_____
_____

In a corm, food is stored in stem tissue. Corms look like bulbs, but if you cut a corm in half you see a mass of homogenous tissue rather than concentric rings of fleshy leaves as in a bulb. Cyclamen, gladiolus, and crocus (right) are corms.

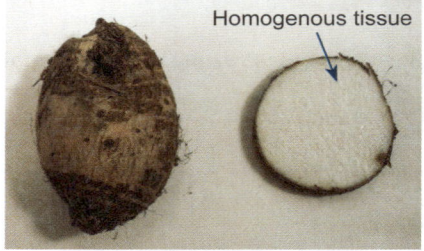

Homogenous tissue

3. Explain how these structures perform the function of reproduction:
   _____
   _____
   _____

Some plants produce copies of themselves (tiny plantlets called bulbils) from axillary buds. In time, these fall off as independent plants. Examples include the hen and chicken fern (*Asplenium bulbiferum*) (right) and kalanchoe (*Bryophyllum*).

Parent plant
Bulbils

_____
_____
_____
_____

Stolons (runners) are horizontal stems that grow above the ground. At certain points along the stolon, where it touches the ground, roots may form and a small plantlet appear. If the stolon breaks, the new plant becomes independent. Strawberry plants produce stolons.

_____
_____
_____

98 Investigating Plant Propagation

Key Question: Does the type of growing medium affect the ability of plants to grow from cuttings?

▸ Many plants can be grown from cuttings, although some plants are easier to propagate in this way than others. The growth medium can affect the cutting's ability to produce roots and start growing.

▸ In the experiment below, you will investigate the effect of different growing mediums on the production of roots from cuttings of the plant coleus (*Coleus scutellarioides*), which is common in garden centers.

Investigation 4.5 Plant propagation

See appendix for equipment list.

1. You will need nine containers of equal size or a nine chambered planter, three planting mediums: washed sand, commercial potting mix, fine bark (or three mediums provided by your teacher), rooting hormone, and nine ice block sticks.
2. You will need to take nine cuttings from the coleus plant provided (or the plant provided by your teacher). Each cutting should have the same number of leaves (2-3) that are roughly the same size. Stems should be cut to the same length (about 10cm).
3. Fill three planting containers with one type of planting medium (e.g. sand), three more with the second type, and the last three containers with the third type.
4. For each container, add water so that the medium is slightly damp. Add the same amount of water to each of the nine containers and allow it to drain through if necessary.
5. Place a small amount of rooting hormone in a container and dip the cut end of each cutting into it, to the same depth.
6. Carefully push one cut stem into each container so that it can stand up on its own. For each container, use the ice block stick to label the container with your name/group and the planting medium.
7. Place the cuttings in a sunny position for several weeks, continuing to water the plants daily with the same amount of water in each container.
8. After 4-5 weeks, carefully remove one of the cuttings from a container. Be careful not to damage the roots. Carefully wash the planting medium off the cutting.
9. Count the number of roots emerging from the stem and record in the table below. Measure the length of each root and record in the table below.
10. Calculate the median number of roots and the median root length for each planting medium.

Sabina Bajracharya CC 4.0

NEED HELP? See Activity 267

	Planting medium											
Pot	1	2	3	Median	1	2	3	Median	1	2	3	Median
Number of roots												
Length of roots (mm)												

1. (a) Is there a difference between the median number of roots and length of the roots between each planting medium?

   _____

   (b) Explain your answer to 1(a): _____

   _____

   _____

   _____

99 Insect Pollinated Flowers

Key Question: How do the structures of insect pollinated flowers carry out the function of reproduction?

- Flowering plants (angiosperms) are highly successful organisms. The egg cell is retained within the flower of the parent plant and the male gametes (contained in the pollen) must be transferred to it by pollination before fertilization can occur. In most angiosperms, the male and female parts occur on the same plant. Some of these plants can self-pollinate, but most have mechanisms that make this difficult or impossible. This means the pollen must be carried to another flower or plant for fertilization to occur. This helps to ensure cross pollination and therefore reduces inbreeding.
- To make this happen, many angiosperms have developed flowers to attract animal couriers, including insects, birds, and mammals. Insects are the most effective pollinators. Flowers attract insects with brightly colored petals, scent, and offers of food such as nectar and pollen.

Petals guide insects towards the pollen or nectar at the center of the flower. This ensures pollen is transferred in the most efficient way.

Many flowers produce pigments that reflect ultraviolet light. This helps them to be seen by many insects, thus attracting specific pollinators.

Cross section of an insect pollinated flower

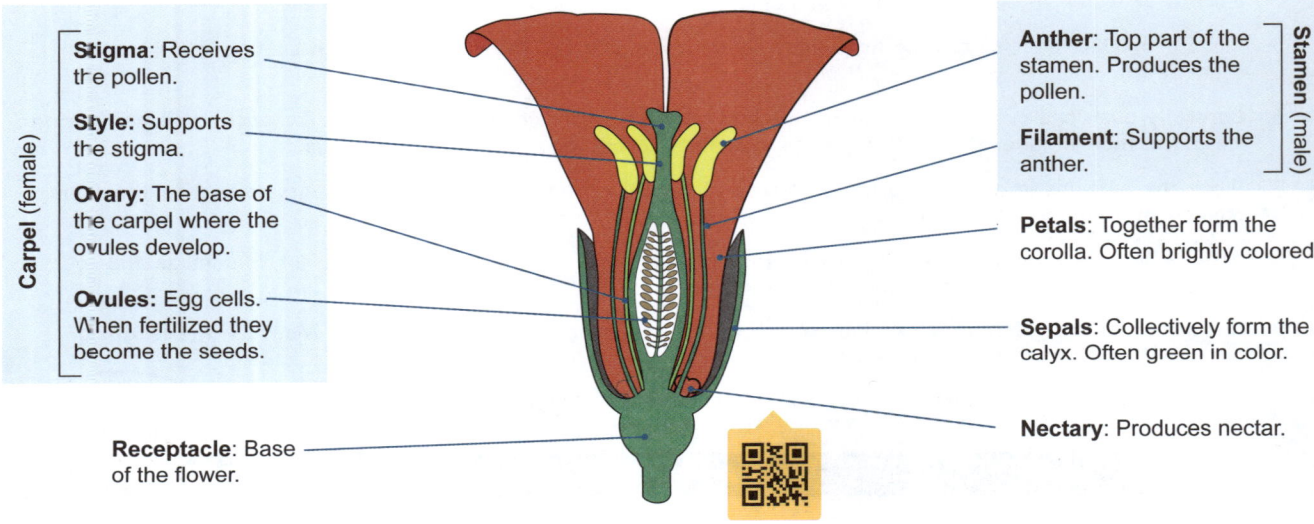

Carpel (female)
- **Stigma:** Receives the pollen.
- **Style:** Supports the stigma.
- **Ovary:** The base of the carpel where the ovules develop.
- **Ovules:** Egg cells. When fertilized they become the seeds.

Receptacle: Base of the flower.

Stamen (male)
- **Anther:** Top part of the stamen. Produces the pollen.
- **Filament:** Supports the anther.

Petals: Together form the corolla. Often brightly colored.

Sepals: Collectively form the calyx. Often green in color.

Nectary: Produces nectar.

1. (a) What is the function of the anther? _____

   (b) What is the function of the stigma? _____

   (c) i. The male part of the flower is the: _____ ii. The female part of the flower is the: _____

2. (a) Why do flowering plants need to attract pollinators? _____

   (b) Name two adaptations that angiosperms use to attract pollinators to their flowers: _____

   (c) How do angiosperms guide insects towards the pollen or nectar? _____

100 Wind Pollinated Flowers

Key Question: How do the structures of wind pollinated flowers carry out the function of reproduction?

The structure of wind pollinated flowers

▸ Wind pollinated plants do not need to attract animal pollinators, so they do not invest energy in producing large, brightly colored flowers or nectar. Wind pollinated flowers typically have many small, drab flowers, and may lack petals altogether, while the anthers and stigma are large and hang clear of the surrounding structures.

▸ Wind pollination is very inefficient and most pollen falls to the ground within 100 m of the flower. The anthers must therefore produce large amounts of pollen to ensure cross-pollination with other plants.

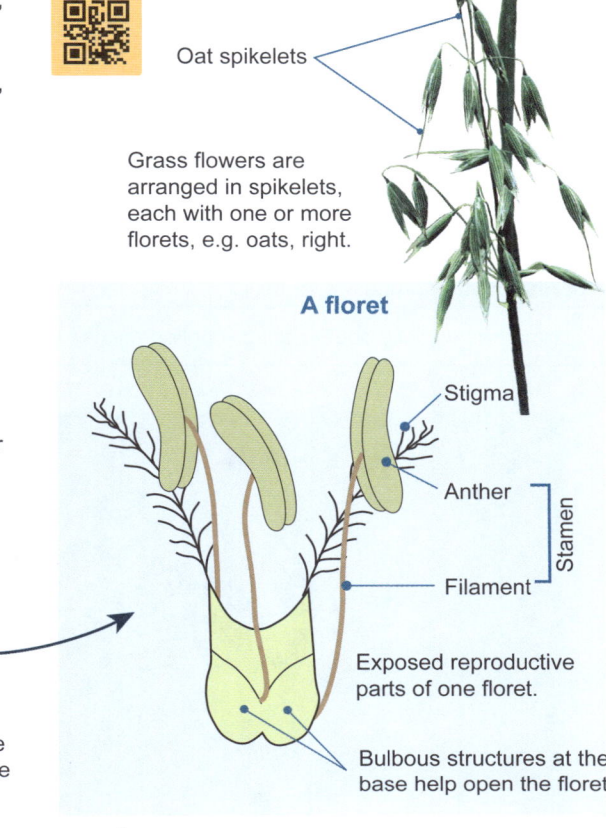

Oat spikelets

Grass flowers are arranged in spikelets, each with one or more florets, e.g. oats, right.

A spikelet

Awn is a bristle-like extension of the floret. It is typical of grasses.

Anther hangs well clear of the leaf scale and produces light-weight pollen which is easily carried away

Large quantities of smooth, dry pollen are produced but there is no scent or nectar.

The large, feathery **stigma** hangs clear of the flower.

Modified leaves enclose both the spikelet and the individual florets.

A floret

Stigma

Anther

Stamen

Filament

Exposed reproductive parts of one floret.

Bulbous structures at the base help open the floret

Only around 10% of flowering plants are wind pollinated, but this 10% includes the grasses, one of the most successful of all angiosperm groups.

Wind pollinated angiosperms range from relatively small, such as this grass plant, to very large, such as oak trees. They are often found in temperate regions.

The flowers of wind pollinated trees are small and numerous, and often have green petals. They often open before the leaves so that pollen can be dispersed without obstruction by leaves.

1. Describe the features of the flowers of wind pollinated plants: _____

_____

_____

2. Identify one advantage and one disadvantage of wind pollination: _____

_____

_____

_____

Comparing insect and wind pollinated plants

▶ The structures of insect and wind pollinated plants reflect the function they perform.

Insect pollinated flowers	Wind pollinated flowers
Large, brightly colored petals to attract insect pollinators.	Small petals, usually pale or dull green.
Often sweetly scented.	No scent.
Contain nectaries to produce nectar as a reward for pollinators.	No nectaries.
Pollen is often sticky, to ensure it is securely attached to pollinators.	Pollen is small and light to be carried by the wind.
Moderate to small amounts of pollen produced.	Pollen produced in vast amounts to ensure at least some grains reach another plant of the same species.
Anthers and stigma inside the petals are placed to ensure pollinators brush against them.	Anthers and stigma extend beyond the petals.
Stigma has a sticky coat to ensure pollen sticks to it.	Stigma is feathery or net-like.

Insect pollinated plant

Wind pollinated plant

3. Using the photographs and the table above to help you, contrast wind and insect pollinated flowers with respect to each of the following characteristics. For each, give reasons for the differences observed:

   (a) Appearance of the flowers: _____

   _____

   _____

   (b) Production of scent and nectar: _____

   _____

   _____

   (c) Position of the reproductive parts (stigma, stamens): _____

   _____

   _____

4. How do the features of flowers carry out their function of reproduction? _____

   _____

   _____

   _____

101 Pollination and Fertilization

Key Question: How do the structures of flowers ensure pollination and fertilization occur efficiently?

- Pollination is the transfer of pollen grains from the male reproductive structures to the female reproductive structures of plants. This must happen before fertilization (the joining of the egg and sperm) can occur.
- Adaptations to ensure cross-pollination (pollination between different plants) include structural and physiological mechanisms associated with the flowers or cones themselves, and reliance on wind and animal pollinators.

Mechanisms to ensure cross pollination

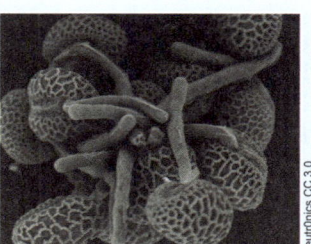

Male willow catkin | Male and female pine cones | Tulip anthers and stigma | Germinating pollen grains

An effective way of ensuring cross-pollination is to have separate male and female plants. This occurs in about 6% of angiosperms, including willows and holly.

Other plants produce separate male and female flowers or cones on the same plant. They may develop at different times to ensure that pollen does not fertilize the same plant.

Some angiosperms have flowers with both male and female structures. They can ensure cross-pollination if the anthers and stigma mature at different times.

In many plants, pollen landing on the stigma of the same plant will not even germinate. This ensures that the egg **cells** are not fertilized by sperm from the same plant.

- Pollen grains are immature male gametophytes, formed by mitosis of haploid microspores within the pollen sac. Pollination is the actual transfer of the pollen from the stamens to the stigma, or from the male cone to the female cone. Pollen grains cannot move independently. They are usually carried by wind or animals.

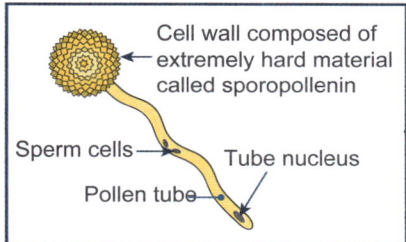

Fertilization in angiosperms

- In angiosperms, when pollen lands on the sticky stigma, it then completes its development by germinating and growing a pollen tube that extends down to the ovary (shown right). The pollen tube is directed by chemicals to the ovule. It enters through the micropyle, a small gap in the ovule.
- Two sperm cells travel down the pollen tube. One sperm nucleus fuses with the egg to form the zygote. A second sperm nucleus fuses with the two polar nuclei within the embryo sac to produce the endosperm tissue (3N), which will provide nutrients for the seed. This is called double fertilization.
- There are usually many ovules in an ovary, so many pollen grains (and fertilizations) are needed before the entire ovary can develop. Many fruits contain multiple seeds.

1. Explain how an egg cell in an angiosperm plant is fertilized: _____

102 Seed Structure and Germination

Key Question: What is the structure of plant seeds and how does the structure help germination of the seed?

Seed structure

- After fertilization has occurred in angiosperms, the ovary develops into the fruit and the ovules within the ovary become the seeds.
- A seed is an entire reproductive unit, containing the embryonic plant in a state of dormancy. During the last stages of maturing, the seed dehydrates until its water content is only 5-15% of its weight. The embryo stops growing and remains dormant until the seed germinates (begins to grow and develop).
- Every seed contains an embryo comprising a rudimentary shoot (plumule), root (radicle), and one or two seed leaves (cotyledons). The embryo and its food supply are encased in a tough, protective seed coat or testa.
- In monocots, the endosperm provides the food supply, whereas in most dicot seeds, the nutrients from the endosperm are transferred to the large, fleshy cotyledons.

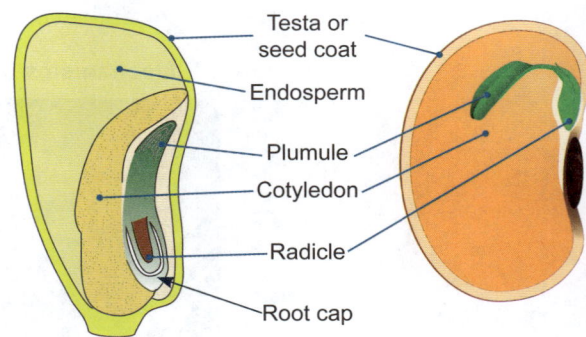

Seed structure in a monocot and dicot

Monocot seed (maize: *Zea mays*)

Dicot seed (garden bean: *Phaseolus vulgaris*)

Seeds only germinate in favorable conditions

- Seeds only germinate when the environmental conditions are favorable to growth. If plants germinate during unfavorable conditions, e.g. too dry or too cold, they are unlikely to survive. As a result, seeds have several adaptations to provide them with an adaptive advantage. These include:
  - Layers of hardened **tissue** to prevent the seed drying out.
  - Tissue to provide nutrients for embryonic growth.
  - Ability to remain dormant until environmental conditions become favorable for germination and growth.
- Germination requires rehydration of the seed and reactivation of the metabolism. The seed absorbs water through the seed coat (testa) and micropyle. As the dry substances in the seed take up water, the **cells** expand, metabolism is reactivated, and embryonic growth begins. Activation stimulates the synthesis of various **hormones** and **enzymes**. Starch in the cotyledons or endosperm is hydrolysed to produce sugars. The mobilized food stores are then delivered to the developing roots and shoots.

Germinating seed

Seed coat

1. Identify the parts of the monocot seed (a) - (f) right:

2. (a) What is the adaptive advantage of germination being triggered by favorable environmental conditions?

   _____
   _____
   _____
   _____
   _____

   (b) What adaptation ensures that a seed does not germinate until there is enough water in the environment?

   _____
   _____
   _____

(a) _____
(b) _____
(c) _____
(d) _____
(e) _____
(f) _____

Word list: Cotyledon, plumule, radicle, testa, root cap, endosperm

Investigating the effect of water on germination

▸ Many factors affect the germination of a seed. In general, there are three requirements for seed germination: water (absorption and reactivation of metabolism), oxygen (for cell respiration), and a temperature that allows metabolism to proceed. Light may or may not be required for germination, depending on the species, although light is required very soon after germination.

▸ Water is essential for the germination process. It enables expansion of the growing cells and activates the enzymes needed for germination. It is also needed for the hydrolysis of stored starch and the mobilization of food molecules.

Investigation 4.6 Germination investigation

See appendix for equipment list.

1. You will need four planting containers and four sets of tomato seeds, (or other small, relatively quickly germinating seeds e.g. mustard seeds), and suitable sterilized growing medium. There should be about 20 seeds in each set (numbers may vary, as the result can be pooled as a class, but around 100 seeds in total per set should be used).

2. Put one set of seeds aside. Soak the other three sets of seeds in water at room temperature for 12 hours, 24 hours, or 36 hours (or other equally spaced time, based on lab availability). Begin the soaking process so that all three sets of seeds finish their soaking at the same time.

3. Plant each of the four sets of seeds in one of the four containers, spacing the seeds in a grid at equal distances from each other. Place the seeds in their containers, in a sunny position, at room temperature.

4. For 12 days, water each set of seeds with 500 mL of water per day, allowing to the water to drain thoroughly.

5. Count the number of seeds germinated each day, and record.

Soaking duration	Days after soaking													Total germinated
	1	2	3	4	5	6	7	8	9	10	11	12	
0													

3. Plot the result on the grid above:

4. At which length of soaking time did the greatest number of seeds germinate? _____

5. Identify one way in which the experiment could be made more accurate: _____
_____

6. Identify one other experiment that could be done to extend this initial experiment: _____
_____

103 Seed Dispersal

Key Question: What features of plants help them disperse their seeds?

▸ Plants disperse their seeds to expand their range and reduce competition. In some cases, the seed is dispersed by itself, but often the fruit or an associated structure helps to disperse seed. Seeds are mainly dispersed by wind, water, and animals.

▸ Wind dispersed seeds have wing-like or feathery structures to catch air currents that carry the seeds. Plants that rely on animals to spread their seeds may have hooks or barbs that attach to hair or sticky secretions that adhere to the skin or hair. Some plants have fleshy fruits that are eaten and the seed is deposited in feces, away from the parent plant. Other dispersal mechanisms rely on explosive discharge or shaking from pods or capsules.

For each example below, research and describe the method of dispersal and the adaptive features associated with it:

1. Dandelion seeds are held in a puff-like cluster:

   (a) Dispersal mechanisms: _____

   (b) Adaptive features: _____
   _____
   _____

2. Acorns are heavy fruits in which the fleshy seeds are encased in a resistant husk:

   (a) Dispersal mechanisms: _____

   (b) Adaptive features: _____
   _____
   _____

3. Coconuts are heavy, buoyant fruits with a thick husk:

   (a) Dispersal mechanisms: _____

   (b) Adaptive features: _____
   _____
   _____

4. Red yucca seeds are held in pods that dry out releasing the small, thin, flat, wedge-shaped seeds:

   (a) Dispersal mechanisms: _____

   (b) Adaptive features: _____
   _____
   _____

5. Yaupon holly seeds are surrounded by a fleshy red fruit:

   (a) Dispersal mechanisms: _____

   (b) Adaptive features: _____
   _____
   _____

104 Responses in Plants

Key Question: How do plants respond to their surroundings?

▸ Plants generally respond to changes in their external environment through changes in patterns of growth. These responses may involve relatively sudden physiological changes, as in flowering, or a steady growth response, such as a **tropism**. Many of these responses involve annual, seasonal, or circadian (daily) rhythms.

Life cycle responses
Plants use seasonal changes (such as falling temperatures or decreasing day length) as cues for starting or ending particular life cycle stages. Such changes are regulated by plant growth factors, such as phytochrome and gibberellin. They enable the plant to avoid conditions unfavorable to growth or survival. Examples include flowering, dormancy and germination, and leaf fall.

Rapid responses to external stimuli
Plants are capable of quite rapid responses. Examples include the closing of stomata in response to water loss (below), opening and closing of flowers in response to temperature, and nastic responses. These responses may follow a circadian rhythm and are protective. They reduce the plant's exposure to abiotic stress or grazing pressure.

Tropisms
Tropisms are growth responses made by plants to external stimuli. The direction of the stimulus determines the direction of the growth response. A tropism may be positive (towards the stimulus), or negative (away from the stimulus). Common stimuli for plants include light, gravity, touch, and chemicals.

Plant competition and allelopathy
Although plants are rooted in the ground, they can still compete with other plants to gain access to resources. Some plants produce chemicals that inhibit the growth of neighboring plants. Such chemical inhibition is called allelopathy. Plants also compete for light and may grow aggressively to shade out slower growing competitors.

Plant responses to herbivory
Many plant species have responded to grazing or browsing pressure with evolutionary adaptations. These enable them to survive cropping or deter browsers. Examples include rapid growth to counteract the constant loss of biomass (grasses), sharp spines or thorns to deter browsers (acacias, cacti), or toxins in the leaf tissues (eucalyptus).

1. Describe the adaptive advantage of a plant responding appropriately to the environment: _____
_____
_____

2. Describe one adaptive response of plants to each of the following stressors in the environment:

   (a) Low soil water: _____

   (b) Falling autumn air temperatures: _____

   (c) Browsing animals: _____

   (d) Low air temperatures at night: _____

105 Tropisms and Growth Responses

Key Question: How do plants respond to external stimuli?

- **Tropisms** are plant growth responses to external stimuli. The stimulus direction determines the direction of the growth response. Tropisms are identified according to the stimulus involved and direction (positive or negative).

- Stimuli are identified as photo- (light), gravi- (gravity), hydro- (water), chemo- (chemicals), and thigmo- (touch). Tropisms act to position the plant in the most favorable growth environment.

Plant growth responses are adaptive. They position the plant in a suitable growing environment, within the limits of the position in which it germinated. The responses to stimuli reinforce the appropriate growth behavior, e.g. shoots grow towards light and away from dark.

Roots grow towards the ground

Tendril wrapping around twig

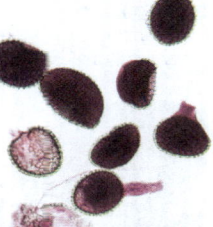
Germinating pollen

Seedlings bending to the light

(a) _____
A positive growth response to a chemical stimulus. Example: Pollen tubes grow towards a chemical, possibly calcium ions, released by the ovule of the flower.

(b) _____
Stems, grow away from the direction of the Earth's gravitational pull. Coleoptiles (the sheath surrounding the young grass shoot) show the same response.

(c) _____
Growth response to water. Roots are influenced primarily by gravity, but will also grow towards water.

(d) _____
Growth responses to light, particularly directional light. Coleoptiles, young stems, and some leaves show a positive response.

(e) _____
Roots respond positively to the Earth's gravitational pull, and curve downward after emerging through the seed coat.

(f) _____
Growth responses to touch or pressure. Tendrils (modified leaves) have a positive coiling response stimulated by touch.

1. Identify each of the plant tropisms described in (a)-(f) above. State whether the response is positive or negative.

2. Describe the purpose of the following tropisms:

   (a) Positive gravitropism in roots: _____

   (b) Positive phototropism in coleoptiles: _____

   (c) Positive thigmomorphogenesis in weak stemmed plants: _____

   (d) Positive chemotropism in pollen grains: _____

3. Explain the purpose of positive phototropism in a seedling: _____

106 Auxins, Gibberellins, and ABA

Key Question: What is the function of plant hormones?

▶ Phytohormones, such as **auxin**, gibberellin, and ABA, play important roles in plant responses to stimuli. They have a wide range of roles in the growth and developmental responses of vascular plants.

Abscisic acid (or ABA) was originally thought to be involved in accelerating abscission (dropping) of leaves and fruit. Now, we believe that it does this in only a few plants. In fact, ABA has roles in seed and bud dormancy, stomal closure, and responses to environmental stress.

Gibberellins are involved in stem and leaf elongation, as well as breaking dormancy in seeds. Specifically, they stimulate **cell** division and cell elongation.

Auxin (indole-acetic acid or IAA) is produced by the shoot tips. It has a role in suppressing growth of lateral buds. This inhibitory influence of a shoot tip or apical bud on the lateral buds is called apical dominance.

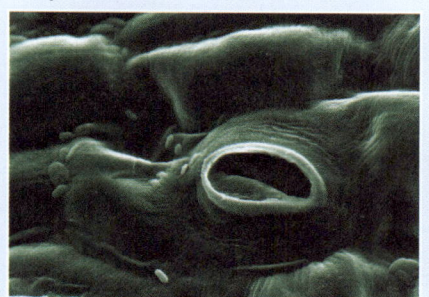

ABA stimulates the closing of stomata in most plant species. Its synthesis is stimulated by water deficiency (water stress). ABA also promotes seed dormancy. It is concentrated in aging leaves, but it is probably not involved in leaf abscission except in a few species.

Gibberellins are responsible for breaking dormancy in seeds and promote the growth of the embryo and emergence of the seedling. Gibberellins are used commercially to hasten seed germination.

Differential auxin transport, and therefore differential growth, is responsible for phototropism. Auxins also produce apical dominance in shoots. Auxin is produced in the shoot tip and diffuses down the stem to inhibit the development of the lateral buds.

ABA is produced in ripe fruit and induces fruit fall. The effects of ABA are generally opposite to those of cytokinins (chemicals that influence plant growth).

Gibberellins cause stem and leaf elongation by stimulating cell division and cell elongation. They are responsible for plant bolting e.g. brassicas (above).

Auxin is required for fruit growth. The maturing seed releases auxin, inducing the surrounding flower parts to develop into fruit.

1. How do gibberellins promote stem elongation? _____

2. (a) What is apical dominance? _____

   (b) How would this affect a tree's growth? _____

107 Plant Hormones as Signal Molecules

Key Question: What is the role of auxin in promoting apical growth in plants?

- **Auxins** are plant **hormones** that have a key role in a range of growth and developmental responses in plants. **Indole-acetic acid (IAA)** is a powerful auxin in plants.
- Gradients in auxin concentration during growth prompt differential responses in specific **tissues** and contribute to directional growth.
- Light is an important growth requirement for all plants. Most plants show an adaptive response of growing towards the light. This growth response is called phototropism.
- The bending of the plants shown on the right is a phototropism in response to light shining from the left and is caused by the plant hormone auxin. Auxin causes the elongation of **cells** on the shaded side of the stem, causing it to bend (photo right).

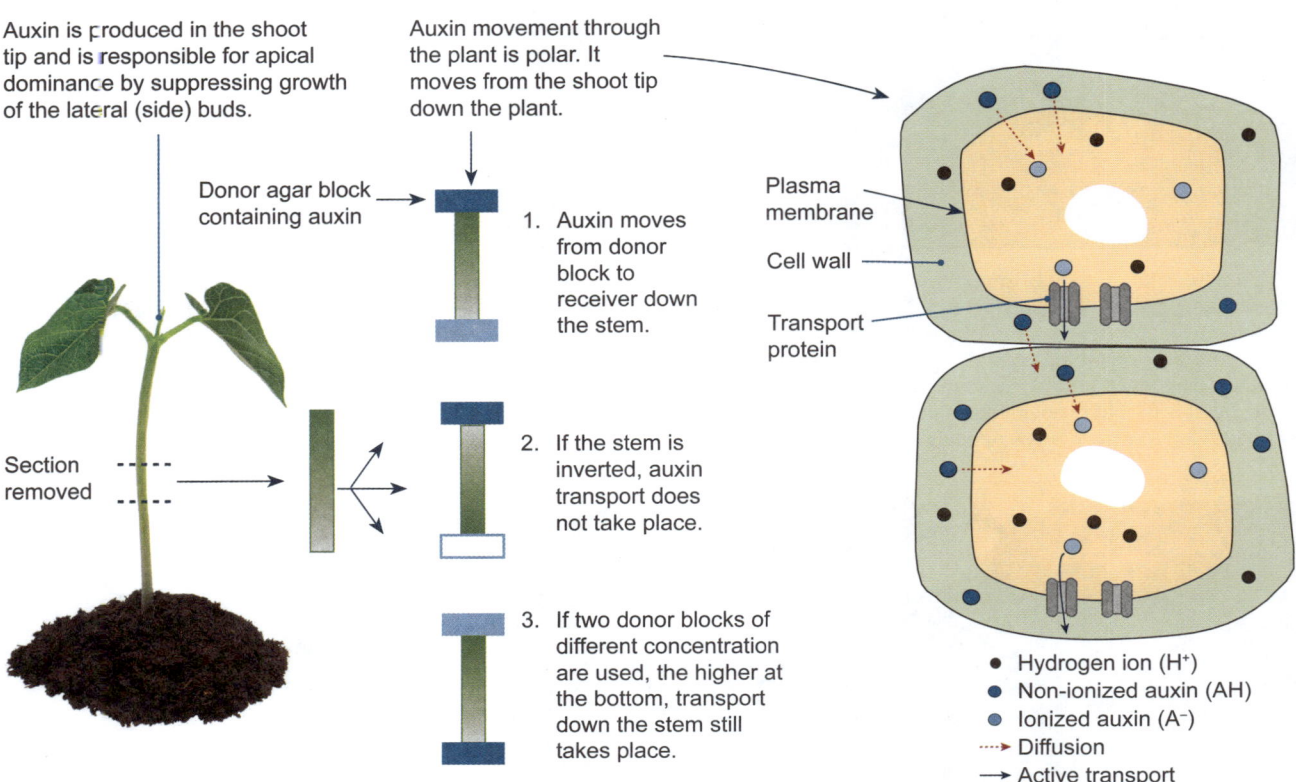

- Hydrogen ion (H⁺)
- Non-ionized auxin (AH)
- Ionized auxin (A⁻)
- ----▶ Diffusion
- ⟶ Active transport

- Under dark conditions, auxin moves evenly down the stem. It is transported cell to cell by diffusion and transport proteins (above right). Outside the cell, auxin is a non-ionized molecule (AH) which can diffuse into the cell. Inside the cell, the pH of the cytoplasm causes auxin to ionize, becoming A⁻ and H⁺. Transport proteins at the basal end of the cell then transport A⁻ out of the cell where it regains an H⁺ ion and reforms AH. In this way auxin is transported in one direction through the plant.
- When plant cells are illuminated by light from one direction, transport proteins in the plasma membrane on the shaded side of the cell are activated and auxin is transported to the shaded side of the plant.

1. What is the term given to the tropism being displayed in the photo (top right)? _____

2. Explain how systems in the plant shoot produced the tropism in question (1) above: _____

_____

_____

_____

_____

Auxin and apical dominance

Auxin is produced in the shoot tip and diffuses down to inhibit the development of the lateral (side) buds. The effect of auxin on preventing lateral bud development can be demonstrated by removing the auxin source and examining the outcome (below).

- In many plants the growth of the shoot apex inhibits the growth of side (lateral) buds. As a result, plants tend to grow a single main stem upwards, which dominates over lateral branches. This response is called apical dominance.
- The hormone responsible for this response is auxin. It acts by stimulating cell elongation.

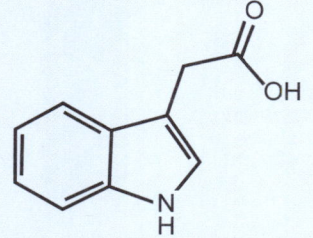

Indole-acetic acid (above) is the only known naturally occurring auxin. It is produced in the apical shoot and young leaves.

No treatment
Apical bud is left intact.

Treatment one
Apical bud is removed. No auxin is applied.

Treatment two
Apical bud is removed. Auxin is applied.

 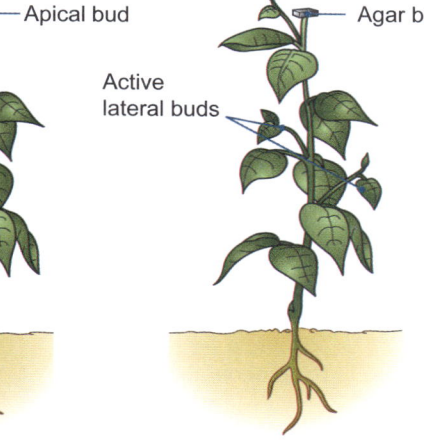 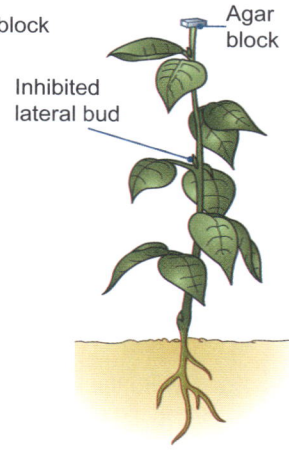

In an intact plant, the plant stem elongates and the lateral buds remain inactive. No side growth occurs.

The apical bud is removed and an agar block without auxin is placed on the cut surface. The seedling begins to develop lateral buds.

The apical bud is removed and an agar block containing auxin is placed on the cut surface. Lateral bud development is inhibited.

3. In the experiment described above, identify:

    (a) The independent variable: _____

    (b) The dependent variable: _____

    (c) The control: _____

4. Draw two conclusions from the experiment described above: _____

5. Describe the role of auxins in apical dominance: _____

6. Study the photo (right) and then answer the following questions:

    (a) Which letter is the apical bud? ____ (b) Which letter is the lateral bud(s)? ____

    (c) Which buds are the largest? _____

    (d) Why would this be important? _____

7. When might lateral buds become important in a plant's growth? _____

108 Investigating Phototropism

Key Question: What is the effect of light on the growth of a plant?

▶ Phototropism is an important plant growth response, as it helps young plants find light when germinating in leaf litter or shaded areas

1. **Directional light:** A pot plant is exposed to direct sunlight near a window and, as it grows, the shoot tip turns in the direction of the Sun. When the plant was rotated, it adjusted by growing towards the Sun in the new direction.

    (a) What hormone regulates this growth response?
    _____

    (b) What is the name of this growth response?
    _____

    (c) How do the cells behave to bring about this change in shoot direction at:

    Point A? _____

    Point B? _____

    (d) Which side (A or B) would have the highest hormone concentration and why?
    _____
    _____

    (e) In the rectangle on the right, draw a diagram of the cells as they appear across the stem from point A to B.

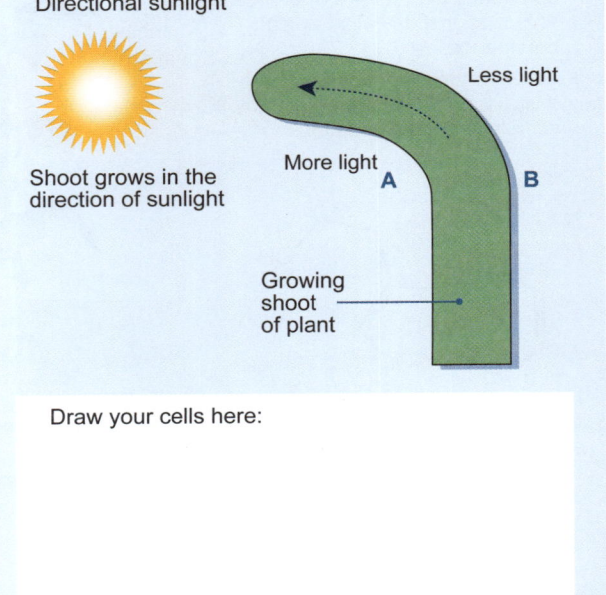

Draw your cells here:

2. **Light excluded from shoot tip:** When a foil cap is placed over the top of the shoot tip, light is prevented from reaching the shoot tip. When growing under these conditions, the direction of growth does not change towards the light source, but grows straight up. State what conclusion you can come to about the source and activity of the hormone that controls the growth response:
    _____
    _____
    _____
    _____

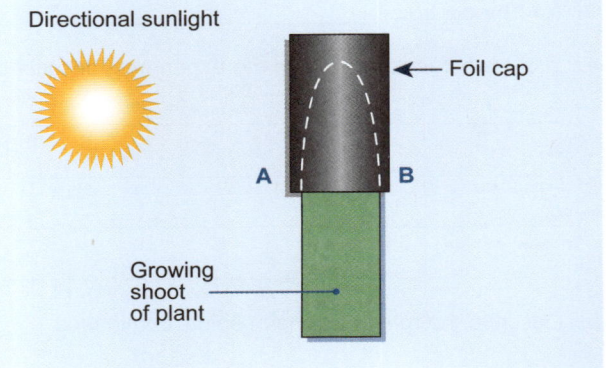

3. **Cutting into the transport system:** Two identical plants were placed side-by-side and subjected to the same directional light source. Razor blades were cut half-way into the stem, thereby interfering with the transport system of the stem. Plant A had the cut on the same side as the light source, while Plant B was cut on the shaded side. Predict the growth responses of:

    Plant A: _____
    _____

    Plant B: _____
    _____

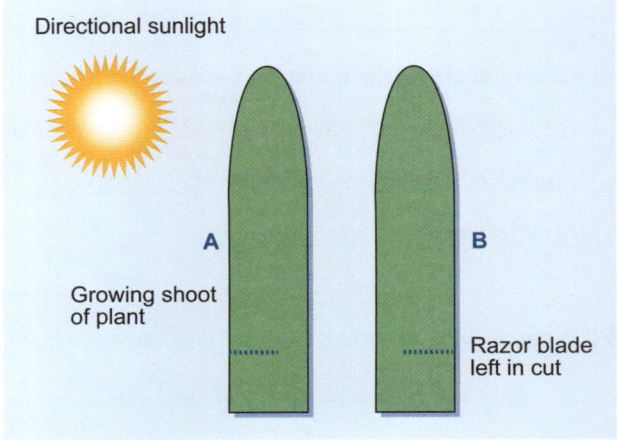

109 Gibberellins and Stem Elongation

Key Question: What is the effect of gibberellins of plant growth responses?

▸ Dwarf pea plants are a selected variety with a mutation that results in impaired gibberellic acid (GA) synthesis. By applying GA to them, the effect of GA on plant growth can easily be demonstrated.

The aim and hypothesis
To investigate the effect of gibberellic acid on stem growth in dwarf pea plants.

If gibberellins are responsible for stem elongation, genetically dwarfed pea plants treated with gibberellic acid will grow taller than untreated dwarf pea plants.

Background
Mendel first described the gene for stem length in pea plants in 1865. In 1997, more than 130 years later, the gene controlling this phenotype was isolated. Called *Le*, it encodes an enzyme called gibberellin 3b-hydroxylase, which converts precursors of gibberellin to an active form. The recessive allele *le* is inactive. Plants that are homozygous for the *le* gene have short stems as they do not produce active gibberellin. This can be verified by giving dwarf peas artificial gibberellins. Instead of remaining short, the plants grow to a normal size.

Method
Students soaked 10 dwarf pea seeds (*Pisum sativum*) overnight in distilled water. The seeds were divided into two groups of five seeds each. The test group received gibberellin treatment and the control group did not. The seeds were planted into two containers filled with potting mix. Once the seeds had germinated, the test group had gibberellic acid paste (500 ppm) painted on them. Both seed groups were watered daily with distilled water. The heights of the germinating shoots were recorded every few days for 20 days (except for one longer break in recording between days 11 and 18 when the students were away).

Normal pea plant

Dwarf pea plant

▸ The results from the experiment are described in the table below.

1. For the control (C) and treatment (T), calculate the mean seedling height and standard deviation for each day and record it in the space provided. Calculate sample standard deviation (s) using a spreadsheet or the equation provided, right. The denominator n - 1 is used for small samples.

$$s = \sqrt{\frac{\sum(x - \bar{x})^2}{n-1}}$$

where $\bar{x}$ = mean

Height (in cm) of dwarf pea plants

Seed #	Days after germination													
	2		4		6		9		11		18		20	
	C	T	C	T	C	T	C	T	C	T	C	T	C	T
1	1.2	1.1	3.1	5.5	4.4	12.1	6.1	22.4	8.3	25.9	12.3	37.5	16.1	38.5
2	1.8	0.6	3.6	5.4	5.1	14.6	7.1	24.7	8.8	28.1	15.9	35.8	19.0	30.1
3	1.3	1.1	3.3	6.6	4.6	15.0	6.4	24.5	8.9	26.8	15.4	34.8	18.1	30.0
4	0.4	0.9	2.9	6.7	4.2	14.2	6.3	21.7	9.5	26.5	12.0	30.3	13.4	38.2
5	1.2	0.4	3.7	4.8	5.1	14.1	6.9	23.6	7.2	25.9	10.8	29.0	12.3	39.2
Mean														
SD														

2. (a) Using a spreadsheet or a separate piece of paper, plot a line graph showing the mean plant heights for the two groups of dwarf peas. Plot the standard deviation (SD) for each mean as error bars either side of the mean.

(b) Describe the effect of gibberellin on the growth of dwarf pea plants: _____

_____

(c) Do the results support the hypothesis? _____

(d) What could have been done to improve this experiment? _____

_____

110 Investigating Gravitropism

Key Question: What is the mechanism that allows plants to response to gravity?

▶ It is believed that **auxin** is the primary regulator in the gravitropic response. The mechanism is shown below but is debated by scientists because it is based on coleoptiles (the sheath surrounding the young grass shoot). This is a specialized, short-lived structure and is probably not representative of plant **tissues** generally.

The role of auxins in gravitropic responses

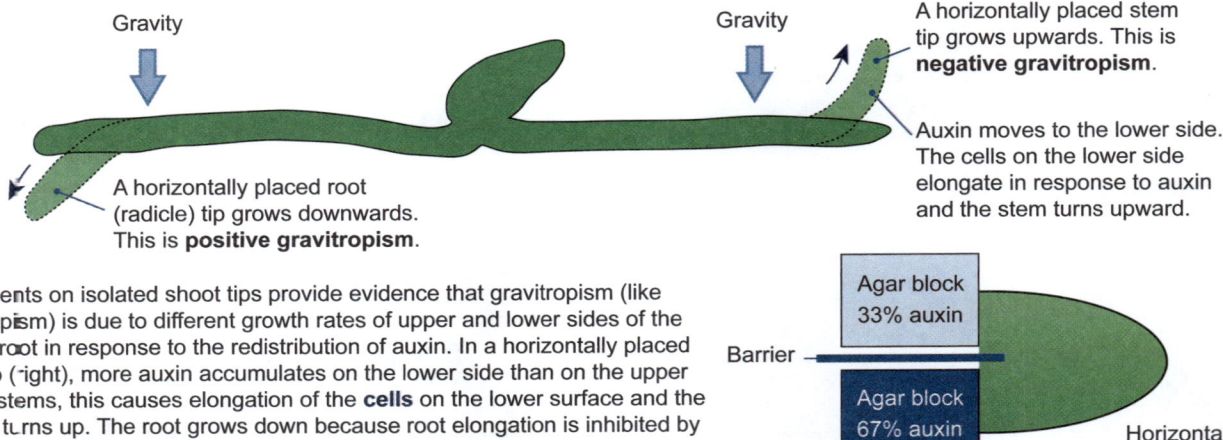

A horizontally placed root (radicle) tip grows downwards. This is **positive gravitropism**.

A horizontally placed stem tip grows upwards. This is **negative gravitropism**.

Auxin moves to the lower side. The cells on the lower side elongate in response to auxin and the stem turns upward.

Experiments on isolated shoot tips provide evidence that gravitropism (like phototropism) is due to different growth rates of upper and lower sides of the stem or root in response to the redistribution of auxin. In a horizontally placed shoot tip (right), more auxin accumulates on the lower side than on the upper side. In stems, this causes elongation of the **cells** on the lower surface and the stem tip turns up. The root grows down because root elongation is inhibited by high levels of auxin on the lower surface (graph below).

Agar block 33% auxin
Barrier
Agar block 67% auxin
Horizontal shoot tip

Auxin concentration and root growth

In a horizontally placed seedling, auxin moves to the lower side in stems and roots. The stem tip grows upwards and the root tip grows down. Root elongation is inhibited by the same level of auxin that stimulates stem growth (graph left). The higher auxin levels on the lower surface cause growth inhibition there. The longest cells are then on the upper surface and the root turns down.

This simple explanation for gravitropism has been criticized because the concentrations of auxins measured in the upper and lower surfaces of horizontal stems and roots are too small to account for the growth movements observed. Other studies indicate that growth inhibitors may interact with auxin in gravitropic responses.

1. Explain the mechanism that facilitates gravitropic response in:

    (a) Shoots (stems): _____

    _____

    (b) Roots: _____

    _____

    _____

2. (a) From the graph above, state the auxin concentration at which root growth becomes inhibited: _____

    (b) State the response of stem at this concentration: _____

3. Explain why the gravitropic response in stems or roots is important to the survival of a seedling:

    (a) Stems: _____

    (b) Roots: _____

111 Investigating Gravitropism in Seeds

Key Question: What is the role of auxin in a plant's response to gravity?

▸ Using the information below, analyze results and draw conclusions about the effect of gravity on the directional growth of seedling roots.

The aim
To investigate the effect of gravity on the direction of root growth in seedlings.

Hypothesis
Roots will always grow towards the Earth's gravitational pull, even when the seedling's orientation is changed.

Method
A damp kitchen paper towel was folded and placed inside a clear plastic sandwich bag. Two pea seeds were soaked in water for five minutes and then placed in the center of the paper towel. The bag was sealed. The plastic bag was then placed on a piece of cardboard which was slightly larger than the plastic bag. The plastic bag was stretched tightly so the plastic held the seeds in place, and secured with staples to the cardboard.

The cardboard was placed upright against a wall. Once the first root from each seed reached 2 cm long, the cardboard was turned 90° degrees.

Daily observations and photographs were made of the root length and direction throughout the duration of the experiment. Photos of one seedling from days 5 and 11 are shown right.

Results
The students took photographs to record changes in growth during the course of the experiment. One seedling at day 5 and 11 is shown below.

Day 5

Day 5, rotated 90° clockwise

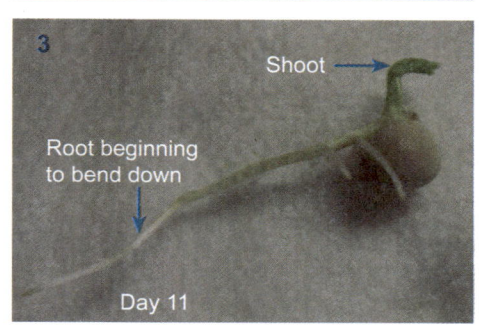
Day 11

Photo 1: This photo was 5 five days after the seed began to germinate.
Photo 2: After photo 1 was taken, the cardboard was rotated 90°.
Photo 3: This photo was taken 6 days after the seed was rotated 90°.

1. (a) In what direction did the root first begin to grow? _____

   (b) Describe what happened to the root when the students rotated the cardboard 90°: _____

   (c) Explain why this occurred: _____

   (d) Predict the result after six more days growth if the students rotated the seedling in photo 3 90° clockwise. Draw your answer in the space right:

2. During the course of the experiment a shoot developed.

   (a) In what direction did the shoot grow at first? _____

   (b) In what direction did the shoot grow after rotation 90° (photo 3)? _____

   (c) Why did this occur? _____

112 Nastic Responses

Key Question: How do plants respond to non-specific stimuli?

- Nastic responses are plant responses in which the direction of the plant response is independent of the stimulus direction. They are reversible and often rapid movements.

Movements of the sensitive Mimosa plant

- The sensitive plant (*Mimosa pudica*) has long leaves composed of small leaflets. When a leaf is touched, it collapses and its leaflets fold together. Strong disturbances cause the entire leaf to droop from its base. This response takes only a few seconds and is caused by a rapid loss of turgor pressure from the **cells** at the bases of the leaves and leaflets.
- The message that the plant has been disturbed is passed quickly around the plant by electrical signals (changes in membrane potential), not by plant **hormones** (as occurs in tropisms). The response can be likened to the nerve impulses of animals, but it is much slower. After the disturbance is removed, turgor is restored to the cells, and the leaflets slowly return to their normal state.
- The adaptive value of these responses is uncertain but may relate to deterring browsers or reducing water loss during high winds.

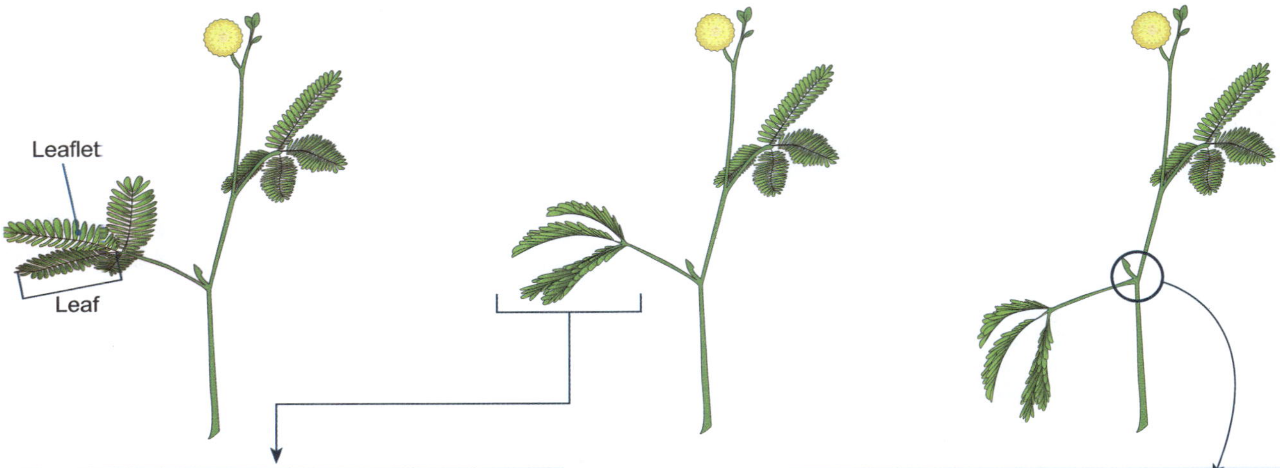

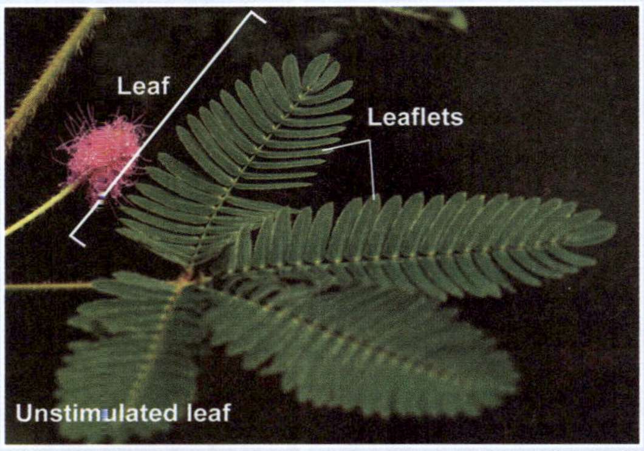

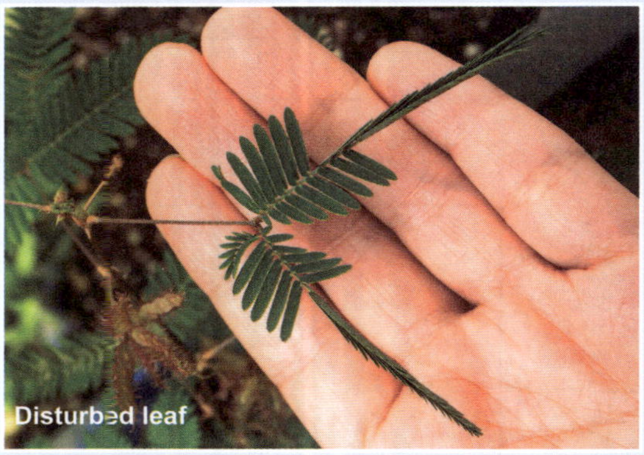

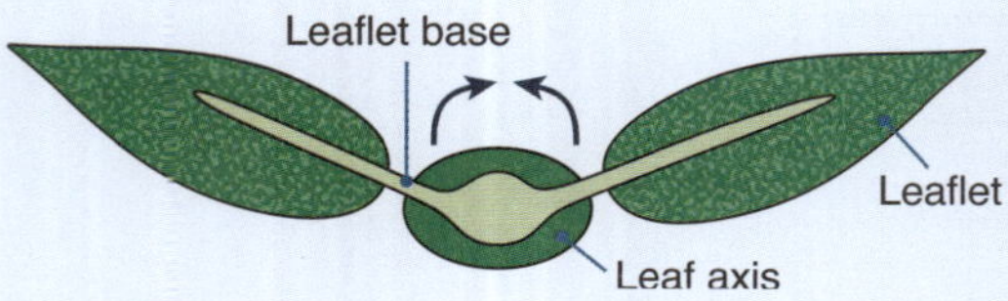

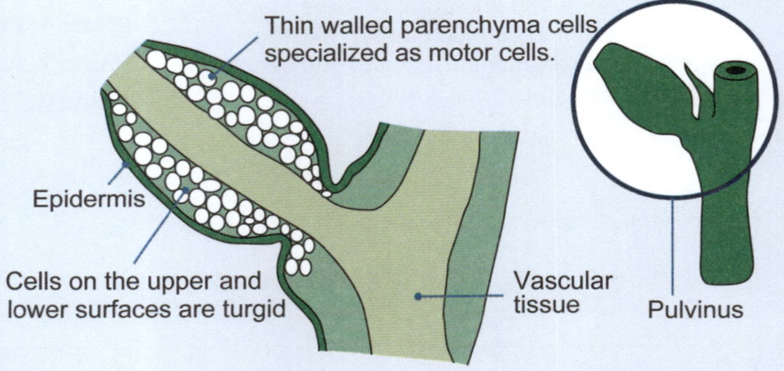

The leaves of Mimosa have joint-like thickenings, the pulvini (sing. pulvinus) at the bases of the petioles and at the bases of each leaflet. The pulvini contain specialized motor cells, which are involved in the rapid leaf movements.

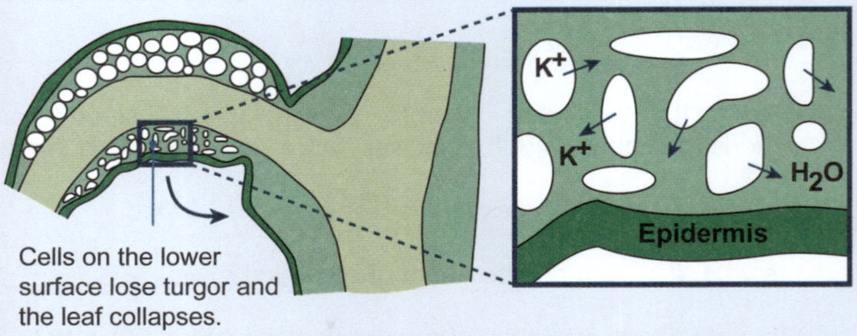

When disturbed, a change in membrane potential of the leaf cells is transmitted to the cells of the pulvinus. These cells respond by actively pumping potassium ions out of the cytoplasm (see inset above). Water follows osmotically and there is a sudden loss of turgor.

This mechanism also operates at the leaflet bases, except that the cells on the upper surface of the pulvinus lose turgor, and the individual leaflets fold up, rather than down (left).

Sleep movements in plants

Many plants show movements in relation to light and dark. *Oxalis* (right) spreads it leaves out during the day to capture sunlight. During the night the leaves are lowered and bend slightly along the midline. This helps to prevent dew accumulating on the leaf and minimizes the risk of damage while the leaf is not being used to capture light.

Day

Night

Thigmonastic responses in carnivorous plants

Some small, specialized plants obtain most of their nitrogen (but not energy) from trapping and digesting small animals, e.g. insects. This allows them to grow in nutrient-poor (particular low nitrogen), high light environments, such as acidic bogs and rocky outcrops.

Venus flytrap (*Dionaea*)

When an insect touches the hairs on a leaf of a Venus flytrap (right), the two lobes of the leaf snap shut, trapping the insect. Once the insect has been digested, the empty leaves reopen. The hairs on the leaf must be touched twice in quick succession for the leaf to close. This means false alarms, such as a twig falling onto the leaf, do not set it off.

Sundew (*Drosera*)

Sundews also show a thigmonastic response. An insect landing on a leaf quickly becomes trapped in the sticky hairs. The hairs fold around the insect and in some species the leaf may curl over, completely enclosing the insect. Carnivory has evolved independently nine times, in five different orders of plants.

1. How is a nastic response different from a tropism? _____

2. (a) Describe the basic mechanism behind the sudden leaf movements in Mimosa: _____

   (b) Explain how the structure of the pulvini helps the Mimosa respond to touch: _____

3. How could the sleep movements of plants (lowering the leaves at night) benefit a plant? _____

4. (a) What structures on the Venus flytrap are used to sense and trap insects? _____

   (b) How does the Venus flytrap ensure the closing of the trap is not falsely triggered? _____

113 Complex Interactions Revisited

Content Anchor Revisited: How do organisms maintain their internal environment and what interactions between organ systems occur to assist this?

▶ At the start of this chapter your were asked to identify the **organ systems** in plants and animals and the interactions between some of those systems.

▶ You should now be able to explain how the organ systems in animals interact and how plant organ systems carry out their functions.

1. Review your answers to questions 3 (a) and (b) in Activity 69 at the start of this chapter: Revise them and add any information you now have to produce a more complete answer:

   (a) _____

   _____

   _____

   _____

   (b) _____

   _____

   _____

   _____

2. Recall the photo of a Venus fly trap that had responded to a person touching it. Review your answer to the organ systems involved in the person touching the plant, and the responses involved in the Venus fly trap catching the person's finger. Were you correct? Revise your answer if needed:

   _____

   _____

   _____

   _____

   _____

   _____

3. Organs and organisms have a hierarchical structure. Describe the hierarchical structure of the nervous system and then describe how it interacts with the muscular and skeletal systems to allow a person to catch a ball:

   _____

   _____

   _____

   _____

   _____

   _____

   _____

   _____

   _____

114 Summing Up

Read each question carefully. Place a cross in the box beside the best answer to the question from the four answer choices provided.

1. Which is the correct order of size, smallest to largest:
   - (a) Cell nucleus, bird, muscle tissue, muscle, muscle cell
   - (b) Bird, muscle tissue, muscle, cell nucleus, muscle cell
   - (c) Bird, muscle, muscle cell, cell nucleus, muscle tissue
   - (d) Cell nucleus, muscle cell, muscle tissue, muscle, bird

2. Which of the following statements is true about exercise?

   A During exercise, breathing rate increases in order to supply more oxygen to the body and remove excess carbon dioxide

   B During exercise, heart rate decreases so that oxygen from the blood has a longer time to be transferred to the muscles

   C During exercise, heart rate increases in order to move more oxygenated blood from the lungs to the muscles and carbon dioxide from the muscles to the lungs

   D During exercise, breathing rate decreases but heart rate increases to move more carbon dioxide to the lungs and gives time for deeper breaths to draw in more oxygen to the lungs

   - (a) A and B
   - (b) A and C
   - (c) B and C
   - (d) B and D

3. The diagram below shows:

   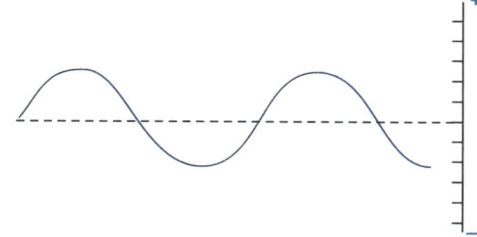

   - (a) Negative feedback because there is a stabilizing effect that returns the internal environment to an optimum set point
   - (b) Positive feedback because the stabilizing systems return the body to the optimum point and this is good for the body
   - (c) Negative feedback because the internal environmental is fluctuating and this is not good for the body
   - (d) Positive feedback because there is no stabilizing effect returning the internal environment to an optimum set point

4. A small cut on the skin could lead to infection. Which two body systems interact to prevent infection in the case of a bleeding cut?
   - (a) Immune and nervous systems
   - (b) Circulatory and immune systems
   - (c) Respiratory and circulatory systems
   - (d) Immune and respiratory systems

5. A dog responds to sound by turning to look in the direction of the sound. Which two body systems are the most important in this response?
   - (a) Muscular and respiratory systems
   - (b) Nervous and circulatory systems
   - (c) Circulatory and nervous systems
   - (d) Nervous and muscular systems

6. The urinary and circulatory systems interact to:
   - (a) Maintain blood volume and pressure
   - (b) Remove metabolic wastes from the body by filtering the blood
   - (c) Control ion levels in the blood by reabsorbing useful ions from filtrate
   - (d) All of the above

7. Blood glucose levels are most likely to be high:
   - (a) Just after eating a meal
   - (b) Just before eating a meal
   - (c) Bloods glucose levels are regulated by insulin so always remain steady
   - (d) About midway between meal times

8. Which of the plant organ systems is most likely involved in detecting and responding to light?
   - (a) The roots
   - (b) The stem
   - (c) The leaves
   - (d) Plants don't detect light

9. In which option below are all the columns correct?

	Leaves and shoots	Stems	Roots
1	Carry out photosynthesis	Anchor plant in soil	Transport nutrients and water
2	Transport nutrients	Produce and store sugars	Absorb water and nutrients from soil
3	Manufacture sugars	Transport nutrients and water	Absorb water and nutrients from soil
4	Exchange gases with the environment	Produce and store sugars	Exchange nutrients with the environment

   - (a) 1
   - (b) 2
   - (c) 3
   - (d) 4

10. Test your knowledge about feedback mechanisms by studying the two graphs below, and answering the questions about them. In your answers, use biological terms appropriately to show your understanding.

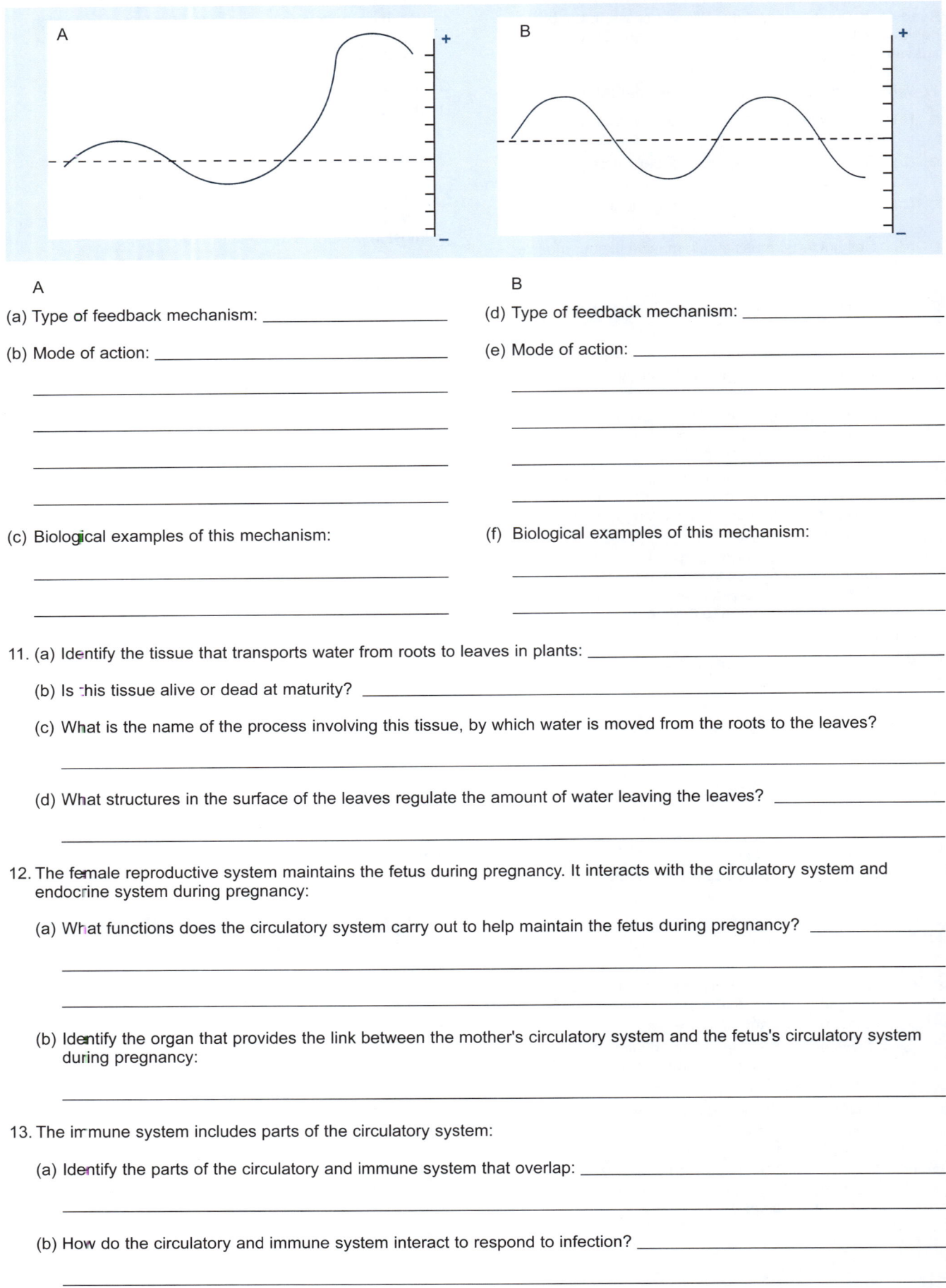

A

(a) Type of feedback mechanism: _____

(b) Mode of action: _____

(c) Biological examples of this mechanism: _____

B

(d) Type of feedback mechanism: _____

(e) Mode of action: _____

(f) Biological examples of this mechanism: _____

11. (a) Identify the tissue that transports water from roots to leaves in plants: _____

(b) Is this tissue alive or dead at maturity? _____

(c) What is the name of the process involving this tissue, by which water is moved from the roots to the leaves?

(d) What structures in the surface of the leaves regulate the amount of water leaving the leaves? _____

12. The female reproductive system maintains the fetus during pregnancy. It interacts with the circulatory system and endocrine system during pregnancy:

(a) What functions does the circulatory system carry out to help maintain the fetus during pregnancy? _____

(b) Identify the organ that provides the link between the mother's circulatory system and the fetus's circulatory system during pregnancy:

13. The immune system includes parts of the circulatory system:

(a) Identify the parts of the circulatory and immune system that overlap: _____

(b) How do the circulatory and immune system interact to respond to infection? _____

14. (a) What is the name given to a plant growth response to directional light? _____

(b) What is the name given to a plant growth response to gravity? _____

(c) What is the name given to a plant response that is independent of stimulus direction? _____

(d) What plant hormone is principally responsible for the phototropic effect? _____

15. (a) What responses are being shown by the orchid in the photo (right): _____

(b) What is the stimulus involved? _____

(c) How would these responses help the plant survive? _____

16. Using the words *receptor*, *effector*, and *control center*, explain how the body maintains homeostasis: _____

17. How does the structure of a flower, such as the tulip (right), help the plant perform the function of reproduction? _____

18. What is the function of root hairs? _____

CHAPTER 5
DNA and Gene Expression

TEKS
Scientific and Engineering Practices

B.1: Investigation and Inquiry
1.G

B.2: Data and Patterns
2.A

B.3: Communicating in Science
3.A 3.B

B.4: Science as a Human Endeavor
4.B 4.C

TEKS
Science Concepts

B7.A identify components of DNA, explain how the nucleotide sequence specifies some traits of an organism, and examine scientific explanations for the origin of DNA

B7.B describe the significance of gene expression and explain the process of protein synthesis using models of DNA and ribonucleic acid (RNA)

B7.C identify and illustrate changes in DNA and evaluate the significance of these changes

B7.D discuss the importance of molecular technologies such as polymerase chain reaction (PCR), gel electrophoresis, and genetic engineering that are applicable in current research and engineering practices

Learning Outcomes
I know I have achieved this when I can:

	Learning Outcome	Activity number
☐	Relate the structure of DNA to gene expression.	116
☐	Contrast DNA and RNA structure.	117
☐	Evaluate models of DNA and construct a paper DNA model.	118
☐	Analyze evidence from past scientists' research into DNA structure.	119
☐	Create a timeline of DNA structure research.	119
☐	Evaluate evidence for the origins of DNA and RNA.	120
☐	Explain the role of DNA in gene expression.	121
☐	Describe the process of transcription.	122
☐	Explain how mRNA editing allows for many protein types to be produced from few genes.	123
☐	Relate codons to amino acids.	124
☐	Define the term degenerate in the context of gene expression.	124
☐	Describe the role of genetic components in the context of gene expression.	125
☐	Model gene expression using plastic bricks.	126
☐	Using scientific vocabulary to explain how the nucleotide sequence can specify traits in an individual.	127
☐	Link mutations to variation in a population, including reasons for why some are retained, some eliminated, and some remain as silent mutations.	128
☐	Distinguish between different types of mutation.	129
☐	Discuss the effects of mutations, using examples.	130
☐	Describe the value of molecular technology to humans, including genetic modification and PCR.	131-132
☐	Analyze and discuss the purpose of gel electrophoresis.	133
☐	Explain the purpose of using recombinant DNA, and describe its usefulness to human industry.	134
☐	Discuss how CRISPR can benefit humans, and human health.	135
☐	Compare different methods of genetically engineering insulin for human use.	136
☐	Discuss how PCR testing can inform medical professions about presence and severity of COVID-19 disease in humans.	137
☐	Discuss how molecular technology can assist genetic research.	138
☐	Research a STEM based career in the field of genetics.	138

RESOURCE HUB
bit.ly/3KXYAnR

ELPS English Language Proficiency Standards

Reading

DNA and RNA. Read and discuss the key question with a partner then use headings, images, and captions to predict where you will find answers to the key question.

After you read the section, return to the key question, and turn it into an answer such as *"Two differences between DNA and RNA are ____."*

200

Writing

The Origin of DNA. Before answering question 7, work with a partner to create a word map based on each hypothesis. Include words associated with the hypotheses.

For example, you might include the words *organic molecules, amino acids,* and *proteins* around the comet hypothesis. Use these words and ideas as you describe, compare, and evaluate the hypotheses.

211

Speaking

DNA Sequence and Traits. Turn to a partner and describe what you see in the cow images or the eye images.

How is the set of images similar? How is it different? How did DNA contribute to the different traits you see in the pictures? Use notes from your discussion to answer the questions.

220

Learning

Effects of Mutations. As you read through this section, highlight cause-and-effect words such as *affect, lead to, account for, change, alter,* and *result in*.

When you finish reading, use the words in a group discussion about the causes and effects of different mutations. Reuse these words and ideas as you answer the questions on pages 225 and 226.

225

Reading

Molecular Technologies and Research. As you read, ask your teacher and classmates for help understanding difficult words and concepts.

A classmate might give a simple synonym such as *single unit* to help you understand the word *monomer* and *repeating unit* to help you understand the word *polymer*.

Your teacher might draw individual colored boxes as an example of monomers and linked colored boxes as an example of polymers.

237

115 Real-Life Superpowers

Content Anchor: What is the result of changes in DNA and can they produce beneficial results?

- No doubt you have come across superheroes in movies and comic books: people with superpowers, such as super speed or strength. Sometimes these superpowers aren't explained; they just exist. At other times, the superpower comes from a mutation caused by some mishap experienced by the character, such as a spider bite or exposure to radiation etc.
- In real-life, events like this would cause you to get sick, or develop cancer, and in some cases die. But in fact there are a number of documented **mutations** that can cause people to develop real-life super-abilities.
- However, unlike movie super-heros, there are often (although not always) side effects to these amazing abilities.
- Some examples of these mutations are shown below:

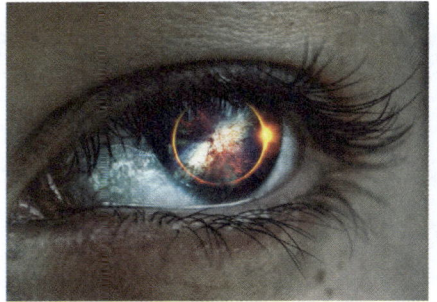

Super vision
Most people have three color detecting cone cells in their retina. Around 20% of women have a mutation in one of their OPN1MW genes that produces cones cells that respond to slightly different wavelengths of light than normal. The result is the ability to see a far greater range of colors than normal.

Super muscles
Muscle development is constrained by the MSTN gene. A mutation to the gene deletes some of the DNA and results in larger muscle growth. The mutation is rare in humans but has been exploited in cattle to produce "double-muscled" breeds, such as the Belgian blue.

Super strong bones
A mutation in the LRP5 gene produces a gain of function mutation. The result is much higher bone density than normal. People with this rare mutation may never break a bone, but there are some negative side effects.

1. If you could, what one superhero power would you have? _____

2. Mutations are a change to a cell's DNA. Why are the majority of mutations harmful to a cell (and organism)?
   _____
   _____
   _____

3. Although some mutations are advantageous (like above), there are always trade-offs for these. Explain why:
   _____
   _____
   _____

4. If mutations alter DNA, how does that actually affect the products cells produce? _____
   _____
   _____
   _____
   _____

116 DNA and Chromosomes

Key Question: How are DNA, genes, and chromosomes related?

▸ In eukaryotes, **DNA** is associated with certain proteins to form **chromatin**. The proteins in the chromatin are responsible for packaging the chromatin into discrete linear structures called **chromosomes**. The extent of packaging changes during the life cycle of the cell, with the chromosomes becoming visible during mitosis.

▸ Each chromosome includes protein-coding regions called **genes**. Segments of DNA called promoters and enhancers are found at the start of eukaryotic genes. These are involved in beginning the process of **transcribing** the gene into mRNA.

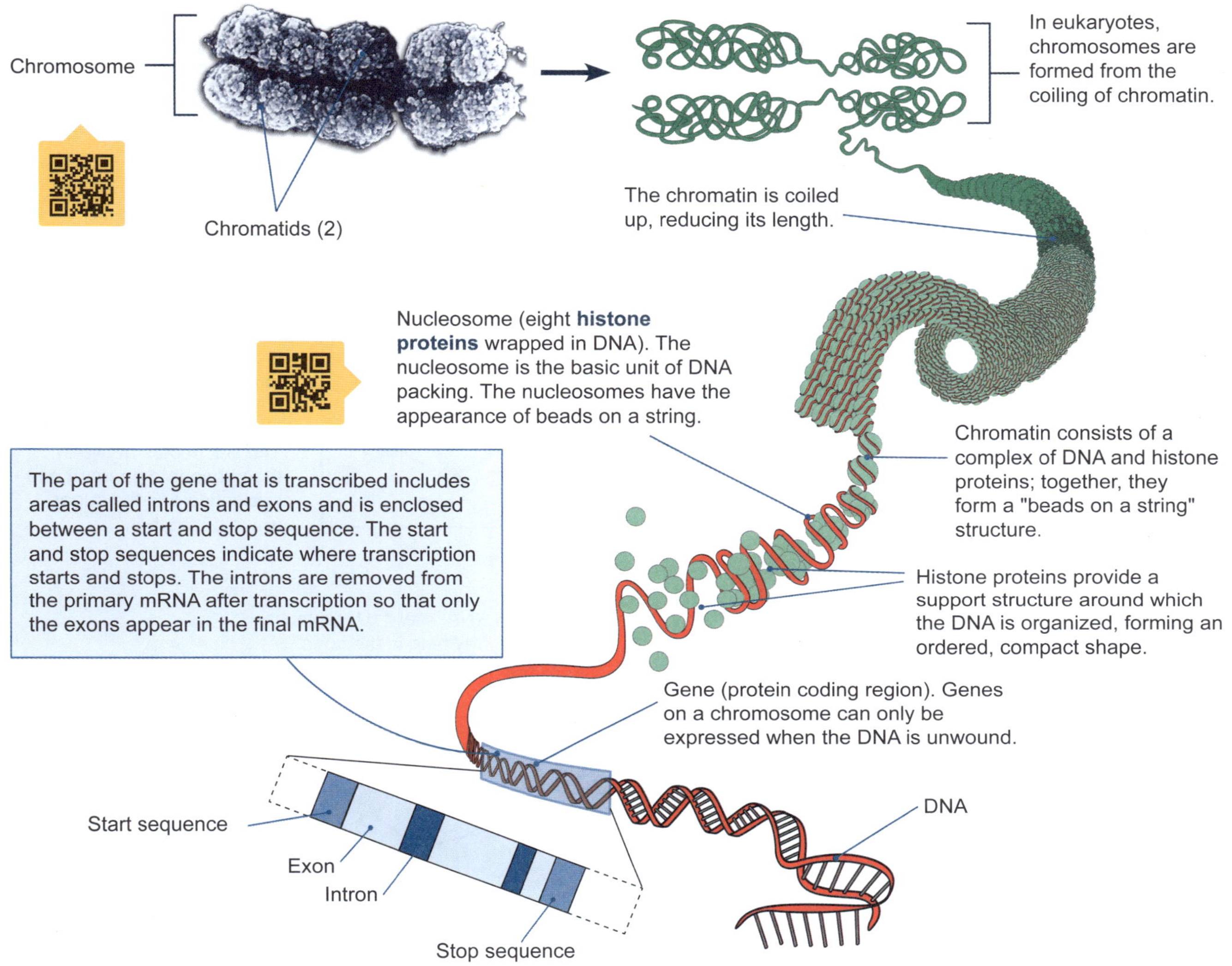

Chromosome — Chromatids (2)

In eukaryotes, chromosomes are formed from the coiling of chromatin.

The chromatin is coiled up, reducing its length.

Nucleosome (eight **histone proteins** wrapped in DNA). The nucleosome is the basic unit of DNA packing. The nucleosomes have the appearance of beads on a string.

Chromatin consists of a complex of DNA and histone proteins; together, they form a "beads on a string" structure.

The part of the gene that is transcribed includes areas called introns and exons and is enclosed between a start and stop sequence. The start and stop sequences indicate where transcription starts and stops. The introns are removed from the primary mRNA after transcription so that only the exons appear in the final mRNA.

Histone proteins provide a support structure around which the DNA is organized, forming an ordered, compact shape.

Gene (protein coding region). Genes on a chromosome can only be expressed when the DNA is unwound.

DNA

Start sequence
Exon
Intron
Stop sequence

1. (a) Suggest a purpose for DNA coiling: _____

   (b) What is the role of histones in this process? _____

2. Identify three parts of the gene that do not appear in the final mRNA: _____

3. What has to happen before a gene can be expressed (made into a protein)? _____

117 DNA and RNA

Key Question: What are the differences between DNA and RNA?

DNA

When long chains of **nucleotides** are joined together, they form nucleic acids. **Deoxyribonucleic acid (DNA)** is a nucleic acid.

DNA consists of a two strands of nucleotides linked together, forming a double helix. A double helix is like a ladder twisted into a corkscrew shape around its length. It is "unwound" in the diagram on the right to show how the bases pair up.

- The DNA backbone is made up of alternating phosphate and sugar molecules.
- Each "rung" of the DNA molecule is made up of two nitrogen bases, joined together by hydrogen bonds between the bases (a hydrogen bond is a bond between hydrogen and an atom such as oxygen).
- Each DNA strand has a direction. Each single strand runs in the opposite direction to the other. This gives the DNA molecule an asymmetrical (uneven) structure.
- The ends of a DNA strand are labeled the 5' (five prime) and 3' (three prime) ends. The 5' end has a terminal phosphate group (off carbon 5), the 3' end has a terminal hydroxyl group (off carbon 3).

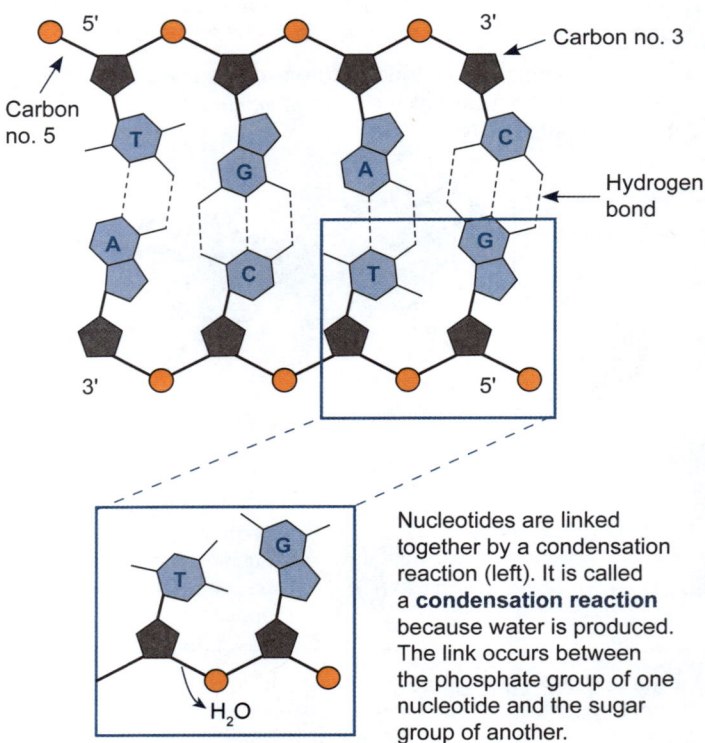

Nucleotides are linked together by a condensation reaction (left). It is called a **condensation reaction** because water is produced. The link occurs between the phosphate group of one nucleotide and the sugar group of another.

RNA

Ribonucleic acid (RNA) is a type of nucleic acid. Like DNA, the nucleotides are linked together through condensation reactions. RNA is single stranded. Its functions include protein synthesis and cell regulation. Three types of RNA are:

- Messenger RNA (mRNA)
- Transfer RNA (tRNA)
- Ribosomal RNA (rRNA)

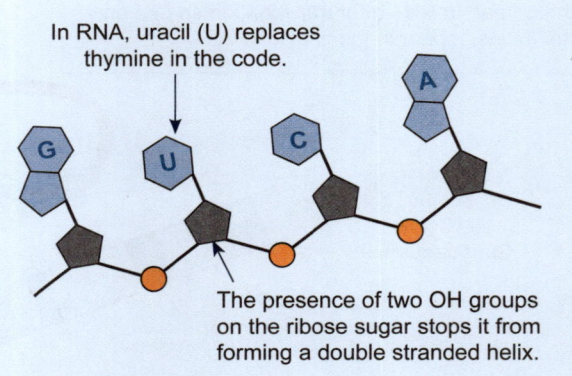

In RNA, uracil (U) replaces thymine in the code.

The presence of two OH groups on the ribose sugar stops it from forming a double stranded helix.

1. Complete the following table, summarizing the differences between DNA and RNA molecules:

	DNA	RNA
Sugar present		
Bases present		
Number of strands		

2. If you wanted to use a radioactive or fluorescent tag to label only the DNA in a cell and not the RNA, what compound(s) would you label?

_____

118 Modeling DNA Structure

Key Question: What is the structure of DNA and how can it be modeled?

Models can be used to show DNA structure

Models of **DNA (deoxyribonucleic acid)** range from simple models (right) to highly sophisticated models constructed with the latest technology. The model of DNA built by Watson and Crick in 1953 was inspired. However, it was also based on information from a number of sources that, together, enabled Watson and Crick to visualize what the molecule would look like. Rosalind Franklin's Photo 51 was crucial to this story. Watson and Crick's model was fundamental to our understanding of inheritance, but it would be another 8 years before researchers determined exactly how the code determined the order of amino acids.

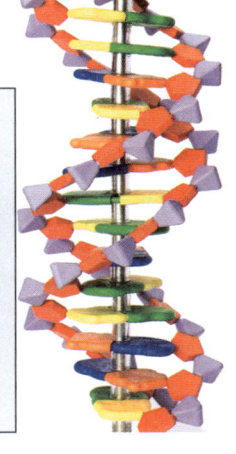

A model needs evidence: Photo 51

The X-ray diffraction patterns of the famous "photo 51" (recreated in the illustration right) provide measurements of different parts of the molecule and the position of different groups of atoms.

The X pattern indicates a helix, but Watson and Crick realized that the apparent gaps in the X (labeled **A**) were due to the repeating pattern of a double helix. The diamond shapes (in blue) indicate the helix is continuous and of constant dimensions and that the sugar-phosphate backbone is on the outside of the helix. The distance between the dark horizontal bands allows the calculation of the length of one full turn of the helix.

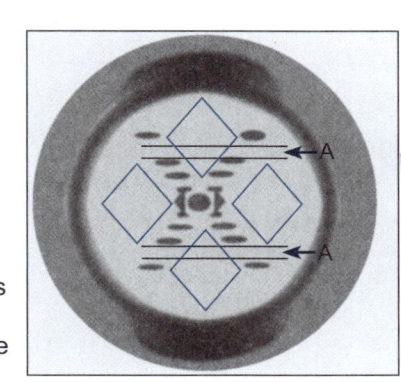

A model can provide different levels of information

The models below show how DNA can be represented in models of different complexity. Each one provides more information about the molecule than the one before it in the sequence.

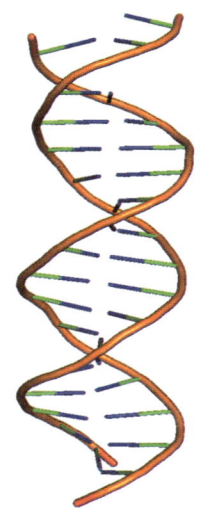
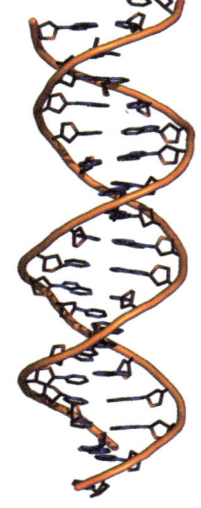
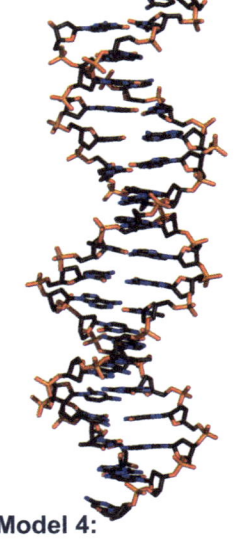

Model 1: DNA has a double helix structure consisting of two strands.

Model 2: Rungs between the two strands join them together.

Model 3: The rungs consist of nitrogen-containing **nucleotide** bases, of which there are four types: cytosine, guanine, adenine, or thymine.

Model 4: Paired nucleotide bases attach to a DNA backbone of alternating deoxyribose sugar molecules joined to a phosphate via phosphodiester bonds.

1. (a) What do all of the DNA models above have in common? _____

   (b) What makes some of the models better than others? _____

Building physical models helps to understand the structure of DNA

The following exercise will help you understand the structure of DNA and learn the base pairing rule for DNA.

The way the nucleotide bases pair up between strands is very specific. The chemistry and shape of each base means they can only bond with one other. Use the information in the table below if you need help, remembering the base pairing rule while you are constructing your DNA molecules.

Chargaff's rules

Before Watson and Crick described the structure of DNA, an Austrian chemist called Chargaff analyzed the base composition of DNA from a number of organisms. He found that the base composition varies between species but that within a species the percentage of A and T bases is equal and the percentage of G and C bases is equal. Validation of Chargaff's rules was the basis of Watson and Crick's base pairs in the DNA double helix model.

DNA base pairing rule

Adenine	always pairs with	Thymine	A	⟷ T
Thymine	always pairs with	Adenine	T	⟷ A
Cytosine	always pairs with	Guanine	C	⟷ G
Guanine	always pairs with	Cytosine	G	⟷ C

Investigation 5.1   Making a DNA model

See appendix for equipment list.

1. Cut out the nucleotides on page 203 by cutting along the columns and rows (see arrows indicating two such cutting points). Although drawn as geometric shapes, these symbols represent chemical structures.

2. Place one of each of the four kinds of nucleotide on their correct spaces below:

   [Place a cut-out symbol for thymine here] — **Thymine**

   [Place a cut-out symbol for cytosine here] — **Cytosine**

   [Place a cut-out symbol for adenine here] — **Adenine**

   [Place a cut-out symbol for guanine here] — **Guanine**

2. Identify and label the following features on the adenine nucleotide above: **phosphate, sugar, base, hydrogen bonds**.

3. Create one strand of the DNA molecule by placing the 9 correct "cut out" nucleotides in the labeled spaces on the following page (DNA molecule). Make sure these are the right way up (with the P on the left) and are aligned with the left hand edge of each box. Begin with thymine and end with guanine.

4. (a) Now create the complementary strand of DNA by using the base pairing rule above.

   (b) The nucleotides have to be arranged upside down. What does this tell you about the DNA molecule?

   _____

5. Once you have checked that the arrangement is correct, glue, paste, or tape these nucleotides in place (next page).

6. Predict what you think would happen to the DNA structure if there was a mistake in the DNA sequence and mismatched base pairing occurred, e.g. A paired with C. You can use your model to visualize this if you want:

   _____

   _____

7. In what way is this simple model deficient as a representation of DNA? _____

   _____

Nucleotides

Tear out this page and separate each of the 24 nucleotides by cutting along the columns and rows (see arrows indicating the cutting points).

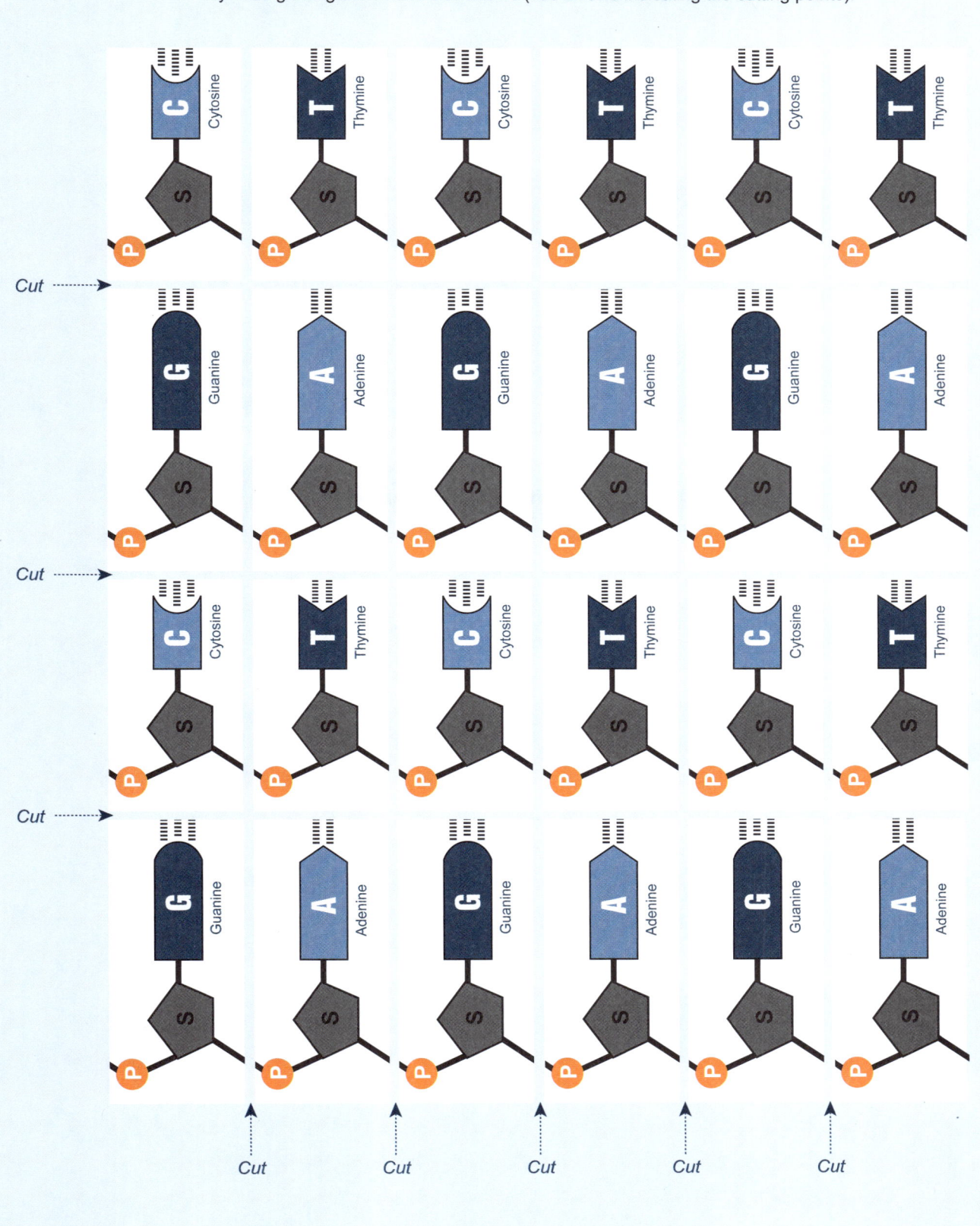

This page is left blank deliberately

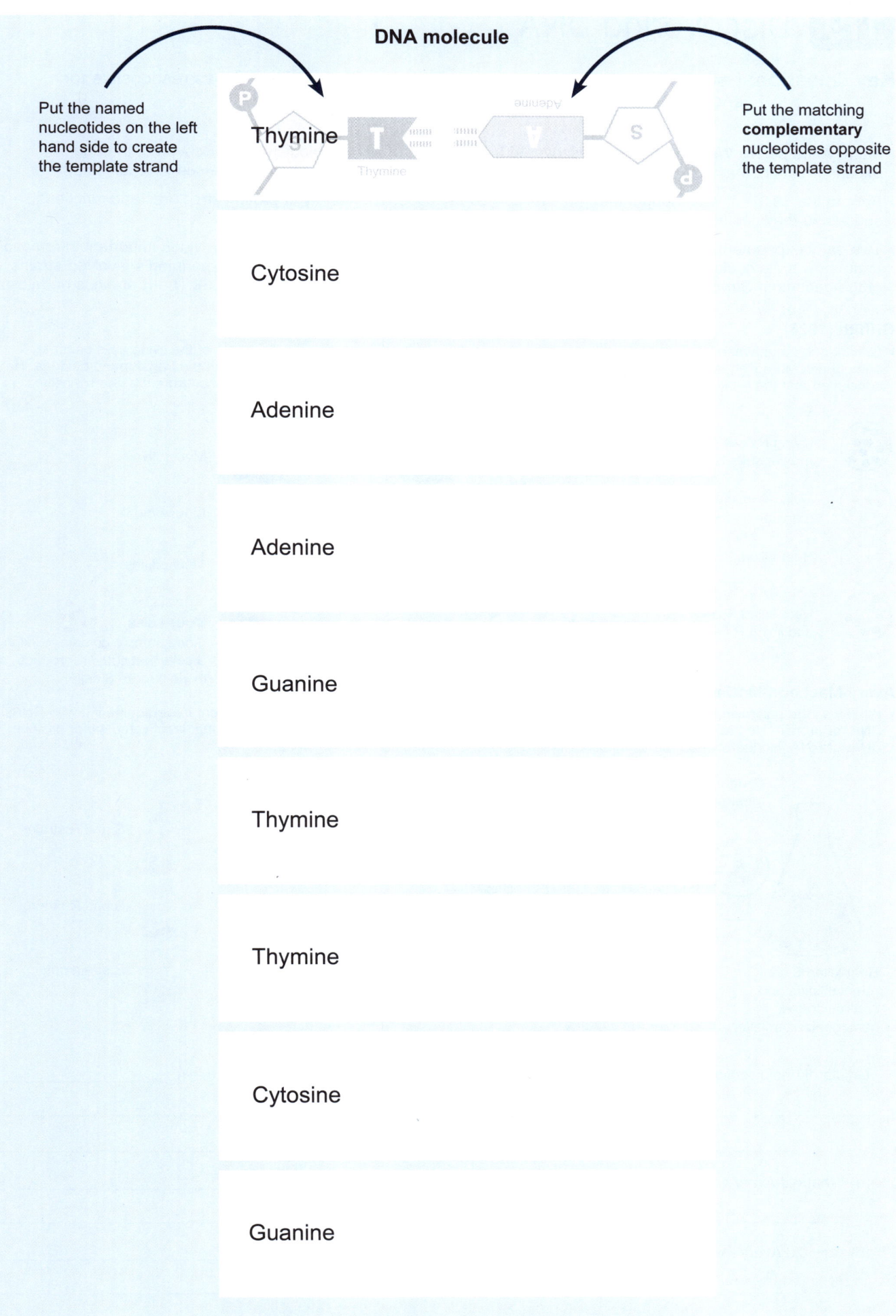

119 Discovering DNA

Key Question: How did scientists figure out the structure of DNA and that it was responsible for encoding information for controlling processes in a cell?

- Many years before Watson and Crick discovered the structure of **DNA**, biologists had discovered, through experimentation, that DNA carried the information responsible for the heritable **traits** we see in organisms.
- Prior to the 1940s, it was thought that proteins carried the code. The variety of protein structures and functions suggested they could account for the many traits we see in organisms.
- Two early experiments, one by Griffith and another by Avery, MacLeod, and McCarty, provided important information about how traits could be passed on and what cellular material was responsible. The experiments involved strains of the bacterium *Streptococcus pneumoniae*. The S strain is pathogenic (causes disease). The R strain is harmless.

Griffith (1928)

- Griffith found that, when he mixed heat-killed pathogenic bacteria with living harmless cells, some of the living cells became pathogenic. Moreover, the newly acquired trait of pathogenicity was inherited by all descendants of the transformed bacteria. He concluded that the living R cells had been transformed into pathogenic cells by a heritable substance from the dead S cells.

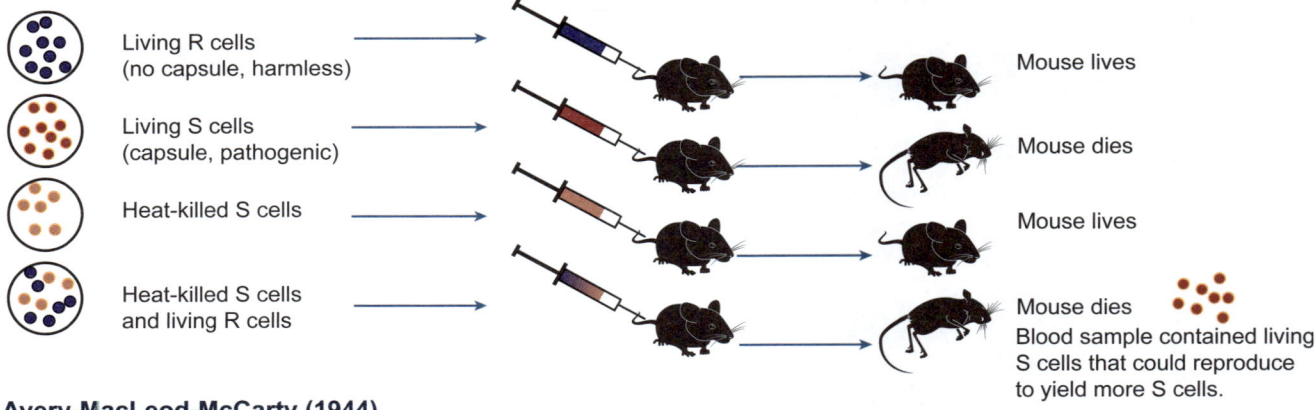

Avery-MacLeod-McCarty (1944)

- What was the unknown transformation factor in Griffith's experiment? Avery designed an experiment to determine if it was RNA, DNA, or protein. He broke open the heat-killed pathogenic cells and treated samples with agents that inactivated either protein, DNA, or RNA. He then tested the samples for their ability to transform harmless bacteria.

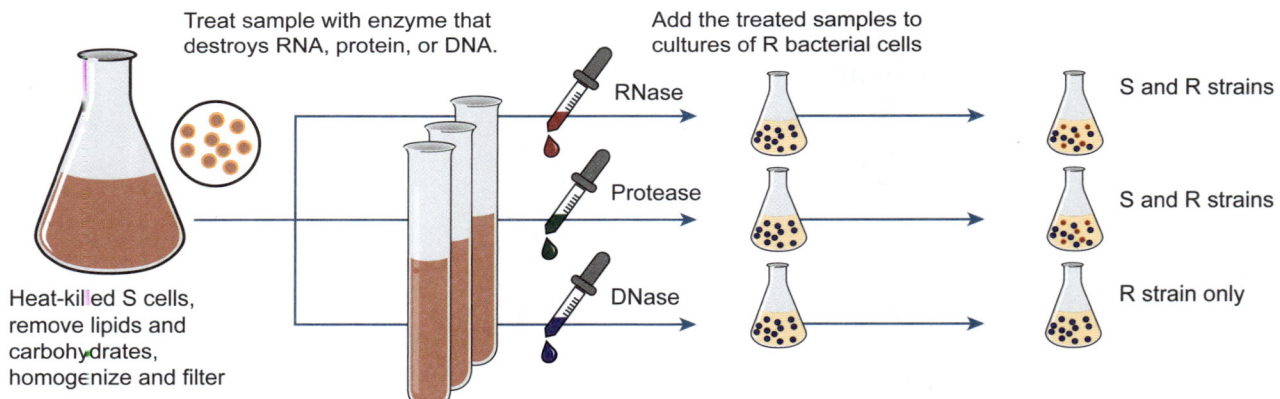

1. Griffith did not predict transformation in his experiment. What results was he expecting? Explain: _____
_____
_____

2. (a) What did Avery's experimental results show? _____
_____

(b) How did Avery's experiment build on Griffith's findings? _____
_____
_____

Hershey and Chase (1952)

▶ Despite the findings of Avery and his colleagues, the scientific community was slow to accept the role of DNA as the carrier of the code. The approaches they used were not fashionable and some scientists criticized the results, saying the procedures they used led to protein contamination. At the time, protein was still favored as the carrier of the code because nucleic acids were not believed to have any biological activity and their structure was not defined. Despite the importance of the work, Avery and his colleagues were overlooked for a Nobel Prize.

▶ The work of Hershey and Chase followed the work of Avery and his colleagues and was instrumental in the acceptance of DNA as the hereditary material. Hershey and Chase worked on viruses called phages, which infect bacteria. Phages are composed of only DNA and protein. When they infect, they inject their DNA into the bacterial cell, leaving their protein coat stuck to the outside.

▶ Hershey and Chase used two batches of phage.
Batch 1 phage were grown with radioactively labeled sulfur, which was incorporated into the phage protein coat.
Batch 2 phage were grown with radioactively labeled phosphorus, which was incorporated into the phage DNA.

▶ The phage were mixed with bacteria, which they then infected. When the bacterial cells were separated from the phage coats, Hershey and Chase looked at where the radioactivity had ended up (right). Hershey and Chase showed conclusively that DNA is the only material transferred from phage to bacteria when bacteria are infected.

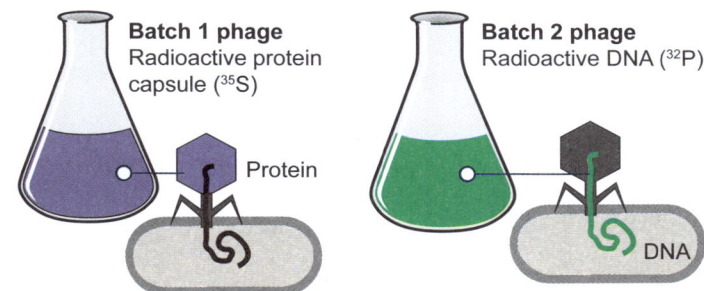

Batch 1 phage
Radioactive protein capsule (^{35}S)

Batch 2 phage
Radioactive DNA (^{32}P)

Homogenization (blending) separates phage outside the bacteria from the cells and their contents. After centrifugation (spinning down), the cells and their DNA form a pellet. Viral protein coats are left in the supernatant (liquid).

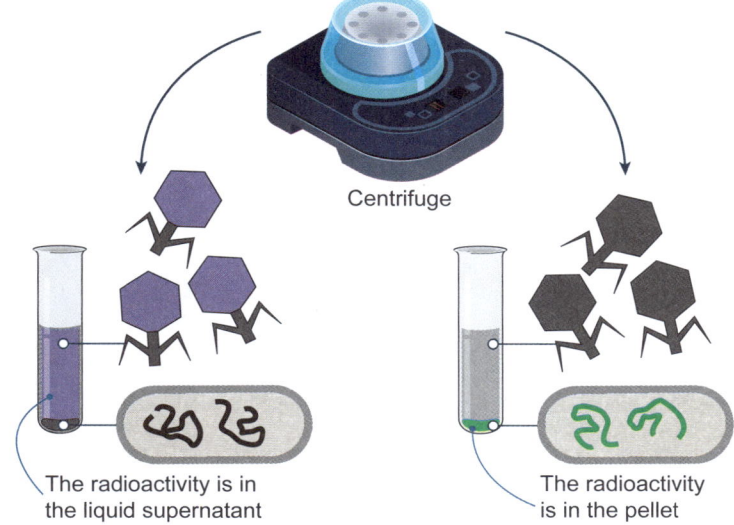

The radioactivity is in the liquid supernatant

The radioactivity is in the pellet

▶ The combined efforts of these three important teams of scientists illustrates that science is an ever evolving process in which teams of researchers build on the work of others, often over decades, or longer.

▶ In 1953, just one year after Hershey and Chase demonstrated that DNA carried hereditary information, James Watson and Frances Crick published their work on the actual structure of DNA: the now familiar double helix.

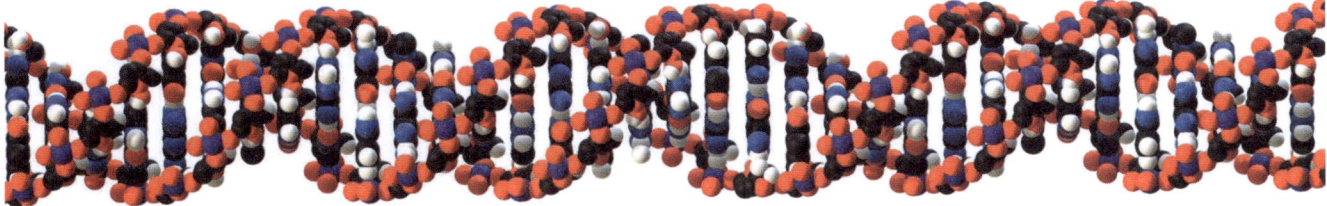

3. (a) How did the Hershey-Chase experiment provide evidence that nucleic acids, not protein, are the hereditary material?

(b) How would the results of the experiment have differed if proteins carried the genetic information?

(c) Why do you think the Hershey-Chase experiment was so successful in convincing the scientific community at the time that DNA was the material that carried the genetic code? Why weren't Avery's experiments equally successful?

How did the work of many scientists contribute to the discovery of DNA's structure?

DNA is easily extracted and isolated from cells. This was first done in 1869, but it took the work of many scientists, working in different areas, many years to determine DNA's structure. In particular, four scientists, Watson, Crick, Franklin, and Wilkins are now recognized as having made significant contributions in determining the structure of DNA. Once the structure of DNA was known, scientists could determine how it was replicated and how it could pass information from one generation to the next.

Discovering the structure of DNA ... a story of collaboration and friction

- Although Watson and Crick are often credited with discovering DNA's structure, the contributions of many scientists were important. This includes not only the contributions from scientists at the time, but also from earlier researchers whose findings contributed to the body of existing knowledge.

- Personal conflicts and internal politics probably prevented DNA's structure being determined earlier. Professional friction between Rosalind Franklin and Maurice Wilkins meant that they worked independently of each other. Watson and Crick analyzed some of Franklin's results, notably "photo 51", without her knowledge or consent and Watson himself recalls that he tended to dismiss her. Photo 51 was crucial to Watson and Crick's model because it showed that DNA was a double helix. Only later, did they acknowledge her considerable contribution.

- Franklin was conservative by nature and opposed to prematurely building theoretical models until there was enough data to guide the model building. However, when she saw Watson and Crick's model, she readily accepted it. Despite her contribution, Franklin did not receive the Nobel prize.

James Watson (left) and Francis Crick (right) in 1953 with their DNA model.

 4. As a class, work together to produce a timeline of the events leading to the discovery of DNA's structure. In small groups, choose one event or researcher and explain its/their significance. You can use the information presented in this activity, as well as any information provided on **BIOZONE's Resource Hub**. Take a photograph of your completed timeline and attach it to this page. How much was the discovery of DNA's structure a collaboration between scientists working in related fields? Use the space below to take notes or attach your timeline.

120 The Origin of DNA

Key Question: What are the current theories for the origin of DNA?

A question of chemistry

- A key question in biology is "How did life get started?" There are various answers to this question which you will explore later in this activity. For now, we might ask a more general question: how do you build complex structures without any real guiding principles?
- Atoms and molecules react in predictable ways. Given specific conditions, specific reactants will always produce the same products. This is how humans are able to reliably produce the many chemical products we use today.
- We can use these reactions to produce products because we know that, under stable conditions, once the product is formed, it will no longer react. If we change the conditions, the product may then react again, but the product of that reaction will then be stable.
- Life is simply a result of the availability of conditions required to carry out certain reactions. Life takes energy from energy-dense molecules and uses it to drive other chemical reactions until the useful energy has been used up. Provided the energy dense molecules are always available, the reactions of life continue.
- The simple molecules from which the more complex molecules of life evolved were simply following the rules of chemistry. So, the question: "How did life start?" is better rephrased as, "What molecules and conditions caused life to start?"

Chemicals always react in the same way under the same conditions. Thus, we know that mixing lead nitrate and potassium iodide in solution at 20°C and 1 atmosphere will produce the yellow precipitate lead iodide. We also know that if the reaction occurs at 80°C the precipitate does not form and lead iodide stays in solution.

1. Consider the Rube Goldberg machine linked in the **BIOZONE Resource Hub**. How is it building a complex object from more simple building blocks?

_____

_____

_____

An RNA world

- Modern life requires complex proteins and other molecules for replication, but these did not exist before life's beginnings. The discovery of ribozymes in 1982 helped to solve the problem of how biological information was stored and replicated.
- Ribozymes are enzymes formed from RNA, which itself can store biological information. The ribozymes can catalyze the replication of the original RNA molecule. This mechanism for self replication has led to the theory of an "RNA world".

Pre-RNA world

- The individual ribonucleotides of RNA are difficult to assemble without enzymes. This has led to proposals of a pre-RNA world, where polymers similar to RNA formed and acted as the very first catalysts.

RNA world

- RNA is able to act as a vehicle for both information storage and catalysis. It provides a way around the problem that genes require enzymes to form, and enzymes require genes to form. The first stage of evolution may have proceeded by RNA molecules folding up to form ribozymes, which could then catalyze the replication of other similar RNAs.
- Over time, RNAs mutated, forming a variety of forms, some of which could assemble proteins using other RNA forms as templates. RNA eventually became double stranded and/or was replaced by DNA as the proteins became more complex.
- The problem with the RNA world hypothesis is that in modern systems, enzymes are needed to detach new RNA copies from their template. This means that, in an RNA world, enzymes (proteins) would already have to exist, which raises the question of how did those enzymes form?

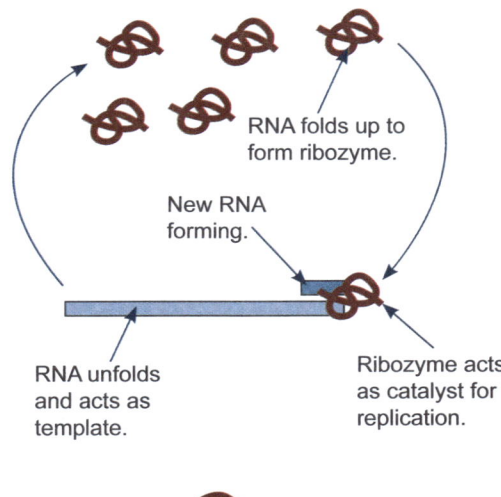

RNA folds up to form ribozyme.

New RNA forming.

RNA unfolds and acts as template.

Ribozyme acts as catalyst for replication.

Mutant ribozyme is able to translate RNA into proteins.

DNA arose much earlier than we thought

▸ The RNA world hypothesis has been the main hypothesis for the evolution for life since the late 1980s. However, in 2020, research showed that the evolution of **DNA** might not have needed RNA to have arisen first. DNA could have been formed in a much more simple process than first thought.

▸ This research is based on the molecule diamidophosphate (DAP). This molecule is a plausible, prebiotic chemical and has been shown to be able to string together deoxynucleotides, forming simple DNA molecules. The new research suggests that RNA and DNA were both present in the prebiotic environment, allowing a much greater range of possible evolutionary pathways for the development of the first fully replicating life.

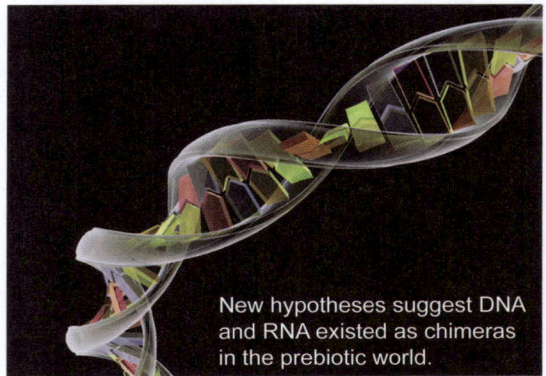

New hypotheses suggest DNA and RNA existed as chimeras in the prebiotic world.

From outer space

▸ A consistently appearing hypothesis for the origin of DNA and life on Earth is that, at the very least, the building blocks of life arrived on comets. This included molecules such as amino acids and proteins.

▸ This hypothesis is supported by the fact that amino acids and other organic molecules have been detected on comets. Evidence exists that Earth was bombarded by multiple, large comets and asteroids (called the late heavy bombardment) around 4 billion years ago, only a few hundred million years before life is believed to have evolved.

Material streaming off comet 67P was found to contain glycine, among other organic molecules.

The Miller-Urey experiment

▸ In the 1950s, Stanley Miller and Harold Urey attempted to recreate the conditions of primitive Earth. They hoped to produce the biological molecules that preceded the development of the first living organisms.

▸ A series of experiments (set up right) varied inputs, based on possible environments for the early Earth, produced a variety of organic molecules. These included amino acids, nucleic acids, several sugars, lipids, adenine, and even ATP (if phosphate is added to the flask). Researchers believe that the early atmosphere may be similar to the vapors given off by modern volcanoes: carbon monoxide (CO), carbon dioxide (CO_2), and nitrogen (N_2). Note the absence of free atmospheric oxygen.

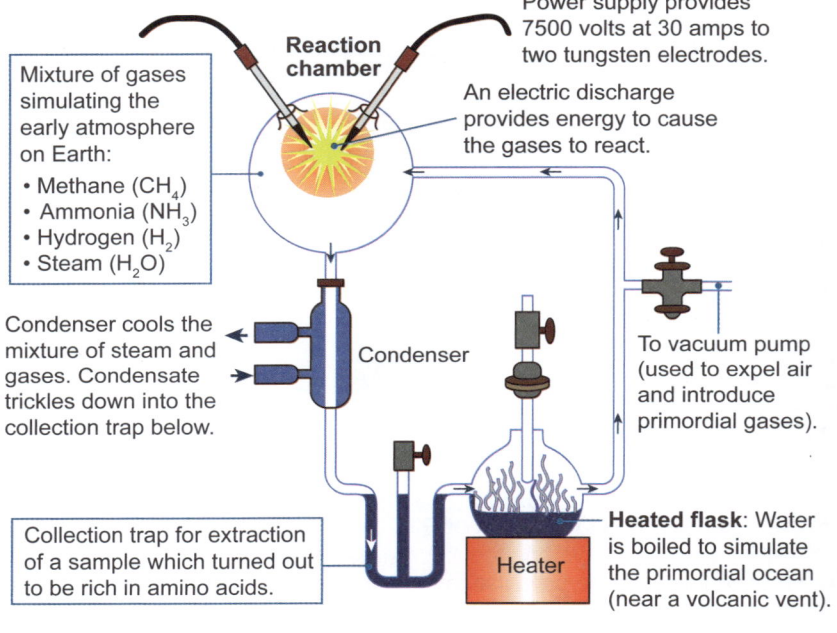

Mixture of gases simulating the early atmosphere on Earth:
- Methane (CH_4)
- Ammonia (NH_3)
- Hydrogen (H_2)
- Steam (H_2O)

Condenser cools the mixture of steam and gases. Condensate trickles down into the collection trap below.

Collection trap for extraction of a sample which turned out to be rich in amino acids.

Power supply provides 7500 volts at 30 amps to two tungsten electrodes.

An electric discharge provides energy to cause the gases to react.

To vacuum pump (used to expel air and introduce primordial gases).

Heated flask: Water is boiled to simulate the primordial ocean (near a volcanic vent).

2. How did the discovery of ribozymes provide evidence for the RNA world hypothesis? _____

_____

_____

_____

3. Describe the two alternative prebiotic world hypotheses: _____

_____

_____

_____

_____

4. State a hypothesis for the Miller-Urey experiment: _____

_____

_____

5. Discuss if the Miller-Urey experiment supported the idea that conditions on early Earth could have produced simple organic molecules:

_____

_____

_____

_____

6. Why is it highly unlikely that life could ever begin again on present day Earth? _____

_____

_____

Comparing hypotheses

7. (a) Evaluate the competing hypotheses for the origin of DNA. If you wish to research more information, the links in the **BIOZONE Resource Hub** will help you. Describe the evidence for and against each hypothesis. Which do you think is more likely to have occurred and why?

_____

_____

_____

_____

_____

_____

_____

_____

_____

_____

_____

_____

_____

_____

(b) Can you come up with a different hypothesis for the origin of DNA? How does your hypothesis solve the problems encountered by the current hypotheses? What problems of its own does it have?

_____

_____

_____

_____

_____

_____

121 Introduction to Gene Expression

Key Question: How is the information in DNA turned into proteins?

Using codes

- You have probably used codes at some time in your life, perhaps as a simple exercise in writing or mathematics. A simple code might be swapping letters for numbers so that 1 = A and 2 = B and so on.
- A more complicated code could be that multiple digits code for a letter. For example: a set of three digits codes for a letter. The code 11_ = A, 12_ = B, and 36_ = Z. The third digit aids in making the code more complex, as it could be used to designate special characters or be swapped occasionally so that 367 = Z, but 368 = R and 363 = A, to throw off people who might want to break the code.

1. Use the code 11_ = A, 12_ = B, and 36_ = Z to work out the meaning of the number sequence:

   301181192290194297111292156246301157249136151

   _____

DNA and gene expression

- We have seen that **DNA** is a sequence of many millions of the **nucleotides** A, T, C, and G and that it is responsible for storing the instructions for how a cell behaves or what it produces.
- How is this done? How is the information in DNA read by the cell? How is that information used to build cell products? The diagram below shows a summary of this process, called **gene expression**.

A summary of eukaryotic gene expression

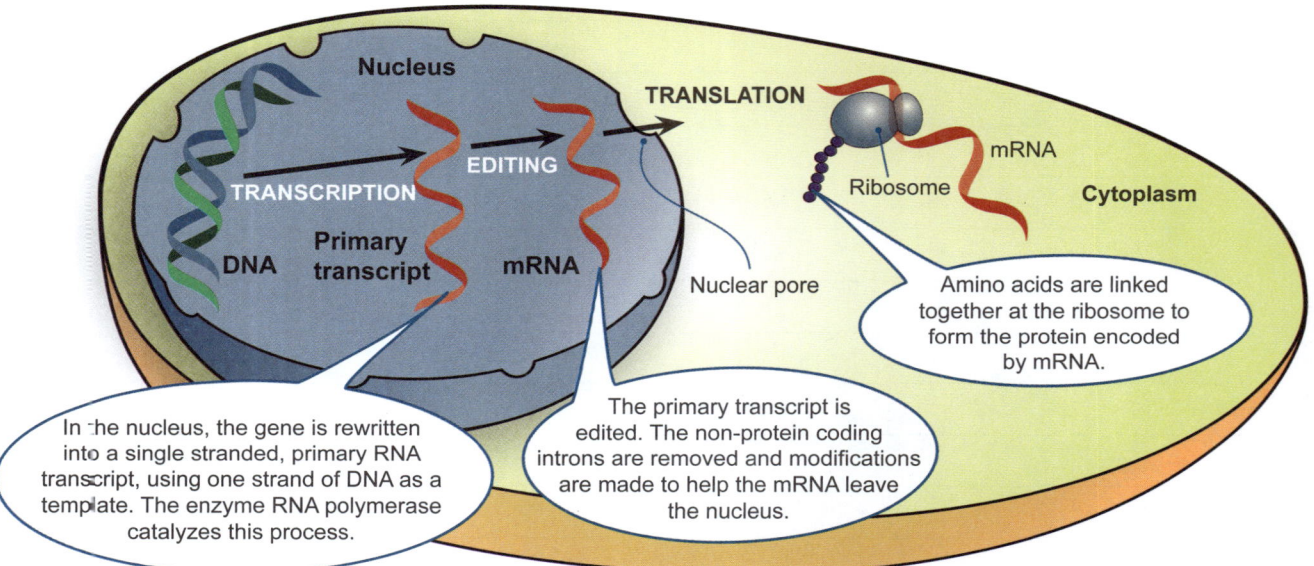

In the nucleus, the gene is rewritten into a single stranded, primary RNA transcript, using one strand of DNA as a template. The enzyme RNA polymerase catalyzes this process.

The primary transcript is edited. The non-protein coding introns are removed and modifications are made to help the mRNA leave the nucleus.

Amino acids are linked together at the ribosome to form the protein encoded by mRNA.

- The DNA sequence contains sections called **genes**. These are the sections of DNA that hold the information for the production of specific protein products. In human DNA, there are between 20,000 to 25,000 genes.
- We can see that there are three main steps in the processing of the information. It involves **transcription** (rewriting) of the DNA (the gene) into mRNA, editing of the mRNA, and the **translation** of the mRNA into an amino acid sequence (proteins).

2. What is the significance of gene expression? _____

   _____

   _____

3. What is the role of DNA in gene expression? _____

   _____

   _____

122 Transcription

Key Question: What is the purpose of transcription, where does it occur, and what are the key steps in the process?

Transcription

- **Transcription** takes place in the nucleus and is carried out by the enzyme RNA polymerase. This enzyme rewrites the **DNA** into a primary RNA transcript using a single template strand of DNA (below).
- A promoter region is found at the start of the **gene** and a terminator region is found at the end. These show the RNA polymerase where to start and stop transcription.

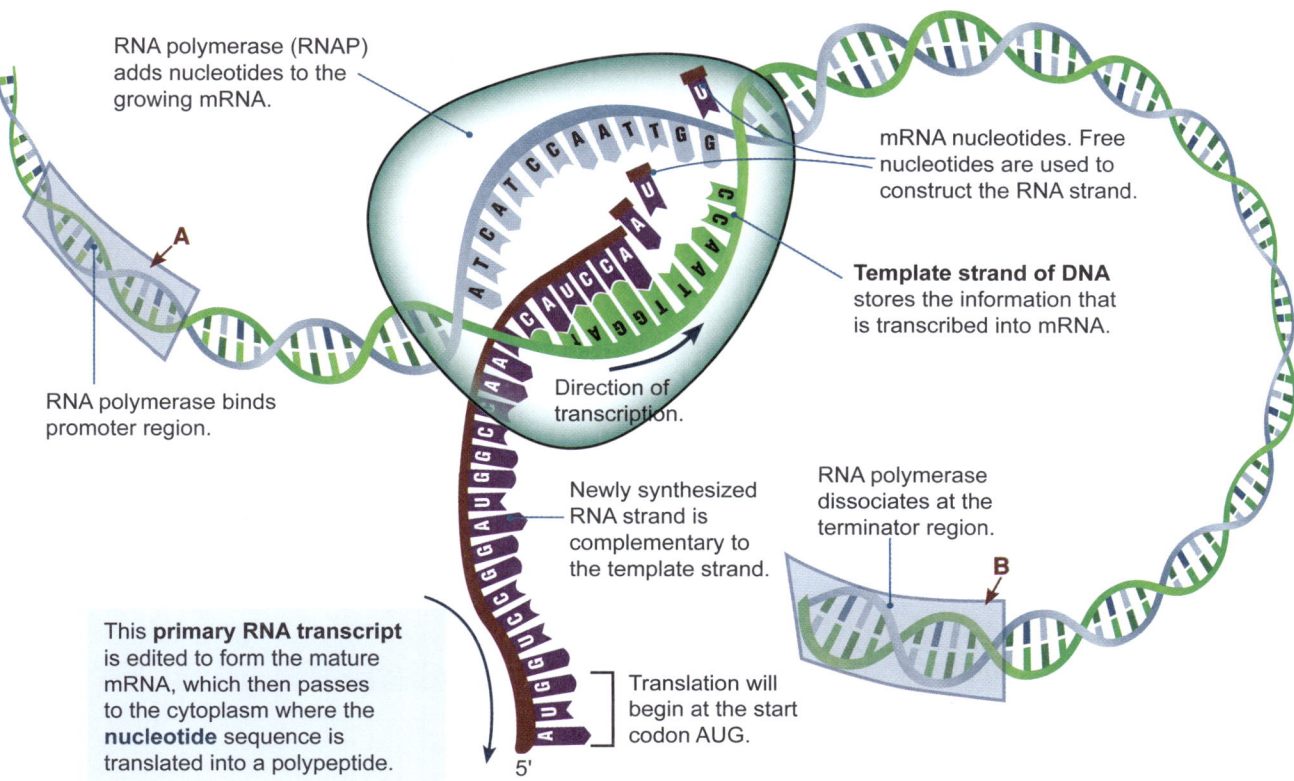

1. (a) Name the enzyme responsible for transcribing the DNA: _____

   (b) What strand of DNA does this enzyme use? _____

   (c) Is the code on this strand the same as or complementary to the RNA being formed: _____

   (d) Which nucleotide base replaces thymine in mRNA? _____

   (e) What is represented by points **A** and **B** on the diagram? _____

2. (a) Why is AUG called the start codon? _____

   (b) What would the three letter code be on the DNA coding strand? _____

3. In your own words, describe the process of transcription: _____

123 mRNA Editing

Key Question: How does mRNA editing affect the primary mRNA or protein produced?

mRNA editing

▶ In eukaryotes, a **gene** is made up of non protein-coding sections called introns and coding sections called exons. After transcription of the DNA, the introns must be removed and the remaining exons spliced together to form the mature mRNA before the gene can be translated into a protein. This editing process also occurs in the nucleus.

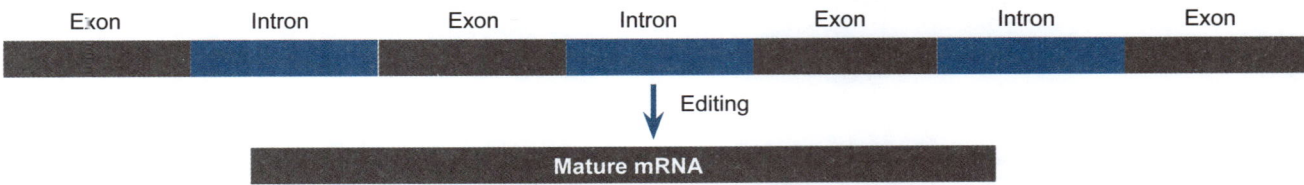

Intron removal can be variable

▶ Human DNA contains 25,000 genes, but produces up to 1 million different proteins. Each gene must therefore produce more than one protein.

▶ Primary mRNA contains exons and introns. Introns are usually removed after transcription and exons are spliced together. However, introns can be removed selectively. Leaving different introns in the final mRNA allows for variation in the final amino acid sequence and, therefore, the protein that is made.

▶ In mammals, the most common method of alternative splicing involves exon skipping, in which not all exons are spliced into the final mRNA (below). Other alternative splicing options create further variants.

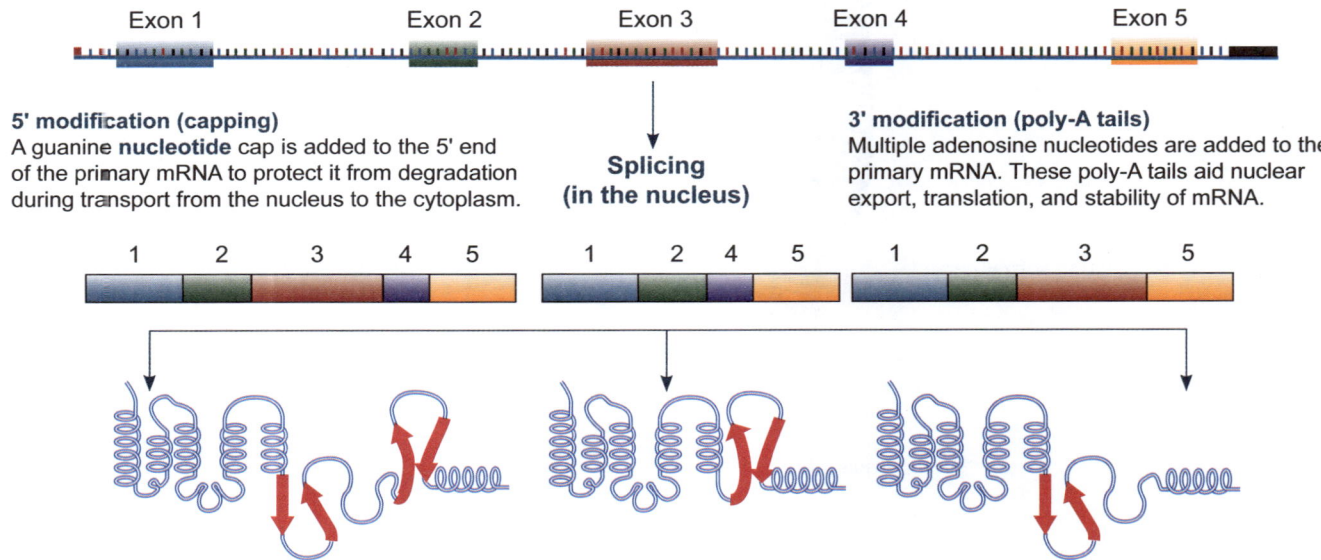

5' modification (capping)
A guanine **nucleotide** cap is added to the 5' end of the primary mRNA to protect it from degradation during transport from the nucleus to the cytoplasm.

Splicing (in the nucleus)

3' modification (poly-A tails)
Multiple adenosine nucleotides are added to the primary mRNA. These poly-A tails aid nuclear export, translation, and stability of mRNA.

1. What is the purpose of the caps and tail on mRNA? _____

2. (a) What happens to the intronic sequences in DNA after transcription? _____

   (b) What is one possible fate for these introns? _____

3. How can so many proteins be produced from so few genes? _____

4. If a human produces 1 million proteins, but human DNA codes for only 25,000 genes, on average, how many proteins are produced per gene?

124 The Genetic Code

Key Question: How is the information in the sequence of nucleotides that make up DNA converted into amino acids?

The genetic code

- The genetic code is the set of rules by which the genetic information in **DNA** (or mRNA) is translated into proteins.
- Just as the example in Activity 121 used a three digit sequence to represent a letter of the alphabet, the genetic information for the assembly of amino acids is stored as a set of three-nucleotide (bases) sequences. These three letter codes on mRNA are called **codons**.
- Each codon represents one of 20 amino acids used to make proteins. The code is effectively universal, being the same in all living things (with a few minor exceptions).
- The code is degenerate, meaning there may be more than one codon for each amino acid. Most of this degeneracy is in the third nucleotide of a codon.

A triplet (three nucleotide bases) codes for a single amino acid. The triplet code on mRNA is called a codon.

	Second letter: U	Second letter: C	Second letter: A	Second letter: G	
U	UUU Phe UUC Phe UUA Leu UUG Leu	UCU Ser UCC Ser UCA Ser UCG Ser	UAU Tyr UAC Tyr UAA STOP UAG STOP	UGU Cys UGC Cys UGA STOP UGG Trp	U C A G
C	CUU Leu CUC Leu CUA Leu CUG Leu	CCU Pro CCC Pro CCA Pro CCG Pro	CAU His CAC His CAA Gln CAG Gln	CGU Arg CGC Arg CGA Arg CGG Arg	U C A G
A	AUU Ile AUC Ile AUA Ile AUG Met	ACU Thr ACC Thr ACA Thr ACG Thr	AAU Asn AAC Asn AAA Lys AAG Lys	AGU Ser AGC Ser AGA Arg AGG Arg	U C A G
G	GUU Val GUC Val GUA Val GUG Val	GCU Ala GCC Ala GCA Ala GCG Ala	GAU Asp GAC Asp GAA Glu GAG Glu	GGU Gly GGC Gly GGA Gly GGG Gly	U C A G

First letter (rows); Third letter (final column)

1. What is the amino acid chain produced from the following mRNA sequences? Use abbreviations for the amino acids.

    (a) GAU CCG UAC GUA CGA ACA AUU ACC: _____

    (b) GGG UUU GCU UGG CAA AAC AGU GCA: _____

    (c) AAA CCC GGG GUA AUU CGC AAU GAU: _____

2. (a) If one RNA base (A, U, C, G) coded for one amino acid, how many amino acids could be coded for in total? _____

    (b) If two RNA bases coded for one amino acid, how many amino acids could be coded for in total? _____

3. (a) What is the effect (in general) of changing the last base in a codon? _____

    (b) How might this affect the chance of a mutation changing the final protein produced after gene expression?

4. Explain what is meant by saying the genetic code is degenerate: _____

125 Translation

Key Question: What is the purpose of translation, where does it occur, and what are the key steps in the process?

Translation

- **Translation** is the process by which the **codons** stored in the mRNA are read by the cell's "machinery" and the information they encode is turned to proteins.
- The important molecules in this process are the mRNA itself, ribosomes, and transfer RNA (tRNA).
- tRNA molecules match amino acids with the appropriate codon on mRNA. The tRNA delivers its amino acid to the ribosome, where enzymes join the amino acids to form a polypeptide chain.
- During translation, the ribosome "wobbles" along the mRNA molecule, joining amino acids together. Enzymes and energy are involved in attaching the tRNA molecules to their amino acids and lengthening the peptide chain.

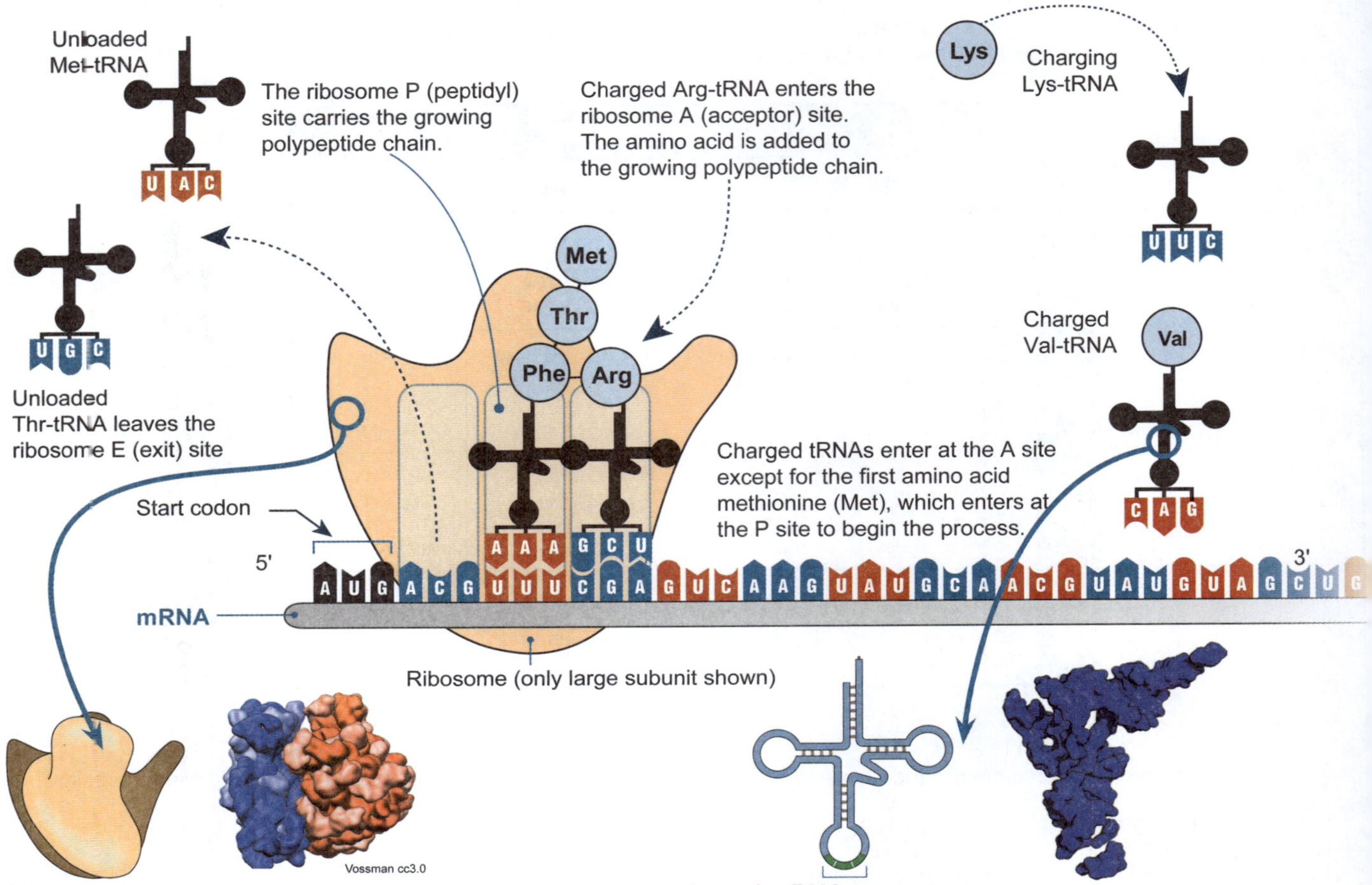

Ribosomes

Ribosomes are composed of ribosomal RNA (rRNA). They exist as two separate sub-units (above) which come together around the mRNA strand. They catalyze the steps for protein synthesis and have specific regions that accommodate transfer RNA (tRNA) molecules loaded with amino acids.

transfer RNA

transfer RNA molecules (tRNA) transfer amino acids. Each tRNA has a 3-base **anticodon**, which is complementary to an mRNA codon. This allows it to deliver the correct amino acid by matching its anticodon to the mRNA codon.

1. What is the role of each of the following components in translation?

    (a) Ribosome: _____

    (b) tRNA: _____

    (c) Amino acids: _____

    (d) Start codon: _____

    (e) Stop codon: _____

2. Many ribosomes can work on one strand of mRNA at a time (a polyribosome system). What would this achieve?

    _____

126 Modeling Gene Expression

Key Question: How can a model be used to explain gene expression?

Models can be used to explain the process of gene expression

The following exercise will help you understand and explain the three important steps in the process of **gene expression**. Using plastic building blocks, you will model how proteins are produced using the information stored in **DNA**.

Investigation 5.2 Modeling gene expression

See appendix for equipment list.

1. Plastic blocks can be used to model a vast number objects and processes. The photographs below, and on the following page, show how they could be used to model the bases that make up DNA, a DNA strand, and tRNA. Using blocks like this, model the process of protein synthesis, including mRNA, ribosomes, transcription and translation.

2. Take photos of your models, print them out, and paste the photos in the spaces on the following pages to show protein synthesis. You could also make a video of the process and share with your class and teacher.

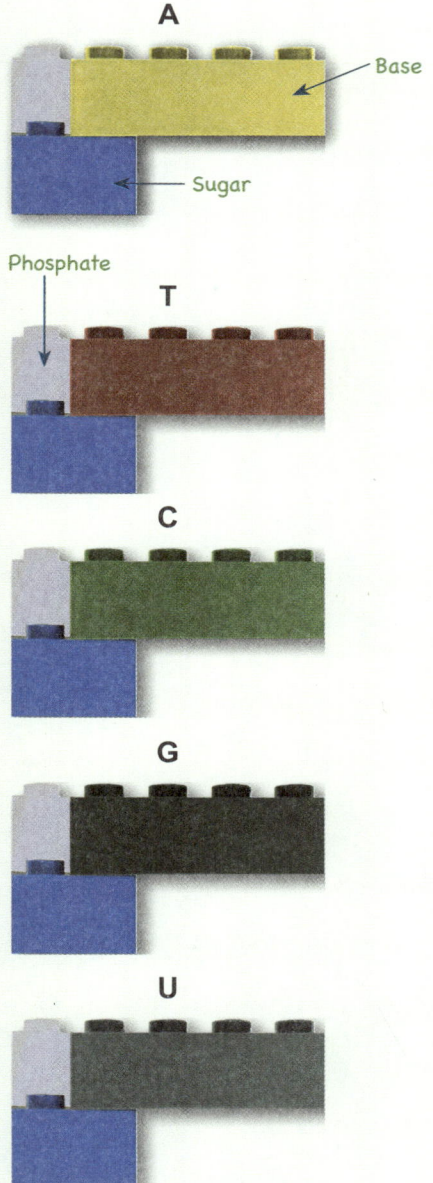

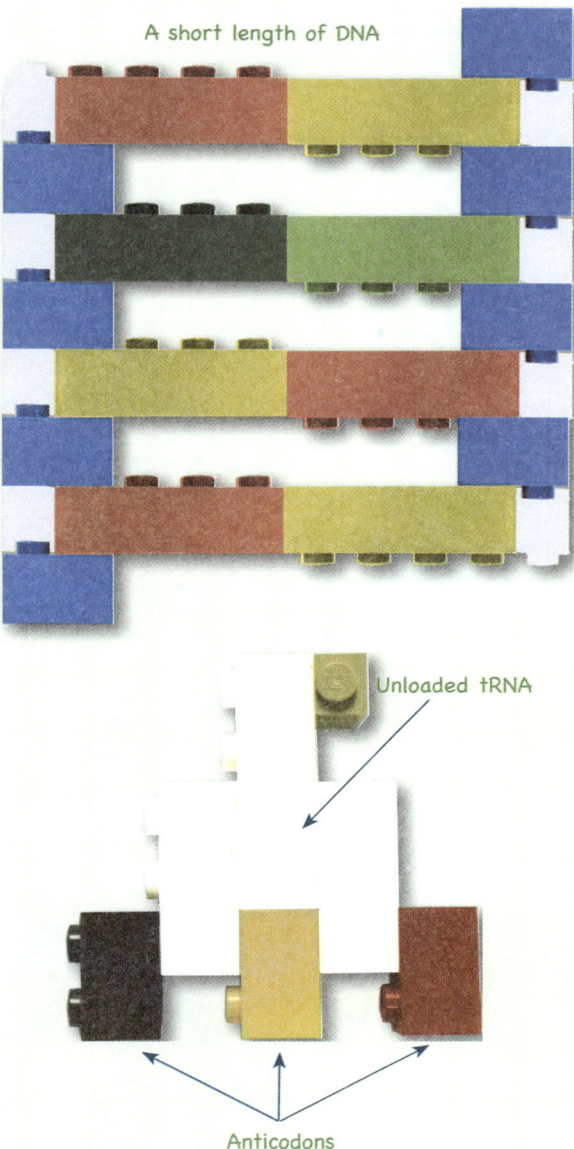

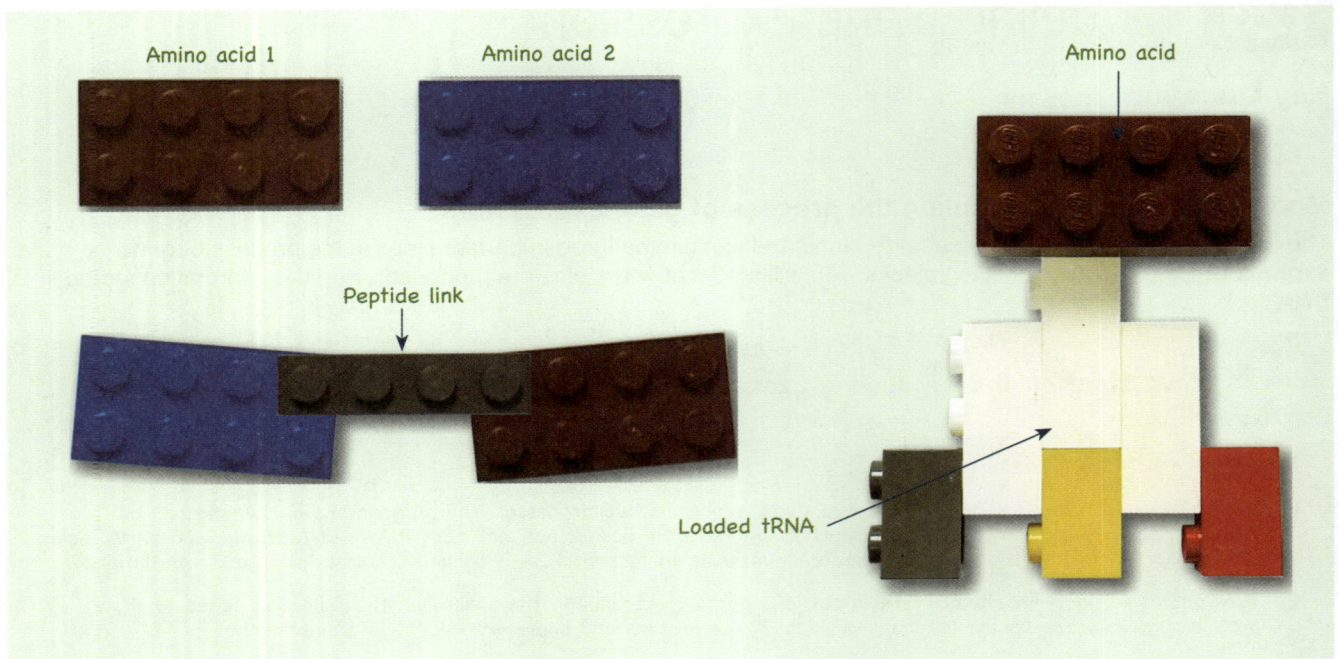

1. Paste your photographs here and annotate them as appropriate:

The diagram below shows a summary of protein synthesis.

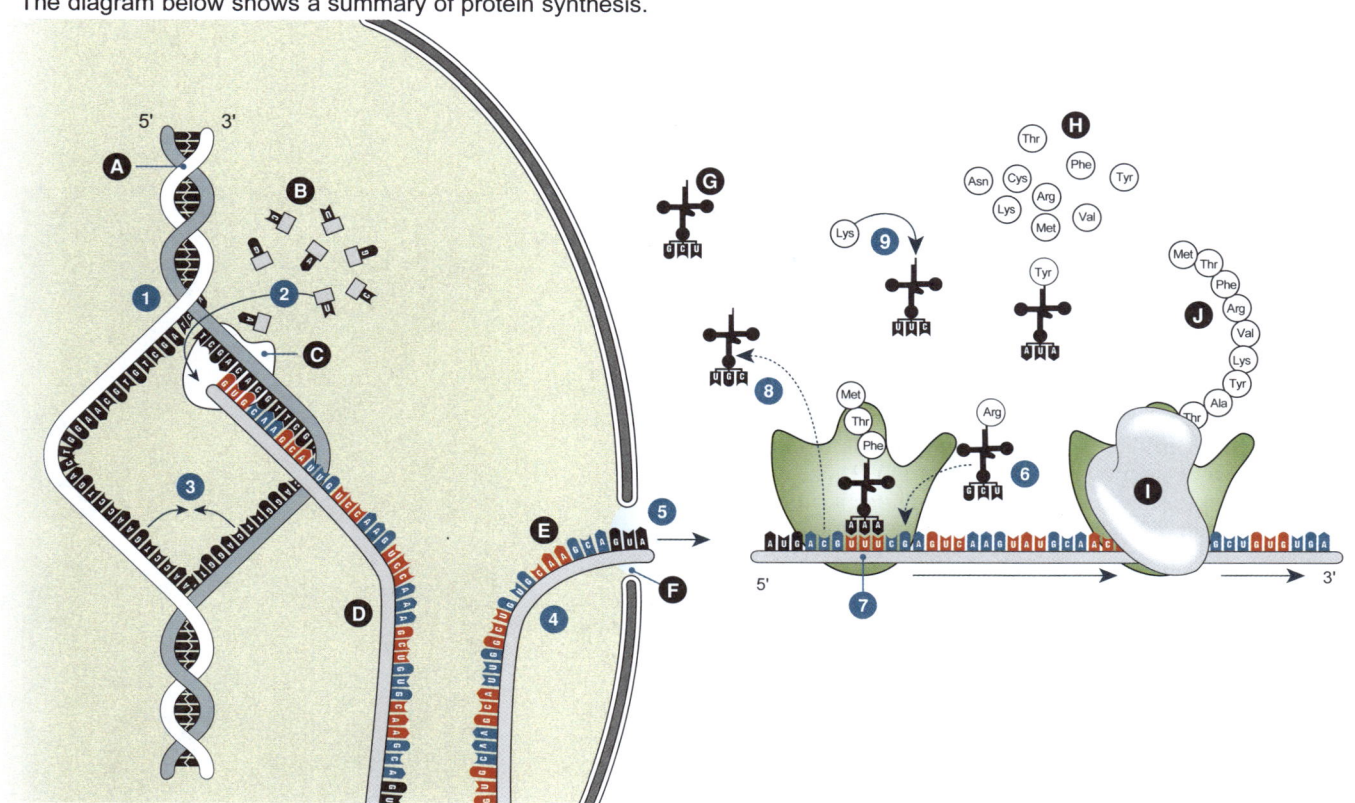

1. Briefly describe each of the numbered processes in the diagram above:

    (a) Process 1: _____

    (b) Process 2: _____

    (c) Process 3: _____

    (d) Process 4: _____

    (e) Process 5: _____

    (f) Process 6: _____

    (g) Process 7: _____

    (h) Process 8: _____

    (i) Process 9: _____

2. Identify each of the structures marked with a letter and write their names below in the spaces provided:

    (a) Structure A: _____   (f) Structure F: _____

    (b) Structure B: _____   (g) Structure G: _____

    (c) Structure C: _____   (h) Structure H: _____

    (d) Structure D: _____   (i) Structure I: _____

    (e) Structure E: _____   (j) Structure J: _____

3. Describe two factors that would determine whether or not a particular protein is produced in the cell:

    (a) _____

    _____

    (b) _____

    _____

127 DNA Sequence and Traits

Key Question: How does the DNA specify the traits of an organism?

The nucleotide sequence affects gene function

- You have seen that the mechanisms for **transcription** and **translation** in the cell are able to take the sequence of **nucleotides** in **DNA** and produce amino acid chains, that then form proteins. These proteins, and importantly enzymes, play important roles in the development of an organism (especially during early growth).

- It becomes apparent then, that specific sequences of nucleotides will produce their own specific amino acid sequences. Those sequences will produce proteins and enzymes that will function in their own specific ways (or may not function at all). This then produces an explanation for the features seen in different organisms, or the different **traits** seen in different individuals of the same species.

- The example below shows part of the nucleotide sequence for the MSTN **gene** found in mammals. The MSTN gene produces an enzyme that inhibits muscle growth. The sequence is specifically from cattle.

    GAT TGT GAT GAA CAC TCC ACA GAA TCT

- Cattle with the nucleotide sequence above will develop in the usual way (top right). However, Belgian blue cattle (bottom right) have a slightly shorter nucleotide sequence in their MSTN gene:

    GAT TGT GAC AGA ATC T

- The result is the distinctive double muscle trait of Belgian blue cattle.

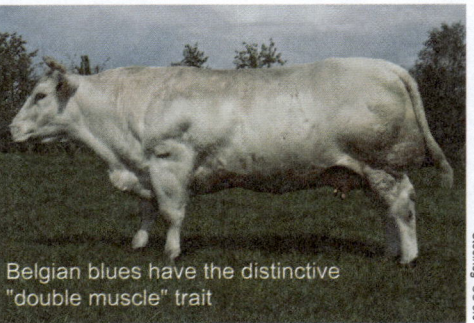
Belgian blues have the distinctive "double muscle" trait

Eye color

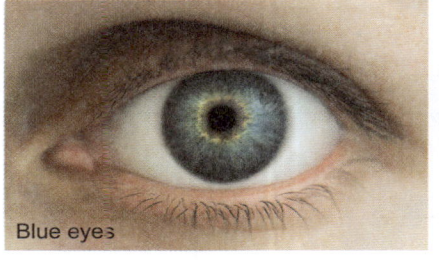
Blue eyes

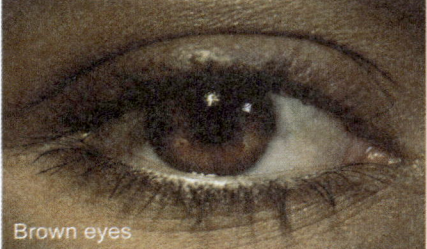
Brown eyes

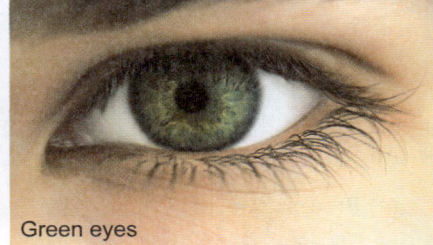
Green eyes

- Eye color is linked to at least eight genes. The genes OCA2 and HERC2 appear to be the most important in determining eye color. An intron in HERC2 contains a promotor for the OCA2 gene. Slight variations in the HERC2 intron will affect how the OCA2 gene is regulated and so will influence eye color. In humans, these genes are very close together on chromosome 15 and so are normally inherited as a single unit.

- Studies have shown that the promotor region for OCA2 is in intron 86 for HERC2. These studies have also shown that a difference in one base in this intron (cytosine instead of thymine) is the difference between having brown eyes and blue eyes.

- Variation in the nucleotide sequence of the OCA2 gene is responsible for other variations in eye color, such as having green eyes.

1. Using examples explain how the nucleotide sequence specifies traits in an organism: _____

    _____

    _____

    _____

    _____

    _____

    _____

128 Mutations

Key Question: What are mutations?

Some mutations are retained, others are eliminated

▸ A **mutation** is a permanent change to the **DNA** sequence of an organism. Mutation is the ultimate source of new variation. Mutations can be small, such the change of a single **nucleotide**, e.g. A to G, or they may be large, such as the duplication of entire **genes** or sometimes entire genomes.

▸ Most mutations are harmful because they disrupt some important cellular process, often by causing a protein to fold incorrectly. Occasionally, they may cause some beneficial change, such as making an enzyme more efficient. Although the DNA replication process is very accurate, it is estimated that, in humans, a mutation occurs once every 30 million base pairs copied during DNA replication prior to meiosis. This means that every person has about 200-300 new mutations that their parents did not have. In most cases, these mutations have little to no effect and are called silent mutations.

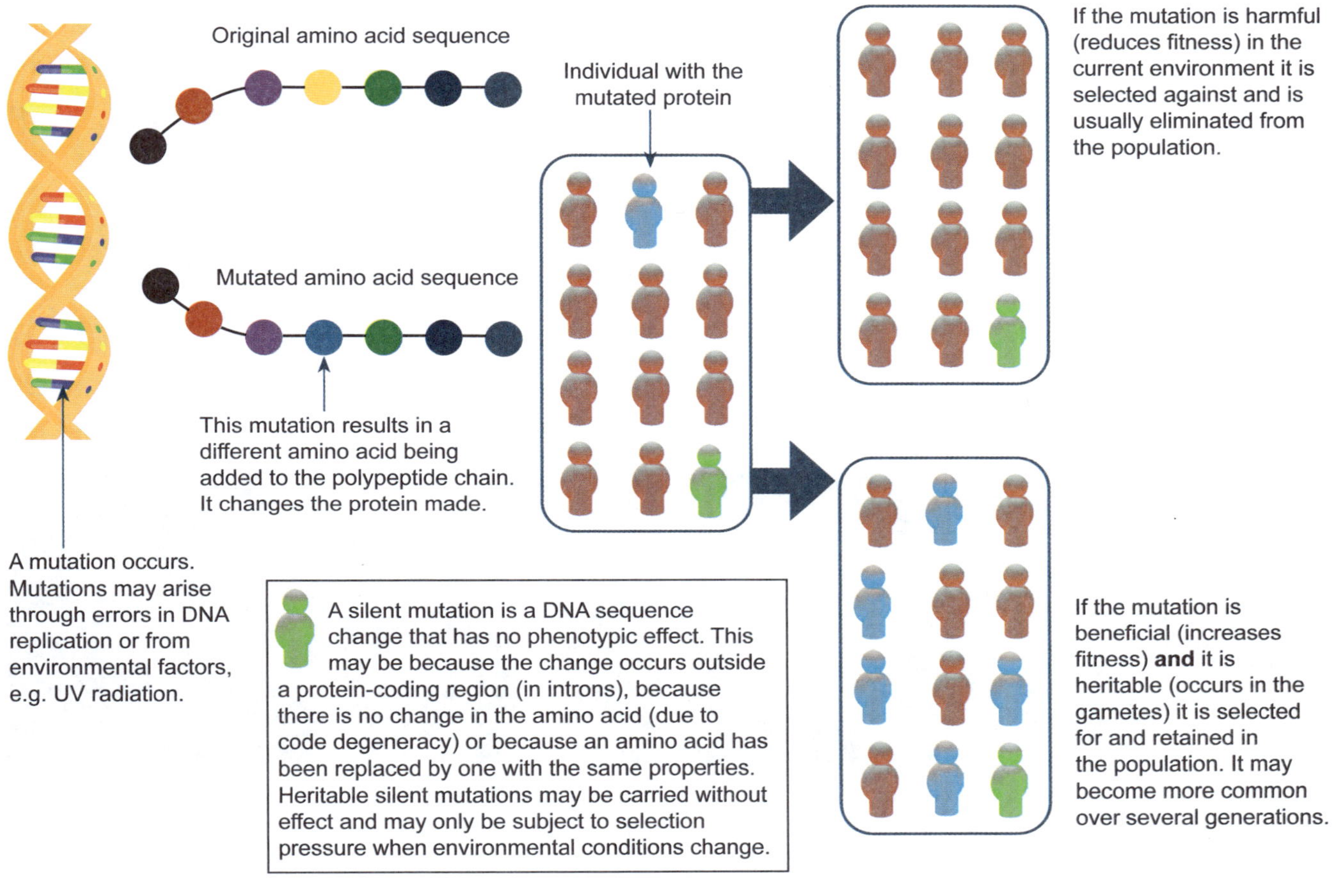

Original amino acid sequence

Individual with the mutated protein

If the mutation is harmful (reduces fitness) in the current environment it is selected against and is usually eliminated from the population.

Mutated amino acid sequence

This mutation results in a different amino acid being added to the polypeptide chain. It changes the protein made.

A mutation occurs. Mutations may arise through errors in DNA replication or from environmental factors, e.g. UV radiation.

A silent mutation is a DNA sequence change that has no phenotypic effect. This may be because the change occurs outside a protein-coding region (in introns), because there is no change in the amino acid (due to code degeneracy) or because an amino acid has been replaced by one with the same properties. Heritable silent mutations may be carried without effect and may only be subject to selection pressure when environmental conditions change.

If the mutation is beneficial (increases fitness) **and** it is heritable (occurs in the gametes) it is selected for and retained in the population. It may become more common over several generations.

1. What is a mutation? _____

2. How does a mutation lead to new variation in a population? _____
   _____
   _____

3. Why are some mutations retained within a population and others eliminated? _____
   _____
   _____
   _____

4. What is a silent mutation? _____
   _____

Somatic and gametic mutations

▸ Gametic cells are the reproductive (sex) cells of an organism (the egg and sperm). Mutations occurring in these cells are called gametic mutations.

▸ Somatic cells (body cells) are all the remaining cells. Mutations to these cells are called somatic mutations.

▸ Only gametic mutations will be inherited. Somatic mutations are not inherited but may affect an organism in its lifetime, e.g. a cancer.

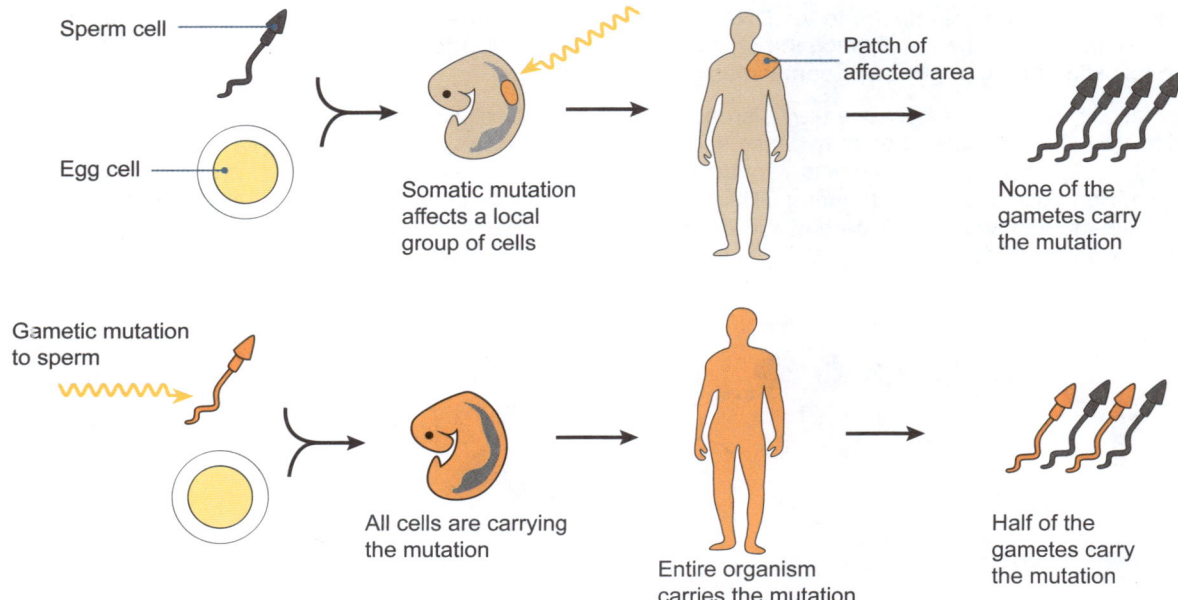

Cause of mutations

▸ Mutations occur spontaneously in all organisms. The natural rate at which a gene will undergo change is normally very low, but this rate can be increased by environmental factors such as ionizing radiation and mutagenic chemicals.

Ionizing radiation

High energy radiation e.g. ultraviolet radiation from the sun, and particle emission from radioactive isotopes, can penetrate tissue and cause DNA damage. Skin cancer from high exposure to ultraviolet is a major health concern in light skinned people.

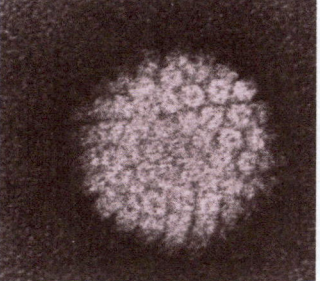

Viruses and microorganisms

Some viruses integrate into the human chromosome, upsetting genes and triggering cancers. Examples include hepatitis B virus (liver cancer), HIV (Kaposi's sarcoma), and HPV (above) which is implicated in cervical cancer.

Poisons and irritants

Many chemicals interact directly with DNA to trigger cancer (they are carcinogenic). Synthetic and natural examples include organic solvents, e.g. benzene and tobacco tar. Those most at risk include workers in the chemicals industries.

Diet, alcohol and tobacco

Diets high in fat, especially fatty, highly preserved meat, can irritate the lower gut, causing bowel cancer. High alcohol intake increases the risk of some cancers. Tobacco tars contain at least 17 known carcinogens. Lung cancer is a common result of smoking.

5. Distinguish between somatic and gametic mutations: _____

6. Explain how mutagens cause mutations: _____

129 Changes to DNA

Key Question: How do mutations change the DNA sequence?

Gene mutations

▸ **Mutations** can change the **DNA** sequence in many different ways. These include deletions, insertions, and substitutions of bases. These may involve just a single **nucleotide**, a few, or many. The type of mutation can have an important effect on the outcome of translation and the amino acid chain (below).

▸ Mutations may occur during replication or when the DNA is repaired after being damaged, e.g. from exposure to ultraviolet (UV) radiation.

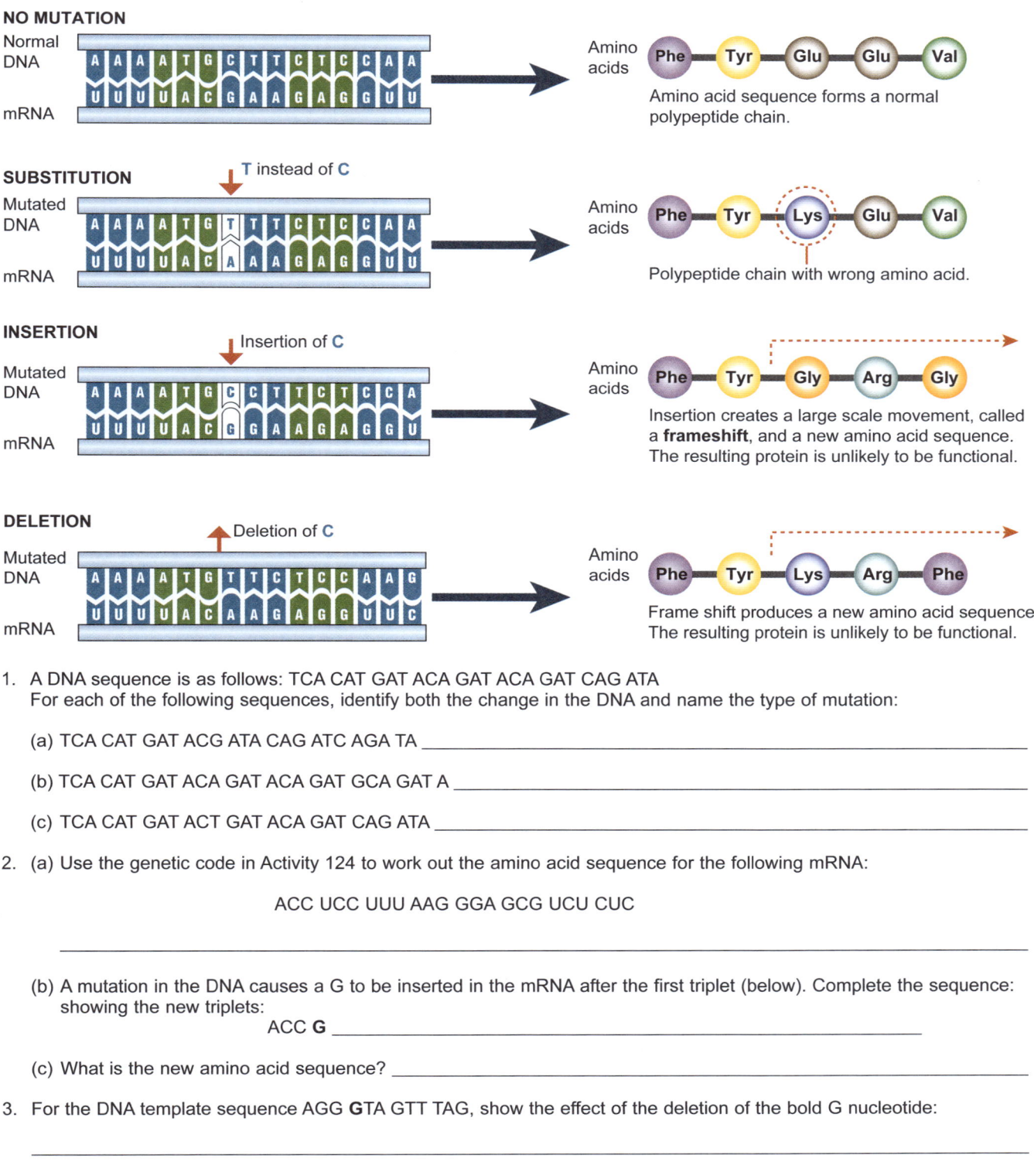

1. A DNA sequence is as follows: TCA CAT GAT ACA GAT ACA GAT CAG ATA
   For each of the following sequences, identify both the change in the DNA and name the type of mutation:

   (a) TCA CAT GAT ACG ATA CAG ATC AGA TA _____

   (b) TCA CAT GAT ACA GAT ACA GAT GCA GAT A _____

   (c) TCA CAT GAT ACT GAT ACA GAT CAG ATA _____

2. (a) Use the genetic code in Activity 124 to work out the amino acid sequence for the following mRNA:

   ACC UCC UUU AAG GGA GCG UCU CUC

   _____

   (b) A mutation in the DNA causes a G to be inserted in the mRNA after the first triplet (below). Complete the sequence: showing the new triplets:
       ACC **G** _____

   (c) What is the new amino acid sequence? _____

3. For the DNA template sequence AGG **G**TA GTT TAG, show the effect of the deletion of the bold G nucleotide:

   _____

   _____

Chromosome mutations

▶ Chromosome mutations (also called block mutations) involve the rearrangement of whole blocks of **genes** involving many bases, rather than individual bases within a gene. They commonly occur during meiosis and they alter the number or sequence of whole sets of genes on the **chromosome**.

Deletion

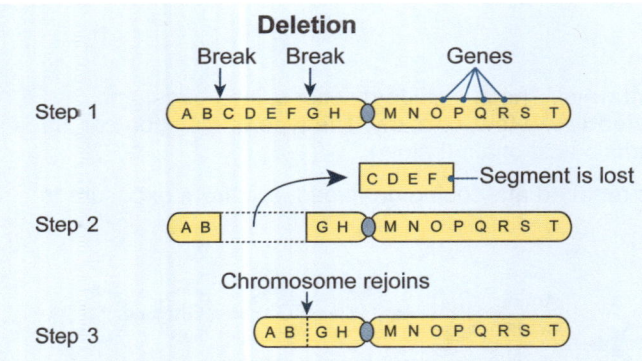

A break may occur at two points on the chromosome, and the middle piece of the chromosome falls out. The two ends then rejoin to form a chromosome that is deficient in some genes. Alternatively, the end of a chromosome may break off and be lost.

Inversion

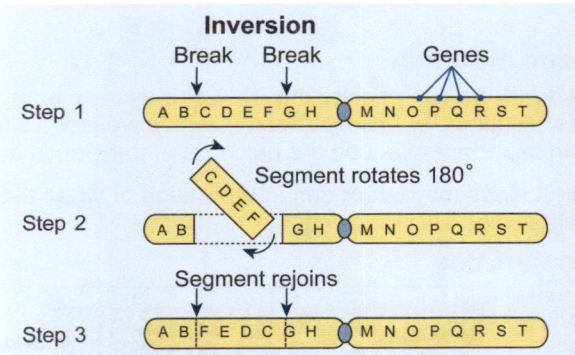

The middle piece of the chromosome falls out and rotates through 180° and then rejoins. There is no loss of genetic material. The genes will be in a reverse order for this segment of the chromosome.

Translocation

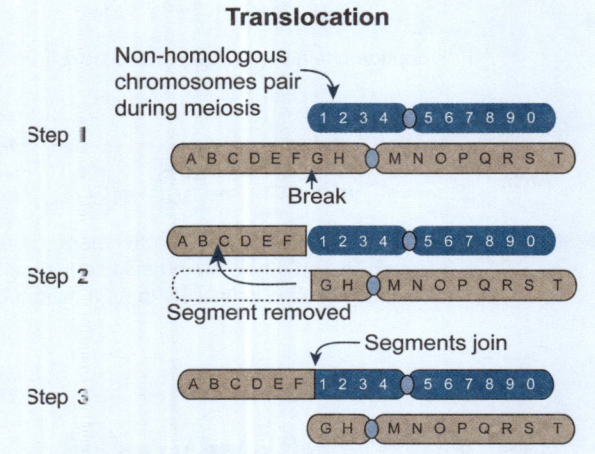

In translocation mutations, a group of genes moves between different chromosomes. The large chromosome (brown) and the small chromosome (blue) are not homologous. A piece of one chromosome breaks off and joins to the other. When the chromosomes are passed to gametes, some gametes will receive extra genes, while some will be deficient.

Duplication

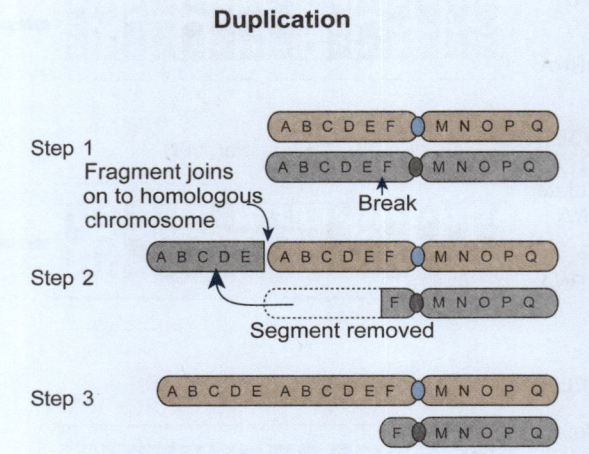

A segment is lost from one chromosome and is added to its homologue. In this diagram, the darker chromosome on the bottom is the 'donor' of the duplicated piece of chromosome. The chromosome with the segment removed is deficient in genes. Some gametes will receive double the genes while others will have no genes for the affected segment.

4. Which of the chromosome mutations above results in a loss of genetic information? _____

5. For the chromosome (block) mutations above, write the new gene sequence after the mutation has occurred:

                   Original sequence   **Mutated sequence**

   (a) Inversion: A B C D E F G H M N O P Q R S T   _____

   (b) Translocation: 1 2 3 4 5 6 7 8 9 0   _____

6. Which type of block mutation is likely to be the least damaging to the organism? Explain your answer:

   _____
   _____

7. Why do translocations sometimes reduce the total number of chromosomes? _____
   _____
   _____

130 Effects of Mutations

Key Question: What types of effects are caused by mutations, and are they always harmful?

Beneficial or harmful?

▸ Most **mutations** have a harmful effect on the organism. This is because changes to the **DNA** sequence of a **gene** can potentially change the amino acid chain encoded by the gene. Proteins need to fold into a precise shape to function properly. A mutation may change the way the protein folds and prevent it from carrying out its usual biological function.

▸ However, sometimes a mutation can be beneficial. The mutation may result in a more efficient protein, or produce an entirely different protein that can improve the survival of the organism.

Beneficial mutations

▸ Beneficial mutations are mutations that increase the fitness of the organisms that possess them. Whether or not the mutation is beneficial often depends on the environment.

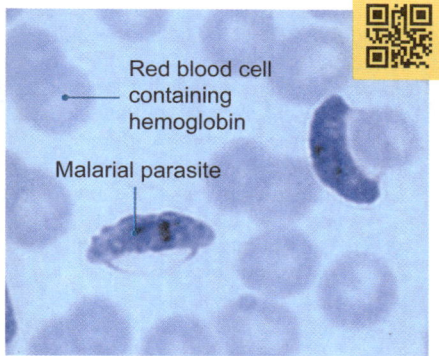

Lactose is a sugar found in milk. All infant mammals produce an enzyme called lactase that breaks the lactose into the smaller sugars, glucose and galactose. As mammals become older, their production of lactase declines and they lose the ability to digest lactose. As adults, they become lactose intolerant and feel bloated after drinking milk. About 10,000 years ago, a mutation appeared in humans that maintained lactase production into adulthood. This mutation is now carried in people of mainly European, African, and Indian descent.

Malaria resistance results from a mutation to the hemoglobin gene (Hb^S) that also causes sickle cell disease. This mutation is beneficial in regions where malaria is common. A less well known mutation (Hb^C) to the same gene, discovered in populations in Burkina Faso, Africa, results in a 29% reduction in the likelihood of contracting malaria if the person has one copy of the mutated gene, and a 93% reduction if the person has two copies. In addition, the anemia that person suffers as a result of the mutation is much less pronounced than in the Hb^S mutation.

Apolipoprotein A1-Milano is a mutation affecting apolipoprotein A1, which helps transport cholesterol through the blood. The mutation causes a change to one amino acid and increases the protein's effectiveness by ten times, dramatically reducing incidence of heart disease. The mutation can be traced back to its origin in Limone, Italy, in 1644. Another mutation, to a gene called PCSK9, has a similar effect, lowering the risk of heart disease by 88%.

1. Why is it that many of the recent beneficial mutations recorded in humans have not spread throughout the entire human population?

_____

_____

_____

2. What selection pressure could act on apolipoprotein A1-Milano to help it spread through a population? _____

_____

_____

_____

3. Why would it be beneficial to be able to digest milk in adulthood? _____

_____

_____

_____

Harmful mutations

▶ Many mutations cause numerous harmful effects because the mutant gene is involved in many different processes in the body.

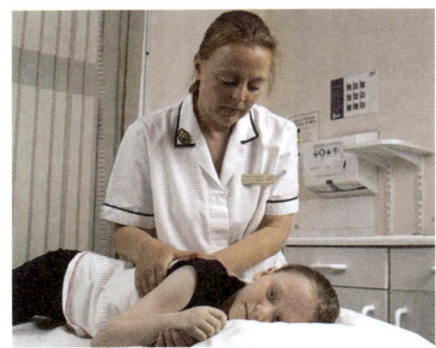

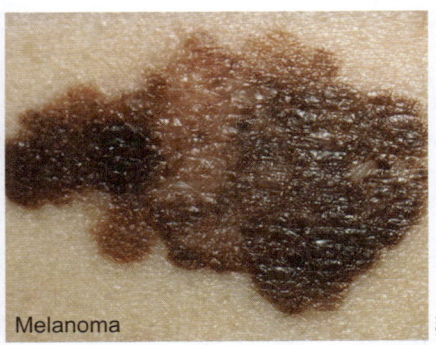

Melanoma

Cystic fibrosis (CF) is an inherited disorder caused by a mutation of the CFTR gene. The CFTR gene's protein product is a membrane-based protein that regulates chloride transport in cells. A mutation, accounting for more than 70% of all defective CFTR genes, causes the loss of a single amino acid. This mutation leads to an abnormal CFTR, which cannot take its proper position in the membrane, nor perform its transport function. Effects of this mutation include disruptions of all glands including pancreas, bronchial glands (chronic lung infections), and sweat glands.

Huntington's disease is a progressive genetic disorder in which nerve cells in certain parts of the brain waste away, or degenerate. Symptoms include shaky hands and an awkward gait. Huntington's disease is caused by a mutation on chromosome 4. A base sequence, CAG, normally repeats between 10 and 28 times on this chromosome, but in people with Huntington's disease, this sequence repeats between 36 to 120 times. Woody Guthrie (above) was an influential folk singer-songwriter who died in 1967 due to complications related to Huntington's disease.

Skin cancer is commonly caused by exposure to ultraviolet light. It is an example of a non-heritable mutation, as skin cells are not involved in the production of reproductive cells (gametes). It has been estimated that there might be 10^{12} mutations in a tumor, occurring both before and after its formation. After exposure to UV light, adjacent thymine bases in DNA become cross-linked to form a "thymine dimer". This disrupts the normal base pairing and affects gene function, usually disrupting the genes regulating the cell cycle.

4. What is the effect of a specific mutation on the CFTR protein? _____

5. The CFTR mutation is a lethal mutation, even more so before modern medicine. However, 1 in 30 white Americans are carriers for the mutation (they have it on one of a pair of chromosomes and don't show any symptoms). The mutation is thought to have appeared about 50,000 years ago. What might the persistence and commonness of such a lethal mutation tell us about it?

6. (a) What causes Huntington's disease? _____

   (b) How does the extent of the mutation affect the symptoms and onset of the disease? _____

7. Whether or not an inherited mutation is harmful or beneficial often depends on the environment. Use an example to show how a single mutation can be either harmful or beneficial under the right conditions:

131 Molecular Technologies and DNA

Key Question: How does DNA manipulation alter an organism's DNA either by adding new DNA or editing the existing DNA?

How are genetically modified organisms produced?

▸ **DNA** manipulation (also called genetic engineering) involves the direct manipulation of an organism's genome using biotechnology. This can be achieved by introducing new DNA into an organism or by editing its existing DNA.

▸ Genetic engineering has wide applications in food technology, industry, agriculture, environmental clean up, pharmaceutical production, and vaccine development. Organisms that have had their DNA altered are called genetically modified organisms (GMOs).

Add a foreign gene

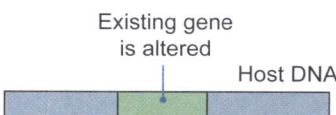

A novel gene (foreign to the recipient) is inserted from another species. This will enable the GMO to express the trait encoded by the new **gene**. Organisms genetically altered in this way are referred to as transgenic.

Alter an existing gene

An existing gene may be altered to make it express at a higher level, e.g. growth hormone, or in a different way, in tissue that would not normally express it. The technique may provide a way to fix a malfunctioning gene.

Delete or 'turn off' a gene

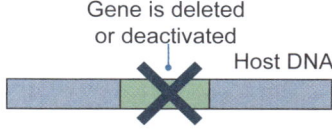

An existing gene may be deleted or deactivated (switched off) to prevent the expression of a trait, e.g. the deactivation of the ripening gene in tomatoes produced the Flavr-Savr tomato.

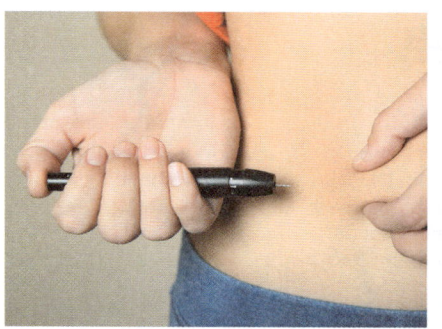

Human insulin, used to treat diabetic patients, is produced by inserting the insulin gene into bacteria or yeast.

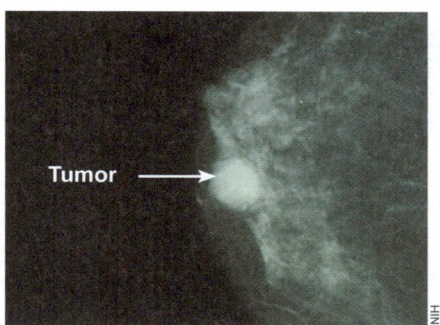

Gene editing technologies, such as CRISPR, are being explored to treat breast cancer (above) and sickle cell disease.

Manipulating gene action is one way in which to control processes such as ripening in fruit so it stays fresher longer.

1. What is DNA manipulation? _____
   _____
   _____

2. For each of the three types of genetic manipulation above, describe why each of the techniques is useful:

   (a) _____
   _____
   _____
   _____

   (b) _____
   _____
   _____
   _____

   (c) _____
   _____
   _____
   _____

132 Polymerase Chain Reaction

Key Question: What are the principles behind the polymerase chain reaction and why is it useful in biotechnology?

▶ The **polymerase chain reaction (PCR)** is a process that can make billions of copies of a target **DNA** sequence of interest so that it can be analyzed. The technique is carried out *in vitro*, i.e. in test tubes / culture dishes, rather than in a living organism. An overview of PCR given below.

▶ PCR's ability to amplify small quantities of DNA means it can be used to identify the presence of organisms in an environment even if they are in very low numbers. Examples include identifying COVID-19 infections (even in people showing no or few symptoms), DNA in ancient bone fragments, and matter from crime scenes.

A single cycle of PCR

DNA polymerase: A thermally stable form of the enzyme is used (e.g. *Taq polymerase*). This is extracted from thermophilic (heat tolerant) bacteria.

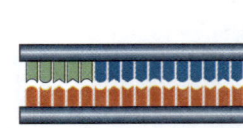

1. Denaturing
A DNA sample (called target DNA) is obtained. It is denatured (DNA strands are separated) by heating at 98°C for 5 minutes.

2. Annealing
The sample is cooled to 60°C. Primers are annealed (bonded) to each DNA strand. In PCR, the primers are short strands of DNA; they provide the starting sequence for DNA extension.

3. Extension/elongation
Free nucleotides and DNA polymerase are added. DNA polymerase binds to the primers and synthesizes complementary strands of DNA, using the free nucleotides.

4. Completed strands
After one cycle, there are now two copies of the original DNA.

Repeat cycle of heating and cooling until enough copies of the target DNA have been produced

1. After only two cycles of replication, four copies of the double-stranded DNA exist. Calculate how much a DNA sample will have increased after:

   (a) 10 cycles: _____   (b) 25 cycles: _____

2. Discuss the importance of PCR to genetic research or industrial bioengineering: _____

   _____

   _____

   _____

   _____

   _____

3. Researchers take great care to avoid DNA contamination during PCR preparation. Explain why: _____

   _____

4. What would happen to the PCR reaction if not enough nucleotides were added to the reaction mix at the beginning of the reaction?

   _____

133 Gel Electrophoresis

Key Question: What is gel electrophoresis, how does it work, and what kind of information does it provide?

What is gel electrophoresis?

- **Gel electrophoresis** is a tool used to isolate fragments of **DNA** for further study. It is also used for DNA profiling, i.e. comparing individuals based on their unique DNA banding profiles.

- DNA has an overall negative charge, so when an electrical current is run through a gel, the DNA moves towards the positive electrode (right).

- The rate at which the DNA molecules move through the gel depends primarily on their size and the strength of the electric field. The gel they move through is full of pores (holes). Smaller DNA molecules move through the pores more quickly than larger ones. The DNA molecules can be stained and visualized as a series of bands. Each band contains DNA molecules of a particular size. The bands furthest from the start of the gel contain the smallest DNA fragments.

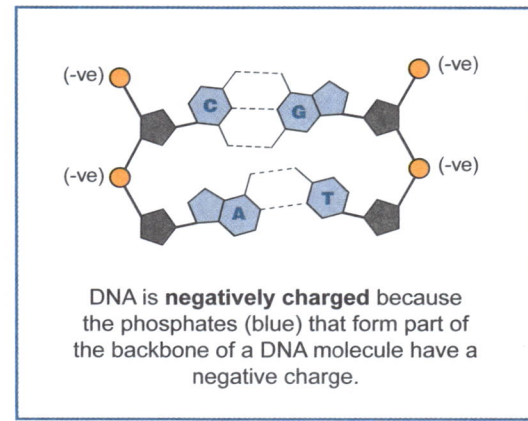

DNA is **negatively charged** because the phosphates (blue) that form part of the backbone of a DNA molecule have a negative charge.

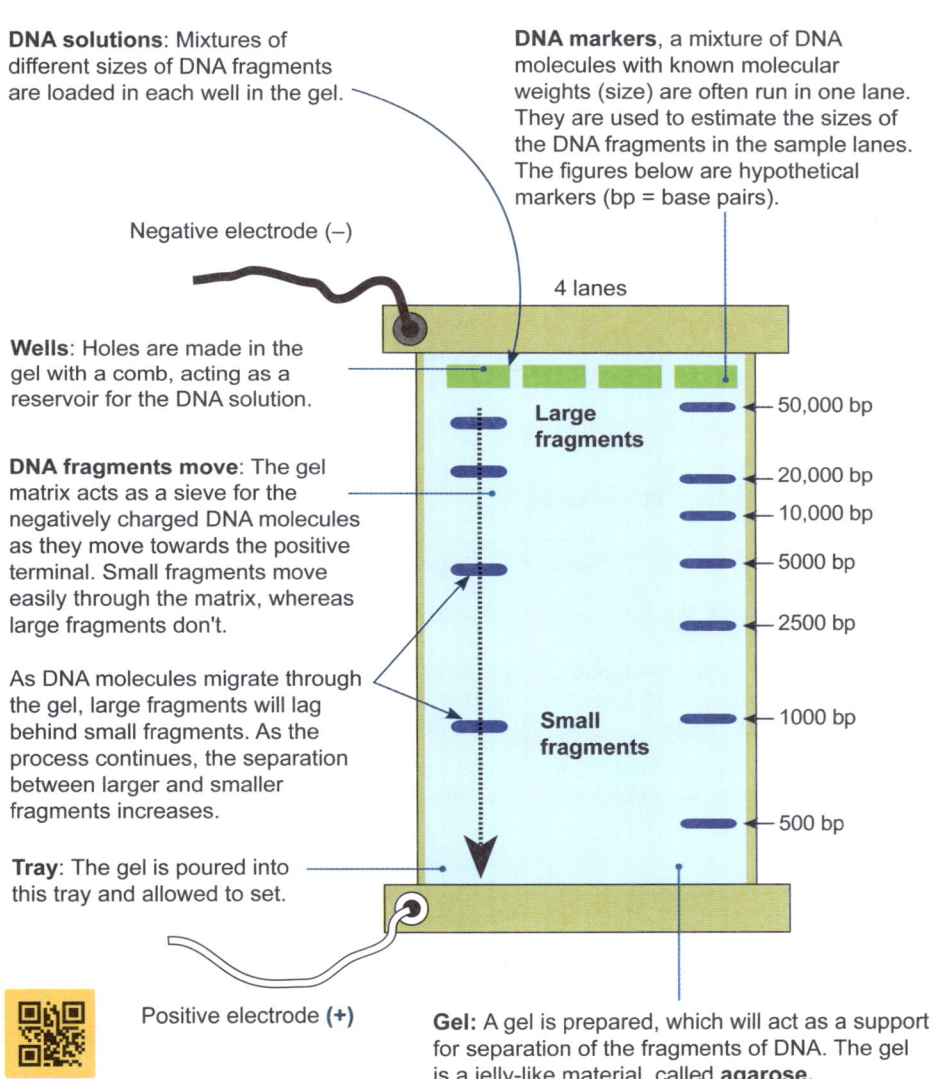

DNA solutions: Mixtures of different sizes of DNA fragments are loaded in each well in the gel.

DNA markers, a mixture of DNA molecules with known molecular weights (size) are often run in one lane. They are used to estimate the sizes of the DNA fragments in the sample lanes. The figures below are hypothetical markers (bp = base pairs).

Negative electrode (−)

4 lanes

Wells: Holes are made in the gel with a comb, acting as a reservoir for the DNA solution.

Large fragments — 50,000 bp, 20,000 bp, 10,000 bp, 5000 bp, 2500 bp

DNA fragments move: The gel matrix acts as a sieve for the negatively charged DNA molecules as they move towards the positive terminal. Small fragments move easily through the matrix, whereas large fragments don't.

As DNA molecules migrate through the gel, large fragments will lag behind small fragments. As the process continues, the separation between larger and smaller fragments increases.

Small fragments — 1000 bp, 500 bp

Tray: The gel is poured into this tray and allowed to set.

Positive electrode (+)

Gel: A gel is prepared, which will act as a support for separation of the fragments of DNA. The gel is a jelly-like material, called **agarose**.

Steps in the process of gel electrophoresis of DNA

1. The gel is placed in an electrophoresis chamber and the chamber is filled with buffer, covering the gel. This allows the electric current from electrodes at either end of the gel to flow through the gel.

2. DNA samples are mixed with a "loading dye" to make the DNA sample visible. The dye also contains glycerol or sucrose to make the DNA sample heavy so that it will sink to the bottom of the well.

3. The gel is covered, electrodes are attached to a power supply and turned on.

4. When the dye marker has moved through the gel, the current is turned off and the gel is removed from the tray.

5. DNA molecules are made visible by staining the gel with methylene blue or ethidium bromide which binds to DNA and will fluoresce in UV light.

6. The band or bands of interest are cut from the gel and dissolved in chemicals to release the DNA. This DNA can then be studied in more detail, e.g. its **nucleotide** sequence can be determined.

1. What is the purpose of gel electrophoresis? _____

_____

_____

Analyzing an electrophoresis gel

▸ Once made, an electrophoresis gel must be interpreted. Different fragment lengths of DNA spread out over the gel, producing a banding pattern.

▸ DNA can be cut into fragments by enzymes called restriction enzymes. Different restriction enzymes cut the DNA at their own specific nucleotide sequence, called a recognition site. This means different restriction enzymes produce different lengths of DNA, creating their different banding patterns for the same piece of DNA.

Recognition sites for selected restriction enzymes

Enzyme	Source	Recognition sites
EcoRI	Escherichia coli RY13	G ^A A T T C
HaeIII	Haemophilus aegyptius	G G ^C C
HindIII	Haemophilus influenzae Rd	A ^A G C T T
HpaI	Haemophilus parainfluenzae	G T T ^A A C
HpaII	Haemophilus parainfluenzae	C ^C G G
MboI	Moraxella bovis	^G A T C
TaqI	Thermus aquaticus	T ^C G A

DNA fragments for gel electrophoresis are produced by "cutting" DNA using restriction enzymes. Restriction enzymes are produced by bacteria as a method of eliminating foreign DNA. About 3000 different restriction enzymes have been isolated. Around 600 are commonly used in laboratories.

Restriction enzymes are named according to the species they were first isolated from, followed by a number to distinguish different enzymes isolated from the same organism.

The symbol ^ shows where the cut is made.

2. (a) The DNA strand below can be "cut" with restriction enzymes. For each of the restriction enzymes listed, identify the cut site and the lengths of DNA produced (in number of bases). List the DNA length in the table below:

GTGACCTTCCGGAGGGCCAAGGGCTACCCCATCGACCTGTACTACCTGATCCGGGGACCTCTTCGG

(b) The diagram below shows an electrophoreses gel. On the gel, draw in the bands that would be seen for each restriction enzyme, based on the lengths of DNA you recorded in your table.

Lengths of DNA

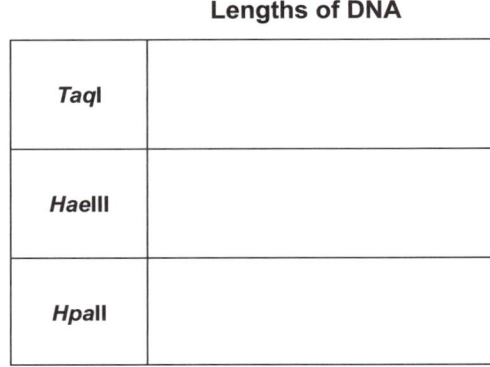

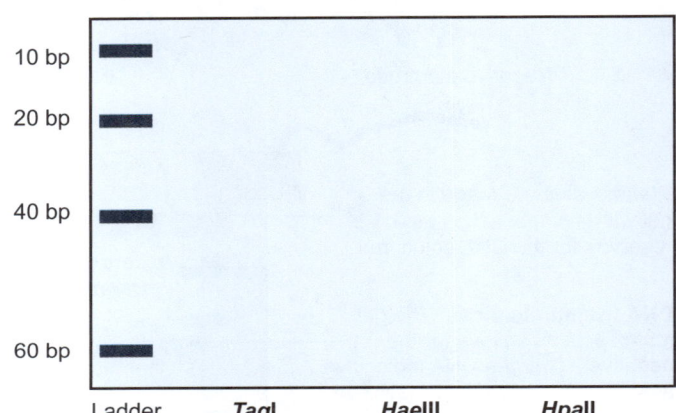

Using electrophoresis to determine relatedness

▸ By using restriction enzymes on the same gene in different species, it is possible to produce banding patterns in electrophoresis gels that can indicate the relatedness of the species. Those with similar banding patterns are more closely related than those with dissimilar banding patterns.

3. (a) Determine the relatedness of each individual (A-E) using each banding pattern on the set of DNA profiles below. When you have done this, complete the phylogenetic tree by adding the letter of each individual.

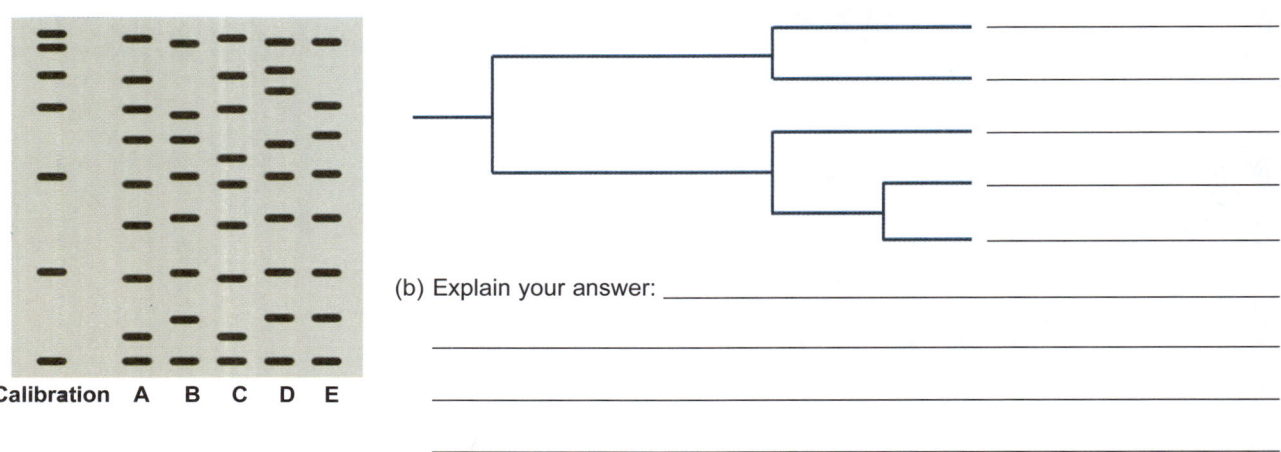

(b) Explain your answer: _____
_____
_____
_____

134 Making Recombinant DNA

Key Question: How can DNA from one species be inserted in the DNA of another species?

▸ Recombinant DNA (rDNA) is produced by combining genetic material from two or more different sources. The production of rDNA is possible because the DNA of every organism is made of the same building blocks (**nucleotides**).

▸ rDNA allows a **gene** from one organism to be moved into, and expressed in, a different organism. Two important tools are used to create rDNA. Restriction enzymes or the CRISPR-Cas9 system cut the DNA and the enzyme DNA ligase is used to join the sections of DNA together.

Overview: How is recombinant DNA made?

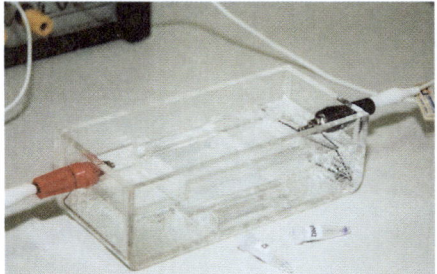

The target DNA is amplified by PCR. It is then cut with restriction enzymes (below). The DNA into which it will be inserted is cut with the same enzymes. The DNA fragments are then run on an **electrophoresis gel** to separate them so they can be identified.

Once the DNA fragments are separated, the gel is placed on a UV viewing platform. The areas of the gel containing the target DNA and the DNA for it to be placed into are cut out and placed in a solution that dissolves the gel releasing the DNA into the solution.

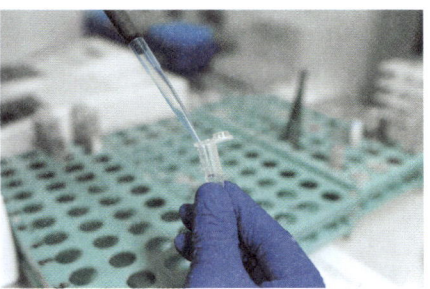

The target DNA and the DNA into which it will be inserted are mixed together. They were cut with the same restriction enzymes so the cut sites will have matching DNA overhangs and bond together to produce a recombinant plasmid (below and next page).

Cutting fragments of DNA

▸ A restriction enzyme is an enzyme that cuts a double-stranded DNA molecule at a specific recognition site (a specific DNA sequence). There are many different types of restriction enzymes and each has a unique recognition site.

▸ Some restriction enzymes produce DNA fragments with two sticky ends (right). A sticky end has exposed nucleotide bases at each end. DNA cut in such a way can be joined to other DNA fragments with matching sticky ends. Such joins are specific to their recognition sites.

▸ Some restriction enzymes produce a DNA fragment with two blunt ends (ends with no exposed nucleotide bases). The piece it is removed from is also left with blunt ends. DNA cut in such a way can be joined to any other blunt end fragment. Unlike sticky ends, blunt end joins are non-specific because there are no sticky ends to act as specific recognition sites.

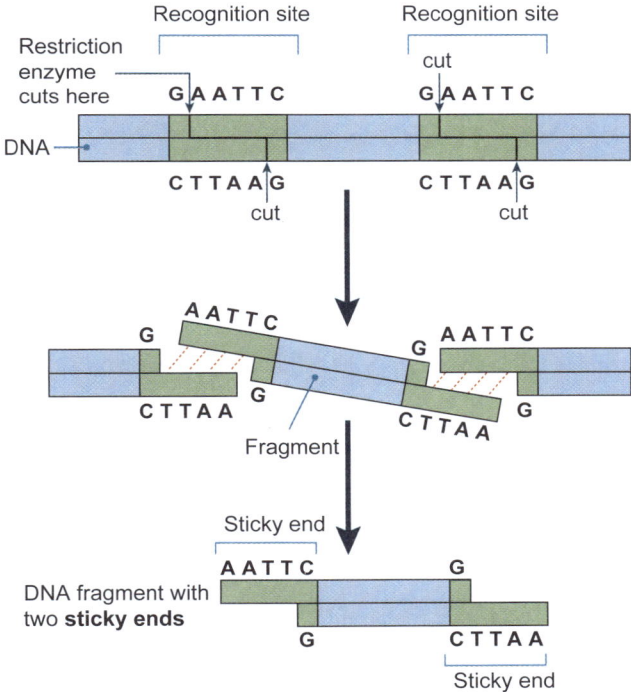

1. What is recombinant DNA? _____

2. What is the difference between sticky end and blunt end fragments? _____

Creating recombinant DNA

1. Two pieces of DNA are cut by the same restriction enzyme (they will produce fragments with matching sticky ends).

2. Fragments with matching sticky ends can be joined by base-pairing. This process is called annealing. This allows DNA fragments from different sources to be joined.

   The DNA fragments are joined by the enzyme DNA ligase. This produces a molecule of recombinant DNA.

3. The joined fragments will usually form either a linear or a circular molecule, as shown here (right) as recombinant plasmid DNA.

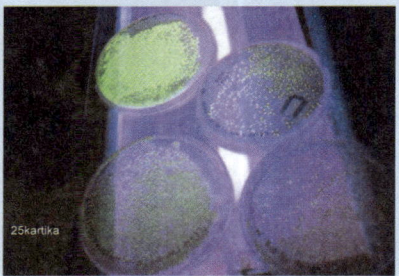

pGLO is a plasmid engineered to contain Green Fluorescent Protein (*gfp*). pGLO has been used to create fluorescent organisms, including the bacteria above (bright green patches on agar plates).

Plasmids are short circular DNA molecules that are separate from the main **chromosomes**.

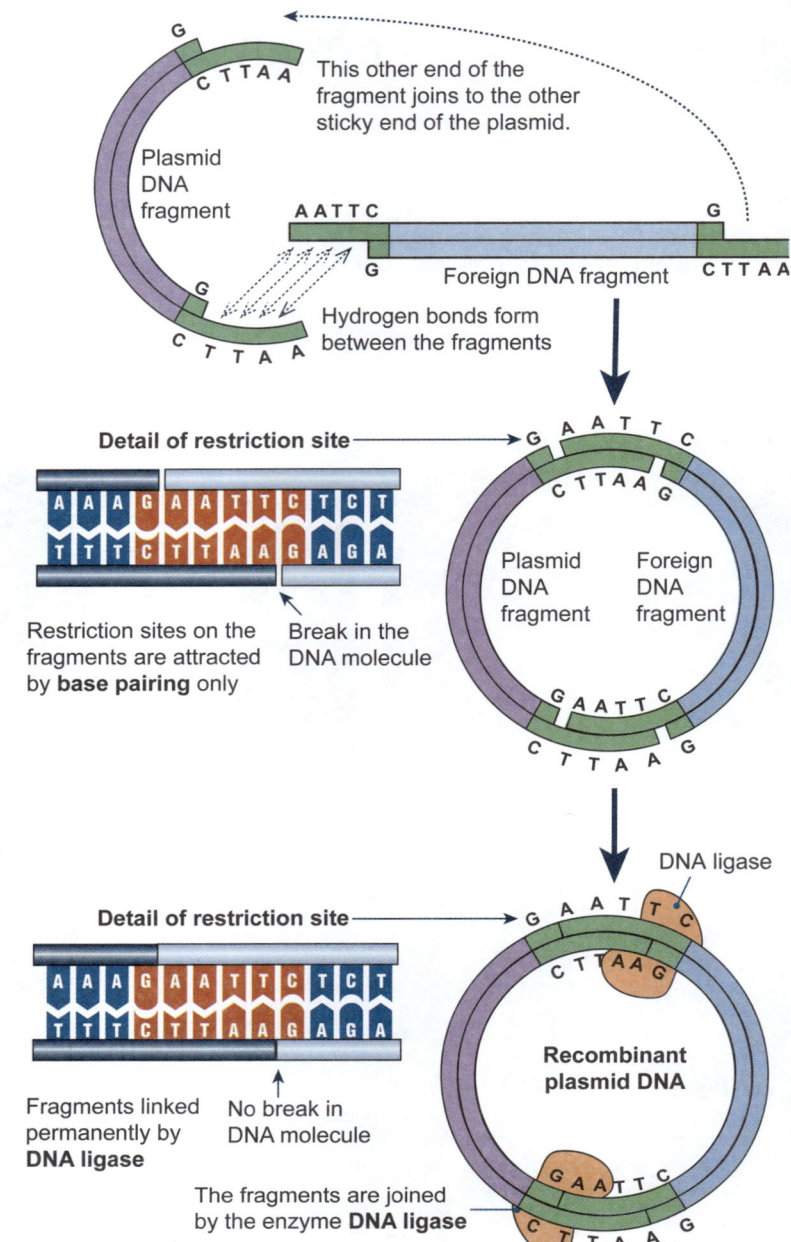

3. Explain in your own words the process of joining two DNA fragments together: _____
_____
_____

4. Explain why ligation can be considered the reverse of the restriction digestion process: _____
_____

5. (a) Why can recombinant DNA be expressed in any kind of organism, even if it contains DNA from another species?
_____
_____
_____
_____

   (b) How can this principle be useful for human industry? _____
_____
_____

135 Gene Editing with CRISPR

Key Question: What is CRISPR/Cas9 and how does it edit DNA?

CRISPR

▸ In 2020, for the first time ever, the Nobel prize in chemistry was awarded to two women scientists, Jennifer Doudna and Emmanuelle Charpentier, for their key discoveries of a molecular tool called CRISPR (pronounced crisper). CRISPR-Cas9 is a form of gene editing technology that occurs naturally in bacteria, who use it to edit the **DNA** of invading viruses. CRISPR is able to target specific stretches of DNA and edit it at very precise locations.

▸ Two key components are required for CRISPR to work. First, an RNA guide locates and binds to the target piece of DNA, then the Cas9 unwinds and cuts the DNA. The technology has potential applications in correcting **mutations** responsible for disease, switching faulty **genes** off, adding new genes to an organism, or studying the effect of specific genes. It represents a major advance because it allows more precise and efficient gene editing at a much lower cost than ever before.

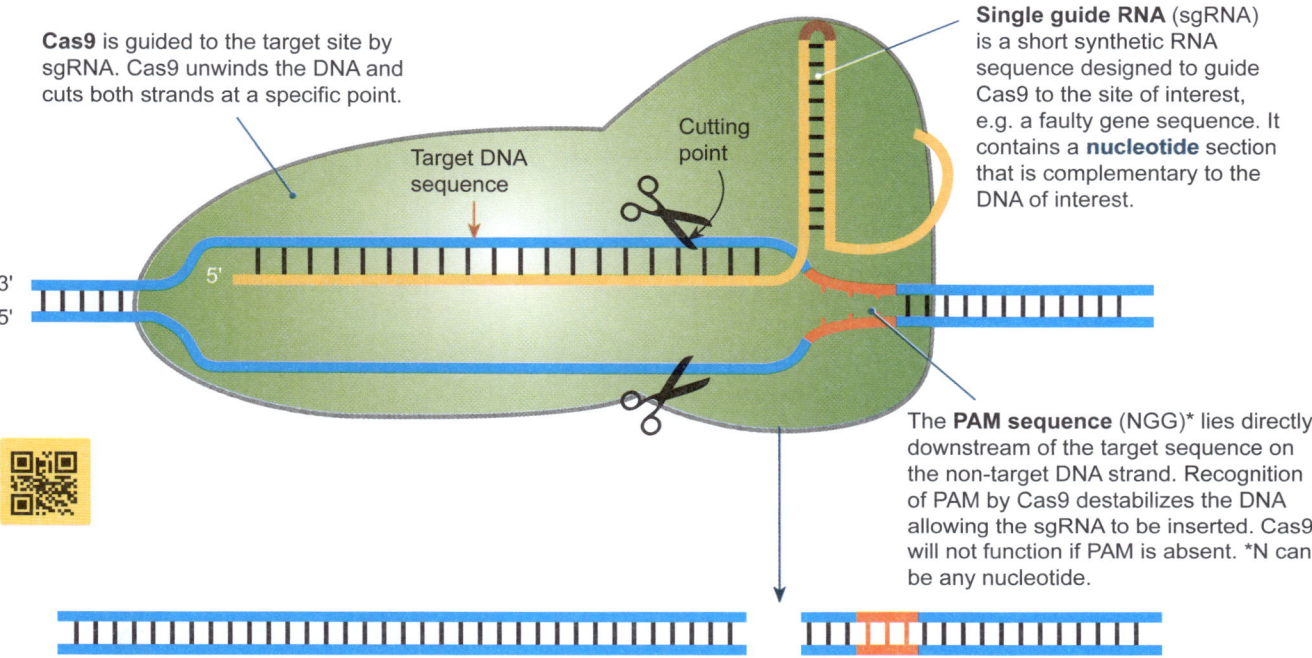

Cas9 is guided to the target site by sgRNA. Cas9 unwinds the DNA and cuts both strands at a specific point.

Single guide RNA (sgRNA) is a short synthetic RNA sequence designed to guide Cas9 to the site of interest, e.g. a faulty gene sequence. It contains a **nucleotide** section that is complementary to the DNA of interest.

The **PAM sequence** (NGG)* lies directly downstream of the target sequence on the non-target DNA strand. Recognition of PAM by Cas9 destabilizes the DNA allowing the sgRNA to be inserted. Cas9 will not function if PAM is absent. *N can be any nucleotide.

The cut DNA can be repaired using one of the following methods:

Gene knock in "gene editing"
A new DNA sequence is inserted into the DNA break. For example, replacing a faulty gene sequence with the correct sequence to restore normal gene function.

Gene knock out "gene silencing"
Errors occur as the cell's normal repair mechanisms mend the broken DNA, causing the insertion or deletion of bases. The resulting frame-shift mutation changes the way the nucleotide sequence is read, either disabling gene function or producing a STOP signal. This technique can be used to silence a faulty gene.

1. What are the roles of the following in CRISPR gene editing:

   (a) Cas9: _____

   (b) sgRNA: _____

2. Research the impact of Doudna and Charpentier's discovery of the CRIPSR system. Describe some uses of the technology and what other scientists believe it should or should not be used for:

   _____
   _____
   _____

3. What benefits are offered by CRISPR technology? _____
   _____

136 Genetic Engineering for Insulin

Key Question: How is genetic engineering used to produce insulin?

- Type I **diabetes mellitus** is a metabolic disease caused by a lack of insulin. Around 25 people in every 100,000 suffer from type I diabetes. It is treatable only with injections of insulin.
- In the past, insulin was taken from the pancreatic tissue of cows and pigs and purified for human use. The method was expensive and some patients had severe allergic reactions to the foreign insulin or its contaminants.

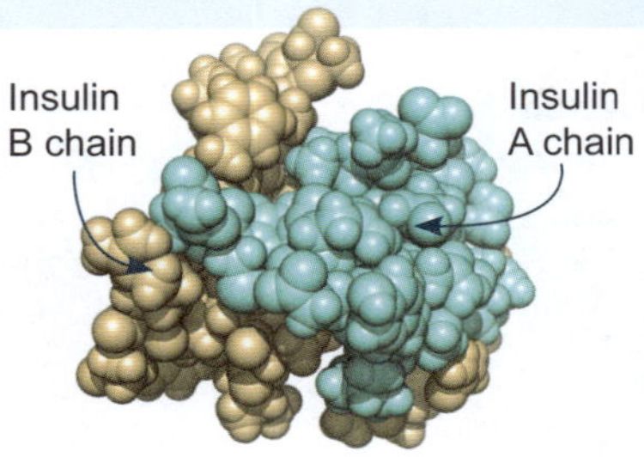

Using bacteria to produce insulin

- Recall that **DNA** can be cut with restriction enzymes and joined with DNA ligase. Any fragments of DNA cut with the same restriction enzymes can be joined together to produce recombinant DNA. Bacteria can pick up plasmids and express the genes that are in them.
- These two ideas are important in the production of insulin using the bacteria *E. coli* (below).

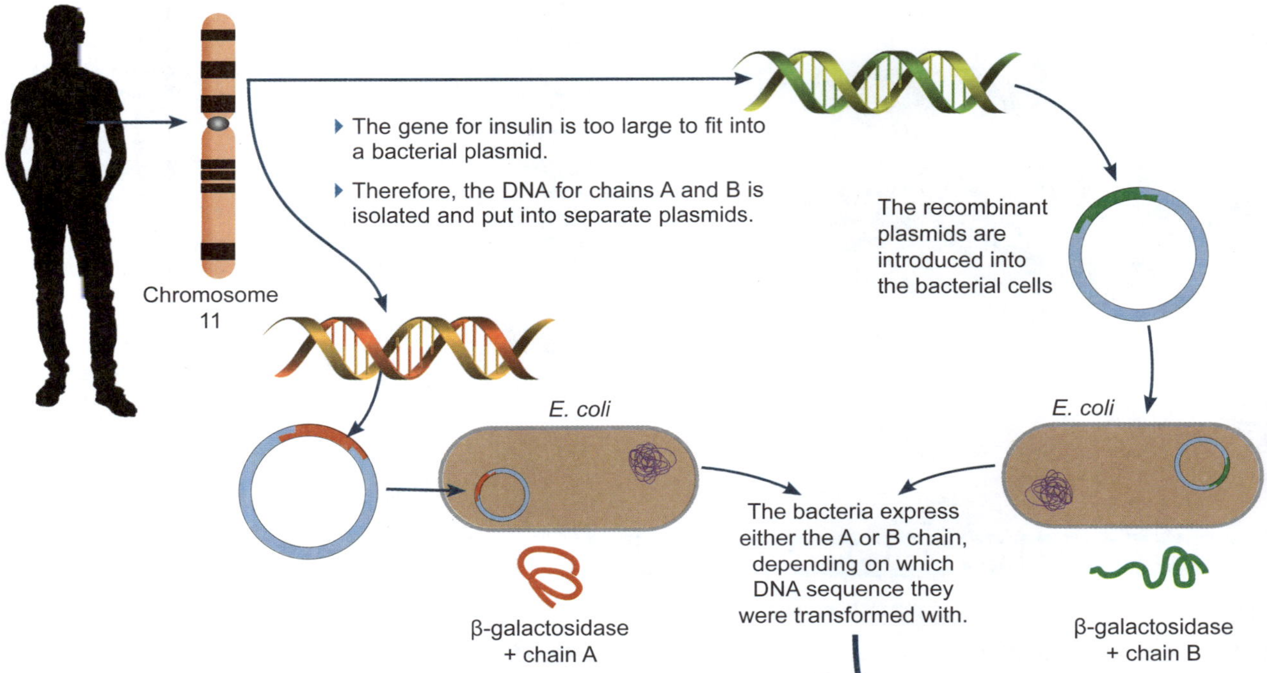

- The gene for insulin is too large to fit into a bacterial plasmid.
- Therefore, the DNA for chains A and B is isolated and put into separate plasmids.

The recombinant plasmids are introduced into the bacterial cells

The bacteria express either the A or B chain, depending on which DNA sequence they were transformed with.

β-galactosidase + chain A

β-galactosidase + chain B

- The **nucleotide** sequence for each insulin chain is isolated and prepared in such a way that bacteria can **transcribe** and **translate** them correctly. The two sequences are small enough to be inserted into a plasmid.
- Plasmids are extracted from *Escherichia coli*. Each nucleotide sequence for the insulin chain is linked to the β-galactosidase gene in the plasmid. This carries a promoter for transcription.
- Restriction enzymes are used to cut plasmids at the appropriate site and the A and B insulin sequences are inserted. The sequences are joined with the plasmid DNA using DNA ligase.
- The recombinant plasmids are inserted back into the bacteria by placing them together in a culture that favors plasmid uptake by bacteria.
- The transgenic bacteria are then grown and multiplied in vats under carefully controlled growth conditions.

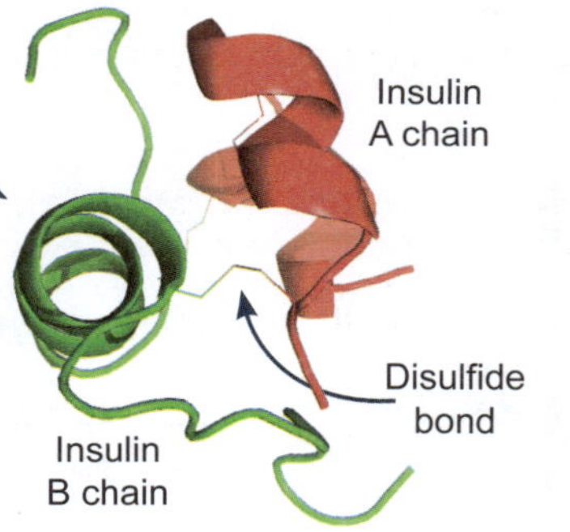

The amino acid chains produced are purified and joined together to produce the final insulin protein.

1. Explain why, when using *E. coli*, the insulin gene is synthesized as two separate A and B chain nucleotide sequences:

_____

_____

_____

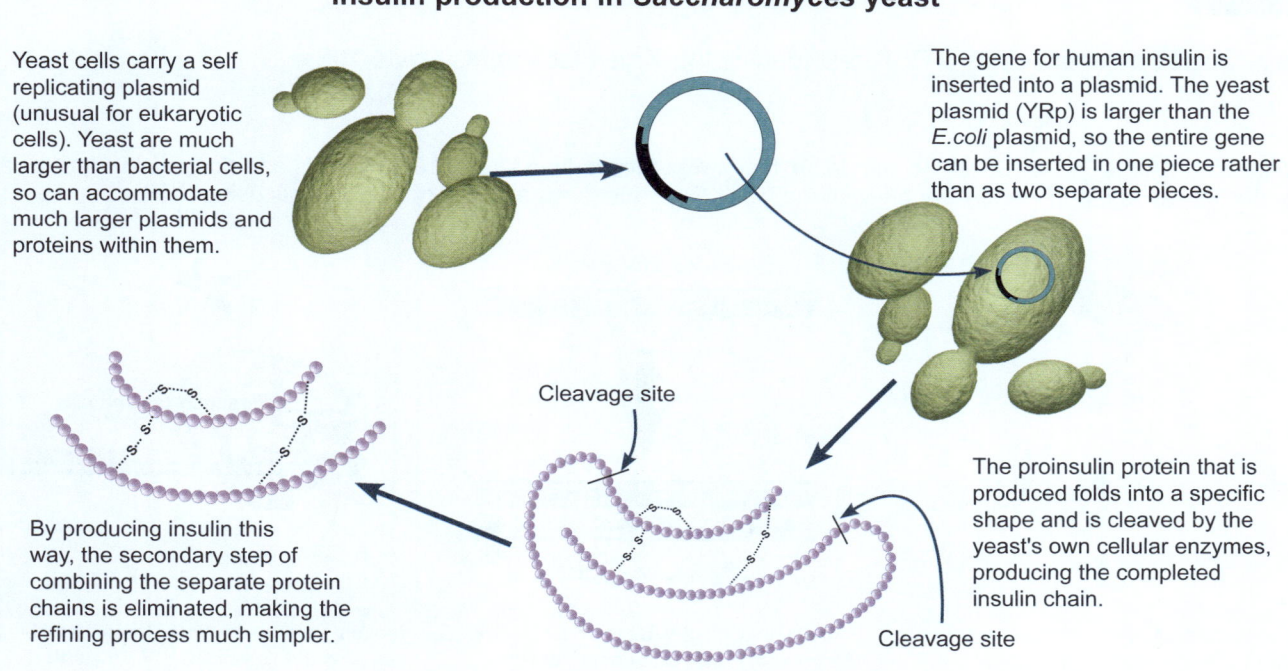

Insulin production in *Saccharomyces* yeast

Yeast cells carry a self replicating plasmid (unusual for eukaryotic cells). Yeast are much larger than bacterial cells, so can accommodate much larger plasmids and proteins within them.

The gene for human insulin is inserted into a plasmid. The yeast plasmid (YRp) is larger than the *E.coli* plasmid, so the entire gene can be inserted in one piece rather than as two separate pieces.

Cleavage site

The proinsulin protein that is produced folds into a specific shape and is cleaved by the yeast's own cellular enzymes, producing the completed insulin chain.

By producing insulin this way, the secondary step of combining the separate protein chains is eliminated, making the refining process much simpler.

Cleavage site

2. Describe the three major problems associated with the traditional method of obtaining insulin to treat diabetes:

   (a) _____

   (b) _____

   (c) _____

3. Explain the reasoning behind using *E. coli* to produce insulin and the benefits that GM technology has brought to people with diabetes:

   _____

   _____

   _____

   _____

4. Why are the synthetic nucleotide sequences ('genes') 'tied' to the β-galactosidase gene? _____

   _____

   _____

5. Yeast (*Saccharomyces cerevisiae*) is also used in the production of human insulin. Discuss the differences in the production of insulin using yeast and *E. coli* with respect to:

   (a) Insertion of the gene into the plasmid: _____

   _____

   _____

   (b) Secretion and purification of the protein product: _____

   _____

   _____

   _____

   _____

137 Testing for Covid-19

Key Question: How does a PCR test identify the virus that causes Covid-19?

▶ When the SARS-CoV-2 (Covid-19) pandemic first swept the world in late 2019, scientists and doctors needed a way to definitively test and identify people with the virus. The most accurate way of doing this is using PCR.

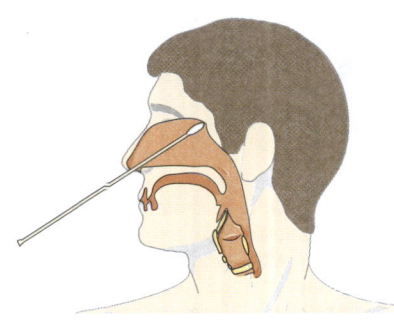

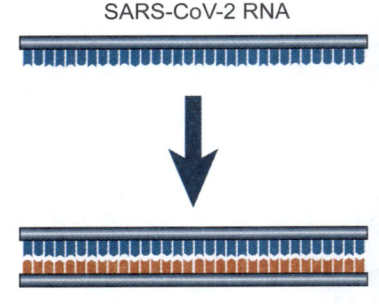

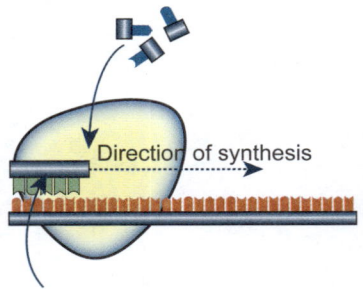

Taking a nasopharyngeal swab from a patient collects some of the genetic material of the SARS-CoV-2 virus, but also genetic material from the patient's own cells, from bacteria, and other viruses.

The SARS-CoV-2 virus is an RNA virus (its genetic material is coded as RNA). The enzyme reverse transcriptase is used to convert the RNA in cDNA (complementary DNA).

Isolating and amplifying Covid-19 DNA requires the use of PCR. Primers that anneal to a DNA sequence found only in the SARS-CoV-2 virus are used. A process called real-time PCR (q-PCR) is used.

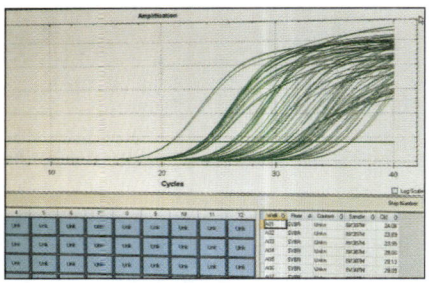

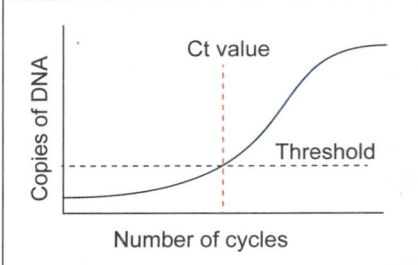

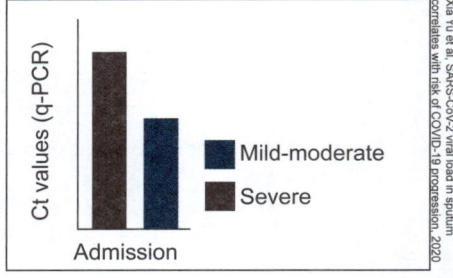

q-PCR uses primers that fluoresce each time a copy of DNA is made. A sensor in the PCR cycler detects this and sends a signal to a computer which plots the number of DNA copies over time, producing an S-shaped graph.

When the level of floresence exceeds a set threshold, it indicates a significant difference between the copied DNA and the background level. The number of PCR cycles taken to reach this threshold is called the cycle threshold value, or Ct value. The lower this value the more DNA was present at the start of the PCR process.

Studies have shown that the lower the Ct value of a patient when Covid-19 is first detected, the more severe their symptoms are likely to be. Studies also show that the Ct value correlates with the probability of disease progression. That is, the lower the Ct value, the greater the probability of the patient progressing to a severe stage of Covid-19.

Xia Yu et al, SARS-CoV-2 viral load in sputum correlates with risk of COVID-19 progression, 2020

1. Discuss the importance of PCR in detecting the SARS-CoV-2 virus and informing doctors on the severity of the infection:

   _____
   _____
   _____
   _____
   _____
   _____

2. Why does the Ct value decrease as the viral load in a patient increases? _____
   _____
   _____
   _____

138 Molecular Technologies and Research

Key Question: How are molecular technologies used to progress research into genetic systems and help produce useful crops?

▸ Many molecular technologies such as PCR and **DNA** recombination have been in use for decades. They have allowed researchers to investigate a wide range of questions and possibilities involving genetics. They have also allowed the development of techniques for the industrial production of important biological molecules or enhanced organisms used in wider industry.

▸ Newer technologies, such as CRISPR, are allowing researchers to expand this research and more easily and accurately develop genetically engineered organisms for use in industry.

Modifying Yukon potatoes

▸ Potatoes are the number one most important vegetable and third most important crop in the world (after rice and wheat). Research into improving a potato plant's output is therefore extremely important for the world's food production.

▸ The potato tuber is mostly just starch. Starch is a large polymer molecule made up of two monomers, amylose and amylopectin. Starches high in amylopectin are more useful in the food (and other) industries because they produce better products such as stabilizers, thickeners, and emulsifiers. Higher amylopectin levels in starch are also better for ethanol production.

Yukon gold potatoes

▸ In 2022, Stephany Toinga-Villafuertes, Maria Isabel Vales and Keerti S. Rafore, working at Texas A&M University, published studies showing the use of CRISPR to stop amylose production in the Yukon gold potato variety. They used *Agrobacterium* to insert the CRISPR/Cas9 system into potato plants. The CRISPR system edited the gbss **gene** in the potato cells. The cells were grown into potato plants that showed normal growth and production quality, but the tubers they produced had very high levels of amylopectin and no amylose in the starch.

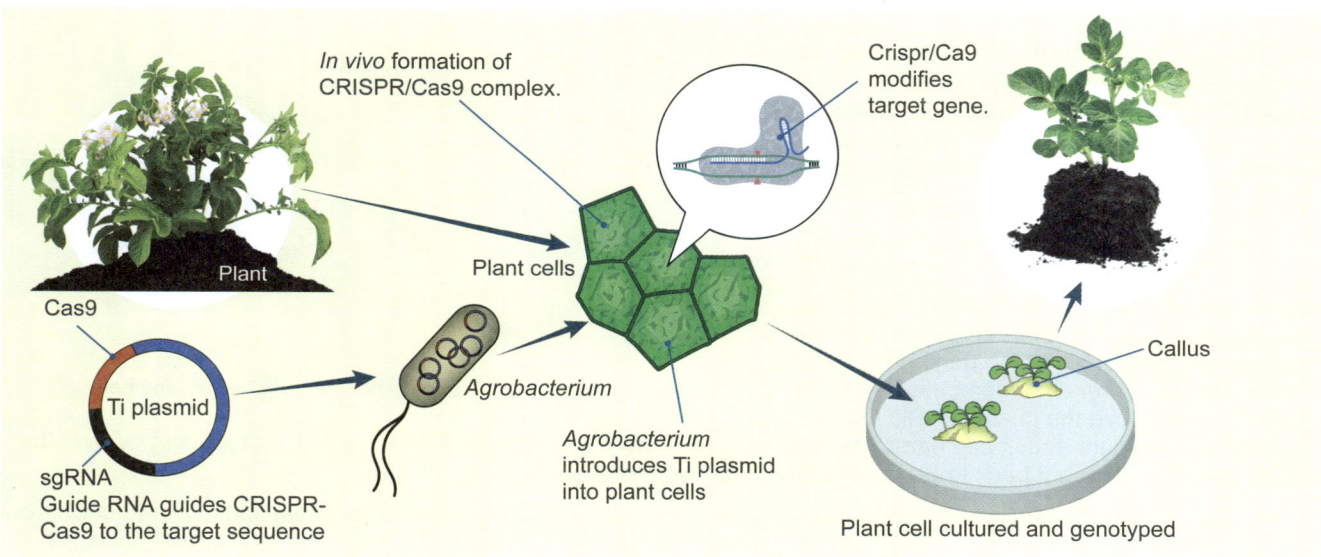

The *in vivo* method uses *Agrobacterium* to deliver a modified Ti plasmid into the plant cells. The Ti plasmid inserts the genes required to introduce CRISPR-Cas9 into the plant DNA. The cell's machinery then produces the CRISPR-Cas9 protein and RNA, which then edits the cell's DNA.

1. Discuss the use of various molecular technologies in researching genetics, including any progress in these techniques:

___
___
___
___
___
___

Tracing the ancestry of the Texas Longhorn

- The Texas Longhorn is possibly one of the most distinctive of cattle breeds. As its name suggests, this cattle breed has immensely long horns compared to most other breeds.
- The horns can have a span of well over 2 meters and the breed holds the world record for the longest horn span in cattle (3.2 meters).
- A wide range of coat colors is possible.
- The ancestors of the modern Longhorns have been in the Americas for centuries, but their exact ancestry was not worked out until 2013 when changes in their DNA were studied using SNPs (single nucleotide polymorphisms).

Using SNPs

- A single nucleotide polymorphism is a substitution of a single base nucleotide at a specific position in the DNA. For example, the two DNA sequences shown below are identical except for one SNP.

CCTGTCGTAATGAATG   CCTGTCGTACTGAATG
GGACACCATTACTTAC   GGACACCATGACTTAC

- These polymorphisms may occur in any place in the genome. If they occur at the specific location at a significant frequency in the population (usually more than 1%), then the SNPs can be considered alleles (different versions of the DNA sequence).
- Certain SNPs may occur more often in one population than another, or only be found in a very specific group. By studying many of these SNPs in a population or species, the ancestry of the population or species can be worked out.

Studying the ancestry of the Texas Longhorn

- Research from the University of Texas by Emily McTavish and David Hillis analyzed 47,506 SNPs from 1461 cattle across 58 different breeds of cattle from all the major continents.
- The study found Longhorns are direct descendents of cattle brought from Spain by Columbus to the New World in the late 1400s.
- Approximately 85% of the Longhorn genome is taurine (the European branch of domesticated cattle). The other 15% of the genome is indicine (the India/Asian branch of domesticated cattle). Indicine cattle have the characteristic hump above the shoulders and floppy ears. The two branches from Europe and India met in Africa producing hybrids that were eventually brought back to Spain and then onto the Americas.
- After release in the Americas, the cattle went feral and evolved to the new conditions. These cattle went out of favor in the 1800s and the population decreased. However, their tolerance to heat, tick parasites, and water stress has brought the Longhorn back into favor.

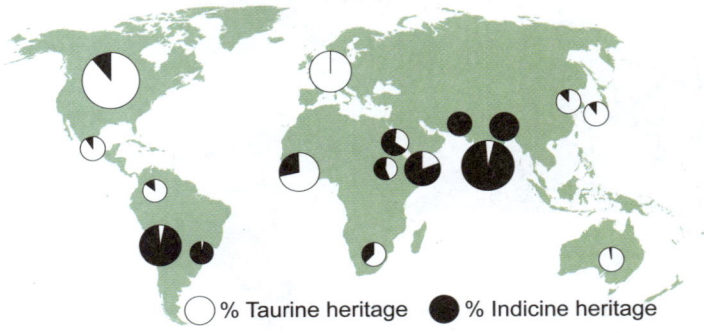

Taurine / Indicine heritage in cattle
○ % Taurine heritage   ● % Indicine heritage

2. What is an SNP? _____

3. How can it be used to compare populations? _____

4. Compare the ratio of taurine to indicine genes in cattle across the globe. Comment on any discrepancies:
_____

5. Like many science careers, genetic engineering brings together various different areas of biology and other sciences. Individually, or in groups of four, carry out research into careers in genetics and related STEM subjects.

   - Resources for research could include museums, libraries, private organizations and companies, online resources, and in-person interviews with people working in relevant areas.
   - Your research should include pathways to a particular career (including qualifications and areas of study) and the subject matter or research area covered by a particular career.
   - Each member of your group should research a different aspect and report back to the group.
   - Once each member of the group has reported back, decide as a group how to present your ideas. This could be by a written report, slideshow, poster, or presentation.
   - Use the space below to organize and summarize your research and notes. This is not your final report.

STEM career		
Description of research and/or data source (written/graphs/tables/images?)	Where was the data sourced? URL/organization	Career information
1		
2		
3		

Importance of this career or job to science	
Importance of this career or job to society	
Pathways to career or job	
Subject areas covered	
Evidence-based summary statements on area of research	

139 Real-Life Superpowers Revisited

Content Anchor Revisited: What is the result of changes in DNA and can they produce beneficial results?

This chapter has studied the structure of **DNA**, the effects of changing its sequence, and how DNA can be investigated and manipulated.

▸ At the start of the chapter you were given some examples of **mutations** that could potentially enhance a person's abilities in some aspects of their biology. You should now be able to explain how mutations occur and why. You should also be able to explain how we can manipulate DNA to work out which parts of the DNA carry a mutation.

▸ The three mutations given as examples at the start of the chapter are show below:

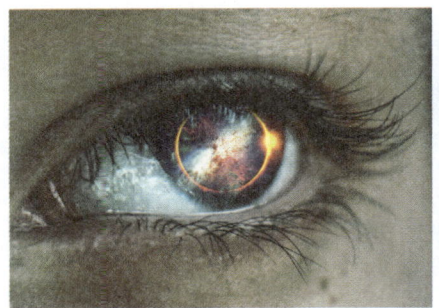  

Tetrachromatic vision
Instead of three types of color sensitive cones in the retina, a few women have four. It results in having one normal OPN1MW gene and one with a mutation. People with this mutation see colors somewhat differently from others.

Increased muscle growth
A mutation that results in a nonfunctional MSTN gene causes increased muscle growth. Studies in mice show no increased strength overall. In cattle the mutation can make giving birth very difficult. Most births need a C-section.

Increased bone density
A mutation in the LRP5 gene produces higher density bones. However, side effects of this are bony growths on the roof of the mouth and thickened jaw. People with this mutation have also noted difficulty staying afloat in water.

1. Describe a disadvantage for each of the mutations above:

    (a) _____

    (b) _____

    (c) _____

2. Part of the coding strand DNA sequence for the LRP5 mutation is shown: GAC TGG GGT GAG ACG. This is the sequence which changes to cause the LRP5$_{V171}$ mutation that results in increased bone density.

    (a) What is the amino acid sequence for the DNA sequence shown? _____

    (b) A mutation from GGT to GTT causes the mutation. What is the change in amino acid? _____

    (c) What type of mutation is this? _____

3. Consider a scenario where a scientist wanted to genetically engineer a person to have the LRP5$_{V171}$ mutation to give them unbreakable bones. Describe the steps they would have to take to genetically engineer this person:

    _____
    _____
    _____
    _____
    _____

140 Summing Up

Read each question carefully. Place a cross in the box beside the best answer to the question from the four answer choices provided.

1. Which is correct in order for the structures from largest to smallest:
   - (a) Gene, chromatid, exon, chromosome
   - (b) Gene, exon, chromosome, chromatid
   - (c) Exon, gene, chromatid, chromosome
   - (d) Exon, chromatid, chromosome, gene

2. During mRNA editing:
   - (a) The primary mRNA is translated into protein
   - (b) Exons are removed to produce the mature mRNA
   - (c) Introns are removed to produce the mature mRNA
   - (d) The primary mRNA is translated into a poly-A tail

3. A chromosome has the following gene sequence (shown as letters for reference) ABCDEFGHIJKLMNOP. A mutation occurs during meiosis so that the sequence becomes ABCDJIHGEKLMNOP. This mutation is a:
   - (a) Deletion
   - (b) Inversion
   - (c) Duplication
   - (d) Translocation

4. Which is the correct sequence for gene expression?
   - (a) DNA → editing → primary transcript → mRNA → translation
   - (b) DNA → editing → primary transcript → translation
   - (c) DNA → primary transcript → editing → mRNA → translation
   - (d) DNA → primary transcript → translation → mRNA editing

5. [Amino acid codon table]

   Using the amino acid table above the what are the amino acids produced by the following mRNA sequence?

   AAA UCC GGA UUU

   - (a) Lys Ser Gly Phe
   - (b) Lys Gly Ser Phe
   - (c) Iie Asp Ser Ala
   - (d) Val Leu Lys Phe

6. According to Chargaff's rules, in a DNA molecule:
   - (a) The amount of A will equal the amount of T
   - (b) The amount of A will equal the amount of G
   - (c) The amount of G will equal the amount of T
   - (d) The amount of C will equal the amount of A

7. The three components of a DNA molecule are:
   - (a) Base, protein, phosphate
   - (b) Phosphate, nucleotide, sugar
   - (c) Phosphate, sugar, base
   - (d) Base, sugar, lipid

8. Three RNA molecules are involved in the production of proteins from DNA:

   X: Delivers amino acids to the ribosome. Matches amino acid molecules with the appropriate codon.

   Y: Primary component of ribosomes which catalyze the production of proteins. Y makes up about 80% of cellular RNA.

   Z: Transcribed from DNA the mature form of Z is produced after the removal of introns and is translated into proteins.

   X, Y, and Z are:
   - (a) X: rRNA, Y: tRNA, Z: mRNA
   - (b) X: rRNA, Y: mRNA, Z: tRNA
   - (c) X: mRNA, Y: tRNA, Z: rRNA
   - (d) X: tRNA, Y: rRNA, Z: mRNA

9. Which of the following single nucleotide mutations is least likely to affect the final protein?
   - (a) A substitution mutation
   - (b) A deletion mutation
   - (c) An insertion mutation
   - (d) None of the mutations will affect the protein

10. Transgenic organisms can produce proteins from foreign DNA inserted into them because:
    - (a) Extra molecules are inserted into the transgenic organism to help produce the proteins.
    - (b) The foreign DNA transforms proteins in the organism into new proteins.
    - (c) DNA is universal to all life and so all cells can translate the DNA into proteins.
    - (d) None of the above.

11. A restriction enzyme, polymerase enzyme, and DNA ligase:
    - (a) Cuts DNA at a specific site, copies a section of DNA, joins cut sections of DNA.
    - (b) Copies a section of DNA, joins cut sections of DNA, cuts DNA at a specific site.
    - (c) Cuts DNA at a specific site, joins cut sections of DNA, copies a section of DNA
    - (d) Joins cut sections of DNA, cuts DNA at a specific site, copies a section of DNA.

12. Put the numbered items in the correct order for producing a transformed bacterial cell and provide a description of each step in the process:

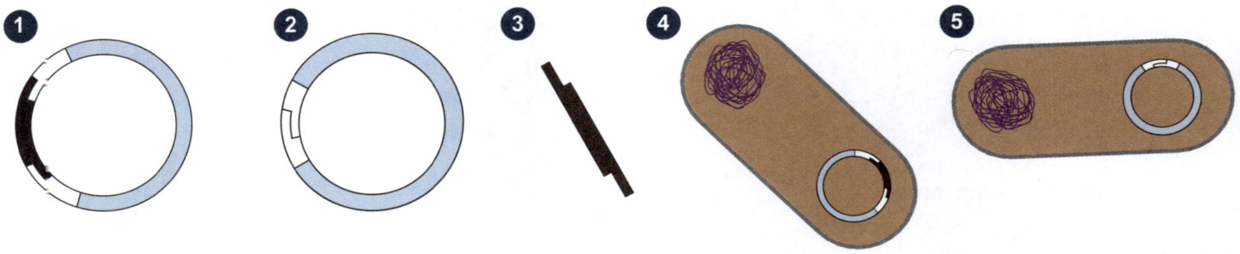

13. Explain why crops such as Bt corn that have been engineered to produce their own natural insecticides or repellent have the potential to decrease biodiversity in the wider ecosystem:

14. An original DNA sequence is as follows: **GCG TGA TTT GTA GGC GCT CTG**

    Point deletion   **Point insertion**   **Point substitution**   **Block deletion**

    Block insertion   **Inversion**   **Duplication**

    From the selection write down which type of mutation has occurred in each of (a) to (d)

    (a) **GCG TGT TTG TAG GCG CTC TG** _____

    (b) **GCG TGA TTT GTA AGG CGC TCT G** _____

    (c) **GCG TGA TTT GGA GGC GCT CTG** _____

    (d) **GCG TGA GTA GGC GCT CTG** _____

15. Define each of the following and explain when it might be used:

    (a) Restriction enzyme: _____

    (b) Recombinant DNA: _____

    (c) CRISPR/cas9 system: _____

    (d) Polymerase chain reaction: _____

16. (a) The schematic below shows the levels of control in gene expression. Fill in the boxes indicating the structures and processes, choosing from the following word list. **Word list**: *5' cap, mRNA in the nucleus, polypeptide, mRNA in the cytoplasm, DNA packing, exon, intron, functional protein, folding and assembly, poly A tail, gene, cleavage or chemical modification, primary mRNA, protein degradation, translation, transcription, exon splicing, nuclear export.*

(b) On the diagram indicate processes with a P and structures with an S.

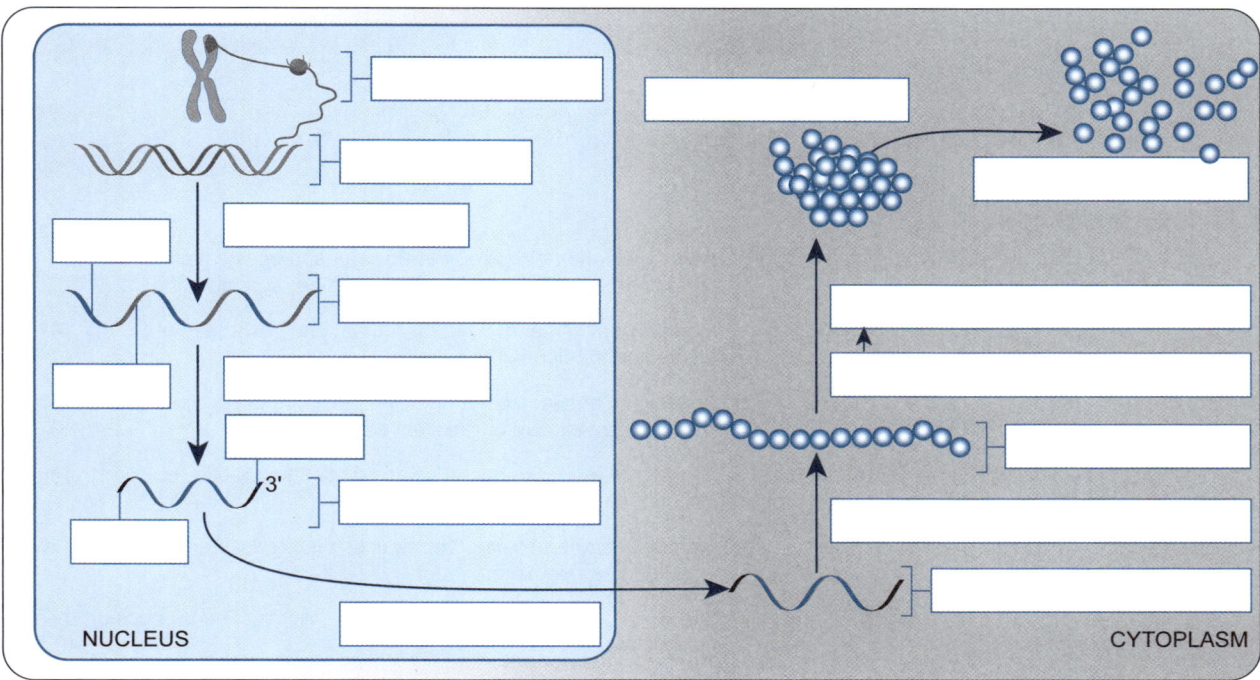

17. Complete the following paragraph by inserting the correct word from the list. Words may be used more than once or not at all:
Word list: *Carbohydrate, cytoplasm, mRNA, nucleus, polypeptide, rRNA, transcription, translation, tRNA*

In eukaryotes, gene expression begins with _____ which occurs in the _____. _____ is the copying of the DNA code into _____. The _____ is then transported to the _____ where _____ occurs. Ribosomes attach to the _____ and help match the codons on _____ with the anticodons on _____. The _____ transports the amino acids to the ribosome where they are added to the growing _____ chain.

18. The electrophoresis gel (below, right) shows four profiles containing five STR sites: the mother (A), her daughter (B), and two possible fathers (C and D). Which of the possible fathers is the biological father?

(a) The biological father is: _____

(b) Explain your answer: _____

(c) Why do profiles B and D only have 9 bands?

CHAPTER 6
Patterns of Inheritance

TEKS
Scientific and Engineering Practices

B.1: Investigation and Inquiry
1.B 1.D 1.F 1.G

B.2: Data and Patterns
2.B 2.C

B.3: Communicating in Science
3.A 3.B

B.4: Science as a Human Endeavor
4.A 4.B

TEKS
Science Concepts

B8.A analyze the significance of chromosome reduction, independent assortment, and crossing-over during meiosis in increasing diversity in populations of organisms that reproduce sexually

B8.B predict possible outcomes of various genetic combinations using monohybrid and dihybrid crosses, including non-Mendelian traits of incomplete dominance, codominance, sex-linked traits, and multiple alleles

Learning Outcomes
I know I have achieved this when I can:

	Learning Outcome	Activity number
☐	Define the genetic terms: trait, true breeding, homologous pairs, allele, homozygous, heterozygous, recessive, dominant, phenotype, genotype, sexual reproduction, and mutation.	142–144
☐	Explain what is meant by the term trait and how these are passed on to offspring via alleles.	142–144
☐	Distinguish between continuous and discontinuous traits, linked to quantitative or qualitative data.	145
☐	Investigate the incidence of selected continuous or discontinuous traits in your class.	145
☐	Discuss the role of sexual reproduction in producing genetic variation in a population.	146
☐	Sequence and recall the names of the stages in the process of meiosis that occurs in sex cells.	147
☐	Explain how recombination during meiosis increases variation in the chromosomes of gametes.	148
☐	Link the number of chromosomes present in gametes or cells at different stages of meiosis.	148
☐	Model the process of meiosis in a classroom activity.	149
☐	Discuss the relationship between gene linkage and variation.	150
☐	Predict the outcome of genetic crosses in Mendelian genetics.	151
☐	Use Punnett squares to calculate offspring genotype and phenotype ratios in a monohybrid cross.	152, 154
☐	Describe the steps required for a test cross, and the expected outcomes for a heterozygous and homozygous parent.	152
☐	Calculate probability of genotypes and phenotypes of offspring.	153
☐	Use Punnett squares to calculate offspring genotype and phenotype frequencies in a dihybrid cross.	155
☐	Contrast between Mendelian and non-Mendelian inheritance.	156
☐	Distinguish between complete and incomplete dominance.	157
☐	Calculate genotype and phenotype probabilities in offspring from incomplete dominance crosses.	157
☐	Define and give examples of codominance.	158
☐	Calculate probability of phenotypes and genotypes in offspring resulting from codominance crosses.	158
☐	Use Punnett squares to calculate offspring phenotype and genotype frequencies from multiple allele crosses.	158
☐	Explain the connection between sex-linked genes and their effect on the phenotype of offspring.	159
☐	Calculate the probability of genotype and phenotypes in offspring from crosses of various sex-linked genes.	159
☐	Compare the outcome of genetic crosses against predicted outcomes, using the chi-square test.	160

RESOURCE HUB
bit.ly/3mtl8T4

ELPS English Language Proficiency Standards

		Page number
Learning	**Whole Chapter.** Read through the chapter. Identify 10 unfamiliar words and record them in your notebook. Working with a partner, use prior knowledge to write brief definitions for the words on each of your lists. Use context clues to guess the meanings of the words. If necessary, look up the words in a glossary or dictionary.	246
Reading	**Meiosis and Variation.** Look at the graphics while you listen to your teacher explain meiosis and variation. As you listen, jot down notes. You can use the text features in the book, such as headings, to help you organize your notes in outline form. What are the two boldfaced headings on the page? Can you use those headings as the main ideas in your outline?	254
Speaking	**Modeling Meiosis.** Working with a partner, carry out steps 1-7 of the investigation. In each step, discuss what to do next and what information to record. For example, you might tell your partner "I will mark the initials and you toss the sticks." Describe your results to the group using the sentence frames: *Our child's phenotype for _____ was ___. Their genotype for _____ was ____.*	256
Speaking	**Mendelian Genetics.** Complete the table. Then work with a partner, taking turns describing each row of the table. Tell your partner which trait each row describes. Describe the possible phenotypes, which are dominant, which are recessive, and how you came to that conclusion. Then give the Mendelian ratio. When you read ratios, use the word "to" in the place of the colon. For example, when you see 3:1, say "3 to 1."	261
Listening	**Monohybrid Crosses.** Listen to your teacher explain the monohybrid crosses. Look at the diagram while you listen and try to identify what part of the process the teacher is talking about. Point to that part of the diagram with your finger. If you are unsure, ask questions. Which Monohybrid Cross diagram uses a Punnett square?	262
Learning	**Probability.** Read this section on probability. Find words you already knew and words you learned earlier in the chapter. Use these words as clues to understand. Use the glossary if necessary. Next, talk with a partner about unfamiliar words. Work together to brainstorm synonyms for challenging words. For instance, synonyms for probability include *chance, likelihood,* and *odds*. Write the synonyms in a concept map and practice using them in sentences. For example, say: "There's a 50% chance that the coin will land heads up." Then write answers to questions 1–4 with your partner.	264

141 Anyone for Chocolate?

Content Anchor: Can we get a chocolate Labrador puppy from black parents?

- Labrador dogs are popular family pets and come in three main colors: black, brown (usually called chocolate), and yellow. There are various shades of each of these three base colors.
- You might expect that if two black Labradors mate, you would always get black puppies, but that's not necessarily the case. You can cross two black animals and get different colored puppies in the litter.
- Puppy color is determined by a number of different **genes**. If you know what the adults' genotypes are, you will know what color of puppies they might be able to produce.
- The gene for yellow color is in a different location from the black/brown gene.

1. What is the likelihood of crossing two black adults and producing a different colored puppy? _____

2. Suggest whether it might be possible to cross a black labrador with a chocolate Labrador and produce yellow puppies:

3. Labrador retrievers can sometimes suffer from a hip disorder. How do you think we could tell if this condition was caused by genetic factors, or dietary and environmental factors?

4. Some Labrador breeders import dogs from other regions or even countries. Discuss in groups why they might do this?

142 What is a Trait?

Key Question: What are traits, and how are they inherited and passed from one generation to the next?

Traits and phenotypes

- **Traits** are particular variants of phenotypic (observed) characteristics, e.g. eye color. Traits may be controlled by one **gene** or many genes and can show **continuous variation**, e.g. height in humans, or **discontinuous variation**, e.g. flower color in pea plants. **Phenotype** and trait are often used interchangeably, but specifically, for example, phenotype refers to the characteristic "eye color", the trait refers to the actual color of the eyes.

- Also, phenotype may refer to one characteristic or a set of characteristics, e.g. eye color, hair color and type, height, etc., that makes up an individual, e.g. every person has a phenotype made up of many traits. In another example, a dog may have the phenotype of one of the major dog groups, e.g. a hound. In order to be this phenotype, it must have various hound traits, such as an acute sense of smell, stamina, and the ability to bay (produce a hunting call).

Traits are inherited

- Phenotypes and their traits are the result of **genotypes**, i.e. the genetic combination of an individual. To see how traits are inherited consider the cross of pea plants on the right.

- A pea plant that is true breeding for green peas is crossed with a pea plant that is true breeding for yellow peas.

- True breeding individuals are those that, when crossed, produce offspring with the same phenotypes and genotypes as the parents. For example, if two true breeding green pea plants are crossed together, all the offspring will produce green peas.

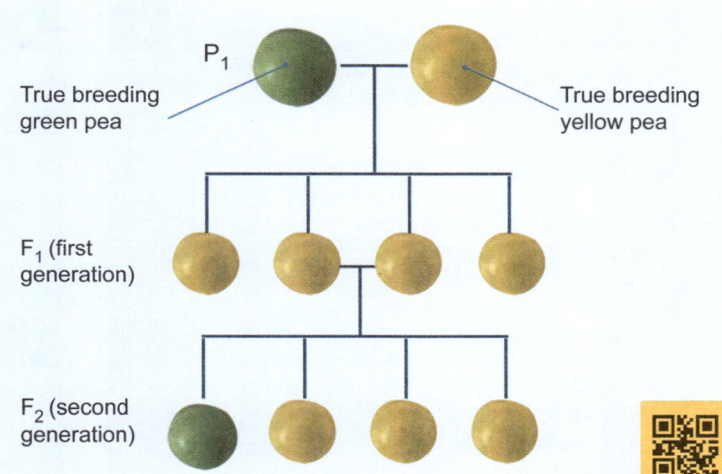

1. Define a trait: _____

2. Define true breeding: _____

3. (a) What was the ratio of yellow seeds to green seeds in the F_2 (second) generation? _____

   (b) Were the pea plants used for the second cross true breeding? _____

   (c) How do you know? _____

   (d) Suggest why the green seed trait did not appear in the F_1 (first) generation: _____

143 Different Alleles For Different Traits

Key Question: What are alleles, and what determines whether a trait will be passed on to offspring?

Homologous chromosomes

In sexually reproducing organisms, chromosomes are generally found in pairs in their cell's nucleus. One of each pair of chromosomes came from the original gametes, formed through meiosis in the parents, and brought together at fertilization. The pairs are called homologues or homologous pairs. Each homologue carries an identical assortment of **genes**, but the version of the gene, known as the **allele**, from each parent may differ. This diagram shows the position of three different genes on the same chromosome that control three different **traits** (A, B, and C).

Having two different versions of gene A is a **heterozygous** condition. Only the dominant allele (A) will be expressed. The dominant allele will determine the phenotype, even if only one copy is present.

When both chromosomes have identical copies of the dominant allele for gene B the organism is **homozygous dominant** for that gene.

When both chromosomes have identical copies of the recessive allele for gene C the organism is said to be **homozygous recessive** for that gene. The recessive allele will only produce the phenotype if two copies are present.

Maternal chromosome originating from the egg of the female parent.

The diagram above shows the complete chromosome complement for a hypothetical organism. It has a total of ten chromosomes, as five, nearly identical pairs (each pair is numbered).

A gene is a unit of heredity. Genes occupying the same position or locus on a chromosome code for the same characteristic, e.g. ear lobe shape.

Paternal chromosome originating from the sperm of the male parent.

1. Define the following terms, describing the allele combinations of a gene in a sexually reproducing organism:

    (a) Heterozygous: _____

    (b) Homozygous dominant: _____

    (c) Homozygous recessive: _____

2. For a gene given the symbol 'A', write down the alleles present in an organism that is:

    (a) Heterozygous: _____ (b) Homozygous dominant: _____ (c) Homozygous recessive: _____

3. What is a homologous pair of chromosomes? _____
   _____
   _____

©2024 **BIOZONE** International
ISBN: 978-1-99-101405-4
Photocopying Prohibited

144 Sources of Variation

Key Question: What are some of the ways in which variation arises?

Mutations
changes to the DNA
Changes to the DNA modifies existing **genes**. Mutations can create new **alleles**.

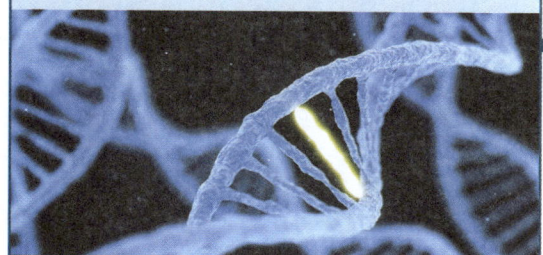

Sexual reproduction
fertilization, and mate selection
Sexual reproduction rearranges and reshuffles the genetic material into new combinations.

Phenotype
The **phenotype** describes the physical characteristics we see in an organism. It is the result of the expression of the **genotype** in a particular environment.

Genes interact with each other and with the environment to influence the phenotype.

Genotype
An individual's genetic makeup. It determines that individual's genetic potential.

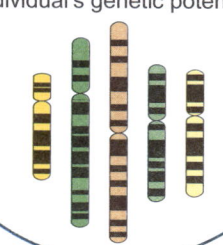

Environmental factors
Environmental factors influence expression of the genotype:
- The external environment includes physical factors, e.g. temperature, and biotic factors, e.g. competition.
- The internal environment, e.g. presence or absence of hormones during development, may also affect expression of the genotype.

1. Define the following terms:

   (a) Mutation: _____

   (b) Genotype: _____

   (c) Phenotype: _____

2. What factors determine the phenotype? _____
   _____
   _____

3. How could two individuals with the same genotype have a different phenotype? _____
   _____
   _____

145 Examples of Genetic Variation

Key Question: What is continuous and discontinuous variation, and what is the difference between quantitative and qualitative traits?

- Individuals show particular variants of phenotypic characters called **traits**, e.g. eye color.
- Traits that show **continuous variation** are called quantitative traits.
- Traits that show **discontinuous variation** are called qualitative traits.

Quantitative traits

Quantitative traits are determined by a large number of **genes**. For example, skin color has a continuous number of variants from very pale to very dark. Individuals fall somewhere on a normal distribution curve of the phenotypic range. Other examples include height in humans for any given age group, length of leaves in plants, grain yield in corn, growth in pigs, and milk production in cattle. Most quantitative traits are also influenced by environmental factors.

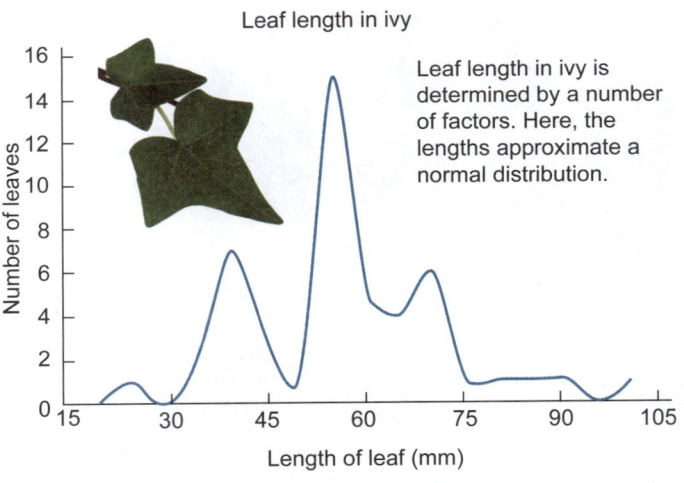

Leaf length in ivy is determined by a number of factors. Here, the lengths approximate a normal distribution.

Grain yield in corn

Growth in piglets

Qualitative traits

Qualitative traits are determined by one or two genes with a very limited number of variants present in the population. For example, blood type in humans has four discontinuous traits A, B, AB, or O. Individuals fall into separate categories. Comb shape in poultry (right) is a qualitative trait and birds have one of four **phenotypes** depending on which combination of four alleles they inherit. The dash (missing allele) indicates that the **allele** may be **recessive** or **dominant**. Albinism is the result of the inheritance of recessive alleles for melanin production. Those with the albino phenotype lack melanin pigment in the eyes, skin, and hair.

Single comb rrpp

Walnut comb R_P_

Pea comb rrP_

Rose comb R_pp

1. What is the difference between continuous and discontinuous variation? _____
   _____
   _____
   _____

2. Identify each of the following phenotypic traits as continuous (quantitative) or discontinuous (qualitative):

   (a) Wool production in sheep: _____

   (b) Hand span in humans: _____

   (c) Blood groups in humans: _____

   (d) Albinism in mammals: _____

   (e) Body weight in mice: _____

   (f) Flower color in snapdragons: _____

Investigation 6.1 Phenotypic variation in your class

See appendix for equipment list.

1. Choose a phenotypic variable in your class you would like to investigate. This could be a quantitative variable (one that can be measured, such as height) or a qualitative (categorical) variable to which you can assign a ranking, e.g. eye color.

2. Record a value for the variable for each person in your class. Tabulate the data in a spreadsheet, e.g. Excel, and graph the result using a histogram; alternatively, use the space below to produce a table and use the grid to produce a plot.

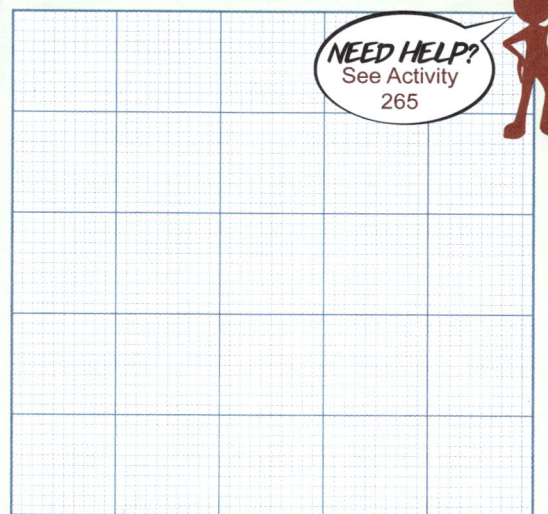

NEED HELP? See Activity 265

3. The data in the table below show foot length for 20 adults.

   (a) In the space, construct a tally chart for the data.

   (b) Plot the data as a histogram on the grid below.

Tally chart

Adult foot length (mm)			
265	272	257	315
300	320	250	250
215	330	240	270
252	270	265	350
315	300	290	310

©2024 BIOZONE International
ISBN: 978-1-99-101405-4
Photocopying Prohibited

146 Sexual Reproduction Produces Genetic Variation

Key Question: Why is variation in a population or species important, and what strategies do sexually reproducing species have to increase variation?

Why do some species show more variation than others?

▶ The photo, below left, shows aphids feeding on a plant. You will notice that all the aphids look the same. Now study the photo of the snails (*Cepaea nemoralis*). Note the variations in the shells. No two appear to be the same!

▶ How is it that the aphids look very similar yet there is huge **variation** between the snail shells? The aphids have reproduced asexually and the offspring are clones of a single parent. This explains why all the individuals look so similar (they are!). The snails are the result of sexual reproduction, in which sex cells (eggs and sperm) combine to produce new individuals.

1. Suggest reasons why variation is important in populations or species: _____

_____

_____

Variation in a population

▶ Most sexually reproducing organisms are diploid (2N), meaning they have two full sets of chromosomes. One set comes from the mother (maternal) and one set from the father (paternal). Sexual reproduction involves a diploid organism producing haploid gametes (cells with one set of chromosomes). When gametes fuse in fertilization to produce a zygote (fertilized egg), the diploid number of chromosomes is restored.

▶ The diagram on the right shows a lineage or pedigree chart of a hypothetical sexually reproducing population. It shows how variation can occur and spread in the population.

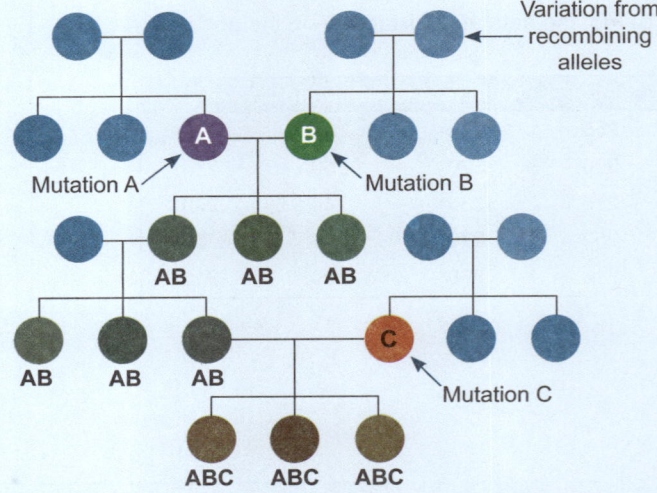

Variation by sexual reproduction
The diagram models how three beneficial mutations could be combined through sexual reproduction.

During meiosis, alleles are recombined in new combinations. Some combinations of alleles may be better suited to a particular environment than others. This variability is produced without the need for mutation. Beneficial mutations in separate lineages can be quickly combined through sexual reproduction.

2. (a) If the population was reproducing asexually, would it be possible for mutations A and B to be combined in one individual?

   _____

   _____

   _____

   (b) How would this affect the development of the asexually breeding population?

   _____

   _____

3. Apart from mutations, what other process is providing variation in the population shown above? _____

_____

147 Meiosis

Key Question: What is meiosis, and how does it produce haploid cells for the purposes of sexual reproduction?

Meiosis produces variation in sex cells

- **Meiosis** is a special type of cell division necessary for the production of gametes (sex cells) for the purpose of sexual reproduction.
- Before meiosis can take place, DNA must be replicated. If mutations (genetic errors) occur at this stage, they will be passed on and inherited by the offspring.
- Meiosis involves duplication of the chromosome, followed by two successive nuclear divisions. It halves the diploid chromosome number.
- An overview of meiosis is shown on the right. In plants and animals, meiosis occurs in the sex organs.
- During meiosis, a process called crossing over may occur. At this stage, homologous chromosomes may exchange **genes**. This further adds to the **variation** in the gametes.
- Meiosis is an important way of introducing genetic variation. The assortment of chromosomes into the gametes (the proportion from the father or mother) is random and can produce a huge number of possible chromosome combinations.

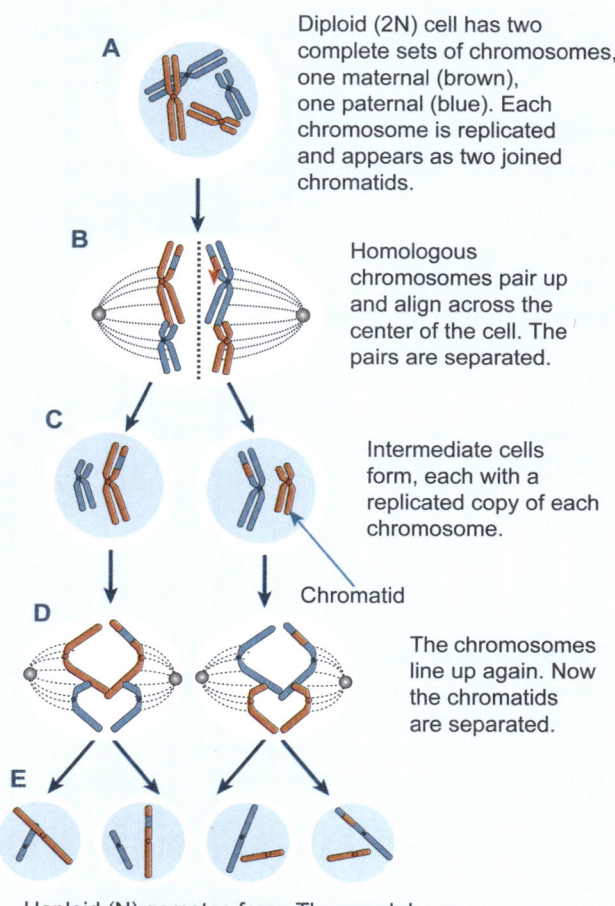

A: Diploid (2N) cell has two complete sets of chromosomes, one maternal (brown), one paternal (blue). Each chromosome is replicated and appears as two joined chromatids.

B: Homologous chromosomes pair up and align across the center of the cell. The pairs are separated.

C: Intermediate cells form, each with a replicated copy of each chromosome. — Chromatid

D: The chromosomes line up again. Now the chromatids are separated.

E: Haploid (N) gametes form. They each have one complete set of chromosomes.

1. What is the purpose of meiosis?

   _____

2. How can variation arise during meiosis? _____

   _____

   _____

3. The diagram above is a simplified model of meiosis. Specific stages in the process have been left out. In pairs or on your own, use the information below to label the spaces in the diagram on the right to produce a more complete model of meiosis. Stages A to E refer to stages in the diagram above. Label A has been done for you.

Meiosis I

Metaphase I
- Random alignment of homologous chromosomes at the equator.
- Chromatids still held at centromere.

Prophase I
- Chromosomes condense.
- Homologs pair up.
- Recombination of alleles occurs as homologous chromosomes exchange DNA in crossing over.

Telophase I
- Nuclear membranes reform, the cell divides and two cells are formed, each with N = 2 chromosomes.
- The spindle fibers disassemble.

Anaphase I
- The pairs of chromosomes separate and move to opposing poles.

Meiosis II

Anaphase II
- The chromatids split at the centromere and are moved by the spindle fibers to opposite poles of the cell.

Metaphase II
- Individual chromosomes line up along the equator of the cell.

Telophase II
- Nuclear membranes reform.
- There are 4 new haploid daughter cells.

Prophase II
- There are now two cells.
- DNA does not replicate again.

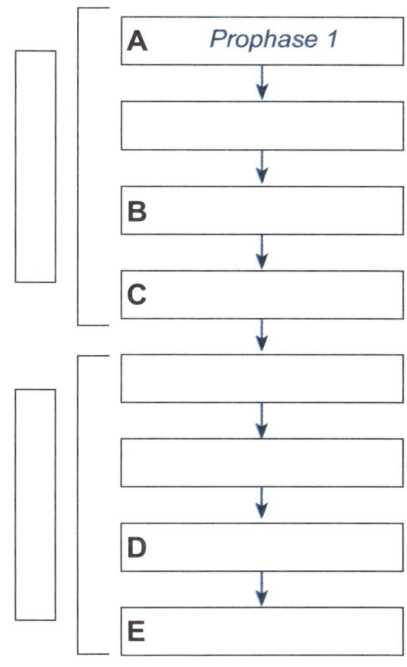

A: Prophase 1

B:

C:

D:

E:

148 Meiosis and Variation

Key Question: What are the important ways of introducing variation into the gametes formed during meiosis?

▶ **Independent assortment** and crossing over lead to **recombination** of **alleles** and are mechanisms that occur during **meiosis**. They increase the genetic **variation** in the gametes, and therefore the offspring.

Independent assortment

Independent assortment is an important mechanism for producing variation in gametes. The law of independent assortment states that allele pairs separate independently during meiosis. This results in the production of 2^x different possible combinations, where x is the number of chromosome pairs. For the example (right), there are two chromosome pairs. The number of possible allele combinations in the gametes is therefore $2^2 = 4$. Only two possible combinations are shown.

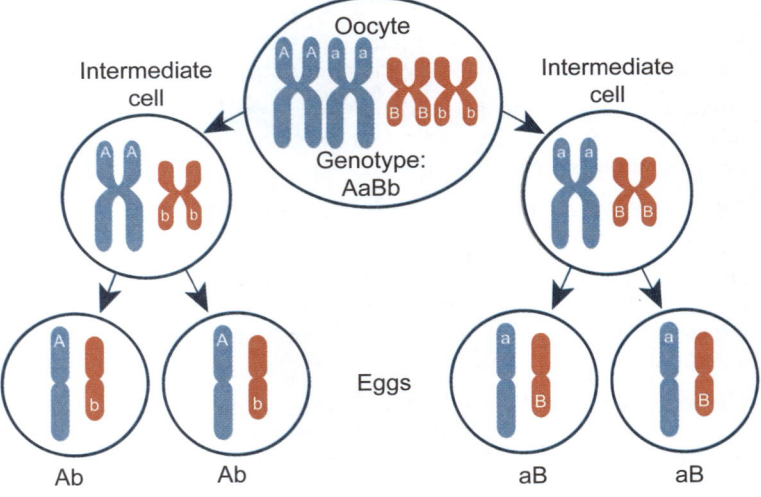

Crossing over and recombination

While they are paired during the first stage of meiosis, the non-sister chromatids of homologous chromosomes may become tangled and segments may be exchanged in a process called crossing over.

Crossing over results in the recombination of alleles, producing greater variation in the offspring than would otherwise occur. Alleles that are linked, i.e. on the same chromosome, may be exchanged and become unlinked.

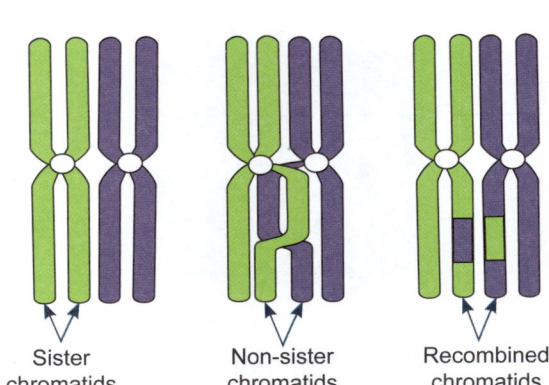

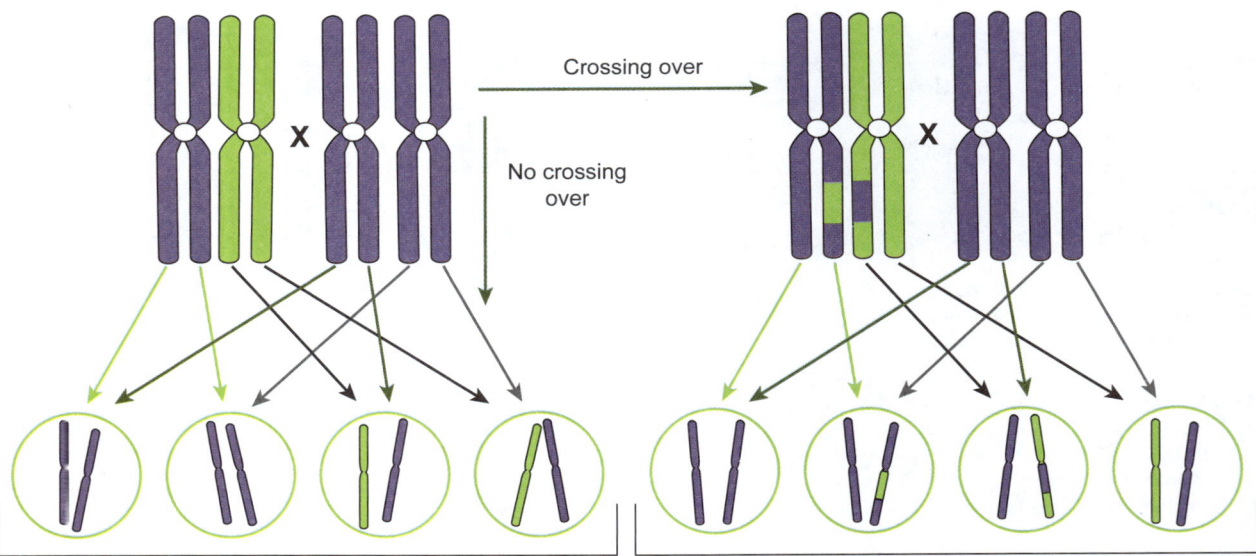

No crossing over, therefore no recombination in this cross results in all the offspring having the same genotypes as the parents.

Crossing over results in recombination. Although half of the offspring are the same as the parents, half have a new genetic combination.

1. (a) Using the diagram on independent assortment (previous page), draw the other two gamete combinations not shown in the diagram:

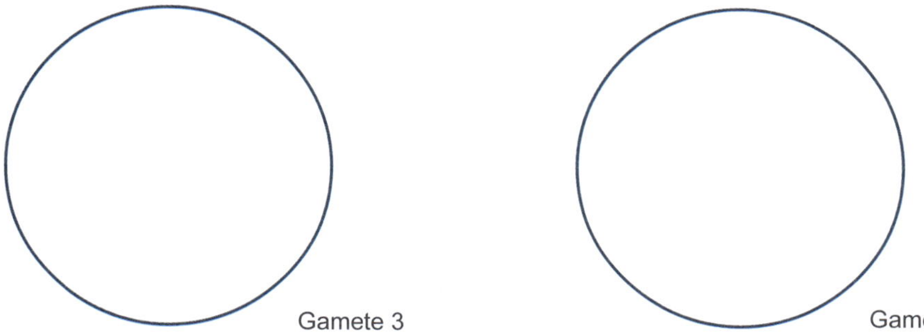

   Gamete 3   Gamete 4

   (b) For each of the following chromosome numbers, calculate the number of possible gamete combinations:

   i. 8 chromosomes: _____

   ii. 24 chromosomes: _____

   iii. 64 chromosomes: _____

2. What are sister and non-sister chromatids? _____

3. (a) What is crossing over? _____

   (b) How does crossing over increase the variation in the gametes (and hence the offspring)? _____

4. Crossing over occurs at a single point between the chromosomes below.

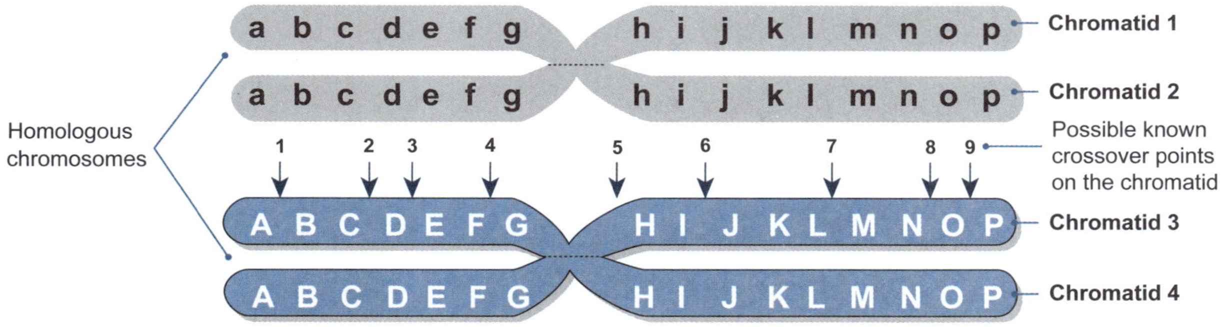

   Homologous chromosomes:
   - Chromatid 1: a b c d e f g | h i j k l m n o p
   - Chromatid 2: a b c d e f g | h i j k l m n o p
   - Possible known crossover points on the chromatid: 1 2 3 4 5 6 7 8 9
   - Chromatid 3: A B C D E F G | H I J K L M N O P
   - Chromatid 4: A B C D E F G | H I J K L M N O P

   (a) Draw the gene sequences for the four chromatids (above), after crossing over has occurred at crossover point 2:

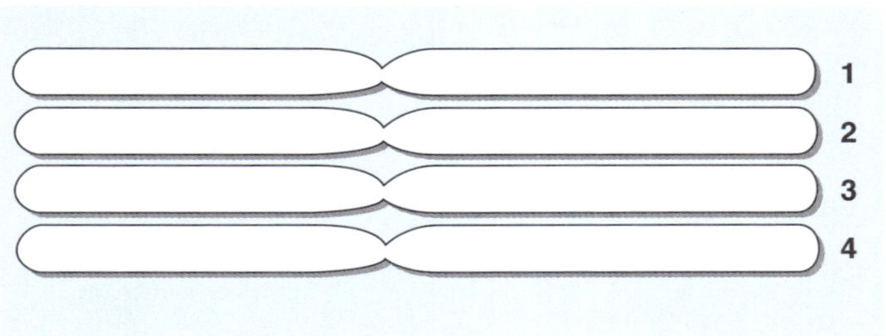

   (b) Which genes have been exchanged between the homologous chromosomes?

   _____

149 Modeling Meiosis

Key Question: How is variation introduced into the gametes formed during meiosis?

Modeling **meiosis** using ice block sticks can help to understand how meiosis creates **variation**. Each of your somatic (body) cells contains 46 chromosomes: 23 maternal and 23 paternal. Therefore, you have 23 homologous pairs. For simplicity, the number of chromosomes studied in this exercise has been reduced to four, i.e. two homologous pairs.

Investigation 6.2 Modeling meiosis using iceblock sticks

See appendix for equipment list.

To study the effect of crossing over on genetic variation, you will work in pairs to simulate the inheritance of two of your own traits: ability to tongue roll and handedness. This activity will take 25-45 minutes.

1. Record your phenotype and genotype for each trait in the table (right). If you have a dominant trait, you will not know if you are heterozygous or homozygous for that trait, so you can choose either genotype.

Chromosome number	Phenotype	Genotype
10	Tongue roller	TT, Tt
10	Non-tongue roller	tt
2	Right handed	RR, Rr
2	Left handed	rr

Step 1

Trait	Phenotype	Genotype
Handedness		
Tongue rolling		

2. Before you start the simulation, partner up with a classmate. Your gametes will combine with theirs (fertilization) at the end of the activity to produce a "child". Decide who will be female, and who will be male. You will need to work with this person again at step 7.

3. Collect four ice block sticks. These represent four chromosomes. Color two sticks blue or mark them with a P. for paternal chromosomes. The plain sticks are the maternal chromosomes. Write your initials on each of the four sticks. Label each chromosome with its number. Label four sticky dots with alleles to describe your phenotype and stick each onto the appropriate chromosome. In the example shown (right), the person is heterozygous for tongue rolling so sticky dots with alleles T and t are placed on chromosome 10. The person is also left handed, so alleles r and r are placed on chromosome 2.

Step 2

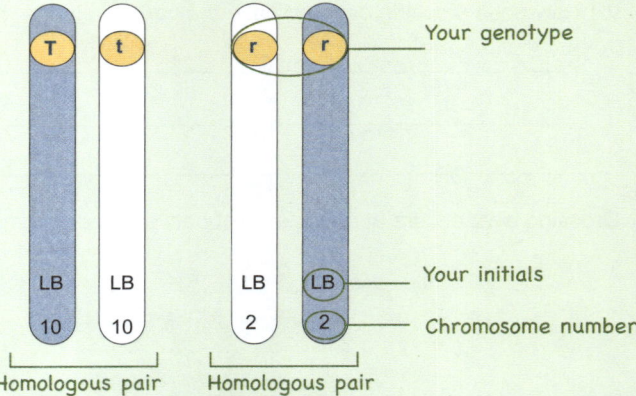

4. Randomly drop the chromosomes onto a table. This represents a cell in either the testes or ovaries.
   Duplicate your chromosomes by adding four more identical ice block sticks to the table (right).
   What are you simulating with this action?

   Simulate the first stage of meiosis by lining the duplicated chromosome pair with their homologous pair (below). For each chromosome number, you will have four sticks touching side-by-side (A, below).
   At this stage, crossing over occurs. Simulate this by swapping sticky dots from adjoining homologues (B, below).

Step 3

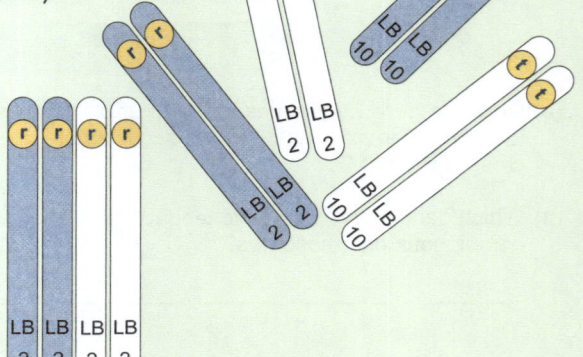

Step 4

(A) 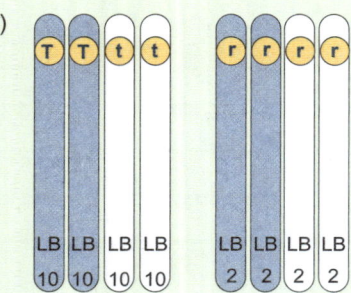 (B)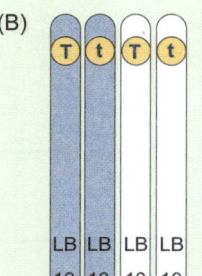

©2024 **BIOZONE** International
ISBN: 978-1-99-101405-4
Photocopying Prohibited

5. Randomly align the homologous chromosome pairs to simulate alignment across the cell's equator (center), as occurs in the next phase of meiosis. Simulate the separation of the chromosome pairs.
For each group of four sticks, two are pulled to each pole of the cell.

Step 5

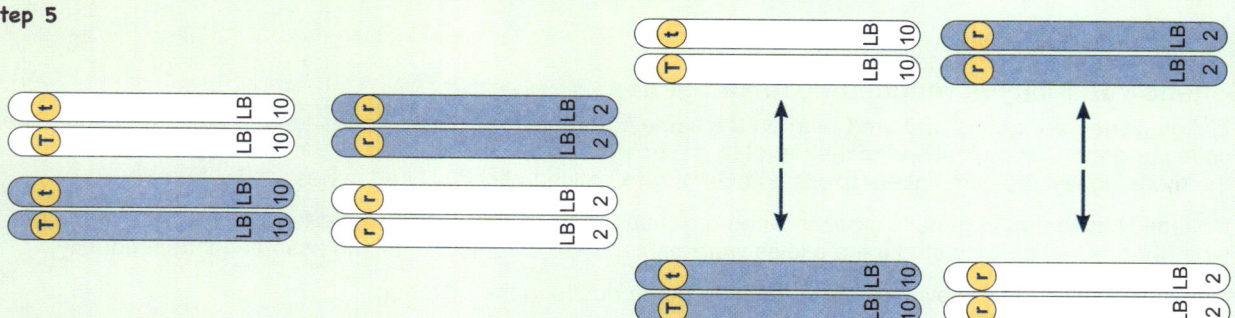

6. Two intermediate cells are formed. If you have completed step 5 correctly, each intermediate cell will be haploid (half the diploid chromosome number shown in step 3) with a mixture of maternal and paternal chromosomes. This is the end of the first division of meiosis. Your cells now need to divide for a second time. Repeat steps 4 and 5 but this time there is no crossing over and you are now separating replicated chromosomes, not homologues. At the end of this process each intermediate cell will have produced two haploid gametes. Each will have a maternal chromosome (white) and a paternal chromosome (blue) (below).

Step 6

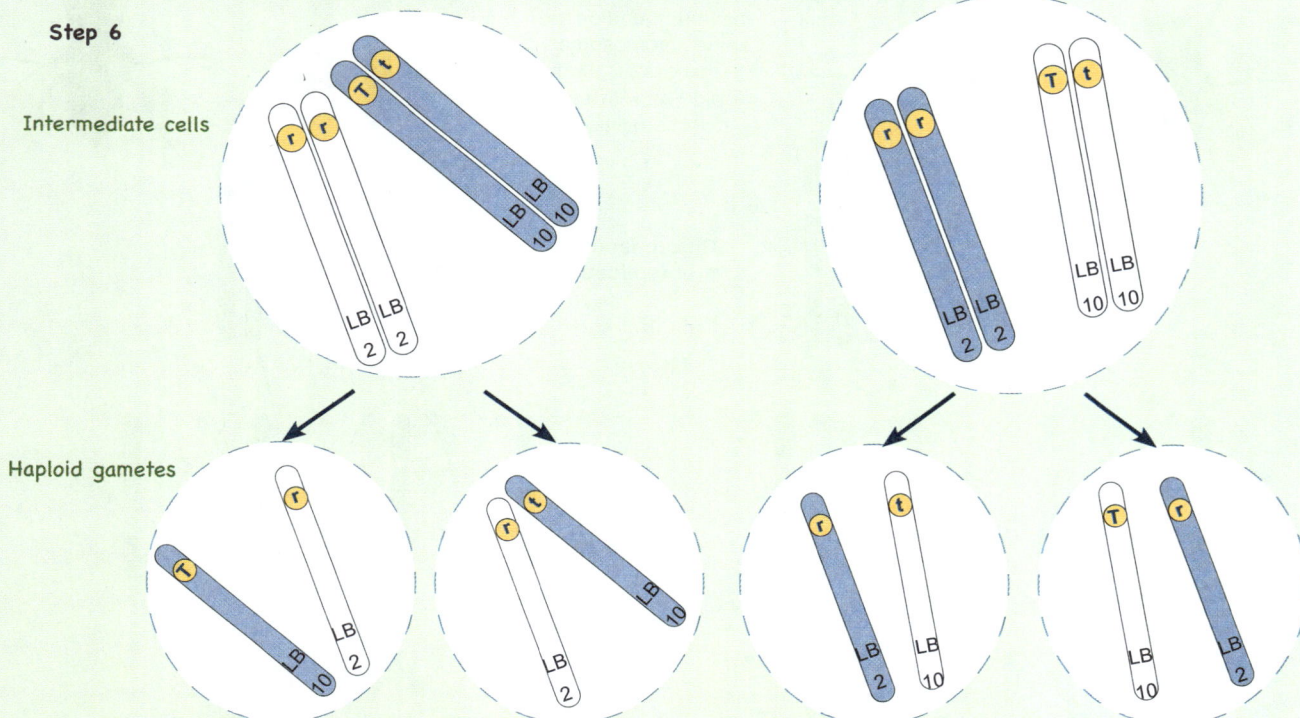

Intermediate cells

Haploid gametes

7. Pair up with the partner you chose at the beginning of the exercise to carry out fertilization. Randomly select one sperm and one egg cell. The unsuccessful gametes can be removed from the table. Combine the chromosomes of the successful gametes. You have created a child! Fill in the following chart to describe your child's genotype and phenotype for tongue rolling and handedness.

Trait	Phenotype	Genotype
Handedness		
Tongue rolling		

8. As a class, create a spreadsheet in which you each enter if you are a tongue roller or non roller and left or right handed. Use the spreadsheet to produce a graph of roller vs non rollers, left vs right handers. Analyze the data to see if there is a correlation between handedness and tongue rolling..

1. What is the significance of each parent contributing half the number of chromosomes to the zygote in terms of producing variation in a population?

_____

_____

_____

150 Linked Genes and Variability

Key Question: How is variation affected when genes and linked?

Genetic variability is reduced by linked genes

- Linked **genes** are genes that are found on the same chromosome. Linked genes tend to be inherited together and so fewer genetic combinations of their **alleles** are possible. The closer genes are to each other on the chromosome, the more 'tightly' they are linked. Crossing over is rare between tightly linked genes.
- Linkage is indicated in genetic crosses when a greater proportion of the offspring from a cross are of the parental type (than would be expected if the alleles were on separate chromosomes and assorting independently).
- Linkage reduces the genetic **variation** that can be produced in the offspring.

Inheritance of linked genes

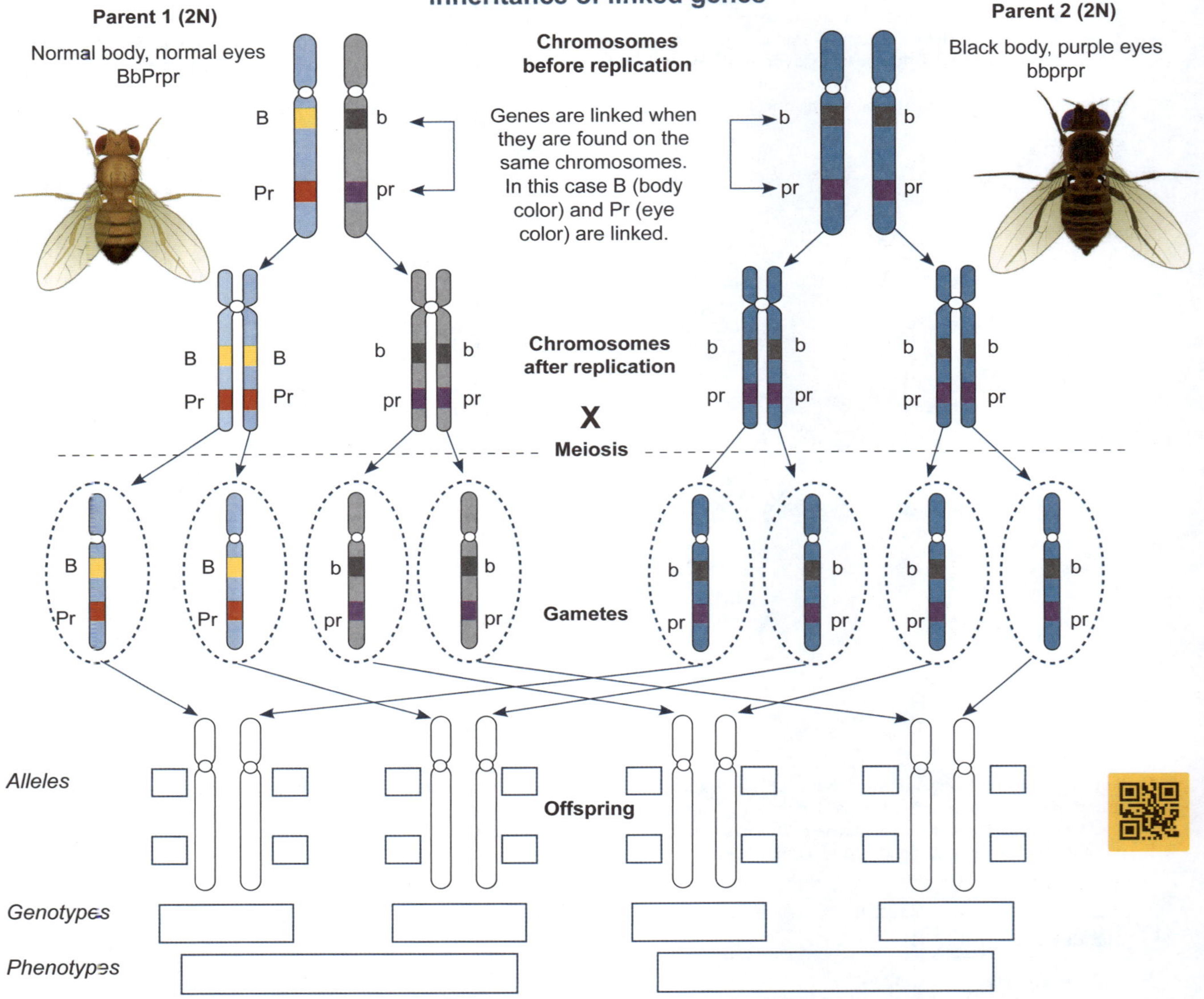

Parent 1 (2N)
Normal body, normal eyes
BbPrpr
Parent 2 (2N)
Black body, purple eyes
bbprpr
Chromosomes before replication
Genes are linked when they are found on the same chromosomes. In this case B (body color) and Pr (eye color) are linked.
B
Pr
b
pr
Chromosomes after replication
X
Meiosis
Gametes
Alleles
Offspring
Genotypes
Phenotypes

1. (a) In the boxes above, write the alleles, the genotypes, and the phenotypes of the offspring:

   (b) How many different genotypes/phenotypes are possible for the cross pictured? ______

   (c) Are they the same as the parental genotypes/phenotypes? YES or NO: ______

2. Explain why linkage reduces genetic variability: ______

______

______

______

8.A

ISBN: 978-1-99-101405-4

151 Mendelian Genetics

Key Question: How can we use Mendelian genetics to predict the outcome of genetic crosses?

Can you roll your tongue?

Are you able to roll your tongue or not? If you are a non-roller, no matter how hard you try, your tongue will not roll up to form a tube? The ability to roll your tongue is often used as an example of simple Mendelian inheritance, meaning it is controlled by one **gene** and the **phenotype** is either present or not present.

In the 1940s, geneticist, Alfred Sturtevant proposed that tongue rolling was inherited according to simple Mendelian rules. Under this scenario, the **allele** for tongue rolling is **dominant**, so if you have at least one dominant allele (T) you will be a tongue roller. Non-rollers have two copies of the **recessive** allele (tt).

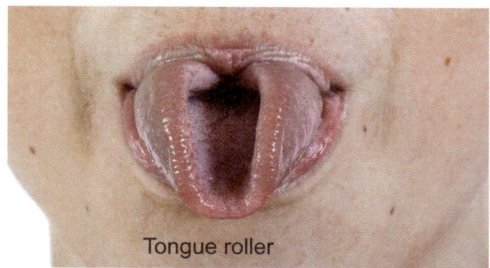
Tongue roller

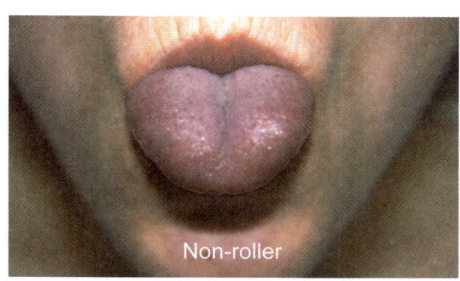
Non-roller

1. (a) Can you roll your tongue? _____

   (b) If you can, state whether each of your parents is able to roll their tongue: _____

   (c) Using the T and t notation for the tongue rolling alleles, decide what your genotype could be and record it here:

   (d) When you compare your phenotype with that of your parents, does it make sense? Explain why or why not:

You may have found in your own investigation above that both of your parents are non-rollers but you (or one of your siblings) can actually roll your tongue. How can this be?

Since Sturtevant's work in the 1940s, we have learned that the genetics of tongue rolling is not as simple as first thought. A study by Komai in 1951 obtained the results presented in the table on the right.

2. (a) Calculate the percentage of rollers from each cross and enter these values on the table.

   (b) What percentage of R offspring were produced by the NR x NR crosses?

   _____

   (c) Does this result suggest tongue rolling is inherited in a classic Mendelian inheritance pattern? Explain your answer:

Parents	R offspring	NR offspring	% R
R x R	928	104	
R x NR	468	217	
NR x NR	48	92	

R = roller, NR = non-roller.

So it's not that simple! Further studies involving identical twins showed that, in some cases, one of the pair could tongue roll while the other could not. Given the (nearly) identical nature of their DNA, we would expect both to have the same phenotype for tongue rolling. We now know that tongue rolling is not determined solely by genetics and that there are also probably environmental factors at play. You may still see many references using tongue rolling as an example of Mendelian inheritance, but keep in mind that it is not that simple.

Mendel's pea experiments

▶ Some of the best known experiments on phenotypes are the experiments carried out by Gregor Mendel on pea plants in the 1800s. Mendel did breeding experiments to study seven phenotypic characters of the pea plant.

▶ During one of the experiments (below) he noticed how **traits** expressed in one generation disappeared in the next, but then reappeared in the generation after that.

▶ In his experiments, Mendel used true breeding plants. When self-crossed (self-fertilized), true breeding plants are **homozygous** for the gene in question and therefore produce offspring with the same phenotypes as the parents. In the experiment described below, Mendel crossed plants true breeding for round seeds with plants true breeding for wrinkled seeds.

Peas show noticeably different traits for pod shape and color, seed shape and color, flower color, height, and position of the flowers on the stem.

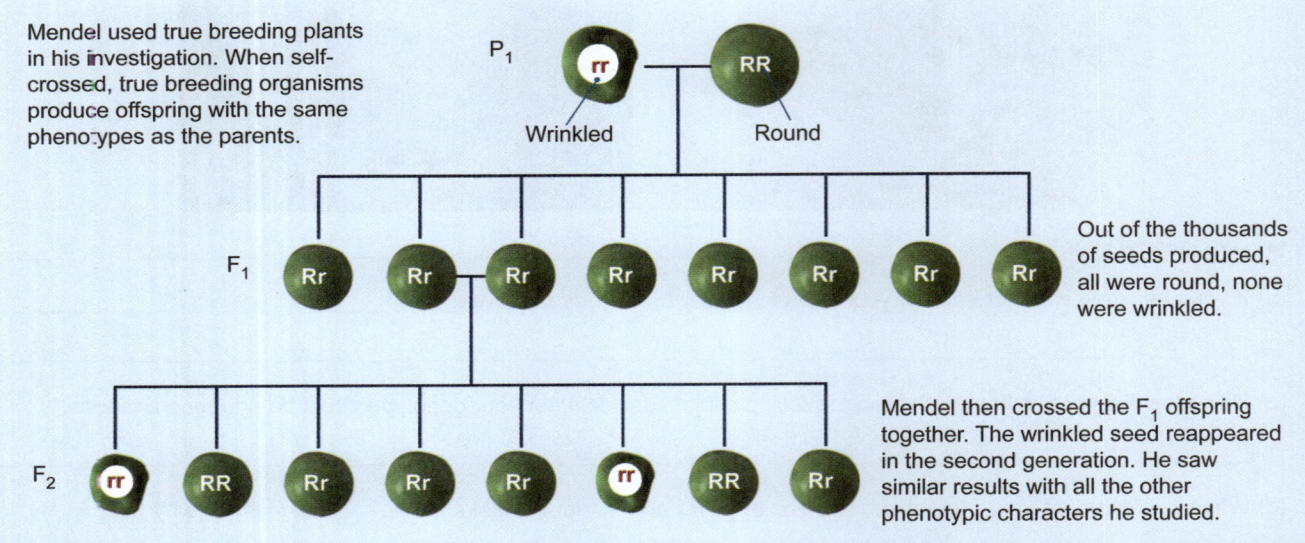

Mendel used true breeding plants in his investigation. When self-crossed, true breeding organisms produce offspring with the same phenotypes as the parents.

Out of the thousands of seeds produced, all were round, none were wrinkled.

Mendel then crossed the F_1 offspring together. The wrinkled seed reappeared in the second generation. He saw similar results with all the other phenotypic characters he studied.

How can these results be explained? Mendel was able to explain his observations in the following way:
▶ Traits are determined by a unit, which passes unchanged from parent to offspring (we now know that these units are genes).
▶ Each individual inherits one unit (gene) for each trait from each parent (each individual has two units).
▶ Traits may not physically appear in an individual, but the units (genes) for them can still be passed to its offspring.

3. In Mendel's pea experiments investigating wrinkled versus round peas above:

   (a) What was the dominant trait? _____

   (b) Which genotypes produced this trait: _____

   (c) If the cross between the F_1 plants produced 76 peas, how many would you expect to be wrinkled? _____

4. Another of Mendel's experiments involved crossing true breeding parents with green seed pods and yellow seed pods together. Green seed pod color is dominant over yellow seed pod color. Based on Mendel's experiment on seed shape above:

   (a) What seed pod color(s) would you expect to observe in the F_1 offspring?

   _____

   (b) What seed pod color(s) would you expect to observe in the F_2 offspring?

   _____

   (c) You can test your prediction by drawing a diagram in the box (right) to illustrate the results of the experiment.

5. Mendel examined seven phenotypic traits (table right). Some of his results from crossing plants heterozygous for the gene in question are shown below. The numbers in the results column represent how many offspring had those particular traits.

(a) Study the results for each of the six experiments below. Determine which of the two phenotypes is dominant, and which is recessive. Place your answers in the spaces in the dominance column in the table below.

(b) Calculate the ratio of dominant phenotypes to recessive phenotypes (to two decimal places). The first one has been done for you (5474 ÷ 1850 = 2.96). Place your answers in the spaces provided in the table below:

Phenotypic characters of the pea plant:
- Flower color (violet or white)
- Pod color (green or yellow)
- Height (tall or short)
- Position of the flowers on the stem (axial or terminal)
- Pod shape (inflated or constricted)
- Seed shape (round of wrinkled)
- Seed color (yellow or green)

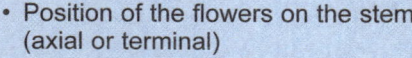

Trait	Possible phenotypes	Results	Dominance	Ratio
Seed shape	Wrinkled / Round	Wrinkled 1850 / Round 5474 / TOTAL 7324	Dominant: Round Recessive: Wrinkled	2.96:1
Seed color	Green / Yellow	Green 2001 / Yellow 6022 / TOTAL 8023	Dominant: Recessive:	
Pod color	Green / Yellow	Green 428 / Yellow 152 / TOTAL 580	Dominant: Recessive:	
Flower position	Axial / Terminal	Axial 651 / Terminal 207 / TOTAL 858	Dominant: Recessive:	
Pod shape	Constricted / Inflated	Constricted 299 / Inflated 882 / TOTAL 1181	Dominant: Recessive:	
Stem length	Tall / Dwarf	Tall 787 / Dwarf 277 / TOTAL 1064	Dominant: Recessive:	

6. Mendel's experiments identified that two heterozygous parents should produce offspring in the ratio of 3 of the dominant phenotype to 1 of the recessive phenotype.

(a) Which three of Mendel's experiments provided ratios closest to the theoretical 3:1 ratio? _____

(b) Suggest why these results deviated less from the theoretical ratio than the others: _____

152 Monohybrid Crosses

Key Question: How can we predict the outcome of genetic crosses?

- Monohybrid crosses are used to show single **gene** inheritance patterns between two individuals. **Monohybrid crosses** can be used to determine **allele** dominance. A simple square grid called a Punnett square is used to determine all of the possible outcomes of a genetic cross.

- The diagram on the right shows how to draw a Punnett square for a monohybrid cross for a **trait** defined by the alleles A and a. Both parents have the allele combination Aa. Crosses can also be drawn using circles and arrows or lines to illustrate segregation of the gametes and fertilization as below.

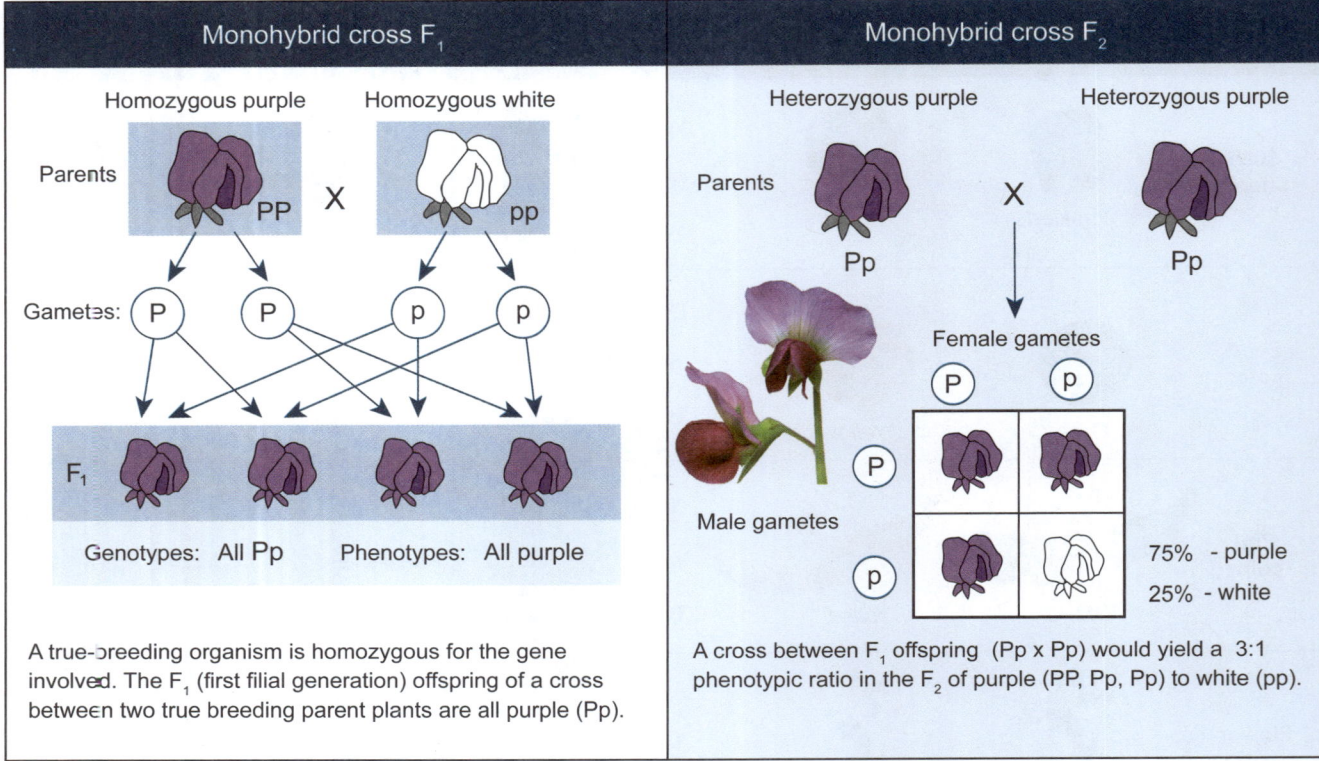

A true-breeding organism is homozygous for the gene involved. The F_1 (first filial generation) offspring of a cross between two true breeding parent plants are all purple (Pp).

A cross between F_1 offspring (Pp x Pp) would yield a 3:1 phenotypic ratio in the F_2 of purple (PP, Pp, Pp) to white (pp).

1. Study the diagrams above and identify the dominant allele and the recessive allele. Explain your reasoning: _____
_____
_____

2. Complete the crosses below:

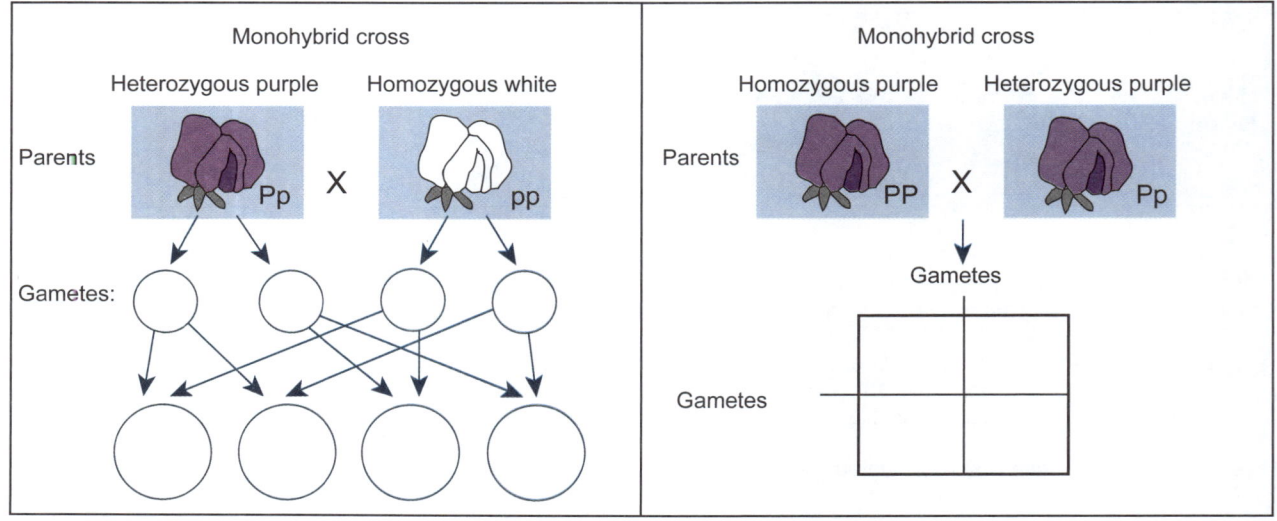

The test cross

▸ It is not always possible to determine an organism's **genotype** by its appearance because gene expression is complicated by patterns of dominance and by gene interactions. The test cross is a special type of back cross used to determine the genotype of an organism with the **dominant phenotype** for a particular trait.

▸ The principle is simple. The individual with the unknown genotype is bred with a **homozygous recessive** individual for the trait(s) of interest. The homozygous recessive can produce only one type of allele (recessive), so the phenotypes of the offspring will reveal the genotype of the unknown parent (below). The test cross can be used to determine the genotype of single genes or multiple genes.

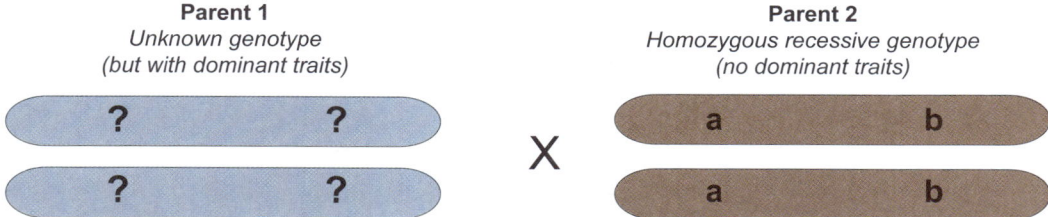

Parent 1
Unknown genotype (but with dominant traits)

X

Parent 2
Homozygous recessive genotype (no dominant traits)

The common fruit fly (*Drosophila melanogaster*) is often used to illustrate basic principles of inheritance because it has several easily identified phenotypes, which act as genetic markers. One such phenotype is body color. Wild type (normal) *Drosophila* have yellow-brown bodies. The allele for yellow-brown body color (E) is dominant. The allele for an ebony colored body (e) is recessive. The test crosses below show the possible outcomes for an individual with homozygous and heterozygous alleles for ebony body color.

A. A homozygous recessive female (ee) with an ebony body is crossed with a homozygous dominant male (EE).

B. A homozygous recessive female (ee) with an ebony body is crossed with a heterozygous male (Ee).

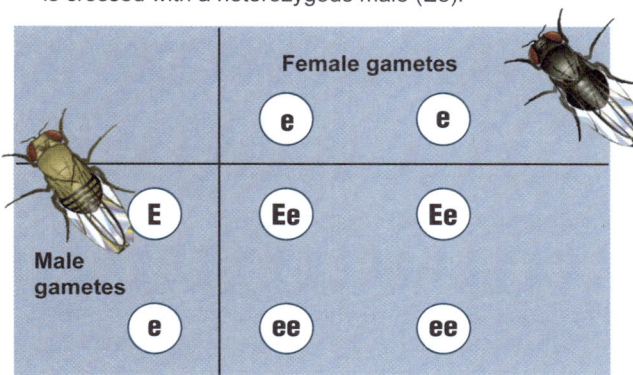

Cross A:
(a) Genotype frequency: _100% Ee_
(b) Phenotype frequency: _100% yellow-brown_

Cross B:
(a) Genotype frequency: _50% Ee, 50% ee_
(b) Phenotype frequency: _50% yellow-brown, 50% ebony_

▸ In crosses involving *Drosophila*, the wild-type alleles are dominant but are given an upper case symbol of the mutant phenotype. This is because there are many alternative mutant phenotypes to the wild type, e.g. vestigial wings and curled wings are both mutant phenotypes, whereas the wild type is straight wing. This can happen in other crosses too, e.g. guinea pigs, if a letter is already used for another character.

3. In *Drosophila*, the allele for brown eyes (b) is recessive, while the red eye allele (B) is dominant. Explain (using text or diagrams) how you would carry out a test cross to determine the genotype of a male with red eyes:

4. Of the test cross offspring, 50% have red eyes and 50% have brown eyes.

    What is the genotype of the male *Drosophila*? _____

153 Probability

Key Question: How can we determine how likely an event is to happen?

Understanding probability

Many events cannot be predicted with absolute certainty; however, we can determine how likely it is that an event will happen. This calculated likelihood of an event occurring is called probability. The probability of an event ranges from 0 to 1. The sum of all probabilities equals 1.

In biology, probability is used to calculate the statistical significance of a difference between means or the probability of an event occurring, e.g. getting an offspring with a certain **genotype** and **phenotype** in a genetic cross.

$$\text{Probability of an event happening} = \frac{\text{Number of ways it can happen}}{\text{Total number of outcomes}}$$

- Tossing a coin and predicting whether it will land heads (H) up or tails (T) up is a good example to illustrate probability.
- There are two possible outcomes; the coin will either land heads up or tails up, and only one outcome can occur at a time. Therefore the probability of a coin landing heads up is 1/2. The likelihood of a coin landing tails up is also 1/2.
- Remember, probability is just an indication of how likely something will happen. Even though we predict that heads and tails will come up 50 times each if we toss a coin 100 times, it might not be exactly that.

1. Calculate the probability that a 6 will occur when you roll a single dice (die):

   _____

The rules for calculating probability

Probability rules are used when we want to predict the likelihood of two events occurring together or when we want to determine the chances of one outcome over another. The rules are useful when we want to determine the probably of certain outcomes in genetic crosses, especially when large numbers of alleles are involved. The probability rule used depends on the situation.

PRODUCT RULE for independent events	SUM RULE for mutually exclusive events
For independent events, A & B, the probability (P) of them both occurring (A&B) = P(A) X P(B)	**For mutually exclusive events, A & B, the probability (P) that one will occur (A or B) = P(A) + P(B)**
Example: If you roll two dice at the same time, what is the probability of rolling two sixes?	**Example**: A single die is rolled. What are the chances of rolling a 2 **or** a 6?
Solution: The probability of getting six on two dice at once is 1/6 x 1/6 = 1/36.	**Solution**: P(A or B) = P(A) + P(B). 1/6 + 1/6 = 2/6 (1/3). There is a 1/3 chance that a 2 or 6 will be rolled.

2. Use the product rule to determine the probability of a first and second child born to the same parents both being boys?

   _____

3. In a cross of rabbits both heterozygous for genes for coat color and length (BbLl x BbLl), determine the probability of the offspring being BbLl. HINT: Calculate probabilities for Bb and Ll separately and then use the product rule. Test your calculation using the Punnett square (right).

   _____

4. In a cross of two individuals with various alleles of four unlinked genes: AaBbCcdd x AabbCcDd, explain how you would calculate the probability of getting offspring with the dominant phenotype for all four traits?

   _____

   _____

   _____

154 Practicing Monohybrid Crosses

Key Question: How can we use a monohybrid cross to study the inheritance pattern of one gene, and what are the predictable ratios in the offspring from this cross?

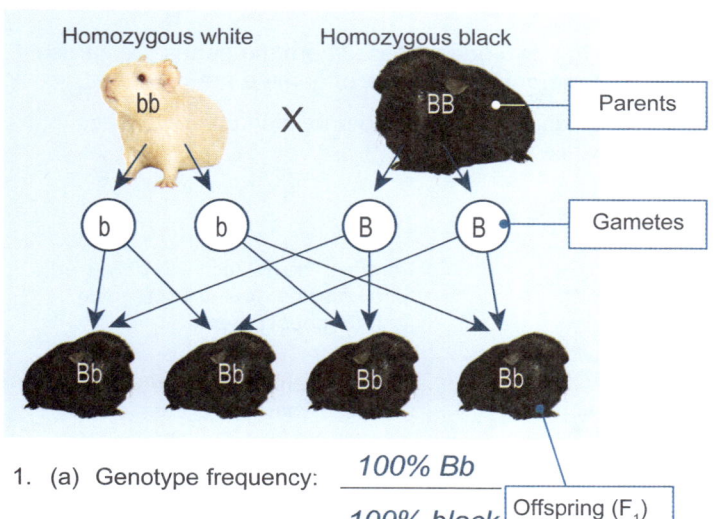

Monohybrid crosses can be used to determine the **genotype** and **phenotype** outcomes for coat color in guinea pigs. Complete the monohybrid crosses below by determining the gametes and phenotypic and genotypic frequencies of the offspring. Question one has been done for you.

1. (a) Genotype frequency: __100% Bb__
   (b) Phenotype frequency: __100% black__

2. (a) Which coat color is dominant? _____
   (b) Which is the dominant allele? _____
   (c) Which coat color is recessive? _____
   (d) Which is the recessive allele? _____

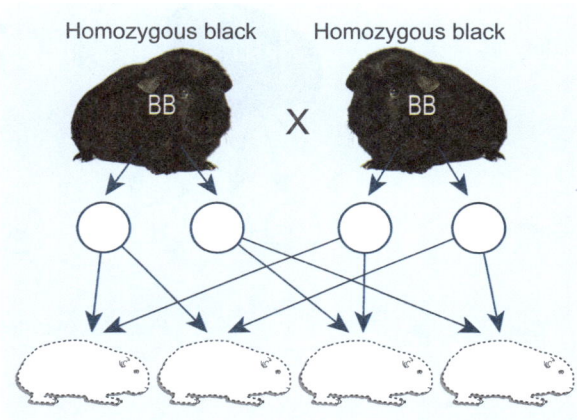

3. (a) Genotype frequency: _____
   (b) Phenotype frequency: _____

4. (a) Genotype frequency: _____
   (b) Phenotype frequency: _____

5. Two parent guinea pigs with the genotypes Bb and BB are crossed:
   (a) What is the probability that any one offspring is BB? _____
   (b) What is the probability that any one offspring is black? _____

6. Two parent guinea pigs with genotypes bb and Bb are crossed:
   (a) What is the probability that any one offspring is Bb? _____
   (b) What is the probability that any one offspring is black? _____

7. A white guinea pig and a black guinea pig are crossed. All of the guinea pigs that are born are white.
   (a) What is the genotype of the black guinea pig? _____
   (b) Explain the result: _____
   _____
   _____

155 Dihybrid Crosses

Key Question: How can we use dihybrid crosses to study the inheritance pattern of two unlinked genes, and what are their predictable ratios?

Dihybrid cross

▸ A **dihybrid cross** studies the inheritance pattern of two **genes**. In a two gene cross, where the genes are carried on separate chromosomes and are sorted independently during **meiosis**, four types of gamete are produced.

▸ The two genes in the example below are on separate chromosomes and control two unrelated characteristics, hair color and coat length. Black (B) and short (L) are **dominant** to white (b) and long (l).

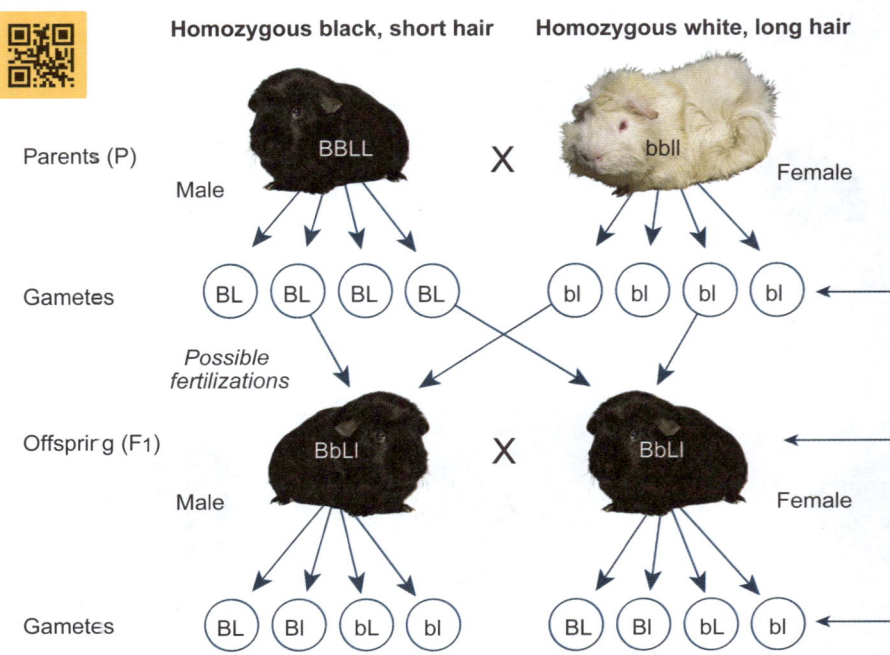

Parents: The notation P is only used for a cross between true breeding (homozygous) parents.

Gametes: Only one type of gamete is produced from each parent (although they will produce four gametes from each oocyte or spermatocyte). This is because each parent is homozygous for both traits.

F_1 offspring: There is only one kind of gamete from each parent, therefore only one kind of offspring produced in the first generation. The notation F_1 is only used to denote the heterozygous offspring of a cross between two true breeding parents.

F_2 offspring: The F_1 were mated with each other (selfed). Each individual from the F_1 is able to produce four different kinds of gamete.

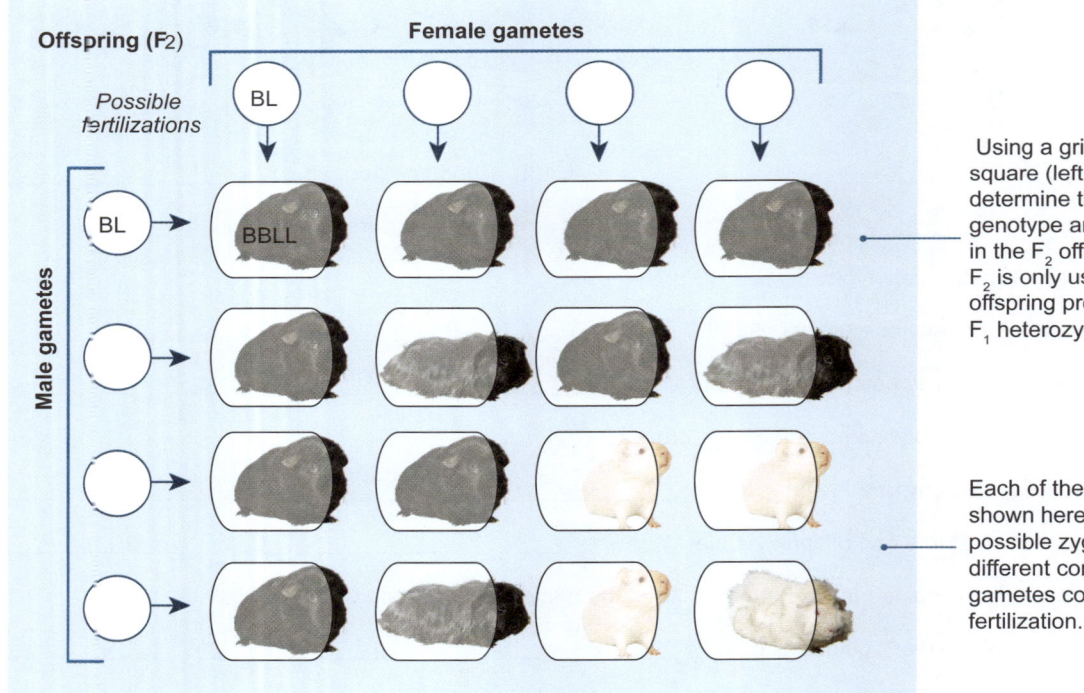

Using a grid called a Punnett square (left), it is possible to determine the expected genotype and phenotype ratios in the F_2 offspring. The notation F_2 is only used to denote the offspring produced by crossing F_1 heterozygotes.

Each of the 16 animals shown here represents the possible zygotes formed by different combinations of gametes coming together at fertilization.

1. Fill in the gametes and complete the Punnett square above.

2. Use the Punnett square to identify the number of each phenotype in the offspring: _____
_____

3. In guinea pigs, rough coat **R** is dominant over smooth coat **r** and black coat **B** is dominant over white **b**. The genes are not linked. A homozygous rough black animal was crossed with a homozygous smooth white:

   (a) State the genotype of the F_1: _____

   (b) State the phenotype of the F_1: _____

   (c) Use the Punnett square to show the outcome of a cross between the F_1 (the F_2):

   (d) Using ratios, state the phenotypes of the F_2 generation:

   _____

   _____

   _____

   _____

   (e) Use the working space on the right to show the outcome of a back cross of the F_1 to the rough, black parent:

   (f) Using ratios, state the phenotype of the offspring of this back cross:

   _____

   (g) A rough black guinea pig was crossed with a rough white guinea pig and produced the following offspring: 3 rough black, 2 rough white, and 1 smooth white. What are genotypes of the parents? Explain your reasoning:

   _____

   _____

   _____

4. In humans, two genes affecting the appearance of the hands are the gene for thumb hyperextension (curving) and the gene for mid-digit hair. The allele for curved thumb (right), **H**, is dominant to the allele for straight thumb, **h**. The allele for mid digit hair (far right), **M**, is dominant to that for an absence of hair, **m**.

   (a) Give all the genotypes of individuals who are able to curve their thumbs, but have no mid-digit hair:

   _____

   (b) Complete the Punnett square to show the possible genotypes from a heterozygous cross between both individuals:

   (c) State the phenotype ratios of the F_1 progeny:

   _____

   _____

   _____

   (d) What is the probability that one of the offspring would have mid-digit hair (show your working)?

   _____

5. In rabbits, spotted coat **S** is dominant to solid color **s**, while for coat color, black **B** is dominant to brown **b**. A brown spotted rabbit is mated with a solid black one and all the offspring are black spotted (the genes are not linked).

   (a) State the genotypes:

   Parent 1: _____

   Parent 2: _____

   Offspring: _____

   (b) Use the Punnett square (top right) to show the outcome of a cross between the F_1 (i.e. the F_2):

   (c) Using ratios, state the phenotypes of the F_2 generation:

   _____

   _____

   _____

6. The Himalayan color-pointed, long-haired cat is a breed developed by crossing a pedigree (true-breeding), uniform-colored, long-haired Persian with a pedigree color-pointed (darker face, ears, paws, and tail), short-haired Siamese.

   Persian   Siamese   Himalayan

   The genes controlling hair coloring and length are on separate chromosomes: uniform color **U**, color pointed **u**, short hair **S**, long hair **s**

   (a) State the genotype of the F_1 (Siamese X Persian): _____

   (b) State the phenotype of the F_1: _____

   (c) Use the Punnett square (right) to show the outcome of a cross between the F_1 (the F_2):

   (d) What ratio of the F_2 will be Himalayan? _____

   (e) State whether the Himalayan would be true breeding:

   _____

   (f) What ratio of the F_2 will be color-point, short-haired cats?

   _____

7. In cats, the following alleles are present for coat characteristics: black (**B**), brown (**b**), short (**L**), long (**l**). The genes are not linked. Use the information to complete the dihybrid crosses below:

   A black short haired (**BBLl**) male is crossed with a black long haired (**Bbll**) female. Determine the genotypic and phenotypic ratios of the offspring:

   Genotype ratio: _____

   _____

   Phenotype ratio: _____

   _____

156 Non Mendelian Genetics

Key Question: How can the patterns that do not fit Mendelian genetics be explained?

Calico cats

▶ Examine the image of the cat on the right. The cat has a coat pattern called calico.

1. Is the cat male or female? _____

2. How do you know? _____
_____

▶ Now consider the following statements:

- Around 75%-80% of orange tabby cats are male.
- All calico cats are female.
- In cats, orange coat color (O) is codominant to black coat color (o).
  If an orange male cat breeds with a black female cat, only calico female and black male offspring will be produced.

3. What genetic rule or rules might be governing these kinds of genetic inheritance patterns? _____
_____
_____

Non Mendelian genetics

▶ When Mendel carried out his famous pea plant breeding experiments, he was either lucky to choose, or had good enough observational skills to initially notice, **traits** that were linked to **genes** that sorted independently and produced only two **phenotypes** per gene.

▶ Had he chosen a different trait or organism, he may have obtained results that made little sense, and it may have taken a lot longer to work out the rules governing them.

▶ We know in Mendelian genetics that genes sort independently and that **alleles** segregate into separate gametes. This results in predictable ratios of **genotypes** and phenotypes in offspring. We know for example, that a monohybrid cross of two **heterozygous** individuals will produce a genotype ratio of 1:2:1 and a phenotype ratio of 3:1. We also know that a **dihybrid cross** of two individuals heterozygous for both genes will produce nine different genotypes and a phenotype ratio of 9:3:3:1.

▶ However, as with many biological rules, there are exceptions. Recall that crossing linked genes results in unexpected gene ratios. Crosses of genes not following Mendel's laws do not produce the expected genotypic and phenotypic ratios.

Examine the crosses below showing the phenotypes of the parent plants and offspring:

4. Can you work out what appears to be the inheritance pattern occurring with the roses shown above? Discuss your ideas in small groups. Write down your ideas below:
_____
_____
_____

5. What is a limitation of Mendel's model of inheritance? _____
_____

157 Incomplete Dominance

Key Question: What inheritance patterns occur when neither allele is dominant over the other?

Crosses involving incomplete dominance

▸ In incomplete dominance, the **heterozygous** offspring are intermediate in **phenotype** between the contrasting **homozygous** parental phenotypes. In crosses involving incomplete dominance, the phenotype and genotype ratios are identical.

▸ The phenotype of heterozygous offspring results from the partial influence of both **alleles**. Examples of incomplete dominance include flower color in snapdragons (*Antirrhinum*) and four o'clocks (*Mirabilis*) (below).

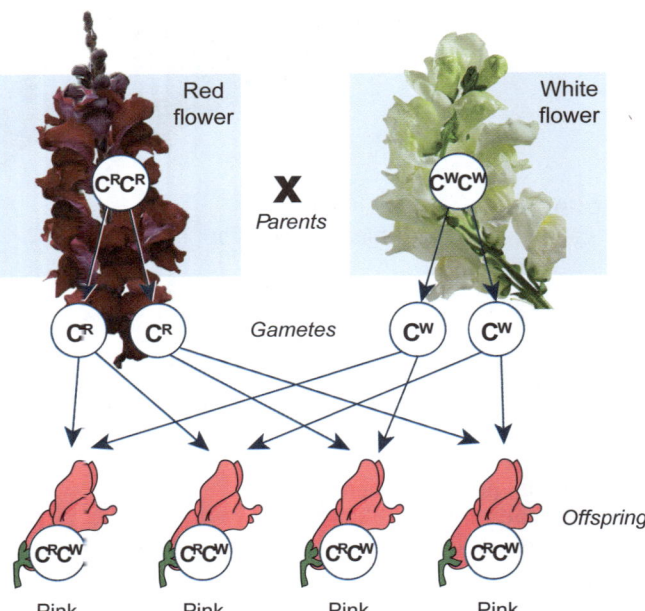

▸ Pure breeding snapdragons produce red or white flowers (left). When red and white-flowered parent plants are crossed, pink-flowered offspring are produced. If the offspring (F_1 generation) are then crossed together, all three phenotypes (red, pink, and white) are produced in the F_2 generation.

▸ Four o'clocks (right) are also known to have flower colors controlled by alleles that show incomplete dominance. Pure breeding four o'clocks produce crimson, yellow, or white flowers. Crimson flowers (right) crossed with yellow flowers produced reddish-orange flowers, while crimson flowers crossed with white flowers produce magenta flowers.

1. A plant breeder wanted to produce snapdragons for sale that were only pink or white, i.e. no red. Determine the phenotypes of the two parents necessary to produce these desired offspring. Use the Punnett square (right) to help you:

   _____

   _____

   _____

2. Another plant breeder crossed two four o'clocks, known to have its flower color controlled by alleles that show incomplete dominance. Pollen from a magenta flowered plant was placed on the stigma of a crimson flowered plant.

   (a) Predict the genotype and phenotype for parents and offspring by filling in the spaces on the diagram on the right.

   (b) State the phenotype ratio of the offspring:

   _____

   _____

   _____

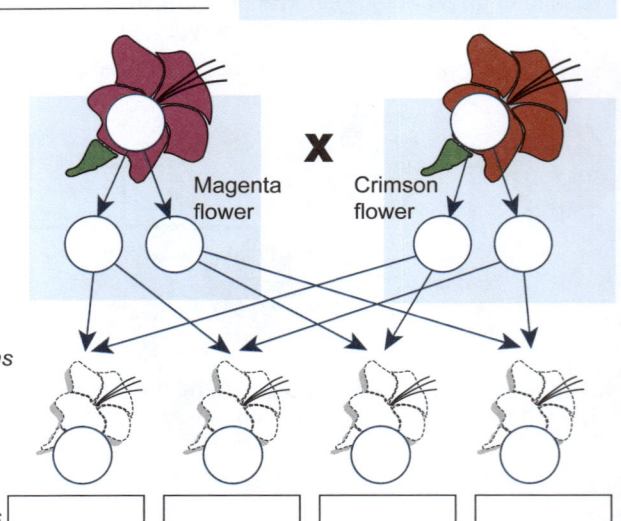

3. Explain why magenta colored snapdragons will never breed true (multiple phenotypes always appear in the offspring):

   _____

   _____

- In its normal form, the MSTN **gene** produces an enzyme that inhibits muscle growth. The mutated form does not operate correctly and mammals with this mutation exhibit enhanced muscle development (called double muscling).
- This **trait** has been used to breed Belgian blue cattle (right) for beef production.
- The MSTN gene displays incomplete dominance. Heterozygous cattle also have enhanced muscle development, while cattle homozygous for the mutant gene show even greater muscle development, to the extent that giving birth is difficult for female cattle.
- Because of this, many breeders opt for cross breeding full Belgian blues with other cattle breeds to produce a manageable increase in muscle growth.

4. (a) A cattle breeder crosses a bull homozygous for double muscling ($M^D M^D$) with a heterozygous cow ($M^D M$). What is the probability that the calf will be heterozygous?

___

(b) What should the breeder do if a 100% chance of producing heterozygous calves is required?

___

- A dog's coat color is governed by at least eight genes and coat texture by at least four genes, most of which have multiple alleles. These genes produce all the coat types and colors seen in dogs.
- The gene producing the merle (M) pattern shows incomplete dominance and is associated with a mottled coat pattern. The **genotype** MM produces a very light, almost white merle pattern, but can produce serious health defects including problems with sight and hearing. The homozygous genotype mm produces a normal (not merle) coat pattern. Mm produces the merle pattern, but no serious health effects.

mm   Mm   MM

5. The image on the right shows a blue merle goldendoodle. This dog results from a cross between a golden retriever and a poodle. As such, it is not considered a true breed, rather the goldendoodle is a designer or mixed breed.

A breeder wants to produce dogs with merle coats but wants to avoid both using dogs and producing puppies that are homozygous for the merle gene (MM). Explain how the breeder can do this in one cross, and provide the genotype/phenotype probabilities of the puppies:

___

6. Explain how incomplete dominance of alleles differs from complete dominance: ___

158 Codominance

Key Question: What inheritance patterns occur when two alleles are both dominant?

Crosses involving codominance

▶ Codominance is an inheritance pattern in which both **alleles** in a heterozygote contribute to the **phenotype** and both alleles are independently and equally expressed.

▶ Examples include the human blood group AB and certain coat colors in horses and cattle. Reddish coat color is equally **dominant** with white. Animals that have both alleles have coats that are roan (both red and white hairs are present).

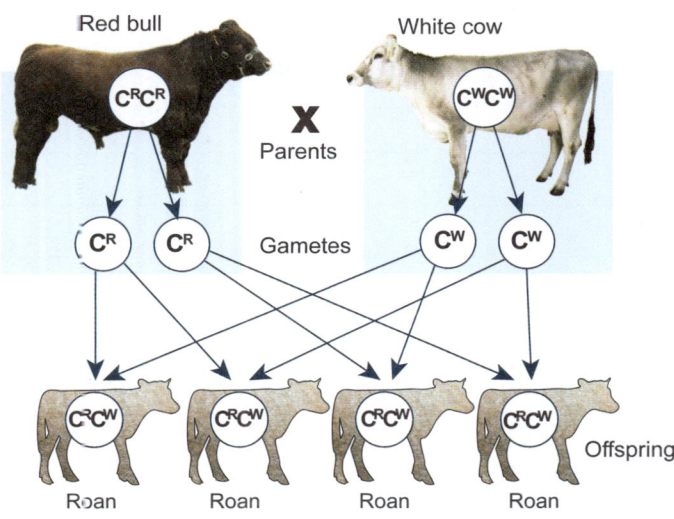

A roan shorthorn heifer

In the shorthorn cattle breed, coat color is inherited. White shorthorn parents always produce calves with white coats. Red parents always produce red calves. When a red parent mates with a white one, the calves have a coat color that is different from either parent; a mixture of red and white hairs, called roan. Use the example (left) to help you to solve the problems below.

1. Explain how codominance of alleles can result in offspring with a phenotype that is different from either parent:

   _____

   _____

2. A white bull is mated with a roan cow (right):

   (a) Fill in the spaces to show the genotypes and phenotypes for parents and calves:

   (b) What is the phenotype ratio for this cross?

   _____

   (c) How could a cattle farmer control the breeding so that the herd ultimately consisted of only red cattle?

   _____

   _____

   _____

   _____

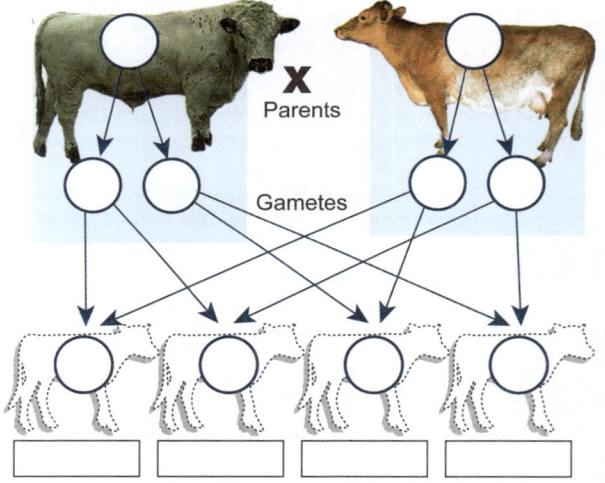

3. In cats the gene for white spotting has three alleles. Two of the alleles, w^s and w, are codominant. A ww cat will not have any white spots, its coat will be a solid color (but not white). A w^sw cat has white spots covering less than half of its body. A w^sw^s cat will have white spots covering more than half of its body.

   (a) If a w^sw cat mates with a ww cat, what will the phenotypic ratio of the offspring be? _____

   _____

   (b) What is the probability of any of the offspring having white spots covering less than half their body? _____

   _____

Codominance and multiple alleles

▸ The four common blood groups of the human 'ABO blood group system' are determined by three alleles: A, B, and O. Where more than two alleles are present for a **gene** it is called a multiple allele system.

▸ The ABO antigens consist of sugars attached to the surface of red blood cells. The alleles code for enzymes (proteins) that join these sugars together.

Recessive allele:	**O**	produces a non-functioning protein
Dominant allele:	**A**	produces an enzyme which forms **A antigen**
Dominant allele:	**B**	produces an enzyme which forms **B antigen**

▸ The allele O is **recessive**. It produces a non-functioning enzyme that cannot make any changes to the basic sugar molecule.

▸ The other two alleles (A, B) are codominant and are expressed equally. They each produce a different functional enzyme that adds a different, specific sugar to the basic sugar molecule.

▸ The blood group A and B antigens are able to react with antibodies present in the blood of other people so blood must always be matched for transfusion.

4. Write down the possible genotypes for each of the possible blood groups: A, AB, B, and O:

_____

_____

5. Complete each of the crosses below to predict the outcome. State the phenotypic ratio for the offspring for each cross:

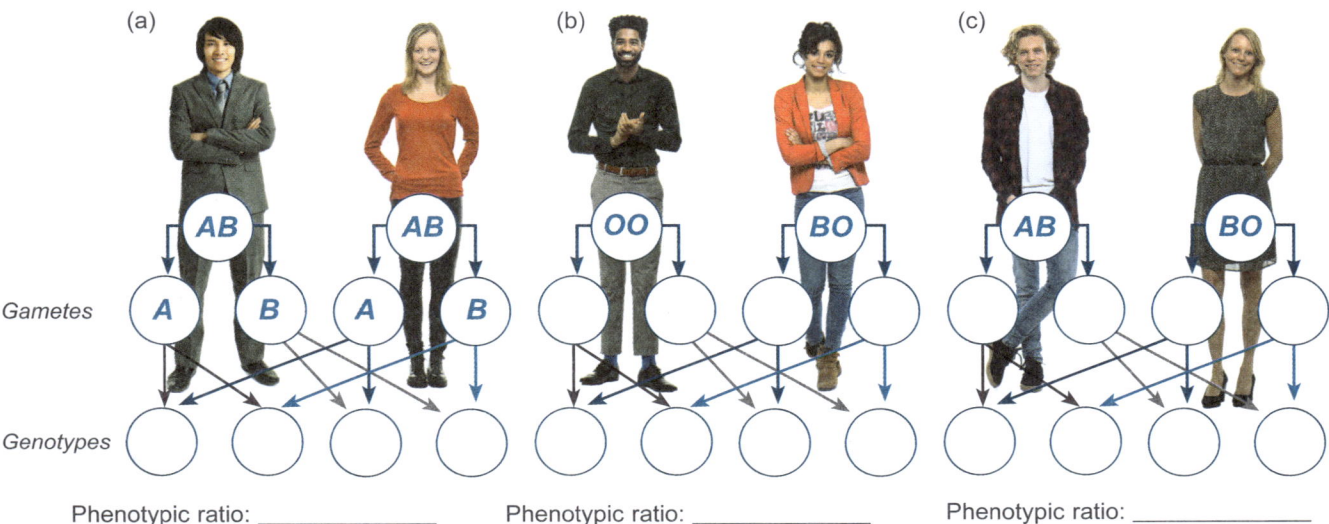

(a) AB × AB

(b) OO × BO

(c) AB × BO

Phenotypic ratio: _____   Phenotypic ratio: _____   Phenotypic ratio: _____

(d) The couple from cross 5 (a) have two children, both are blood group B. Does this fit with your prediction?

_____

_____

6. (a) A man with blood group A has a child with a woman with blood group B. What is the probability of the child being blood group AB? Use the space below for drawing Punnett squares:

(b) Probability of child being blood group AB: _____

159 Sex Linkage

Key Question: How does having genes on the sex chromosomes affect the possible phenotypes of genetic crosses?

Sex linkage

▶ Sex linkage refers to the way **genes** on the sex chromosomes are inherited and expressed. In humans, the sex chromosomes are X and Y, but sex linkage usually involves genes on the X chromosome, which has many more genes than the Y chromosome.

▶ X-linked **recessive traits** are usually seen only in males (XY) and occur rarely in the females (XX) because females may be **heterozygous** (carriers). X-linked **dominant** traits do not necessarily affect males more than females. In humans, recessive sex linked genes are responsible for a number of heritable disorders in males. Y-linked disorders are rare and usually associated with infertility.

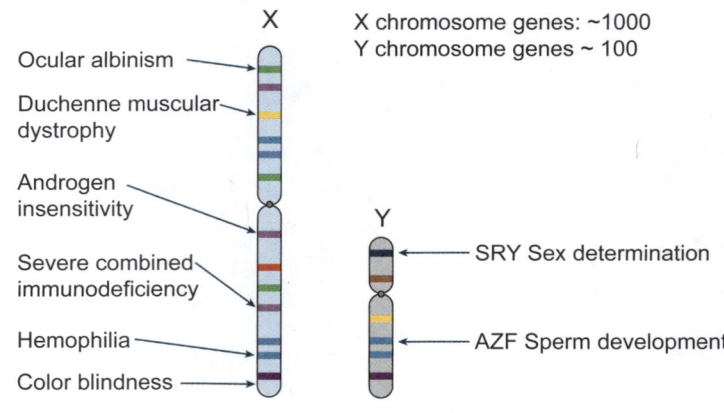

Genetic comparison of X and Y chromosomes

1. Explain why sex linked conditions are more likely linked to alleles on the X chromosome than on the Y chromosome:

___

▶ Hemophilia is a **recessive** disorder linked to the X-chromosome that results in ineffective blood clotting when a blood vessel is damaged. The most common type, hemophilia A, occurs in 1 in 5000 male births. Any male who carries the gene will express the phenotype. Hemophilia is extremely rare in women.

2. A couple wish to have children. The woman knows she is a carrier for hemophilia. The man is not a hemophiliac. Use the notation X^h for hemophilia and X^H for the dominant allele to complete the diagram on the right including the parent genotypes, gametes and possible fertilizations. Write the genotypes and phenotypes in the table below.

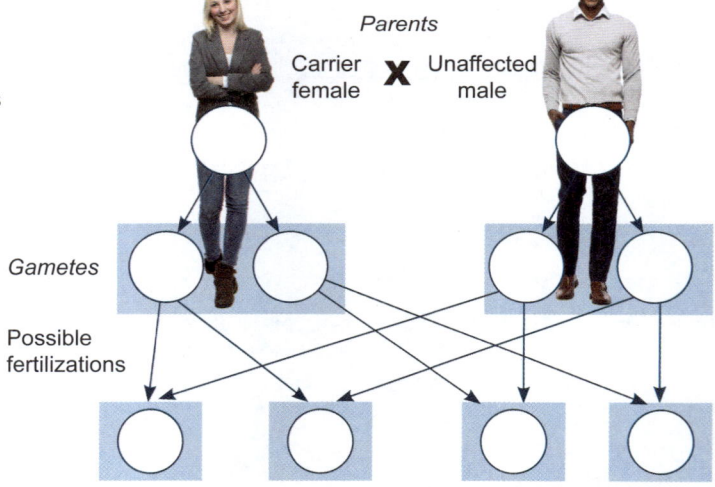

	Genotypes	Phenotypes
Male children		
Female children		

3. A second couple also wish to have children. The woman knows her maternal grandfather was a hemophiliac, but neither her mother or father were. Determine the probability she is a carrier ($X^H X^h$) Use the Punnett squares, right, to help you:

___

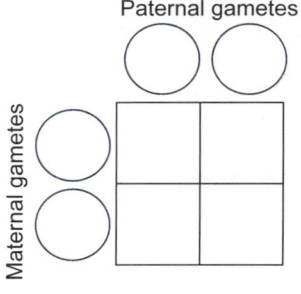
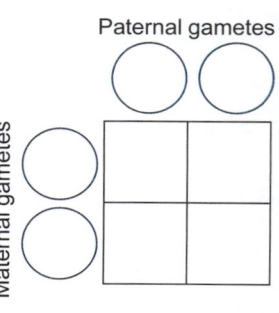

▶ A rare form of rickets in humans is determined by a **dominant allele** of a gene on the X chromosome (it is not found on the Y chromosome). This condition is not successfully treated with vitamin D therapy. The allele types, **genotypes**, and **phenotypes** are as follows:

Allele types		Genotypes		Phenotypes
X^R	= affected by rickets	$X^R X^R$, $X^R X$	=	Affected female
X	= unaffected	$X^R Y$	=	Affected male
		XX	=	Unaffected female,
		XY	=	unaffected male

▶ As a genetic counsellor, you are presented with a couple where one of them has a family history of this disease. The male is affected by this disease and the female is unaffected. The couple, who are thinking of starting a family, would like to know what their chances are of having a child born with this condition. They would also like to know what the probabilities are of having an affected boy or affected girl. Use the symbols above to complete the diagram (right) and determine the probabilities stated below (expressed as a proportion or percentage).

Parents: Unaffected female × Affected male

Gametes

Possible fertilizations

Children

4. Determine the probability of having:

   (a) Affected children: _____

   (b) An affected girl: _____

   (c) An affected boy: _____

▶ Another couple with a family history of the same disease also come in to see you to obtain genetic counselling. In this case, the male is unaffected and the female is affected. The female's father was not affected by this disease. Determine what their chances are of having a child born with this condition. They would also like to know what the probabilities are of having an affected boy or affected girl. Use the symbols above to complete the diagram (right) and determine the probabilities stated below (expressed as a proportion or percentage).

Parents: Affected female (unaffected father) × Unaffected male

Gametes

Possible fertilizations

Children

5. Determine the probability of having:

   (a) Affected children: _____

   (b) An affected girl: _____

   (c) An affected boy: _____

6. Why are males much more likely to inherit X-linked recessive disorders than females?

_____

_____

_____

7. From what you know about sex linkage, what two features could you use to detect a Y-linked disorder in a pedigree?

   (a) _____

   (b) _____

160 Testing the Outcomes of Genetic Crosses

Key Question: How does having genes on the sex chromosomes affect the possible phenotypes of genetic crosses?

Testing the outcome of genetic crosses against predicted ratios

The chi-squared test for goodness of fit (χ^2) can be used for testing the outcome of **dihybrid crosses** against an expected (predicted) Mendelian ratio.

Using χ^2 in Mendelian genetics

- In genetic crosses, certain ratios of offspring can be predicted based on the known **genotypes** of the parents. The chi-squared test is a statistical test to determine how well observed offspring numbers match (or fit) expected numbers. Raw counts should be used and a large sample size is required for the test to be valid.
- In a chi-squared test, the null hypothesis predicts the ratio of offspring of different **phenotypes** is the same as the expected Mendelian ratio for the cross, assuming independent assortment of **alleles** (no linkage, i.e. the **genes** involved are on different chromosomes).
- Significant departures from the predicted Mendelian ratio indicate linkage (the genes are on the same chromosome) of the alleles in question.
- In a *Drosophila* genetics experiment, two individuals were crossed (the details of the cross are not relevant here). The predicted Mendelian ratios for the offspring of this cross were 1:1:1:1 for each of the four following phenotypes: gray body-long wing, gray body-vestigial wing, ebony body-long wing, ebony body-vestigial wing.
- The observed results of the cross were not exactly as predicted. The following numbers for each phenotype were observed in the offspring of the cross:

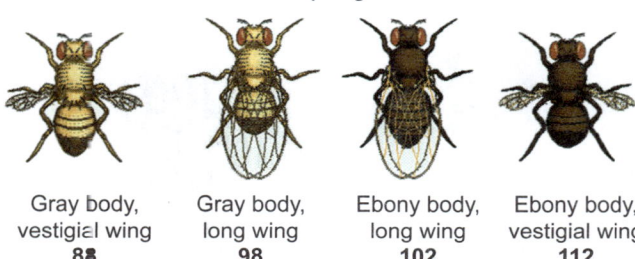

Gray body, vestigial wing	Gray body, long wing	Ebony body, long wing	Ebony body, vestigial wing
88	98	102	112

Table 1: Critical values of χ^2 at different levels of probability. By convention, the critical probability for rejecting the null hypothesis (H_0) is 5%. If the test statistic is less than the tabulated critical value for $P = 0.05$ we cannot reject H_0 and the result is not significant. If the statistic is greater than the tabulated value for $P = 0.05$ we reject (H_0) in favor of the alternative hypothesis.

Degrees of freedom	Level of probability (P)					
	0.50	0.20	0.10	0.05	0.02	0.01
1	0.455	1.64	2.71	3.84	5.41	6.64
2	1.386	3.22	4.61	5.99	7.82	9.21
3	2.366	4.64	6.25	7.82	9.84	11.35
4	3.357	5.99	7.78	9.49	11.67	13.28
5	4.351	7.29	9.24	11.07	13.39	15.09
	Do not reject H_0 ←			Reject H_0 →		

Steps in performing a χ^2 test

1. Enter the observed value (O)
Enter the values of the offspring into the table (below) in the appropriate category (column 1).

2. Calculate the expected value (E)
In this case the expected ratio is 1:1:1:1. Therefore the number of offspring in each category should be the same (i.e. total offspring/ no. categories). 400 / 4 = 100 (column 2).

3. Calculate O−E and (O−E)²
The difference between the observed and expected values is calculated as a measure of the deviation from a predicted result. Since some deviations are negative, they are all squared to give positive values (columns 3 and 4).

4. Calculate χ^2
For each category, calculate $(O - E)^2 / E$. Then sum these values to produce the χ^2 value (column 5).

$$\chi^2 = \Sigma \frac{(O - E)^2}{E}$$

5. Calculate degrees of freedom
The probability that any particular χ^2 value could be exceeded by chance depends on the number of degrees of freedom. This is simply one less than the total number of categories (this is the number that could vary independently without affecting the last value) In this case 4 − 1 = 3.

6. Use χ^2 table
On the χ^2 table with 3 degrees of freedom, the calculated χ^2 value corresponds to a probability between 0.2 and 0.5. By chance alone a χ^2 value of **2.96** will happen 20% to 50% of the time. The probability of 0.0 to 0.5 is higher than 0.05 (i.e 5% of the time) and therefore the null hypothesis cannot be rejected. We have no reason to believe the observed values differ significantly from the expected values.

	1	2	3	4	5
Category	O	E	O−E	(O−E)²	(O−E)²/E
GB, LW	98	100	−2	4	0.04
GB, VW	88	100	−12	144	1.44
EB, LW	102	100	2	4	0.04
EB, VW	112	100	12	144	1.44
				$\chi^2 \rightarrow$	2.96

The following problems examine the use of the chi-squared (χ^2) test in genetics.

1. In a tomato plant experiment, two **heterozygous** individuals were crossed (the details of the cross are not relevant here). The predicted Mendelian ratios for the offspring of this cross were 9:3:3:1 for each of the four following phenotypes: purple stem-jagged leaf edge, purple stem-smooth leaf edge, green stem jagged leaf edge, green stem-smooth leaf edge.

   The observed results of the cross were not exactly as predicted. The number of offspring with each phenotype are provided:

   Observed results of the tomato plant cross
   Purple stem-jagged leaf edge   12   Green stem-jagged leaf edge   8
   Purple stem-smooth leaf edge   9   Green stem-smooth leaf edge   0

   (a) State your null hypothesis for this investigation: _____

   _____

   (b) State your alternative hypothesis: _____

   _____

2. Use the chi-squared (χ^2) test to determine if the difference between the observed and expected phenotypic ratios are significant. Use the table of critical values of χ^2 at different P values on the previous page.

   (a) Enter the observed values (number of individuals) and complete the table to calculate the χ^2 value:

Category	O	E	O – E	(O – E)²	$\frac{(O-E)^2}{E}$
Purple stem, jagged leaf					
Purple stem, smooth leaf					
Green stem, jagged leaf					
Green stem, smooth leaf					
	Σ				Σ

   (b) Calculate χ^2 value using the equation:

   $$\chi^2 = \Sigma \frac{(O-E)^2}{E} \qquad \chi^2 = _____$$

   (c) Calculate the degrees of freedom: _____

   (d) Using the χ^2 table, state the P value corresponding to your calculated χ^2 value:

   _____

   (e) State your decision:
       reject: H_0 / do not reject H_0:

   _____

3. Students carried out a plea plant experiment, where two heterozygous individuals were crossed. The predicted Mendelian ratios for the offspring were 9:3:3:1 for each for the four following phenotypes: round-yellow seed, round-green seed, wrinkled-yellow seed, wrinkled-green seed.

   The observed results were as follows: Round-yellow seed 441   Wrinkled-yellow seed 143
                                           Round-green seed 159   Wrinkled-green seed   57

   Use a separate piece of paper to complete the following:

   (a) State the null and alternative hypothesis (H_0 and H_A): _____

   (b) Calculate the χ^2 value: _____

   (c) Calculate the degrees of freedom and state the P value corresponding to your calculated χ^2 value: _____

   (d) State whether or not you reject your null hypothesis: _____

4. Comment on the whether the values obtained above are similar. Suggest a reason for any difference:

   _____

   _____

   _____

161 Anyone for Chocolate? Revisited

Content Anchor Revisited: Can we get a chocolate Labrador puppy from black parents?

At the beginning of the chapter you were asked whether a pair of black Labradors could be crossed to give a chocolate (brown) colored puppy.

▸ The **gene** for black coat color is represented by 'B' and is **dominant** over the **recessive** form, 'b', which gives a chocolate brown coat color.

▸ Yellow coat color is determined by a different, unlinked gene known as 'E'. This time, the recessive form of the gene 'e' can interfere with the black/brown coat color and mask it, resulting in a yellow puppy. If the dominant form of the gene 'E' is present, black/brown coloration is not masked and dogs will always be black or chocolate colored.

1. Complete the following crosses for Labradors, including genotype and phenotype ratios: Black dogs (B) are dominant to brown dogs (b):

    (a) Bb x Bb:

    (b) Bb x bb

2. Two chocolate colored Labradors with genotypes bbEe are mated. Fill in the Punnett square below and circle any yellow puppies that might be expected in the litter:

3. Explain whether mating two yellow Labradors would result in any black or chocolate colored puppies:

    _____
    _____
    _____

4. Two slight variations are possible in yellow Labradors. They can have either brown noses, lips, and eye rims, or black noses, lips, and eye rims. Which of the genes, B or E, might be causing this and why?

    _____
    _____
    _____

162 Summing Up

Read each question carefully. Place a cross in the box beside the best answer to the question from the four answer choices provided.

1. Alternate forms of the same gene are known as:
   - (a) Gametes
   - (b) Alleles
   - (c) Genotypes
   - (d) Homozygotes

2. The information within a Punnett square is used to predict:
   - (a) The phenotypes of the offspring
   - (b) The phenotypes of the parents
   - (c) The genotypes of the parents
   - (d) The genotypes of the offspring

3. The law of independent assortment:
   - (a) Describes how alleles of different genes sort independently of one another during meiosis
   - (b) Holds true only for unlinked genes
   - (c) Was first described by Gregor Mendel
   - (d) All of the above

4. Which of the following is not a homozygous genotype?
   - (a) AACC
   - (b) AaBaCc
   - (c) bbcczz
   - (d) bbdd

   This information refers to Question 5 and 6

5. In a species of mouse, fur color is caused by the following alleles: B = black and b = brown

   Which genotype for the organism will have brown fur?
   - (a) BB
   - (b) Bb
   - (c) bb
   - (d) All of the above

6. If two heterozygous mice were crossed, what would be the phenotypic ratio of black to brown in the offspring?
   - (a) 1:1
   - (b) 1:2:1
   - (c) 3:1
   - (d) 9:3:3:1

7. In humans, a gene involved in a sex-linked trait is located on:
   - (a) The X chromosome
   - (b) The Y chromosome
   - (c) Any autosome
   - (d) Either the X or Y chromosome

8. Two flowers, heterozygous for a single gene, are crossed. The probability any of the F_1 generation will be homozygous recessive is:
   - (a) 0.25
   - (b) 0.5
   - (c) 0.75
   - (d) 1

9. Recombination occurs when:
   - (a) Unlinked genes cross over during meiosis
   - (b) Linked genes cross over during mitosis
   - (c) The full 2N chromosome number is restored during fertilization
   - (d) Genes on homologous non-sister chromatids cross over during meiosis

10. If a male is color blind, from whom did he receive the recessive allele?
    - (a) Mother
    - (b) Father
    - (c) Either mother or father
    - (d) Not enough information

11. Human females are much less likely to be color blind than males because:
    - (a) Female eyes are more sensitive than male eyes
    - (b) Females need two recessive alleles two become color blind, rather than one in males
    - (c) Females only ever have the dominant allele
    - (d) None of the above

12. A cross between homozygous purple-flowered and homozygous white-flowered pea plants results in offspring with purple flowers. This demonstrates
    - (a) The blending model of genetics
    - (b) True-breeding
    - (c) Dominance
    - (d) A dihybrid cross

13. In a hypothetical gene, the alleles F and G are codominant. For the cross below:

	F	G
F	FF	FG
G	FG	GG

    How many phenotypes will there be?
    - (a) 2
    - (b) 3
    - (c) 4
    - (d) 5

14. Describe the process of meiosis and the type of cells produced. Compare the cells produced to the parental cell. You may use diagrams to illustrate your answer:

15. Provide an explanation of how meiosis increases genetic variation in reproductive cells:

16. A plant with orange-striped flowers was cultivated from seeds. The plant was self-pollinated and the F_1 progeny appeared in the following ratios: 89 orange with stripes, 29 yellow with stripes, 32 orange without stripes, 9 yellow without stripes. Call orange O and stripes S.

    (a) Determine which alleles responsible for the phenotypes observed are dominant and which are recessive:

    (b) Determine the genotype of the original plant with orange striped flowers:

17. In cats, the following alleles are present for coat characteristics: black (B), brown (b), short (L), long (l), tabby (T), blotched tabby (tb). Use the information to complete the dihybrid crosses below:

    A black, short haired (BBLl) male is crossed with a black long haired (Bbll) female. Determine the genotypic and phenotypic ratios of the offspring:

    Genotype ratio: _____

    _____

    Phenotype ratio: _____

    _____

    _____

18. Snapdragons have red, pink, or white flowers. Roan coats in Shorthorn cattle is a mix of red and white hairs.

    (a) Define monohybrid cross: _____

    _____

    (b) Use the inheritance patterns of codominance and incomplete dominance to explain these results (you may use extra paper if required):

    _____

    _____

    _____

    _____

    _____

    _____

    _____

    _____

    _____

19. In a dispute over parentage, the mother of a child with blood group O identifies a male with blood group A as the father. The mother is blood group B. On a separate sheet of paper draw Punnett squares to show possible genotype/phenotype outcomes to determine if the male is the father and the reasons (if any) for further dispute:

    _____

    _____

    _____

20. You may have heard the expression that a person's blood type may be A positive (or A+) or AB negative (AB–). The + or – refers to the presence of antigens associated with the Rh gene (Rhesus factor). There are two Rh alleles: D (Rh factor present (+)) and d (Rh factor not present (–)).
    The ABO gene is found on chromosome 9, while the Rh gene is found on chromosome 1 and they are therefore inherited independently.

    (a) Given that the ABO blood system produces four phenotypes (A, AB, B, and O), what phenotypes are possible if the Rh blood system is included in blood phenotypes?

    _____

    (b) A woman is O–. Her husband is B+. Write down the genotypes of the woman and man:

    _____

    (c) What is the probability of the couple having a child with a blood type of B–?

    _____

    _____

21. In groups of four, carry out research into the impact of past and current genetics research and knowledge on both scientific thought and society.

   - This could include aspects of how the results of studies into genetic and heredity impacted the scientific community at the time of the research, of its on going impact, or how current advances in genetics impact society today.
   - It could also include aspects such as the cost and benefit to society of scientific knowledge and research in genetics and the contributions of different scientists or aspects of the methodology they used, to the overall understanding of genetics and heredity.
   - You could also describe a solution to a genetic disease that has been solved by current genetic research and knowledge.
   - Each member of your group should research a different aspect and report back to the group.
   - Once each member of the group has reported back, decide as a group how to present your ideas. This could be by a written report, slideshow, poster, or presentation.
   - Use the space below to organize and summarize your research and notes. This is not your final report.

Area of research	
Scientists involved, research they were involved in,	
Impact of their findings (advances in knowledge, reaction by society etc)	

Description of research and/or data source (written/graphs/tables/images?)	Where was the data sourced? URL/organization	What evidence-based claim(s) does the source/data make? (summary of source/data)

Positive impacts of area of research	
Negative impacts of area of research	
Evidence-based summary statements on area of research	

CHAPTER 7
Common Ancestry

TEKS
Scientific and Engineering Practices

B.1: Investigation and Inquiry
1.A 1.F 1.G 1.H

B.2: Data and Patterns
2.B

B.3: Communicating in Science
3.A 3.B

B.4: Science as a Human Endeavor
4.A 4.B

TEKS
Science Concepts

B9.A analyze and evaluate how evidence of common ancestry among groups is provided by the fossil record, biogeography, and homologies, including anatomical, molecular, and developmental

B9.B examine scientific explanations for varying rates of change such as gradualism, abrupt appearance, and stasis in the fossil record

Learning Outcomes
I know I have achieved this when I can:

	Outcome	Activity number
☐	Explain the link between new scientific evidence and changes in scientific understanding about common ancestry.	163
☐	Identify and elaborate on examples of scientific evidence for common ancestry of organisms on earth.	164
☐	Link the process of fossil formation to the likely presence or absence of fossilized species.	165
☐	Discuss the link between the fossil record and layered rock strata.	166
☐	Interpret rock profiles containing fossils and rocks of different ages.	167
☐	Define transitional fossils, and discuss how they are able to provide evidence for common ancestry.	168
☐	Identify differences in homologous structures due to adaptation for different functions.	169
☐	Discuss how the pentadactyl limb in different mammals can provide evidence for common ancestry.	169
☐	Evaluate the process of biogeography in providing evidence for common ancestry, using information from provided examples.	170
☐	Analyze data from phylogenetic trees to discuss how DNA differences can provide evidence for common ancestry.	171
☐	Analyze amino acid sequences from different species to identify the extent of common ancestry.	172
☐	Analyze data of embryo development from different species to discuss how it can be used as evidence for common ancestry.	173
☐	Suggest explanations for the abrupt appearance of new fossils in the Cambrian explosion.	174
☐	Link the presence of 'living fossils' to apparent stasis in the fossil record.	175
☐	Evaluate punctuated equilibrium and gradualism as models that can provide scientific explanations for stasis and abrupt appearance of fossils in the fossil record.	175
☐	Explore the time-line of scientific discoveries leading to Darwin's theory of evolution through natural selection.	176

RESOURCE HUB
bit.ly/3SNa27M

ELPS English Language Proficiency Standards

		Page number
Speaking	**Transitional Fossils.** With a partner, discuss transitional fossils before answering the activity question. Develop your own definition of the term transitional fossil, then take turns asking and answering questions about them. What were they? Why were they important? Asking and answering your own questions about transitional fossils will prepare you to write about them in question 1(a) and 1(b).	290
Reading	**Biogeography and Common Ancestry.** Use the arrows on the map to help you understand the migration of the camel family. Find the place where camels originated, then follow the arrows. How did early camels get to Asia? Now look at the green shaded areas. Which color green shows where members of the camel family are today? .	295
Speaking	**Changes in the Fossil Record.** For question 1, summarize your ideas as key points (not whole sentences) on index cards, and use the cards to help you present your ideas to the class. Connect your facts to the Cambrian Explosion using phrases like *The Cambrian Explosion was significant because _____ and calcium and phosphate ions allowed _____.* Practice with a partner before you present to the class, taking turns giving each other feedback	299
Writing	**Dinosaur or Bird revisited.** Think back to what you learned about dinosaurs early in the chapter. Then revisit the evolution graphic on page 286. Can any of these branches of science be used to prove that birds are part of the avian dinosaur group? Why or why not? Make sure to use the names for specific types of evidence in your answer.	306

163 Dinosaur or Bird?

Content Anchor: How does scientific evidence allow us to continually build ideas about the common ancestry between different groups of organisms?

The "old" velociraptor model

Thanks to movies such as Jurassic Park, the Velociraptor genus of dinosaurs is well known. These sleek, hairless hunters lived in packs and seemed to have distinctly scaly, reptilian skin. What evidence allowed us to form an impression of these dinosaurs? The first piece of the puzzle showed up in 1923, when the fossilized remains of a damaged skull and a toe claw were found in the Mongolian desert.

Many models used in the past, such as the one to the right, were based on the limited evidence available to scientists at the time.

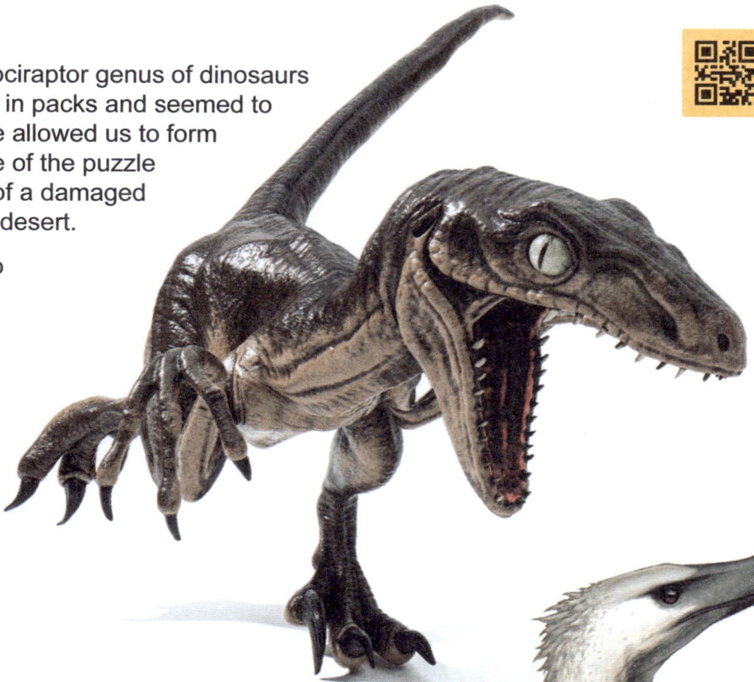

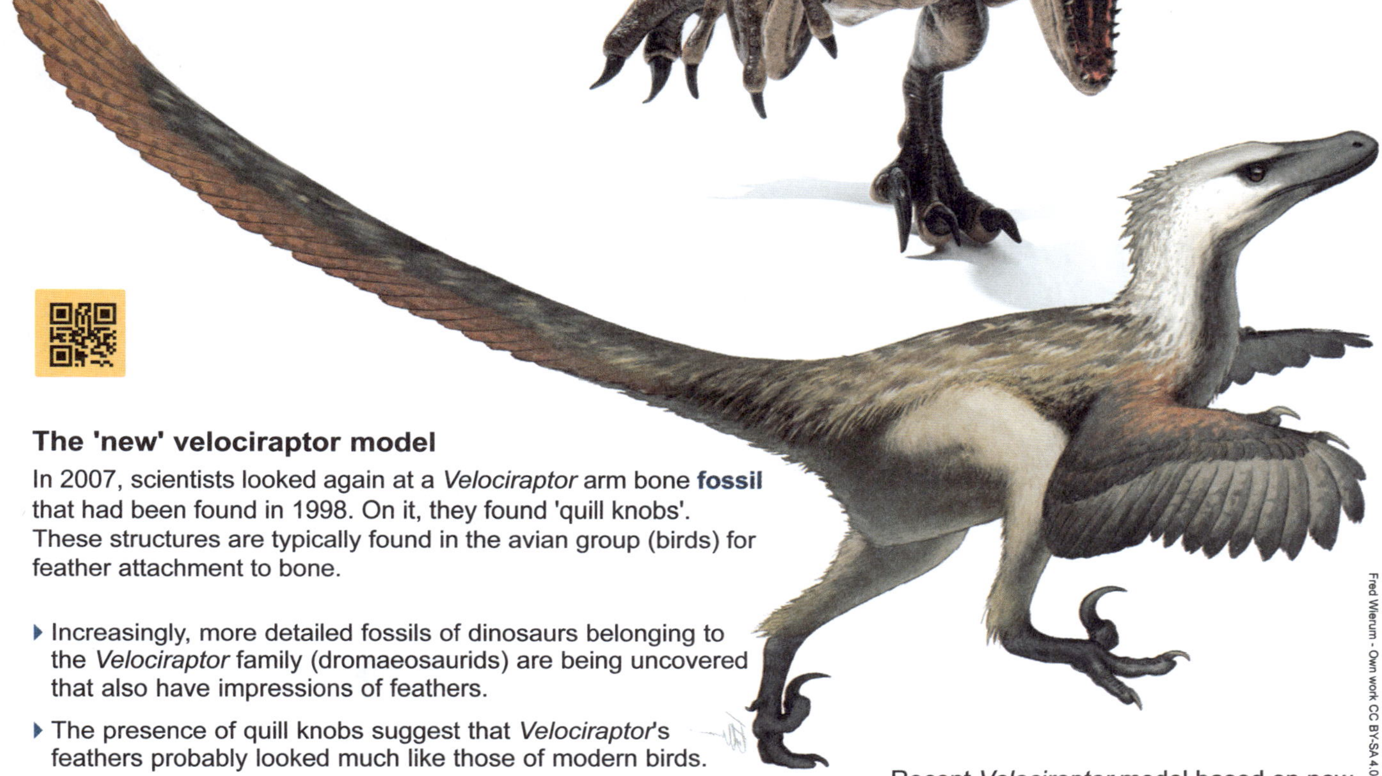

The 'new' velociraptor model

In 2007, scientists looked again at a *Velociraptor* arm bone **fossil** that had been found in 1998. On it, they found 'quill knobs'. These structures are typically found in the avian group (birds) for feather attachment to bone.

▸ Increasingly, more detailed fossils of dinosaurs belonging to the *Velociraptor* family (dromaeosaurids) are being uncovered that also have impressions of feathers.

▸ The presence of quill knobs suggest that *Velociraptor*'s feathers probably looked much like those of modern birds.

▸ Fossils of dinosaurs in the same group as *Velociraptor* show similar respiratory systems to those found in birds today.

Recent *Velociraptor* model based on new 'avian' discoveries from fossil records.

1. Why does new evidence cause a change in scientific ideas? _____
_____
_____
_____

2. Some scientists believe that birds should be classified as dinosaurs. What evidence might they use for their claim?
_____
_____
_____
_____

164 Evolution and Common Ancestry

Key Question: Where can evidence for evolution, and therefore common ancestry, be found?

How are evolution and common ancestry linked?

▶ **Evolution** describes the heritable changes in a population's gene pool over time. It is the mechanism by which we can explain how different groups are linked to a **common ancestor**. The evidence for evolution, and therefore common ancestry, comes from many branches of science (below) and includes evidence from living populations, as well as from the past. New methods have enabled scientists to clarify the origin of the eukaryotes and to recognize two prokaryote domains, Bacteria and Archaea. The universality of the genetic code and the similarities in the molecular workings of all cells provide powerful evidence for a common ancestor to all life on Earth.

Comparative anatomy
Comparative anatomy examines the similarities and differences in the anatomy of different species. Similarities (**homology**) in anatomy, e.g. the bones forming the arms in humans and the wings in birds and bats, indicate descent from a common ancestor.

Geology
Geological strata (the layers of rock, soil, and other deposits such as volcanic ash) can be used to determine the relative order of past events, and therefore the relative dates of fossils. Fossils in lower strata are older than fossils in higher (newer) strata, unless strata have been disturbed.

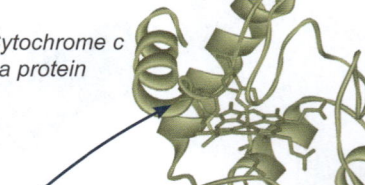

Cytochrome c - a protein

DNA comparisons
DNA can be used to determine how closely organisms are related to each other. More closely related species have greater homology in their DNA sequences.

Protein evidence
Similarities (and differences) between proteins provide evidence for determining shared ancestry. Fewer differences in amino acid sequences reflects closer genetic relatedness.

EVOLUTION

Fossil record
Fossils, like this shark's tooth (left), are the remains of long-dead organisms. They provide a record of the appearance and extinction of organisms.

Developmental evidence
The study of developmental processes and the genes that control them gives insight into evolutionary processes. This field of study is called evolutionary developmental biology (evo-devo).

Biogeography
The geographical distribution of living and extinct organisms provides evidence of common ancestry and can be explained by speciation, extinction, and continental drift. The biogeography of islands, e.g the Galápagos Islands, provides evidence of how species evolve when separated from their ancestral population on the mainland.

Chronometric dating
Radiometric dating techniques, such as carbon dating, allow scientists to determine an absolute date for a fossil by dating it or the rocks around it. Absolute dating has been used to assign ages to strata, and construct the **geologic time scale**.

1. Working in groups of 2-4, in turn, select one of the pictures above and, in words, explain to the rest of your group how the image can be linked to the specific category of evidence for common ancestry. For this activity, you may draw on prior knowledge, or you may be given some time by your teacher to find out more information from books or online.

165 Fossil Formation

Key Question: How do fossils form and preserve a record of once-living organisms?

▸ **Fossils** are the remains of long-dead organisms that have become preserved in the Earth's crust. They can be of plants or animals. Fossilization is more likely to be a 'snapshot' event that happens as a result of a storm or infrequent flood, rather than a gradual accumulation of specimens. Many fossils contain no original remains of the organism's body. Instead, they are composed of replacement minerals where the body or impression once was.

▸ Fossils provide a record of the appearance and extinction of organisms, from species to whole taxonomic groups. The **fossil record** can be used to establish the relative order of past events, and links to common ancestors of related groups.

How are fossils formed?

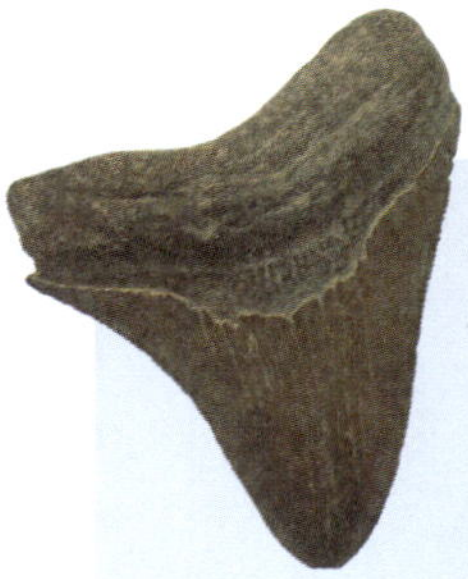

Fossilization occurs best when an organism dies in a place where sediment can be laid down relatively quickly. This is often an aquatic environment, e.g. an estuary, but it can be caused by rapid burial such as by a landslide or volcanic ash.

After death, the flesh may rot or be scavenged, but hard materials, usually bones and teeth, are able to remain long enough for burial.

Soft material, such as the cartilaginous skeletons of sharks, doesn't fossilize well. Often, the only remains are their teeth (above), one of the most commonly found fossils in Texas.

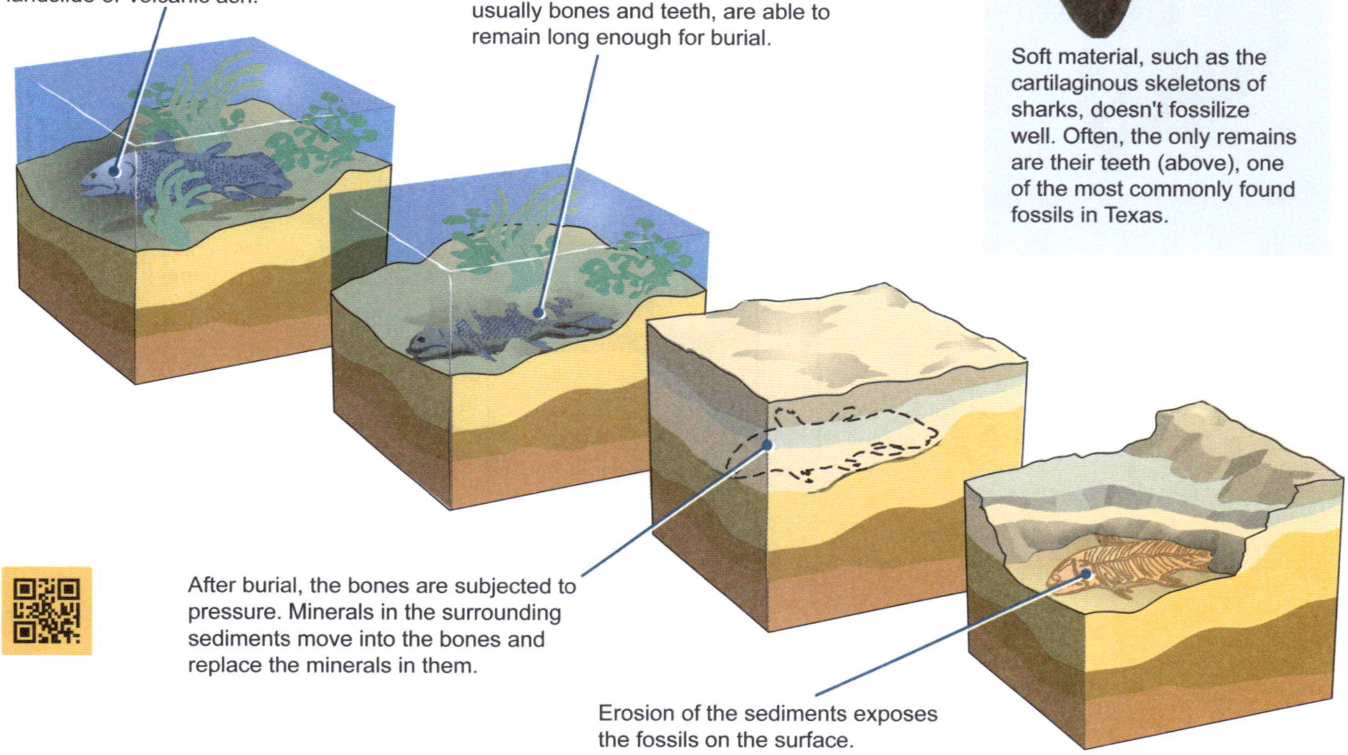

After burial, the bones are subjected to pressure. Minerals in the surrounding sediments move into the bones and replace the minerals in them.

Erosion of the sediments exposes the fossils on the surface.

1. Fossils are usually rare, with many past species not yet, or possibly ever, represented. Why is this the case?

   _____

   _____

2. The vast majority of organisms are fossilized in water bodies. Why might this be? _____

   _____

   _____

   _____

3. What types of organisms would you expect to see most represented in the fossil record, and why? _____

   _____

   _____

166 The Fossil Record

Key Question: How can the fossil record be used as evidence for common ancestry?

Rock strata are layered through time

▶ Rock strata are arranged in the order in which they were deposited, unless they have been disturbed by geological events. Most recent layers are near the surface and the oldest are at the bottom. **Fossils** can be used to establish the sequential order of past events in a rock profile.

▶ The **fossil record** can be calibrated against a time scale, using dating techniques, to build up a picture of the evolutionary changes that have taken place.

Profile with sedimentary rocks containing fossils

Fossilized fern frond

New fossil types mark changes in environment
In the strata at the end of one geologic period, it is common to find many new fossils that become dominant in the next. Each geologic period had a different environment from the others. Their boundaries coincided with drastic environmental changes and the appearance of new niches. These produced new selection pressures, resulting in new adaptive features in the surviving species as they responded to the changes.

Recent fossils are found in more recent sediments
The more recent the layer of rock, the more resemblance there is between the fossils found in it and living organisms.

Fossil types differ in each stratum
Fossils found in a given layer of sedimentary rock are generally significantly different from fossils in other layers.

Extinct species
The number of extinct species is far greater than the number of species living today.

More primitive fossils are found in older sediments
Fossils in older layers tend to have quite generalized forms. In contrast, organisms alive today have specialized forms.

Gaps in the fossil record

▶ The fossil record contains gaps. Without a complete record, it can sometimes be difficult to determine an evolutionary sequence. Scientists must also use additional information to produce an order of events that best fits all the evidence.

▶ Gaps in the fossil record can occur because fossils are destroyed due to tectonic earth processes, the fact that some organisms do not fossilize well, and fossils of particular species have not yet been found.

1. Discuss the importance of fossils as a record of evolutionary change over time: _____

_____

_____

_____

_____

_____

_____

167 Interpreting the Fossil Record

Key Question: How can we use fossils in rock strata to order past events, from oldest to most recent?

The diagram below shows a hypothetical rock profile from two locations, separated by a distance of 67 km. Differences exist in the rock layers at the two locations. Apart from layers D and L, which are volcanic ash deposits, all other layers are composed of sedimentary rock. Use the information in the diagram to answer the questions below.

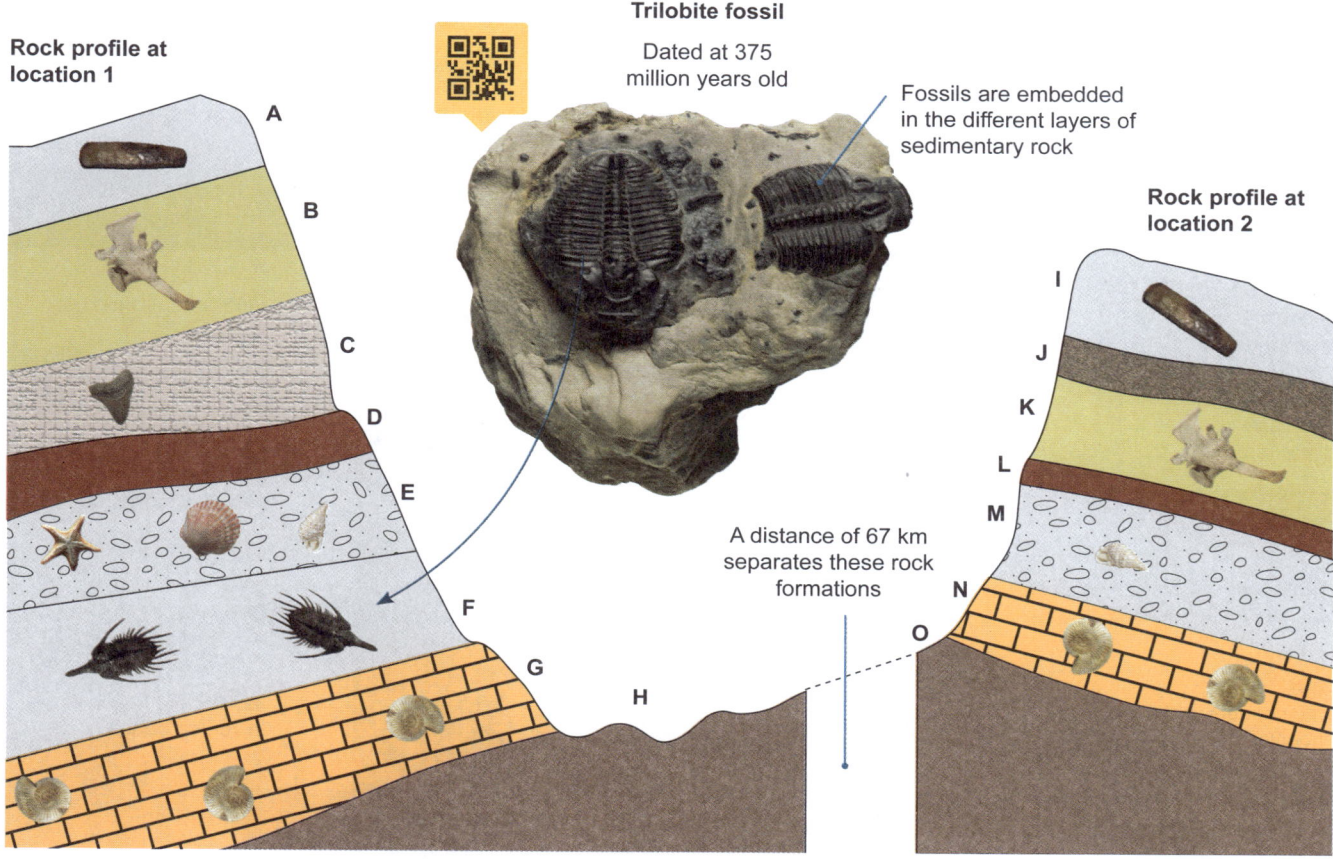

1. Assuming there has been no geological activity to disturb the order of the rock layers, state in which rock layer (A-O) you would find:

   (a) The youngest rocks at Location 1: _____   (c) The youngest rocks at Location 2: _____

   (b) The oldest rocks at Location 1: _____   (d) The oldest rocks at Location 2: _____

2. (a) State which layer at location 1 is of the same age as layer M at location 2: _____

   (b) Explain the reason for your answer in 2 (a): _____

   _____

   _____

   _____

3. (a) State which layers present at location 1 are missing at location 2: _____

   (b) State which layers present at location 2 are missing at location 1: _____

4. The rocks in layer H and O are sedimentary rocks. Why are there no visible fossils in these layers? _____

   _____

   _____

   _____

168 Transitional Fossils

Key Question: How do transitional fossils provide important links in the fossil record?

- **Transitional fossils** have a mixture of features that show intermediate states found in two different, but related, groups. Transitional fossils provide important links in the **fossil record** and give evidence to support how one group may have given rise to the other by evolutionary processes.
- Important examples of transitional fossils include horses, whales, and *Archaeopteryx* (below), a transitional form between birds and non-avian dinosaurs.
- *Archaeopteryx* was crow-sized (50 cm length) and lived about 150 million years ago. It had a number of birdlike (avian) features, including feathers. However, it also had many non-avian features, which it shared with theropod dinosaurs. Although not a direct ancestor of birds, *Archaeopteryx* and birds shared a **common ancestor**.

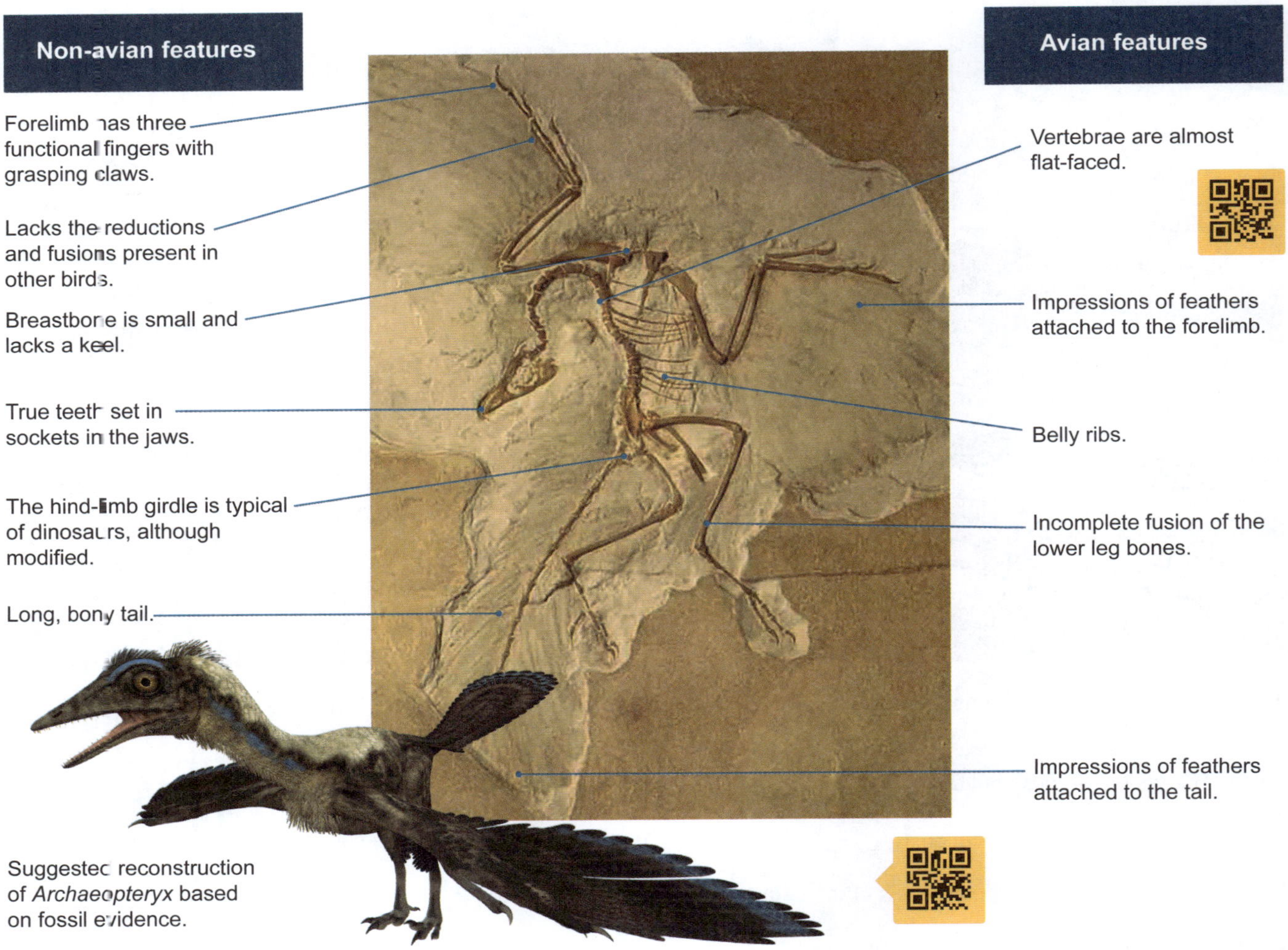

Non-avian features

- Forelimb has three functional fingers with grasping claws.
- Lacks the reductions and fusions present in other birds.
- Breastbone is small and lacks a keel.
- True teeth set in sockets in the jaws.
- The hind-limb girdle is typical of dinosaurs, although modified.
- Long, bony tail.

Avian features

- Vertebrae are almost flat-faced.
- Impressions of feathers attached to the forelimb.
- Belly ribs.
- Incomplete fusion of the lower leg bones.
- Impressions of feathers attached to the tail.

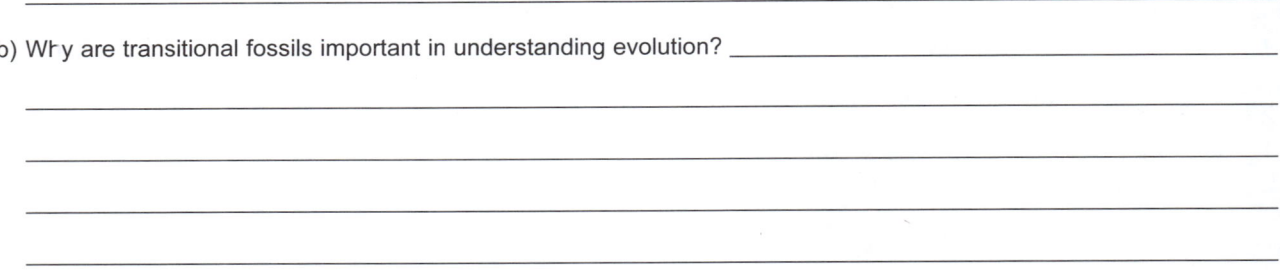

Suggested reconstruction of *Archaeopteryx* based on fossil evidence.

1. (a) What is a transitional fossil? _____

   _____

   _____

   (b) Why are transitional fossils important in understanding evolution? _____

   _____

   _____

   _____

   _____

   _____

Evolution of the Horse

Horse **evolution** is well documented. Many fossils exist that document stages in its evolution from a three toed animal (about the size of a medium sized dog) to the large, single toed horse (and related species) we know today. Although each fossil type is a separate species, the similarity of skeletal structures and features derived from earlier species clearly connects them in a evolutionary sequence, and links them to a common ancestor.

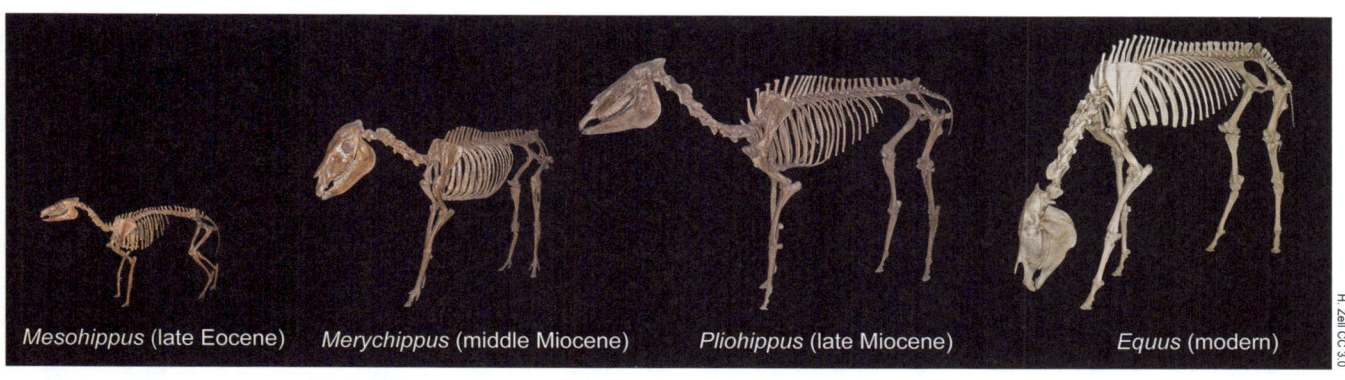

Mesohippus (late Eocene)   *Merychippus* (middle Miocene)   *Pliohippus* (late Miocene)   *Equus* (modern)

2. (a) How many modern species of horse (genus *Equus*) can you think of? _____

   (b) What feature connects these species as *Equus*? _____

3. The evolution of modern whales from an ancestral land mammal is also well documented in the fossil record. How could scientists hypothesize, from examining the fossils, that the common ancestor of whales and mammals lived on land?
   _____
   _____

4. How might relatively complete fossil records, such as that of the horse, provide evidence for evolution and common ancestry of organisms?
   _____
   _____
   _____
   _____
   _____
   _____
   _____
   _____
   _____
   _____
   _____
   _____
   _____
   _____

Equus — 1.6 m
Merychippus — 1.25 m
Mesohippus — 0.6 m
Hyracotherium (Eohippus) — 0.4 m

Millions of years ago

169 Anatomical Homology

Key Question: What is anatomical homology, and how can it provide evidence for common ancestry?

Anatomical homologous structures

▶ Anatomically homologous structures are found in different organisms and are the result of their inheritance from a **common ancestor**. Their presence indicates an evolutionary relationship between organisms. Homologous anatomical structures have a common origin, but they may have different functions.

▶ For example, the forelimbs of birds and sea lions are homologous anatomical structures. They have the same basic skeletal structure, but have different functions. A bird's wings have been adapted for flight and a sea lion's flippers are modified as paddles for swimming. The diagrams below show some homologous structures.

Duck wing (top) and sea lion flipper.   Snail foot (top) and squid tentacles.   Moth forewing (top) and beetle elytron.

1. Describe the purpose of the homologous structures above and the adaptations to provide for their purpose:

(a) Duck wing	(c) Snail foot	(e) Cecropia moth forewing

(b) Sea lion flipper	(d) Squid tentacles	(f) Beetle elytron

2. What might the anatomical homology of the sea lion's flipper and the duck's wing tell you about the features of the common ancestor of the two groups?

_____

_____

_____

3. Why might looking at anatomical homology be a particularly good method to investigate common ancestry in fossils?

_____

_____

_____

The pentadactyl limb

A pentadactyl limb has five fingers or toes, with the bones arranged in a specific pattern (below, left). Scientists can identify and use the presence of the same pentadactyl limb bones in quite different species of organisms as useful evidence for **evolution** from a past common ancestor between species.

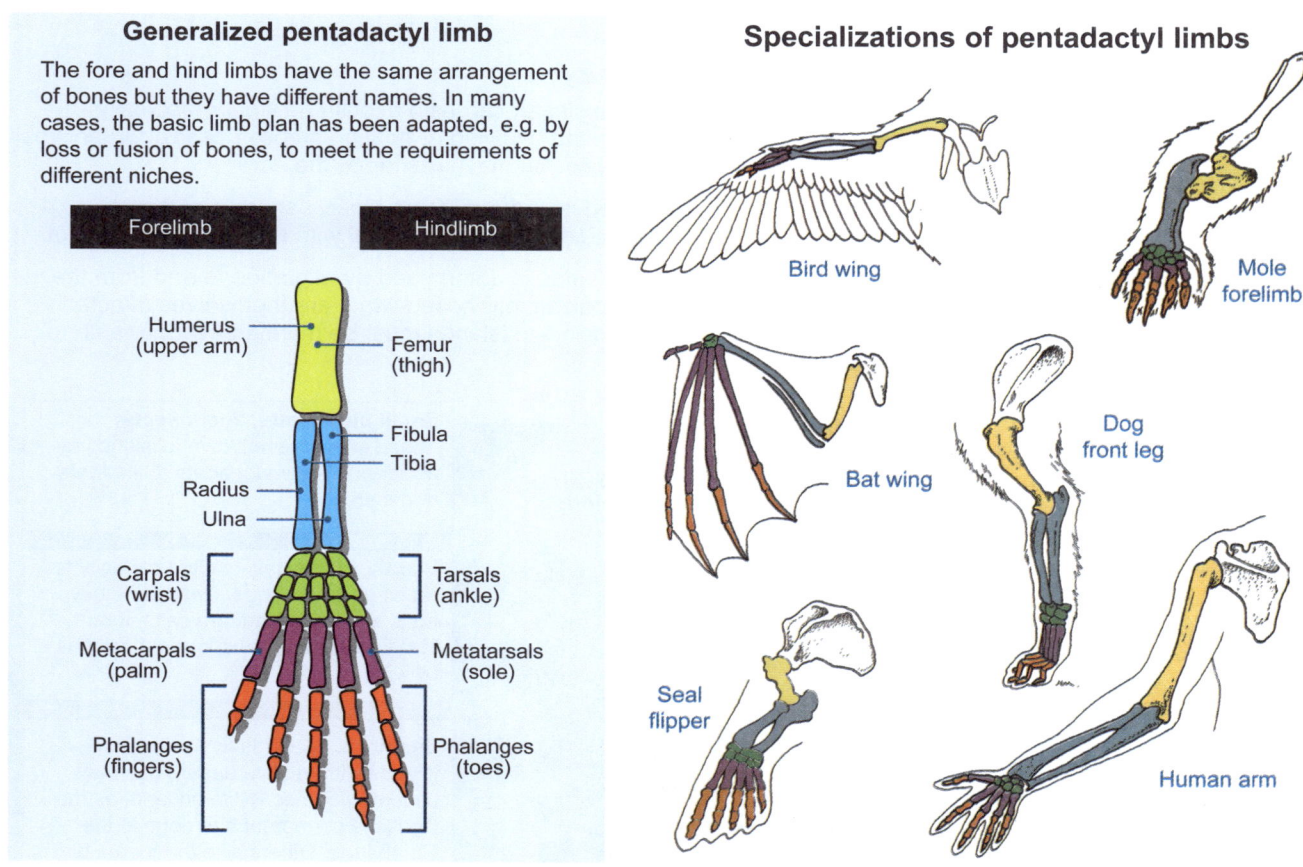

4. Briefly describe the purpose of the major anatomical change that has taken place in each of the limb examples above:

   (a) Bird wing: _____

   (b) Human arm: _____

   (c) Seal flipper: _____

   (d) Dog front leg: _____

   (e) Mole forelimb: _____

   (f) Bat wing: _____

5. How could anatomical homology in the pentadactyl limb provide evidence for common ancestry? _____
   _____
   _____
   _____

6. Bats, birds, and some insects all have wings that enable flight. Why do you think this is not an example of anatomical homology?
   _____
   _____

7. Why is a winged mammal fossil not likely to be a common ancestor of bats and another four legged group of mammals?
   _____
   _____

170 Biogeography and Common Ancestry

Key Question: How does biogeography provide evidence for common ancestry, especially species of isolated island biota?

Biogeography provides evidence for common ancestry

▶ Biogeography is the study of the geographical distribution of species. It can help explain how geographically diverse species shared a **common ancestor** before they dispersed by flying or floating across oceans. Organisms also crossed land bridges that no longer exist, as geological processes have reshaped the earth.

▶ Biogeography can also account for why a species on an isolated island may more closely share a common ancestor with a species on the nearby mainland, than with species on a distant island with a similar environment.

▶ The diversity and uniqueness of plants and animals found on islands is determined by migration to and from the location as well as extinctions and diversifications following colonization. These events are themselves affected by a number of other factors. The organisms that successfully colonize islands must be marine based, or able to survive long periods at sea or in the air.

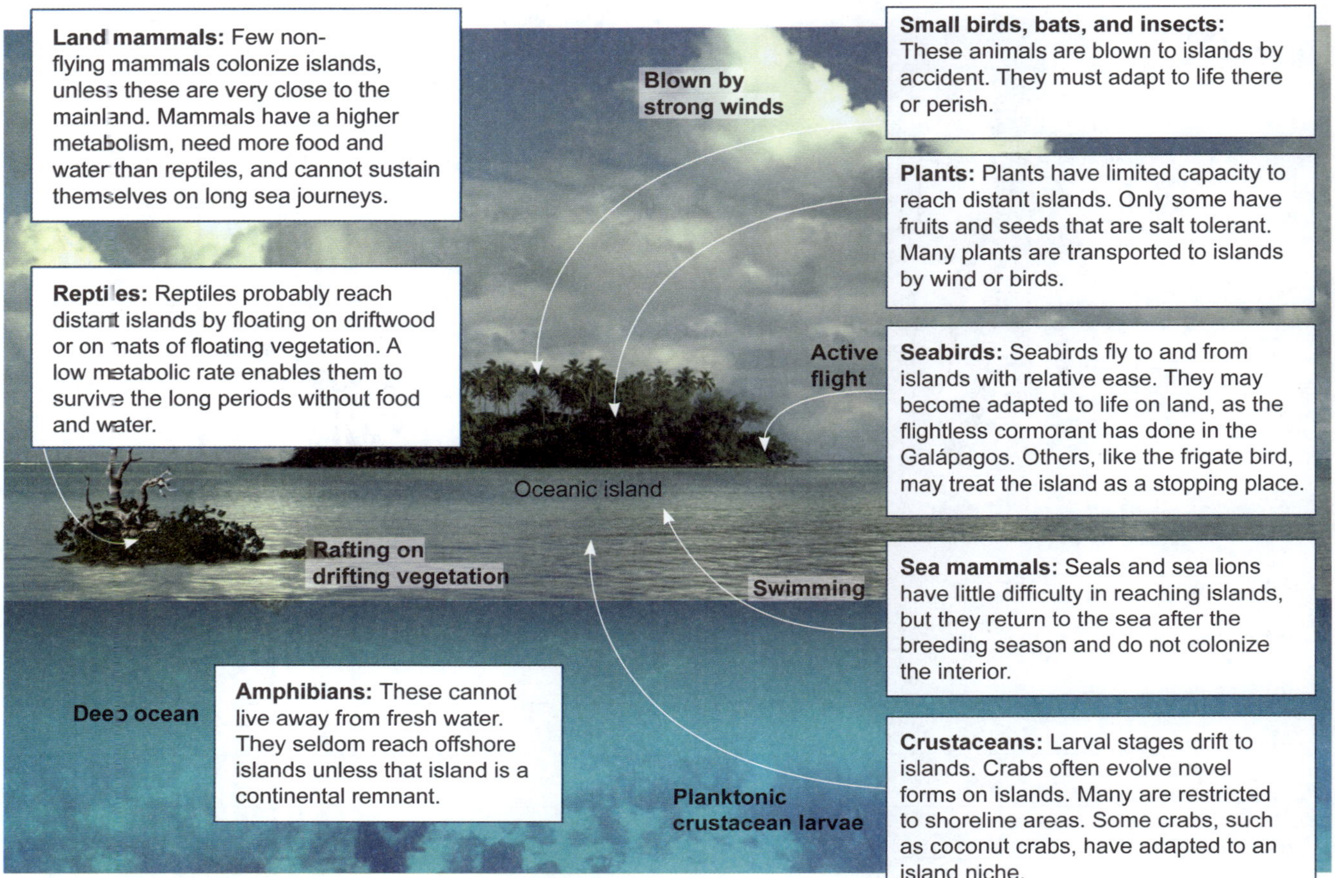

Land mammals: Few non-flying mammals colonize islands, unless these are very close to the mainland. Mammals have a higher metabolism, need more food and water than reptiles, and cannot sustain themselves on long sea journeys.

Reptiles: Reptiles probably reach distant islands by floating on driftwood or on mats of floating vegetation. A low metabolic rate enables them to survive the long periods without food and water.

Amphibians: These cannot live away from fresh water. They seldom reach offshore islands unless that island is a continental remnant.

Small birds, bats, and insects: These animals are blown to islands by accident. They must adapt to life there or perish.

Plants: Plants have limited capacity to reach distant islands. Only some have fruits and seeds that are salt tolerant. Many plants are transported to islands by wind or birds.

Seabirds: Seabirds fly to and from islands with relative ease. They may become adapted to life on land, as the flightless cormorant has done in the Galápagos. Others, like the frigate bird, may treat the island as a stopping place.

Sea mammals: Seals and sea lions have little difficulty in reaching islands, but they return to the sea after the breeding season and do not colonize the interior.

Crustaceans: Larval stages drift to islands. Crabs often evolve novel forms on islands. Many are restricted to shoreline areas. Some crabs, such as coconut crabs, have adapted to an island niche.

1. The Galápagos and the Cape Verde Islands are both tropical island groups found close to the equator. Their plants and animals are quite different, and more closely linked to different, distant mainland sources. Explain why this is the case:

2. Using one or more examples, describe how biogeography provides support for common ancestry:

Biogeographical evidence for the common ancestor of the camel group

▶ The current fragmented distribution of the camel family can be explained by natural phenomena such as migration, plate tectonics, glacial cycles, and changes in sea level. The camel family, Camelidae, consists of six modern-day species on three continents: Asia, Africa, and South America. Biogeographical evidence can be used to evaluate the common ancestry of the camel group. Three principles about the dispersal and distribution of land animals are:

- When very closely related animals were present at the same time in widely separated parts of the world, it is highly probable that there was no barrier to their movement in between the localities in the past.
- The most effective barrier to movement of land animals (especially mammals) was a sea between continents.
- The scattered distribution of living species can be explained by migration out of their original range or extinction in regions between the current populations.

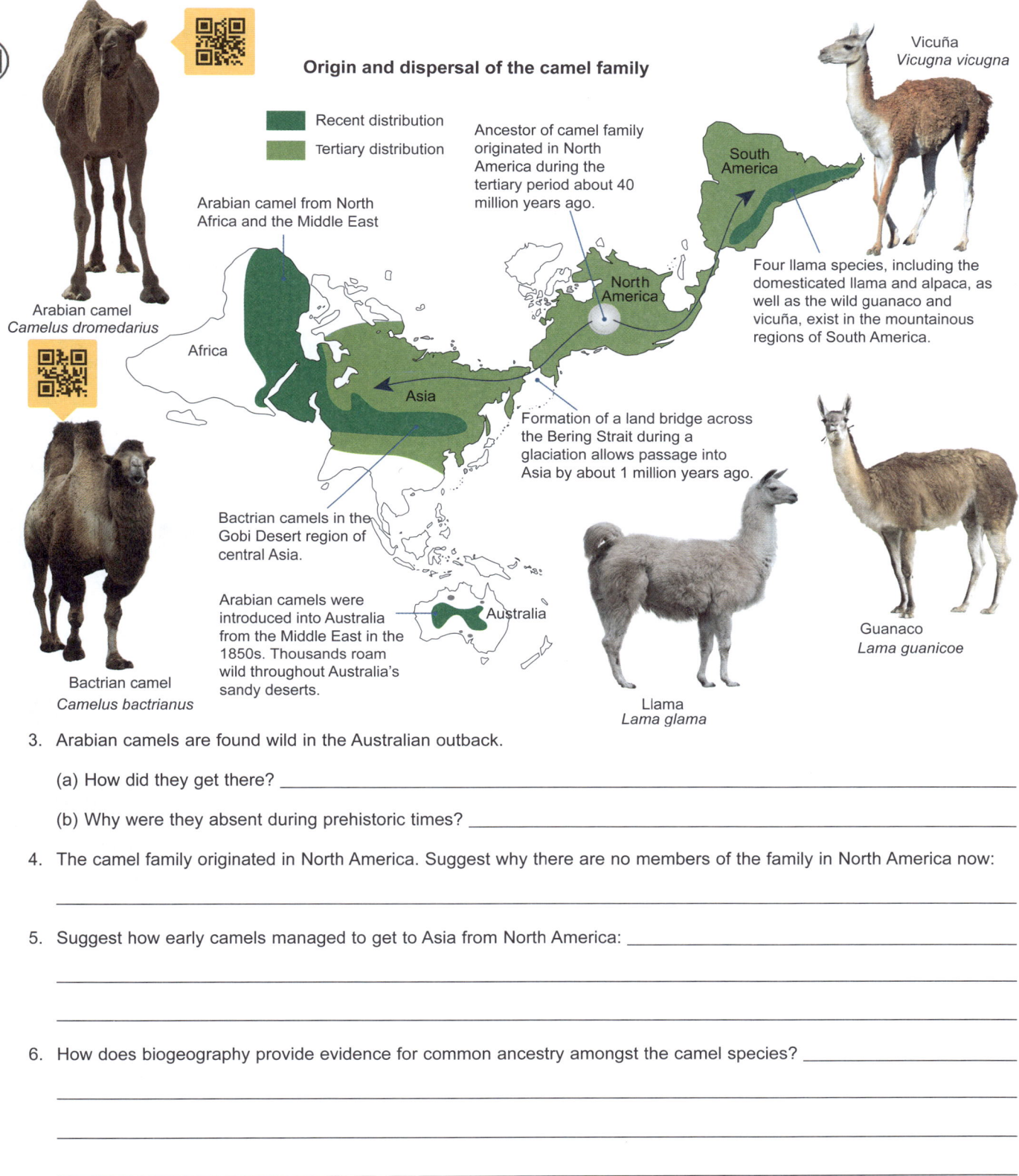

3. Arabian camels are found wild in the Australian outback.

   (a) How did they get there? _____

   (b) Why were they absent during prehistoric times? _____

4. The camel family originated in North America. Suggest why there are no members of the family in North America now:
   _____

5. Suggest how early camels managed to get to Asia from North America: _____
   _____
   _____

6. How does biogeography provide evidence for common ancestry amongst the camel species? _____
   _____
   _____
   _____

171 DNA Evidence for Common Ancestry

Key Question: How can comparison of DNA sequences be used as evidence for common ancestry?

- Molecular **homology** looks at similarities between species at a molecular level. These could be DNA sequences or protein structure. Similarities at a molecular level imply a recent **common ancestry**. More differences indicate that species share a common ancestor further back in evolutionary history.
- The advancement of techniques in molecular biology is providing increasingly large amounts of information about the genetic makeup of organisms. DNA sequencing and comparison of sequences, along with the use of computer databases, are now frequently used when analyzing the evolutionary histories and relationships of different species.
- Bioinformatics allows biological information to be stored in databases where it can be easily retrieved, analyzed, and compared. Comparison of DNA or protein sequences between species enables researchers to investigate and better understand their evolutionary relationships.

An overview of the bioinformatics process

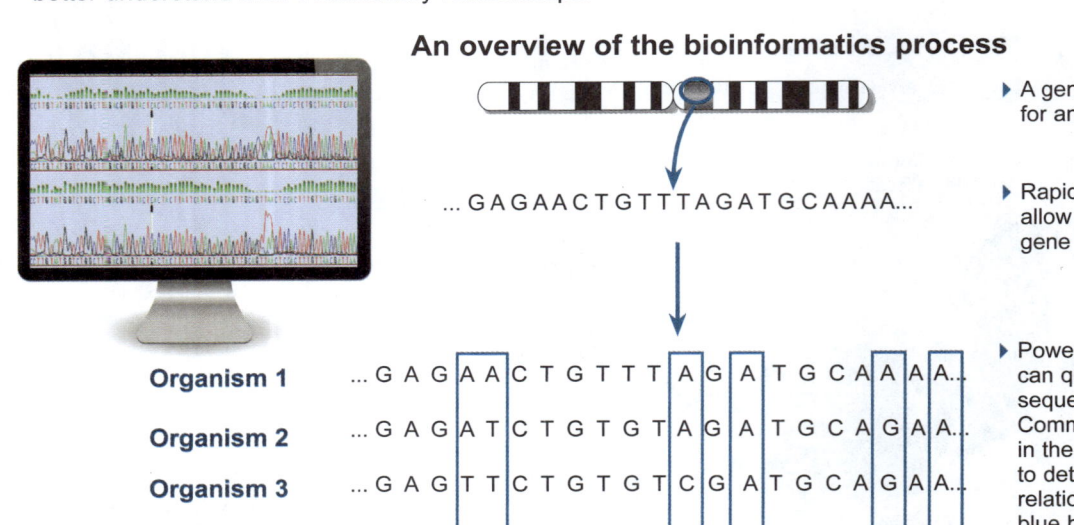

- A gene of interest is selected for analysis.
- Rapid sequencing technologies allow the DNA sequence of the gene to be quickly determined.
- Powerful computer software can quickly compare the DNA sequences of many organisms. Commonalities and differences in the DNA sequence can help to determine the evolutionary relationships of organisms. The blue boxes indicate differences in the DNA sequences.

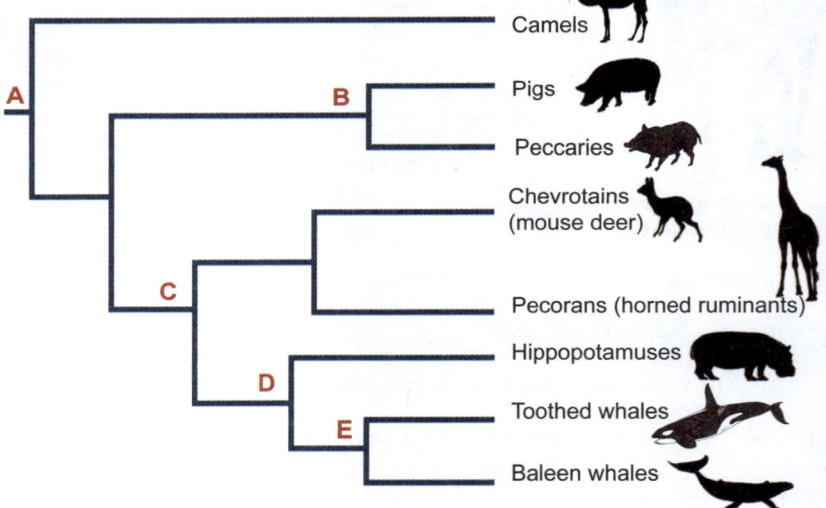

- Once sequence comparisons have been made, the evolutionary relationships can be displayed as a **phylogenetic tree**. The example (right) shows the evolutionary relationships of the whales to some other land mammals.
- Bioinformatics has played an important role in determining the origin of whales and their transition from a terrestrial (land) form to a fully aquatic form.
- This phylogenetic tree was determined by comparing repetitive DNA fragments that are inserted into chromosomes after they have been reverse transcribed from an mRNA molecule. The locations of these repetitive fragments are predictable and stable, so they make reliable markers for determining species relationships. If two species have the same repeats in the same location, they are very likely to share a close common ancestor.

1. The diagram above shows the relatedness of several mammals, as determined by DNA sequencing of 10 genes:

    (a) Which land mammal are whales most closely related to? _____

    (b) Which letter shows where whales and the organism in (a) last shared a common ancestor? _____

    (c) Pigs were once considered to be the most closely related land ancestor to the whales. Use the phylogenetic tree above to describe the currently accepted relationship:

    _____

    _____

172 Protein Evidence for Common Ancestry

Key Question: How can comparisons of protein homology determine common ancestry between living species?

Protein homology

- The amino acid sequence of proteins can be used to establish molecular **homologies** (similarities) between organisms. Any change in the amino acid sequence reflects changes in the DNA sequence. As genetic relatedness increases, the number of amino acid differences due to mutation decreases.

- Some proteins are common to many different species. These proteins are often highly conserved, meaning they mutate (change) very little over time. This is because they have critical roles, e.g. in cellular respiration, and mutations are likely to be detrimental to their function.

- Evidence indicates that these highly conserved proteins are homologous and have been derived from a **common ancestor**. Because they are highly conserved, changes in the amino acid sequence are likely to represent major divergences between groups during the course of **evolution**.

The Pax-6 protein provides evidence for evolution

- The Pax-6 protein regulates eye formation during embryonic development.

- The Pax-6 gene is so highly conserved that the gene from one species can be inserted into another species, and still produce a normally functioning eye.

- This suggests the Pax-6 proteins are homologous, and the gene has been inherited from a common ancestor.

An experiment inserted mouse Pax6 gene into fly DNA and turned it on in a fly's legs. The fly developed fly eyes on its legs!

Hemoglobin homology

- Hemoglobin is the oxygen-transporting blood protein found in most vertebrates. Hemoglobin DNA sequences from different organisms can be compared to determine evolutionary relationships and common ancestry.

- As genetic relatedness decreases, the number of amino acid differences between the hemoglobin chains of different vertebrates increases (below). For example, there are no amino acid differences between humans and chimpanzees, indicating they recently shared a common ancestor. Humans and frogs have 67 amino acid differences, indicating they had a common ancestor a very long time ago.

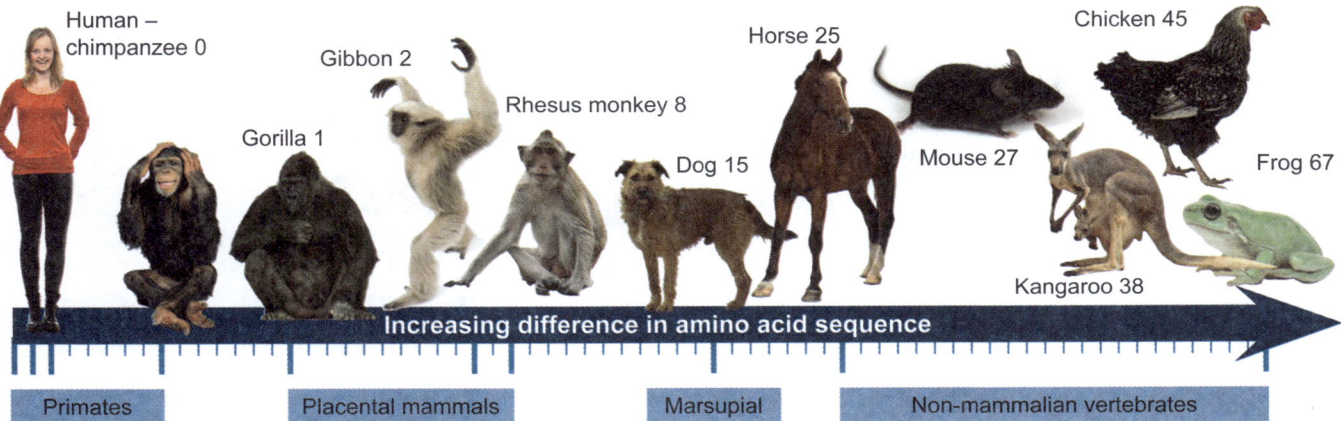

1. Compare the differences in the hemoglobin sequence of humans, rhesus monkeys, and horses. What do these tell you about the relative relatedness of these organisms?

   _____
   _____

2. The fossil record shows that amphibians diverged from the lineages that evolved into mammals before the birds did. How useful are protein molecular homologies in providing evidence for common ancestry?

   _____
   _____
   _____

173 Developmental Homology

Key Question: How can similar stages of embryonic development in different species provide evidence for common ancestry?

Shared stages of development reflect common ancestry

- Developmental biology studies the process by which organisms grow and develop. Today, it focuses on the genetic control of development and its role in producing the large differences we see in the adult appearance of different species.
- During development, vertebrate embryos pass through the same stages, in the same sequence, regardless of the total time period of development.
- The developmental **homology** is strong evidence of their shared **common ancestry**, where embryonic features are determined by the same genes.
- The stage of embryonic development is identified using a standardized system, e.g. Carnegie stages, and is based on the development of structures, not by size or the number of days of development.

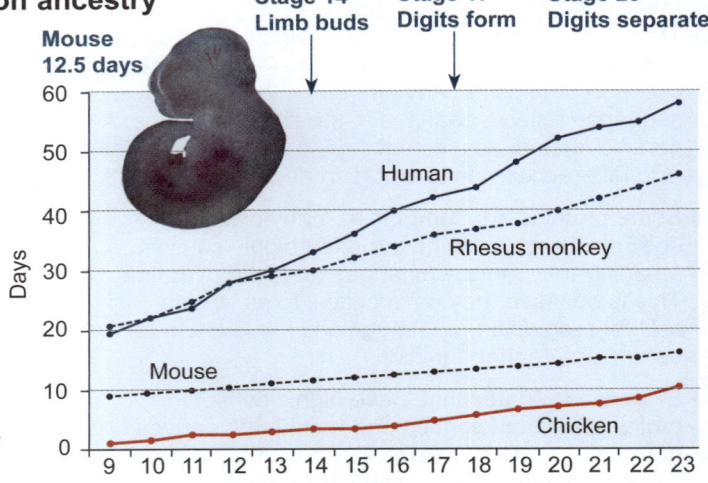

Limb homology and the control of development

- As we have seen, homology, e.g. in limb structure, is evidence of shared ancestry. How do these **homologous structures** become so different in appearance? The answer lies in the way the same genes are regulated during development.
- All vertebrate limbs form as buds at the same stage of development. At first, the limbs resemble paddles, but apoptosis (programmed cell death) of the tissue between the developing bones separates the digits to form fingers and toes.
- Like humans, mice have digits that become fully separated by apoptosis of the tissue between the bones during development. In bat forelimbs, this controlled destruction of the tissue between the forelimb digits is inhibited. The developmental program is the result of different patterns of expression of the same genes in the two types of embryos.

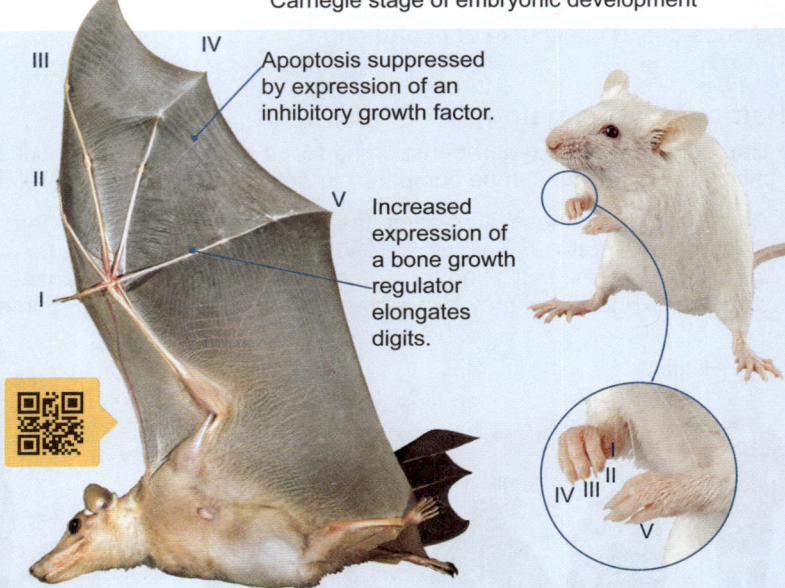

Bat wings are highly specialized structures with unique features, such as elongated wrist and fingers (I-V) and membranous wing surfaces. The forelimb structures of bats and mice are homologous, but how the limb looks and works is quite different.

1. Describe a feature of vertebrate embryonic development that supports evolution from a common ancestor:

   _____
   _____

2. Why might features that are seen in embryos no longer be present in adults? _____
   _____
   _____
   _____

3. How might vertebrate embryonic development provide evidence for common ancestry that anatomical homology cannot?
   _____
   _____

174 Changes in the Fossil Record

Key Question: Why is there often an abrupt appearance in the fossil record of many new groups?

What was the Cambrian Explosion?

▶ Earth's past and present can be organized into a **geologic time scale**, each section with its own name, period of time, and recognizable set of expected **fossils** found in rocks laid down at the same time. This geologic time scale model, showing time periods and sequences, can be represented below:

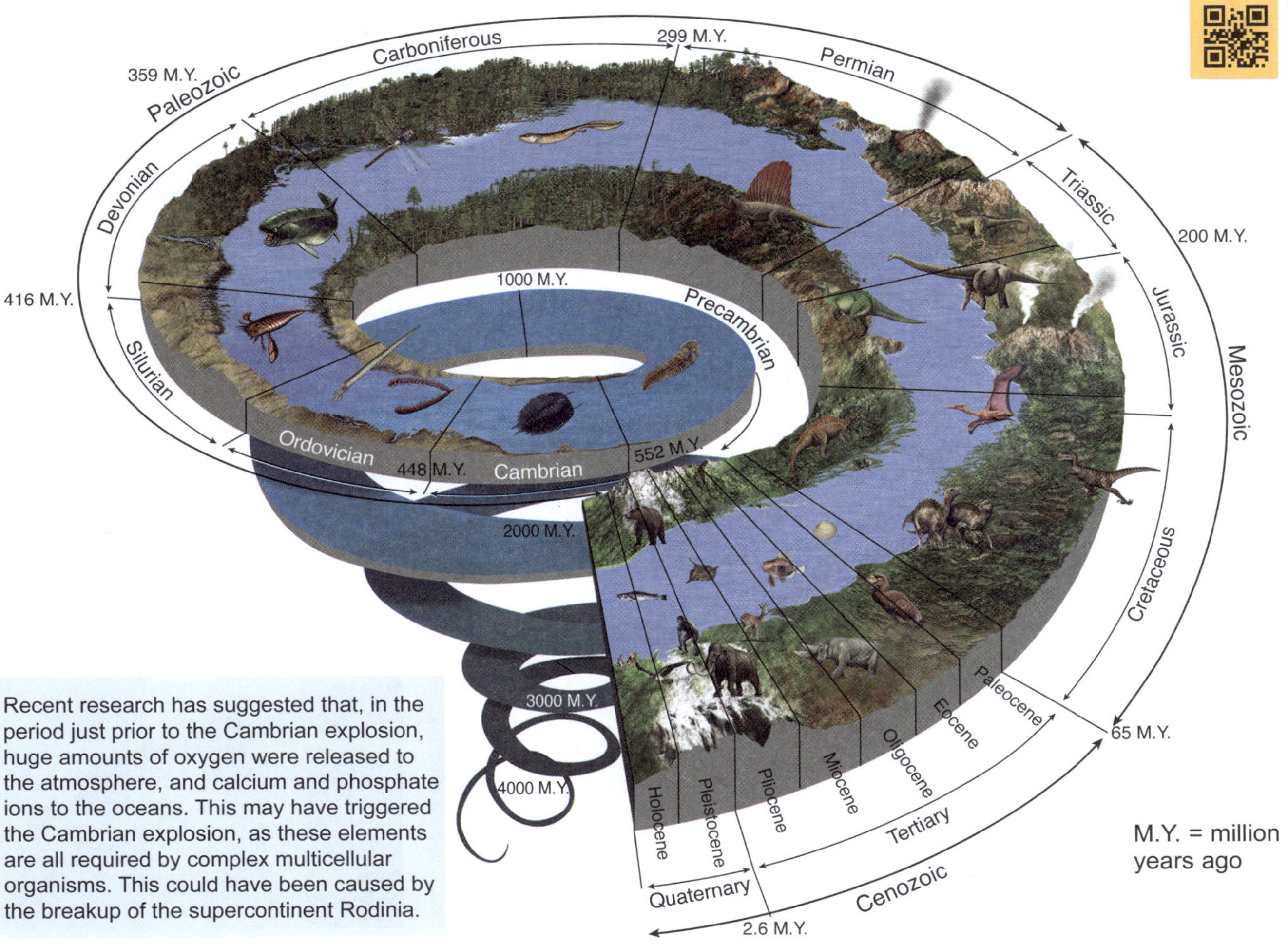

Recent research has suggested that, in the period just prior to the Cambrian explosion, huge amounts of oxygen were released to the atmosphere, and calcium and phosphate ions to the oceans. This may have triggered the Cambrian explosion, as these elements are all required by complex multicellular organisms. This could have been caused by the breakup of the supercontinent Rodinia.

M.Y. = million years ago

▶ An event known as the 'Cambrian explosion' has been revealed by the **fossil record**. It occurred over a relatively short geologic time period of around 20 million years, during the Early Cambrian Period (538 million years ago (mya)).

▶ Only sparse numbers of fossils of unicellular or simple multicellular organisms have been found in the fossil record for the period prior to the Cambrian explosion. The abrupt appearance in the fossil record of complex multicellular organisms represent most of the major phylum groups present on Earth today.

1. In small groups, discuss some ideas about what may have been the cause(s) for the sudden appearance of complex organisms in the fossil record during the Cambrian explosion. Record your group's ideas below:

_____
_____
_____
_____
_____

What is an abrupt appearance?

▸ Fossils of new organisms sometimes seem to appear abruptly in the fossil record, sometimes with little indication of their evolutionary ancestry.

▸ However, this is often due to: 1) the fossil record is limited or disjointed and the vast majority of organisms will not, or have not, fossilized, and 2) often fossil lineage is difficult to identify because fossils are partial, and missing important identifying features.

▸ The abrupt appearance of new species usually occurs due to the rapid evolution of ancestral species into many species (adaptive radiation) when conditions provide many new niches to be exploited.

The appearance of trilobites (left) define the beginning of the Cambrian period. *Parvancorina* (right) is an early arthropod from about 558 mya that has similarities to early trilobites. Although the first true trilobites seem well developed, they can be traced back to earlier organisms, such as *Parvancorina*.

Solving the puzzle

▸ Life on Earth can be traced back to at least 3.5 billion years ago (bya). Stromatolites, mounds built up from cyanobacteria secretions, are some of the earliest fossils known. Examples can be seen at places such as Glacier National Park, in Montana, and the Great Slave Lake in Canada. Stromatolites peak in the fossil record at around 1.3 bya. By the start of the Cambrian period, they had declined to about 20% of their peak abundance. It is suggested that this is due to the earlier appearance of sufficiently complex grazers.

▸ Indeed, the fossil record now shows that, although there is a rapid appearance of new types of organisms during the Cambrian, increasingly complex organisms were evolving in the long periods of time before this. The question then becomes, "What were the conditions that caused this rapid expansion of organisms?", rather than "Where did they come from?"

Stromatolites appeared from around 3.45 bya and peaked in abundance about 1.3 bya.

Spriggina is an early bilateral animal from the Ediacaran period between 635 and 538 mya. Its modern affiliations are still debated.

Ottoia is a soft bodied worm from the Cambrian period. It shows greater complexity than earlier life forms.

2. Did the organisms of the Cambrian period really abruptly appear? Discuss your ideas: _____

_____

_____

_____

3. What were the conditions that caused the sudden appearance of many new complex organisms in the fossil record at the start of the Cambrian period?

_____

_____

_____

_____

175 Punctuated Equilibrium and Gradualism

Key Question: How can models of the rate of change in evolution be used to explain the abrupt appearance or stasis of species in the fossil record?

'Living fossils' and stasis in the fossil record

▶ Scientists developed the punctuated equilibrium model (following page) to explain the often abrupt appearance of organisms in the fossil record.

▶ Some species, once appearing in the fossil record, appear to undergo little to no structural change for vast periods of time. In some cases they become extinct or, like the coelacanth, remain relatively unchanged for long periods of time. The coelacanth appeared to become extinct about 65 million years ago, but was rediscovered as still surviving 'living fossils'. This phenomena is known as stasis. The punctuated equilibrium model can be used to explain how little 'apparent' change has taken place in fossils over a time period.

▶ Fossils can usually only reveal morphological (structural) features of organisms. Scientific studies have shown that molecular level evolutionary changes are continuously occurring at typical rates in all organisms. These can be observed in the genetic material, not seen in fossils, even if this is not reflected in any visible changes to the organism's structure.

Coelacanth: rediscovered alive after 80 million years

▶ Fossils of the coelacanth (left), dating back to 410 million years ago, have been found. The last of the species was assumed to have become extinct 65 million years ago, during the time that the dinosaurs died out. Coelacanths are thought to be more closely related to the common ancestor of four legged tetrapods (amphibians, reptiles, and mammals) than typical ray-finned fish.

Coelacanth fossil found in sandstone

▶ Living coelacanths (right) were rediscovered in 1938, in deep waters off the coast of South Africa. They bore a remarkable, physical similarity to the fossilized specimens but not to any other closely related living species. Scientists called the fish species a 'living fossil'.

Latimeria chalumnae / Coelacanth

1. How could the coelacanth have remained almost morphologically unchanged for over 400 million years?

   _____
   _____
   _____

2. How might the fossil record alone be unable to show if any evolutionary change has occurred in 'living fossils'?

   _____
   _____
   _____

3. Some scientists prefer to call seemingly unchanged organisms 'durable species', rather than living fossils. Why might this be a more accurate term for describing the coelacanth?

   _____
   _____

Models of rate of change in evolution

▶ Two main models have been proposed for the rate at which evolution occurs: phyletic gradualism and punctuated equilibrium. It is likely that both mechanisms operate at different times and in different situations. Interpretations of the fossil record vary, depending on the time scales involved.

▶ During its initial speciation event, a species may have accumulated changes gradually, e.g. over 50,000 years. If that species survives for 5 million years, the evolution of its defining characteristics would have been compressed into just 1% of its evolutionary history. In the fossil record, the species would appear quite abruptly.

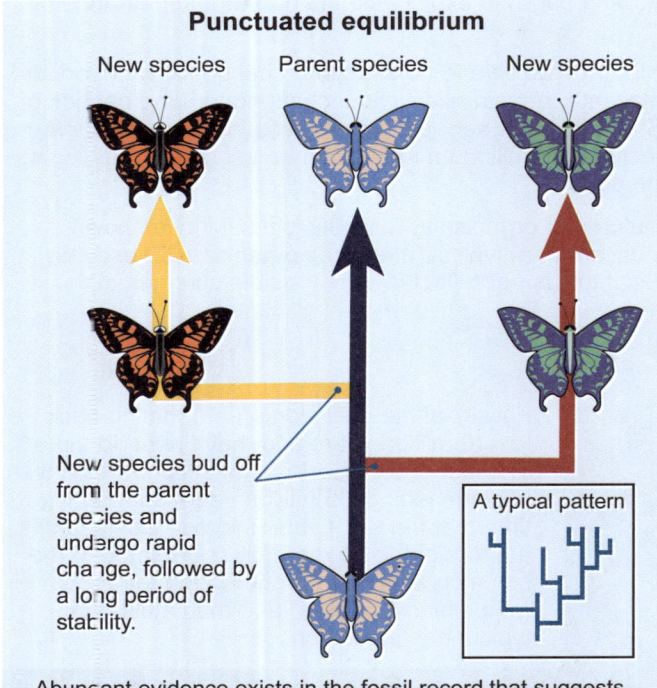

Punctuated equilibrium

New species bud off from the parent species and undergo rapid change, followed by a long period of stability.

Abundant evidence exists in the fossil record that suggests, many species stayed much the same for long periods of time (called stasis) instead of gradually changing. These periods were punctuated by short bursts of **evolution** which produce new species quite rapidly. According to the punctuated equilibrium theory, most of a species' existence is spent in stasis and little time is spent in active evolutionary change. The stimulus for evolution occurs when some crucial aspect of the environment changes.

Phyletic gradualism

Each species undergoes gradual changes in its genetic makeup and phenotype.

New species diverges from the parent species.

Phyletic gradualism assumes that populations slowly diverge by accumulating adaptive characteristics in response to different selective pressures. If species evolve by gradualism, there should be transitional forms seen in the fossil record, as is seen with the evolution of the horse. Trilobites, an extinct marine arthropod, are another group of animals that have exhibited gradualism. In a study published in 1987, a researcher found that they actually changed gradually over a three million year period.

4. How would the fossil record appear as supporting evidence for each of the models of evolutionary change?

   (a) Punctuated equilibrium: _____

   _____

   (b) Gradualism: _____

   _____

5. In the fossil record of early human evolution, species tend to appear suddenly and linger for often very extended periods, before disappearing suddenly. There are few examples of smooth inter-gradations from one species to the next. Which of the above models best describes the rate of human evolution?

   _____

6. Recap: Some species show apparently little evolutionary change over long periods of time (100+ million years).

   (a) Name two examples of such species: _____

   (b) What term is given to this lack of evolutionary change? _____

   (c) Such species are often called 'living fossils'. Why might they have been described in this way? _____

   _____

   (d) Why would such species change so little over evolutionary time? _____

   _____

Testing a model

▸ Does a group follow phyletic gradualism, or punctuated equilibrium? From each model we can make predictions. If a group follows phyletic gradualism, it follows that there should be a series of **transitional fossils** in the fossil record (provided the transitional organisms fossilized).

▸ It also follows that each transitional fossil should be more or less within the range of features of the related species, on either side of it in the fossil record.

▸ In 1987 Peter Sheldon carried out a study of the number of pygidial ribs of eight genus of trilobites using nearly 3500 specimens covering a period of 3 million years. From the graph below it can be seen that in each genera the number of ribs gradually changes over time (below).

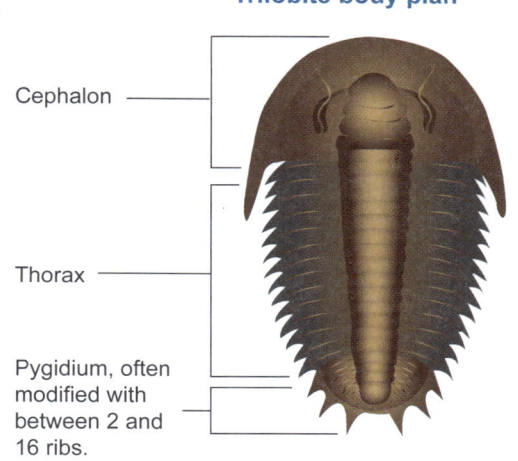

Trilobite body plan

Cephalon

Thorax

Pygidium, often modified with between 2 and 16 ribs.

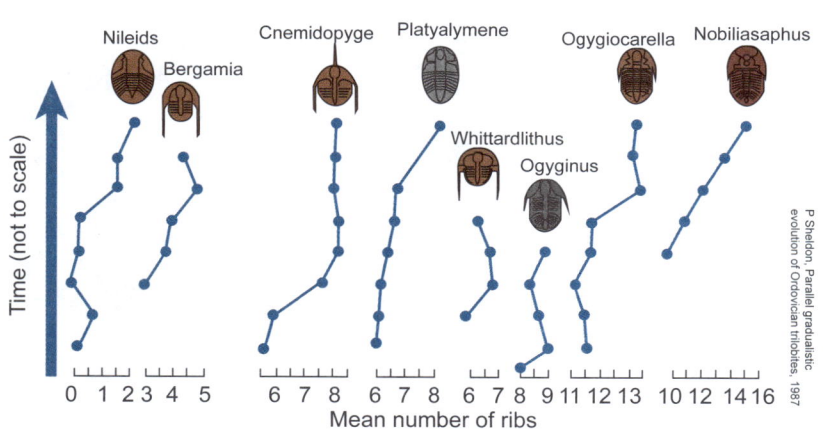

▸ A separate study by Alan Cheetham looked at the marine invertebrate genus *Metrarabdotus* (a Byrozoan) which has an extensive fossil record. He measured 46 features in around 1000 specimens, covering a period of about 10 million years. He was able to produce a graph showing the morphological differences between specimens over time (right).

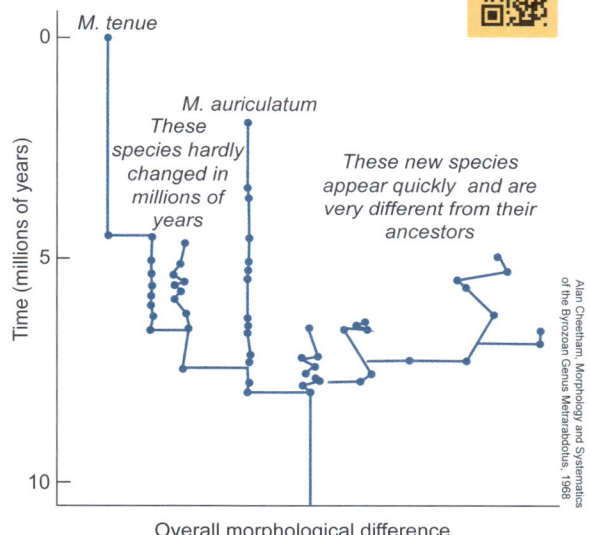

7. What do the results of Sheldon and Cheetham's investigations tell us about phyletic gradualism and punctuated equilibrium as models of evolution?

___

8. Why might some groups of organisms experience phyletic gradualism but others punctuated equilibrium?

___

9. Consider the diagram (right) showing the evolution of a hypothetical fish species, A, over time. Does the appearance of new species over time resemble phyletic gradualism or punctuated equilibrium? Explain:

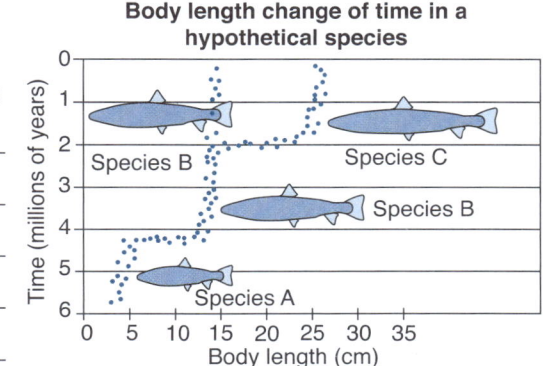

Body length change of time in a hypothetical species

___

176 Developing the Theory of Evolution

Key Question: How did a succession of scientists' observations, ideas, and hypotheses advance the development of the theory of evolution?

Living organisms appear to be different from those that existed in the past

▶ Fossilized remains of organisms have been discovered and puzzled over by humans for thousands of years. Some remains could be identified as being the same, or very similar, to currently living organisms. Other **fossils** seemed to be 'mystery' organisms that no-one could identify.

1. Why did scientists in the first half of the 1800s and earlier have difficulty interpreting newly discovered dinosaur fossil remains to determine what type of organism or species that were from, or how they were related to modern species?

   _____
   _____
   _____
   _____
   _____

Recreated iguanodon dinosaur models (1853)

2. What types of scientific evidence, not available in the early 1800s, are available to today's scientists to help them understand how dinosaurs are related to each other, and to current living species?

   _____
   _____
   _____

In the mid 1800s, sculptures that attempted to recreate the recently discovered dinosaurs, made famous by scientist Richard Owen, were installed in Crystal Palace, London, England. Scientists at the time had little fossil evidence on which to base their assumptions of dinosaur form on, so modeled them after current species of living reptiles. Future fossil evidence would lead to completely changing what they knew about dinosaurs.

▶ Around 1800, French scientist, **Jean-Baptiste Lamarck**, proposed a solution to account for how differences between current species, specifically the giraffe, and those found in ancestral fossils developed. Lamarck reasoned that, as food became more scarce for each generation of giraffe ancestor, the animal had to stretch higher. The longer necked giraffe then passed this physical characteristic onto the next generation. This meant that, eventually, those short-necked giraffe ancestors were replaced, in time, with the long necked giraffes we have today.

3. If you were able to go back in time and tell Lamarck what you know about genetics, to help him with his ideas on giraffe neck length 'growing over time', what would you explain to him?

   _____
   _____
   _____
   _____
   _____
   _____
   _____

Lamarck observed giraffe ancestors had shorter necks
Time →

4. Although Lamarck's ideas about how change occurred over time in organisms were later proven wrong, why were his ideas still considered to have contributed to understanding how organisms change, while sharing common ancestors?

   _____
   _____
   _____

The voyage of the Beagle

▶ In 1831, an aspiring young English naturalist, **Charles Darwin**, took a job on a British Royal Navy ship, the HMS Beagle. He was to collect and catalog samples of living and non-living specimens, including any fossils.

▶ Darwin was able to observe a wide range of life-forms, from many diverse and isolated regions. He found that island groups, such as the Galápagos Islands, were particularly rich in biodiversity. After his five year journey, he returned home. He began to hypothesize about how similar, yet different, species could have formed over time, and may have descended from a **common ancestor**. Darwin used his observational evidence to support his ideas, eventually forming the basis of his developing **theory of evolution by natural selection**.

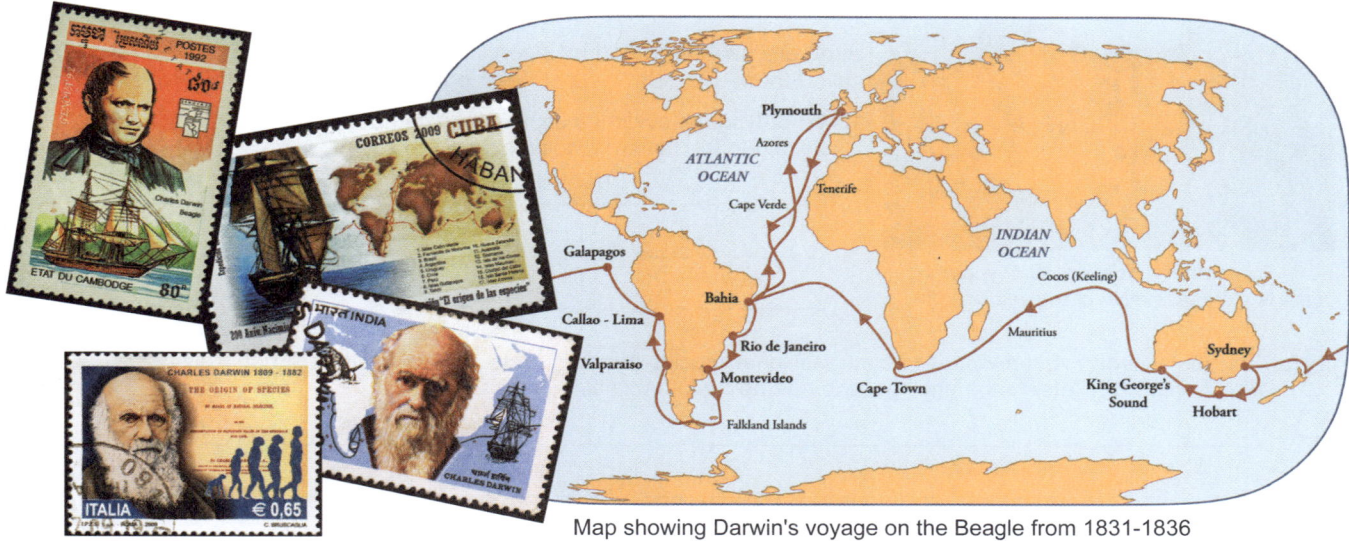

Map showing Darwin's voyage on the Beagle from 1831-1836

5. How might the five year voyage around the world on the Beagle have contributed to Darwin's ideas about evolution?

_____
_____
_____

Battle of the evolutionary theories

▶ Recall that, in science, a theory is a term used when an explanation of a phenomenon has been developed after significant amounts of evidence and observations support it, as well as being scientifically testable (see ch10, p426).

▶ Charles Darwin formulated his theory of evolution privately between 1837-1839. He spent many more years gathering evidence before receiving a letter from Alfred Russel Wallace that covered many of his own ideas. This prompted him to finally publish his theory of evolution by natural selection in 1859, 20 years after he had first thought of it.

Charles Darwin (1809-1882)   Alfred Wallace (1823-1913)

▶ **Alfred Russel Wallace**, a contemporary of Darwin's, spent many years collecting and studying species from tropical regions, especially South America and Indonesia. From his studies, he realized that new species arose by adapting to their environment. He expressed this in a letter to Darwin although he did not use the term natural selection.

NEED HELP? See Activity 248

6. Define the terms scientific hypothesis, a scientific theory, and a scientific law, comparing and contrasting the concepts:

_____
_____
_____
_____

177 Dinosaur or Bird? Revisited

Content Anchor Revisited: How does scientific evidence allow us to continually build ideas about the common ancestry between different groups of organisms?

125 MYA feathered fossil of *Zhenyuanlong* fossil is a close relative of the velociraptor.

A dinosaur and a bird?

Scientific theory is developed by repeated collection and rigorous examination of data by the scientific community to confirm validity. However, if new evidence arises, then a theory must be changed to accommodate it.

Birds are now classified by scientists as belonging to a group of dinosaurs called theropods. Birds have been on Earth for at least 150 million years.

1. What types of evidence could scientists have used to determine that birds belong to the (avian) dinosaur group?

   _____
   _____
   _____

2. (a) Scientists identify *Archaeopteryx* as a transitional species between birds and dinosaurs, but not likely a direct ancestor. Refer to the phylogenetic tree below and explain why this might be:

   _____
   _____

   (b) At what point in the phylogenetic tree would we expect to find the last common ancestor between birds and velociraptors? Explain your reasoning:

   _____
   _____

   (c) Which of the above points are fossils likely to share some features of birds? Explain your reasoning:

   _____
   _____

   (d) Why is the ability to fly not an appropriate indicator of bird and dinosaur classification?

   _____
   _____
   _____

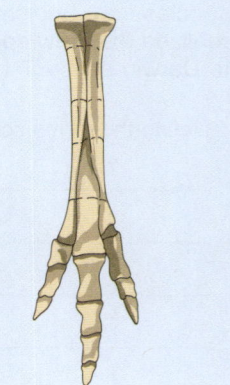

3. The foot bones of *Tyrannosaurus rex* and a chicken look similar (right). What is a probable explanation for this?

   _____
   _____
   _____

©2024 BIOZONE International
ISBN: 978-1-99-101405-4
Photocopying Prohibited

178 Summing Up

Read each question carefully. Place a cross in the box beside the best answer to the question from the four answer choices provided.

1. Mammals share a common ancestor with bacteria. In which **order** did the following features arise in ancestors of the mammals since splitting from the bacteria group?

   ☐ (a) Eukaryote cell, tissue, lungs, skeleton
   ☐ (b) Tissue, eukaryote cell, skeleton, lungs
   ☐ (c) Eukaryote cell, skeleton, lungs, fur
   ☐ (d) Eukaryote cell, lungs, skeleton, fur

2. Some rocks aged between 600-538 million years ago contain a diverse range of unusual species, known as the Ediacaran biota. Many of these are not represented by modern day groups. Most disappeared from the fossil record at the start of the Cambrian explosion, from 538 million years onwards. Which model or theory **best describes** this pattern?

   ☐ (a) Stasis
   ☐ (b) Punctuated equilibrium
   ☐ (c) Gradualism
   ☐ (d) Natural selection

3. *Archaeopteryx* is considered a transitional fossil. Which statement is the **most correct** to complete the following sentence?

   Archaeopteryx is a transitional species between theropod dinosaurs and birds because —

   ☐ (a) It was the common ancestor of theropods and birds
   ☐ (b) It had wings and feathers
   ☐ (c) Fossils of the *Archaeopteryx* could be dated to the time period between theropods and birds
   ☐ (d) It had features of both theropods and birds

4. This phylogenetic tree shows the further evolutionary relationships among some ratite (flightless) bird groups based on homologous structures found in both living species and fossils. Which **statement is true** based on the data.

   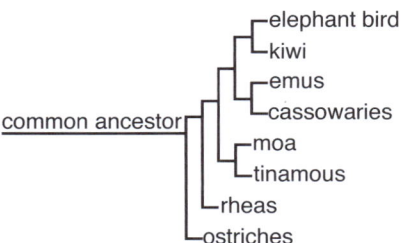

   ☐ (e) Moa are more closely related to tinamous than to kiwi.
   ☐ (f) Ostriches are more closely related to moa than to rheas.
   ☐ (g) Rhea and ostriches are more closely related than emu and cassowaries.
   ☐ (h) Elephant birds are those most closely related to the common ancestor.

5. The opossums of South America and similar looking possums of Australia both have long tails that assist them moving around and living in trees. Both have a pouch where they rear undeveloped young and both groups are marsupials.

   Molecular homology links a South American species of small marsupial, the monito del monte (pictured), found in Argentina and Chile, as being more closely related to the Australian marsupials than all the other South American marsupials.

   José Luis Bartheld CC BY 2.0

   What does this evidence imply about the common ancestry of the Australian marsupials?

   ☐ (a) The monito del monte is the common ancestor to the Australian possums
   ☐ (b) All Australian marsupials originated from a South American common ancestor
   ☐ (c) The common ancestor of the South American opossums originated in Australia
   ☐ (d) Opossums and possums are likely to have had a possum-like common ancestor

6. Scientists originally assumed that all aquatic mammals, such as whales, dolphins, walruses, and manatees, were closely related. We now know that dolphins are more closely related to cows and walruses are more closely related to dogs and bears, than they are each other. What most influenced scientists changing ideas about the common ancestry of aquatic mammals?

   ☐ (a) The theory of natural selection
   ☐ (b) Better observation and understanding of aquatic mammals
   ☐ (c) Newly discovered transitional fossils, combined with new molecular homology technologies
   ☐ (d) Similar features identified between dogs, bears and walrus, and between dolphins and cows

7. Humans are more closely related to mice than to sheep or cows. How is this relevant to why mice are often more useful as substitutes for human medical investigations than sheep or cows? Select the most correct answer:

   ☐ (a) Human and mice are more alike at a molecular level than to the other species, so share a greater proportion of DNA sequences
   ☐ (b) Humans and mice have a more similar diet to each other than cows or sheep, so they are likely to have the same metabolic reaction to medications
   ☐ (c) Cows and sheep are herd animals, and have different behavior responses than humans and mice would to investigations.
   ☐ (d) Mice are smaller and easier to work with in labs

8. Test your vocabulary by matching each term to its correct definition, as identified by writing the letter in the correct box.

(i) fossil record ☐

(ii) gradualism ☐

(iii) abrupt appearance ☐

(iv) stasis ☐

(v) punctuated equilibrium ☐

A A species or fossil record of an organism or group that shows no major anatomical change over long periods of time

B A fossil of a new species or group appearing without transitional fossils apparent in older rocks

C A model that shows evolution occurring at a steady and consistent pace

D A model that shows evolution occurring in 'bursts' alternating with 'dormant' periods

E A sequential collection of fossils showing change from one species or group over time

9. The diagram (left) shows the evolutionary relationship of a group of birds based on DNA similarities:

(a) Place an X at the last common ancestor of all the birds:

(b) How many years ago did storks diverge from vultures?

_____

(c) What are the most closely related birds? _____

_____

(d) What is the difference in DNA (score) between:

i: Storks and vultures: _____

ii: Ibises and shoebills: _____

(e) Which of the birds is the least related to vultures?

_____

10. Insects are extremely adaptable and have a wide range of body forms. Consider the wing structure of the insects below:

Butterfly — A. Forewing, B. Hindwing
Fly — C. Wing, D. Haltere
Beetle — E. Hindwing, F. Elytron

(a) Use the letters to identify the wing structures that are homologous on the images above: _____

(b) What does the homology of these structures indicate? _____

_____

_____

11. Compare and contrast anatomical and molecular homology comparison as methods for identifying common ancestry:

_____

_____

_____

_____

_____

Common ancestry of the Tiktaalik

In 2004, a fossil of an unknown vertebrate was discovered in northern Canada. It was subsequently named *Tiktaalik roseae*, after the Inuit word for large freshwater fish. The *Tiktaalik* fossil was well preserved and many interesting features could be identified. These are shown in the photograph of the fossil below.

Tiktaalik's head is flattened horizontally like that of a crocodile with the eyes on top, looking up.

The shoulder bones are not attached to the skull, allowing its neck to turn independently of the body.

Rod-like bones that help pump water over gills are present, but the presence of ribs indicates that lungs were also present.

The bones of the limbs have a primitive **pentadactyl** arrangement, similar to tetrapods, which allowed it to support its body weight.

Fish-like fins are clearly visible.

The fossil of *Tiktaalik* was covered with scales much like those of fish.

12. Use the anatomical information above to place *Tiktaalik* on the time line of vertebrate evolution. Discuss the evidence for your decision.

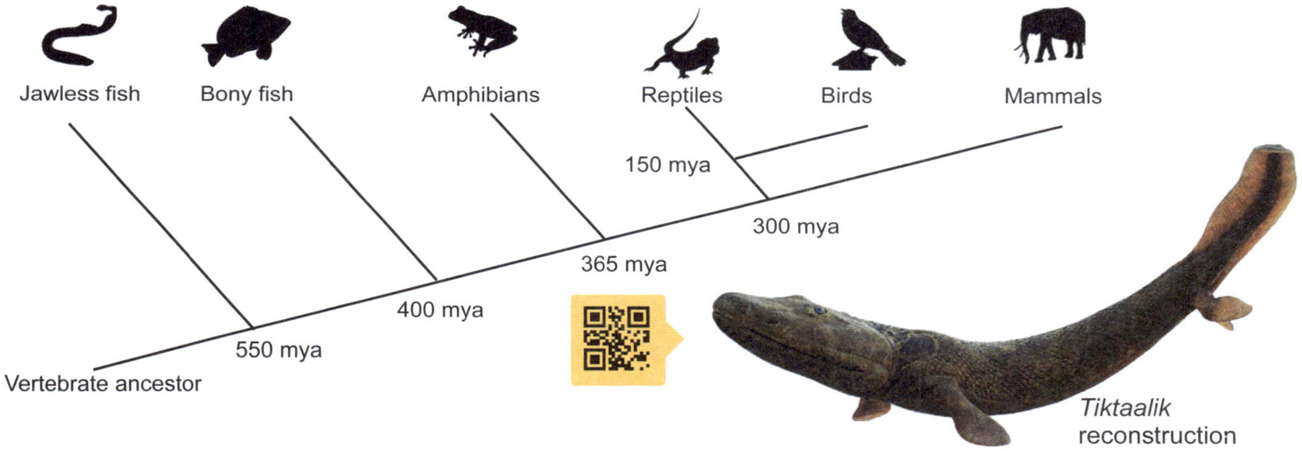

Jawless fish — Bony fish — Amphibians — Reptiles — Birds — Mammals

550 mya — 400 mya — 365 mya — 300 mya — 150 mya

Vertebrate ancestor

Tiktaalik reconstruction

_____
_____
_____
_____
_____
_____
_____
_____
_____
_____

CHAPTER 8
Evolution and Natural Selection

TEKS
Scientific and Engineering Practices

B.1: Investigation and Inquiry
`1.B` `1.C` `1.E` `1.F` `1.G`

B.2: Data and Patterns
`2.A` `2.B` `2.C` `2.D`

B.3: Communicating in Science
`3.A` `3.B`

B.4: Science as a Human Endeavor
`4.A` `4.B`

TEKS
Science Concepts

`B10.A` analyze and evaluate how natural selection produces change in populations and not in individuals

`B10.B` analyze and evaluate how the elements of natural selection, including inherited variation, the potential of a population to produce more offspring than can survive, and a finite supply of environmental resources, result in differential reproductive success

`B10.C` analyze and evaluate how natural selection may lead to speciation

`B10.D` analyze evolutionary mechanisms other than natural selection, including genetic drift, gene flow, mutation, and genetic recombination, and their effect on the gene pool of a population

Learning Outcomes
I know I have achieved this when I can:

		Activity number
☐	Identify the factors involved in the process of natural selection.	180
☐	Evaluate how factors that result in differential reproductive success can cause a change of inherited characteristics in a population over time.	180
☐	Investigate the process of natural selection using a model.	181
☐	Discuss the importance of variation in populations as a required factor needed for natural selection to occur.	182
☐	Evaluate how natural selection acts upon the beak phenotype in Galápagos finches to provide evidence for evolution by natural selection.	183
☐	Analyze and evaluate the effect of selection pressures on populations that can result in directional selection, disruptive selection, and stabilizing selection, giving examples of each.	184
☐	Analyze data related to directional selection of peppered moth populations of different colors in industrial areas of the UK.	185
☐	Measure the change of allele frequency in a theoretical gene pool, linking to evidence for natural selection.	186
☐	Analyze data on the relationship between the rock pocket mice coat color phenotype and the selection pressure of rock color in the environment.	187
☐	Carry out a spreadsheet simulation activity to investigate the effect of gene pool changes on rock pocket mice.	188
☐	Define the term species, using both BSC and PSC concepts.	189
☐	Link isolating mechanisms to speciation, giving examples.	190
☐	Compare and contrast patterns of evolution: divergent and convergent evolution, and adaptive radiation.	191
☐	Explain and differentiate between the terms gene flow and genetic drift, as evolutionary mechanisms.	192
☐	Analyze how lack of gene flow creates reduced diversity in gene pools, using examples.	193
☐	Research the cost-benefit of wildlife corridors as a means to increase gene flow between populations.	193
☐	Analyze changes in gene pools due to genetic drift, from data provided.	194
☐	Calculate allele frequency change in populations due to the founder effect.	195
☐	Analyze the impact of the bottleneck effect on Texan red wolf populations.	196
☐	Research the impact of a beneficial mutation on the gene pool of a population, using a selected example.	197
☐	Analyze the relationship between genetic recombination and the addition of variation to a population's gene pool.	198
☐	Discuss the changes over time due to selection pressures in the tusk phenotype of an African elephant population.	199

RESOURCE HUB
bit.ly/3yaOp7Z

ELPS English Language Proficiency Standards

Page number

Learning

How Does an Elephant Lose its Tusks? Use the question words in question numbers 1 and 2 to decide how to start your answers. For example, question 1 (a) begins "What do you think…?" Begin your answer with "I think…" Use the words: *might*, *advantage*, and *disadvantage* in your answer to question 1(b). What two different ways can you begin your answer to question 2?

312

Learning

Modeling Natural Selection with M&M's®. As you carry out the investigation, practice describing the results in each round. Use the sentence frame: *In round _____, the proportion of _____ [color] was _____.* To answer the questions, use and reverse the wording of the questions: *Over time, the blue M&M's _____. This model is useful because _____.* If you have trouble describing, ask your partner how they might say it.

315

Speaking

Modeling Natural Selection with M&M's®. Carry out the M&M's® modeling activity with a partner. At each stage, discuss your results. What is happening to the color distribution of the M&M's®? At the end of the activity, discuss your results. Together, answer the questions: Why did this happen? How does this represent the process of evolution? Optionally, explain your results to another pair.

315

Listening

Selection Pressure in Populations. Listen as your teacher explains the term *selection pressure* and make a note of its meaning. Using the graphs on page 319 as a guide, practice explaining the difference between types of selection when your classmates ask questions. Use the words *directional*, *disruptive*, and *stabilizing* in your answers."

319

Reading

How Species Form. Work independently or with a partner. Before reading about species formation, examine the diagram on Ancestral Population. What changes does it represent? Now read the text about species formation above the diagram, using a glossary if needed. As you read each paragraph, compare its content to the diagram. When you have finished reading, try to answer the question: *How do species develop?*

329

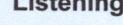
Listening

Mutations and the Gene Pool. Discuss the questions with a group. What words do you recognize from earlier chapters? Try to use them in your response. Pay attention to the way others use the words: *allele*, *recessive*, *generation*, and *beneficial*. Use some of these words in your responses. When others disagree, make notes of the reasons for disagreement.

340

179 How does an Elephant Lose its Tusks?

Content Anchor: How does poaching cause African elephants to be born without tusks?

Tusks and African elephants

- African elephants can be found across 23 countries in Africa and have adapted to a wide range of habitats.
- Both male and female African elephants typically have tusks, which are adapted from extended teeth. The tusks can vary in length.
- The tusks are made of ivory, a material much sought after by humans. Poaching (illegal hunting) of elephants has greatly reduced the total **population**, with around 14,000 individuals still being killed every year, mainly for their tusks.
- Tusklessness is a **phenotype** that is controlled by genes. Typically, around 2-4% of an African elephant population are born without tusks. Almost all are female.
- In some heavily poached areas, scientists have observed up to 60% of the elephants having the tuskless phenotype.

Carved ivory tusk

1. (a) What might be the evolutionary advantage of having large tusks? Discuss in groups, and write your ideas below:

___

(b) Why might that advantage now be a disadvantage to African elephants? ___

___

2. How does the term, "Survival of the fittest" relate to the phenomenon of increasing numbers of African elephants being born without tusks in a population?

___

180 How Does Natural Selection Work?

Key Question: How does natural selection act as a mechanism for evolution?

The factors of natural selection

▶ **Evolution** is the change in inherited characteristics, or phenotypes, in a **population** over generations.

▶ Evolution is the result of interaction between three factors that affect reproductive success:
(1) the potential for populations to increase in numbers, producing more offspring than can survive and reproduce.
(2) inheritable genetic **variation** in **phenotype** as a result of mutation and sexual reproduction.
(3) competition for a finite supply of environmental resources.

▶ When better adapted organisms survive to produce a greater number of viable offspring, we call it **natural selection**. This has the effect of increasing their proportion in the population so that they become more common. It is the basis of Darwin's theory of evolution by natural selection (see activity 176).

We can demonstrate the basic principles of evolution using the model analogy of a "population" of M&M's® candy.

#1

In a bag of M&M's®, there are many colors. This represents the variation in a population. As you and a friend eat through the bag of candy, you both leave the blue ones. Neither of you like this color so you return them to the bag.

#2

The blue candy becomes more common...

#3

Eventually, you are left with a bag of blue M&M's®. Your selective preference for the other colors changed the make-up of the M&M's® population. This is the basic principle of selection that drives evolution in natural populations.

Darwin's theory of evolution by natural selection

Darwin's theory of evolution by natural selection is outlined below. It is widely accepted by the scientific community today and is one of the founding principles of modern science.

Overproduction
Populations produce too many young: many must die.

Populations generally **produce more offspring than are needed to replace the parents**. Natural populations normally maintain constant numbers. A certain number will die without reproducing.

Variation
Individuals show variation: some variations are more favorable than others.

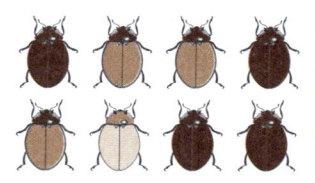

Individuals in a population have **different phenotypes** and therefore, genotypes. Some **traits** are better suited to the environment, and individuals with these have better survival and reproductive success.

Natural selection
Natural selection favors the individuals best suited to the environment at the time.

Individuals in the population **compete for limited resources**. Those with favorable variations will be more likely to survive. Relatively more of those without favorable variations will die.

Inherited
Variations are inherited: The best suited variants leave more offspring due to **increased reproductive success**.

Variation, selection, and population change

1. Variation through mutation and sexual reproduction:
In a population of brown beetles, mutations independently produce red coloration and 2 spot marking on the wings. More individuals are born than survive to maturity, and the individuals in the population compete for limited resources.

Red   Brown mottled   Red 2 spot

2. Selective predation:
Brown mottled beetles are eaten by birds but red ones are avoided.

3. Change in the genetics of the population:
Red beetles have better survival and reproductive success and become more numerous with each generation. Brown beetles have poor **fitness** and become rare.

Natural populations, like the ladybug population above, show genetic variation. This is a result of mutation, which creates new **alleles**, and sexual reproduction, which produces new combinations of alleles. Some variants are more suited to the environment of the time than others. These variants will leave more offspring, as described in the hypothetical population model (right).

1. What produces the genetic variation in populations? _____

2. (a) Define evolution: _____
   _____

   (b) Identify the three factors that interact to bring about evolution in populations: _____
   _____
   _____
   _____
   _____

3. Using your answer to 2(b) as a basis, evaluate how the genetic make-up of a population could change over time:
   _____
   _____
   _____
   _____
   _____
   _____

181 Modeling Natural Selection with M&M's

Key Question: How can modeling be used to investigate the process of natural selection?

Investigation 8.1 Investigating natural selection with M&M's®

See appendix for equipment list.

Modeling natural selection with M&M's®

You can easily demonstrate **natural selection** using M&M's. M&M's have several colors, but they all taste the same. Therefore the only basis for selection is color. We call this **selection pressure**. We can see the effect of this when people act as M&M's "predators" and preferentially select their favorite color candy to eat.

1. Work in pairs. Wash hands well, then open a bag of M&M's and pour them onto a clean surface.

2. Count the total "**population**" number of M&M's and record it here: _____

3. First, calculate the proportion of each color in the original population using:
Proportion = number of each color ÷ total "population" number. Enter these values below (row START).

4. Each person chooses 10 M&M's, <u>avoiding the blue ones</u>. Pick your favorite color first. If your favorite color is not available, e.g. blue, choose your second favorite. Discard candy if handled by another student.

5. Calculate the number of M&M's <u>remaining</u> and the proportion of each color in this population. Proportion = number of each color ÷ total number remaining. Record these proportions below (Round 1).

6. Return the remaining M&M's to the bag (including the blue ones). From a second bag of M&M's restore the original "population" number using the proportions you calculated above. This represents the surviving M&M's reproducing. Calculate the number of each color of M&M's needed using the equation:
Proportion × number of M&M's you both ate (20). Round up or down to get whole numbers.

7. Repeat steps 3-5 three more times. Each round represents a generation.
Record the proportions left after each round (take photos after each round if you want).

	Blue	Green	Yellow	Orange	Red	Brown
START						
Round 1						
Round 2						
Round 3						
Round 4						

1. What happens to the blue M&M's® over time? Explain: _____
_____
_____

2. How is this model useful due to similarities to the process of natural selection? _____
_____
_____

3. What are the limitations of using M&M's to model natural selection? _____
_____
_____
_____

182 The Role of Variation in Populations

Key Question: Why is variation an important factor to enable natural selection to act on populations?

▶ **Variation** refers to the diversity of **phenotypes** within a **population**. Variation is important in **evolution** because it is the raw material for selecting Favorable phenotypes. Those individuals with phenotypes well suited to the current environment have a greater chance of surviving and reproducing to pass on their genes to the next generation, i.e. they will have higher **fitness**. Individuals with less favorable phenotypes are less likely to survive and reproduce and their genes will have a lower representation in the next generation. These elements of the scientific theory of evolution are modeled below:

Mutations
- Gene (point) mutations
- Chromosome rearrangements

Provides the source of all new genetic information (all new **alleles**).

Selection pressures
- Competition
- Predation
- Climatic factors
- Disease and parasitism

Favorable phenotypes
Phenotypes well-suited to the prevailing environment have better survival and greater reproductive success (i.e. higher fitness). They produce many offspring with the favorable **traits**.

Sexual reproduction
- Independent assortment
- Crossing over
- Recombination
- Mate selection

Rearrangement and shuffling of the genetic material into new combinations.

Favor some phenotypes more than others

The phenotype is the product of the many complex interactions between the genotype, the environment, and the chemical tags and markers that regulate the expression of the genes (epigenetic factors).

Genotype

Determines the genetic potential of an individual.

Environmental factors influence the expression of the genotype in producing the phenotype.

Phenotype

Each individual in the population is a 'TEST CASE' for its combination of alleles.

Selection pressures on the phenotype will affect an individual's fitness. Selection pressures are those factors in the environment that determine whether an organism will be more or less successful at surviving and reproducing.

Unfavorable phenotypes
Phenotypes poorly suited to prevailing environment have lower fitness and produce few offspring with the unfavorable traits.

Environmental factors
- Diet/nutrients • pH
- Temperature • Wind exposure
- Sunlight

1. (a) What is variation? _____
   _____

   (b) Identify the sources of variation in sexually reproducing organisms: _____
   _____

2. What is the importance of variation to evolutionary change? _____
   _____
   _____

3. (a) What is meant by fitness and why is it important? _____
   _____

   (b) Define the term selection pressure and explain how it relates to fitness: _____
   _____
   _____

183 Natural Selection in Galápagos Finches

Key Question: How did studying Galápagos finch beaks provide evidence for evolution by natural selection?

- **Natural selection** acts on the **phenotypes** of a **population**. Individuals with phenotypes that increase their **fitness** produce more offspring, increasing the proportion of the genes corresponding to that phenotype in the next generation.
- Numerous population studies have shown that natural selection can cause phenotypic changes in a population relatively quickly. The effect of natural selection on a population can be verified by making quantitative measurements of phenotypic **traits**.

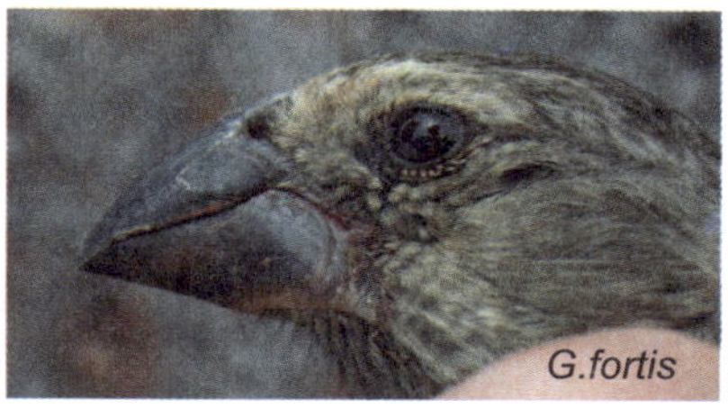

Beak adaptations and feeding in Galápagos finches (Darwin's finches)

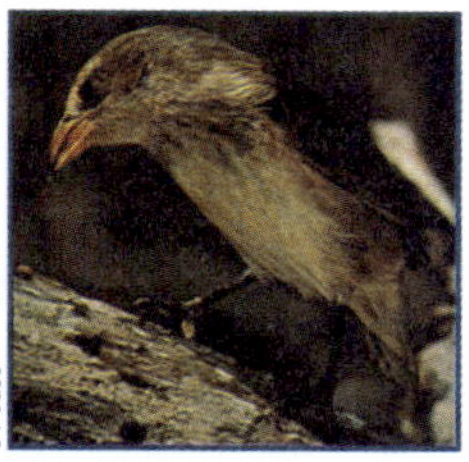

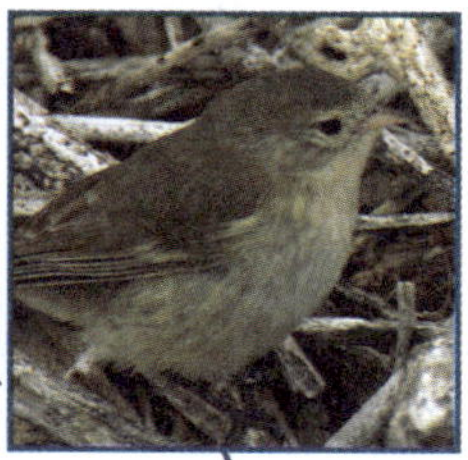

The beaks of the **warbler finches** are the thinnest of the Galápagos finches. They are used to spear insects and probe flowers for nectar.

The **cactus ground finches** have evolved probing beaks to extract seeds and insects from cacti.

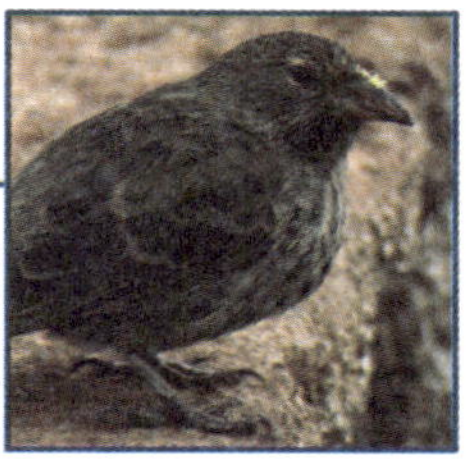

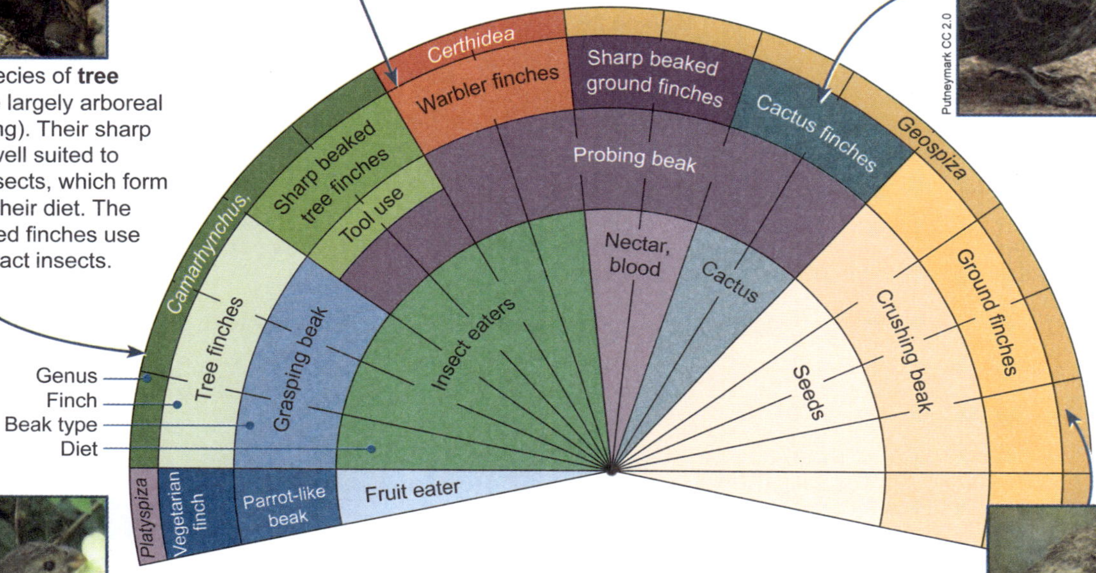

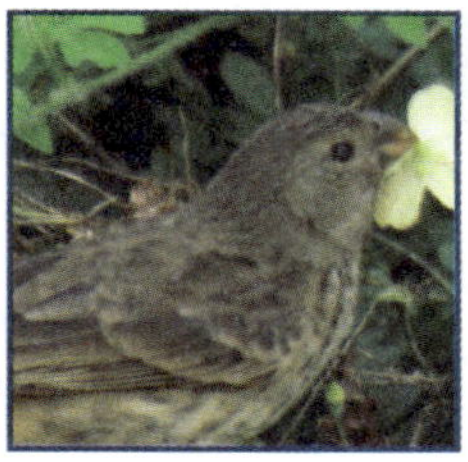

The five species of **tree finches** are largely arboreal (tree dwelling). Their sharp beaks are well suited to grasping insects, which form the bulk of their diet. The sharp beaked finches use tools to extract insects.

Vegetarian finches feed on buds, leaves, flowers, and fruit. Their parrot-like beaks are evolved for food manipulation.

Ground finches have crushing type beaks for seed eating. The finches differ mainly in body size and in the size of their beaks.

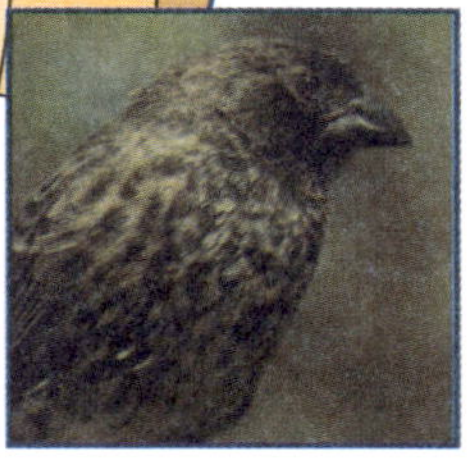

1. What are the main factors that contributed to the natural selection seen in the Galápagos finches? _____

2. Using information from Darwin's finches above, evaluate how differential reproduction of the vegetarian finches would be affected in a year when fruit became extremely scarce, but there was still adequate insect food available:

Adaptations are heritable

▶ The finches on the Galápagos island (Darwin's finches) are famous as they are commonly used examples of how **evolution** produces new species.

▶ Of all the Galápagos Island finches, the medium ground finch (*Geospiza fortis*) is particularly well studied. A population on the Daphne Major island had their beak depth measured shortly before the island experienced a severe drought. Researchers were interested in how the drought had affected the birds, and measured the beak depth of the survivors and their offspring.

Medium ground finch

In this activity, you will analyze data from the measurement of beak depths of the medium ground finch (*Geospiza fortis*) on the island of Daphne Major, near the center of the Galápagos Islands. The measurements were taken in 1976, before a major drought hit the island, and in 1978, after the drought (survivors and survivors' offspring).

Note: In table (right) No. = number of.

Beak depth (mm)	No. 1976 birds	No. 1978 survivors	Beak depth of offspring (mm)	Number of birds
7.30-7.79	1	0	7.30-7.79	2
7.80-8.29	12	1	7.80-8.29	2
8.30-8.79	30	3	8.30-8.79	5
8.80-9.29	47	3	8.80-9.29	21
9.30-9.79	45	6	9.30-9.79	34
9.80-10.29	40	9	9.80-10.29	37
10.30-10.79	25	10	10.30-10.79	19
10.80-11.29	3	1	10-80-11.29	15
11.30+	0	0	11.30+	2

3. Use the data above to draw two separate sets of histograms:

   (a) On the left hand grid, draw side-by-side histograms for the number of 1976 birds per beak depth and the number of 1978 survivors per beak depth.

   (b) On the right hand grid, draw a histogram of the beak depths of the offspring of the 1978 survivors.

NEED HELP? See Activity 265

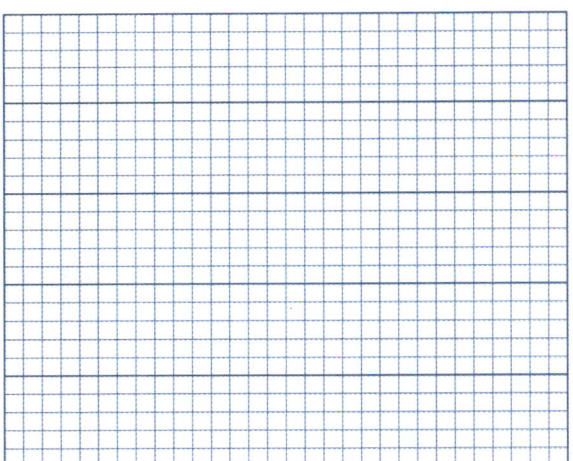
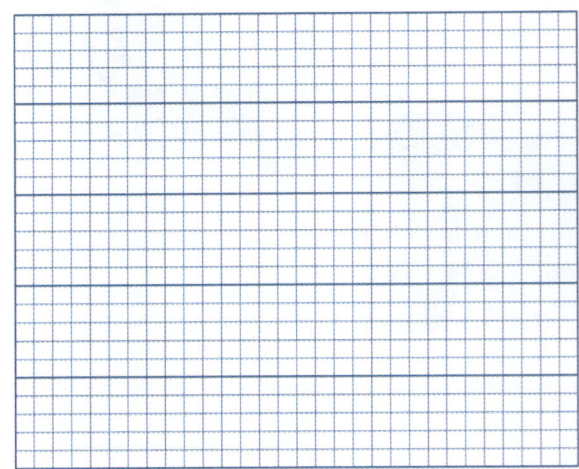

4. (a) On both of your graphs, mark the approximate mean beak depth.

   (b) How much has the average moved from 1976 to 1978? _____

   (c) Is beak depth heritable? What does this mean for the process of natural selection in the finches? _____

184 Selection Pressure in Populations

Key Question: How do environmental factors create selection pressure on populations?

- Environmental factors act as a **selection pressure**, favoring survivability of some **traits** over others. Individuals with **phenotypes** better suited to the environment have better reproductive success and there are more of them. More of these successful **alleles** will exist in the **population**.

- Over time, **natural selection** may lead to a permanent change in the genetic makeup of a population. Natural selection is always linked to the suitability of the phenotype in the current environment, so it is a dynamic process. It may favor existing phenotypes or shift the phenotypic median, as demonstrated in the models shown below.

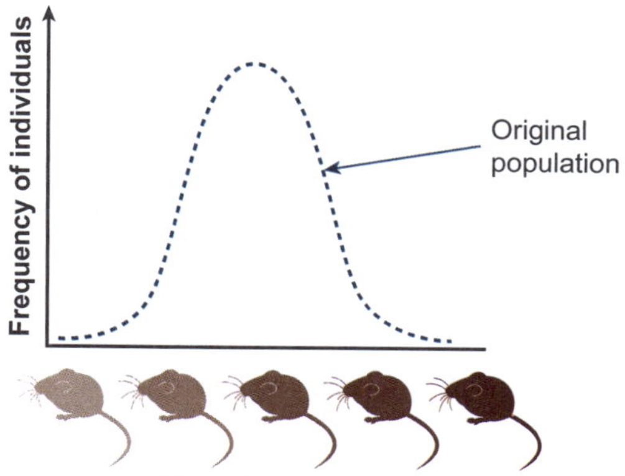

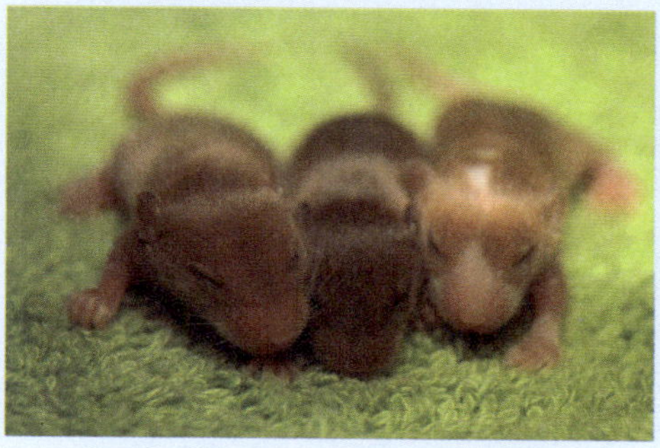

Natural selection acts on phenotypic variation. Even slight variations may be enough for selection to occur. The white streak on the mouse on the right may make it stand out to predators. The darker mouse may be able to more easily hide in the shadows.

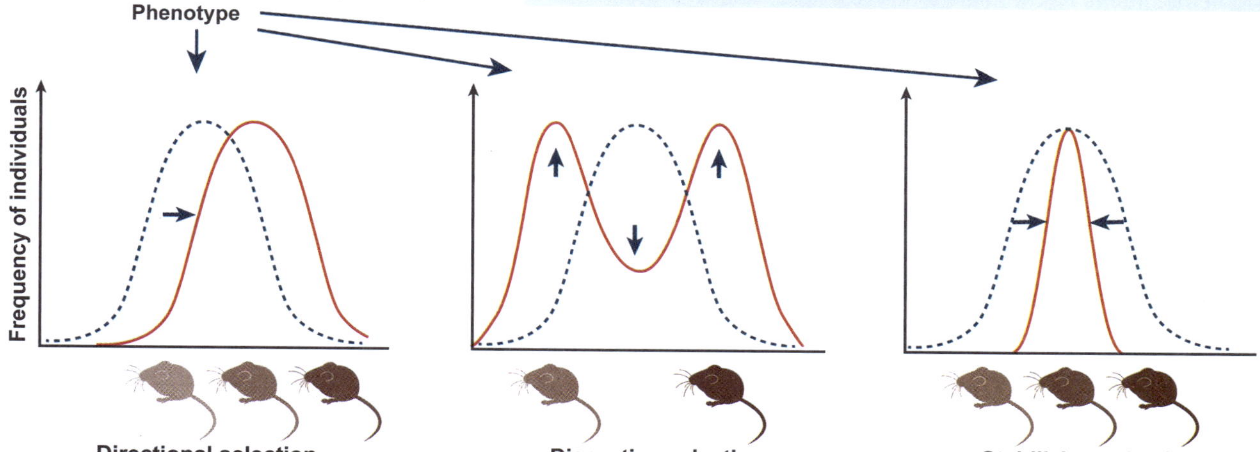

Directional selection
An environmental pressure, e.g. predation, or higher temperatures, selects against one of the phenotypic extremes. The adaptive phenotype is shifted in one direction and one phenotype is favored over others.

Disruptive selection
Disruptive selection favors two phenotypic extremes at the expense of intermediate forms. Disruptive selection may occur when environments or resources are fluctuating or distinctly divergent.

Stabilizing selection
Extreme variations are selected against and the middle range (most common) phenotypes are retained in greater numbers. Stabilizing selection decreases **variation** for the phenotypic character involved.

1. Analyze how fluctuating (as opposed to stable) environments favor disruptive (diversifying) selection: _____

_____

_____

_____

2. Evaluate the likely effect of rapid environmental change on a population with very low phenotypic variation: _____

_____

_____

_____

Examples of selection types

Directional selection

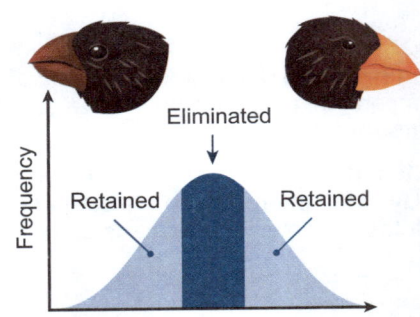

Fossil evidence shows European black bears experienced directional selection during the most recent ice age. During the glacials, larger sized back bears predominated, due to an advantage in cold conditions with limited food resources. Inter-glacial periods favored smaller bears.

Disruptive selection

A prolonged drought on Santa Cruz Island in the Galápagos resulted in a population of ground finches that was bimodal for beak size. Competition for the usual medium-sized seed sources was so intense that selection favored birds able to exploit either small or large seeds.

Stabilizing selection

Stabilizing selection operates most of the time, in most populations, and acts to prevent divergence from the adaptive phenotype, e.g. birth weight of human infants such that they fit through the woman's pelvic bones or number of eggs laid in a nest.

3. Which of the graphs below relate to the examples above?

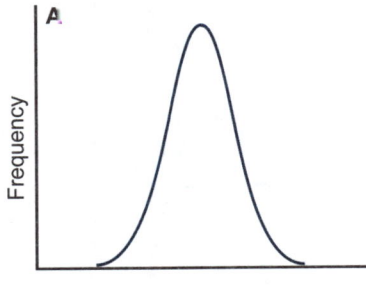
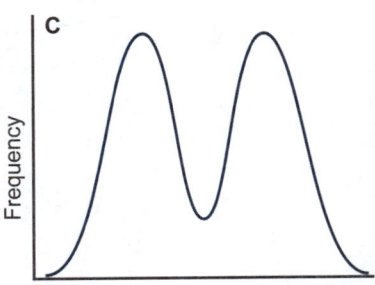

A: _____

B: _____

C: _____

4. Disruptive selection can be important in the formation of new species:

   (a) Describe the evidence from the ground finches on Santa Cruz Island that provides support for this statement:
   _____
   _____

   (b) The ground finches on Santa Cruz Island are one interbreeding population with a strongly bimodal distribution for the phenotypic character beak size. Suggest what conditions could lead to the two phenotypic extremes diverging further:
   _____

   (c) Predict the consequences of the end of the drought and an increased abundance of medium sized seeds as food:
   _____

5. Explain why the number of eggs in a bird's nest is most likely governed by stabilizing selection: _____
_____
_____
_____

185 Directional Selection in Moth Populations

Key Question: How did environmental selection pressure create change in the peppered moth populations in the past?

- **Selection pressures** on the peppered moth during the Industrial Revolution shifted the common **phenotype** from the gray form to the melanic (dark) form in **populations**.
- Genetically determined melanism is relatively common in animals and leads to different forms existing in the population. During the Industrial Revolution in the UK, high levels of pollution from burning of coal caused trees to become dark with soot. Selection Favored the proliferation of dark (melanic) forms over the pale (non-melanic) forms of the peppered moth (*Biston betularia*). Melanic forms were better camouflaged against predatory birds. This shift in phenotype is an example of directional selection.

The gene controlling color in the peppered moth, is located on a single locus. The **allele** for the melanic (dark) form (M) is dominant over the allele for the gray (light) form (m).

Melanic form
Genotype: MM or Mm

Gray form
Genotype: mm

The peppered moth, *Biston betularia*, has two forms, shown right: a gray, mottled form, and a dark, melanic form. During the Industrial Revolution, the gray form was more visible to birds in industrial areas where the trees were dark with sooty deposits. Birds preyed upon them more often, so more of the dark forms survived. In rural areas, more of the lighter, gray forms persisted.

Museum collections of the peppered moth over the last 150 years show a marked change in the frequency of the melanic form. Moths collected prior to the major onset of the Industrial Revolution in England, were mostly the gray form. The frequency of the darker melanic forms increased in industrial times and regions.

In the 1940s and 1950s in the UK, coal burning was still at intense levels around the industrial centers of Manchester and Liverpool. During this time, the melanic form of the moth was dominant. With the decline of coal burning factories and the introduction of the Clean Air Act in cities, air quality improved between 1960 and 1980. Sulfur dioxide and smoke levels dropped to a fraction of their previous levels. This coincided with a sharp fall in the relative numbers of melanic moths (right).

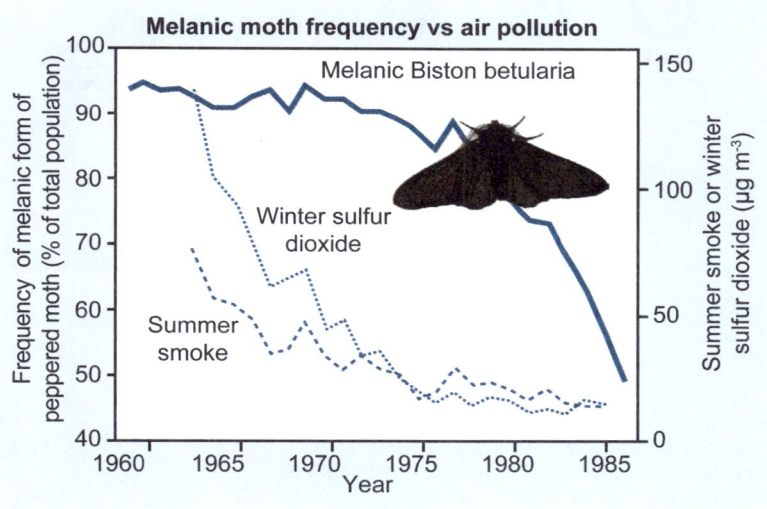

Melanic moth frequency vs air pollution

1. The populations of peppered moth in England have undergone changes in the frequency of an obvious phenotypic character over the last 150 years. What is the phenotypic character?

   _____

2. Describe how the selection pressure on the gray form has changed with change in environment over the last 150 years:

   _____
   _____
   _____

3. The level of pollution dropped around Manchester and Liverpool between 1960 and 1985. How could scientists ensure that there was a valid correlation between this data and a decrease in melanic moths?

   _____
   _____

186 Measuring Gene Pool Change

Key Question: What is a gene pool, and how is it significant for evolutionary change?

▸ A **gene pool** is defined as the sum total of all the genes present in a **population** at any one time:
- Not all the individuals will be breeding at a given time.
- The population may have a distinct geographical boundary which prevents or limits flow of genes.
- Each individual is a carrier of part of the total genetic complement of the population.

▸ **Natural selection** can alter the **allele** frequencies in gene pools.

▸ The model below shows a hypothetical population of beetles undergoing changes as it is subjected to natural selection. The phases shown below represent the same gene pool, undergoing change over time. The beetles have two **phenotypes** (dark and pale) determined by the amount of pigment deposited in the cuticle. The gene controlling this character is represented by two alleles, A and a. Your task is to analyze the gene pool as it changes.

1. For each phase in the gene pool below fill in the following tables (the first columns have been done for you):

   (a) Count the number of A and a alleles separately. Enter the count into the top row of the table (left hand columns).
   (b) Count the number of each type of allele combination (AA, Aa and aa) in the gene pool. Enter the count into the top row of the table (right hand columns).
   (c) For each of the above, work out the frequencies as percentages (bottom row of table):

   Allele frequency = No. counted alleles ÷ Total no. of alleles x 100

Phase 1: Initial gene pool

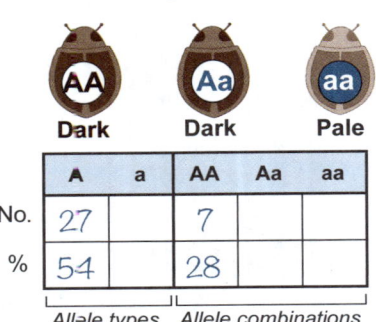

	A	a	AA	Aa	aa
No.	27		7		
%	54		28		

Allele types | Allele combinations

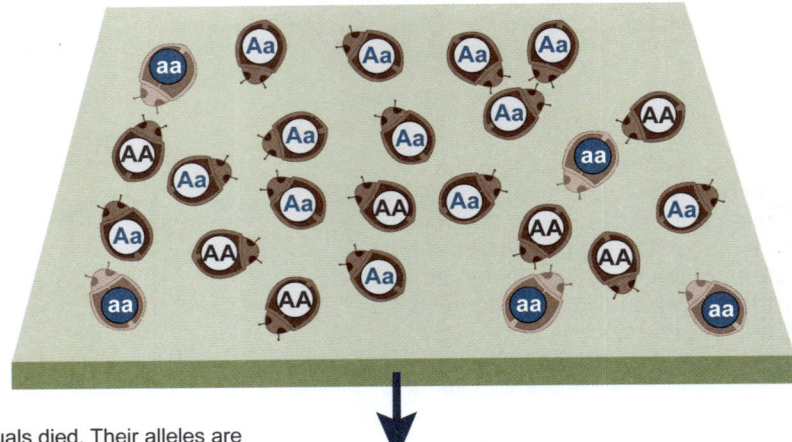

Two pale individuals died. Their alleles are removed from the gene pool.

Phase 2: Natural selection

In the same gene pool at a later time there was a change in the allele frequencies. This was due to the loss of certain allele combinations due to natural selection. Some of those with a genotype of aa were eliminated (poor **fitness**).
These individuals (surrounded by small red arrows) are not counted for allele frequencies; they are dead!

	A	a	AA	Aa	aa
No.					
%					

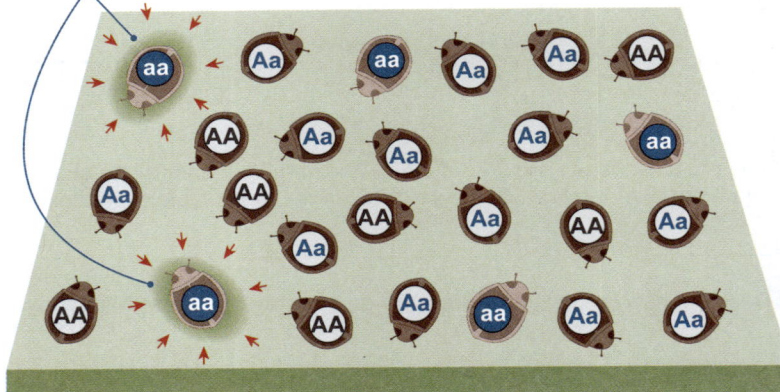

2. What evidence demonstrates that evolution has occurred in the beetle population after phase 2 in the scenario above?

_____

_____

187 Natural Selection in Rock Pocket Mice

Key Question: How does natural selection act upon the coat color of rock pocket mice?

Rock pocket mice are found in the deserts of southwestern United States and northern Mexico. They are nocturnal, foraging at night for seeds, while avoiding owls (their main predator). During the day, they shelter from the desert heat in their burrows. The coat color of the mice varies from light brown to very dark brown. Throughout the desert environment in which the mice live, there are outcrops of dark volcanic rock. The presence of these outcrops and the mice that live on them provide an excellent study in **natural selection**. The need to blend into their surroundings to avoid predation is an important **selection pressure** acting on the coat color of rock pocket mice.

▸ The coat color of the Arizona rock pocket mice is controlled by the Mc1r gene (a gene that in mammals is commonly associated with the production of the pigment melanin). Homozygous dominant (AA) and heterozygous mice (Aa) have dark coats, while homozygous recessive mice (aa) have light coats.

▸ The coat color of mice in New Mexico is not related to the Mc1r gene.

▸ 107 rock pocket mice from 14 sites were collected and their coat color and the rock color they were found on were recorded by measuring the percentage of light reflected from their coat (low percentage reflectance equals a dark coat). The data are presented on the right:

Site	Rock type (V volcanic)	Percent reflectance (%) Mice coat	Percent reflectance (%) Rock
KNZ	V	4	10.5
ARM	V	4	9
CAR	V	4	10
MEX	V	5	10.5
TUM	V	5	27
PIN	V	5.5	11
AFT		6	30
AVR		6.5	26
WHT		8	42
BLK	V	8.5	15
FRA		9	39
TIN		9	39
TUL		9.5	25
POR		12	34.5

1. (a) What is the genotype(s) of the dark colored mice? _____

   (b) What is the genotype of the light colored mice? _____

2. Using the data in the table above and the grids below and on the facing page, draw column graphs of the percent reflectance of the mice coats and the rocks at each of the 14 collection sites.

NEED HELP? See Activity 265

3. (a) What do you notice about the reflectance of the rock pocket mice coat color and the reflectance of the rocks they were found on?

   _____

   _____

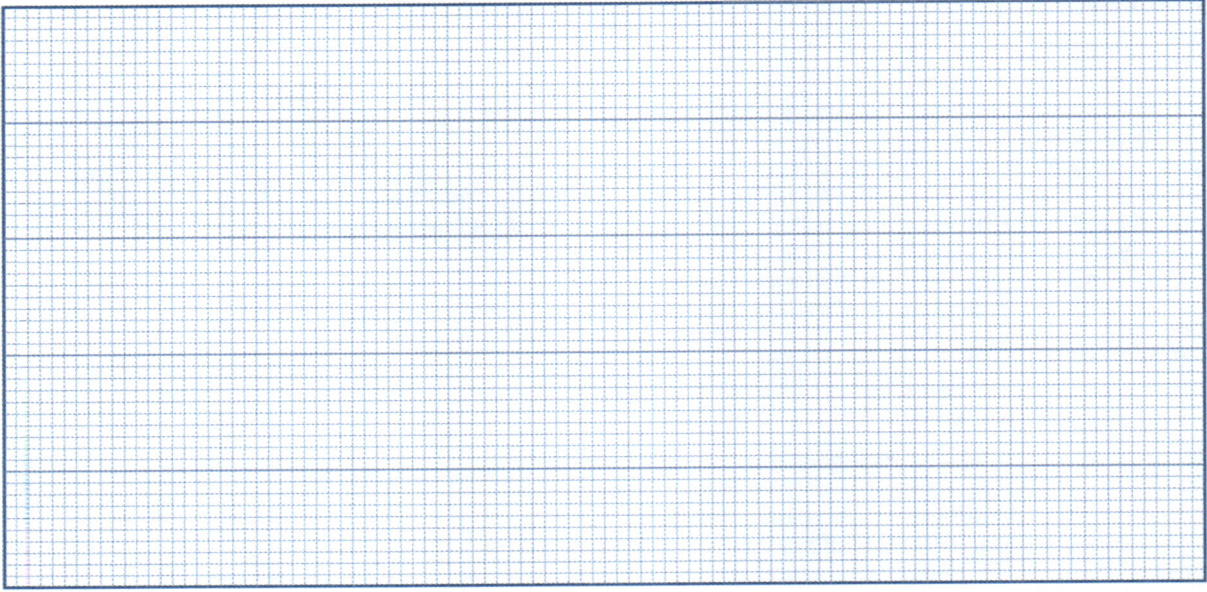

Barn owls hunt by sight, and are the main predator of the rock pocket mouse

   (b) Suggest a cause for the pattern in 3(a). How do the phenotypes of the mice affect where the mice live?

   _____

   _____

   _____

   _____

   (c) What are two exceptions to the pattern you have noticed in 3(a)? _____

   _____

   _____

   _____

   (d) How might these exceptions have occurred? _____

   _____

   _____

4. The rock pocket mice populations in Arizona use a different genetic mechanism from the New Mexico populations to control coat color. What does this tell you about the evolution of the genetic mechanism for coat color?

   _____

   _____

188 Modeling Natural Selection in Rock Pocket Mice

Key Question: How can we use a computer model to simulate changes in the gene pool due to natural selection?

Investigation 8.2   Investigating gene pool changes

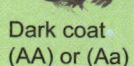

Light coat (aa)   Dark coat (AA) or (Aa)

Changes in **gene pools** are often modeled using physical representations of the **genotypes** in a **population**. This type of modeling is tedious and subject to human error. Modeling genotypic changes using a spreadsheet is quicker and allows the model to be changed to simulate different scenarios occurring in the gene pool. Some **natural selection** labs work by manually placing tokens representing alleles into a bag and withdrawing them randomly to make genotypes. A certain genotype is then selected against by not returning it to the bag and the next generation is drawn from the remaining **alleles**. This exercise can be long and difficult to manipulate, and it reduces the population numbers over time so that an accurate simulation is not entirely possible.

These problems can be solved by modeling natural selection using a spreadsheet to compute allele changes over time. Once the formulae are in place, the spreadsheet can be manipulated in different ways to produce a more accurate, yet still simple, simulation. Download the spreadsheet from the **BIOZONE Resource Hub bit.ly/3yaOp7Z** or use the notes and screen shots below to recreate the spreadsheet yourself.

PART 1: Setting Up the spreadsheet to investigate natural selection in rock pocket mouse coat colors

1. Open a new spreadsheet. First, switch off automatic calculation. This makes calculation of future **allele** frequencies simpler and under manual control, so that you can calculate them when you're ready. Each spreadsheet program will have slightly different ways of doing this. For Microsoft Excel, click on the **Formulas** tab then on the **Calculation Options** menu and click **Manual**. Calculations can then be made using the **Calculate Now** button beside the Calculation Options menu or by using the **F9** button.

2. The headings A and B represent the alleles A and a. This is necessary because the COUNTIF formula used later in the spreadsheet is not case sensitive (it does not recognize the difference between A and a).

3. 0.5 is the frequency of the A allele in the initial population (generation 0). The frequency of the B allele in the population is equal to 1–A. In our initial population, 50% of the alleles will be A and 50% will be B (A and a in the population).

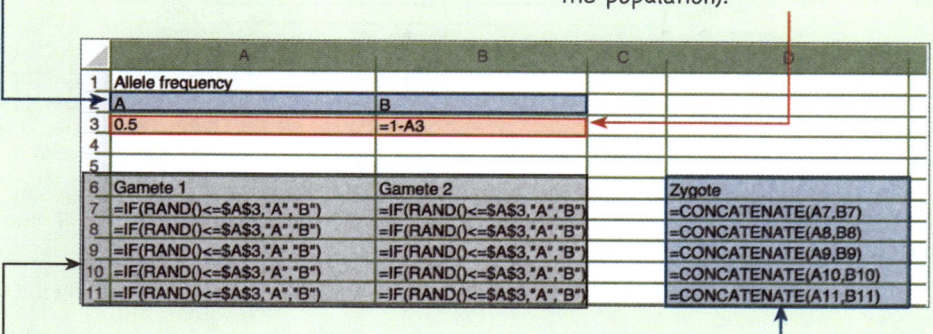

4. The RAND formula produces a random number between 0 and 1 and compares it to the number in cell A3. If the random number is less than or equal to the number in cell A3, then an A is displayed in the cell. If the random number is greater, a B is displayed. The $ symbol tells the spreadsheet that cell A3 is a reference cell and must not change.

5. The CONCATENATE formula takes gametes A and B and puts them together to make the zygote.

6. Highlight cells and copy down all formulas to row 56 to produce 100 random gametes containing alleles A or B and 100 zygotes.

7. You now need to count up the number of AA, AB, and BB genotypes

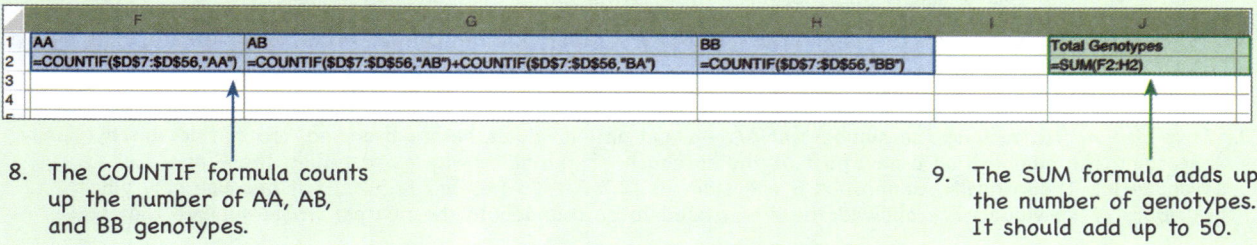

8. The COUNTIF formula counts up the number of AA, AB, and BB genotypes.

9. The SUM formula adds up the number of genotypes. It should add up to 50.

10. Now you must calculate the number of A and B alleles present in this generation (Generation 1). In cell **F4**, type the heading **A** and in cell **G4**, type the heading **B**. In cell **I4**, type the heading **Total Alleles**.

11. Cell F5 adds up all the alleles from the AA genotype and the A alleles from the AB genotype. Cell **G5** adds up all the B alleles. Cell I5 adds up all the alleles. Click **Calculate Now** and you should see the number 100 appear.

12. Cells F9 and G9 calculate the frequency of As and Bs in Generation 1. Cell I9 adds up cell F9 and G9. This should add to 1.

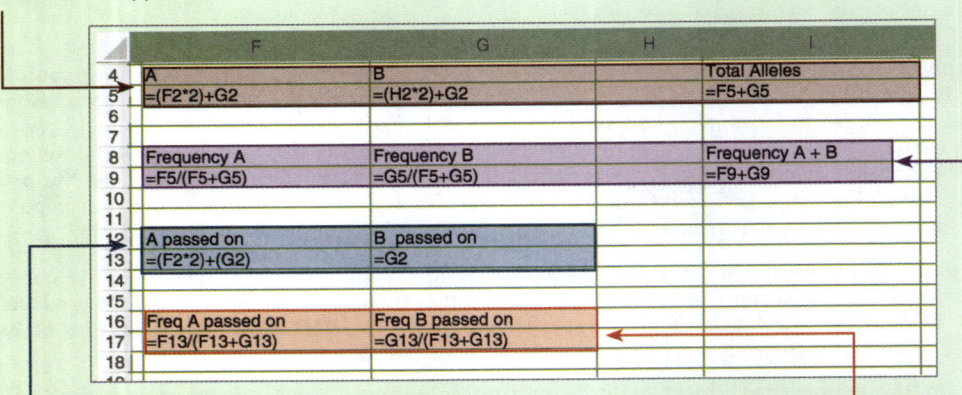

13. The **selection pressure** is against the recessive phenotype. The alleles in the recessive genotype (BB) will not be passed on, so the frequency of alleles in the population is different from the frequency of alleles that will be passed on. These cells calculate the **number** of alleles that will be passed on (excluding those in genotype BB).

14. These cells calculate the frequency of alleles that will be available to the next generation.

15. Finally, you must keep a record of each generation's allele frequencies before mating, i.e. before BB is excluded.

16. Gen 0 was your starting population. Copy down to cell **F31** to get ten generations.

17. The frequency of A in Gen 0 was 0.5

18. The frequency of a is simply 1–A. Copy down to **H31**.

19. Click **Calculate Now**. Note the numbers that appear in cells **F9** and **G9**. Type the number in **F9** into **G22**. This is the frequency of the A allele in the first generation.

20. Now type the number in cell **F17** into cell **A3** and click **Calculate Now** to produce the second generation of alleles in cells F9 and G9. Again enter the number in F9 into G23 and the number in F17 into A3 before clicking **Calculate Now**.

21. Each time you do this, the spreadsheet calculates a new generation of genotypes and their alleles based in the number you enter into A3.

22. Save your spreadsheet.

PART 2: Natural Selection lab

Now that you have built the spreadsheet and are familiar with it, you can begin the natural selection lab.

1. To do this, you will select against the recessive light color coat **phenotype** (and hence the aa genotype, represented as BB in the spreadsheet). In this scenario, any BB individuals never get to breed (it is irrelevant what the phenotype is, simply that no BB individuals will enter their alleles into the next generation).

2. To start the lab, make sure **0.5** is entered into cell **A3**. Enter **0.5** into cell **G21** and make sure the cells below them are clear. Highlight cells **F17** and **G17** and under the **Format** menu click **Cells**, then click the **Number** category and set it to **2** decimal places. Click **OK**.

3. It is also worth tracking the numbers of AA, Aa and aa individuals before breeding. You can do this by simply recording the numbers on a new part of the spreadsheet, the same way as recording the A and a allele frequencies. Theoretically, Generation 0 will start as 12.5 AA, 25 Aa, and 12.5 aa, but because only whole numbers of individuals are allowed these will need to be rounded to the nearest whole number that still

produces a total of 50 (12, 26, 12). In cell **F35**, type the heading **Gen0**. **Highlight** the cell and copy it down to cell **F45**. In cell **G34**, type the heading **AA**, Cell **H34** type **Aa** and in **I34** type **aa**. Into cell **G35** type **12**, in **H35** type **26**, and in **I35** type **12**.

	F	G	H	I
35	Gen 0	12	26	12
36	Gen 1			
37	Gen 2			

4. Click **Calculate Now**.

5. Enter the results in cell **F9** into Generation 1 A (cell **G22**). Enter the numbers in **F2**, **G3**, and **H2** into **G36**, **H36**, and **I36**. Enter the number from **F17** into **A3** and click **Calculate Now** again.

6. Repeat this until you have ten generations of alleles.

7. This data can be used to develop explanations consistent with the scientific theory of **evolution**.

PART 3: Graphing the data

1. You can now produce a graph of the results. Highlight the cells **F20** to **H31** and click **Insert** then click on a **line graph** with markers.

2. The graph should automatically produce two lines for A and a. Give the graph appropriate titles and axes labels by clicking **Add Chart Element** (depending on your spreadsheet program) and selecting **title** and **axes** labels.

3. Repeat this for the AA, Aa, and aa individuals.

4. Print the graphs and staple them to this page.

1. (a) What happens to the frequency of the **a** alleles in rock pocket mice over ten generations when the aa genotype is totally excluded from passing its alleles to the next generation?

___

(b) What happens to the frequency of the **A** alleles in rock pocket mice over ten generations when the aa genotype is totally excluded from passing its alleles to the next generation?

___

(c) Why do your observations from (a) and (b) happen? ___

___

(d) What is the effect on the rock pocket mouse coat color phenotypes over time? (Assume AA and Aa produce the same dominant dark coat phenotype and aa is the recessive light coat phenotype).

___

(e) In summary, the rock pocket mice can be prolific breeders and produce excess offspring, but not all will survive. Evaluate how this situation has led to differential reproduction success:

___

189 What is a Species?

Key Question: How can we define a species?

▶ The **species** is the basic unit of taxonomy, how scientists classify and group organisms. The **biological species concept** (BSC) defines a species as a group of organisms capable of interbreeding to produce fertile offspring.

▶ There can be difficulties in applying the biological species concept (BSC) in practice as some closely related species are able to interbreed to produce fertile hybrids, e.g. species of *Canis*, which includes wolves, coyotes, domestic dogs, and dingoes. These breeds of domestic dog, above, may look and act differently, but they can all breed with each other and with the wolf, a different species, to produce viable offspring.

▶ The BSC is more successfully applied to animals than to plants and organisms that reproduce asexually. Plants hybridize easily and can reproduce vegetatively. For some, e.g. cotton and rice, first generation hybrids are fertile, but second generation hybrids are not.

▶ The BSC cannot be applied to extinct organisms. Some organisms, such as bacteria, can transfer genetic material to unrelated species. Increasingly, biologists are using DNA analysis to clarify relationships between closely related populations.

Another way of defining a species is by using the **phylogenetic species concept** (PSC). Phylogenetic species are defined by their evolutionary history. This is determined on the basis of shared, derived characteristics. These are characteristics that evolved in an ancestor and are present in all its descendants.

▶ The PSC defines a species as the smallest group that all share a derived character state. It is useful in paleontology, because biologists can compare both living and extinct organisms.

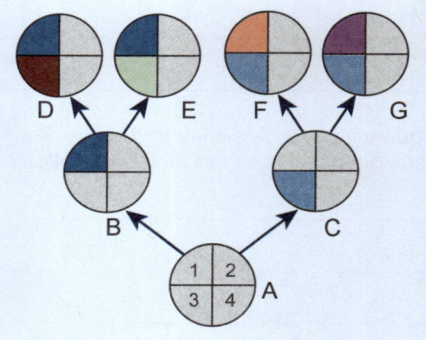

Species B and C are related to species A as they share three of four characteristics with it. However, they only share two characteristics with each other.

D and E share characteristics with B, while F and G share characteristics with C.

1. Why is it difficult to fully define a species in practice? _____

_____

_____

2. How is the biological species concept different from the phylogenetic species concept? _____

_____

_____

3. There often appear to be greater differences between different breeds of dog than there are between different species of the *Canis* genus. Why are dogs all considered one species?

_____

_____

_____

190 How Species Form

Key Question: How do isolating mechanisms lead to the formation of new species?

Species formation

- **Species** evolve in response to **selection pressures** from the environment. These may be naturally occurring or caused by humans.

- The diagram below represents a possible sequence for the **evolution** of two hypothetical species of butterfly from an ancestral **population**. As time progresses (from top to bottom of the diagram below), the amount of genetic difference between the populations increases, with each group becoming increasingly isolated from the other. **Gene flow** is reduced when populations are separated (see activity 193). Continual reduction in gene flow by isolating mechanisms may eventually lead to the formation of new species.

The Monarch butterfly, *Danaus plexippus*, is the official insect of Texas. Six different subspecies have been found around the world.

- The isolation of two **gene pools** from one another may begin with geographical barriers. This may be followed by isolating mechanisms that occur before the production of a **zygote**, e.g. behavioral changes, and isolating mechanisms that occur after a zygote is formed, e.g. hybrid sterility. As the two gene pools become increasingly isolated and different from each other, they are progressively labeled: population, race, and subspecies. Finally, they attain the status of separate species.

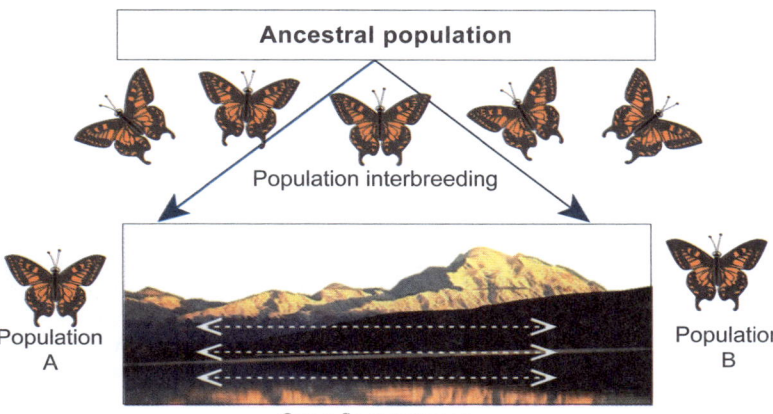

Ancestral population

Population interbreeding

Population A — Population B

Gene flow common

A species of butterfly lives on a grassland plateau scattered with boulders. During colder weather, some butterflies rest on the sun-heated boulders to absorb the heat, while others move to the lower altitude grassland to avoid the cold.

Continued mountain building raises the altitude of the plateau, separating two sub-populations of butterflies, one in the highlands, the other in the lowlands.

Race A — Race B

Gene flow uncommon

In the highlands, boulder-sitting butterflies (BSBs) do better than grass-sitting butterflies (GSBs). Darker BSBs have greater **fitness** than light BSBs. In the lowlands, the opposite is true. BSBs only mate on boulders with other BSBs. In the lowlands, light GSBs blend in with the grass and survive better than darker butterflies.

Subspecies A — Subspecies B

Gene flow very rare

Over time, the boulder-sitting butterflies occur only in the highlands and grass-sitting butterflies in the lowlands. Occasionally, wind brings members of the two populations together, but if they happen to mate (left), the hybrid offspring are not viable or have a much lower fitness.

Species A — Species B

Separate species

Eventually, gene flow between the separated populations stops, as **variation** between the populations increases. They fail to recognize each other as members of the same species.

Time →

Geographic isolation

▸ Geographic isolation describes the isolation of a species population (gene pool) by some kind of physical barrier, e.g. mountain range, water body, desert, or ice sheet. Geographic isolation is a frequent first step in the subsequent reproductive isolation of a species.

▸ An example of geographic isolation leading to speciation is the large variety of cichlid fish in the rift lakes of East Africa (right). Geological changes to the lake basins have been important in the increase of cichlid fish species.

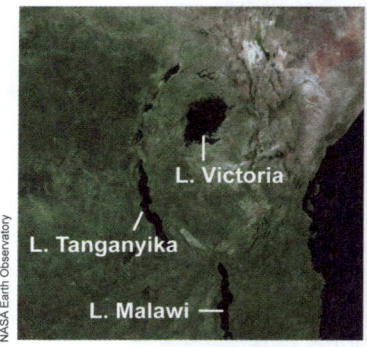

Reproductive isolating mechanisms

▸ Reproductive isolating mechanisms (RIMs) are reproductive barriers that are part of a species' biological physiology and, therefore, do not include geographic isolation.

▸ RIMS prevent interbreeding and, therefore, gene flow between species. Single barriers may not completely stop gene flow, so most species commonly have more than one type of barrier.

▸ Some RIMs (**prezygotic** - before fertilization) include differences in breeding season, differences in mating behaviors, and differences in physical copulatory structures.

▸ Other RIMS (**postzygotic** - after fertilization) occur after fertilization (formation of the zygote) has occurred. They are important in maintaining the integrity of closely related species. There are several different postzygotic mechanisms operating at different stages. The first prevents development of the zygote. Even if the zygote develops into a viable offspring there are further mechanisms to prevent long term viability. These include premature death or (more commonly) infertility.

The white-tailed antelope squirrel (left) and the Harris's antelope squirrel (right) in the southwestern United States and northern Mexico, are separated by the Grand Canyon (center) and have evolved to occupy different habitats.

Mules are a cross between a male donkey and a female horse. The donkey contributes 31 chromosomes while the horse contributes 32, making 63 chromosomes in the mule. This produces sterility in the mule as meiosis cannot produce gametes with an even number of chromosomes.

1. Identify some geographical barriers that could separate populations: _____

2. Why is a geographical barrier not considered a reproductive isolating mechanism? _____

3. Identify the two types of reproductive isolating mechanisms, and evaluate how each leads to speciation: _____

4. Why is more than one reproductive isolation barrier needed to completely isolate a species? _____

191 Patterns of Evolution

Key Question: What particular patterns of evolution might be seen in populations moving into a new environment?

- The diversification of one **species** into one or more separate species can follow particular patterns.
- Divergent **evolution** occurs when two species diverge from a common ancestor. Divergence is common in evolution and is responsible for evolutionary radiations. When divergent evolution involves the formation of a large number of species to occupy different niches, this is called an adaptive radiation. When unrelated species evolve similar forms as a result of similar **selection pressures**, it is called convergent evolution (convergence).

Divergent evolution
A lineage splits and evolves independently due to different selection pressures in different environments. Species may later occupy the same environment, e.g. black swan and mute swan.

Convergent evolution
Unrelated or distantly related species in similar environments and under similar selection pressures evolve similar features, e.g. streamlined swimming form in aquatic birds and mammals.

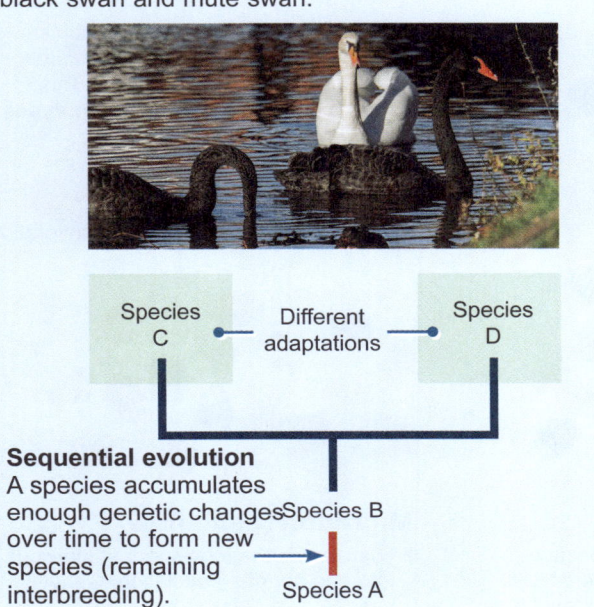

Sequential evolution
A species accumulates enough genetic changes over time to form new species (remaining interbreeding).

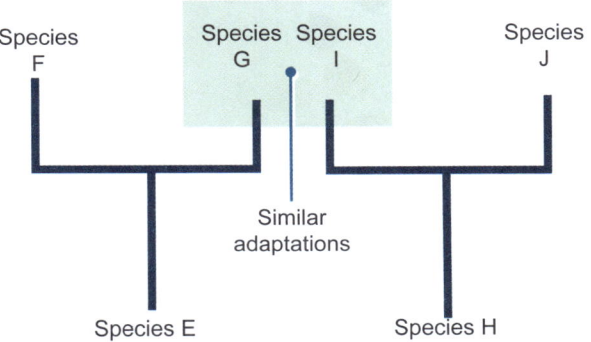

Adaptive radiation
- The earliest true mammals evolved about 195 million years ago, long before they underwent their major adaptive radiation some 65-50 million years ago. Most ancestors to the modern forms were very small (12 cm) with a similar form to modern shrews. Many were nocturnal and fed on insects and other invertebrates. *Megazostrodon* is a typical example. This animal is known from fossil remains in South Africa and first appeared in the Early Jurassic period, about 195 million years ago.
- Climatic change, as well as the extinction of the dinosaurs and their related forms, suddenly left many niches vacant for exploitation by such adaptable 'generalists'. All modern mammal orders developed relatively quickly.

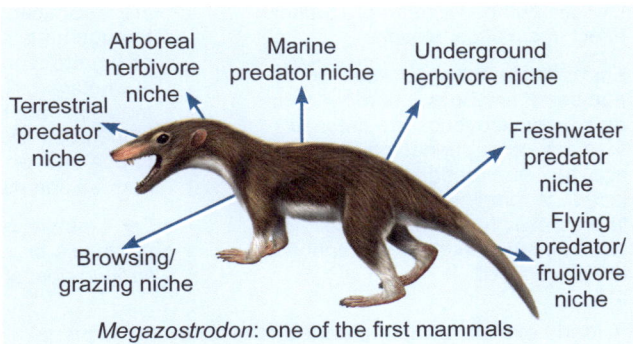

Megazostrodon: one of the first mammals

1. What is the difference between divergent evolution and adaptive radiation? _____

_____

2. What is the difference between divergence and sequential evolution? _____

_____

3. Penguins and dolphins have converged on a streamlined body form for moving through the water. What other groups of animals have also converged on this body shape?

_____

192 Evolutionary Mechanisms in Gene Pools

Key Question: Aside from natural selection, what other evolutionary mechanisms can change a gene pool over time?

▸ **Mutations**, **gene flow**, **genetic drift**, and **recombination** all contribute to changes in the genetic makeup of a **population**. Recall that the total genetic material of a population is its **gene pool**. Changes to a population's gene pool over time is known as **evolution**.

▸ Four microevolutionary processes can contribute to genetic change in populations: **1.** Mutation alters the genetic material and produces new genetic variations. **2.** Recombination 'shuffles' the combinations of parent **alleles** in each offspring. Both mutation and recombination occur during sexual reproduction. **3.** Migration creates gene flow as genetic material enters or leaves a population. **4.** Genetic drift alters the frequency of genetic variants randomly; its effects are due to chance events. Increasingly, genetic drift is being recognized as an important agent of change, especially in small, isolated populations, e.g. island colonizers.

Recombination

▸ Genetic **variation** arises through recombination of genetic material as a result of sexual reproduction. Offspring inherit a random assortment of alleles from both parents and show phenotypic variation from one another.

Genetic variation produces phenotypic variation e.g. color of ladybugs. This phenotypic variation is the raw material for **natural selection.** This ladybug population has five different **phenotypes**: black, dark brown, tan, brick red, and pale.

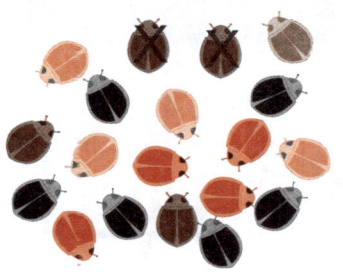

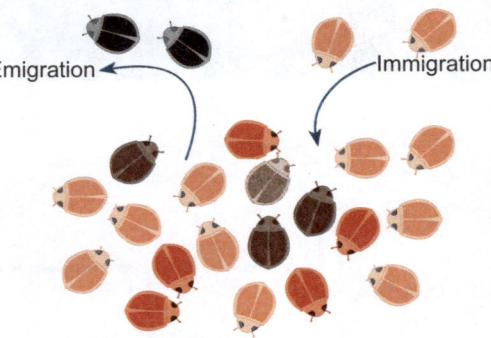

Emigration ← → Immigration

Genetic drift

▸ Genetic drift is the change in the frequency of specific genetic variations due to random events. Genetic drift has a more pronounced effect in small populations.

▸ For example, falling rocks kill a number of ladybugs, but more of the dark brown ladybugs are crushed than any other phenotype (image above). The proportion of dark brown ladybugs remaining in the population is drastically reduced, and their representation in the next generation is also reduced.

Mutation

▸ A population's gene pool is a collection of all its alleles. New alleles, and therefore phenotypes, are introduced to the gene pool through mutation. The gradual change to a gene pool over time is what we call evolution. Some mutations are beneficial and stay in the population, being passed on through generations. Others are harmful and may die out, over time.

▸ For example, a mutation could produce a larger ladybug with a new spotted phenotype (image above).

Migration (gene flow)

▸ Migration is the movement of individuals into and out of a population. Genetic variants can enter or leave the population via immigration or emigration. Gene flow tends to decrease the genetic differences between populations because genetic material is being exchanged.

▸ In the example above, several black ladybugs have left and some very pale lady birds have arrived, changing the proportion of remaining phenotypes in the population.

1. Clearly explain what is meant by the following terms:

   (a) Gene flow: _____

   _____

   _____

   (b) Genetic drift: _____

   _____

   _____

   (c) Sexual reproduction: _____

   _____

   _____

193 Gene Flow

Key Question: What is the effect of gene flow on the allele frequencies of a population, and how does population size affect its influence?

▶ **Gene flow** is the movement of genes into or out of a **population** (immigration and emigration). A population may gain or lose **alleles** through gene flow. Gene flow tends to reduce the differences between populations because the **gene pools** become more similar. The model below graphically represents the elements of gene flow.

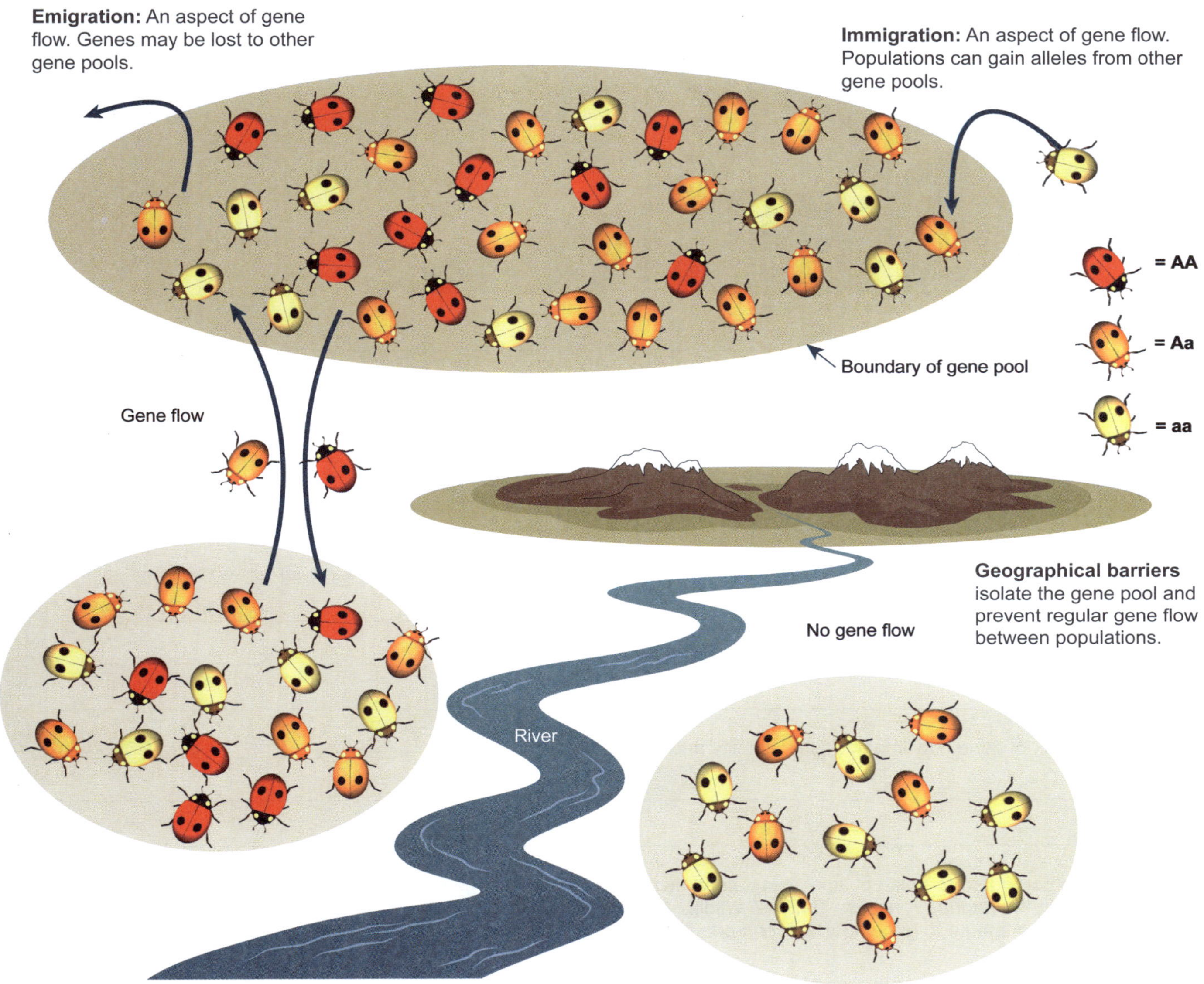

Emigration: An aspect of gene flow. Genes may be lost to other gene pools.

Immigration: An aspect of gene flow. Populations can gain alleles from other gene pools.

= AA

= Aa

= aa

Boundary of gene pool

Gene flow

Geographical barriers isolate the gene pool and prevent regular gene flow between populations.

No gene flow

River

Gene flow: Genes are exchanged with other gene pools as individuals move between them. Gene flow is a source of new genetic **variation** and tends to reduce differences between populations that have accumulated because of **natural selection** or **genetic drift**. Recall that lack of gene flow can lead to speciation (new **species** forming) in isolated populations, over time.

▶ The allele frequencies of large populations are more stable because there is a greater reservoir of variability and they are less affected by changes involving only a few individuals.

▶ Small populations have fewer alleles to begin with and so the severity and speed of changes in allele frequencies are greater when gene flow occurs.

▶ Endangered species with very low population numbers or restricted distributions, such as the Texas ocelot and Florida panther, may experience severe and rapid allele changes.

▶ Human intervention to save endangered populations with low diversity often involves artificially creating gene flow by introducing individuals from different populations, even similar subspecies. This has happened in the example of the Texas puma. Migratory corridors can also be created, such as those helping the Texas ocelot.

Lack of gene flow and the Texas ocelot

▸ The Texas ocelot, *Leopardus pardalis*, is an endangered wild cat that was once found across Texas. Presently, the only Texan populations of ocelots that remain are in two small areas in the deep south. Both small populations, in Laguna Atascosa National Wildlife Refuge and in Willacy County, are isolated from each other due to surrounding farm and human occupied land that acts as a geographical barrier and prevents gene flow. These isolated areas are also called 'genetic islands'.

▸ Small numbers in both populations, under 100 ocelots in total, and the prevention of gene flow, mean that the reduced diversity in the Texan ocelot gene pool is too small to maintain a healthy and viable wild population.

▸ Human-created migratory corridors, where connecting land is reforested and protected, can be used to increase gene flow between gene pools of isolated populations.

1. How can gene flow be defined? _____

2. In general, how is gene flow likely to positively impact the gene pool of a population? _____

3. Why are smaller populations more affected by a lack of gene flow? _____

4. How has lack of gene flow impacted the viability and 'fitness' of the Texas ocelot? _____

5. (a) Wildlife corridors are being built in Texas, such as the Tobin Land Bridge in San Antonio, and the under-highway tunnels in the Laguna Atascosa National Wildlife Refuge, home of a small Texas ocelot population. How do these corridors contribute to gene flow between populations?

   (b) The wildlife corridors can be expensive to build, but can contribute to a species survival. Use the **Biozone Resource Hub** and your own research to complete the cost-benefit analysis of the wildlife corridor as a means of species conservation of the Texas ocelot below:

Costs of wildlife corridor	Benefits of wildlife corridor

Texas puma, *Puma concolor*

Gene flow from the Texas puma saves the Florida panther

- American pumas, panthers, and mountain lions all belong to the same species, *Puma concolor*. A number of subspecies, including the Florida panther, can be found in the Americas.
- The Florida panther population was reduced to fewer than 30 wild individuals in the mid-1990s. The low genetic diversity of the population's gene pool resulted in an accumulation of genetic disorders, including heart failure, and a lowered immunity to diseases and parasites.
- Scientists enabled gene flow from a closely related subspecies, the Texan Puma, in the form of eight female panthers.
- The addition of genes from the Texan puma into the Florida panther gene pool improved the population's health. The incidence of heart defects in the panthers dropped from 21% to 6% over three decades. Fertility also increased, and visual defects in the population also declined.

6. Why do rare, recessive mutations tend to have more of a negative impact in the gene pools of small populations of species with limited or no gene flow?

7. What features are likely to create a geographical barrier to gene flow between populations?

8. Speciation can result from a long-term lack of gene flow. Why might this be?

9. (a) A subspecies of *Puma concolor*, the Texas puma, was used to save the Florida panther. What is a subspecies, and how can it be distinguished from a species?

(b) The solution that was designed by scientists to increase gene flow in the Florida panther could also increase the risk of the subspecies disappearing altogether. Evaluate the introduction of genes from the Texas Puma into the Florida panther, while also considering the consequences of not introducing new genes:

10. Texas Longhorns originated from imported African and Portuguese cattle and were allowed to roam freely for nearly 400 years around uncultivated Texan wildland. How did this contribute to the breed's ability to tolerate drought, and provide greater immunity to tick fever, which negatively effects many other cattle breeds?

194 Genetic Drift

Key Question: How does the evolutionary mechanism of genetic drift, including the founder and bottleneck effect, create change in the gene pool of a population?

▸ **Genetic drift** is the change in **allele** frequencies in a **gene pool** due to random (chance) events. It may result in the loss (or fixation) of any allele, including beneficial ones, within a gene pool. Genetic drift has a greater effect when a **population** is small and it is an important agent of genetic change. In natural systems, small populations are often the result of the founder effect or a genetic bottleneck (see following activities). Both of these mechanisms are well documented in natural populations.

How does genetic drift reduce variation in populations?

▸ The change in allele frequencies within a population through genetic drift can be illustrated using the random sampling of marbles from a jar. The different alleles are represented by blue and orange marbles.

▸ The starting population contains an equal number of blue and orange marbles (10 of each). Random mating is represented by selecting 10 marbles at random. Twenty marbles representing the new allele proportions are placed into a new jar to represent the second generation, and the process is repeated for subsequent generations.

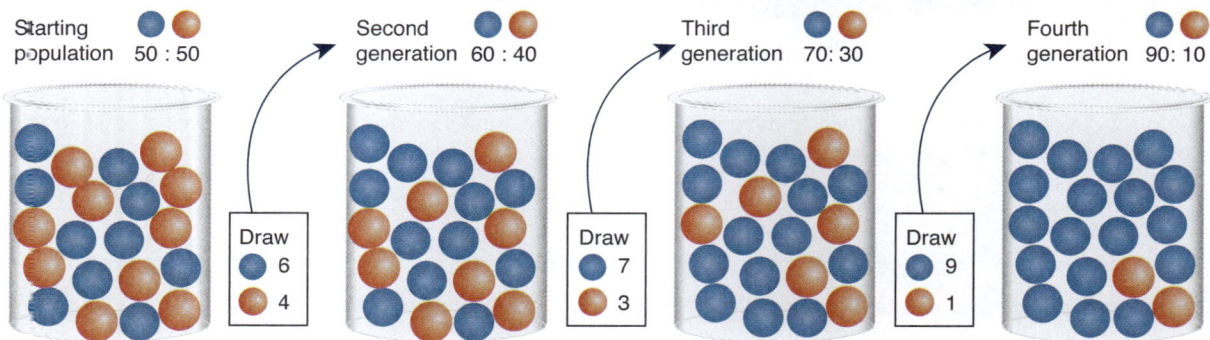

▸ In the example above, the orange marbles are becoming less frequent within the population and the amount of genetic **variation** within the population is reducing. Unless the proportion of orange marbles increases, it will eventually be lost from the population altogether and the allele for the blue marble becomes fixed (the only variant).

▸ If environmental conditions change so that the blue allele becomes detrimental, the population may become extinct (the potentially adaptive orange allele has been lost).

▸ In small populations, genetic drift can be a major agent of rapid change because the loss of any one individual represents a greater proportion of the total population.

The graph on the right shows a simulation of the effect of genetic drift on populations of various sizes.

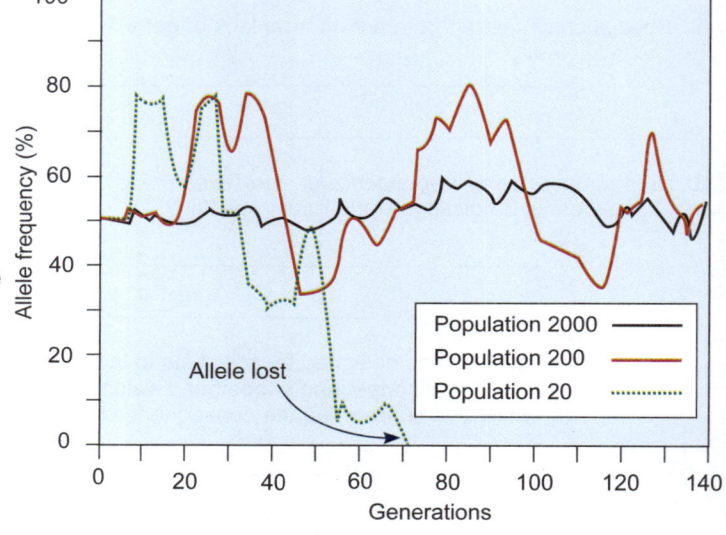

1. Explain why the large population shows less allele frequency fluctuation: _____

   _____

2. Use the diagram above to predict the shape of the line if a population of 200,000 was used in the simulation: _____

3. (a) Does genetic drift in a small population increase or decrease the number of heterozygotes? _____

   (b) How could this affect a population's long term viability? _____

   _____

195 The Founder Effect

Key Question: How does the founder effect result in differences between the proportions of alleles found in a parent and founder population?

▶ If a small number of individuals from a large **population** becomes isolated from the original, parent population, their sample of **alleles** is unlikely to be in the same proportion as the alleles of the parent population. This is called the **founder effect**. It can result in the colonizing (founder) population evolving in a different direction from the parent population. This is particularly so if the founder population comes under different **selection pressures** in a new environment and is missing alleles that are present in the original parent population.

Mainland population — Some individuals from the mainland population are carried at random to the offshore island by natural forces such as strong winds.

Island population

Founders can become isolated by migration, e.g. to an island, but also by geological events, such as the formation of mountains or straits.

This population may not have the same allele frequencies as the mainland population.

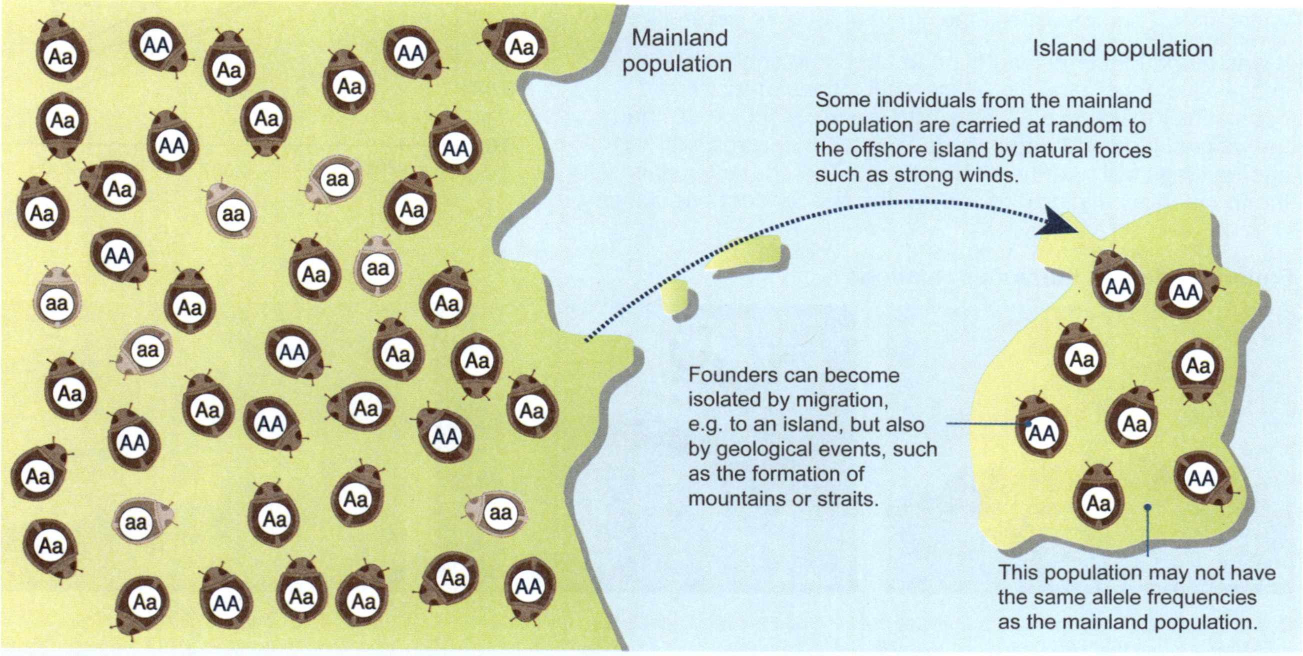

Mainland population

	Allele frequencies		Phenotype frequencies	
	Actual numbers	Calculate %	Dark	Pale
Allele A				
Allele a				
Total				

AA = Dark, aa = Pale, Aa = Dark

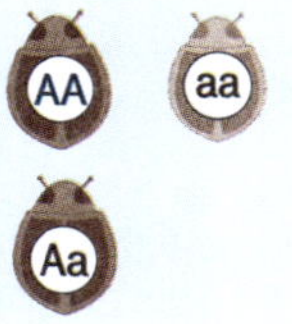

Colonizing island population

	Allele frequencies		Phenotype frequencies	
	Actual numbers	Calculate %	Dark	Pale
Allele A				
Allele a				
Total				

AA = Dark, aa = Pale, Aa = Dark

NEED HELP? See Activity 252

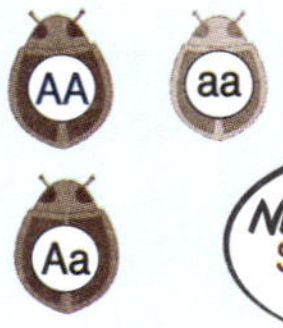

1. Compare the mainland population to the population which ended up on the island (use the spaces in the tables above):
   (a) Count the phenotype numbers for the two populations (i.e. the number of dark and pale beetles).
   (b) Count the allele numbers for the two populations: the number of dominant alleles (A) and recessive alleles (a). Calculate these as a percentage of the total number of alleles for each population.

2. How are the allele frequencies of the two populations different? _____

3. (a) What changes are likely when a founder population is isolated in a new environment? _____

   (b) What factors might influence the end result or the speed of the changes? _____

Founder effect in brown anole lizards

▶ In 2004, Hurricane Francis wiped out the brown anole lizard (*Anolis sagrei*) populations on several cays (small sandy islands) around the Bahamas. Scientists used this as a chance to study the founder effect. They took pairs of lizards from the mainland and placed them on different cays.

▶ The vegetation on the cays is much smaller and scrub-like compared to the much larger trees of the mainland. On the mainland, scientists noted that the lizards use their long limbs to climb around the trees. They hypothesized that the populations isolated on the cays would eventually evolve shorter limbs to adapt to the scrub-like, less supportive vegetation. They measured the limb length over several years.

▶ It was found that limb length indeed became shorter over successive generations in all the populations. Importantly, populations founded by lizards with the longest legs still had the longest legs and populations founded by lizards with the shortest legs still had the shortest legs. The characteristics of the founder populations influenced the descendant populations.

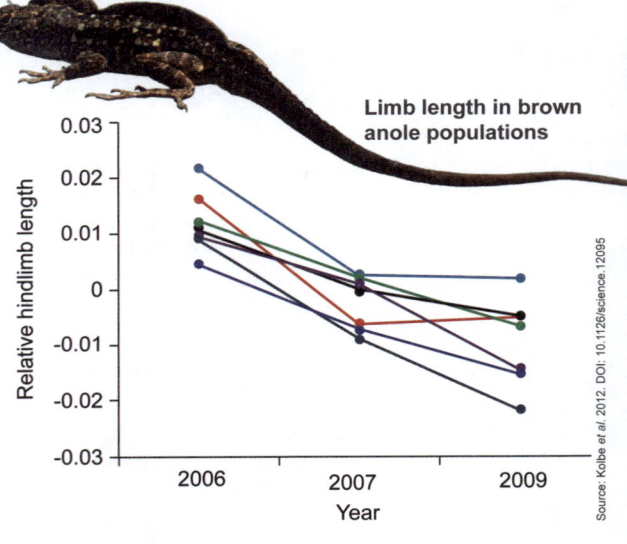

Limb length in brown anole populations

Source: Kolbe et al. 2012. DOI: 10.1126/science.12095

Founder effect in human populations

Pitcairn Island

Norfolk Island
thinboyfatter CC 2.0

Tristan da Cunha
Brian Gratwicke CC 2.0

Due to the frequent episodes of human migration around the world there are many instances of the founder effect in human populations. In 1790, nine mutineers from the ship HMS Bounty along with six Tahitian men, eleven Tahitian women, and a baby girl settled on Pitcairn island. The population eventually grew to 193 by 1856.

In 1856 the entire population of Pitcairn Island resettled on Norfolk Island after it was decided Pitcairn was over populated. The effect of this can still be seen in genetic studies of the Norfolk Island population. In 1859, 16 people returned to Pitcairn Island and founded a new population, that eventually reached 250 people by 1936. The population is now around 50.

Tristan da Cunha sits 2,400 km from Africa and more than 3,500 km from South America. The current settlement was founded by the English in 1817. In 1961, a genetic study traced 14% of all genes in the population of 300 to one founding couple. Around 47% of the population are now affected by asthma. From the 15 original settlers at least three had asthma.

4. (a) Why were conditions good for setting up an experiment on the founder effect on the cays around the Bahamas?

(b) Describe how the founder effect was demonstrated in the brown anole lizards: _____

5. (a) The rate of asthma in the UK is about 8%. Calculate the rate of asthma in the original Tristan da Cunha settlers:

(b) How has genetic drift affected the current population of Tristan da Cuhna? _____

196 Genetic Bottlenecks

Key Question: How do genetic bottlenecks affect the genetic diversity of populations?

- Populations may sometimes be reduced to low numbers by predation, disease, or periods of climatic change. These large-scale reductions are called genetic (or population) bottlenecks. The sudden **population** decline is not necessarily selective and it may affect all phenotypes equally. Large scale catastrophic events, such as fire or volcanic eruptions, are examples of such non-selective events. Affected populations may later recover, having squeezed through a 'bottleneck' of low numbers. However, although a population's numbers may recover, its genetic diversity often does not.

- The models below illustrate how population numbers may be reduced as a result of a catastrophic event. Following such an event, the **gene pool** of the surviving remnant population may be markedly different from that of the original gene pool. The small population may return to previous levels, but with a reduced genetic diversity.

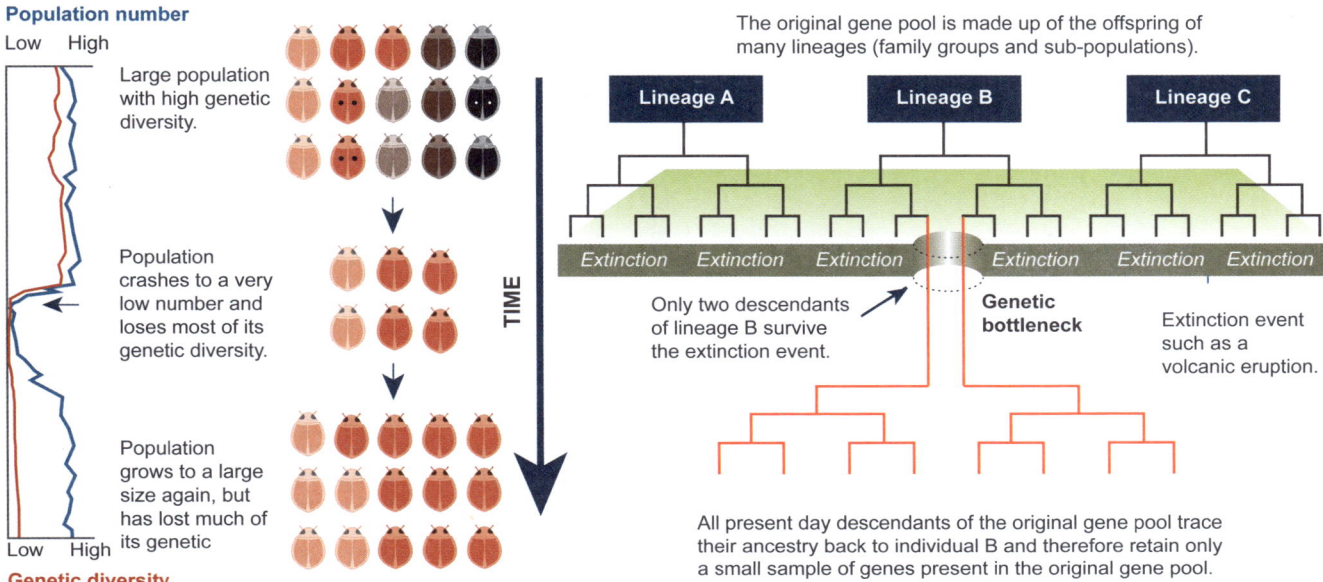

Bottlenecks and low allelic diversity in red wolves

The American red wolf, *Canis rufus*, was once found in large populations that stretched across Eastern United States. It is smaller than the gray wolf and has been impacted to a greater extent by hunting and loss of habitat.

In the 1980s, the last remaining wild individuals were taken into captivity in an effort to begin rebuilding numbers. The Fossil Rim Wildlife Center in Texas is having success in breeding pups as part of a **species** survival plan. However, using such low numbers for breeding has resulted in a genetic bottleneck and the new breeding population has extremely low genetic diversity. Inbreeding, due to limited fertile adults, has also compounded the problem.

Coyotes can crossbreed with red wolves and produce viable offspring. Some Texan coyotes have had up to 60% of their genetic material identified as coming from red wolves. These genes have been called 'ghost genes', so named as they were once presumed lost when the red wolves died. Scientists hope to breed the ghost genes from the coyotes back into the red wolf population to restore some of its lost genetic diversity.

Red wolf, *Canis rufus*

1. Endangered species are often subjected to genetic bottlenecks. Explain how genetic bottlenecks affect the ability of a population of an endangered species to recover from its plight:

2. What has been the genetic consequence of bottleneck events in the Texas red wolf population?

197 Mutations and the Gene Pool

Key Question: How does the evolutionary mechanism of mutation affect the gene pool in a population?

▸ **Mutations** are the source of all new **alleles**. Therefore, mutations can change the frequency of existing alleles by competing with them.

▸ A mutation that is beneficial to the organism is very rare. However, beneficial mutations can give the organism a greatly increased survival advantage, such as hiding from predators or finding food, so that the new allele can increase in frequency in the **gene pool** very quickly.

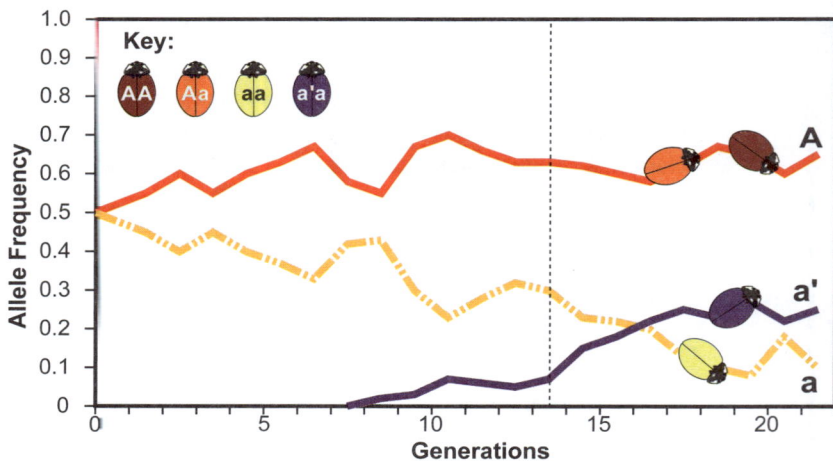

▸ Some mutations are so beneficial, like that producing the 'anti-freeze' protein found in species of Antarctic ice-fish (above), that they allow species to occupy completely new habitats.

▸ Typically, any mutation is likely to be small in impact. Multiple mutations result in cumulative changes to the gene pool over many generations.

▸ Mutations that are non-beneficial to survival are rapidly removed from the gene pool, and known as deleterious.

▸ Mutations that do not cause a consequential change in phenotype, are known as silent mutations. **Selection pressures** do not act on them to remove mutant alleles any faster than other alleles in the gene pool. They are useful for understanding common ancestry between **species**, and for DNA identification.

▸ In the graph above, a random mutation creates a new recessive allele: **a'** and causes a purple **phenotype.**

▸ The frequency of this new allele increases when environmental conditions change, giving it a competitive advantage over the other recessive allele: **a**

▸ The frequency of **A** remains relatively stable.

▸ Eventually, the **a** allele may be lost from the **population** altogether.

1. What is a beneficial mutation? _____
_____

2. From the graph above, describe how the gene pool appear (name alleles) after another 10 generations has passed:
_____
_____
_____

3. Paraphrase an explanation for why beneficial mutations are likely to rapidly increase in a gene pool: _____
_____
_____

4. Either individually, or in pairs, research an animal or plant, preferably a local example, that has had one or more beneficial mutations added to the gene pool in the past. Describe the species, the type of beneficial mutation, and how intervention has increased the organism's 'fitness' or reproductive success. Create a report, either written, or digital. Attach the report here, or provide a link, if digital. Include text, images, and a list of information sources.
_____
_____

198 Genetic Recombination and the Gene Pool

Key Question: How does the evolutionary mechanism of genetic recombination affect the gene pool in a population?

Recombination increases genotype diversity

▸ Sexual reproduction results in **allele** combinations in offspring that are different from their parents. You will recall that each gene consists of two alleles, each on a separate homologous chromosome. Each offspring will receive one chromosome from each parent during fertilization via the maternal (female) or paternal (male) gametes. The alleles are separated in a random shuffling process, as homologous pairs of chromosomes "line up" during meiosis. Gamete formation is the result. This process is modeled below.

▸ Furthermore, sections of genes on a chromosome can be 'swapped' during the meiotic **crossing over** phase with close by homologous chromosomes, introducing even more **variation**.

▸ The random meiotic separation of chromosomes, along with shuffling of genetic material by exchanging equivalent sections of maternal and paternal chromosomes, followed by random pairing of chromosomes within the gametes during fertilization is known as genetic **recombination**. This process increases the variation in a **gene pool**.

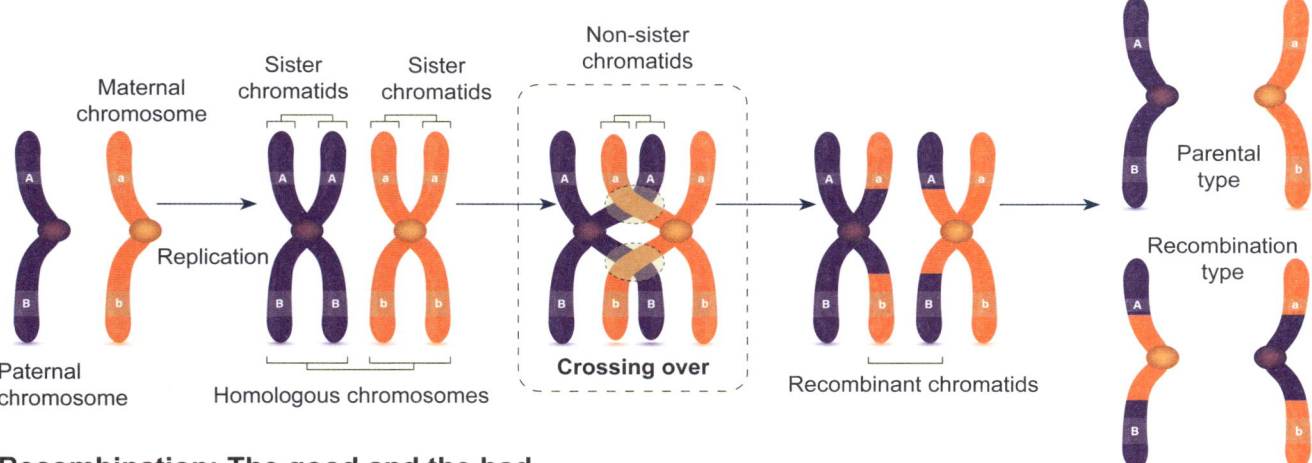

Recombination: The good and the bad

▸ As the **phenotype** is determined by the **genotype**, offspring will appear slightly different from the parents, and from each other. Genetic recombination is an evolutionary mechanism that creates variation in a **population** by creating new genetic combinations. This increases diversity in a gene pool.

▸ However, due to the random nature of recombination, the favorable combinations of genes that enhanced survival and reproduction of the parents can be lost in offspring.

▸ Recombination rates vary between groups of organisms. The rate is higher in fungi and invertebrates, lower in vertebrates, and lower again in some types of plant, such as conifers, where linkage of genes reduces variation.

1. Summarize how genetic recombination is able to add diversity into a gene pool: _____
   _____
   _____

2. How does increased genetic diversity affect the process of microevolution (changes in the gene pool over time)?
   _____
   _____
   _____

3. (a) Some gene combinations on chromosomes are more resistant to recombination than others, i.e. there is more chance that they are passed onto offspring together. Why might this be an advantage?
   _____
   _____

   (b) What two or more combined features (phenotypes) do you share with your family, and/or brothers and sisters?
   _____

199 How does an Elephant Lose its Tusks? Revisited

Content Anchor Revisited: How does poaching cause African elephants to be born without tusks?

1. Critique this statement: an elephant is more likely to lose its tusks if poaching is occurring?

   _____
   _____

2. Tuskless females will pass the tuskless genes to 50% of their daughters. The tuskless gene is carried on the X (female) chromosome. Why is the increase in tuskless African elephants over time an example of natural selection?

   _____
   _____
   _____
   _____

3. (a) A long term study of a Mozambique population of African elephants showed changing proportions of 2, 1, and no tusk elephants. What most likely caused an increase in tuskless elephants from 1970-2000?

   Population size vs presence of tusks in elephants

   Number of tusks: 2, 1, 0

   (n = 54) 18.5%, 9.3%, 68.5%
   no data available
   (n = 108) 50.9%, 8.3%, 40.7%
   (n = 91) 33%, 7.7%, 58.4%

   Data from Campbell-Station et al. (2021)

   _____
   _____
   _____
   _____
   _____
   _____

   (b) Three factors are relevant to observations seen in the study of the Mozambique African elephant population:
   i. The presence of the tuskless gene is fatal to males, hence the lack of tuskless male elephants in the population.
   ii. Tusks are used by elephants to perform many tasks and are an important evolutionary adaptation.
   iii. Conservation efforts were made after 2000 that reduced poaching of the elephants.
   Relate the factors above to the changes in the proportion of tuskless elephants in the Mozambique population since 2000, and link them to the process of evolution by natural selection:

   _____
   _____
   _____
   _____
   _____
   _____
   _____
   _____

200 Summing Up

Read each question carefully. Place a tick or cross in the box beside the best answer to the question from the four answer choices provided.

1. The mean birth weight for babies born in the US is 3.5 kg but the range of typical birth weights is quite narrow. This is an example of:

   ☐ (a) Directional selection
   ☐ (b) Disruptive selection
   ☐ (c) Stabilizing selection
   ☐ (d) None of the above

2. The statements (A)-(D) below describe scenarios in hypothetical populations:

   A A forest fire burns through an area containing black and gray squirrels. It kills 70% of the black squirrels and 30% of the gray squirrels.
   B Dark colored mice are able to avoid being eaten by owls more easily than light colored mice.
   C Deer from a population become isolated from another population because of a huge slip blocking a hilltop pass.
   D A plant nursery is infected with a fungal infection that affects the leaves of plants. Plants with thick leaf cuticles are more likely to survive.

   In which of the scenarios is genetic drift likely to have a significant effect on allele frequencies?

   ☐ (a) A
   ☐ (b) B and D
   ☐ (c) A and C
   ☐ (d) All of the above

3. Humans, birds, and whales all use limbs for locomotion. The skeletal structures are pictured below.

   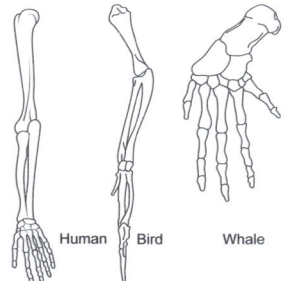

   The structures pictured above arose through:

   ☐ (a) Mutation
   ☐ (b) Extinction
   ☐ (c) Convergent evolution
   ☐ (d) Adaptive radiation

4. The four important concepts of Darwin's theory of natural selection are:

   ☐ (a) Finite resources, overpopulation, variation, inheritance
   ☐ (b) Overproduction, variation, artificial selection, inheritance
   ☐ (c) Overproduction, natural selection, increased fitness, inheritance
   ☐ (d) Finite resources, variation, natural selection, inheritance

5. The image below shows a representative collection of peppered moth, *Biston betularia,* from 1850 to 1950. The image shows how the color of moths has changed over time. This is most probably the result of:

   ☐ (a) Stabilizing selection
   ☐ (b) Artificial selections
   ☐ (c) Directional selection
   ☐ (d) Disruptive selection

6. Which of the following is likely to have a greater effect on the allele frequencies in a small population than the allele frequencies of a large population:

   ☐ (a) Genetic drift
   ☐ (b) Natural selection
   ☐ (c) Gene flow
   ☐ (d) All of the above

7. Red wolves and coyotes are able to breed with each other and produce living hybrid offspring. What might be the best method of defining them both as separate species?

   ☐ (a) Common ancestry
   ☐ (b) Biological species concept
   ☐ (c) Genetic mapping
   ☐ (d) Phylogenetic species concept

8. What type of evolutionary pattern do these two unrelated desert species of reptiles exhibit?

   Greater short-horned lizard (*Phrynosoma hernandesi*). North America.

   Thorny devil (*Moloch horridus*). Australia.

   ☐ (a) Common ancestry
   ☐ (b) Convergent evolution
   ☐ (c) Divergent evolution
   ☐ (d) Natural selection

9. Test your vocabulary by matching each term to its correct definition, as identified by writing the letter in the correct box.

(i) Gene pool [ ]

(ii) Genetic drift [ ]

(iii) Variation [ ]

(iv) Mutation [ ]

(v) Natural selection [ ]

A A change in the base sequence of DNA; the ultimate source of new alleles.

B The differences between individuals in a population as a result of genes and environment.

C Change in allele frequencies in a population due to random (chance) events.

D The process by which favorable heritable traits become more common in successive generations.

E The collective group of genes in a population.

10. The graph on the right shows the difference in escape acceleration in two populations of mosquitofish (*Gambusia*) in the Bahamas. One population is subject to heavy predation, the other is not. Predict how this difference might lead to speciation over time if the populations remain separate:

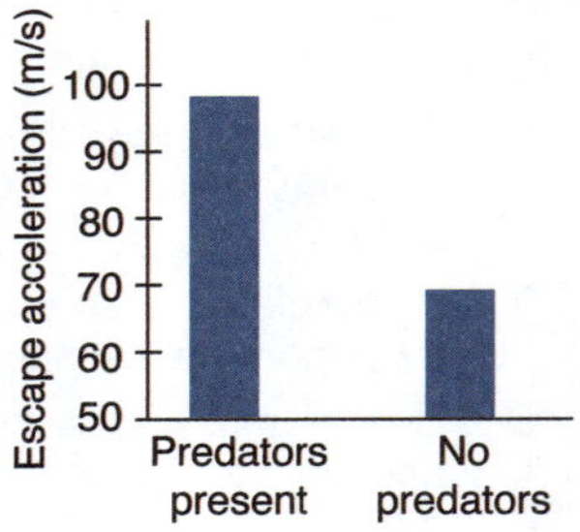

___

11. (a) Describe the features of stabilizing selection: ___

(b) Using an example, explain what might cause selection to shift from stabilizing to directional: ___

12. Until 1992, the Illinois prairie chicken was destined for extinction. The population had fallen from millions to 25,000 in 1933, and then to 50 in 1992. The dramatic decline in the population in such a short time resulted in a huge loss of genetic diversity, which led to inbreeding. This resulted in a decrease in fertility and an ever-decreasing number of eggs hatching successfully (right). In 1992, a translocation program began, bringing in 271 birds from Kansas and Nebraska. Use the data to discuss the trend seen, linking to changes in the gene pool:

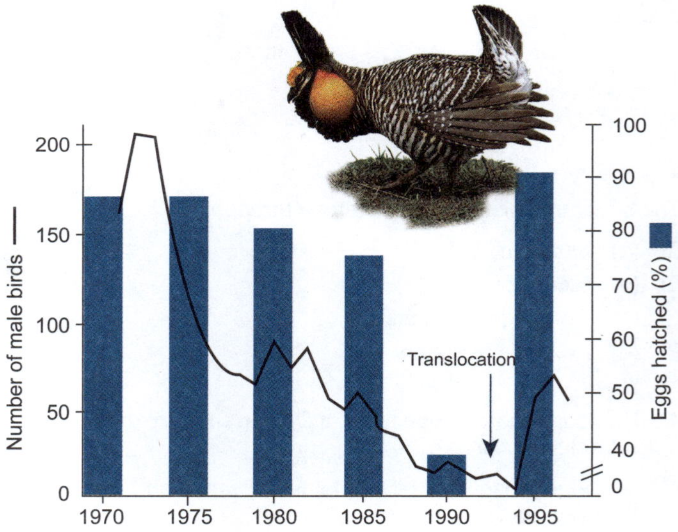

___

13. The model below shows the divergence of an ancestral population into two new species. In the boxes below, describe the level of gene flow, presence of reproductive isolating mechanisms, and its effect for each of the labelled points (a-c):

(a)

(b)

(c)

14. The two plants shown right are unrelated. The left hand image shows a cactus from North America, while the right hand image shows a Euphorbia from Africa. Both these plants live in deserts.

   (a) Identify the pattern of evolution displayed by these plants:

   _____

   _____

   (b) Describe the environments associated with the adaptations.

   _____

   _____

15. Compare and contrast the role of genetic drift and natural selection in changing the genetic makeup of a population:

_____
_____
_____
_____
_____
_____
_____
_____

16. Fill in the missing gaps using the word list provided (note there are more words provided than will be needed):
    genetic recombination, adaptive advantage, DNA, beneficial, harmful, variation, alleles, eliminated, crossing-over,

   Mutations are changes in an organism's _____ and they are the source of new _____ in a

   population. Most mutations are _____ and so are _____ from the population.

   Inherited beneficial mutations provide an _____ so are retained in the population. Genetic

   recombination is another evolutionary mechanism also increases _____ into a gene pool. One way it

   does this is by a process called _____, separating groups of alleles from each other.

CHAPTER 9
Ecological Interactions

TEKS
Scientific and Engineering Practices

B.1: Investigation and Inquiry
1.B 1.C 1.D 1.E 1.F 1.G

B.2: Data and Patterns
2.A 2.B 2.C 2.D

B.3: Communicating in Science
3.A 3.B 3.C

B.4: Science as a Human Endeavor
4.A 4.B

TEKS
Science Concepts

B13.A investigate and evaluate how ecological relationships, including predation, parasitism, commensalism, mutualism, and competition, influence ecosystem stability

B13.B analyze how ecosystem stability is affected by disruptions to the cycling of matter and flow of energy through trophic levels using models

B13.C explain the significance of the carbon and nitrogen cycles to ecosystem stability and analyze the consequences of disrupting these cycles

B13.D explain how environmental change, including change due to human activity, affects biodiversity and analyze how changes in biodiversity impact ecosystem stability

Learning Outcomes
I know I have achieved this when I can:

	Learning Outcome	Activity number
☐	Define and use the following terms in context: ecosystem, biotic factor, abiotic factor, habitat, and niche.	202-204
☐	Link ecosystem stability and resilience to high biodiversity in ecosystems, using case studies and data.	205-207
☐	Identify the type of species interactions from given examples, and describe whether each species benefits, is harmed, or is not impacted by the relationship.	208
☐	Investigate, using provided case studies and data, how each of the following type of ecological relationship can influence ecosystem stability: predator-prey, competition, parasitism, commensalism, and mutualism.	209-218
☐	Evaluate how predator-prey, competition, parasitism, commensalism, and mutualism can influence ecosystem stability.	209-218
☐	Relate the sequence of energy and matter moving through an ecosystem to the trophic levels of a species.	219-221
☐	Analyze the consequences of disruption to matter cycles that impact ecosystem stability.	223
☐	Calculate energy efficiency in an ecosystem, applying 10% rule.	224
☐	Evaluate the use of ecological pyramids to represent the amount of energy, organisms, or biomass in each trophic level.	225
☐	Collect and analyze data from a virtual river simulation to construct biomass and energy pyramids.	226
☐	Analyze the changes in ecosystem stability due to disruption to biomass and energy flow.	227
☐	Use and develop models to demonstrate the importance of functioning carbon and nitrogen cycles to ecosystem stability.	228-234
☐	Analyze consequences of carbon cycle and nitrogen cycle disruptions, including climate change, ocean acidification, and aquatic eutrophication.	231, 234
☐	Investigate the link between increased atmospheric carbon, in the form of carbon dioxide, and increased ocean acidification.	232
☐	Analyze the effect of permanent ecosystem change, either natural or human-caused, on biodiversity.	235
☐	Calculate biodiversity values in provided examples using the Simpson's Index of Diversity method.	236
☐	Link the loss of keystone species to ecosystem destabilization.	237
☐	Analyze and investigate, using group simulations, the impact of human activity on biodiversity.	238
☐	Explain and model how marine harvesting, specifically overfishing, has led to a decrease in fish biodiversity, and marine ecosystem stability in our oceans.	239
☐	Debate the pros and cons of prioritizing protection of endangered forest species to maintain biodiversity.	240
☐	Research the impacts of human activity on a local species.	241
☐	Explain the impacts of damming on ecosystem biodiversity.	242

RESOURCE HUB
bit.ly/3KXYlnw

ELPS English Language Proficiency Standards

Speaking

A Mammoth Task. Working with a partner, read about mammoth ecosystems and then respond to the questions. Read each question aloud and agree on what is being asked. Then work with your partner to find evidence in the text to support an answer. Work together to write the answer in one or two sentences. When a question calls for you to make inferences, recall what you learned in previous chapters.

348

Reading

The Ecological Niche. Boldface terms are key vocabulary that you need to learn and understand. Often, clues to the term's meanings can be found in the same or in nearby sentences. For instance, the first point on this page includes the boldface term niche. The rest of the sentence tells you that the term refers to the organism's position in its environment. The following points refine the definition. Note that e.g. is the abbreviation for the Latin phrase *exempli gratia*, meaning "for example."

351

Reading

A Case Study in Ecosystem Resilience. Both the text at the top of the page and the graph tell the same story. Read the text, then trace the cycles on the chart. Notice that the lines on the chart show the rise and fall of different populations. The solid black line represents the spruce budworm population. Note also that each number in the graph corresponds with a numbered explanation. For instance, number 1 in the graph marks a high point in the spruce budworm population. Work with a partner to summarize the cause-and-effect relationships described by the graphic.

355

Listening

Ocean Acidification. Collaborate with a lab partner to complete Investigation 9.5. First, make sure you both understand the purpose of the investigation. Then, discuss with your partner the best way to carry out the investigation. During your discussion, confirm understanding of what your partner says. Take notes to describe each step in the process, as well as you and your final partner's observations. Finally, discuss your responses to questions 3(a-c) and 4 with your partner before answering. Agree on the answers and write them down.

393

Reading

Human Activity and Biodiversity. With a partner, read the bulleted sentences at the top of page 404. Note the boldfaced key terms: biodiversity, habitats, climate change. Discuss with your partner the meaning of these terms and write brief definitions in your own words. If needed, refer to the glossary. As you read the rest of this section and perform the investigation, use the definitions you wrote to aid comprehension. Check understanding with peers and your teacher as often as needed.

404

201 A Mammoth Task

Content Anchor: How could bringing back the woolly mammoth help restore the stability of a lost ecosystem?

Mammoth ecosystems

▸ Woolly mammoths belonged to the same family as modern Asian and African elephants. They lived on Earth from about 300,000 years ago to around 10,000 years ago. Alongside other large, grazing herbivores, they occupied an **ecosystem** of treeless grasslands. In winter, they scraped off snow with their tusks, grazing and trampling the grassland. This maintained the landscape, keeping the ground compacted and frozen and preventing trees and shrubs from establishing.

▸ Evidence suggests human hunting activity, in conjunction with climatic warming may have contributed to the extinction of the large grazers, including mammoths, in these grasslands. Without the trampling effect of the grazers, the ground grew softer and other plants were able to establish themselves. What was formerly grassland changed such that small shrubs and trees grew. The ground began to thaw, melting the permafrost cover, and changing the ecosystem.

1. What was the effect of large grazers on the tundra?

   _____

   _____

   _____

2. Frozen tundra prevented millions of tonnes of carbon, contained within trapped methane molecules, being released back into the atmosphere. What would be the effect of the tundra melting and how might this impact ecological stability?

   _____

   _____

   _____

Frozen tundra, much as it would have appeared for mammoths   Modern, thawing tundra

3. A conservation group is attempting to recreate the mammoth's ecosystem in Siberia, Russia. They have called it Pleistocene Park. Over 100 species of animal have been brought in, including bison, reindeer, and moose. There is also future hope that eventually, they may be able to bring in a genetically modified "mammoth".

   (a) Why must other species be introduced alongside the mammoth? _____

   _____

   _____

   _____

   (b) How might bringing back the mammoth affect the modern tundra? What effect might this have on climate change?

   _____

   _____

   _____

   _____

202 Components of an Ecosystem

Key Question: What are the abiotic and biotic components that comprise an ecosystem?

▸ **Ecosystems** are natural units made up of all the living organisms (**biotic factors**) and the physical (**abiotic factors**) in an area.
▸ Abiotic factors are non-living physical factors including soil, water, atmosphere, temperature, and sunlight (SWATS).
▸ Biotic factors are all the living organisms in an area, including plants, animals, fungi, protists, and microorganisms.
▸ The interactions of living organisms with each other and with the physical environment help determine the features of an ecosystem and maintain its stability. The components of an ecosystem are linked to each other, and to other ecosystems, through nutrient cycles and energy flows.

BIOTIC FACTORS	ABIOTIC FACTORS		
The living organisms in the environment, including their interactions, e.g. as competitors, predators, or symbionts. ▸ Plants ▸ Animals ▸ Microorganisms ▸ Fungi ▸ Protists, e.g. algae, protozoa	**Atmosphere (air)** ▸ Wind speed ▸ Wind direction ▸ Humidity ▸ Light intensity/quality ▸ Precipitation ▸ Temperature	**Hydrosphere (water)** ▸ Dissolved nutrients ▸ pH ▸ Salinity ▸ Dissolved oxygen ▸ Precipitation ▸ Temperature	**Geosphere (rock/soil)** ▸ Nutrient availability ▸ Soil moisture ▸ pH ▸ Composition ▸ Temperature ▸ Depth

1. (a) What are biotic factors? _____
   _____

   (b) Give an example of a biotic factor: _____

2. (a) What is an abiotic factor? _____
   _____

   (b) Give an example of an abiotic factor: _____

3. How are the components of an ecosystem linked? _____
   _____

4. How do the particular characteristics of an ecosystem arise? _____
   _____

203 Habitat and Tolerance Range

Key Question: What is a habitat? How does the tolerance range of an organism determine its optimum position in a habitat?

The habitat is where an organism lives

▶ A **habitat** is the natural environment in which an organism lives. It includes all the **abiotic** and **biotic factors** in that occupied area. Habitats vary widely in scale, as do the abiotic factors that influence them.

A habitat may be vast and relatively homogeneous, e.g. the open ocean. Predatory barracuda (above) occur around reefs and in the open ocean.

For sessile organisms like this fungus, a suitable habitat may be defined by a small area, such as a decaying log.

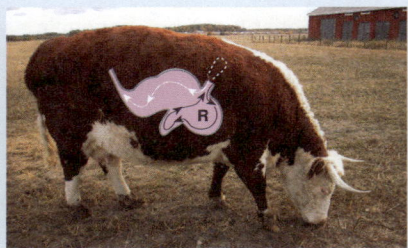

For microbial organisms, such as those in the ruminant gut, habitat is defined by the chemical environment within the rumen (R) of the host, in this case, a cow.

Tolerance range determines distribution in the habitat

▶ Each species has a tolerance range for factors in its environment (below). However, members of a population are individually different and vary in their tolerance. Organisms are usually most abundant where the conditions for their survival and reproduction are optimal. Outside optimum conditions, the environment is less favorable for survival and fewer individuals exist.

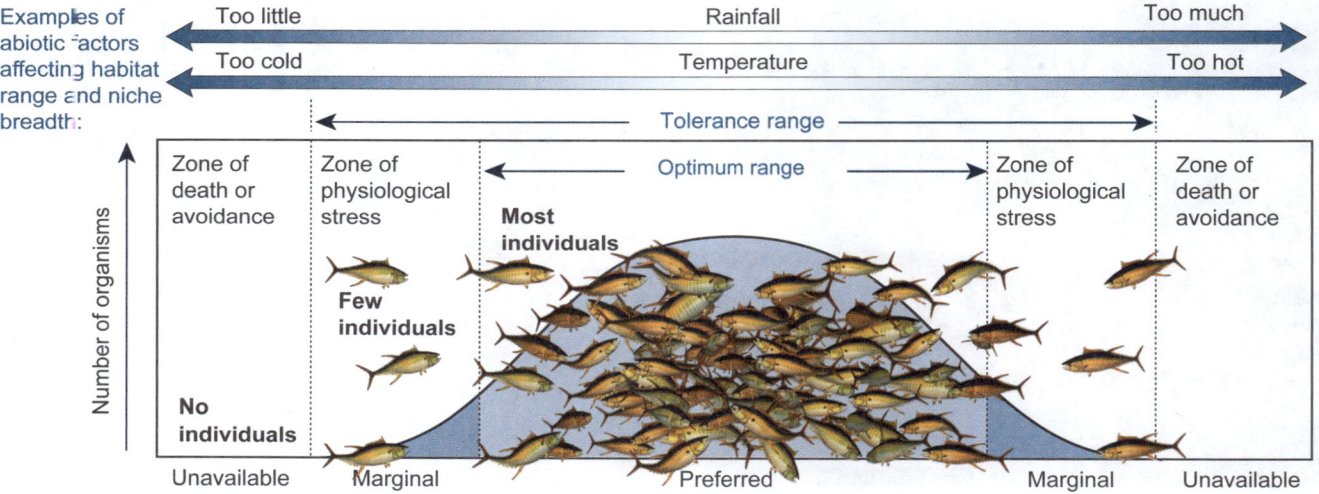

1. In which part of an organism's range will competition for resources be most intense, and why? _____

_____

Species tolerant of large environmental variations tend to be more widespread than organisms with a narrow tolerance range. The Atlantic blue crab (left) is widespread along the Atlantic coast, from Nova Scotia to Argentina. Adults tolerate a wide range of water salinity, ranging from almost fresh to highly saline. This species is an omnivore and has a varied diet, including shellfish, carrion, and animal waste.

2. Suggest an advantage to being able to tolerate variations in a wide range of environmental factors?

_____

_____

_____

204 The Ecological Niche

Key Question: What is an organism's niche? How is it influenced by interactions with other species?

The niche is the functional role of an organism

▸ The ecological **niche** (or niche) of an organism describes its functional position in its environment. The fundamental niche describes the full range of environmental conditions in which an organism can exist.

▸ The realized niche describes the actual conditions in which an organism lives. This is usually narrower than the fundamental niche, as interactions with other species, e.g. **competition**, affect conditions.

▸ Central to the niche concept is the idea that two species with exactly the same niche cannot coexist because they would compete for the same resources, and one would exclude the other. More often, species compete for only some of the same resources. These competitive interactions limit population sizes and influence distributions.

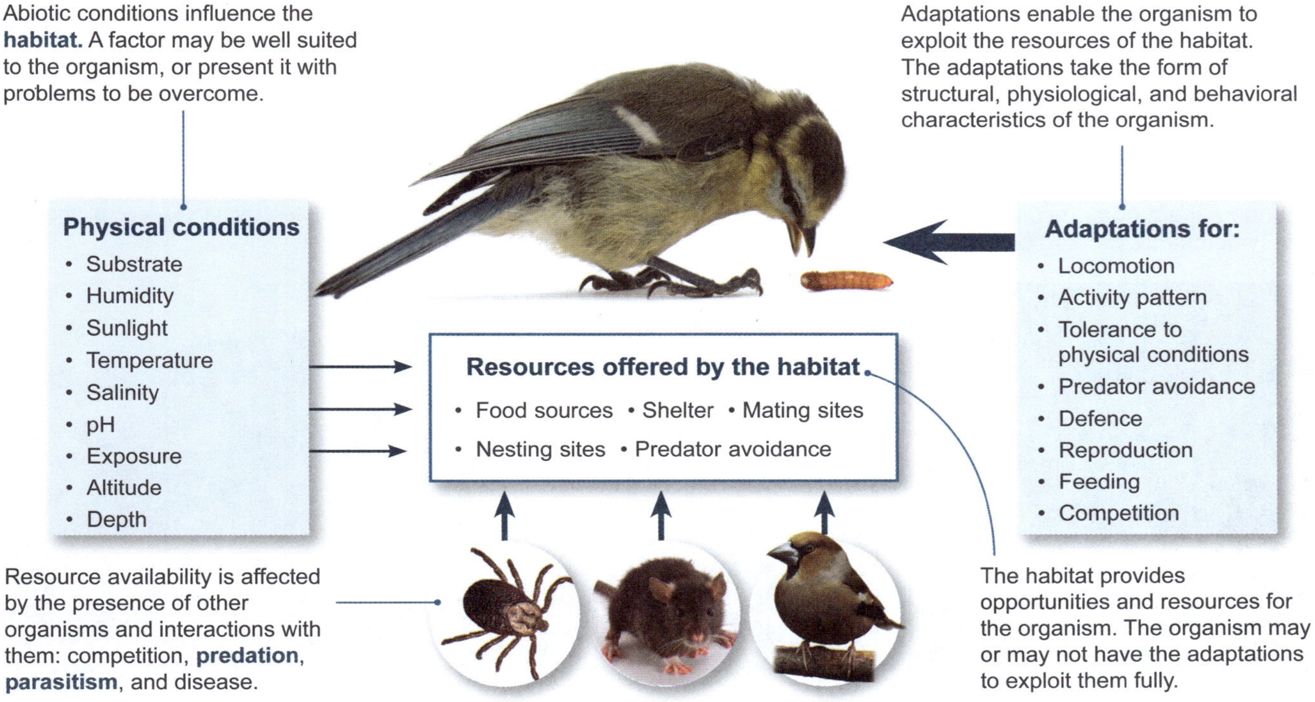

Abiotic conditions influence the **habitat.** A factor may be well suited to the organism, or present it with problems to be overcome.

Physical conditions
- Substrate
- Humidity
- Sunlight
- Temperature
- Salinity
- pH
- Exposure
- Altitude
- Depth

Resources offered by the habitat
- Food sources • Shelter • Mating sites
- Nesting sites • Predator avoidance

Resource availability is affected by the presence of other organisms and interactions with them: competition, **predation**, **parasitism**, and disease.

Adaptations enable the organism to exploit the resources of the habitat. The adaptations take the form of structural, physiological, and behavioral characteristics of the organism.

Adaptations for:
- Locomotion
- Activity pattern
- Tolerance to physical conditions
- Predator avoidance
- Defence
- Reproduction
- Feeding
- Competition

The habitat provides opportunities and resources for the organism. The organism may or may not have the adaptations to exploit them fully.

1. What is the niche of an organism, and how is it different from the organism's habitat? _____

2. (a) In what way is the size of the realized niche flexible? _____

   (b) How does competition with another species affect the size of an organism's niche? Explain: _____

3. The diagram (right) shows the resource-use curves of two bird species. They overlap in the size of the food items they exploit. On the diagram, use A, B, and C to mark:

   (a) The food item sizes most exploited by species A.
   (b) The food item sizes most exploited by species B.
   (c) The regions where competition is most intense.

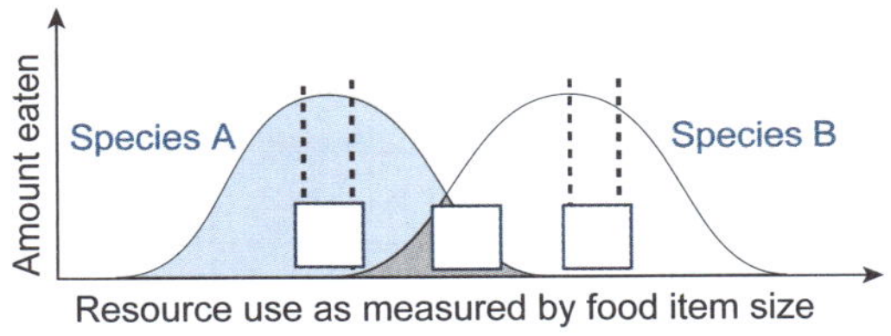

©2024 **BIOZONE** International
ISBN: 978-1-99-101405-4
Photocopying Prohibited

205 Ecosystem Dynamics

Key Question: How do ecosystems remain relatively stable over the long term, while responding to short-term and cyclical changes?

The dynamic ecosystem

- **Ecosystems** are dynamic and are constantly changing. Many ecosystem components, including the seasons, predator-prey cycles, and disease cycles, are cyclical. Some cycles may be short term, such as the change of seasons. Others are long term, such as the growth and retreat of deserts.

- Although ecosystems may change constantly over the short term, they may be relatively stable over longer periods. For example, some tropical areas have wet and dry seasons, but over hundreds of years the ecosystem as a whole remains unchanged.

- Change can be introduced by a shift of any **biotic** or **abiotic factors** away from the seasonal normal. Human impact has been responsible for many ecosystem changes, both on an ecosystem specific scale and a global scale. Examples are over exploitation of organisms, or human-induced **climate change**. These changes affect numerous abiotic factors across multiple ecosystems, creating instability.

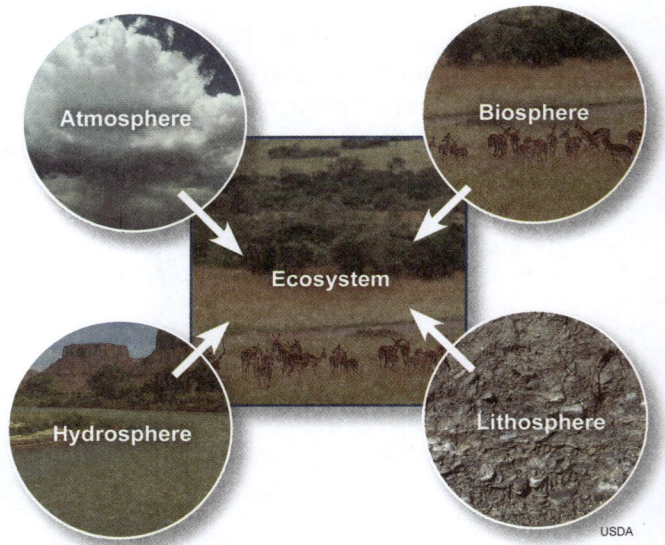

Ecosystems are a result of the interactions between biological (biotic) and physical (abiotic) factors.

An ecosystem may remain stable for many hundreds or thousands of years, provided that the components interacting within it remain stable.

Small scale changes usually have little effect on an ecosystem. Fire or flood may destroy some parts, but enough is left for the ecosystem to return to its original state relatively quickly.

Large scale disturbances such as volcanic eruptions or large scale open cast mining destabilize and ecosystem and change it forever.

1. What is meant by the term dynamic ecosystem? _____
   _____
   _____
   _____

2. (a) Describe a small scale event occurring in your local area that an ecosystem recovered from: _____
   _____
   _____

   (b) Describe two large scale events that an ecosystem may not recover from: _____
   _____
   _____

Ecosystem stability

▸ Ecosystem stability is affected by its inertia (the ability to resist disturbance) and resilience (the ability to recover from external disturbances).

▸ Ecosystem stability is closely linked to **biodiversity**. More biodiverse systems are more stable. Researchers hypothesize that greater diversity causes a greater number of biotic interactions, and few, if any, vacant niches. The system is resistant to invasions and enough species are present to protect ecosystem functions if one is lost. This hypothesis is supported by experimental evidence. However, there is uncertainty over what level of biodiversity provides stability or what factors will stress a system beyond its tolerance.

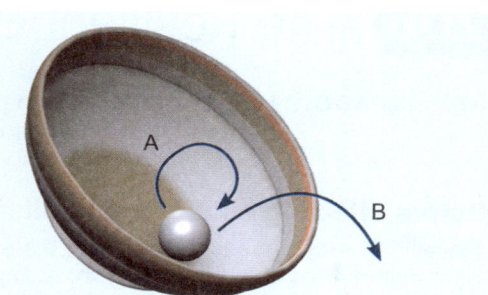

The stability of an ecosystem can be illustrated by a ball in a tilted bowl. Given a slight disturbance the ball will eventually return to its original state (**line A**). However, given a large disturbance the ball will roll out of the bowl and the original state with never be restored (**line B**).

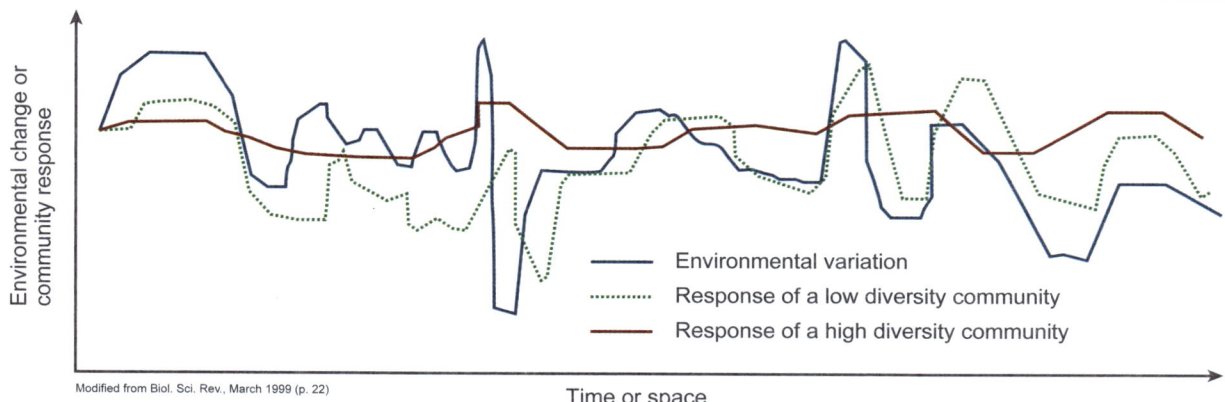

Response to environmental change

Modified from Biol. Sci. Rev., March 1999 (p. 22)

▸ In models of ecosystem function, higher species diversity increases the stability of ecosystem functions, such as productivity and nutrient cycling. In the graph above, note how the low diversity system varies more consistently with variations in the environment, whereas changes in the high diversity system are more gradual.

▸ In any one ecosystem, some species have a disproportionate effect on ecosystem stability due to their key role in an ecosystem function, e.g. nutrient recycling. These species are called **keystone species** (activity 237).

3. Why is ecosystem stability higher in ecosystems with high biodiversity than in those with low biodiversity?

_____

_____

4. The effect of changes in ecosystems can be difficult to measure in the field, so researchers often build small scale simulations of an ecosystem. The graph (right) shows the effect of adding nutrients to a marine ecosystem e.g. nutrient runoff from land into the sea. Algal growth-promoting medium was added at 2, 10, or 20% to seawater, together with 0.1 mL of an algal mix. Two days after adding the growth medium and algae, six copepods were added to each chamber. The chambers were sealed and the population size in each chamber was measured over time.

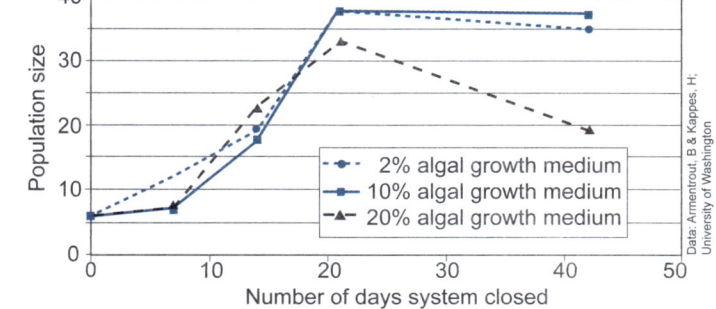

(a) Which chamber had the greatest environmental disturbance?

_____

(b) Which chamber(s) were able to withstand the environmental disturbance? _____

(c) What does this tell us about the stability and resilience of the system being studied? _____

_____

206 The Resilient Ecosystem

Key Question: How is the resilience of an ecosystem affected by its biodiversity, health, and the frequency with which it is disturbed?

Factors affecting ecosystem resilience

▶ Recall that **resilience** is the ability of the **ecosystem** to recover after disturbance. It is affected by three important factors: **biodiversity**, ecosystem health, and frequency of disturbance. These factors are summarized below:

Ecosystem biodiversity
▶ The greater the biodiversity of an ecosystem, the greater the chance that all the roles (niches) in an ecosystem will be occupied. This makes it harder for invasive species to establish and easier for the ecosystem to recover after a disturbance.

Ecosystem health
▶ Intact ecosystems are more likely to be resilient than ecosystems suffering from species loss or disease.

Disturbance frequency
▶ Single disturbances to an ecosystem can be survived, but frequent disturbances make it more difficult for an ecosystem to recover. However, some ecosystems depend on frequent natural disturbances for their maintenance, e.g. grasslands rely on natural fires to prevent shrubs and trees from establishing. The various grass species in this ecosystem evolved to survive frequent fires.

▶ A study of coral and algae cover at two locations in Australia's Great Barrier Reef (right) showed how ecosystems recover after a disturbance. At Low Isles, frequent disturbances, e.g. from cyclones, made it difficult for corals to reestablish, while at Middle Reef, infrequent disturbances made it possible for coral to reestablish its dominant position in the ecosystem.

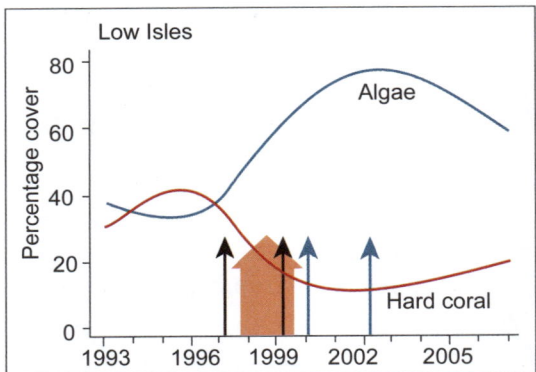

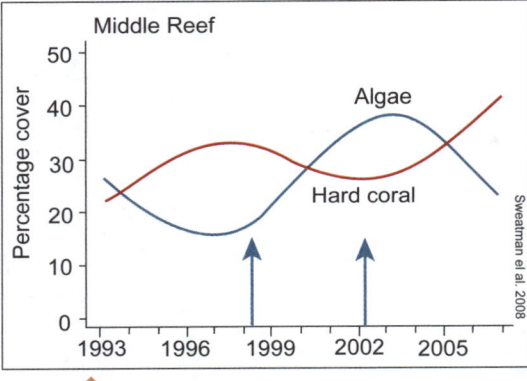

➔ Crown of thorns starfish outbreak
➔ Cyclones   ➔ Bleaching event

Resilience and harvesting

It is important to consider the resilience of an ecosystem when harvesting resources from it. For example, logging and fishing remove organisms from an ecosystem for human use. Most ecosystems will be resilient enough to withstand the removal of a certain number of individuals. However, excessive removal may go beyond the ecosystem's ability to recover. Examples include the overfishing of North Sea cod and deforestation in the Amazon basin.

Fishing

Deforestation

1. Define ecosystem resilience: _____

2. Why did the coral at Middle Reef remain abundant from 1993 to 2005 while the coral at Low Isles did not?
_____
_____
_____

3. How might over-harvesting affect an ecosystem's resilience? _____
_____
_____

207 A Case Study in Ecosystem Resilience

Key Question: How are resilient ecosystems able to recover from moderate fluctuations?

Spruce budworm and balsam fir

A case study of **ecosystem resilience** is provided by the spruce/fir forest community in northern North America. Organisms in the community include the spruce budworm, and balsam fir, spruce, and birch trees. Despite its name, the spruce budworm is a greater pest to balsam fir than spruce. The community fluctuates between two extremes:

▶ During budworm outbreaks, the environment favors the spruce and birch species.

▶ Between spruce budworm outbreaks, the environment favors the balsam fir.

Balsam fir

Spruce budworm

1. Under certain environmental conditions, the spruce budworm population grows so rapidly it overwhelms the ability of predators and parasites to control it.

2. The spruce budworm feeds on balsam fir (despite its name), killing many trees. The spruce and birch trees are left as the major species.

3. The population of budworm eventually collapses because of a lack of food.

4. Balsam fir saplings grow back in thick stands, eventually out-competing the spruce and birch. Evidence suggests that these cycles have been occurring for possibly thousands of years.

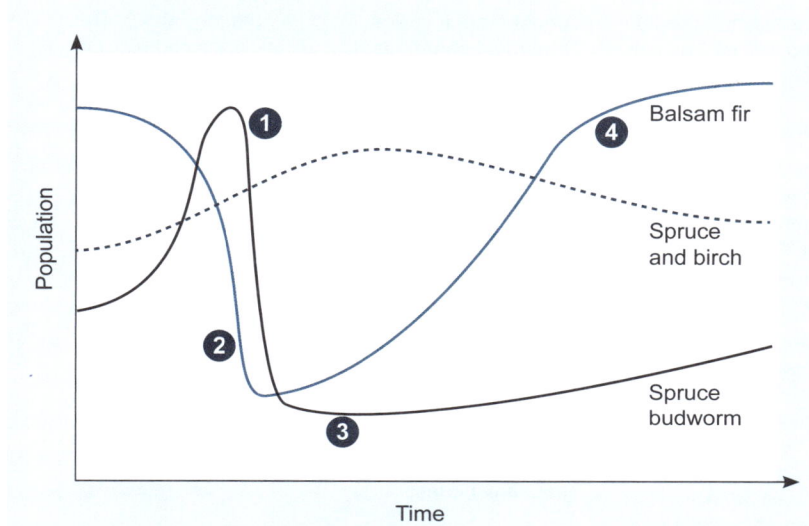

1. (a) Why could predators not control the budworm population? _____

    _____

    (b) What was the cause of the budworm population collapse after its initial rise? _____

2. Under what conditions does the balsam fir out-compete the spruce and birch? _____

3. In what way is the system resilient in the long term? _____

208 Species Interactions

Key Question: How do interactions and relationships between species influence the size and distribution of their populations?

- Within **ecosystems**, each species interacts with others. In many of these interactions, one of the parties in the relationship is disadvantaged. Predators eat prey, parasites and pathogens exploit their hosts, and species compete for limited resources. These interactions limit the number of organisms in a population and prevent any one population from becoming too large.
- Not all relationships involve exploitation. Some species form relationships that are mutually beneficial. Mutualistic relationships can enable two species to exist in greater numbers than either would alone.
- The interactions can be grouped into symbiotic relationships between two organisms of different species: **mutualism**, **parasitism**, and **commensalism**; and community interactions: **predation** and **competition**.

Type of interaction between species				
Mutualism	**Commensalism**	**Parasitism**	**Predation**	**Competition**
Both species benefit. If both species depend on the symbiosis for survival the mutualism is called obligate. Mutualism can involve more than two species. **Examples**: Flowering plants and their insect pollinators. The flowers are pollinated and the insect gains food. Ruminants and their rumen protozoa and bacteria. The microbes digest the cellulose in plant material and produce short-chain fatty acids, which the ruminant uses as an energy source.	One species benefits and the other is unaffected. It is likely that most commensal relationships involve some small benefit to the apparently neutral party. **Example**: The squat anemone shrimp, lives among the tentacles of sea anemones, where it gains protection and scavenges scraps of food from the anemone. The anemone appears to be neither harmed, nor gain any benefit.	The parasite lives in or on the host, taking all its nutrition from it. The host is harmed but not usually killed directly. Parasites may have multiple hosts; their transmission is often linked to **food webs**. **Example**: Parasitic isopods cut the blood supply to the tongue of the host fish, causing it to fall off. The parasite attaches to what is left of the tongue, feeding on blood or mucus.	A predator kills the prey and eats it. Predators may prey on a range of species or they may prey exclusively on one species. Predation is a type of exploitation. Herbivory is an equivalent exploitation between herbivores and plants. **Examples**: Praying mantis consuming insect prey. Canada lynx eating snowshoe hare. Starfish eating mussels and limpets.	Individuals of the same or different species compete for the same, limited resources. Both parties are detrimentally affected. **Examples**: Neighboring plants of the same and different species compete for light and soil nutrients. Vultures compete for the remains of a carcass. Insectivorous birds compete for suitable food in a forest. Tree-nesting birds with similar requirements compete for nest sites.
	(a)	(b)	(c)	(d)
A ⇄ B Benefits Benefits				

1. In the spaces above, draw a simple model to show whether each species/individual in the interaction described is harmed or benefits. The first one has been completed for you.

2. (a) What is a key difference between community interactions and symbiotic relationships? _____

    _____

    (b) Provide an example of a species interaction from your local area: _____

    _____

Examples of interactions between different species are illustrated below. For each example, identify the type of interaction, and explain how each species in the relationship is affected.

3. The Monarch butterfly is the official insect of Texas. It is the key pollinator of milkweed plants and, in return, the monarch lays its eggs on the plants. Its caterpillars rely on the milkweed for food.

   (a) Identify this type of interaction: _____

   (b) Describe how each species is affected (benefits/harmed/no effect):

   _____

   _____

4. The American alligator can be found in freshwater ecosystems in parts of Texas. They capture food by ambush, lying submerged in water until animals, such as tortoise, are close enough to lunge at.

   (a) Identify this type of interaction: _____

   (b) Describe how each species is affected (benefits/harmed/no effect):

   _____

   _____

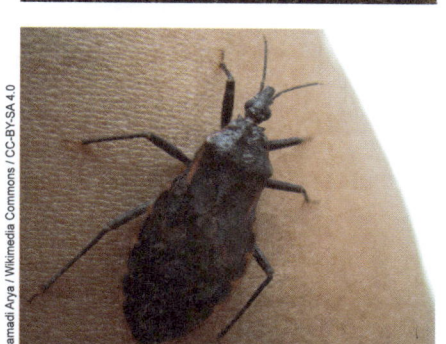

5. The Kissing bug consists of 11 species that can be found across Southern United States, including Texas. It bites the skin of animals and sucks the blood for nourishment. The kissing bug is the vector responsible for Chagas disease in humans.

   (a) Identify this type of interaction: _____

   (b) Describe how each species is affected (benefits/harmed/no effect):

   _____

   _____

   _____

6. Large herbivores expose insects in the vegetation as they graze. A bird called the cattle egret is widespread in tropical and subtropical regions. It follows the herbivores as they graze, feeding on the disturbed insects when the herbivore moves away.

   (a) Identify this type of interaction: _____

   (b) Describe how each species is affected (benefits/harmed/no effect):

   _____

   _____

7. Ruby-throated hummingbirds are nectar feeders found in Texas. These small, active birds target plants with orange and red flowers in particular. They use their long beaks to obtain nectar from the flowers. In return, the birds pollinate the flowers, carrying pollen to other plants.

   (a) Identify this type of interaction: _____

   (b) Describe how each species is affected (benefits/harmed/no effect):

   _____

   _____

8. Explain the similarities and differences between a predator and a parasite: _____

_____

209 Predator-Prey Relationships

Key Question: How are populations of predators and prey related, and how do they change over time?

Do predators limit prey numbers?

- It was once thought that predators always limited the numbers of their prey populations. While this is often true for invertebrate predator-prey systems, prey species are often regulated more by factors other than **predation**, including climate and the availability of food.
- In contrast, predator populations can be strongly affected by the availability of prey, especially when there is little opportunity for prey switching, i.e hunting another prey if the preferred one becomes scarce.
- Predator and prey populations may settle into a stable pattern, where predator numbers follow those of the prey, with a time lag (right).

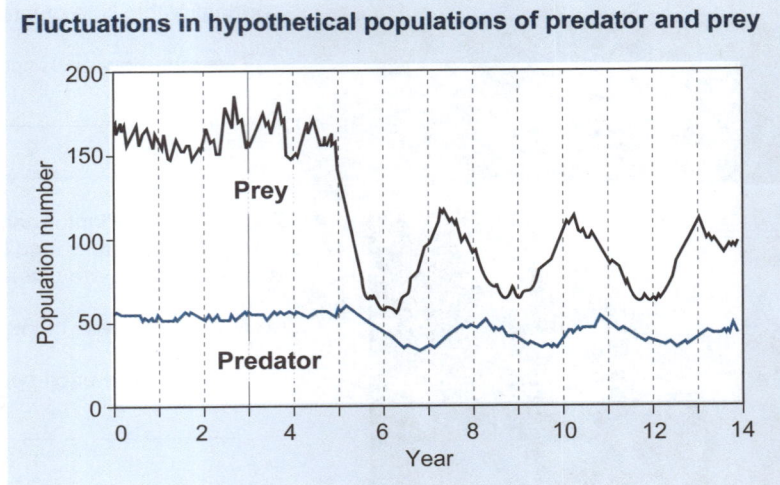

Fluctuations in hypothetical populations of predator and prey

Population changes over time: predator-prey cycles

- Predators hunt and eat other animals. As a result, their population growth can be closely linked to that of their prey.
- In many herbivore-carnivore systems, prey populations fluctuate seasonally with changes in vegetation growth. Predators may respond to increases in prey by increasing their rate of reproduction. This type of numerical response shows a time lag, associated with the time it takes for the predator population to respond by producing more young. One of the most famous of these time-lagged predator-prey cycles is that recorded for the Canada lynx and its prey, the snowshoe hare (below).

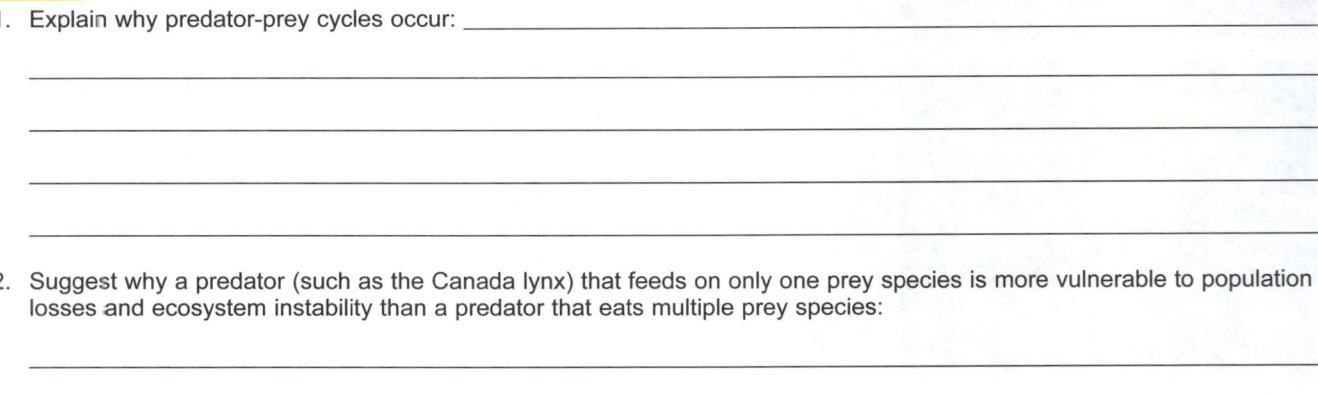

Changes in snowshoe hare and Canadian lynx populations

1. Explain why predator-prey cycles occur: _____

_____

_____

_____

2. Suggest why a predator (such as the Canada lynx) that feeds on only one prey species is more vulnerable to population losses and ecosystem instability than a predator that eats multiple prey species:

_____

_____

_____

A case study in predator-prey numbers

▸ In some areas of Northeast India, a number of woolly aphid species colonize and feed off bamboo plants. The aphids can damage the bamboo so much that it can no longer be utilized by local people for construction and textile production.

▸ Giant ladybug beetles (*Anisolemnia dilatata*) feed exclusively on the woolly aphids of bamboo plants. These beetles could be used as biological control agents to reduce woolly aphid numbers, and limit the damage to the bamboo plants.

▸ The graph below shows the relationship between the giant ladybug beetle and the woolly aphid, when grown in controlled laboratory conditions.

Bamboo plants are home to many insect species, including ladybugs and aphids.

Aphids feed on the bamboo sap. Ladybugs are predators of the aphids (below).

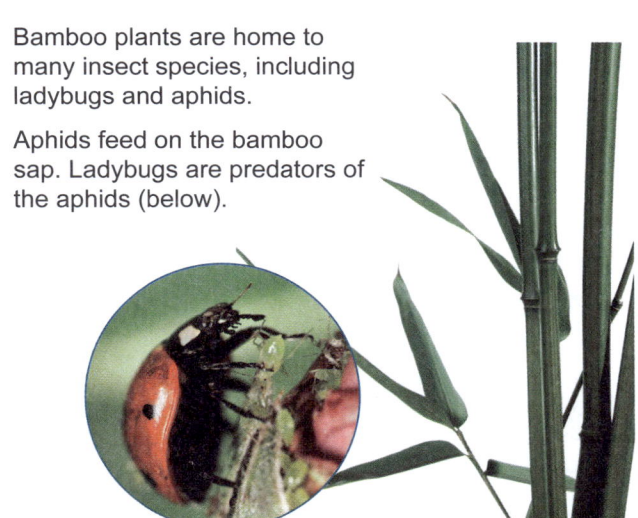

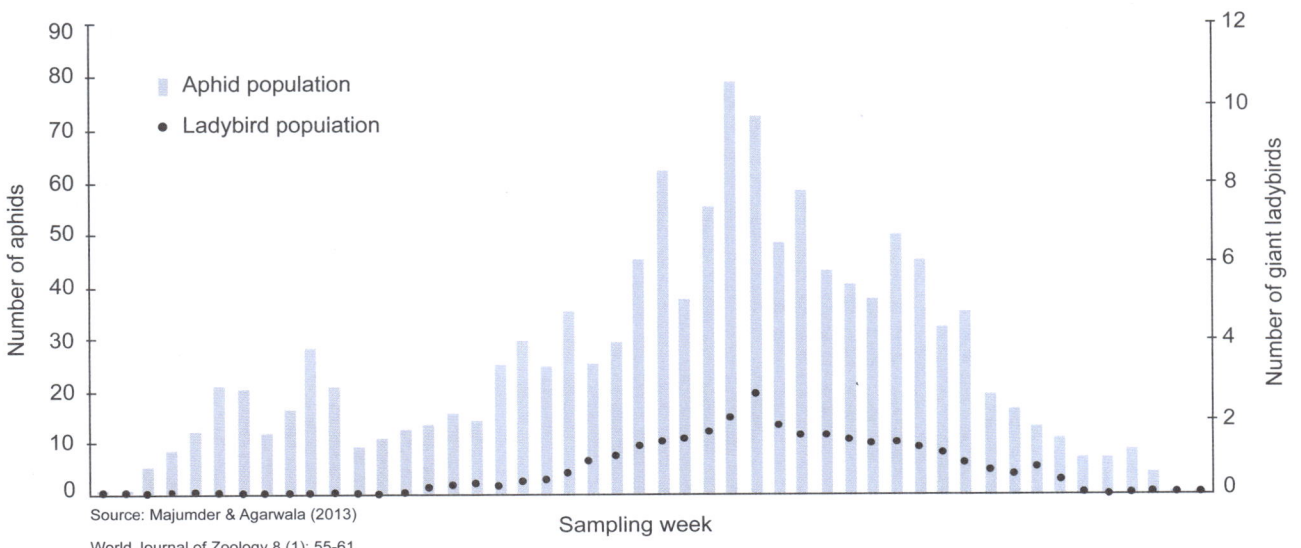

Source: Majumder & Agarwala (2013)
World Journal of Zoology 8 (1): 55-61

3. (a) Using different colored pens, mark the two points on the graph above where the peak numbers of woolly aphids and giant ladybugs occur.

   (b) Do the peak numbers for both species occur at the same time? _____

   (c) Why do you think this is? _____
   _____
   _____

4. (a) What is the response of the ladybug population when their prey decline? _____
   _____
   _____

   (b) Although this was a laboratory situation, what features of the ladybug predator suggest it would be a good choice to control woolly aphids?
   _____
   _____
   _____
   _____

210 Predation and Destabilized Ecosystems

Key Question: How can predator-prey interactions destabilize an ecosystem?

What happened when wolves were introduced to Coronation Island?

▸ Coronation Island is a small island (116 km^2) off the Alaskan coast. A resident black tailed deer population had overgrazed the island. As a result, very little forest understorey remained and many common plant species were absent. The forest was open and park-like, not dense like a typical South Alaskan forest. Researchers noted that the deer on the island were smaller than those in other populations and several died each year from malnutrition because there was not enough food.

▸ In 1960, the Alaska Department of Fish and Game released two breeding pairs of timber wolves (below right) onto the island. Their aim was to control the black-tailed deer (top right), which had been overgrazing the land.

▸ Initially, wolves preyed on deer and bred successfully. Deer numbers began to fall and the island's vegetation began to return. By 1964, the vegetation was quite abundant. However, within a few years, the deer population crashed. The wolves ran out of food (deer) and began eating each other, causing a drop in wolf numbers. Within 8 years of the wolves being introduced, only one wolf remained on the island. By 1983, wolves were absent and the deer were once again abundant.

▸ The changes in the predator-prey system on Coronation Island are represented by the model below:

Black tailed deer

Timber wolf

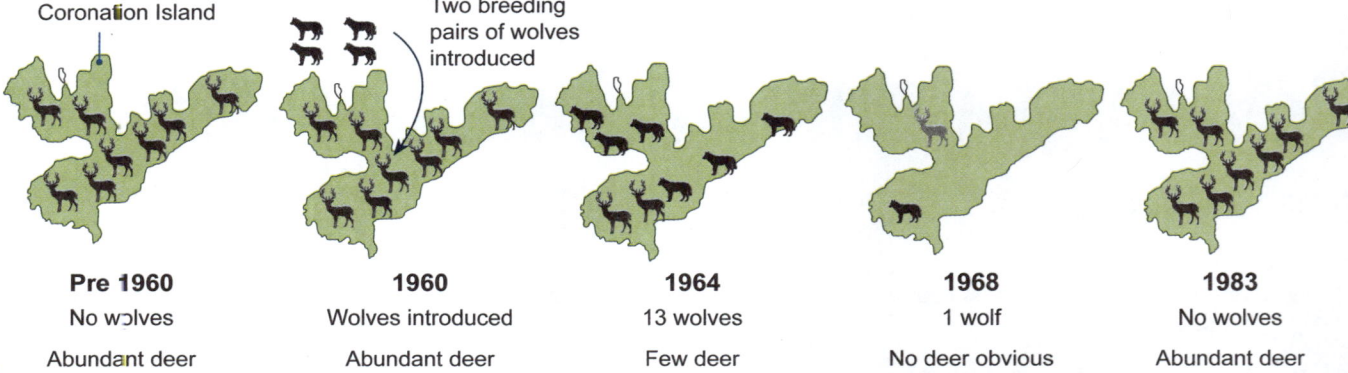

Pre 1960	1960	1964	1968	1983
No wolves	Wolves introduced	13 wolves	1 wolf	No wolves
Abundant deer	Abundant deer	Few deer	No deer obvious	Abundant deer

1. Working in pairs, evaluate the introduction of wolves to Coronation Island. What did it show? Was it a success?

   _____
   _____
   _____
   _____

2. If the experiment was carried out at a larger scale on a bigger island with more resources, do you think the outcome would have been different? What is your reasoning?

   _____
   _____
   _____
   _____
   _____

211 Investigating Predator-Prey Stability

Key Question: Does a predator-prey system stabilize over time?

▸ Any environment has limited resources, such as water, food, or nesting sites. Some of these will be continuously renewed, but at a limited rate. The prey population is considered a resource for predators.

▸ If a new predator species moves into an **ecosystem** with a small initial population, at first it may find excess prey resources and be able to expand quickly, but as prey are used up population growth will eventually slow down.

Investigation 9.1 Investigating predator-prey populations

See appendix for equipment list.

1. This activity is best done outside for ease of movement. Do not place beans in mouth or throw them around.
2. Each student needs a spoon and a cup. For the class, divide 100 dried beans (or peas, beads, or small marbles, etc.) into two sets of 50. The beans **represent the prey**.
3. Place each set of 50 dried beans onto a separate tray with an upturned edge (so the beans don't roll off), wide enough so that the beans are single layered and have a small amount of space between them.
4. The teacher chooses a student (or has a volunteer) to be the data recorder. They will need a pen or pencil and a copy of Table 1 below.
5. Students arrange themselves in a circle about 10 m in diameter. Place the trays in the center of the circle, but about a meter apart.
6. The teacher will choose one student, **representing the predator**, to start the first round of "feeding". The student places their cup on the ground to mark their position in the circle. The teacher starts the timer and tells the student. This student has 15 seconds to run to either tray and using only their spoon, pick up as many beans as they can. They then run back to their place and tip the beans into their cup (if beans fall off the spoons, they are not picked up until the end of this step). They keep doing this until the 15 seconds is up.
7. The number of beans in the cup are then counted. For each 7 beans in the cup, another student will participate in the next round. The teacher will choose this student. The data recorder notes down the generation (1) and the population of feeding students.
8. All the beans are returned to the trays so that there are again 50 beans on each tray.
9. The second generation of students is then given 15 seconds to collect beans as before. Again, for each 7 beans in a cup another student to added to the next generation (it is not the total number of beans that is counted but the number in each cup e.g. if there are 2 cups and one has 11 beans and the other has 17 beans then 3 extra students are added, not 4). Again, the generation and student number is recorded.
10. This continues for 10 generations (teacher may decide longer if time and beans allow).
11. Repeat the experiment, but this time the teacher will decide on a **different value than 7** to raise or lower the number of beans needed to spawn another feeding student, e.g. 5 or 9 beans.

Table 1

Generation	Number of students
0	1
1	
2	
3	
4	
5	
6	
7	
8	
9	
10	

Table 2

Generation	Number of students
0	1
1	
2	
3	
4	
5	
6	
7	
8	
9	
10	

1. Plot a line graph of the results for both table 1 and 2 on the grid below. Include a key:

2. Describe the shape of the graph: _____

3. Which predator population grows the fastest? _____

4. What factor is limiting the rate of growth of the predator populations? _____

5. The population growth rate can be calculated by: **Growth rate = change in population number ÷ change in time** (in this case, generations).

    (a) For table 1, what is the growth rate of the population from generation 0 to 3? _____

    (b) For table 1, what is the growth rate of the population from generation 4 to 7? _____

    (c) For table 1, what is the growth rate of the population from generation 8 to 10? _____

6. (a) What is the approximate maximum number of predator for the first experiment? _____

    (b) What is the approximate maximum number of predator of the second experiment? _____

7. Explain why the population growth rates for the graphs change the way they do: _____

212 Competition for Resources

Key Question: How does competition for limited resources negatively affect species?

- No organism exists in isolation. Each organism interacts with other organisms, and with the physical (**abiotic**) components of its environment.
- **Competition** occurs when two or more organisms are competing for the same limited resource, e.g. food or space, or if the organisms occupy the same niche in the same **habitat**.
- Competition harms all competitors. The negative effects of competition limit population numbers because resources are limited, and growth, reproduction, and survival are affected.
- Populations with smaller numbers, due to intense competition, are less resilient to change and therefore make an **ecosystem** less stable.
- Competition can occur between members of the same species (**intraspecific competition**), or between members of different species (**interspecific competition**).

A complex system of interactions occurs between the different species living on this coral reef in Hawaii. Population numbers will be limited by competition for limited resources, such as food and space on the reef.

Examples of limited resources

Space can be a limited resource
These sea anemones are competing for space in a tidal pool. Some species defend areas, called territories, which contain the resources they need.

Suitable mates can be hard to find
Within a species, individuals may compete for a mate. These male whitetail deer are fighting to determine which one will mate with the females.

Food is usually a limited resource
In most natural systems, competition exists between individuals of the same species for food, and between different species with similar diets.

1. (a) What is competition? _____

    _____

    _____

    (b) Why does competition occur? _____

    _____

    _____

    (c) Why does competition have a negative effect on all competitors? _____

    _____

    _____

    _____

213 Intraspecific Competition

Key Question: Why does intraspecific competition occur, and how does intraspecific competition regulate population size?

- **Intraspecific competition** occurs when individuals of the same species compete for the same limited resources. In addition to food, space, nutrients, and light, intraspecific competition also includes competition for mating partners and breeding sites.
- In most cases, intraspecific competition is more intense than **competition** between different species because individuals are all competing for the same resources, e.g. same food and mates. It is an important factor in limiting the population size of many species.
- Strategies such as territoriality and social hierarchies can reduce the conflict associated with resource competition and can be important in determining which individuals in the population will breed.

How does intraspecific competition limit population size?

- Most resources are limited. This is a major factor in determining how large a population can grow.
- As population numbers increase, demands on resources are higher and they are used up more quickly. This means that some individuals receive fewer resources than others. High population densities may also increase the influence of population-limiting factors such as disease. Populations respond by decreasing their numbers. This occurs by:
- Reduced survival (more individuals die).
- Reduced birth rates (fewer individuals are born).
- If resources increase, e.g. food increases, population numbers can increase. The relationship between resources and population numbers is shown on the right.

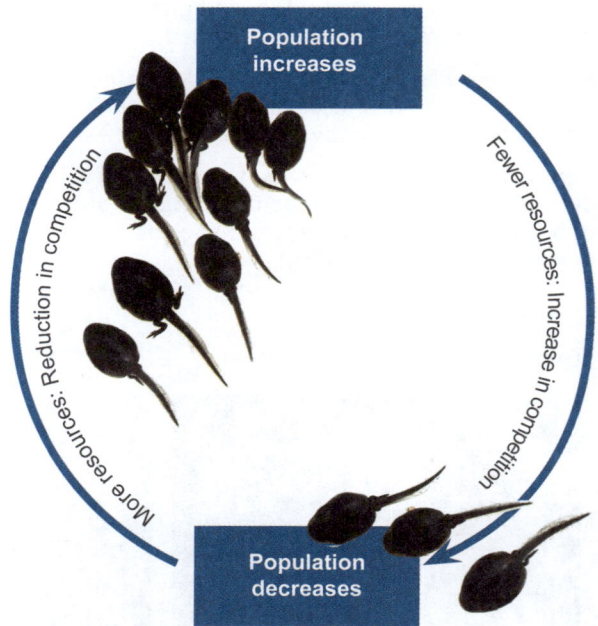

Scramble competition

Direct competition between members of the same species for a finite resource is called scramble competition. These silkworm caterpillars are all competing for the same food. When it is insufficient, none of the individuals may survive.

Contest competition

In contest competition, there is a winner and a loser and resources are obtained completely or not at all. For example, male elephant seals fight for territory and mates. Unsuccessful males may not mate at all.

Competition in social species

In some animals, strict social orders ensure that dominant individuals will have priority access to resources. Lower ranked individuals must rely on what remains. If food is very limited, only dominant individuals may receive enough to survive.

1. (a) What is intraspecific competition? _____

   (b) How does scramble competition differ from contest competition? _____

Territories and limitations of population size

- Territoriality in birds and other animals is usually a result of intraspecific competition. A territory is a defended area containing the resources required by an individual or breeding pair to survive and reproduce. Territories space organisms out in the **habitat** according to the availability of resources. Those animals without territories usually do not breed.
- South American **rufous-collared sparrow** males and females occupy small territories (below). These birds make up 50% of the population. The remaining 50% of the population, called floaters, occupy home ranges (which are undefended areas) within the territory boundaries. They are tolerated by the territory owners, but these floaters do not breed.
- By using tagging studies and removal of birds, researchers found that when a territory owner (male or female) dies or disappears, it is replaced by a floater of the appropriate sex. This is shown for the females in the left diagram as a darker region.

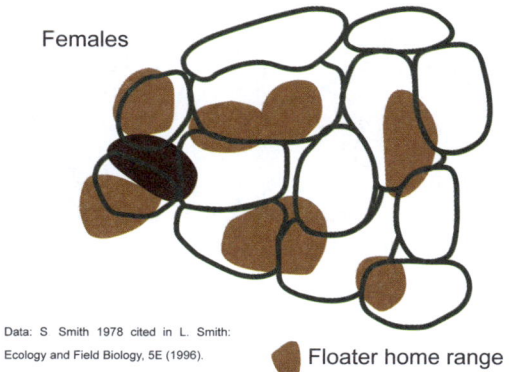
Females

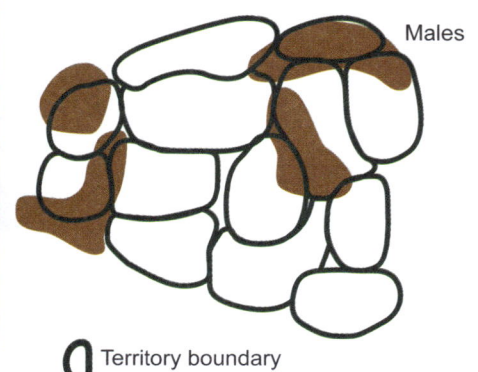
Males

Data: S Smith 1978 cited in L. Smith: Ecology and Field Biology, 5E (1996).

● Floater home range   ◐ Territory boundary

2. (a) Explain the relationship between resource availability and intraspecific competition: _____

   (b) What happens to population numbers as intraspecific competition increases? _____

   (c) Intraspecific competition is considered to be a density-dependent process. What do you think this means?

3. Territoriality is a way to ensure that at least some individuals have the resources to survive and reproduce, thus maintaining ecosystem stability, and is established and maintained by direct conflict and by calls and displays.

   (a) Identify the benefits of possessing a territory: _____

   (b) Can you think of a cost of having a territory? _____

   (c) What evidence is there from the rufous-collared sparrow study to show that territoriality can limit population size?

214 Interspecific Competition

Key Question: Why does interspecific competition occur, and how does it affect the species involved?

Interspecific competition involves individuals of different species competing for the same limited resources. They may do this by:

- Interfering directly with the ability of others to gain access to the resource.
- Exploiting the resource before other individuals can get access to it (exploitative competition).

▸ Interspecific competition is usually less intense than competition between members of the same species. This is because competing species have different requirements for at least some resources, e.g. different **habitat** or food preferences. In other words, their **niches** are different, even if they exploit some of the same resources.

▸ Interspecific competition can have a role in limiting a population's size and determining species occurrence and distribution. However, in naturally occurring populations, it is generally less effective at limiting population size than intraspecific competition, especially in animals. This is because each species usually has alternative resources it can exploit to avoid **competition**.

▸ Interspecific competition in natural plant communities is dependent on nutrient availability and will be greater when soil nutrients are low. Fast growing plants with large, dense root systems can absorb large amounts of nitrogen. This depletes soil nitrogen so that other plants cannot grow close to them. Similarly, fast growing plants may quickly grow tall enough to intercept the available light and prevent the germination of plants nearby. Pest plants often have this strategy and can become very difficult to control.

▸ Sometimes, humans may introduce a species with the same resource requirements as a native species. The resulting competition can lead to the decline of the native species and a destabilization of the **ecosystem**.

In some communities, many different species may be competing for the same resource. This type of competition is called interference competition because the individuals interact directly over a scarce resource. In the example above, three species compete for what remains of a carcass.

The plant species in the forest community above compete for light, space, water, and nutrients. A tree that can grow taller than those around it will be able to absorb more sunlight, and grow more rapidly to a larger size, than the plants in the shade below.

1. (a) What is interspecific competition? Describe an example: _____
   _____
   _____

   (b) Why is interspecific competition usually less intense than intraspecific competition? _____
   _____
   _____
   _____

   (c) Why is interspecific competition generally less effective at limiting population size than intraspecific competition?
   _____
   _____
   _____

215 The Impact of Competing Invasive Species

Key Question: What impact do invasive species have on stability of the ecosystem into which they are brought, either as introduced or invasive species?

- Introduced species have evolved at one place in the world and been transported by humans, either intentionally or inadvertently, to another region. Some of these introductions are directly beneficial to humans and controlled in their impact, e.g. introduced agricultural plants and animals, and Japanese clams and oysters.
- Invasive species are introduced species that have a negative effect on the stability of the **ecosystems** into which they have been imported. Originally, these species may have been brought by humans as pets, food, ornamental specimens, or decoration, but have escaped into the wild.
- In their new environment, they usually lack natural predators, diseases, or other controls to limit their growth. They often out-compete native species for resources such as food and space, or become a predator that prey species have no defence against.
- Other species may have been accidentally transported in cargo shipments or in the ballast water of ships. Some have been deliberately introduced to control another pest species and have themselves become a problem.
- Some of the most destructive of all invasive species are fast growing plants and algae, e.g. mile-a-minute weed, a perennial vine from Central and South America; Miconia, a South American tree invading Hawaii and Tahiti; and Caulerpa seaweed, an aquarium strain of algae now found in the Mediterranean.
- Two introductions, one unintentional and the other deliberate, are described below.

Mile-a-minute weed covering woodland in Maryland, USA

Kudzu: a deliberate introduction

Kudzu (*Pueraria lobata*) is a climbing vine native to south-east Asia. It was first introduced to the United States in 1876 as an ornamental plant. Kudzu was widely planted during the Dust Bowl Era to try to conserve soil and, in 1940, the government paid farmers to plant it. However, by 1953 the payments stopped as kudzu escaped farms and invaded woodlands, outcompeting native plants for light and space. By 1970, it was declared a weed and in 1997 was placed on the noxious weeds list. Huge amounts of money are spent every year to try to control it. Today, investigations indicate that kudzu is estimated to cover 3 million hectares of land in the southeastern US.

Red fire ant: an accidental introduction.

Red fire ants (*Solenopsis invicta*) were accidentally introduced into the southeastern states of the United States from South America in the 1920s and have spread north each year. Red fire ants competitively displace populations of native insects and ground-nesting wildlife. They also damage crops and are very aggressive, inflicting a nasty sting. Investigations from the USDA estimates damage and control costs for red fire ants at more than $6 billion a year.

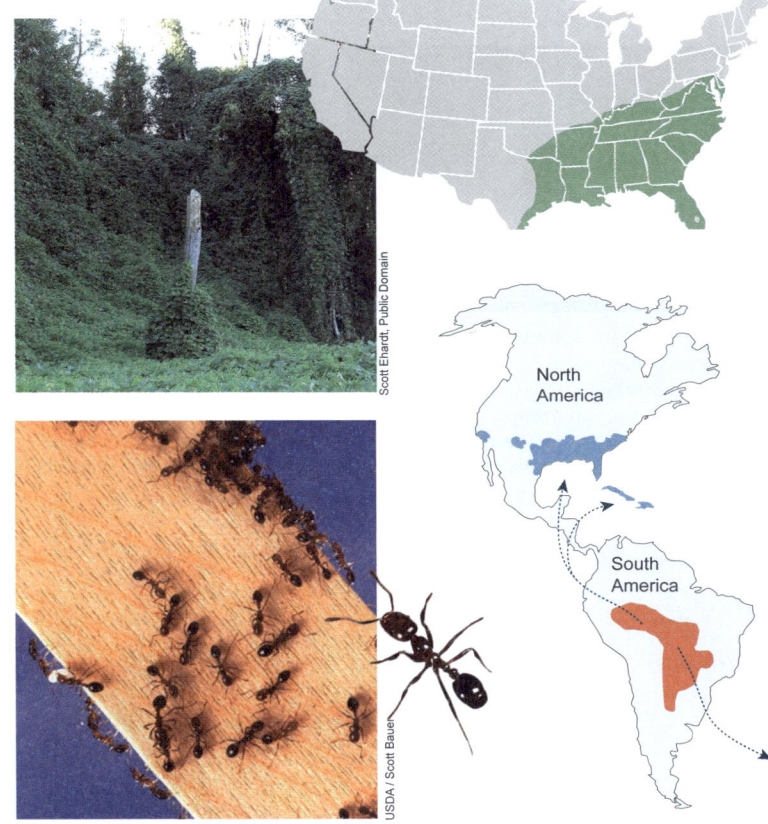

1. Explain why many invasive species become invasive when introduced to a new area: _____

_____

_____

2. Investigate an invasive pest species in your area and discuss the impact it has had on the stability of a local ecosystem. Research valid information from reputable sites, write a short summary of your findings, and attach it to this page.

216 Parasitism: One-sided Benefits

Key Question: How can parasitic outbreaks impact ecosystem stability?

- A symbiosis is any form of very close ecological relationship between two or more species. **Parasitism** is a type of symbiosis in which one organism (the parasite) benefits at the expense of the other (the host).
- Parasites always harm their host but do not usually kill it. The parasite benefits by obtaining nutrition, but often gains other advantages, such as protection.
- Newly introduced parasites can destabilize an **ecosystem** if the hosts do not have adequate adaptations for defense against it, leading to death and a greatly reduced population.
- A parasitic outbreak also impacts humans, such as that caused by an outbreak of the single cell *Cyclospora* parasite in Texas, in 2022.

The **emerald cockroach wasp**: a small parasitoid with a strategy to subdue its host species.

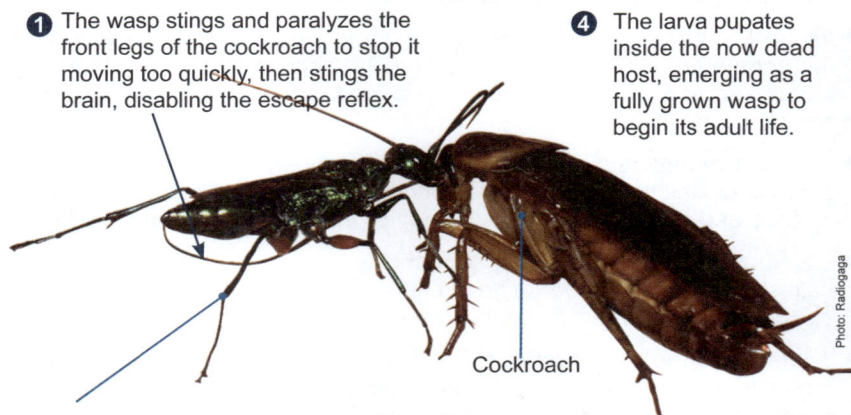

① The wasp stings and paralyzes the front legs of the cockroach to stop it moving too quickly, then stings the brain, disabling the escape reflex.

② The wasp leads the cockroach to its burrow by tugging on the cockroach's antennae.

③ Once inside the burrow, the wasp lays an egg on the cockroach's abdomen. The larva hatches and lives inside the roach, eating its organs in an order that ensures the host remains alive long enough for the larva to complete development.

④ The larva pupates inside the now dead host, emerging as a fully grown wasp to begin its adult life.

Vampire bat (*Desmodus rotundus*)

Male anglerfish

Sea lampreys attached to host fish

Vampire bats are sometimes called the only mammalian parasite. They have many features of parasites: they are specialized for feeding on blood (hematophagous) and do not kill their host. However, they do not live on the host, but return to a roost when they have finished feeding.

Animals are not always parasites of other species. Male deep sea anglerfish seek out the much larger females and attach to them by biting and holding on to their skin. They then slowly regress as they merge with the female, becoming little more than a sperm producing appendage.

Lampreys are jawless fishes that feed by attaching to the flanks of a fish with their rubbery sucker mouthparts. They scrape away the skin with their teeth and then suck the body fluids from the host. Lampreys can swim, but do not need to while attached to the host.

1. (a) How might we define a parasitic outbreak? _____

    _____

    (b) Hosts can evolve concurrently with a resident parasite, so that it possesses adaptations to survive an infestation, such as the Texan longhorn cattle and parasitic ticks. Research has shown that newly introduced parasites can create major disruption to the established food web, and therefore the ecosystem. Why might this be?

    _____

    _____

    _____

2. How might parasitism be affected by a future changing, warmer climate? _____

    _____

    _____

    _____

The Indiana bat and fungal parasitism

▶ The Indiana bat (*Myotis sodalis*) is an insectivorous (insect eating) bat native to North America. It is found in the eastern US, from New Hampshire to northern Florida, and west to Iowa, Missouri, and Oklahoma. In winter, Indiana bats hibernate in caves. In spring, they migrate north and make their home in tree cavities or under loose tree bark.

▶ Winter hibernation allows the bats to conserve energy when there are too few insects available as food. The bats do not eat during hibernation, but live off stored fat laid down before winter when there is ample insect food available. However, each time a bat is woken during hibernation, energy stores are used up more rapidly. If they are disturbed too often, they may not have enough fat reserves to survive the winter.

▶ The Indiana bat has an important ecological role. It is a major predator of night-flying insects and contributes energy to the cave **food web** through guano (excrement) and decomposition of its body when it dies.

The Indiana bat population is estimated to be around 244,000. Very few caves provide suitable hibernation conditions, so the majority of the population hibernates in only a few caves. About 23% hibernate in caves in Indiana. The bats hibernate in tight clusters (above).

How is a parasitic fungus affecting the Indiana bat?

Population numbers of the Indiana bat have declined hugely over time and they are listed as endangered. A parasitic fungus causing white nose syndrome (WNS) has had a large impact on the Indiana bat population.

▶ WNS is a fungal disease that grows around the mouth and wings of hibernating bats. The disease causes abnormal behaviors in the bats, such as daytime flights during winter. Many bats of different species have been killed by WNS.

▶ WNS is spread by direct contact with other infected bats or with infected material in the cave, e.g. guano. The fungus can be infectious for a very long time in the cave and can be spread to new caves by people and bats.

3. Why is the white nose syndrome outbreak considered parasitic? _____
_____

4. Evaluate the impact that a WNS outbreak has on the cave ecosystem stability:

    (a) In the short term: _____
    _____
    _____
    _____

    (b) In the long term: _____
    _____
    _____
    _____

5. Parasitism, causing WNS, has contributed to the severe reduction in the Indiana bat population in the last decade. Investigate factors leading to WNS outbreaks, considering appropriate sources to collect useful data and information. Design a solution to halt the population decline of Indiana bats and ultimately restore their numbers so that they are no longer listed as endangered. You should consider the effect of biological factors and human activities on bat conservation and use both scientific and engineering solutions in your conservation plan.

    Once you have established your plan, you should present the information as an educational resource, e.g. a pamphlet, poster, or slide presentation, designed to teach the general public about the importance of the Indiana bat to ecological stability and how they can help protect it.

217 Commensalism: Free for the Taking

Key Question: How does commensalism support an ecosystem to maintain sufficient populations of particular species?

- Recall that **commensalism** occurs when some species have a symbiotic relationship with another one that provides resources or services that would be difficult for it to acquire otherwise. The other species gains no obvious benefit from the relationship.
- A healthy population of the beneficial partner in this relationship is likely to be important for maintaining **ecosystem** stability. They control populations of some food sources and are, in turn, likely to be a food source for another.

Eastern screech owls and Texas blind snakes

- Both Eastern screech owls and Texas blind snakes are found in many places across the state. The blind snake spends most of its life underground, coming up to the surface infrequently during the night or if it rains.
- Researchers have found live blind snakes in some Eastern screech owl nests, brought there by the owls themselves.
- The owls do not eat the blind snakes. They benefit from the snake eating rotting meat garbage in the nest. This keeps it clean and reduces the chance of disease.
- The blind snake obtains no direct benefit. It can find food equally well without being brought to the nest. The snake will leave the nest and return to the ground if it has the opportunity.
- This type of ecological relationship is called commensalism.
- Data from investigations indicated that owlets (young owls) from the nests that contained blind snakes were healthier, and had a faster growth rate than those from nests with no blind snakes present.

1. What factors in the owl and blind snake commensalism make it likely that this behavior will perpetuate?
   _____
   _____

2. What might be the impact on the eastern screech owl population, and therefore the ecosystem stability, if the blind snake populations disappeared from the owl habitats?
   _____
   _____
   _____

3. Many examples of commensalism exist, such as cattle egrets and cattle, tree frogs and trees, ball moss and trees. Investigate any example, aside from those given in this chapter and gather information to answer the questions below:

   (a) What example have you selected: _____

   (b) What species benefits from the relationship, and how? _____
   _____
   _____

   (c) What role does the species that benefits from the relationship play in the ecosystem: is it a predator, a food source for other species, or does it have another role?
   _____

   (d) What might be some consequences of your selected benefited species being removed or disappearing from the ecosystem? Investigate how might that affect the stability of the ecosystem using several sources, such as those from an online search and library books. Summarize your notes, and attach your report to the page.

218 Mutualism: A Beneficial Dependence

Key Question: How can the disruption of mutualistic relationships result in ecosystem instability?

Mutualistic relationships and access to energy

▶ Mutualistic relationships can provide both parties with access to a resource that would otherwise be unavailable (usually food) and create new pathways for energy flow in **ecosystems**.

▶ Stable ecosystems depend on the presence of both species in sufficient numbers to maintain healthy populations.

Resource-resource relationships: One resource is traded for another (usually food or a nutrient)

Staghorn coral

Worker termite

Reef building corals rely on **mutualism** with algae in their tissues. The algae supply the coral with energy (glucose and glycerol) and, in return, obtain a **habitat** and the compounds they need for photosynthesis and growth (CO_2 and nitrogen). This symbiosis is crucial to recycling nutrients within the reef.

Termites eat wood. They rely on a bacterial community in their gut to break down the cellulose in wood to produce the fatty acids they use for energy. This obligate relationship provides food for both microbes and termites. Termites are responsible for recycling most of the dead wood in tropical ecosystems.

Lichens are a composite organism formed by an alga or cyanobacterium living among the filaments of a fungus. These symbioses are very successful, involving about 20% of all fungal species. Lichens are common pioneer species in ecosystems and make energy available to higher **trophic levels** that would otherwise be lost.

Service-resource relationships: A service is performed in exchange for a resource, e.g. food for protection

Ant guards its aphids

Honeybee pollinating a purple crocus

Some ant species "farm" aphids and protect them from predation by beetles. In return, the ants harvest and consume the honeydew excreted by the aphids. The relationship provides a reliable source of high-energy food for the ants, enabling them to meet their colony's needs.

Many flowering plant species have a mutualistic relationship with their bee pollinators. The bee obtains food (nectar) and the plant receives a service (pollination). The mutualism is normally, but not always, facultative; i.e. neither species depends exclusively on the other for survival.

Many plants have symbiotic mycorrhizal relationships with fungi that live in or on their roots. The fungi help the plant extract nutrients from the soil. This means the plant is much more productive than would it might otherwise be, making more energy and matter available to the community as a whole.

1. Some mutualistic relationships are the result of evolved co-dependence between two species, so that each species survival depends on the presence of the other. Detail the vital role in resource creation that each pair of species has:

   (a) Staghorn coral and algae: _____

   _____

   (b) Worker termite and gut microbes: _____

   _____

   (c) Algae and fungus in a lichen: _____

   _____

The rhizosphere

▸ Some plants deliberately release up to 40% of the carbohydrates they produce from photosynthesis into the soil around their roots. These sugars are often modified into complex, larger molecules by the plant before it releases them into the rhizosphere (the area directly around the roots).

▸ Four key groups of bacteria are attracted to plant roots as they push through the soil. These are similar to those found in the human gut. The bacteria 'fix' nitrogen from the air in a process that plants can't do, and release those nitrogen-containing products.

▸ A plant will release different chemical signaling molecules through its roots, depending on the particular bacteria species it requires. It nurtures a sort of bacterial 'zoo', forming a mutualistic relationship that is beneficial to all parties.

▸ The density of bacteria in the rhizosphere can reach over one billion individuals per gram of soil. This protects the roots from other 'unfriendly' microbes which could enter and attack the plant.

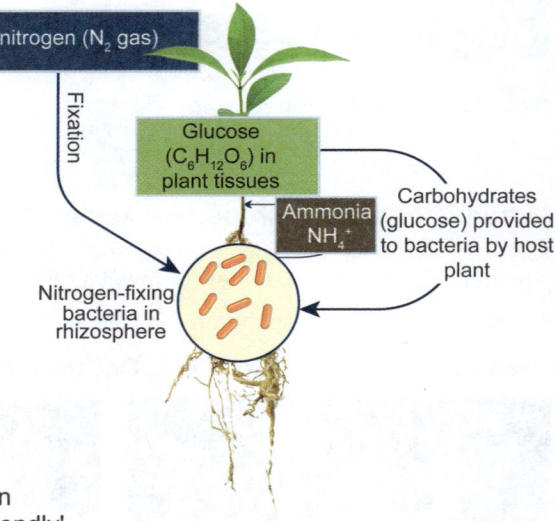

2. The rhizosphere is also known as a plant's 'external gut'. As well as containing similar bacteria to those found in human digestive systems, the rhizosphere displays some similarities in function. Drawing on any prior knowledge, what might be the main mutualistic function of the rhizosphere?

_____

_____

3. What might be some possible consequences to the plant and soil ecosystem stability if the mutualistic rhizosphere bacteria were damaged by the use of chemical fertilizers:

_____

_____

_____

Investigation 9.2   Mutualistic relationships in my community

See appendix for equipment list.

1. Pollination between insects and flowers is a mutalistic relationship that can be investigated and observed easily. This activity could take place close to school if there is a suitable area for a field trip, or near your home if the investigation is assigned for homework.

2. You will need to go to an area that has pollinators and flowers present. You must wear suitable clothing and slip resistant shoes. Be careful during your observations, to prevent any injuries, especially from stinging insects that you may be close to. Do not handle the insects. Be aware if you have allergic reactions to insects, such as bees, and take appropriate precautions and medicines if required. Report to your teacher immediately if stung or injured during observations.

3. Within your class, or student group, stay in one area for at least 10 minutes. Closely observe a flowering plant being pollinated by insects, or birds, if possible, in your location.

4. Record your observations of pollination of flowers using photos, scientific drawings, and notes. Your observations will need to be detailed enough to identify both insect and plant.

4. From your observations, complete a project, either individually or in pairs, in any suitable format and setting. This can be either digital or on paper, and must include both text and images. Any keywords you use should be defined. Your insect and plant must be identified by common and scientific names. Include detailed information about the mutualistic relationship between both species, what type of resource or service each species benefits from, and what might be the consequences to the ecosystem if either species were to decline in number or disappear from the area. Include one or more possible solutions to prevent loss of these species. Present your findings as a class presentation.

219 Eat or be Eaten

Key Question: How does energy and matter move through an ecosystem where the Texas blind salamander is the apex predator?

- A rare type of amphibious salamander, the **Texas blind salamander**, *Eurycea rathbuni* (above), can only be found in a very small region near San Marcos, Texas. This region contains 67 square kilometers of an underground cave system. Scientists estimate that the species shares a common ancestor with other salamanders dating back 10-15 million years before it was isolated.

- Despite its lack of eyes, the Texas blind salamander is the apex (top) predator in the cave aquifer **ecosystem**. It captures its food using a pressure sensing ability, similar to the lateral line sensory organ in some fish. It also has an extra large head and has more teeth than other salamander species.

- The salamander eats shrimps, water snails, and other invertebrates, although it can survive months without eating.

1. (a) Draw a simple diagrammatic model in the box below to represent the feeding relationships between the aquatic plants, herbivorous invertebrates, and the carnivorous Texas blind salamander described above:

   (b) All life on Earth needs energy to survive. Animals obtain energy from the food they eat, either from plants or by eating other animals. Where do plants obtain their energy from?

   _____

2. What might happen to the aquifer ecosystem described above if water levels fell significantly, due to human use?

   _____
   _____
   _____

220 Photoautotrophs and Heterotrophs

Key Question: How can organisms be divided into groups based on their role of matter cycling and energy flow through an ecosystem?

▸ **Autotrophs**, also known as **producers**, make their own food using environmental energy sources. Photoautotrophs use light energy in the process of photosynthesis. Another group, called chemoautotrophs use chemical energy from inorganic chemicals.

▸ **Heterotrophs** consume the food sources made by autotrophs. They are also called **consumers**.

Respiration
Heat given off from metabolic activity.

Growth and new offspring
New offspring as well as new tissues, e.g. branches and leaves.

Eaten by consumers
Some tissue eaten by herbivores and omnivores.

Photoautotrophs

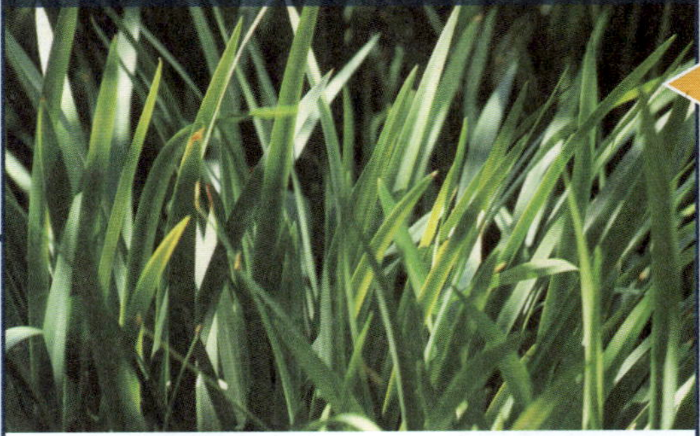

Wastes
Metabolic waste products are released.

SUN

Sunlight is the most common form of energy input for most ecosystems, which rely on photoautotrophs as producers.

Photoautotrophs use **photosynthesis** to fix carbon using the energy in sunlight. Examples: *green plants, algae, some bacteria.*

Reflected light
Solar radiation not utilized by the producer is reflected off the surface of the organism.

Dead tissue
Available to detritivores and decomposers

Respiration
Heat given off from metabolic activity.

Growth and new offspring
New offspring as well as growth and weight gain.

Eaten by carnivores
Some tissue eaten by carnivores and omnivores.

Heterotrophs

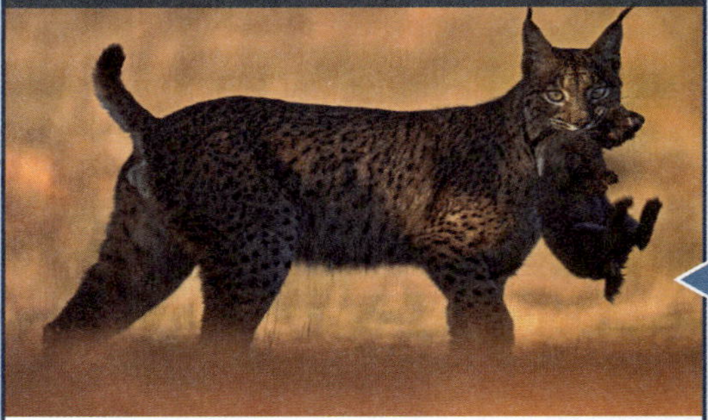

Wastes
Metabolic waste products are released (e.g. as urine, feces, carbon dioxide).

Heterotrophs rely on other living organisms or organic particulate matter for their energy. Heterotrophs include herbivores, carnivores, detritivores, and decomposers.

Examples: *animals, some protists, some bacteria*

Food
Consumers obtain lipids, carbohydrates, and proteins from sources such as plants and other animals. They use these to provide the energy for their metabolic processes and the raw materials they need for growth and tissue maintenance. The organic molecules they obtain are broken down by hydrolysis.

Dead tissue
Available to detritivores and decomposers

What is a photoautotroph?

- The term autotroph means self feeding, and most of the autotrophs on Earth are photoautotrophs. They utilize the energy from the sun to produce glucose, a simple sugar, in a process called photosynthesis.

- Plants are the most visible, and largest, photoautotrophs, but algae, and some bacteria and protists, can also photosynthesize. The color of the chlorophyll substance, a key component used during the photosynthetic process, often distinguishes this group, with 'body' structures appearing green.

- Photosynthesis transforms sunlight energy into chemical energy initially as glucose. This is converted to starch and energy is released when this undergoes further metabolic processes. The inputs and outputs of photosynthesis are shown on the leaf diagram (right).

- Many photoautotrophs have adapted to maximize their ability to obtain the requirements, for photosynthesis, including light.

Sugar (stored energy) and water
Water
Carbon dioxide gas (CO_2)
Sunlight
Oxygen gas (O_2)

Photosynthesis by marine algae provides oxygen and absorbs carbon dioxide. Most algae are microscopic, but some, like this kelp, are large.

On land, vascular plants (plants with transport tissues) are the main producers of food.

Producers, e.g. grasses, make their own food, and are also the ultimate source of food and energy for consumers, such as these cows.

1. (a) Define a photoautotroph in your own words: _____

   (b) In pairs, make a list of at least 10 different species in your local area that are photoautotrophs (producers). Record them below: _____

2. Selecting one of your named photoautotrophs from above, describe its habitat, and name the other organisms that rely on it as a food source: _____

3. Using information from the models on the previous pages, why do you think it is essential for an ecosystem to maintain healthy populations of photoautotrophs (producers)? _____

What is a heterotroph?

▸ Heterotrophs cannot make their own food. They must obtain it by consuming other organisms (by eating or extracellular digestion). Heterotrophs are commonly called consumers. In addition to animals, fungi, and some protists and bacteria are consumers. Heterotrophs (herbivores, carnivores, decomposers) are categorized according to where they get their energy from.

▸ Heterotrophs rely on producers for survival, even if they do not consume them directly. Herbivores, such as the rabbit below, gain their energy by eating plants. Although higher level consumers, such as the eagle, may feed off herbivores, they still ultimately rely on plants to sustain them. Without the plants, the rabbit would not survive, and the eagle could not eat it.

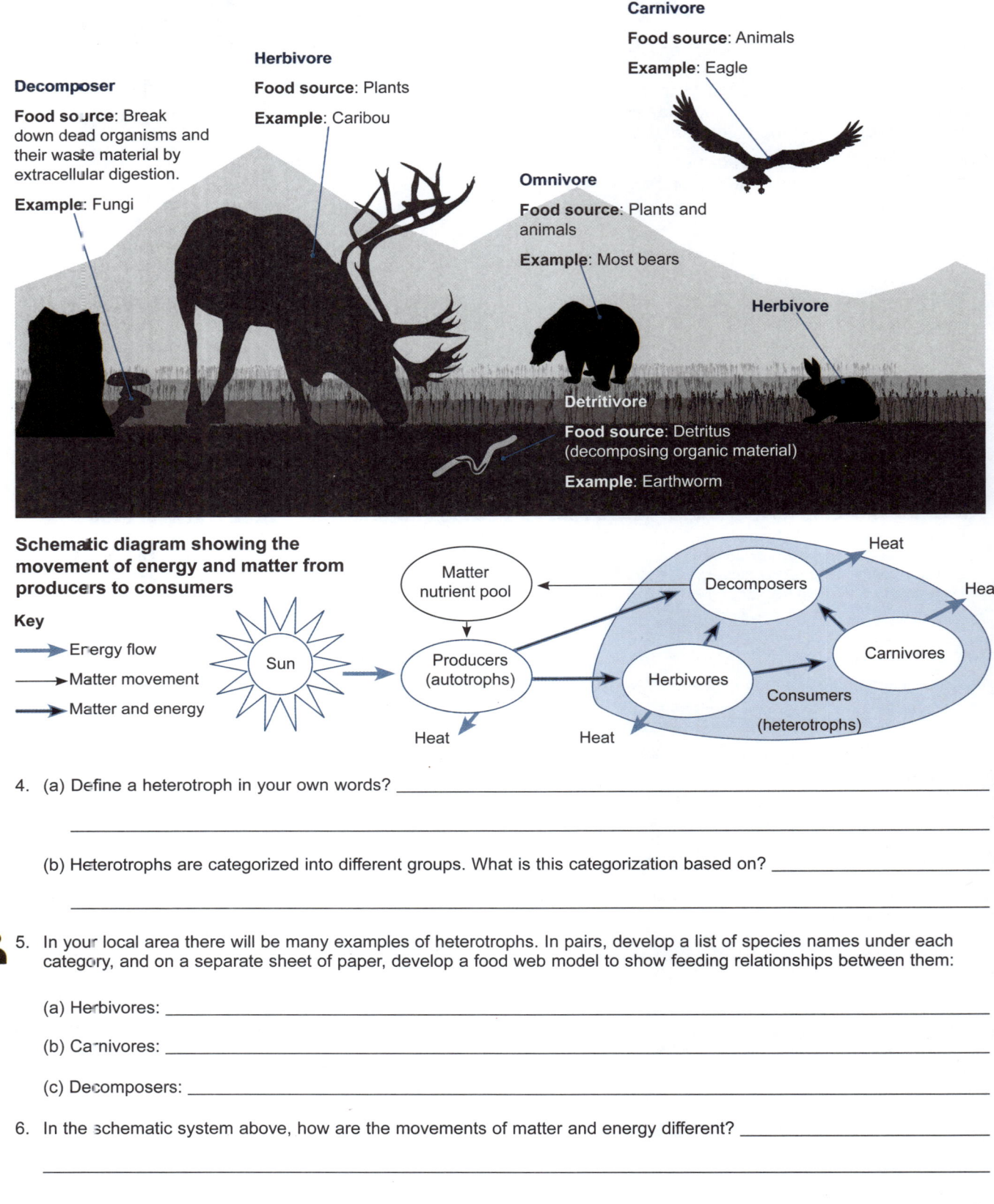

Decomposer
Food source: Break down dead organisms and their waste material by extracellular digestion.
Example: Fungi

Herbivore
Food source: Plants
Example: Caribou

Carnivore
Food source: Animals
Example: Eagle

Omnivore
Food source: Plants and animals
Example: Most bears

Herbivore (rabbit)

Detritivore
Food source: Detritus (decomposing organic material)
Example: Earthworm

Schematic diagram showing the movement of energy and matter from producers to consumers

Key
→ Energy flow
→ Matter movement
→ Matter and energy

4. (a) Define a heterotroph in your own words? _____
   _____

   (b) Heterotrophs are categorized into different groups. What is this categorization based on? _____
   _____

5. In your local area there will be many examples of heterotrophs. In pairs, develop a list of species names under each category, and on a separate sheet of paper, develop a food web model to show feeding relationships between them:

   (a) Herbivores: _____

   (b) Carnivores: _____

   (c) Decomposers: _____

6. In the schematic system above, how are the movements of matter and energy different? _____
   _____
   _____

221 Trophic Levels

Key Question: How do trophic levels determine the sequence in which matter (and energy) move from autotrophs to different types of heterotrophs?

Food chains and food webs

▶ Organisms in **ecosystems** interact through their feeding (or trophic) relationships. These interactions can be shown in a **food chain**, which is a simple model to illustrate how energy and matter, in the form of food, pass from one organism to the next. Each organism in the chain is a food source for the next. A **food web** is used to show the interconnections between organisms in an ecosystem, where an organism often relies on more than one other organism for food.

Trophic levels

▶ The levels of a food chain are called **trophic** (feeding) **levels**. An organism is assigned to a trophic level based on its position in the food chain. Organisms may occupy different trophic levels in different food chains or during different stages of their life.

▶ Arrows link the organisms in a food chain. The direction of the arrow shows the flow of energy and matter through the trophic levels. At each link, energy is lost from the system as heat. This loss of energy limits how many links can exist. Most food chains begin with **producers**, which use the energy in sunlight to make their own food. Therefore, sunlight is the ultimate source of energy for life on Earth. Producers are eaten by primary **consumers** (herbivores). Secondary (and higher level) consumers eat other consumers, as shown below.

Great white heron spears a fish to eat

Autotrophs (Producers)	Primary consumers	Secondary consumers	Tertiary consumers
	Herbivores	Omnivores & carnivores	Omnivores & carnivores
Trophic level: 1	Trophic level: 2	Trophic level: 3	Trophic level: 4

1. In your own words, describe what a food chain is: _____
   _____

2. (a) A simple food chain for a cropland ecosystem is pictured below. Label the organisms with their trophic level and trophic status, e.g. primary consumer.

   Corn → Mouse → Corn snake → Hawk

   Trophic level: _____ _____ _____ _____

   Trophic status: _____ _____ _____ _____

   (b) What is the ultimate energy source for both the food chains pictured on this page? _____

   (c) Why are there rarely more than five or six links in a food chain? _____
   _____

222 Matter Cycles Through an Ecosystem

Key Question: How is matter, that is essential to sustain the life of organisms, retained within an ecosystem?

Matter cycles through the ecosystem

▸ The Earth is, effectively, a closed system and the total amount of matter in a closed system is conserved. This means that Earth's fixed supply of nutrients must be recycled to sustain life. Nutrients move between various parts of the Earth: the atmosphere (air), hydrosphere (water), soils, and living organisms.

▸ Carbon, hydrogen, nitrogen, sulfur, phosphorus, and oxygen move through **ecosystems** in cycles, although the cycles may intersect and interact at various points. Energy drives the cycling of matter within and between systems.

▸ The rate of nutrient cycling can vary widely. Some nutrients, e.g. phosphorus, are cycled slowly. Others, such as nitrogen are cycled more quickly (see model below). The type of environment and **biodiversity** of an ecosystem can also have a large effect on the rate at which nutrients are cycled.

Cycling nitrogen

Nitrogen must be in a form that can be utilized by plants. Some bacteria, called nitrogen fixers, can convert nitrogen gas to nitrate. Other bacteria convert nitrates back into nitrogen gas. Both processes are anaerobic.

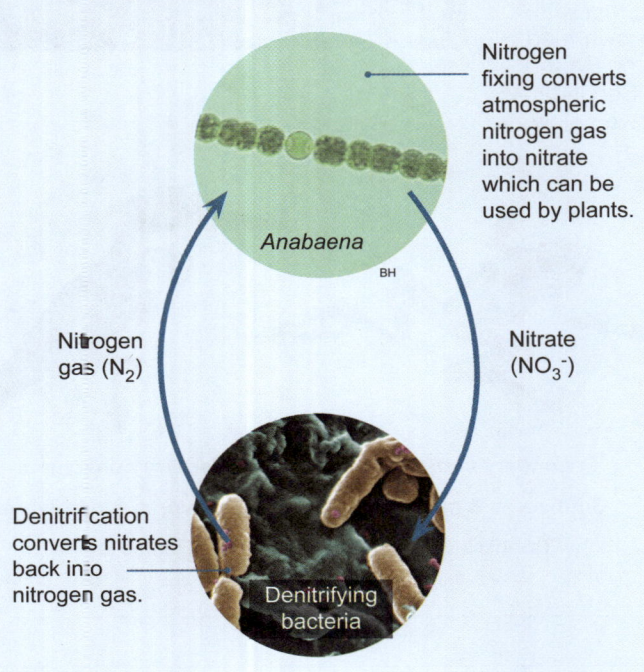

Cycling carbon and oxygen

Photosynthesis and cellular respiration link the cycling of carbon and oxygen. Aerobic respiration, the conversion of glucose and oxygen into carbon dioxide and water, is part of the carbon and oxygen cycles. The carbon dioxide and water are converted back into glucose and oxygen by the anaerobic process of photosynthesis.

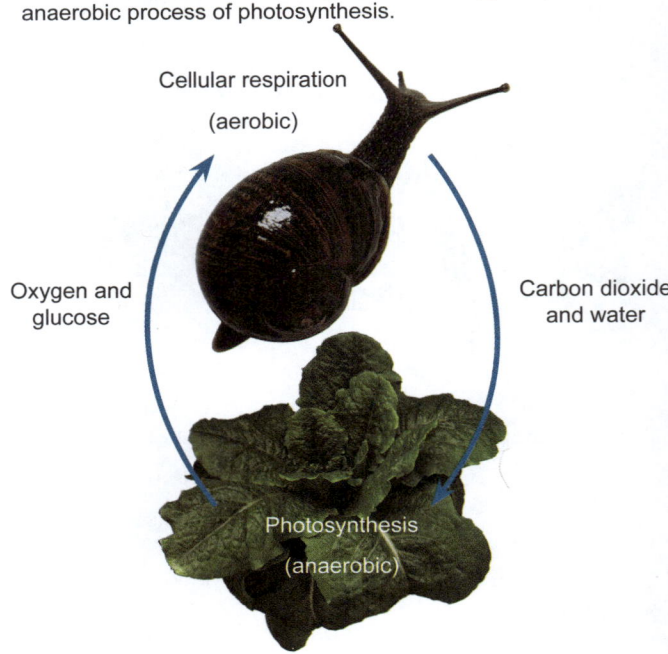

1. Consider a simple ecosystem where a cow eats grass and the cow is consumed by a human in a hamburger. How can the concept of matter cycling be applied to this simple system?

2. How do photosynthesis and cellular respiration interact to cycle matter through an ecosystem?

223 Disruption of Matter Cycles

Key Question: How can the disruption of matter cycles impact an ecosystem?

▸ Due to the finite amount of matter in an **ecosystem**, its stability can be negatively impacted if matter is removed.
▸ Natural events, such as wildfires and volcanic eruptions, can remove large quantities of matter from an ecosystem, as can human activity.

Disrupting matter cycles in the rangelands

▸ Rangelands are large, relatively undeveloped areas populated by grasses, grass-like plants, and scrub. They are usually semi-arid to arid and include grasslands, tundra, scrublands, coastal scrub, alpine areas, and savannah.

▸ Globally, rangelands cover around 50% of the Earth's land surface. The USA has about 3.1 million km² of rangeland, of which 1.6 million km² is privately owned. Rangelands cover 60% of Texas, or 90 million acres.

▸ Rangelands are often used to graze livestock, e.g. sheep and cattle, but they regenerate slowly because of low rainfall. Careful management is required to prevent damage and soil loss as a result of overgrazing.

▸ Grasses grow continuously from a plant meristem (dividing cells) close to the ground, so the leaf can be cropped without causing growth to stop. This allows a field to be grazed in a near-continuous fashion. Grazing by animals stimulates grass to grow and removes dead material. Manure returns material back to the ground and helps fertilize the grass, helping regrowth.

▸ Overgrazing occurs when too many animals are grazed for too long on a section of pasture and the grass does not have enough time between cropping to regrow. Overgrazing may destroy the meristem, in which case plant regeneration stops. Exposed soils may become colonized by invasive species or eroded by wind and rain.

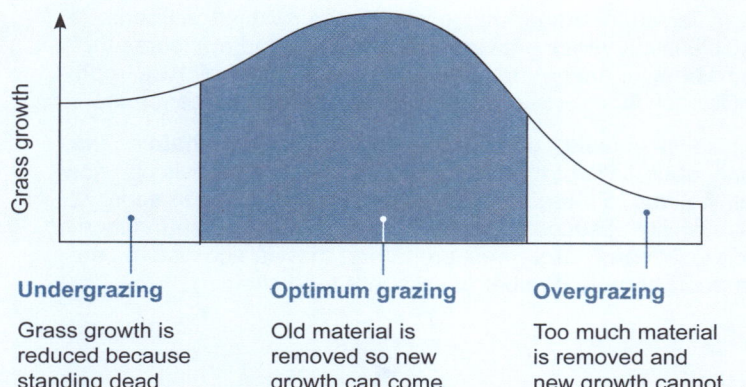

Undergrazing
Grass growth is reduced because standing dead material leaves little room for new growth to come through.

Optimum grazing
Old material is removed so new growth can come through, but enough growing material is left to allow recovery.

Overgrazing
Too much material is removed and new growth cannot become established. Plants die and soil is lost through erosion.

1. (a) Annotate the model below showing a simple matter cycle through grass and Texas Longhorn cattle in the rangelands, showing where the matter cycle is disrupted due to over grazing.
Beside the model, analyze the possible consequences of overgrazing to each species.

(b) Texas Longhorn

(c) Grass

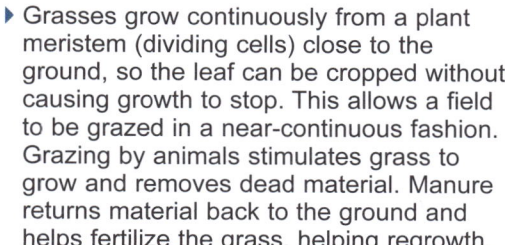

224 Energy Flows Through an Ecosystem

Key Question: How does energy flow through an ecosystem?

Conservation of energy and trophic efficiency

- Energy flows through an ecosystem from one **trophic level** to the next. However, only 5-20% of energy is transferred between each trophic level.
- The Law of Conservation of Energy states that energy cannot be created or destroyed, only transformed from one form to another, e.g. light energy to chemical energy, in the bonds of molecules.
- Each time energy is transferred (as food) from one trophic level to the next, some energy is given out as heat, usually during cellular respiration. This means the amount of energy available to the next trophic level is less than at the previous level.
- Potentially, we can account for the transfer of energy from its input, as solar radiation, to its release, as heat from organisms because energy is conserved. The percentage of energy transferred from one trophic level to the next is the trophic efficiency. It varies between 5% and 20% and measures the efficiency of energy transfer. An average figure of 10% trophic efficiency is often used. This is called the ten percent rule.
- This loss is multiplied over the length of a **food chain** so that only about 0.1% of the energy fixed in photosynthesis can flow all the way to a tertiary **consumer**. The reduction in energy availability at successive trophic levels limits the abundance of top level carnivores (apex predators) that an ecosystem can support and the number of links in a food chain.

Calculating available energy

The energy available to each trophic level will equal the amount entering that trophic level, minus total losses from that level (energy lost as heat + energy lost to detritus).

Heat energy is lost from the ecosystem to the atmosphere. Other losses become part of the detritus and may be utilized by other organisms in the ecosystem.

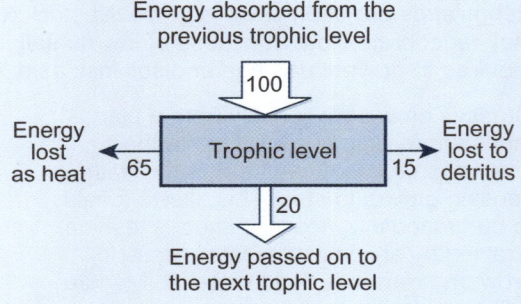

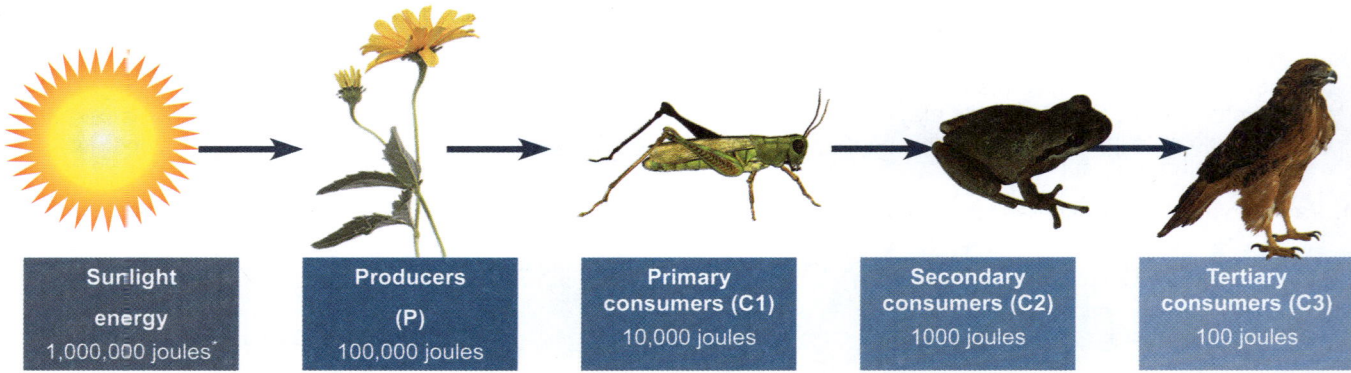

*Note: joules are units of energy

1. Why is the energy available to a particular trophic level less than the energy in the previous trophic level? _____
   _____
   _____
   _____

2. Why must ecosystems receive a continuous supply of energy from the Sun? _____
   _____
   _____

3. (a) What is trophic efficiency? _____
   _____

   (b) In general, how much energy is transferred between trophic levels? _____

Quantifying energy transfer and trophic efficiency in an ecosystem

4. Identify the process occurring at each of points 1-4 on the diagram.

   1: _____

   2: _____

   3: _____

   4: _____

5. Calculate the amount of energy transferred to each trophic level. Write your answers in the spaces labeled (a)-(d):

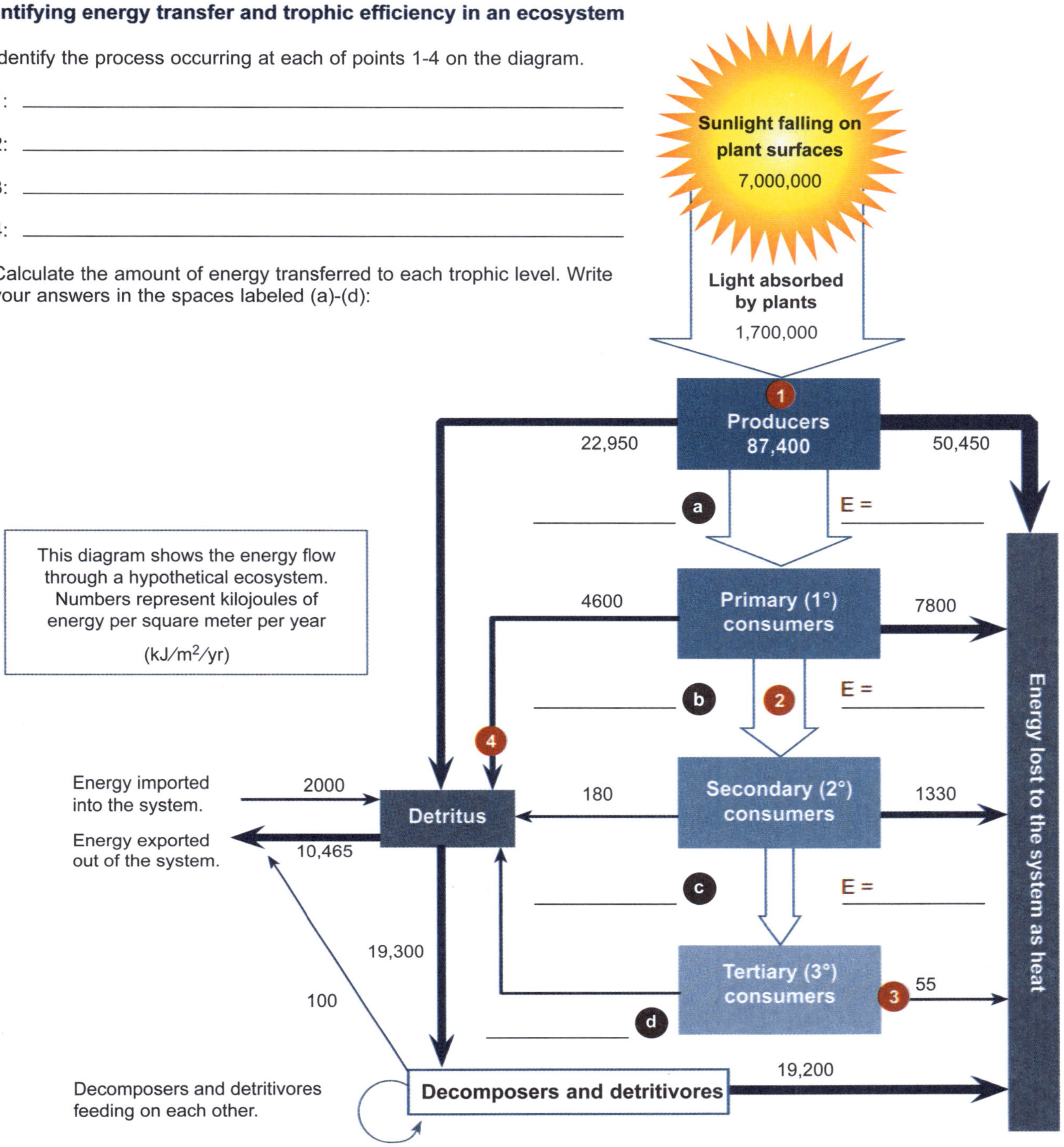

This diagram shows the energy flow through a hypothetical ecosystem. Numbers represent kilojoules of energy per square meter per year (kJ/m^2/yr)

Sunlight falling on plant surfaces: 7,000,000
Light absorbed by plants: 1,700,000

Producers: 87,400
— 22,950 →
— 50,450 →
(a) _____ E = _____

Primary (1°) consumers
— 4600 →
— 7800 →
(b) _____ E = _____

Secondary (2°) consumers
— 180 →
— 1330 →
(c) _____ E = _____

Tertiary (3°) consumers
— 55 →
(d) _____

Energy imported into the system: 2000
Energy exported out of the system: 10,465
Detritus — 19,300 →
100
Decomposers and detritivores feeding on each other.
Decomposers and detritivores — 19,200 →

Energy lost to the system as heat

6. (a) Calculate the percentage of light energy that is absorbed when it falls on the plants:

   Light absorbed by plants ÷ sunlight falling on plant surfaces x 100: _____

   (b) Calculate the percentage of absorbed light energy that is converted (fixed) into producer energy:

   Producers ÷ light energy x 100: _____

   (c) What percentage of light energy is absorbed but not fixed? _____

   (d) Account for the difference between the amount of energy absorbed and the amount fixed by producers: _____

   _____

   _____

7. Calculate the percentage efficiency of transfer between each trophic level. Write each figure on the diagram in the spaces provided (**E =**). Which transfer is the most efficient?

   _____

225 Ecological Pyramids

Key Question: How can the number of organisms, amount of energy, or amount of biomass at each trophic level be represented in an ecosystem?

The energy, biomass, or numbers of organisms at each **trophic level** in any **ecosystem** can be represented by an ecological pyramid. The first trophic level is placed at the bottom of the pyramid and subsequent trophic levels are stacked on top in their "feeding sequence". Ecological pyramids provide a convenient model to illustrate the relationship between different trophic levels in an ecosystem.

▸ Pyramid of numbers shows the numbers of individual organisms at each trophic level.

▸ Pyramid of biomass measures the mass of the biological material at each trophic level.

▸ Pyramid of energy shows the energy contained within each trophic level. Pyramids of energy and biomass are usually quite similar in appearance.

▸ This generalized ecological pyramid (right) shows a conventional pyramid shape, with a large number of producers at the base, and decreasing numbers of **consumers** at each successive trophic level.

▸ Ecological pyramids for this plankton-based ecosystem have a similar appearance, regardless of whether we construct them using energy, or biomass, or numbers of organisms.

▸ Units refer to biomass or energy. The images provide a visual representation of the organisms present.

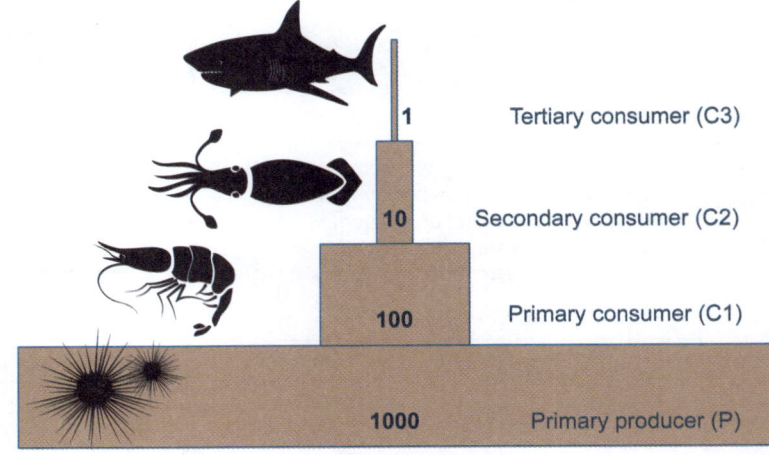

Not to scale

1. (a) What major group is missing from the pyramid above? _____

   (b) Explain the significance of this group in a food web: _____

   _____

   _____

There are benefits and disadvantages to each type of pyramid.

▸ A pyramid of numbers provides information about the number of organisms at each level, but it does not account for their size, which can vary greatly. For example, one large producer might support many small consumers.

▸ A pyramid of biomass is often more useful because it accounts for the amount of biological material at each level. The number of organisms at each level is multiplied by their mass to produce biomass.

▸ While number and biomass pyramids provide information about an ecosystem's structure, a pyramid of energy provides information about function, i.e. how much energy is fixed, lost, and available to the next trophic level.

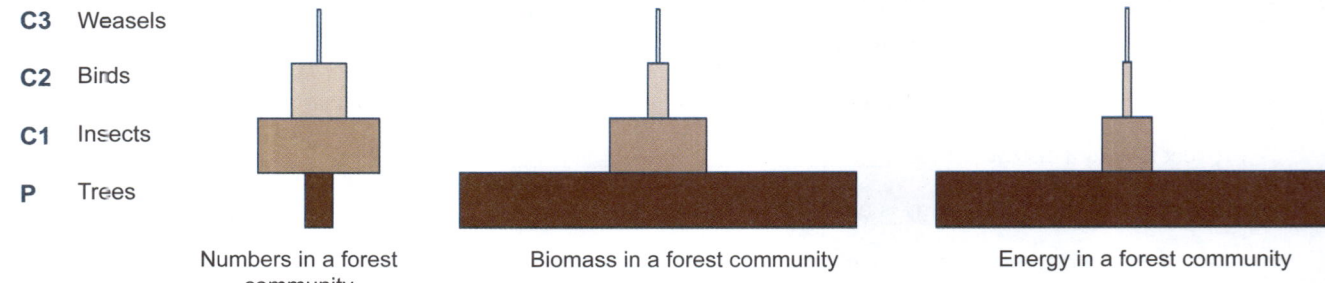

2. What is the advantage of using a biomass or energy pyramid, rather than a pyramid of numbers, to express the relationship between different trophic levels?

   _____

   _____

   _____

A pyramid of numbers can sometimes be inverted

In some ecosystems, e.g. a forest ecosystem (right), a few large producers can support all the organisms at the higher trophic levels. This is due to the large size of the producers, e.g. a large tree can support many individual consumer organisms. A pyramid of energy for this system would be a conventional pyramid shape because the energy of the producers is enough to support the consumer levels.

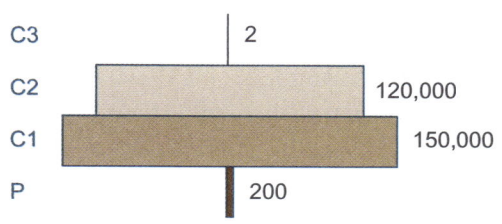

Numbers in 10 m² oak forest, England

3. The forest community above has relatively few producers. How can it support a large number of consumers?

_____

_____

_____

_____

4. The table (below) shows the number of organisms at each trophic level of a grassland community.

   (a) Draw the pyramid of numbers for this data in the space below:

| Numbers in a grassland community ||
Trophic level	Number of organisms
Producer	1,500,000
Primary consumer	200,000
Secondary consumer	90,000
Tertiary consumer	1

   (b) Do you think the pyramid of energy would be similar or different? Explain your answer: _____

   _____

   _____

5. Would a pyramid of energy ever have an inverted (upturned) shape? Explain your reasoning: _____

   _____

   _____

An unusual biomass pyramid

You can see from the previous examples that biomass and energy pyramids usually look very similar. However, in some ecosystems, a single (one-time) measure of biomass may indicate that a smaller, producer biomass is supporting a larger, consumer biomass (figure below right). What this pyramid does not show is the rate at which the producers (algae) are reproducing in order to support the larger biomass of consumers.

Phytoplankton (algae)

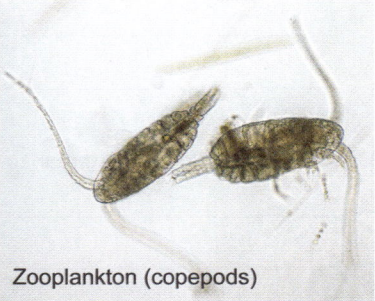
Zooplankton (copepods)

6. Give a possible explanation of how a small biomass of producers (algae) can support a larger biomass of consumers (zooplankton):

   _____

   _____

   _____

Zooplankton and bottom fauna — 21 g/m
Algae — 4 g/m²
Biomass

226 Investigating Ecological Pyramids

Key Question: What patterns do we see in ecological pyramids of real-world examples?

Investigation 9.3 Exploring biomass pyramids

See appendix for equipment list.

1. You can work individually or in pairs for this investigation. It makes use of an online interactive module "Exploring **Biomass Pyramids**". The module is based on real research from an aquatic ecosystem in Panama (Mary Power, 1984). The work examined the ecology of armored catfish (Ancistris sp.) in the Rio Frijoles. These small fish browse algae growing on the substrate. In this investigation, you will collect and analyze data from a virtual river to construct pyramids of energy and biomass. The investigation includes embedded questions, which you will answer in order to proceed.

2. Access the interactive module via **BIOZONE's Resource Hub** or by typing **bit.ly/3LHITkW**

3. Launch the interactive from the button on the left hand corner of the screen. Read through the introduction, then click the LAUNCH FIELD STUDY button.

4. The next screen will invite you to explore the pools of the Rio Frijoles. Once you have done that, you can commit to a pool using COMMIT TO POOL button at the bottom of the screen.

5. Follow the on-screen instructions to make a prediction about the shape of the biomass pyramid for this ecosystem. Once you have done this, move on to sample the algal community and quantify its biomass, and then count the catfish and quantify their biomass.

Aquarium specimen of armored catfish (Ancistris species) showing suckered mouth

---

1. Do your calculations from the investigation support your original prediction? Explain: _____

   _____
   _____

2. Continue with the interactive to run the trophic simulator and examine the productivity of algae over a longer period of time. What does the pyramid of biomass look like now?

   _____
   _____

3. You will be asked to summarize your findings. Paraphrase your summary below: _____

   _____
   _____
   _____
   _____
   _____

4. If you wish, continue the interactive session to explore how algal productivity is affected by the amount of sunlight reaching the pond and how this affects the number of consumers that can be supported. At the end of the interactive session, you can generate a report. Attach your report to this page.

227 Disruption to Biomass and Energy Flow

Key Question: How do different disruption events affect the way biomass and energy move through an ecosystem, and what impact do they have on ecosystem stability?

What about disturbances?

▸ Recall that energy moves through **trophic levels** in the form of biomass, a chemical energy reservoir. Disruptions to **ecosystems** through loss of productivity in the **autotroph** (**producer**) trophic level can alter the abundance of organisms at higher levels. This can lead to a simplification of **food chains** and a loss of **consumer** species, and disrupt ecosystem stability.

▸ In undisturbed communities, producer levels contain more biomass than higher trophic levels. Fires reduce producer biomass, which has an impact on the higher trophic levels from the bottom up. In contrast, an increase in **predation**, such as might occur with the introduction of predator from outside its natural range, affects lower levels from the top down.

▸ Communities already compromised by introduced predators will be more vulnerable to disturbances (such as fire) that reduce producer biomass, because primary consumers will feel the impact of both a reduction in producer biomass and predation.

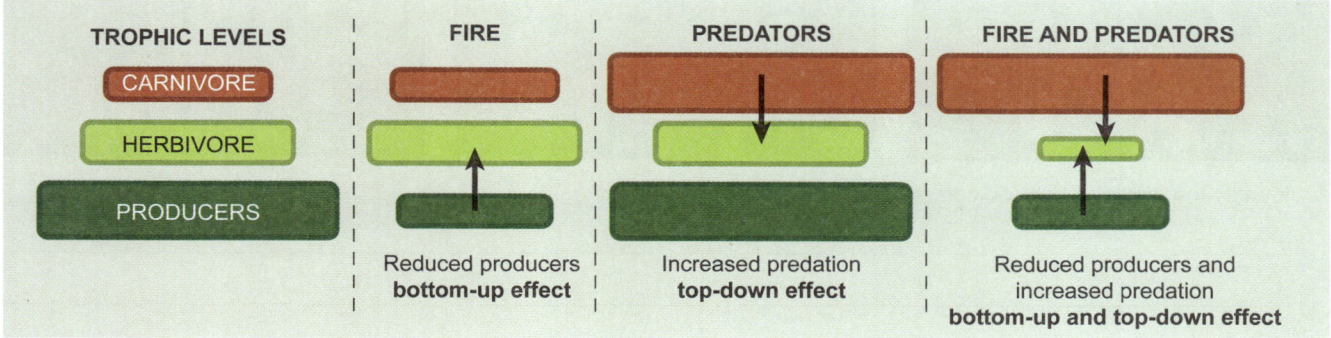

1. (a) Describe examples of events that could lead to a loss of producer biomass: _____
   _____
   _____

   (b) Use the models above to predict the likely the effect of reduced producer biomass on other trophic levels:
   _____
   _____

2. In terms of energy availability, use information from the model to explain the effect of introduced carnivores on a system:
   _____
   _____
   _____

3. The coast of Texas contains habitats that migrating birds rely on for feeding stops along their migration path, e.g. the golden-winged warbler and the ruby-throated hummingbird. Much of the vegetation in the area was destroyed or damaged due to hurricane Harvey, that hit the coast of Texas in 2017. Draw and label a diagram to represent the relative biomass of the two trophic levels before and after the hurricane, using either of these two bird species as examples.

Golden-winged warbler

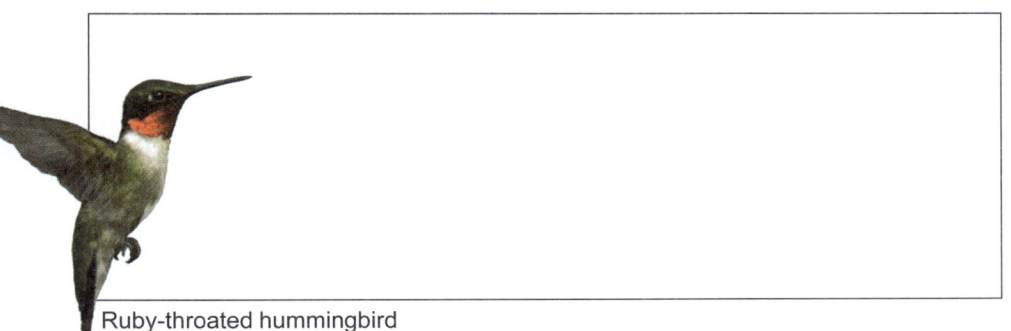
Ruby-throated hummingbird

228 Nutrient Cycles

Key Question: How does matter cycle through the biotic and abiotic compartments of Earth's ecosystems?

Composting food waste models a simple nutrient cycle

▸ All things on Earth are made up of chemical elements, such as carbon, nitrogen, and phosphorus. These elements are transferred through the Earth's systems via a series of biological and chemical processes. In composting, food scraps, lawn clippings, and other green waste are decomposed (broken down). Decomposition produces a rich organic material called humus, and releases minerals in forms that can then be taken up by the roots of plants, e.g. ammonium. Application of humus to soil returns chemical elements to the biosphere as the plants use them to grow. Stable ecosystems rely on carbon and nitrogen, especially, to be recycled for reuse.

Compost heap

Soil

Vegetable garden and hen

1. Identify the biotic and abiotic components in the example above: _____

2. How could the chemical elements from the producers be transferred to the consumer level? _____

3. How could the chemical elements in the consumer be transferred back to the abiotic system? _____

4. (a) The diagram (right) shows a simplified model of the cycling of chemical elements on Earth. You will recall that producers are also known as autotrophs, and consumers and decomposers are classified as heterotrophs. Add arrows to indicate the transfer of energy in this system (including inputs and outputs).

   (b) What are the inorganic substances represented by the green arrow, and how do they enter the producers:

   _____

   _____

   (c) By what process do the nutrients move from one trophic level to the next, as indicated by the red arrows?

   _____

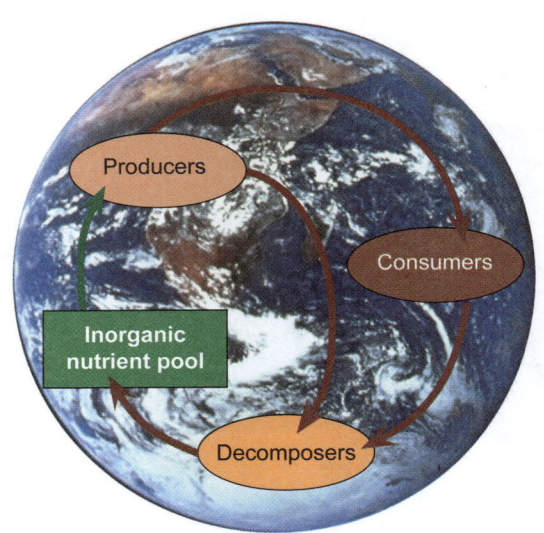

Nutrients cycle through ecosystems

▶ Nutrient cycles move and transfer chemical elements, e.g. carbon, hydrogen, nitrogen, and oxygen, through an **ecosystem**. Because these elements are part of many essential nutrients, their cycling is called a nutrient cycle, or a biogeochemical cycle. The term biogeochemical means that biological, geological, and chemical processes are involved in nutrient cycling.

▶ In a nutrient cycle, the nutrient passes through the **biotic** (living) and **abiotic** (physical) components of an ecosystem. This is modeled in the diagram below. Recall that energy drives the cycling of matter within and between systems. Matter is conserved throughout all these transformations, although it may pass from one ecosystem to another.

Processes in a generalized biogeochemical cycle

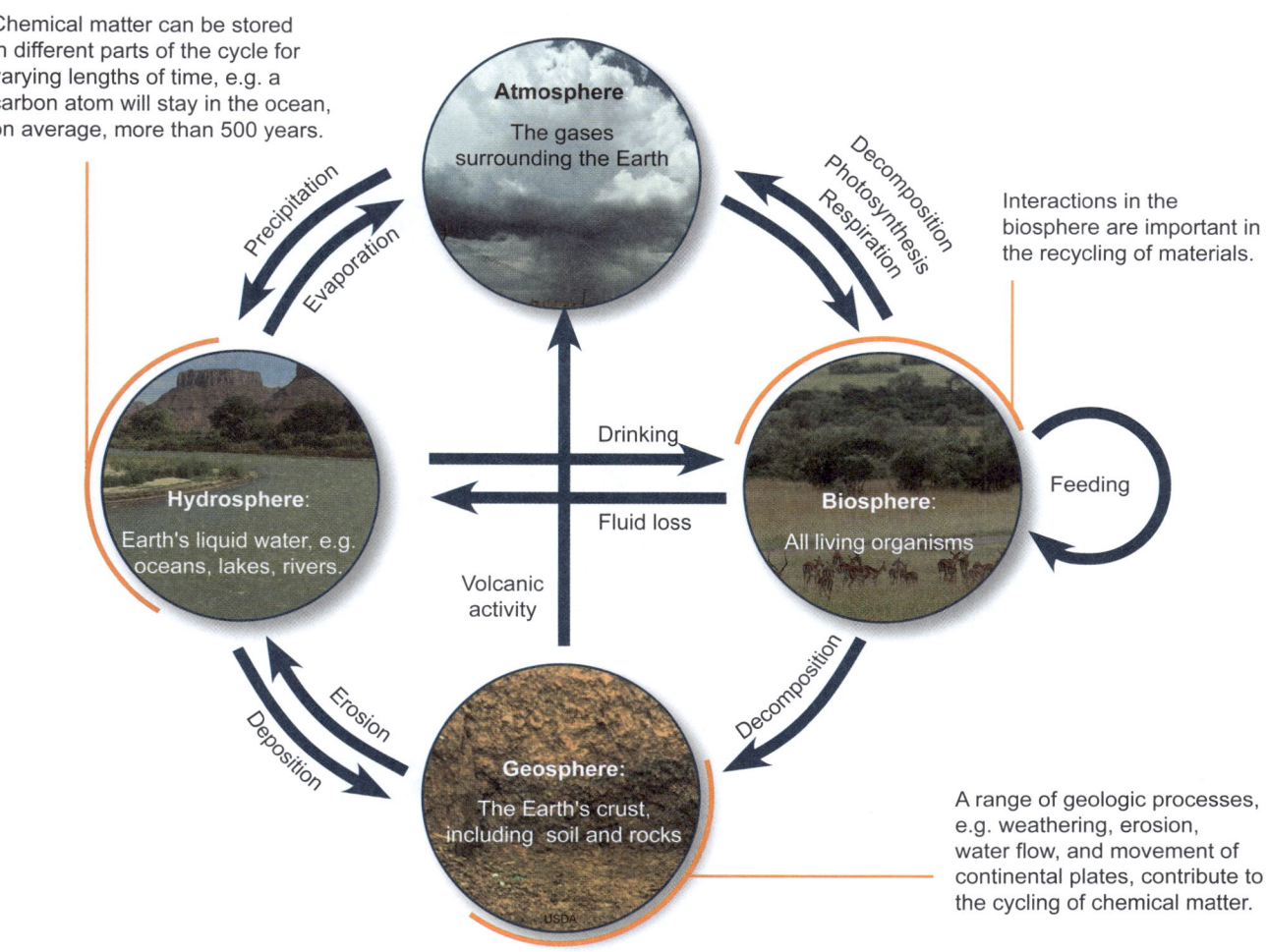

Chemical matter can be stored in different parts of the cycle for varying lengths of time, e.g. a carbon atom will stay in the ocean, on average, more than 500 years.

Interactions in the biosphere are important in the recycling of materials.

A range of geologic processes, e.g. weathering, erosion, water flow, and movement of continental plates, contribute to the cycling of chemical matter.

5. Define the term "nutrient cycle": _____

6. Why is it important that matter containing carbon and nitrogen is cycled through an ecosystem? _____

7. Select one or more components from the diagram above (black text), and provide an example of this process occurring in your local area: _____

229 The Carbon Cycle

Key Question: How does carbon cycle between the atmosphere, biosphere, geosphere, and hydrosphere?

- Carbon is the essential element of life. Its unique properties allow it to form an almost infinite number of different molecules. In living systems, the most important of these are carbohydrates, fats, nucleic acids, and proteins.

- Areas where carbon is stored are known as sinks. Carbon in the atmosphere is found mainly as carbon dioxide (CO_2) and methane (CH_4). In rocks, it is most commonly found as either coal (mostly carbon) or limestone (calcium carbonate). The continual moving of a fixed amount of carbon around the earth is known as the **carbon cycle**.

- The most important processes in the carbon cycle are photosynthesis and respiration. Photosynthesis removes carbon from the atmosphere and converts it into organic molecules. This organic carbon may eventually be returned to the atmosphere as CO_2 through respiration (the oxidation of glucose to produce usable energy for metabolism). The activity of volcanoes also releases CO_2 into the atmosphere.

- Stable **ecosystems** rely on a balanced carbon cycle. Earth and biological processes work in different time scales to move carbon from one sink to another. Human activity has accelerated the shift of carbon into the atmosphere.

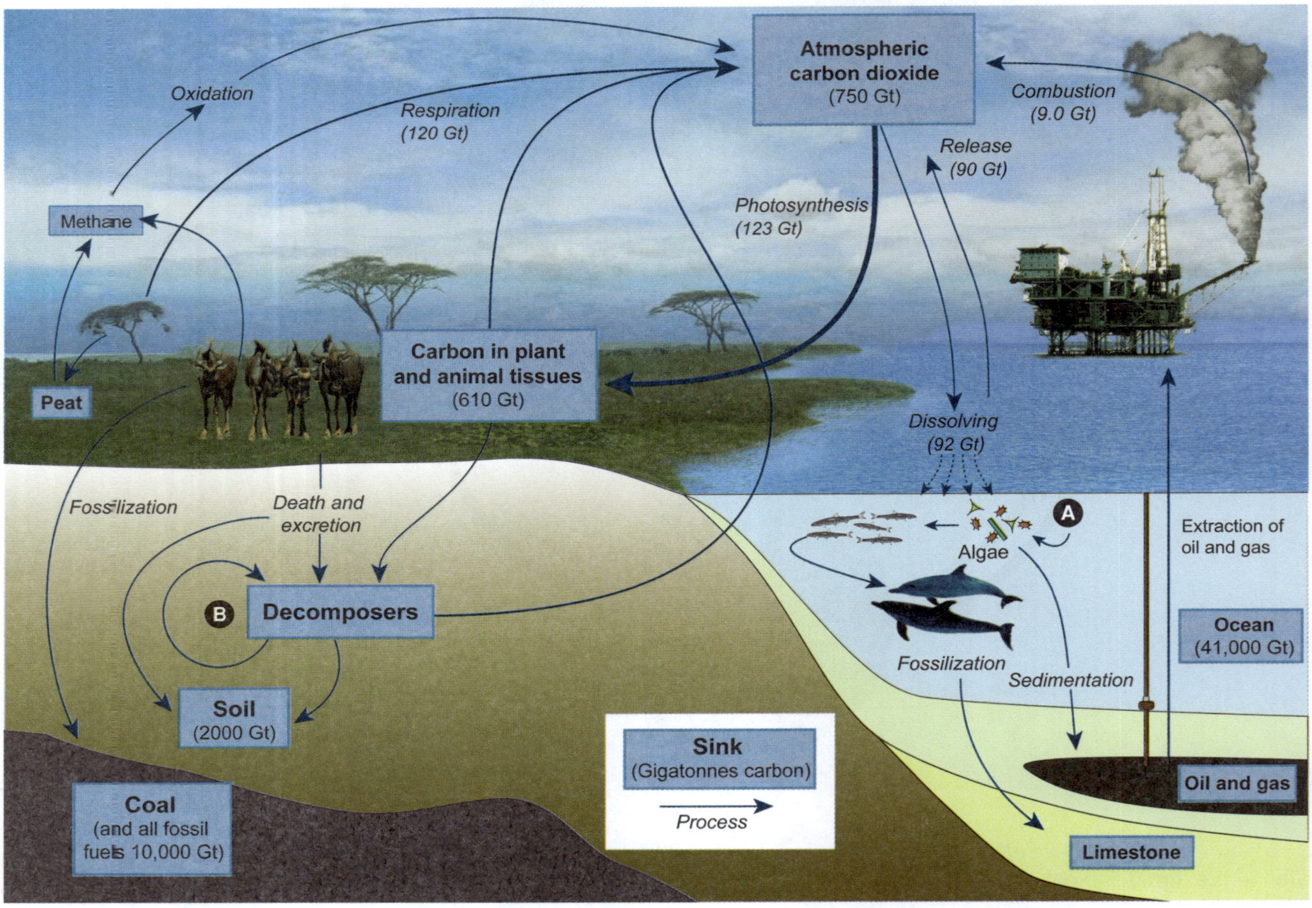

1. (a) In what form is carbon found in the atmosphere? _____
   (b) In what three important molecules is carbon found in living systems? _____
   (c) In what two forms is carbon found in rocks? _____

2. (a) Name two processes that remove carbon from the atmosphere: _____
   (b) Name two processes that add carbon to the atmosphere: _____

3. What is the effect of deforestation, and burning of coal and oil on carbon cycling? _____

230 Modeling the Carbon Cycle

Key Question: How can a simple model be used to represent the carbon cycle?

Investigation 9.4 A model of the carbon cycle

See appendix for equipment list.

⚠ **Living organisms should be handled with care and respect.**

1. In this investigation, you will make a simple ecobottle to model the **carbon cycle** in a small closed ecosystem. Your group will be provided with the following equipment: A large, clear soda bottle with a lid; filtered pond water; aquarium gravel; a source of detritus, e.g. dead leaves; aquatic plant (such as Cabomba); small pond snails.

2. Use the equipment to set up a bottle ecosystem. You will need to think about how long you wait before you close the system off, how much air gap you will have, how much organic material you will add, and where you will put your ecosystem (light/dark).

3. Draw a scientific drawing of your bottle ecosystem. Label the picture to include important design features and a key with the type and total numbers of each organism.

4. Leave your bottle ecosystem for a week. Observe it carefully at various times during the week. After a week, note down any changes since you set it up.

5. Return any living organisms back to the aquarium and dispose of any waste materials.

Cabomba

1. (a) What produces the O_2 in your system?

    _____

   (b) What produces the CO_2 in your system?

    _____

2. The pond water contains small microorganisms. What is their role in this system?

    _____
    _____

3. Was your system stable? Explain why (or why not):

    _____
    _____
    _____
    _____

4. In the space provided, draw a simple diagram to show how carbon cycles between the aquatic plant and the animals:

231 Disruptions to the Carbon Cycle

Key Question: How have the concentrations of carbon dioxide in the Earth's atmosphere changed over time, and what are the implications of this for ecosystem stability?

Carbon in the form of carbon dioxide is increasing in concentration in the atmosphere

▶ The concentration of gases in the Earth's atmosphere fluctuate on small scales over the short term, and on large scales over the long term. Earth's early atmosphere was very different from today's. Even after a nitrogen-oxygen atmosphere was established, the concentrations of gases such as oxygen (O_2) and carbon dioxide (CO_2) changed greatly, causing major changes to the Earth's climate.

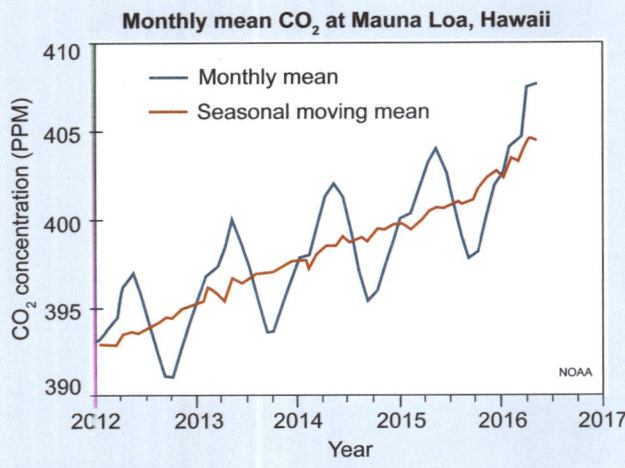

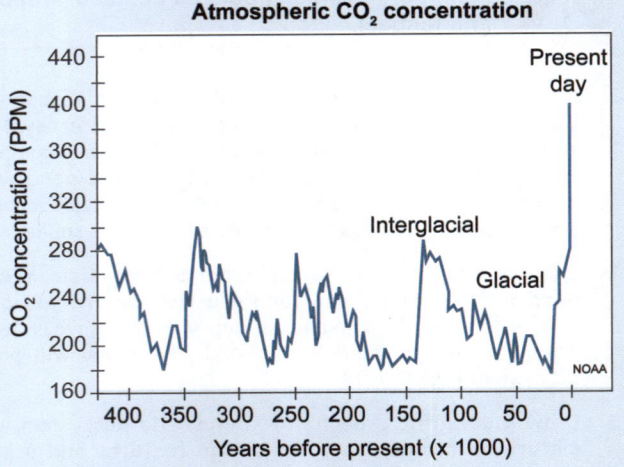

Seasonal CO_2

The vast majority of the land on Earth is situated in the Northern Hemisphere. Across North America and northern Eurasia there are vast tracts of conifer and deciduous forest. During winter, plants in these forests carry out little or no significant photosynthesis. However, the rest of the world keeps on respiring and decomposing, so CO_2 levels rise over the winter season. During spring, deciduous trees and other plants begin photosynthesizing again, removing CO_2 from the atmosphere. Because the winds from the Northern and Southern Hemispheres do not readily mix, this effect is not canceled out by forests in the Southern Hemisphere (which is also mostly covered in oceans).

Long term CO_2

Climate scientists can measure trapped CO_2 in polar ice cores to determine concentrations prior to modern recording. They have found that the concentration of CO_2 has cycled relatively consistently over the last several hundred thousand years. The rise and fall of the atmosphere's CO_2 concentration correlates with the rise and fall of the Earth's surface temperature over at least the last 400,000 years. CO_2 is not the direct cause of the change in temperature, which is actually caused by changes in Earth's orbit. However, it does contribute to a positive feedback effect that helps warm the planet. The steep rise in CO_2 concentration over the last 60 years has been attributed to the burning of fossil fuels.

1. Atmospheric carbon, in the form of carbon dioxide, has fluctuated in the past. What is different about the most recent increase in atmospheric carbon dioxide?

2. What factors do you think might account for this difference in increasing carbon dioxide in the atmosphere, and where in the carbon cycle has the carbon come from?

3. Why can we not easily shift the excess atmospheric carbon back into other carbon sinks or reservoirs (storage areas)?

Carbon dioxide is a greenhouse gas

▸ The Earth's atmosphere comprises a mix of gases including nitrogen, oxygen, and water vapor. Small quantities of carbon dioxide (CO_2), methane, and a number of other trace gases are also present. Together, water, CO_2, and methane produce a greenhouse effect that moderates the surface temperature of the Earth.

▸ The term greenhouse effect describes the natural process by which heat is retained within the atmosphere by these greenhouse gases letting in sunlight, but trapping the heat that would normally radiate back into space. The greenhouse effect results in the Earth having a mean surface temperature of about 15°C, 33°C warmer than it would have without an atmosphere. About 75% of the natural greenhouse effect is due to water vapor, however CO_2 is a more significant factor.

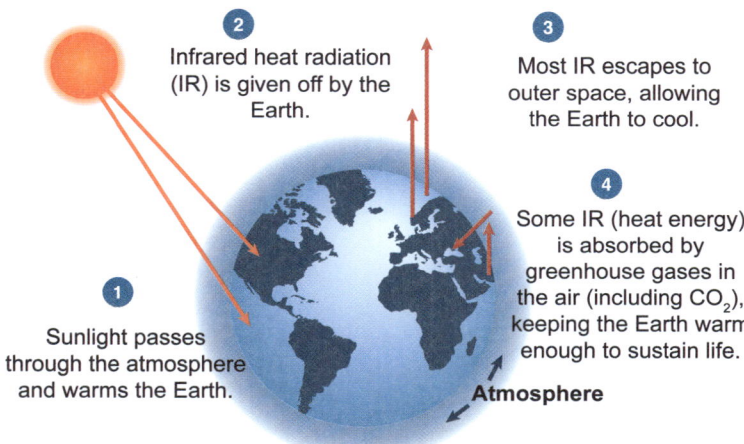

1. Sunlight passes through the atmosphere and warms the Earth.
2. Infrared heat radiation (IR) is given off by the Earth.
3. Most IR escapes to outer space, allowing the Earth to cool.
4. Some IR (heat energy) is absorbed by greenhouse gases in the air (including CO_2), keeping the Earth warm enough to sustain life.
5. **Enhanced greenhouse effect:** Increasing levels of CO_2 increase the amount of heat retained, causing the atmosphere and Earth's surface to heat up.

▸ Fluctuations in the Earth's surface temperature as a result of climate shifts, albeit slow and gradual, that are normal and have occurred throughout the Earth's history. However, since the mid 20th century, Earth's surface temperature has been increasing at a rapid rate. This phenomenon is called global warming and leads to **climate change**. Scientific evidence shows that it can be directly attributed to increasing levels of CO_2 and other greenhouse gases emitted into the atmosphere.

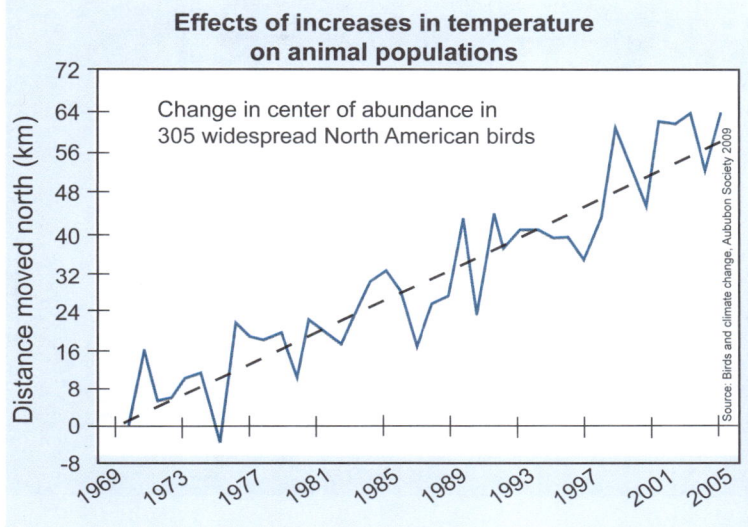

Animals living at altitude are also affected by warming climates and are being forced to shift their normal range. As temperatures increase, the snow line increases in altitude, pushing alpine animals to higher altitudes. In some areas of North America, this has resulted in local extinction of the North American pika (*Ochotona princeps*), a herbivore.

A number of studies indicate that animals are beginning to be affected by increases in global temperatures. Data sets from around the world show that birds are migrating up to two weeks earlier to summer feeding grounds and are often not migrating as far south in winter.

4. How might the disappearance of the North American pika disrupt the stability of the alpine ecosystem? _____

_____

_____

5. What consequences to ecosystem stability might result from migrating birds moving north each year? _____

_____

_____

_____

6. Investigate any local species that may be affected by current or future temperature rises. Record them below:

_____

_____

232 Ocean Acidification

Key Question: How does the increasing amount of carbon dioxide in the atmosphere affect the pH, and therefore stability of the marine ecosystem in the oceans?

- The pH of the oceans has fluctuated throughout geological history, but has always remained at around pH 8.1 - 8.2. Recent studies have measured current ocean pH at around 8.0.
- The oceans act as a carbon sink, absorbing much of the CO_2 produced from burning fossil fuels. When CO_2 reacts with water, it forms carbonic acid (H_2CO_3), which decreases the pH of the oceans.
- H_2CO_3 dissociates into HCO_3^- and H^+ ions. CO_3^{2-} ions from the ocean waters react with the extra H^+ ions to form more HCO_3^- ions. This process lowers the CO_3^{2-} ions available to shell-making organisms, leading to thinner and deformed shells.

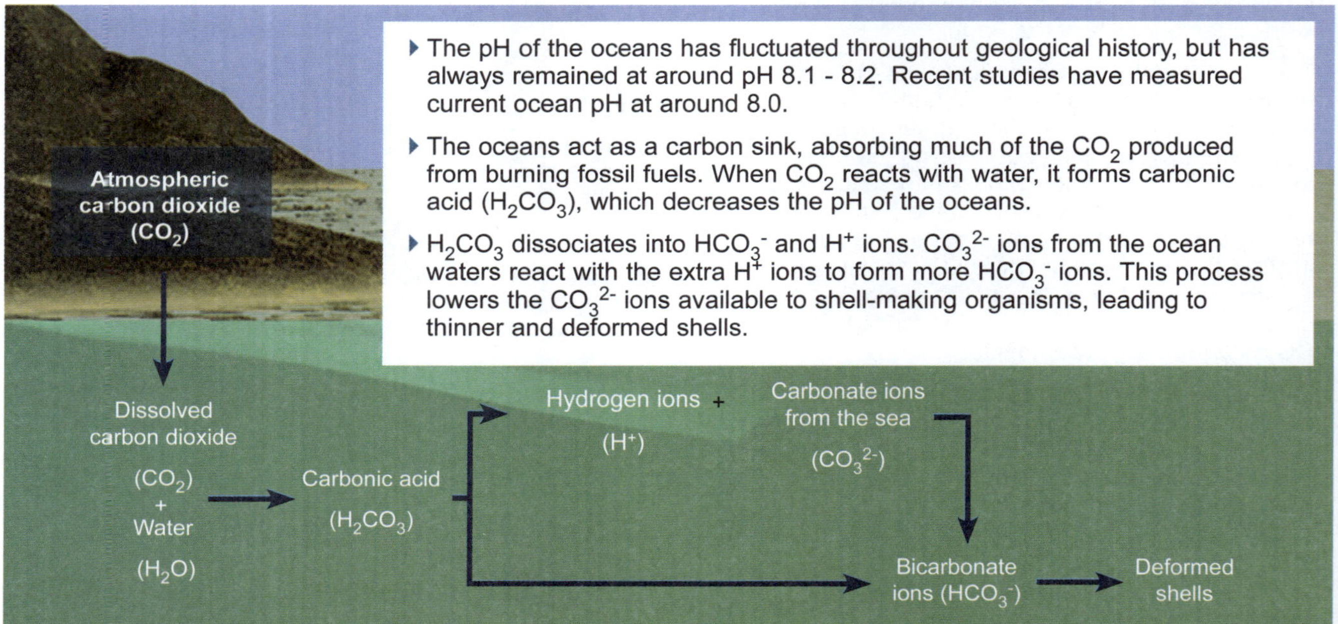

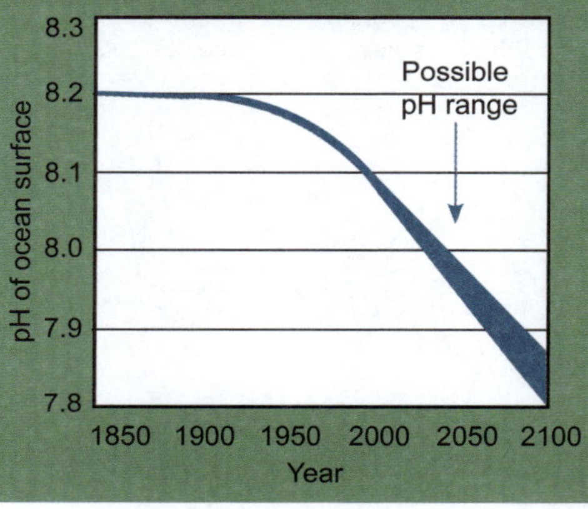

Atmospheric carbon dioxide (CO_2) → Dissolved carbon dioxide (CO_2) + Water (H_2O) → Carbonic acid (H_2CO_3) → Hydrogen ions (H^+) + Carbonate ions from the sea (CO_3^{2-}) → Bicarbonate ions (HCO_3^-) → Deformed shells

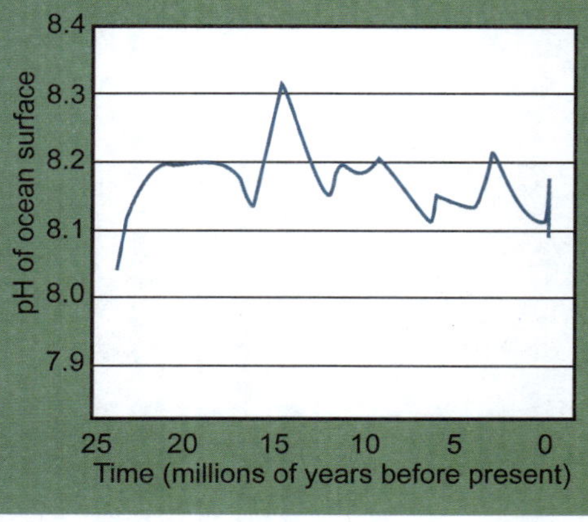

- pH is a logarithmic scale, so even a small change in pH represents a large change in H^+ concentration. Some areas of the ocean, e.g. areas of increased human activity or underwater volcanic eruptions are more affected by pH change than others.

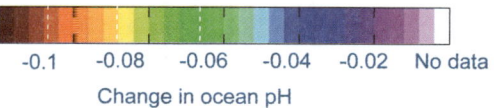

Change in ocean pH: -0.1, -0.08, -0.06, -0.04, -0.02, No data

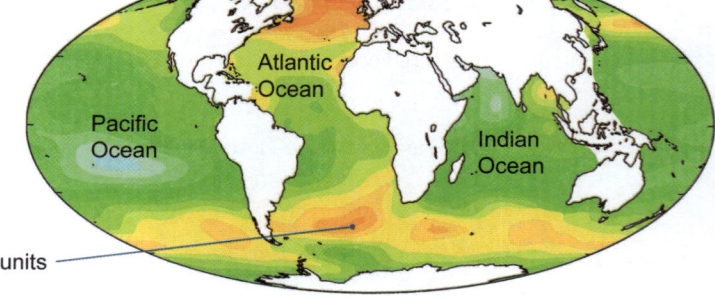

Change of -0.09 pH units

1. (a) What does the term "ocean acidification" mean? _____

   (b) Describe the trend in ocean pH since the 1850s: _____

2. Summarize how a disrupted carbon cycle is leading to ocean acidification: _____

CO_2 and water

▶ In this activity, you will use dry ice (solid CO_2) to determine the effect of CO_2 on pH. The change in pH will be measured using universal indicator (a solution that changes color as pH changes - see the color guide below).

< 3 Strong acid	3-6 Weak acid	7 Neutral	8-11 Weak base	> 11 Strong base

Investigation 9.5 Investigating how dry ice affects pH

See appendix for equipment list.

Wear protective eyewear and insulated gloves when handling dry ice.

Work in pairs or small groups for this investigation.

1. You have dry ice (solid carbon dioxide), 100mL flasks, and universal indicator. Plan an investigation to test if adding carbon dioxide changes the pH of the water.

2. Attach a plan of your investigation – indicating both the independent variable (what you are changing), your dependent variable (what you are measuring), and what you control. Record your hypothesis below (Q3a)

3. Observe what occurs in each flask. Record your observations below (paste photos if you want):

   _____
   _____
   _____
   _____
   _____

3. (a) What is a suitable hypothesis for this investigation? _____
   _____

   (b) What was responsible for this change? Provide evidence: _____
   _____
   _____

   (c) Suggest why this happens? _____
   _____
   _____

4. Evaluate how this experiment could be improved to closer replicate the ocean acidification phenomenon: _____
   _____
   _____
   _____

233 The Nitrogen Cycle

Key Question: What is the importance of nitrogen cycling through the ecosystem?

- Nitrogen is essential for building proteins. Nitrogen gas is converted to nitrates, which are taken up by plants. Animals obtain nitrogen by feeding on plants and/or animals. Nearly 80% of the Earth's atmosphere is made of nitrogen gas. As a gas, nitrogen is very stable and unreactive, effectively having no interaction with living systems. However, nitrogen is extremely important in the formation of amino acids, which are the building blocks of proteins.

- Nitrogen may enter the biosphere during lightning storms. Lightning produces extremely high temperatures in the air (around 30,000°C). At such high temperatures, nitrogen reacts with oxygen in the air to form ammonia and nitrates which dissolve in water and are washed into the soil.

- Some bacteria can fix nitrogen directly from the air. Some of these bacteria are associated with plants (especially legumes) and produce ammonia (NH_3). This can be converted to nitrates (NO_3^-) by other bacteria. Other bacteria produce nitrites (NO_2^-).

- Nitrates are absorbed and used by plants to make amino acids. Animals gain their nitrogen by feeding on plants (or other animals). Nitrogen is returned to the atmosphere by denitrifying bacteria which convert nitrates back into nitrogen gas.

- The movement of the fixed amount of nitrogen around earth is known as the **nitrogen cycle**.

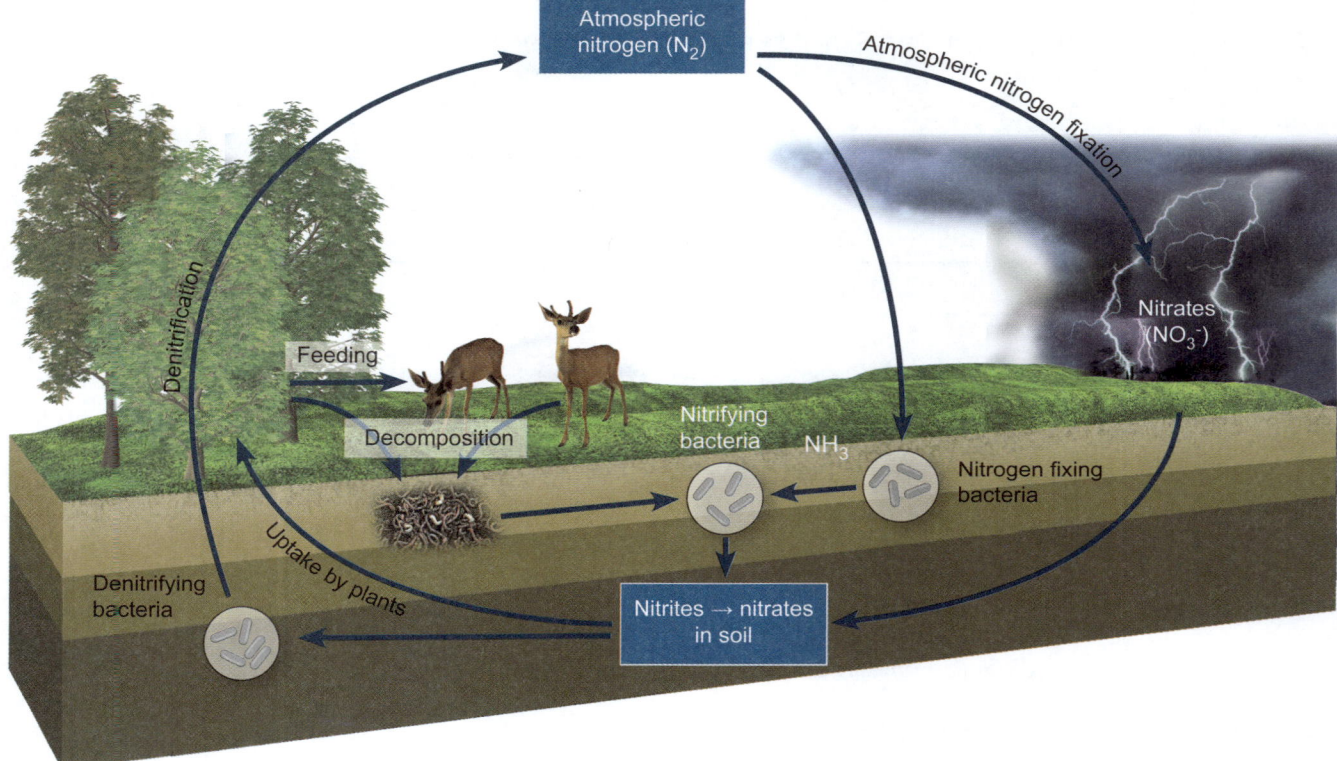

1. Name two processes that fix atmospheric nitrogen: _____

2. What process returns nitrogen to the atmosphere? _____

3. What essential organic molecule does nitrogen help form? _____

4. What role do plants play in the availability of nitrogen to animals? _____

Elements change form as they cycle

▶ As elements cycle, they change their chemical form, which affects their availability to living things. Nitrogen transformations affect its ability to be used by living organisms. Before nitrogen (N) can be incorporated into DNA and proteins, it must be transformed into a form that is available to plants (and therefore to consumers). Bacteria play a very important role in this, transferring nitrogen between the **abiotic** and **biotic** systems.

▶ The Earth's atmosphere is about 78% nitrogen gas. It is very stable and unreactive and is therefore unavailable to most living organisms. Atmospheric nitrogen can be fixed (captured) and transformed into usable forms, e.g. nitrate, by biological or non-biological pathways. Nitrogen can then enter food chains via plant uptake of nitrate nitrogen.

▶ Oxidation of N_2 by lightning forms nitrogen oxides. These dissolve in rain, forming nitrates, which enter soil. In biological pathways, nitrogen fixing bacteria in the soil, water, and in the roots of some plants convert nitrogen gas to ammonia. The ammonia undergoes oxidation by soil bacteria (nitrification) eventually to nitrate. The reverse process, denitrification, is carried out by anaerobic bacteria and returns nitrogen to the atmosphere.

The nitrogen cycle

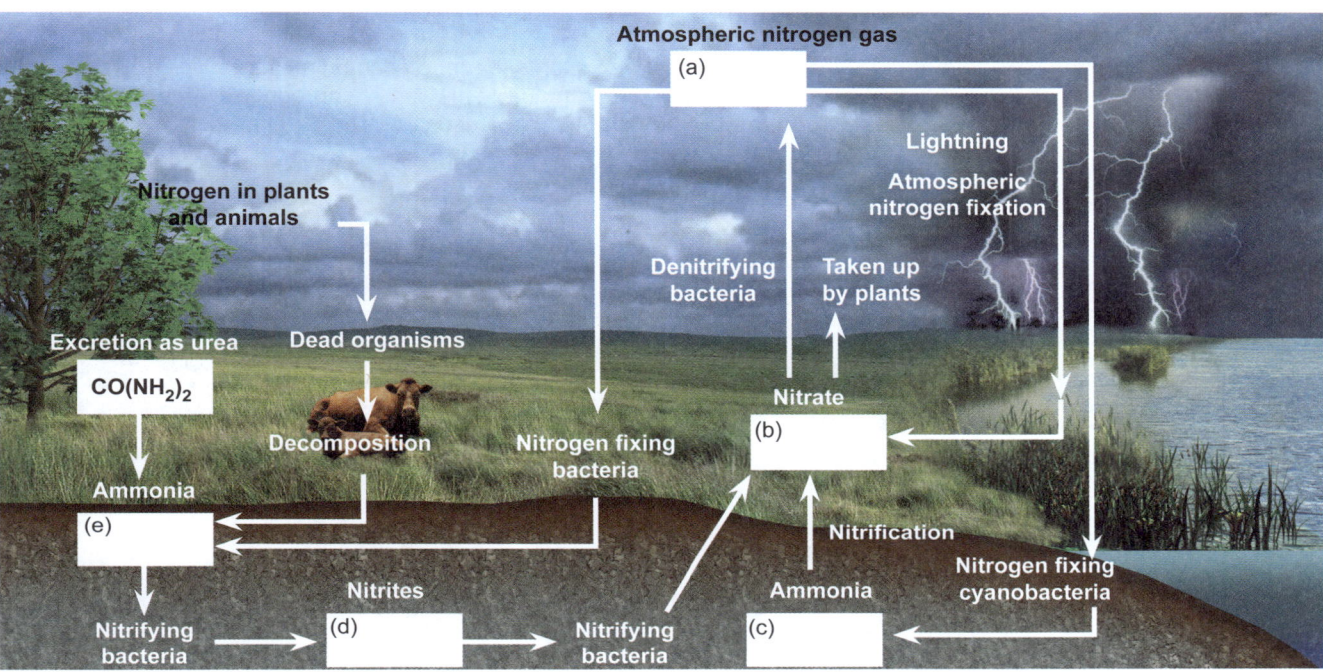

5. **Extension:** In the boxes (a-e) above write the appropriate chemical form of nitrogen (urea has been completed for you). You may need to complete some online research if you are not familiar with the chemical formulae:

6. Identify the form of nitrogen that can be taken up by plants: _____

7. (a) What biological process produces ammonia from atmospheric nitrogen: _____

   (b) What biological process produces nitrate in the soil: _____

   (c) What biological process releases nitrogen to the atmosphere? _____

8. Use the model above to explain the importance of bacteria in the cycling of nitrogen: _____

9. What is the importance of the nitrogen cycle in maintaining ecosystem stability? _____

234 Disruptions to the Nitrogen Cycle

Key Question: What are the consequences of disrupting the nitrogen cycle?

- The largest disruptions to the **nitrogen cycle** occur through farming and effluent (liquid waste or sewage) discharges. Other interventions include burning, which releases nitrogen oxides into the atmosphere, and irrigation and land clearance, which leach nitrate ions from the soil.
- Farmers apply organic nitrogen fertilizers to the land in the form of green crops and manures, replacing the nitrogen lost through harvesting. Until the 1950s, atmospheric nitrogen could not be made available to plants except through microbial nitrogen fixation. During World War II, Fritz Haber developed the Haber process, combining atmospheric nitrogen and hydrogen to form gaseous ammonia. The ammonia is converted into ammonium salts and sold as inorganic fertilizer.
- The Haber process made inorganic nitrogen fertilizers readily available at relatively low cost, increasing crop yields and revolutionizing farming. However, it comes at a large energetic cost: around 1-2% of the world's annual energy supply is used in the Haber process.
- Application of nitrogen fertilizers at rates that exceed uptake by plants leads to accumulation of nitrogen in the soil. Excess soil nitrogen can kill plants that have low nitrogen requirements. This can encourage growth of non-native plants at the expense of native species. **Biodiversity** and community composition can be altered.

The nitrogen cycle is altered through application of nitrogen based fertilizers (left) and removal of biomass by harvesting (right).

1. Study the graph (right).

   (a) Describe the trend in nitrogen fertilizer use since 1850: _____

   _____

   (b) Emissions of nitrogen in various forms (termed NO_x) have increased over time. What could be contributing to this increase?

   _____

   _____

   _____

   _____

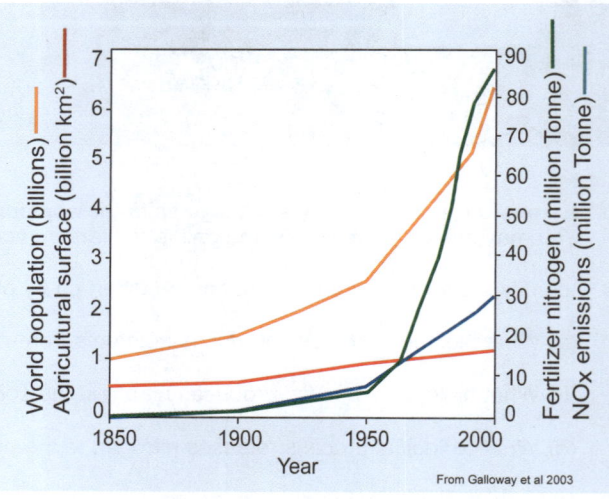

2. What are some consequences of adding excess nitrogen fertilizer to the ecosystem? _____

   _____

   _____

3. What are some consequences of removing biomass from the ecosystem? _____

   _____

   _____

Global changes in nitrogen inputs and outputs between 1860 and 1995 in million tonnes

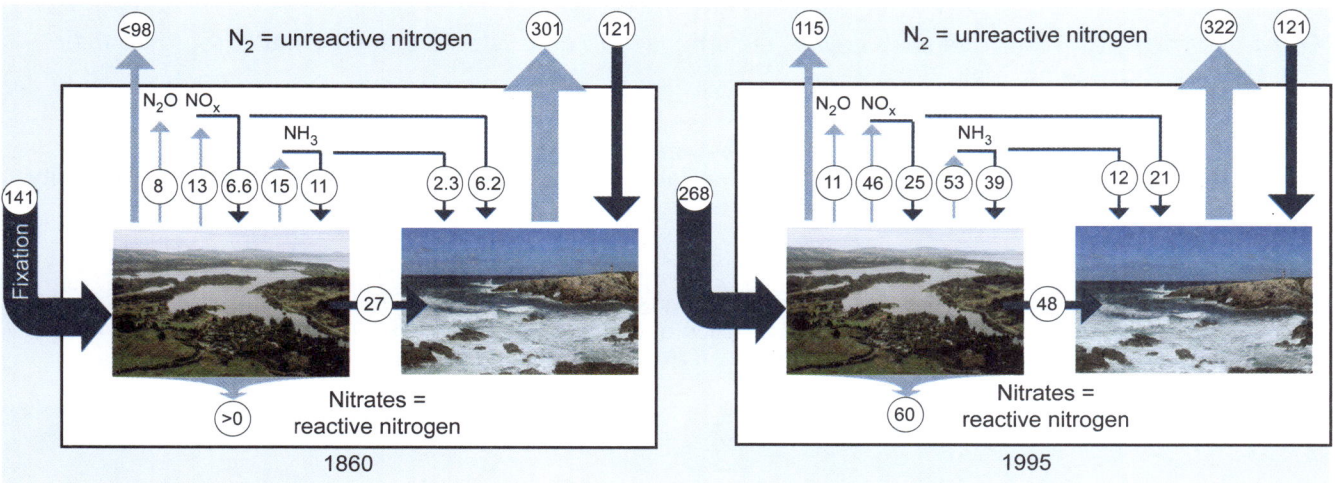

4. Using the quantitative models of nitrogen movement (1860 and 1995) above, calculate the increase in nitrogen deposition in the oceans from 1860 to 1995 and compare this to the increase in release of nitrogen from the oceans:

_____
_____
_____
_____
_____

What are the consequences for aquatic ecosystems of altering nitrogen cycles?

▸ Excess nitrogen from fertilizer use can leach into groundwater and run off into surface waters. This extra nitrogen load is one of the causes of increased enrichment (eutrophication) of lakes and coastal waters. An increase in algal production also increases decomposer activity, depleting oxygen and leading to the death of fish and other aquatic organisms.

▸ At the same time, eutrophic conditions allow undesirable species, tolerant of low oxygen, to increase in numbers. This can permanently disrupt the ecosystem stability, removing food sources for many other animals.

▸ Many aquatic microorganisms also produce toxins. These can accumulate in the water, fish, and shellfish. The rate at which nitrates are added has increased faster than the rate at which nitrates are returned to the atmosphere as N_2 gas.

▸ In Texas, eutrophication has been observed in a number of inland lakes surrounded by farmland. It has been observed in estuaries, rivers, and bays on the southern coast. In built-up urban areas, where fertilizer is used for lawns and garden plants, excess nitrogen can run off from storm water drains and into aquatic **ecosystems**.

5. (a) How does excess nitrogen enter waterways? _____
_____
_____
_____

(b) How do increased levels of nitrogen in water affect the organisms and the biodiversity of the affected environment?
_____
_____
_____
_____

235 Ecosystem Changes can be Permanent

Key Question: Can there be such severe disturbances to ecosystems that they never return to their original state?

- **Ecosystems** are dynamic and constantly fluctuate between particular conditions. However, large scale changes can occasionally occur that completely change and destabilize an ecosystem. This often reduces the **biodiversity**. These include **climate change**, volcanic eruptions, or large scale fires.

Volcanic eruptions change ecosystems

- Volcanic eruptions can cause extreme and sudden changes to local, or even global, ecosystem stability, removing most, if not all, of the previous biodiversity. The eruption of Mount St. Helens in 1980 provides a good example of how the natural event of a volcanic eruption can cause extreme and long lasting changes to an ecosystem.

Mount St Helens, one day before the eruption.

Mount St. Helens, post-eruption.

Before the 1980 eruption, Mount St. Helens had an almost perfect, classic, conical volcano structure. The forests surrounding it were predominantly conifer, including Douglas-fir, western red cedar, and western white pine. Around 35 mammal species and numerous bird species occupied the forests, with amphibians and fish living in the surrounding lakes and streams.

Destruction from the heat, ash, and mudslides was patchwork and random. Surviving plants, animals, and fungi acted as source populations to "reseed" the ecosystem. The volcanic ash provided plentiful nutrients for lakes and the biodiversity of many lakes rapidly returned to normal. However, around 600 square kilometers of forest **habitats** were lost.

Schematic of eruption and recovery

The eruption covered about 600 km^2 (dark blue) in ash, up to 180 m deep in some areas. The schematic shows the general area of bare land per decade since the eruption.

- 1980
- 1990
- 2000
- 2010

Coldwater Lake was formed when Coldwater Creek was blocked by eruption debris.

Spirit Lake was completely emptied in the initial eruption. All life except bacteria was extinguished. The lake has since begun to recover.

The forests on the northern flank of the mountain have vanished, replaced by pumice plains, thinly covered in low growing vegetation.

North Fork Toutle River originated from Spirit Lake before the eruption. It now originates from the mountain's crater. The river itself is now laden with sediment due to erosion.

- Eruption crater
- Original summit
- Mt St Helens

1. Describe the major change to the ecosystem on Mount St Helens' northern flank after the eruption: _____
_____

2. Study the eruption schematic on the previous page. Why has the recovery of the biodiversity in the ecosystem area shown in light blue (2010) been so much slower than elsewhere?
_____
_____

3. Describe the large scale change that occurred at Coldwater Creek after the eruption. What effect would this have had on the biodiversity of the local ecosystem?
_____
_____

4. Describe the large scale change that occurred in the biodiversity of Spirit Lake. Explain why this would cause an almost complete change in the lake ecosystem:
_____
_____
_____

5. (a) What two major changes have occurred in the North Fork Toutle River? _____
_____
_____

   (b) How might this affect the biodiversity in this riverine ecosystem? _____
_____
_____
_____

Deforestation changes ecosystems

▶ Dolly Sods is a rocky high plateau area in the Allegheny Mountains of eastern West Virginia, USA. Originally, the area was covered with spruce, hemlock, and black cherry. During the 1880s, logging began in the area and most commercially viable trees were cut down. The logging caused the underlying humus and peat to dry out. Sparks from locomotives and campfires frequently set fire to this dry peat, producing fires that destroyed almost all the remaining forest. In some areas, the fires were so intense that they burnt everything right down to the bedrock, destroying seed banks. One fire, during the 1930s, destroyed over 100 km^2 of forest. The forests have never recovered. The once forested landscape is now mostly open meadow, resulting in substantial **biodiversity** changes.

The original dense forests of Dolly Sods included spruce, hemlock, white oak and black cherry. These species have been replaced mostly by maple, birch, beech, and low growing scrub.

6. (a) Identify the large scale ecosystem change that occurred at Dolly Sods: _____
_____

   (b) What caused this change in the ecosystem? _____
_____

   (c) Explain why the biodiversity of the ecosystem has not been able to recover quickly after the change: _____
_____
_____

236 Biodiversity in an Ecosystem

Key Question: How is biodiversity measured in an ecosystem?

What is biodiversity?

Biodiversity is the amount of biotic variation within a given group. It could be the number of species in a particular area or the amount of genetic diversity in a species. **Ecosystem** biodiversity refers to the number of ecosystems in a given region and is usually correlated with species diversity. To ensure ecosystem stability, conservation of biodiversity is important.

Biodiversity of Earth	
Type of organism	Estimated number of species
Protozoa	36,400
Brown algae, diatoms	27,500
Invertebrates	7.6 million
Plants	298,000
Fungi	611,000
Vertebrates	60,000

An estimate of the number of eukaryotic species on Earth is 8.7 million, of which only 1.2 million have been formally described. Prokaryotic species are so variable and prolific, the number of species is virtually inestimable and could be hundreds of millions.

Biodiversity tends to be clustered in certain parts of the world, called hotspots, where species diversity is high. Tropical forests and coral reefs (above) are some of the most diverse ecosystems on Earth.

Measuring biodiversity

Biodiversity is measured for a variety of reasons, e.g. to assess the success of conservation work or to measure the impact of human activity. One measure of biodiversity is to simply count all the species present (the species richness). However, this may give an inaccurate impression of the ecosystem's biodiversity. Species evenness gives a measure of relative abundance of species, i.e. how close the numbers of each species in an environment are.

Two ecosystems with quite different biodiversity.

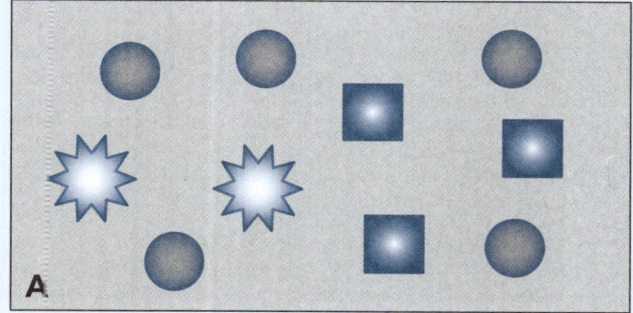
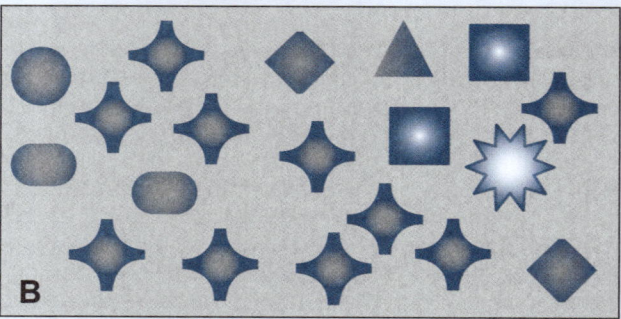

Species richness is a simple method of estimating biodiversity in which the number of species is counted. This does not show if one species is more abundant than others. Species evenness is a method in which the proportions of different species in an ecosystem are estimated.

Diversity Indices use mathematical formulae based on the species abundance and the number of each species to describe the biodiversity of an ecosystem. Many indices produce a number between 0 and 1 to describe biodiversity, 1 being high diversity and 0 being low diversity.

1. Calculate the number and percentage of eukaryotic species that are still to be formally described: _____

2. Describe in words the species richness and species evenness of ecosystem A and B above:

    A _____

    B _____

Comparing the biodiversity of different ecosystems

▶ Simpson's Index of Diversity: $D = 1 - (\sum(n/N)^2)$ is a method of quantitatively measuring diversity. In the equation, n = number of individuals of each species in the sample and N = total number of individuals (of all species) in the sample. It generates a value between 0 and 1. The closer the value is to 1, the higher the biodiversity.

▶ At student observed that, in a conifer plantation, there seemed to be only a few different invertebrate species and wondered if more would be found in a nearby oak woodland. They carried out an investigation and the results are tabled below. The invertebrates are not drawn to scale.

Species	Number of animals (n)	n/N	$(n/N)^2$
Species 1	35		
Species 2	14		
Species 3	13		
Species 4	12		
Species 5	8		
Species 6	6		
Species 7	6		
Species 8	4		
	$\sum n = 98$		$\sum(n/N)^2 =$

Species	Number of animals (n)	n/N	$(n/N)^2$
Species 1	74		
Species 2	20		
Species 3	3		
Species 4	3		
Species 5	1		
Species 6	0		
Species 7	0		
Species 8	0		
	$\sum n = 101$		$\sum(n/N)^2 =$

Species 1 Mite | Species 2 Ant | Species 3 Earwig | Species 4 Woodlice | Species 5 Centipede | Species 6 Longhorn beetle | Species 7 Small beetle | Species 8 Pseudoscorpion

3. (a) Complete the two tables above by calculating the values for n/N and $(n/N)^2$ for the student's two sampling sites:

   (b) Calculate the Simpson's Index of Diversity for site 1: _____

   (c) Calculate the Simpson's Index of Diversity for site 2: _____

   (d) Compare the diversity of the two sites and suggest any reasons for it: _____

4. (a) Species richness is a measure of the number of different species in an area. Which of the two areas sample above has the greatest species richness?

   (b) Why would measuring species richness not be as informative as measuring species diversity? _____

237 Keystone Species and Ecosystem Stability

Key Question: What are keystone species, and why are they important for ecosystem stability?

Keystone species

- **Keystone species** play a crucial role in **ecosystems**. Their actions are key to maintaining the dynamic equilibrium of an ecosystem. Some species have a disproportionate effect on the stability of an ecosystem. These species are called keystone species (or key species). The term keystone species comes from the analogy of the keystone in a true arch. If the keystone is removed, the arch collapses.

- The role of the keystone species varies from ecosystem to ecosystem. However, if they are lost, the ecosystem can rapidly change, or collapse completely. The pivotal role of keystone species is a result of their influence in some aspect of ecosystem functioning, e.g. as predators, prey, or processors of biological material.

An archway is supported by a series of stones, the central one being the keystone. Although this stone is under less pressure than any other stone in the arch, the arch collapses if it is removed.

Keystone species make a difference

- The idea of the keystone species was first hypothesized in 1969 by US scientist, Robert Paine. He studied an area of rocky seashore, noting that diversity seemed to correlate with the number of predators present. To test this, he removed the starfish from an 8 m by 2 m area of seashore. Initially, the barnacle population increased rapidly before collapsing and being replaced by mussels and gooseneck barnacles. Eventually, the mussels crowded out the gooseneck barnacles and the algae that covered the rocks. Limpets that fed on the algae were lost. The number of species in the study area dropped from 15 to 8.

Ochre starfish - Paine removed these in his famous study.

Elephants play a key role in maintaining the savannas by pulling down even very large trees for food. This activity maintains the grasslands.

The burrowing of prairie dogs increases soil fertility and channels water into underground stores. Their continuous grazing promotes grass growth and diversity.

Mountain lions prey on a wide range of herbivores and, to some extent, dictate their home ranges and the distribution of scavenger species.

1. Define the term keystone species: _____

_____

_____

2. Prairie dog colonies are often destroyed by ranchers who believe they compete with cattle for food. How might this lead to the collapse of prairie ecosystems?

_____

_____

_____

_____

Sea otters as keystone species

▸ Sometimes, the significant keystone effects of a species becomes evident when a species declines rapidly to the point of near extinction. This is illustrated by the sea otter example described below.

▸ Sea otters live along the Northern Pacific coast of North America and have been hunted for hundreds of years for their fur. Commercial hunting didn't fully begin until about the mid 1700s, when large numbers were killed and their fur sold to overseas markets.

▸ The drop in sea otter numbers had a significant effect on the local marine environment. Sea otters feed on shellfish, particularly sea urchins, which eat kelp that provides habitat for many marine creatures. Without the sea otters to control the sea urchin population, sea urchin numbers increased and the kelp forests were severely reduced.

▸ Sea otters are critical to ecosystem function. When their numbers were significantly reduced by the fur trade, sea urchin populations exploded and the kelp forests, on which many species depend, were destroyed.

▸ The effect can be seen on Shemya and Amchitka Islands. Where sea otters are absent, large numbers of sea urchins are found, and kelp is almost absent.

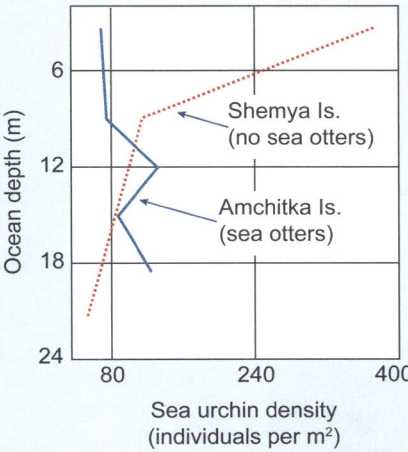
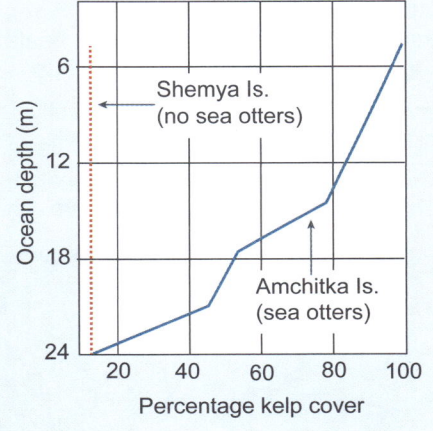

Kelp are large seaweeds, a type of brown algae. There are many forms and species of kelp. Giant kelp can grow to 45 m long.

Similar to how forests on land provide diverse habitats for terrestrial species, kelp provides habitat, food, and shelter for a variety of marine animals.

Sea urchins kill kelp by eating the holdfast that secures the kelp to the seabed. Unchecked urchin populations can quickly turn a kelp forest into an "urchin barren".

3. What is the importance of kelp in the ecosystem? _____

4. (a) What effect do sea otters have on sea urchin numbers? _____

   (b) What effect do sea urchins have on kelp cover? _____

   (c) What evidence is there that the sea otter is a keystone species in these Northern Pacific coastal ecosystems? _____

5. From the new evidence above, if your definition of a keystone species has changed go back to Q1. and edit it.

238 Human Activity and Biodiversity

Key Question: How is human activity affecting biodiversity on Earth?

- The activities of an expanding human population are contributing to an increase in extinction rates above the natural level, and to local and global reductions in **biodiversity**.
- As human demand for resources increases, increasing pressure is placed on **habitats** and their natural populations. The effects of these pressures are often detrimental to ecosystem stability and biodiversity.
- A decline in biodiversity reduces the ability of ecosystems to resist change and recover from disturbance. Humans depend both directly and indirectly on healthy ecosystems, so a loss of biodiversity affects us too.

How does climate change affect biodiversity?

- **Climate change** is the result of increasing amounts of greenhouse gases, such as carbon dioxide, being added into the atmosphere. These trap heat from the Sun, which leads to a continuous rise in the average temperature of the Earth's surface. Climate change is a significant contributor to loss of biodiversity.
- Evidence indicates that human activities causing release of greenhouse gases are responsible for the current rapid climate warming. These are activities such as deforestation, agriculture, industrial use of fossil fuels, and transport.
- Climate change is changing habitats throughout the world. Those organisms that are mobile or able to adapt to the changes are more likely to survive. Those that cannot, are likely to become locally or globally extinct.

Coral bleaching and marine biodiversity

Coral reefs are one of the most biodiverse ecosystems on Earth. They provide a habitat for many marine species to breed and feed. Coral reefs are threatened by human activities such as over-fishing and pollution, but the greatest threat comes from climate change.

Most corals obtain their nutrition from photosynthetic organisms living in their tissues. When coral becomes stressed, e.g. by an increase in ocean temperature, the photosynthetic organisms are expelled. The coral becomes white or bleached (left) and is more vulnerable to disease.

Half of the coral reefs in the Caribbean were lost in one year (2005) due to a large bleaching event. Warm waters centered around the northern Antilles, near the Virgin Islands and Puerto Rico, expanded southward and affected the coral reefs.

Biodiversity on land

The number of quaking aspen trees has declined significantly across the US in recent years. Climate change, which has reduced rainfall and increased drought conditions, is thought to be the cause. Quaking aspen is a **keystone species**, so its loss in some regions has a significant effect on North American biodiversity. Moose, elk, deer, black bear, and snowshoe hare browse its bark, and aspen groves (below) support up to 34 species of birds.

Climate change and the polar bear

Arctic temperatures have been above the 1981-2022 global average every year since 1988, and the extent of Arctic sea ice has also been decreasing (below). This is having a detrimental effect on polar bears, which rely on the sea ice to hunt their prey (seals), mainly in winter. In Canada's Hudson Bay, the sea ice is melting earlier and forming later. As a result, the bears must swim further to hunt and their hunting time is cut short. Survival and breeding success is reduced because they put on less weight during their main feeding time of year.

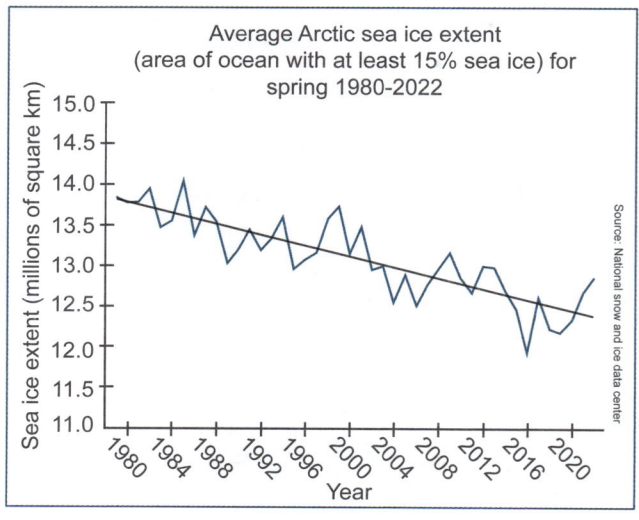

How land use affects biodiversity

Natural grasslands (top photo) are diverse and productive ecosystems. Ancient grasslands may have contained 80-100 plant species, in contrast to currently cultivated grasslands, which may contain as few as three species. Unfortunately, many of the management practices that promote grassland species diversity conflict with modern farming methods. Appropriate management can help to conserve grassland ecosystems, while still maintaining the land's viability for agriculture and food production.

Demand for food increases as the population grows. Modern farming techniques favor monocultures (bottom, photo) to maximize yield and profit. However, monocultures, in which a single crop type is grown year after year, are low diversity systems and food supplies are vulnerable if the crop fails. Rice, maize, and wheat alone make up two thirds of human food consumption and are staples for more than 4 billion people. This creates issues of food insecurity for the human population, currently, and in the future.

- Human activities have had major effects on the biodiversity of Earth.

- Nearly 40% of the Earth's land surface is devoted to agricultural use. In these areas, the original biodiversity, a polyculture of plants and animals, has been severely reduced. Many of these areas are effectively monocultures, where just one type of plant is grown.

- The graph (right) shows that, as the land is more intensively used, the populations and variety of plants and animals fall.

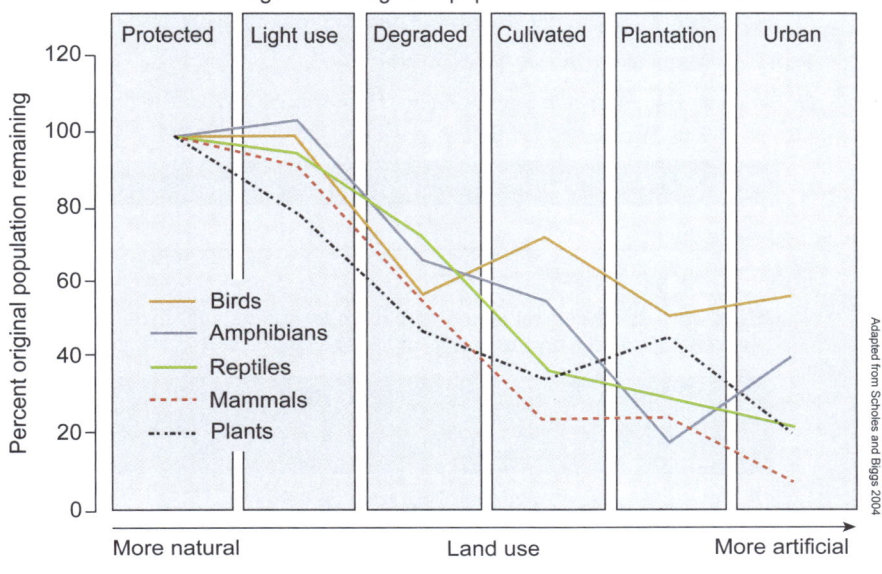

Adapted from Scholes and Biggs 2004

1. How does coral bleaching affect marine biodiversity? _____

2. (a) Describe the trend in Arctic sea ice extent since the 1980s: _____

   (b) Reduced sea ice means polar bears need to swim longer distances to hunt seals. What effect is this likely to have on the polar bear population numbers? _____

3. Study the land use graph above and describe the effect of land development on biodiversity: _____

Investigation 9.6 Investigating biodiversity and human impacts

Biodiversity changes in an ecosystem can be modelled using a simulation. Different species in an ecosystem can be represented by different colored tokens or beads. Individuals in an ecosystem can be removed or added to represent different effects, including effects of human activity. These can be overpopulation, overexploitation, adverse habitat alterations, pollution, invasive species, and changes in climate.

There are many ways to quantify biodiversity. For the purpose of this activity, you will use **Simpson's Index of Diversity,** once more. A simple online calculator can be found at bit.ly/3zhL4od and also accessed through BIOZONE's Resource Hub.

PART 1: Setting Up the simulation

1. Your teacher will mix up enough colored tokens into a large bag or bowl so that every group has enough for 2 or 3 handfuls. The mix has a ratio of Red:4 Blue:3 Green:2 Yellow:1 for every 10 beads or tokens.

2. Working in small groups, your group will take a handful of randomly mixed different colored beads or tokens from a large container that your teacher has already prepared. Spread your handful onto a tray.

PART 2: Measuring initial biodiversity

3. Add up the total number of each color and record in the table. Use the **Simpson's Index of Diversity** calculator to calculate initial diversity. Calculate **Simpson's Index of Diversity**

Species	Red	Blue	Green	Yellow	Simpson's Index of Diversity
Total number					

PART 3: Human impacts

4. After each of these following rounds, add number of each color/species and calculate **Simpson's Index of Diversity.** Develop and perform at least 2 more scenarios and calculate **Simpson's Index of Diversity**

Scenario	Red	Blue	Green	Yellow	Simpson's Index of Diversity
A. Invasive species: Species Red has risen to plague proportions. Add 2 additional red tokens for every red you have in your tray.					
B. Climate change: Species yellow has been impacted significantly and cannot adapt to temperature increase. Remove all yellow tokens.					
C.					
D.					

Biodiversity in an ecosystem can be improved using a wide range of solutions, such as preventing the cutting of trees, restoring the habitats and replanting, banning animal hunting or capture, making areas protected, re-introducing species.

4. (a) How does the simulation represent changes in biodiversity? _____

_____

_____

(b) What trend are you seeing in biodiversity due to human impacts, giving an example from above? _____

_____

_____

_____

PART 4: Biodiversity solutions

5. In your groups, develop at least four different solutions and the action that the solution will create for the different colored tokens, i.e. remove, take away, add a new color. Start with a new handful of tokens.

Species	Red	Blue	Green	Yellow	Simpson's Index of Diversity
Total number					

6. Simulate each solution with your tokens and calculate the **Simpson's Index of Diversity** after each.

Scenario	Red	Blue	Green	Yellow	Simpson's Index of Diversity
A.					
B.					
C.					
D.					

5. (a) How would we know if a solution was effective at increasing biodiversity? _____

_____

_____

(b) Which of your solutions was the most effect at increasing biodiversity and why do you think this might be? _____

_____

_____

6. Selecting one of your solutions from above, what might be any possible constraints of cost, safety, and reliability, as well as cultural, and environmental impacts if it was actioned as a real-world solution?

_____

_____

_____

_____

7. How well do you consider that your biodiversity solution is represented by the simulation and what are its limitations?

_____

_____

_____

_____

239 Human Impacts on Marine Biodiversity

Key Question: What is the impact of unsustainable fishing on fish stocks?

- Fishing is an ancient human tradition. It provides food, and is economically, socially, and culturally important. Today, it is a worldwide resource extraction industry. Decades of overfishing in all of the world's oceans have pushed commercially important species, such as cod (right), into steep decline. Overfishing has caused the collapse of many fisheries. Unsustainable fishing practices continue throughout the world's oceans.

- According to the United Nations Food and Agriculture Organization (FAO), almost half the ocean's commercially targeted marine fish stocks are either heavily or over-exploited. Without drastic changes to the world's fishing operations, many fish stocks will soon be lost.

Grand Banks fishery

Lost fishing gear can entangle all kinds of marine species. This is called **ghost fishing**.

Overfishing has resulted in many fish stocks at historic lows and fishing effort (the effort needed to catch fish) at unprecedented highs.

Huge fishing trawlers are capable of taking enormous amounts of fish. Captures of 400 tonnes at once are common.

Bottom trawls and dredges cause large scale physical damage to the seafloor. Non-commercial, bottom-dwelling species in the path of the net can be uprooted, damaged, or killed. An area of 8 million km² is bottom trawled annually.

The limited selectivity of fishing gear results in millions of marine organisms being discarded for economic, legal, or personal reasons. These organisms are defined as by-catch and include fish, invertebrates, protected marine mammals, sea turtles, and sea birds. Many of the discarded organisms die. Estimates of the worldwide by-catch is approximately 30 million tonnes per year.

Percentage of catch taken

- Rest of world
- Northwest Pacific
- Northeast Atlantic
- Western central Pacific
- Southwest Pacific

The single largest fishery is the Northwest Pacific, taking 26% of the total global catch.

Percentage exploitation of fisheries

- Recovering
- Depleted
- Under exploited
- Over exploited
- Moderately exploited
- Fully exploited

52% of the world's fished species are already fully exploited. Any increase in catch from these species would result in over-exploitation. 7% of the fish species are already depleted and 17% are over-exploited.

Fishing techniques have become so sophisticated, and efforts are on such a large scale, that thousands of tonnes of fish can be caught by one vessel, on one fishing cruise. Fishing vessels can reach over 100 m long. Here, 360 tonnes of Chilean jack mackerel are caught in one gigantic net.

Tuna is a popular fish type, commonly found canned in supermarkets and as part of sushi. However, virtually all tuna species are either threatened or vulnerable. Demand for the fish appears insatiable, with a record price of $1.7 million being paid in 2013 for a 221 kg bluefin tuna.

Illegal fishing was a major problem in the 1990s. Thousands of tonnes of catches were being unreported. International efforts have reduced this by an estimated 95%, helping the recovery of some fish stocks. Naval patrol vessels (above) have helped in targeting illegal fishing vessels.

Over-fishing is only one way that humans affect fish populations. The reduction in quality **habitats** as a result of human activities also has an effect. The bleaching of corals, related to **climate change** and sea water acidification, has the potential to seriously affect fish populations.

It is estimated that 8 million tonnes of plastic finds its way into the sea each year. Plastic can have severe detrimental effects on marine life, especially those that mistake plastic bags for jellyfish or other prey species. Some areas are so polluted it is dangerous to eat fish caught there.

An example of the effect a fishery can have on fish stocks is the Galápagos Island sea cucumber population. The commercial fishery began in 1993 and, by 2004, the sea cucumber population had dropped by 98%.

1. Why is overfishing of a fish species not sustainable? _____

2. Why is returning by-catch to the sea not always as useful as it might appear? _____

3. What percentage of commercial fish stocks are fully exploited, overexploited, or already depleted? _____

4. Identify three ways in which fish stocks are over-exploited: _____

Investigation 9.7 A model of human impacts on fish stocks

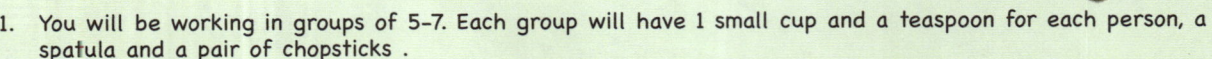

See appendix for equipment list.

1. You will be working in groups of 5-7. Each group will have 1 small cup and a teaspoon for each person, a spatula and a pair of chopsticks.
2. Each group will have 2 bowls, each filled with beads: 60 red beads (representing target fish), 40 yellow beads (representing other fish), and 40 blue beads (representing protected species). One bowl is the ocean bowl, one bowl is the reservoir bowl.
3. For each "fishing season," randomly assign the spatula (representing an ocean trawler) to one student, and a set of chopsticks (representing line and hook traditional fishing) to another. Remaining students use their teaspoon (representing commercial net fishing).
4. Students have 20 seconds all together, as a group to use their "fishing" method to gather as many "target fish" as possible from their ocean bowl, and put them into their cup.
5. Each student counts their "catch" and records the number, and fishing method, on the data chart below.

Fishing season	Fishing method	Target fish (red)	'Other' fish (yellow)	Protected species (blue)
1				
2				
3				
4				
5				

6. Replenish remaining beads/fish in the ocean bowl from the group's reservoir bowl. Each bead/fish remaining is doubled to account for breeding (add one bead of the correct color for each one left in the ocean bowl).
7. Group members with fewer than 5 target fish (red beads) in their cup sit out the next season, as they have gone "bankrupt".
8. Redistribute spatula and chopsticks to 2 random group members. Start season 2.
9. Repeat for 5 seasons, or until all target fish (red beads) have gone.
10. If a group's "ocean" has run out of "target" fish, they can spread out and "fish" in another group's ocean (bowl) by joining in their next round.

5. How did the type of fishing method affect:

   (a) The total number of fish caught in any season? _____

   _____

   _____

   (b) The proportion of "target fish" caught, compared to other and protected species (bycatch)? _____

   _____

   _____

6. Why do the "oceans" still become depleted of target fish eventually, even when they reproduce each year? _____

   _____

   _____

   _____

7. How did students moving to other oceans to fish affect fish stocks? _____

   _____

   _____

8. In your original groups, discuss some ideas about sustainable fishing methods and how they might prevent overfishing and loss of biodiversity. Agree to a set of group rules that incorporates the sustainability ideas. Repeat the activity. Record the new bead/fish counts in the chart on the previous page, but in a different colored pen or pencil.

   (a) Describe one of your new sustainability rules used in round two, and how it reduced overfishing in the activity:

   _____
   _____
   _____

   (b) What new rule was successful in reducing the bycatch of other fish (yellow) and protected species (blue)?

   _____
   _____
   _____

9. Which of your new rules could be applied to the problem of overfishing in the real world? _____
   _____
   _____

10. (a) What parts of the fishing investigation activity do you think represented the "real thing" particularly well? Why?
    _____
    _____

    (b) What parts of the fishing investigation activity do you think misrepresented the "real thing"? Why?
    _____
    _____

11. A fishbone chart helps categorize causes of a problem. Write a major problem that is caused by overfishing at the fish head; list four factors that contribute to the problem as headings, and conduct an online search to find some reasons for each factor, listing these underneath. Some reasons may appear in several categories. One heading has been completed for you. Discuss your findings in a group, or as a class.

e.g. Reduction of fish stocks

Fish being taken before they can reproduce; too many fish being taken every year, reduced reproduction rates.

PROBLEM

e.g. Collapse of the Atlantic salmon fishery

240 Deforestation and Species Survival

Key Question: How does deforestation impact species survival?

Deforestation

▶ At the end of the last glacial period, about 10,000 years ago, forests covered around 45% of the Earth's land surface. Forests now cover about 31% of Earth's surface. These include the cooler temperate forests of North and South America, Europe, China, and Australasia, and the tropical forests of equatorial regions. Over the last 5000 years, the loss of forest cover is estimated at 1.8 billion hectares. A net loss of around 5.2 million hectares has occurred the last 10 years alone. Temperate regions where human civilizations have historically existed the longest, e.g. Europe, have suffered the most, but now the vast majority of deforestation is occurring in the tropics. Intensive clearance of forests during settlement of the most recently discovered lands has extensively altered their landscapes and permanently changed or decreased **biodiversity**.

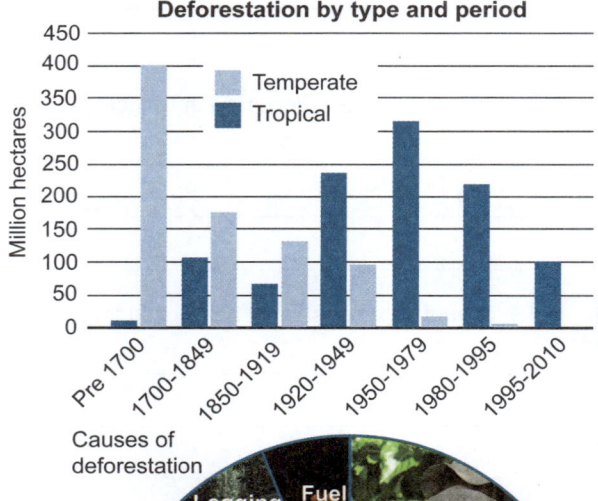

Causes of deforestation

▶ Deforestation is the end result of many interrelated causes, which often center around socioeconomic drivers. In many tropical regions, the vast majority of deforestation is the result of subsistence farming. Poverty and a lack of secure land can be partially solved by clearing small areas of forest and producing family plots. However, huge areas of forests have been cleared for agriculture, including ranching and production of palm oil plantations. These produce revenue for governments through taxes and permits, producing an incentive to clear more forest. Just 14% of deforestation is attributable to commercial logging, although combined with illegal logging it may be much higher.

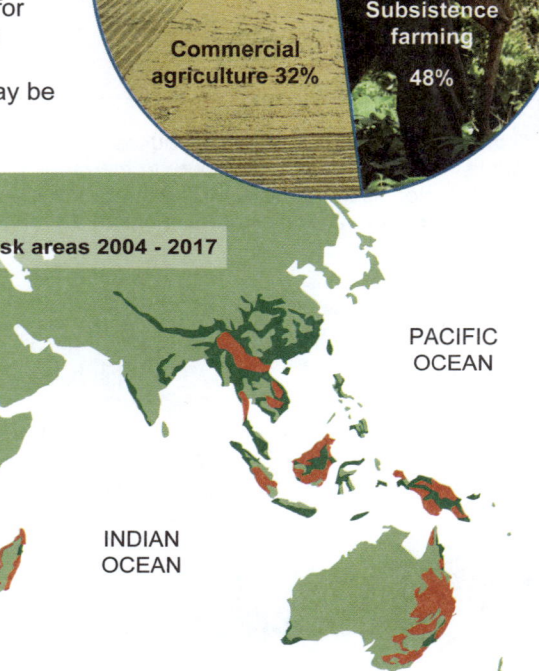

Causes of deforestation: Logging 14%, Fuel 5%, Subsistence farming 48%, Commercial agriculture 32%

Tropical forest deforestation risk areas 2004 - 2017

▶ It is important to distinguish between deforestation involving primary (old growth) forest and deforestation in plantation forests. Plantations are regularly cut down and replaced, and can artificially inflate a country's apparent forest cover or rate of deforestation. The loss of primary forests is far more important as these are refuges of high biodiversity, including rare species, many of which are not found anywhere else in the world.

▶ Temperate deforestation is still a problem. However, in equatorial regions the pace of deforestation is accelerating. This is of global concern as species biodiversity is highest in the tropics and **habitat** loss puts a great number of species at risk. Deforestation fronts (above) are large areas of forests that are under threat, with around 10% of forest cover lost between 2004 -2017, and the remaining areas at significant risk of further future losses.

The Pacific Northwest and the spotted owl

▶ The forests of the Pacific Northwest of the USA include the primary forests of Oregon and Washington states. They include stands of redwood, Douglas-fir, western red cedar, and shore pine, as well as alder and maple. The region is home to hundreds of species of wildlife, many dependent on the primary (old-growth) forest. One of these, the northern spotted owl, is a **keystone species** and an important indicator of healthy old growth forest. Listed as threatened under the Endangered Species Act, it was first protected in 1990 after nearly a century of old growth logging. In 1994, in response to the owl's status, the Northwest Forest Plan (NWFP) was implemented to govern land use on federal lands in the Pacific Northwest. In particular, it called for extensive old growth reserve lands and restrictions on old growth logging.

▶ The northern spotted owl ranges over wide areas of the Pacific Northwest. It requires forests with a dense canopy of mature and old-growth trees, abundant logs, standing snags, and live trees with broken tops. Only old growth forests offer these characteristics. When forced to occupy smaller areas of less suitable habitat, the birds are susceptible to starvation, **predation**, and **competition**.

▶ Although the plight of the northern spotted owl triggered the NWFP, it is the fate of the old growth forests, the species they support, and the ecosystem services they provide that are at stake. Despite protections, the number of northern spotted owls is declining at three times the predicted rate.

1. Describe the trend in temperate and tropical deforestation over the last 300 years: _____

2. What are some of the causes of deforestation? _____

3. Why is it important to distinguish between total forest loss and loss of primary forests when talking about deforestation?

4. Deforestation in temperate regions has largely stabilized and there has been substantial forest regrowth. However, these second growth forests differ in structure and composition to the forests that were lost. Why might this be of concern?

5. **"The Pacific Northwest old growth forest should be fully protected and not logged."**
Work in one of two groups, with one group arguing in support of this statement and one group arguing against it. Present group arguments to the class as a whole. Complete your own research, ensuring validity, using the **Resource Hub** for more information. Present your arguments as a report and summarize the main points in the space below:

241 Can't See the Wood for the Trees

Key Question: How has human activity affected a high biodiversity area, and what are some possible solutions for restoring it, or preventing more loss?

▶ The Madrean Pine-Oak Woodlands is an area of around 500,000 km² in Mexico, with patches that extend into Southern United States. The **ecosystem** has a very high **biodiversity**, but is threatened by human activity.

The Madrean Pine-Oak Woodlands is a biodiversity hotspot containing over 5000 species of flowering plants, many endemic (found nowhere else) to the area, as well as 500 bird, 380 reptile, 200 amphibian, and 330 mammal species.

Biodiversity loss in the Madrean Pine-Oak Woodlands is occurring due to commercial logging, deforestation (clearing forests) to provide agricultural land, **climate change**, and poor forest management leading to fires.

1. Why does having many endemic species make it extra important to conserve the biodiversity in this area?

    _____

    _____

    _____

2. Each of the factors listed below affects the Madrean Pine-Oak Woodlands ecosystem. Work in small groups to discuss each of these and then write what you think some of those effects might be:

    (a) Commercial logging: _____

    _____

    _____

    _____

    (b) Deforestation for agriculture: _____

    _____

    _____

    _____

    (c) Climate change: _____

    _____

    _____

    _____

    (d) Uncontrolled forest fires: _____

    _____

    _____

    _____

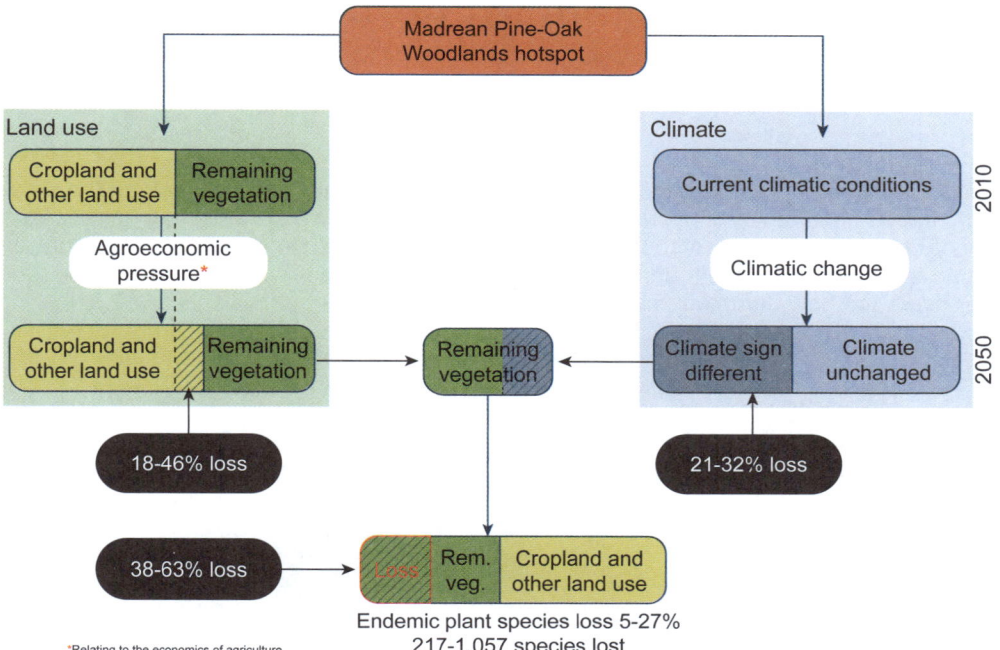

3. Using information from the diagram above, what are the key factors that will affect plant biodiversity loss in the Madrean Pine-Oak Woodlands hotspot?

___

4. Research an endangered species found in your local area and present a one page report of your findings, including features about the species, its habitat and niche, the population size, and how human activity has affected it. What potential conservation action(s) might be effective in reducing the loss of your selected species? Consider the cost-benefits of such actions, including economic, social, and ethical considerations? Attach your report to this page.

___

242 The Effects of Damming on Biodiversity

Key Question: What is the effect of damming on ecosystems?

▶ Rivers are dammed for a variety of reasons, including production of electricity, flood control, and to provide water for irrigation. The increased importance of hydroelectricity and the need for water reservoirs has resulted in many rivers being dammed. Once seen as an environmentally friendly way to produce renewable electricity, the enormous damage dams can do to **ecosystems,** both upstream and downstream of the dam, is often realized too late. Most of the world's major rivers are now dammed; many more than once.

Impacts of dams

Reservoir
Displaces communities, inundates and fragments ecosystems.

Water quality
Water quality can be severely reduced for many years after filling. Aquatic communities in free flowing water are destroyed.

Dam
Fish migration is blocked, separating spawning waters from rearing waters. Sediment accumulates in the reservoir, reducing fertility of downstream soils and leading to erosion of river deltas, which are deprived of their sediment supply.

Downstream impact
Disrupted water flow and low water quality near the dam reduces biodiversity.

U.S. Bureau of Reclamation

Order of environmental impact of dams

▶ First order impacts occur immediately and are **abiotic** (physical). They include barrier effects (blocking water flow), and effects to water quality and sediment load.

▶ Second order effects occur soon after the establishment of the dam. These include effects related to the change in the physical and **biotic** environments. Rates of primary production change and the morphology of the ecosystem changes, e.g. river valley flooding and erosion.

▶ Third order impacts occur after the change in ecosystem and relate to the establishment of new ecosystem and **biodiversity** change: from a flowing water system to a still water system.

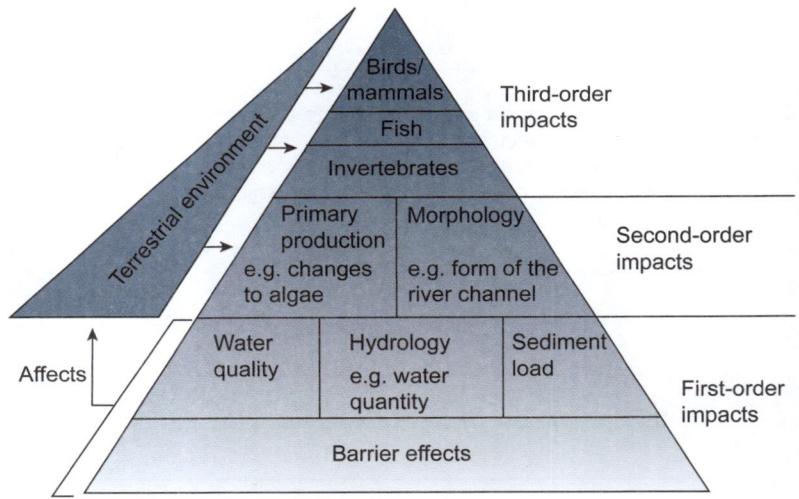

1. Identify three reasons for damming rivers: _____

   _____

2. Describe how dams can affect aquatic ecosystems: _____

   _____

   _____

   _____

   _____

The Yangtze River

1987   2006   Pollution behind the dam   Siberian Crane

The Three Gorges Dam (above) on the Yangtze river, China, is 2.3 km wide and 101 m high, with a reservoir 660 km long. Its construction caused the river water level to rise by 100 m. This rise in water level flooded important wetland **habitats**, where wading birds, including the Siberian crane, over-wintered. It also flooded 13 cities and hundreds of towns. The waste left behind in the flooded cities and towns continues to pollute the reservoir's waters. Dams reduce flood damage by regulating water flow downstream. However, this also prevents deposition of fertile silts below the dam and increases downstream erosion.

Colorado River

A number of dams are found along the Colorado River, which runs from Colorado through to Mexico. The two largest hydroelectric dams on the river are the Glen Canyon Dam and the Hoover Dam. Both dams control water flow through the Colorado River and were controversial even before their construction started.

The construction of the Glen Canyon and Hoover Dams effectively ended the annual flooding of the Colorado River and allowed invasive plants to establish in riparian (riverside) zones. Reduced flow rates have also impacted the populations of many downstream fish species.

The once vast Colorado River delta has been reduced to just 5% of its original size due to the construction of various dams. The river itself no longer reaches the sea. The photo above shows the river just 3 km below Morelos Dam, an irrigation diversion dam that takes almost all of the remaining river water.

3. Using the pyramid diagram, explain why the environmental impacts of dams occur in a specific order: _____

4. Describe the effects of the Three Gorges Dam on the ecosystems of the Yangtze River: _____

5. Describe the effects of the dams on the Colorado River on the river's ecosystems: _____

243 Humans Depend on Biodiversity

Key Question: What are the ecosystem services that humans depend upon?

Ecosystems provide services

- Humans depend on Earth's ecosystems for the services they provide. These **ecosystem** services include resources such as food and fuel, as well as processes such as purification of the air and water. These directly affect human health.
- The **biodiversity** of an ecosystem affects its ability to provide these services.
- Biologically diverse and resilient ecosystems that are managed in a sustainable way are better able to provide the ecosystem services on which we depend.
- The UN has identified four categories of ecosystem services: supporting, provisioning, regulating, and cultural.
- Regulating and provisioning services are important for human health and security (security of resources and security against natural disasters).
- Cultural services are particularly important to the social fabric of human societies and contribute to well being. These are often things we cannot value in monetary terms.

Biodiversity is important in crop development, e.g. promoting disease resistance. Many medical breakthroughs have come from understanding the biology of wild plants and animals.

High biodiversity creates buffers between humans and infectious diseases, e.g. Lyme disease, and increases the efficiency of processes such as water purification.

Biodiversity and ecosystem health are essential for reducing the effects of human activities, e.g. pollution, and the effects of environmental disasters, e.g. eruptions and landslides.

1. What are ecosystem services, and why are they important to humans? _____
   _____
   _____
   _____

2. What is the relationship between biodiversity and the ability of an ecosystem to provide essential ecosystem services?
   _____
   _____
   _____

244 A Mammoth Task Revisited

Content Anchor Revisited: How could bringing back the woolly mammoth restore a lost ecosystem?

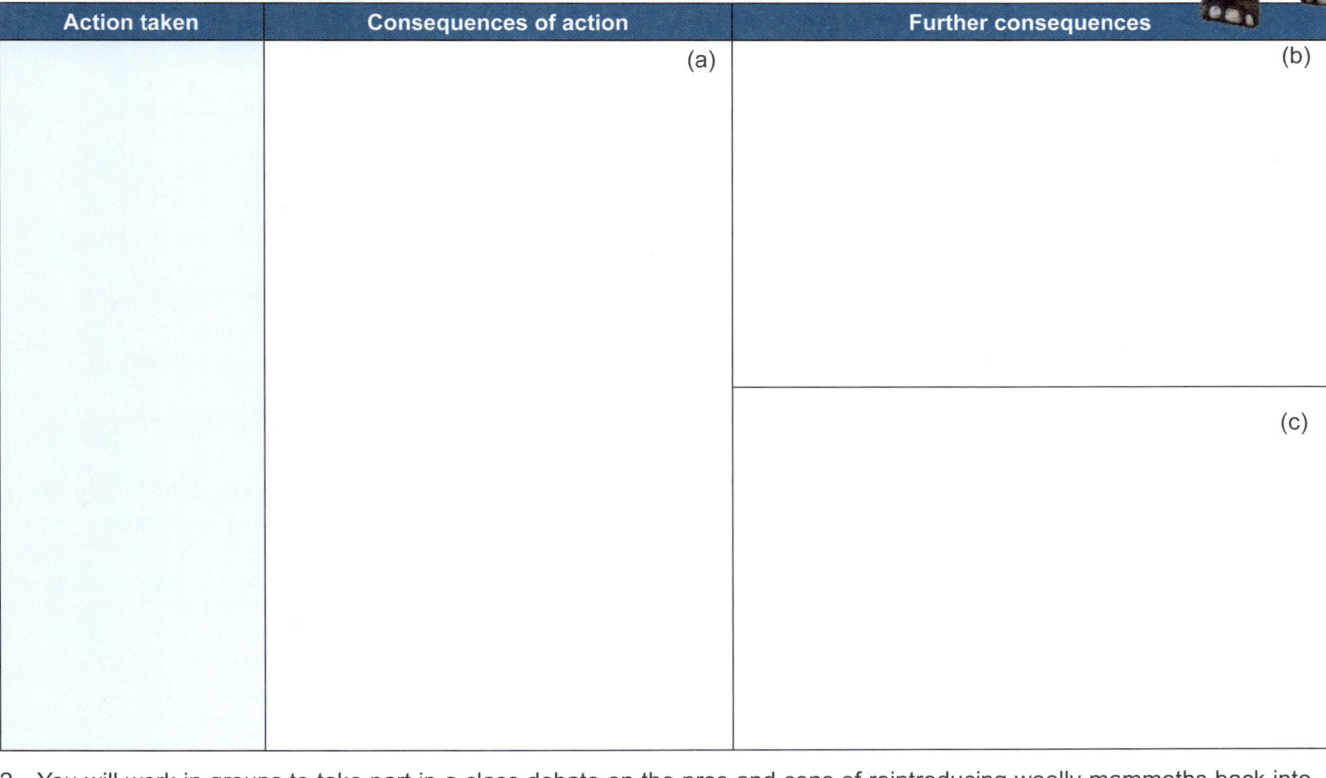

1. Earth's northern **ecosystems** have been without woolly mammoths for around 10,000 years. Fill in the chart below with possible consequences of re-introducing the mammoth as a **keystone species** to the ecosystem. Write two further consequences (knock-on effects) arising from this to the right. In small groups, compare and discuss the various answers.

Action taken	Consequences of action	Further consequences
	(a)	(b)
		(c)

2. You will work in groups to take part in a class debate on the pros and cons of reintroducing woolly mammoths back into the northern European ecosystem as a means to restore stability to the ecosystem. Divide each larger group into two smaller teams of around 3 students.

    (a) Which side of the debate are you arguing for? For (pro) or against (con)? _____

    (b) Work together in your smaller group to decide upon at least three key points that you will use to support your side of the debate. Record those points down below.

    _____
    _____
    _____

    (c) Construct a summary statement linking the introduction of woolly mammoth as a means to reduce **climate change**. Draw on your learning from activities you have completed within this chapter to help craft your answer.

    _____
    _____
    _____
    _____
    _____
    _____
    _____

245 Summing Up

Read each question carefully. Place a cross in the box beside the best answer to the question from the four answer choices provided.

Questions 1-2 relate to the plot below.

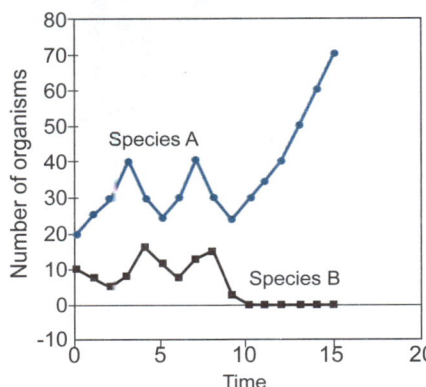

1. Species A and B are found in the same ecosystem. What could be inferred from the data presented?

   ☐ (a) Species B is the prey of species A
   ☐ (b) Species B is the predator of species A
   ☐ (c) Species A is a keystone species
   ☐ (d) Species A is a parasite of species B

2. Which of the following statements would explain why a species would show cyclic population changes?

   ☐ (e) A keystone species in the ecosystem is removed
   ☐ (f) The vegetation on which it depends varies seasonally
   ☐ (g) Climate shifts alter the prevalence of parasites that affect the species
   ☐ (h) A predator of the species is preyed on by a species that is introduced to the ecosystem

3. Pumas prey on a species of deer in a region. The deer feed on vegetation. What is the most likely outcome if the puma population declines markedly as a result of hunting by humans?

   ☐ (a) The deer population will cycle between high and low numbers until the predator returns
   ☐ (b) The deer population will increase and strip the region of vegetation
   ☐ (c) The deer population will decline and vegetation cover will increase
   ☐ (d) The deer population will be stable

4. Which of the following might be expected if a keystone species is eliminated from an ecosystem?

   ☐ (a) Interspecific competition increases.
   ☐ (b) There is one less niche available.
   ☐ (c) The ecosystem biodiversity declines.
   ☐ (d) There is no significant change.

5. The plot below shows the results of the experimental removal of the starfish *Pisaster* from a region of an ecosystem:

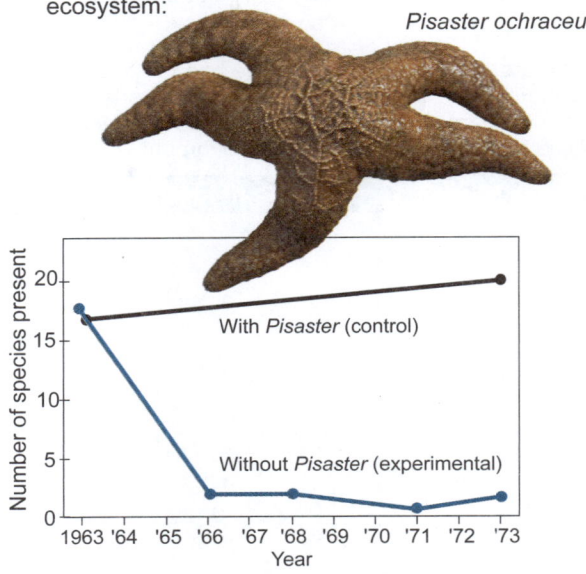

Pisaster ochraceus

From information on the plot it could be said that:

   ☐ (a) *Pisaster* is an effective predator
   ☐ (b) Removing *Pisaster* only causes a short term effect
   ☐ (c) *Pisaster* has an important role in maintaining species diversity
   ☐ (d) Removing *Pisaster* in a benefit to the ecosystem

6. The photograph below depicts:

   ☐ (a) Intraspecific competition and predation
   ☐ (b) Mutualism
   ☐ (c) Intra- and interspecific competition and predation
   ☐ (d) Interspecific competition and predation

7. Increasing carbon dioxide gas in the atmosphere leads to a stronger greenhouse gas effect. What is the main reason for this?

   ☐ (a) Carbon dioxide gas reflects light energy back down to the earth's surface
   ☐ (b) Carbon dioxide gas acts as a insulating blanket to keep heat in
   ☐ (c) Carbon dioxide gas blocks light energy from leaving the earth's atmosphere
   ☐ (d) Carbon dioxide gas is a greenhouse gas that increases the amount of heat energy retained in the atmosphere

8. Describe the species interactions below as mutualism, commensalism, competition, predation, or parasitism, and note which species benefits and which is harmed (if any).

(a) _____

(b) _____

(c) _____

9. A census of a deer population on an island forest reserve indicated a population of 2000 animals in 1960. In 1961, ten wolves (natural predators of deer) were brought to the island in an attempt to control deer numbers. The numbers of deer and wolves were monitored over the next nine years. The results of these population surveys are presented (right).

(a) Plot a line graph for the results. Use one scale (on the left) for numbers of deer and another scale (on the right) for the number of wolves. Use different symbols or colors to distinguish the lines and include a key.

(b) What does the plot show? _____

Island population surveys (1961-1969)

Year	Wolf numbers	Deer numbers
1961	10	2000
1962	12	2300
1963	16	2500
1964	22	2360
1965	28	2244
1966	24	2094
1967	21	1968
1968	18	1916
1969	19	1952

(c) When did the ecosystem show a trend of becoming stable? _____

10. Human activity can have major effects on ecosystems and their biodiversity. Use information from the model below to discuss how each human activity could decrease biodiversity, and affect the ecosystem stability:

Fragmentation of habitats
Building roads and towns in undeveloped habitat separates habitats and makes it difficult for organisms to move between them.

Introduction of new species
Some introductions may have a limited impact, but most are harmful. Introduced species can become pests as they often have few predators in their new environment. They can quickly spread and cause huge environmental damage.

Simplifying natural ecosystems
Plowing diverse grasslands and replanting them with a low diversity pasture affects the interrelationships of thousands of species.

Eliminating competing species
Often, native plant species compete for the same resources as farmed species, e.g. livestock. Native species are often excluded from farmed areas by fences or removal, reducing system biodiversity.

Overharvesting
Overgrazing of native grasslands by livestock or overharvesting of trees from forests or fish from the sea can alter the balance of species in an ecosystem and have unpredictable consequences.

Pollution
Pollution can include the deliberate dumping of waste into the environment or unintended side effects of chemical use, e.g. estrogens in the environment or emission of ozone-depleting gases.

(a) Fragmentation of native habitats: _____

(b) Introduction of new species: _____

(c) Eliminating competing species: _____

(d) Pollution: _____

(e) Intensive agriculture and over fishing: _____

CHAPTER 10
Science Practices

TEKS
Scientific and Engineering Practices

B.1: Investigation and Inquiry

1.A 1.B 1.C 1.D
1.E 1.F 1.G 1.H

The student, for at least 40% of instructional time, asks questions, identifies problems, and plans and safely conducts classroom, laboratory, and field investigations to answer questions, explain phenomena, or design solutions using appropriate tools and models.

B.2: Data and Patterns

2.A 2.B 2.C 2.D

The student analyzes and interprets data to derive meaning, identify features and patterns, and discover relationships or correlations to develop evidence-based arguments or evaluate designs.

B.3: Communicating in Science

The student develops evidence-based explanations and communicates findings, conclusions, and proposed solutions.

B.4: Science as a Human Endeavor

4.A 4.C

The student knows the contributions of scientists and recognizes the importance of scientific research and innovation on society.

Learning Outcomes
I know I have achieved this when I can:

	Learning Outcome	Activity number
☐	Discuss the features of Science, in small groups.	246
☐	Define and link the terms system and model, in a science context.	247
☐	Define and compare the scientific terms hypothesis, law, and theory.	248
☐	Generate a hypothesis from a provided case, and describe the assumptions used.	249
☐	Convert between decimal and standard form in given numerical values.	250-251
☐	Discuss the value of processing raw data.	252
☐	Calculate fractions and ratios from provided numerical values.	253
☐	Evaluate the usefulness of logarithm and semi-log graphs for processing exponential data.	254
☐	Calculate the percentage error for provided measurements.	255
☐	Classify data as quantitative, ranked, or qualitative.	256
☐	Evaluate the suitability of collecting qualitative or quantitative data in different types of investigations.	256
☐	Define independent, dependent, and control variables, describing the purpose of each in an investigation.	257, 274
☐	Discuss the value of accurate data recording, including tables, and the use of dataloggers.	258-259
☐	Plot a line graph, from provided data.	261, 274
☐	Draw a scatter plot, including a line of best fit.	262
☐	Distinguish between correlation and causation in data.	263
☐	Process raw data and draw a bar graph and histogram, from provided data.	264-265
☐	Calculate percentages from provided data and use the values to construct pie graphs.	266
☐	Calculate mean, median, and mode, from provided data.	267
☐	Calculate standard deviations, explaining what this statistical tool indicates about the data and sampling bias of the data.	268-269
☐	Construct a biological drawing from a provided photograph.	271
☐	Identify safety issues and risks in the classroom laboratory, and also in fieldwork settings.	272
☐	Discuss procedures for collecting qualitative data, in a provided case study.	273
☐	Process raw data into a data table.	274
☐	Evaluate an investigation method, from a provided case study.	274

RESOURCE HUB
bit.ly/3ycPUSZ

246 How Do We Do Science?

Key Question: What are the key aspects of the scientific method and how can we use the scientific method to better understand our world?

- Science is a way of understanding the world we live in: how it formed, the rules it obeys and how it changes over time. In brief, the scientific method begins with **observations** and asking questions about those observations. A statement to be tested is produced (a hypothesis). Investigations and experiments gather **data**. The data is analyzed for patterns that will prove or disprove the hypothesis. Based on the results, the hypothesis will be accepted or rejected. If it is rejected, a new hypothesis will need to be produced and tested.
- Science influences, and is influenced by, society and technology. As society's beliefs and needs change, and technology advances, what is or can be researched is also affected. Scientific discoveries advance technology and can change society's beliefs.
- Science can never answer questions with absolute certainty. It can be confident of certain outcomes, but only within the limits of the data. Science might help us predict with 99.9% certainty that a system will behave a certain way, but that still means there's one chance in a thousand it won't.

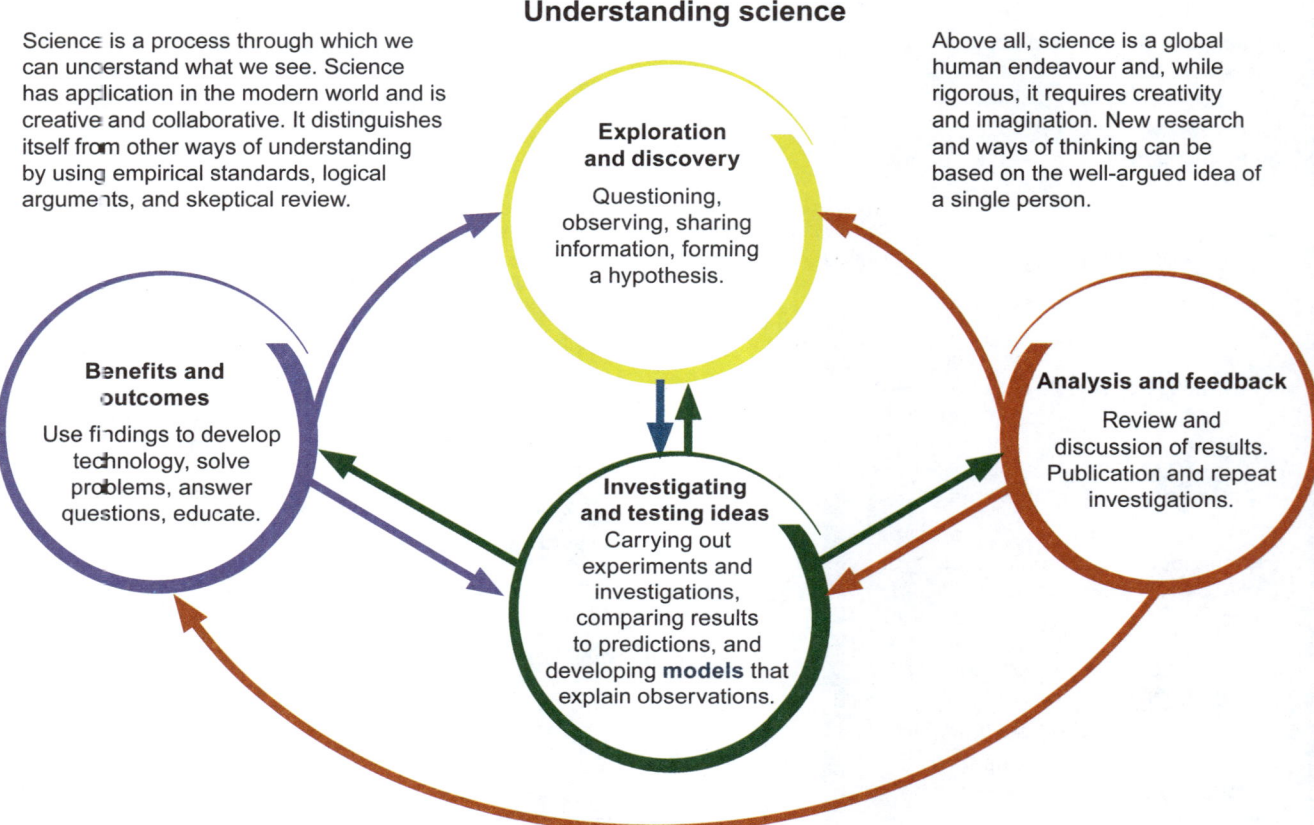

Understanding science

Science is a process through which we can understand what we see. Science has application in the modern world and is creative and collaborative. It distinguishes itself from other ways of understanding by using empirical standards, logical arguments, and skeptical review.

Above all, science is a global human endeavour and, while rigorous, it requires creativity and imagination. New research and ways of thinking can be based on the well-argued idea of a single person.

Exploration and discovery
Questioning, observing, sharing information, forming a hypothesis.

Benefits and outcomes
Use findings to develop technology, solve problems, answer questions, educate.

Investigating and testing ideas
Carrying out experiments and investigations, comparing results to predictions, and developing **models** that explain observations.

Analysis and feedback
Review and discussion of results. Publication and repeat investigations.

1. The buttons below make five statements about science. Your teacher will divide your class into five groups. Each group will address one statement. Discuss the statement as a group and present a brief written summary of what the statement means, what evidence there is to support it, and whether you agree with it. Have one person in each group present the group's views to the whole class.

- Science is a process through which we can understand what we see
- Science has application and relevance in the modern world
- Science is ongoing: it moves in a direction of greater understanding
- Science is a global human endeavour
- Science is exciting, dynamic, creative, and collaborative

2. Science encompasses a vast array of disciplines. In groups of four, research a discipline in which science is used. Each person in the group should research a different discipline and present a short description to your group (oral or written). You should include a brief description of the subject and an example of an investigation or experiment, including the results in that discipline (try to find an interesting experiment). From your group, choose the most interesting experiment to present to the class.

247 Systems and System Models

Key Question: What are models and why do we use them in science?

- A system is a set of interrelated components that work together. Energy flow in ecosystems (such as the one in the image on the right), gene regulation, interactions between organ systems, and feedback mechanisms are all examples of systems studied in biology.
- Scientists often used **models** to learn about biological systems. A model is a representation of a system and is useful for breaking a complex system down into smaller parts that can be studied more easily. Often, only part of a system is modelled. As scientists gather more information about a system, more **data** can be put into the model so that, eventually, it represents the real system more closely.

Modeling data

There are many different ways to model data. Often, seeing data presented in different ways can help us understand it better. Some common examples of models are shown here.

Visual models

Visual models can include drawings, such as these plant cells on the right.

Three dimensional models can be made out of materials such as modeling clay and sticks, like this model of a water molecule (below).

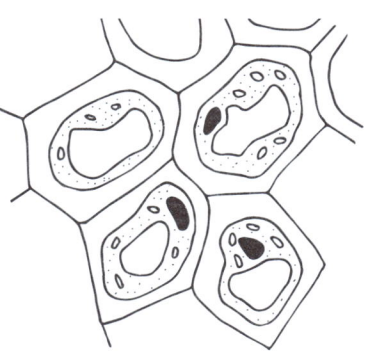

Mathematical models

Displaying data in a **graph** or as a mathematical equation, as shown below for logistic growth, often helps us to see relationships between different parts of a system.

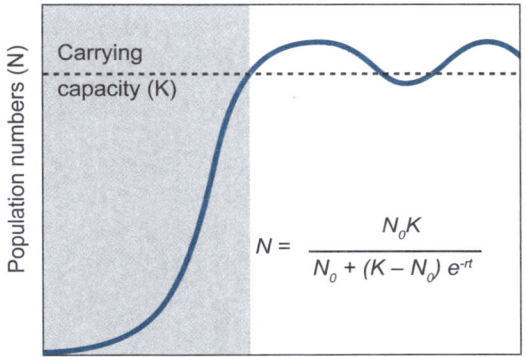

$$N = \frac{N_0 K}{N_0 + (K - N_0) e^{-rt}}$$

Analogy

An analogy is a comparison between two things. Sometimes, comparing a biological system to an everyday object can help us to understand it better. For example, the heart pumps blood in blood vessels in much the same way a fire truck pumps water from a fire hydrant through a hose. Similarly, ATP is like a fully charged battery in a phone.

ATP is like… …a charged phone battery

1. What is a system? _____

2. (a) What is a model? _____

   (b) Identify the advantages and disadvantages of using a model to describe a system: _____

248 Hypotheses, Laws, and Theories

Key Question: In science, what is the difference between a hypothesis, a theory, and a law?

▸ Like many subjects, science has its own use of language. Some words used in science do not mean the same as when they are used by the general public. This can often cause confusion when scientific ideas are discussed by the public.

▸ Three important words used in science are hypothesis, law, and theory.

Hypothesis

A possible answer to a question. A statement that can be tested:

"During meiosis each gamete receives one copy of a gene."

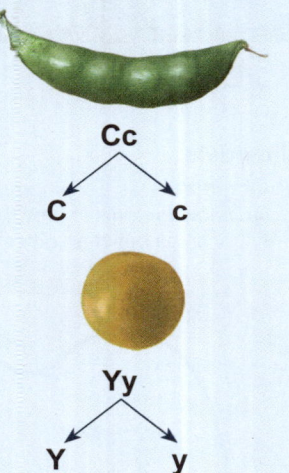

Hypotheses can be accepted or rejected, based on the outcome of the investigation.

Law

A statement based on experiment and observation that predicts the outcome of a situation. Laws are often mathematical, (e.g. $F = ma$) but they can be statements:

"For a diploid individual, during meiosis, alleles at the same locus, i.e. the two copies of a gene, segregate into separate gametes".

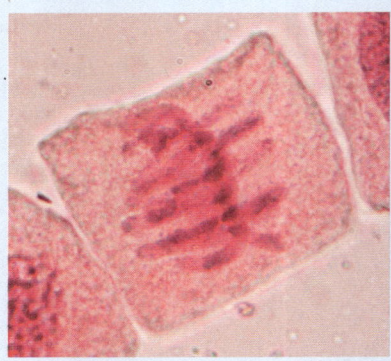

Laws generally resist change because they are designed to fit all the data. They may be revised if new data becomes available.

Theory

A comprehensive explanation of a set of scientific observations, experiments, and data. Like laws they can be used to predict outcomes:

"Segregation is the division of alleles during the first division in meiosis. Homologous chromosomes carrying the genes are pulled apart and move to separate, intermediate cells."

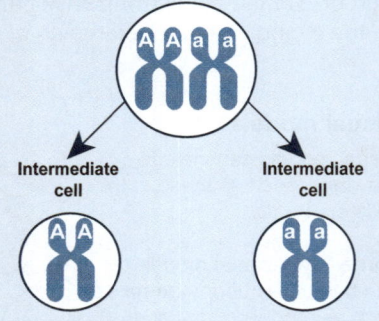

For any set of observations, there may be a number of theories put forward, e.g. Lamarck vs Darwin. Scientists will favor the theory that best explains the observations and predicts new ones.

Scientific laws and theories must be able to be extensively tested and conclusively proven

1. What is the difference in the purpose of a scientific theory and a scientific law? _____

_____

_____

_____

2. Explain why a theory cannot become a law and law cannot become a theory: _____

_____

_____

_____

3. The saying "It's just a theory" does not apply in science. Explain why: _____

_____

_____

_____

249 Observations and Assumptions

Key Question: What is the importance of making observations, producing hypotheses, and recognizing assumptions?

Observations and hypotheses

▶ An **observation** is watching or recording what is happening. Observation is the basis for forming hypotheses and making **predictions**. An observation may generate a number of hypotheses (tentative explanations for what we see). Each hypothesis will lead to one or more predictions, which can be tested by investigation.

▶ A hypothesis is often written as a statement to include the prediction: **"If X is true then, if I do Y (the experiment), I expect Z (the prediction)"**. Hypotheses are accepted, changed, or rejected on the basis of investigations. A hypothesis should have a sound, theoretical basis and should be testable.

Observation 1:
- Some caterpillar species are brightly colored and appear to be highly visible to predators, such as insectivorous birds. Predators appear to avoid these caterpillars.
- These caterpillars are often found in groups, rather than as solitary animals.

Observation 2:
- Some caterpillar species have excellent camouflage. When alerted to danger, they are difficult to see because they blend into the background.
- These caterpillars are usually found alone.

Assumptions

Any investigation requires you to make **assumptions** about the system you are working with. Assumptions are features of the system you are studying that you assume to be true, but that you do not (or cannot) test. Some assumptions about the examples above include:

- Insect eating birds have color vision.
- Caterpillars that look bright to us, also appear bright to insectivorous birds.
- Birds can learn about the taste of prey by eating them.

Read the two observations about the caterpillars above and then answer the following questions:

1. Generate a hypothesis to explain the observation that some caterpillars are brightly colored and highly visible, while others are camouflaged and blend into their surroundings:

   _____
   _____

2. Describe one of the assumptions being made in your hypothesis: _____
   _____
   _____

3. Generate a prediction about the behavior of insect eating birds towards caterpillars: _____
   _____
   _____

250 Accuracy and Precision

Key Question: What do accuracy and precision mean, how are they different, and why are they important when taking measurements?

The terms **accuracy** and **precision** are often used when talking about measurements.
- **Accuracy** refers to how close a measured value is to its true value, i.e. the correctness of the measurement.
- **Precision** refers to the closeness of repeated measurements to each other, i.e. the ability to be exact. For example, a digital device such as a pH meter (right) will give very precise measurements, but its accuracy depends on correct calibration.

Using the analogy of an archery target, repeated measurements are compared to hits. This analogy is useful when distinguishing between accuracy and precision.

Accurate but imprecise

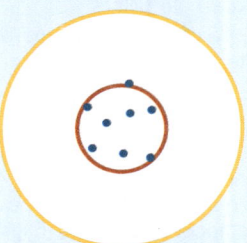

The measurements are all close to the true value but quite spread apart.
Analogy: The arrows are all close to the bullseye.

Inaccurate and imprecise

The measurements are all far apart and not close to the true value.
Analogy: The arrows are spread around the target.

Precise but inaccurate

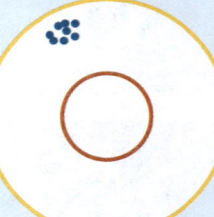

The measurements are all clustered close together but not close to the true value.
Analogy: The arrows are all clustered close together but not near the bullseye.

Accurate and precise

The measurements are all close to the true value and also clustered close together.
Analogy: The arrows are clustered close together near the bullseye.

Significant figures

Significant figures (sf) are the digits of a number that carry meaning contributing to its precision. They communicate how well you could actually measure the data. For example, you might measure the height of 100 people to the nearest cm. When you calculate their mean height, the answer is 175.0215 cm. If you reported this number, it implies that your measurement technique was accurate to 4 decimal places. You would have to round the result to the number of significant figures you had accurately measured. In this instance the answer is 175 cm.

Non-zero numbers (1-9) are always **significant**.

All zeros between non-zero numbers are always **significant**.

$$0.005704510$$

Zeros to the left of the first non-zero digit after a decimal point are not significant.

Zeros at the end of number where there is a decimal place are **significant** (e.g. 4600.0 has five sf).
BUT
Zeros at the end of a number where there is no decimal point are not significant (e.g. 4600 has two sf).

1. Why are precise but inaccurate measurements not helpful in a biological investigation? _____
_____
_____
_____

2. State the number of significant figures in the following examples:

   (a) 3.15985 _____ (d) 1000.0 _____

   (b) 0.0012 _____ (e) 42.3006 _____

   (c) 1000 _____ (f) 120 _____

251 Working With Numbers

Key Question: How is mathematical notation used, and how does converting and manipulating numbers make them easier to understand?

Commonly used mathematical symbols

In mathematics, universal symbols are used to represent mathematical concepts. They save time and space when writing. Some commonly used symbols are shown below.

- $=$ Equal to
- $<$ The value on the left is less than the value on the right
- $<<$ The value on the left is much less than the value on the right
- $>$ The value on the left is greater than the value on the right
- $>>$ The value on the left is much greater than the value on the right
- $\propto$ Proportional to. $A \propto B$ means that $A = $ a constant x B
- $\sim$ Approximately equal to

Decimal and standard form

- Decimal form (also called ordinary form) is the longhand way of writing a number (e.g. 15,000,000). Very large or very small numbers can take up too much space if written in decimal form and are often expressed in a condensed standard form. For example, 15,000,000 is written as 1.5×10^7 in standard form.

- In standard form a number is always written as $A \times 10^n$, where A is a number between 1 and 10, and n (the exponent) indicates how many places to move the decimal point. n can be positive or negative.

- For the example above, A = 1.5 and n = 7 because the decimal point moved seven places (see below).

$$1\,5\,000\,000 = 1.5 \times 10^7$$

- Small numbers can also be written in standard form. The exponent (n) will be negative. For example, 0.00101 is written as 1.01×10^{-3}.

$$0.00101 = 1.01 \times 10^{-3}$$

- Converting can make calculations easier. Work through the following example to solve $4.5 \times 10^4 + 6.45 \times 10^5$.

1. Convert $4.5 \times 10^4 + 6.45 \times 10^5$ to decimal form: _____

2. Add the two numbers together: _____

3. Convert to standard form: _____

Estimates

- When carrying out calculations, typing the wrong number into your calculator can put your answer out by several orders of magnitude. An estimate is a way of roughly calculating what answer you should get, and helps you decide if your final calculation is correct.

- Numbers are often rounded to help make estimation easier. The rounding rule is, if the next digit is 5 or more, round up. If the next digit is 4 or less, it stays as it is.

- For example, to estimate 6.8 x 704 you would round the numbers to 7 x 700 = 4900. The actual answer is 4787, so the estimate tells us the answer (4787) is probably right.

Use the following examples to practise estimating:

4. 43.2 x 1044: _____

5. 3.4 x 72 ÷ 15: _____

6. 658 ÷ 22: _____

Conversion factors and expressing units

- Measurements can be converted from one set of units to another by using a conversion factor. This is a numerical factor that multiplies or divides one unit to convert it into another.

- Conversion factors are commonly used to convert non-SI units to SI units, e.g. converting pounds to kilograms. Note that mL and cm^3 are equivalent, as are L and dm^3.

- In the space below, convert 5.6 cm^3 to mm^3 (1 cm^3 = 1000 mm^3):

7. _____

- The value of a **variable** must be written with its units where possible. SI units or their derivations should be used in recording measurements: volume in cm^3 (mL) or dm^3 (L), mass in kilograms (kg) or grams (g), length in meters (m), time in seconds (s). To denote 'per', you can use a solidus (/) or a negative exponent, e.g. per second is written as /s or s^{-1} and per meter squared is written as /m^2 or m^{-2}.

- For example, the rate of oxygen consumption should be expressed as:

Oxygen consumption (mL/g/s) or
Oxygen consumption mL g^{-1} s^{-1})

252 Tallies, Percentages, and Rates

Key Question: How is unprocessed (raw) data manipulated or transformed to make it easier to understand and to identify important features?

▶ The **data** collected by measuring or counting in the field or laboratory is called **raw data**. Raw data often needs to be processed into a form that makes it easier to identify its important features, e.g. trends, and make meaningful comparisons between samples or treatments. Basic calculations, such as totals (the sum of all data values for a **variable**), are commonly used to compare treatments. Some common methods of processing data include creating tally charts, and calculating percentages and rates. These are explained below.

Tally Chart
Records the number of times a value occurs in a data set

HEIGHT (cm)	TALLY	TOTAL
0 – 0.99	III	3
1 – 1.99	ʰʰʰʰ I	6
2 – 2.99	ʰʰʰʰ ʰʰʰʰ	10
3 – 3.99	ʰʰʰʰ ʰʰʰʰ II	12
4 – 4.99	III	3
5 – 5.99	II	2

- A useful first step in analysis; a neatly constructed tally chart doubles as a simple histogram.
- Cross out each value on the list as you tally it, to prevent double entries.

Percentages
Expressed as a fraction of 100

Men	Body mass (kg)	Lean body mass (kg)	% lean body mass
Athlete	70	60	85.7
Lean	68	56	82.3
Normal weight	83	65	78.3
Overweight	96	62	64.6
Obese	125	65	52.0

- Percentages express what proportion of data fall into any one category, e.g. for pie charts.
- Allows meaningful comparison between different samples.
- Useful to monitor change, e.g. % increase from one year to the next.

Rates
Expressed as a measure per unit time

Time (minutes)	Cumulative sweat loss (mL)	Rate of sweat loss (mL/min)
0	0	0
10	50	5
20	130	8
30	220	9
60	560	11.3

- Rates show how a variable changes over a standard time period, e.g. one second, one minute, or one hour.
- Rates allow meaningful comparison of data that may have been recorded over different time periods.

Example: Height of 6-day old seedlings.

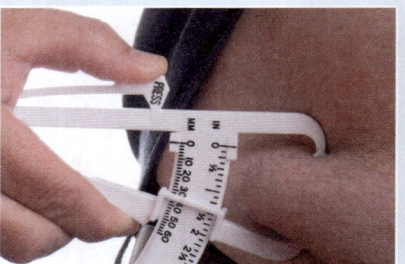
Example: Percentage of lean body mass in men.

Example: Rate of sweat loss during exercise in cyclists.

1. What is raw data? _____

2. Why is it useful to process raw data and express it differently, e.g. as a rate or a percentage? _____

3. Identify the best data transformation in each of the following examples:

   (a) Comparing harvest (in kg) of different grain crops from a farm: _____

   (b) Comparing amount of water loss from different plant species: _____

253 Fractions and Ratios

Key Question: How are fractions and ratios used to provide a meaningful comparison of sample data where the sample sizes are different?

Fractions

▸ Fractions express how many parts of a whole are present.

▸ Fractions are expressed as two numbers separated by a solidus (/). For example, 1/2.

▸ The top number is the numerator. The bottom number is the denominator. The denominator can not be zero.

Simplifying fractions

▸ Fractions are often written in their simplest form (the top and bottom numbers cannot be any smaller, while still being whole numbers). Simplifying makes working with fractions easier.

▸ To simplify a fraction, the numerator and denominator are divided by the highest common number that divides into both numbers equally.

▸ For example, in a class of 10 students, two had blonde hair. This fraction is 2/10. To simplify this fraction 2 and 10 are divided by the highest common factor (2).

$$2 \div 2 = 1 \text{ and } 10 \div 2 = 5$$

▸ The simplified fraction is 1/5.

Adding fractions

▸ To add fractions, the denominators must be the same. If the denominators are the same the numerators are simply added, e.g. 5/12 + 3/12 = 8/12

▸ When the denominators are different one (or both) fractions must be multiplied to give a common denominator, e.g. 4/10 + 1/2. By multiplying 1/2 by 5 the fraction becomes 5/10. The fractions can now be added together (4/10 + 5/10 = 9/10).

Ratios

▸ Ratios give the relative amount of two or more quantities, i.e. it shows how much of one thing there is relative to another.

▸ Ratios provide an easy way to identify patterns.

▸ Ratios do not require units.

▸ Ratios are usually expressed as *a : b*.

▸ In the example below, there are 3 blue squares and 1 gray square. The ratio would be written as 3:1.

Calculating ratios

▸ Ratios are calculated by dividing all the values by the smallest number.

▸ Ratios are often used in Mendelian genetics to calculate phenotype (appearance) ratios. Some examples for pea plants are given below.

882 inflated pod   *299 constricted pod*

To obtain the ratio, divide both numbers by 299.
299 ÷ 299 = 1
882 ÷ 299 = 2.95
The ratio = 2.95 : 1

495 round yellow   152 wrinkled yellow   158 round green   55 wrinkled green

For the example above of pea seed shape and color, all of the values were divided by 55. The ratio obtained was:
9 : 2.8 : 2.9 : 1

1. (a) A student prepared a slide of the cells of an onion root tip and counted the cells at various stages in the cell cycle. The results are presented in the table (right). Calculate the ratio of cells in each stage (show your working):

   _____

   _____

   (b) Assuming the same ratio applies in all the slides examined in the class, calculate the number of cells in each phase for a cell total count of 4800.

Cell cycle stage	No. of cells counted	No. of cells calculated
Interphase	140	
Prophase	70	
Telophase	15	
Metaphase	10	
Anaphase	5	
Total	240	4800

2. Simplify the following fractions:

   (a) 3/9 : _____ (b) 84/90: _____ (c) 11/121: _____

3. In a class, 5/20 students had blue eyes. In another class, 5/12 students had blue eyes. What fraction of students had blue eyes in both classes combined?

   _____

254 Dealing with Large Numbers

Key Question: How does using logarithms or log-linear (semi-log) graphs make large scale changes in numerical data more manageable?

- In biology, numerical data indicating scale can often decrease or increase exponentially. Examples include the exponential growth of populations, exponential decay of radioisotopes, and the pH scale.
- Exponential changes in numbers are defined by a function. A function is simply a rule that allows us to calculate an output for any given input. Exponential functions are common in biology and may involve very large numbers.
- Log transformations of exponential numbers can make them easier to handle.

Exponential function
- Exponential growth occurs at an increasingly rapid rate in proportion to the growing total number or size.
- In an exponential function, the base number is fixed (constant) and the exponent is variable.
- The equation for an exponential function is $y = c^x$.
- Exponential growth and decay (reduction) are possible.
- Exponential changes in numbers are easy to identify because the curve has a J-shape appearance due to its increasing steepness over time.
- An example of exponential growth is the growth of a microbial population in an unlimiting, optimal growth environment.

Log transformations
- A log transformation makes very large numbers easier to work with. The log of a number is the exponent to which a fixed value (the base) is raised to get that number. So $\log_{10}(1000) = 3$ because $10^3 = 1000$.
- Both $\log_{10}$ (common logs) and $\log_e$ (natural logs or $\ln$) are commonly used.
- Log transformations are useful for **data** where there is an exponential increase or decrease in numbers. In this case, the transformation will produce a straight line plot.
- To find the $\log_{10}$ of 32, using a calculator, key in log 32 = . The answer should be 1.51.
- Alternatively, the untransformed data can be plotted directly on a log-linear scale (as below). This is not difficult. You just need to remember that the log axis runs in exponential cycles. The paper makes the log for you.

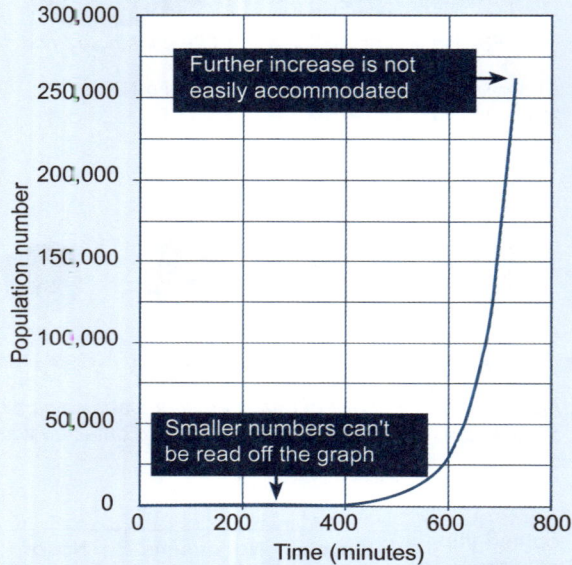

Example: Cell growth in a yeast culture where growth is not limited by lack of nutrients or build up of toxins.

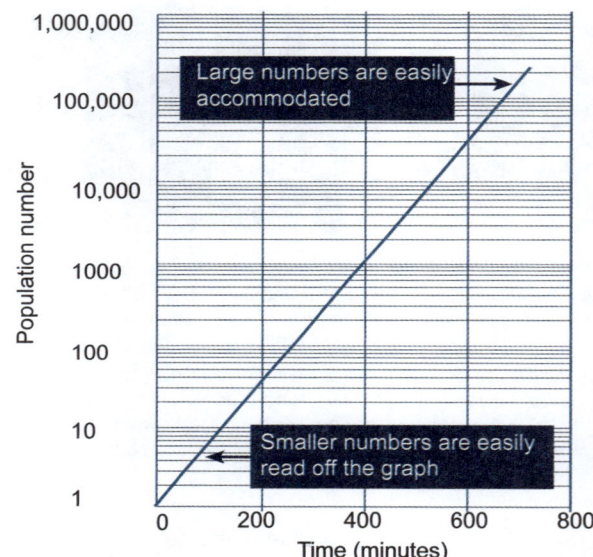

Example: The same yeast cell growth plotted on a log-linear scale. The y axis present 6 exponential cycles.

1. Why is it useful to plot exponential growth using semi-log paper? _____

2. What would you do to show yeast exponential growth as a straight line plot on normal graph paper? _____

3. Log transformations are often used when a value of interest ranges over several orders of magnitude. Can you think of another example of data from the natural world where the data collected might show this behavior? _____

255 Apparatus and Measurement

Key Question: Why must the apparatus used in experimental work be appropriate for the experiment or analysis and be used correctly?

Selecting the correct equipment

It is important that you choose equipment that is appropriate for the type of measurement you want to take. For example, if you wanted to accurately weigh out 5.65 g of sucrose, you need a balance that accurately weighs to two decimal places. A balance that weighs to only one decimal place would not allow you to make an accurate enough measurement.

Study the glassware (right). Which would you use if you wanted to measure 225 mL? The graduated cylinder has graduations every 10 mL, whereas the beaker has graduations every 50 mL. It would be more accurate to measure 225 mL in a graduated cylinder.

Percentage errors

Percentage error is a way of mathematically expressing how far out your result is from the ideal result. The equation for measuring percentage error is:

$$\frac{\text{experimental value - ideal value}}{\text{ideal value}} \times 100$$

For example, to determine the **accuracy** of a 5 mL pipette, dispense 5 mL of water from the pipette and weigh the dispensed volume on a balance. The mass (g) = volume (mL). The volume is 4.98 mL.

$$\frac{\text{experimental value (4.98) - ideal value (5.0)}}{\text{ideal value (5.0)}} \times 100$$

The percentage error = –0.4% (the negative sign tells you the pipette is dispensing **less** than it should).

Recognizing potential sources of error

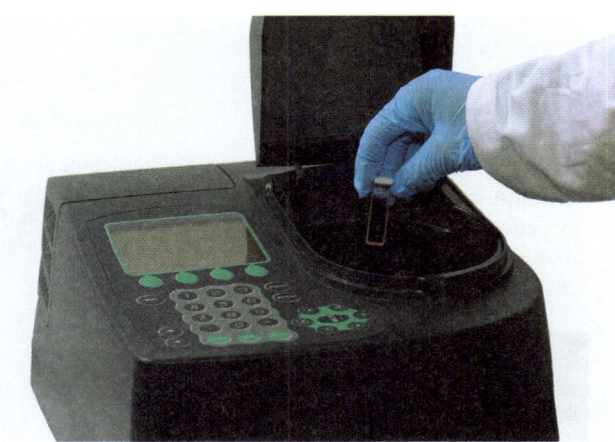

It is important to know how to use equipment correctly to reduce errors. A spectrophotometer measures the amount of light absorbed by a solution at a certain wavelength. This information can be used to determine the concentration of the absorbing molecule, e.g. density of bacteria in a culture. The more concentrated the solution, the more light is absorbed. Incorrect use of the spectrophotometer can alter the results. Common mistakes include incorrect calibration, errors in sample preparation, and errors in sample measurement.

A cuvette (left) is a small, clear tube designed to hold spectrophotometer samples. Inaccurate readings occur when:
- The cuvette is dirty or scratched (light is absorbed giving a falsely high reading).
- Some cuvettes have a frosted side to aid alignment. If the cuvette is aligned incorrectly, the frosted side absorbs light, giving a false reading.
- Not enough sample is in the cuvette and the beam passes over, rather than through the sample, giving a lower absorbance reading.

---

1. Assume that you have the following measuring devices available: 50 mL beaker, 50 mL graduated cylinder, 25 mL graduated cylinder, 10 mL pipette, 10 mL beaker. What would you use to accurately measure:

    (a) 21 mL: _____   (b) 48 mL: _____   (c) 9 mL: _____

2. Calculate the percentage error for the following situations (show your working):

    (a) A 1 mL pipette delivers a measured volume of 0.98 mL: _____

    _____

    (b) A 10 mL pipette delivers a measured volume of 9.98 mL: _____

    _____

256 Types of Data

Key Question: What types of data may be collected during an investigation?

Data is information collected during an investigation. Data may be **quantitative**, **qualitative**, or ranked. When planning a biological investigation, it is important to consider the type of data that will be collected. It is best to collect quantitative or numerical data, because it is easier to analyze it objectively (without bias).

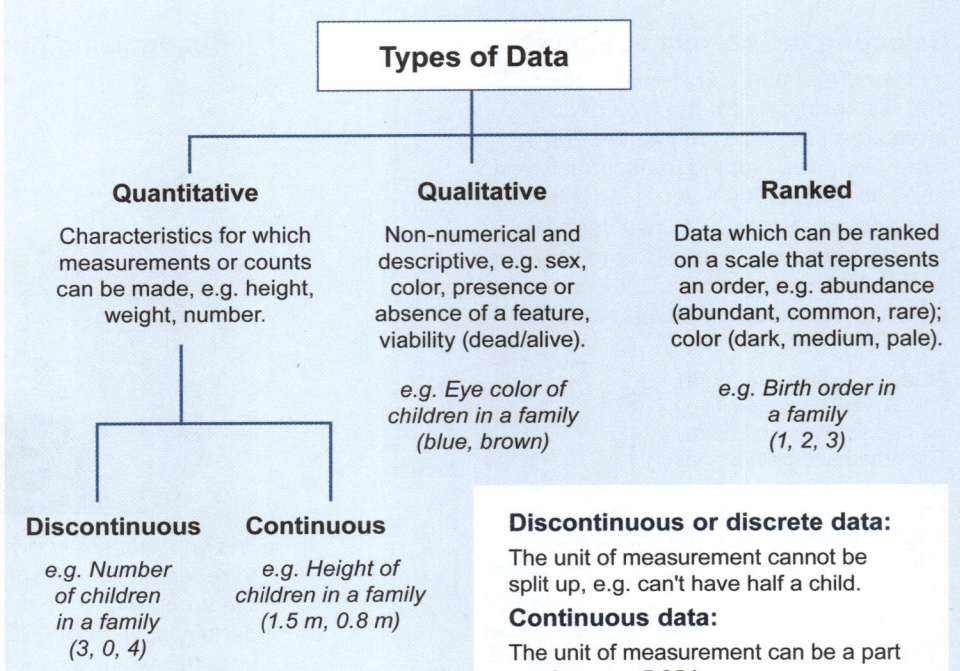

Types of Data

Quantitative
Characteristics for which measurements or counts can be made, e.g. height, weight, number.

Qualitative
Non-numerical and descriptive, e.g. sex, color, presence or absence of a feature, viability (dead/alive).

e.g. Eye color of children in a family (blue, brown)

Ranked
Data which can be ranked on a scale that represents an order, e.g. abundance (abundant, common, rare); color (dark, medium, pale).

e.g. Birth order in a family (1, 2, 3)

Discontinuous
e.g. Number of children in a family (3, 0, 4)

Continuous
e.g. Height of children in a family (1.5 m, 0.8 m)

Discontinuous or discrete data:
The unit of measurement cannot be split up, e.g. can't have half a child.

Continuous data:
The unit of measurement can be a part number, e.g. 5.25 kg.

A: Skin color

B: Eggs per nest

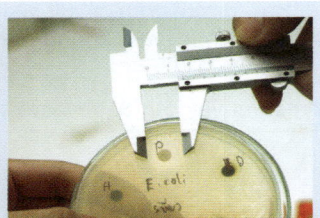

C: Bacterial colony diameter

1. For each of the photographic examples A-C above, classify the data as quantitative, ranked, or qualitative:

   (a) Skin color: _____

   (b) Number of eggs per nest: _____

   (c) Bacterial colony diameter: _____

2. Why is it best to collect quantitative data where possible in biological studies? _____

   _____

   _____

   _____

   _____

3. Give an example of data that could not be collected quantitatively, and explain your answer: _____

   _____

   _____

257 Variables and Controls

Key Question: What are variables in an investigation?

Types of variables

A **variable** is a factor that can be changed during an experiment, e.g. temperature. Investigations often look at how changing one variable affects another.

There are several types of variables:

- Independent
- Dependent
- Controlled

Only one variable should be changed at a time. Any changes seen are a result of the changed variable.

Remember! The **dependent variable** is "dependent" on the **independent variable**. Example: *When heating water, the temperature of the water depends on the time it is heated for. Temperature (dependent variable) depends on time (independent variable).*

Dependent variable
- Measured during the investigation.
- Recorded on the y axis of the graph.

Controlled variable
- Factors that are kept the same.

Independent variable
- Set by the experimenter, it is the variable that is changed.
- Recorded on the graph's x axis.

Experimental controls

▶ A **control** is the standard or reference treatment in an experiment. Controls make sure that the results of an experiment are due to the variable being tested, e.g. nutrient level, and not due to another factor such as the equipment not working correctly.

▶ A control is identical to the original experiment except that it lacks the altered variable. The control undergoes the same preparation, experimental conditions, **observations**, measurements, and analysis as the test group.

▶ If the control works as expected, it means the experiment has run correctly, and the results are due to the effect of the variable being tested.

Test plant (nutrient added)

Control plant (no nutrient added)

An experiment was designed to test the effect of a nutrient on plant growth. The control plant had no nutrient added to it. Its growth sets the baseline for the experiment. Any growth in the test plant greater than that seen in the control plant is due to the added nutrient.

1. What is the difference between a dependent variable and an independent variable? _____
_____
_____

2. Why do we control the variables we are not investigating? _____
_____

3. What is the purpose of the experimental control? _____
_____

258 Recording Results

Key Question: How does accurately recording results (using tables or data loggers) make it easier to understand and analyze your data later?

Ways to record data

▸ Recording your results accurately is very important in any type of scientific investigation. If you have recorded your results accurately and in an organized way, it makes analyzing and understanding your **data** easier. Log books and dataloggers are two methods by which data can be recorded.

Log books

A log book records your ideas and results throughout your scientific investigation. It also provides proof that you have carried out the work.

- An A4 lined exercise book is a good choice for a log book. It gives enough space to write ideas and record results and provides space to paste in photos or extra material such as printouts.
- Each entry must have the date recorded.
- Make sure that you can read what you write at a later date. A log book entry is meaningless if it is incomplete or cannot be read.

Dataloggers

A datalogger (also called a data recorder) is an electronic device that automatically records data over time.

- Dataloggers have a variety of sensors to measure different physical properties. Common sensors include light, temperature, pH, conductivity, and humidity.
- Dataloggers can be used in both field or laboratory experiments, and can be left to collect data without the experimenter being present.
- Information collected by the datalogger can be downloaded to a computer or phone (below) so that the data can be accessed and analyzed.

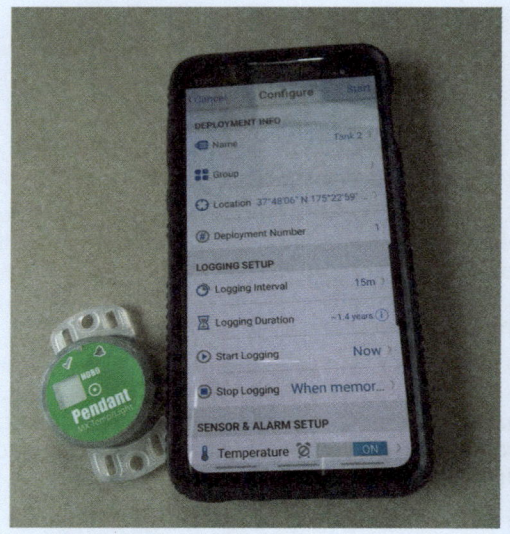

If you are not using a datalogger, a **table** (above) is often a good way to record and present your results as you collect them. Tables can also be useful for showing calculated values (such as rates and means). Recording data in a table as your experiment proceeds, lets you identify any trends early on and change experimental conditions if necessary.

1. Why is it important to accurately record your results? _____

2. Why must log book entries be well organized? _____

3. (a) What is a datalogger? _____

   (b) What are the advantages of using a datalogger over manually recording results? _____

259 Constructing Tables

Key Question: What is the purpose of recording data in an organized table during an experiment?

- **Tables** are used to record **data** during an investigation. Your log book should present neatly tabulated data (right).
- Tables allow a large amount of information to be condensed, and can provide a summary of the results.
- Presenting data in tables allows you to organize your data in a way that allows you to more easily see the relationships and trends.
- Columns can be provided to display the results of any data transformations such as rates. Basic **descriptive statistics** (such as **mean** or standard deviation) may also be included.
- Complex data sets tend to be graphed rather than tabulated.

Features of tables

Tables should have an accurate, descriptive title. Number tables consecutively through a report.

Heading and subheadings identify each set of data and show units of measurement.

Table 1: Length and growth of the third internode of bean plants receiving three different hormone treatments.

Independent variable in the left column.

Treatment	Sample size	Mean rate of internode growth (mm/day)	Mean internode length (mm)	Mean mass of tissue added (g/day)
Control	50	0.60	32.3	0.36
Hormone 1	46	1.52	41.6	0.51
Hormone 2	98	0.82	38.4	0.56
Hormone 3	85	2.06	50.2	0.68

Control values should be placed at the beginning of the table.

Each row should show a different experimental treatment, organism, sampling site etc.

Columns for comparison should be placed alongside each other. Show values only to the level of significance allowable by your measuring technique.

Organize the columns so that each category of like numbers or attributes is listed vertically.

1. What are two advantages of using a table format for data presentation?

    (a) _____

    (b) _____

2. Why might you tabulate data before you presented it in a graph? _____

260 Which Graph to Use?

Key Question: How does the type of data you collected affect the type of graph you should choose to display your data?

What type of data have you collected?

- One **variable** is a category
- One variable is a count

→ **Use a pie graph**

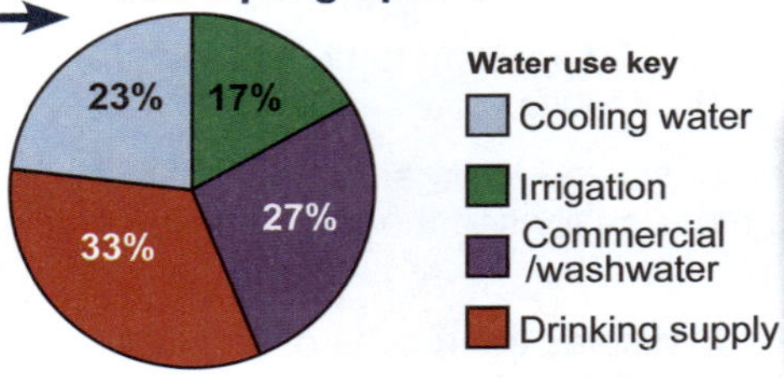
23%
17%
27%
33%
Water use key
Cooling water
Irrigation
Commercial /washwater
Drinking supply

Use to compare proportions in different categories.

- One variable is a category
- One variable is continuous data (measurements)

→ **Use a bar or column graph**

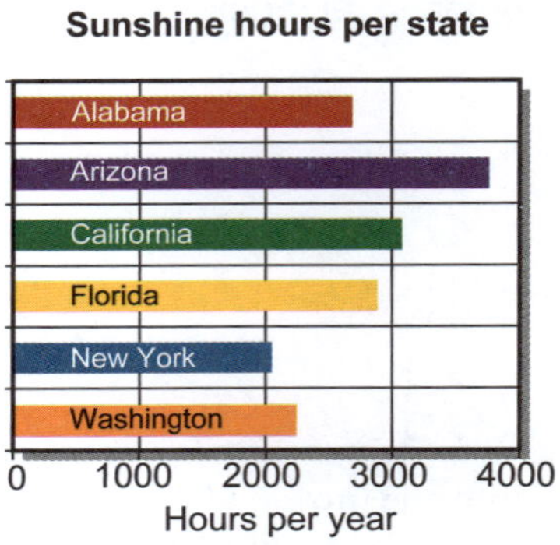
Sunshine hours per state
Alabama
Arizona
California
Florida
New York
Washington
0
1000
2000
3000
4000
Hours per year

Use to compare different categories (or treatments) for a continuous variable.

- One variable is continuous data (measurements)
- One variable is a count

→ **Use a histogram**

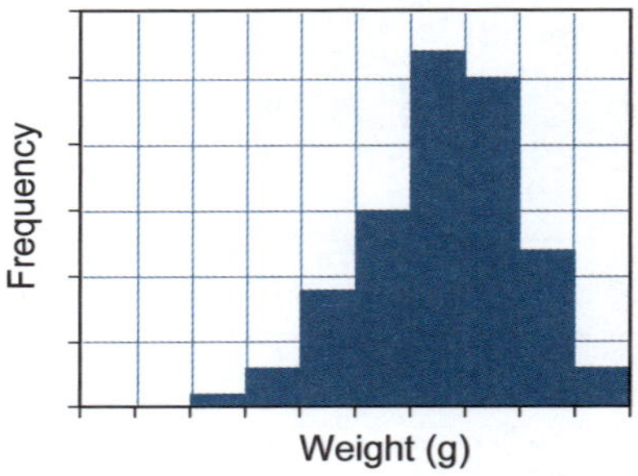
Frequency
Weight (g)

Use to show a frequency distribution for a continuous variable.

- Both variables are continuous
- The response variable is dependent on the **independent** (manipulated) variable

→ **Use a line graph**

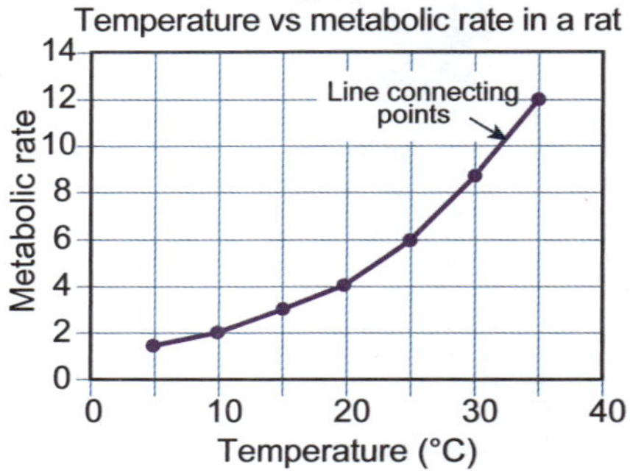
Temperature vs metabolic rate in a rat
Line connecting points
Metabolic rate
0
2
4
6
8
10
12
14
0
10
20
30
40
Temperature (°C)

Use to illustrate the response to a manipulated variable.

- Both variables are continuous
- The two variables are inter-dependent but there is no manipulated variable

→ **Use a scatter plot**

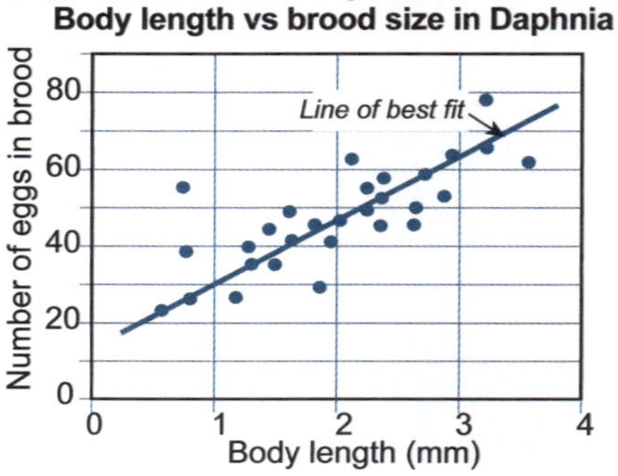
Body length vs brood size in Daphnia
Line of best fit
Number of eggs in brood
0
20
40
60
80
0
1
2
3
4
Body length (mm)

Use to illustrate the relationship between two correlated variables.

ISBN: 978-1-99-101405-4

261 Drawing Line Graphs

Key Question: What kind of data is plotted on line graphs, and how do they show the relationship between the independent variable and the dependent variable?

Graphs provide a way to visually see **data** trends. Line graphs are used when one variable (the **independent variable**) affects another, the **dependent variable**. Important features of line graphs are:

- The data must be continuous for both variables.
- The dependent variable is usually a biological response.
- The independent variable is often time or the experimental treatment.
- The relationship between two variables can be represented as a continuum and the data points are plotted accurately and connected directly (point to point).
- Line graphs may be drawn with a measure of error. The data are presented as points (the calculated **means**), with bars above and below, indicating a measure of variability or spread in the data, e.g. standard deviation.
- More than one curve can be plotted per set of axes. If the two data sets use the same measurement units and a similar range of values for the dependent variable, one scale on the y axis is used. If the two data sets use different units and/or have a very different range of values for the dependent variable, two scales for the y axis are used (see right). Distinguish between the two curves with a key.

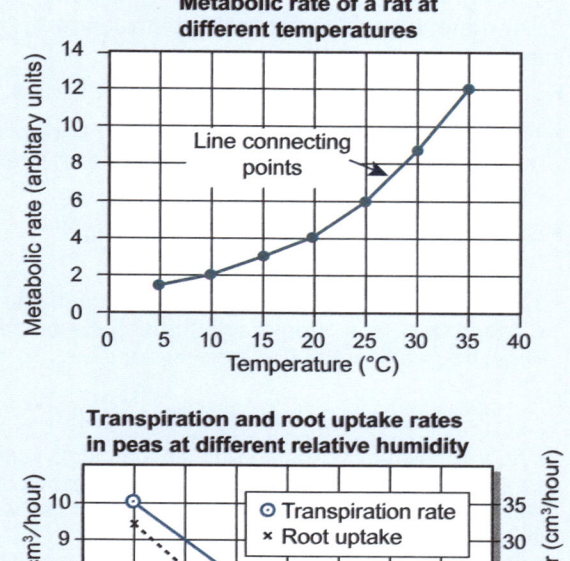

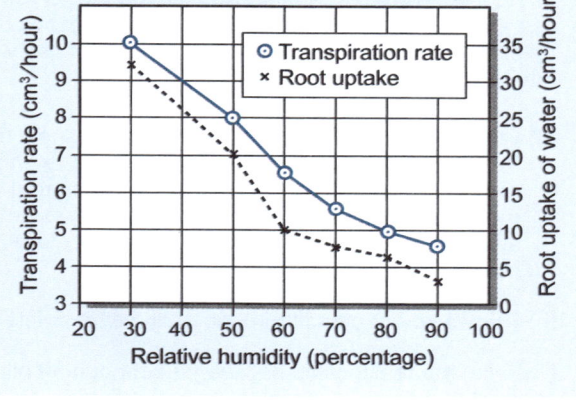

1. The results (shown right) were collected in a study investigating the effect of temperature on the activity of an enzyme.

   (a) Using the results provided, plot a line graph on the grid below:

   (b) Estimate the rate of reaction at 15°C: _____

Lab Notebook

An enzyme's activity at different temperatures

Temperature (°C)	Rate of reaction (mg of product formed per minute)
10	1.0
20	2.1
30	3.2
35	3.7
40	4.1
45	3.7
50	2.7
60	0

©2024 **BIOZONE** International
ISBN: 978-1-99-101405-4
Photocopying Prohibited

262 Drawing Scatter Graphs

Key Question: How does a scatter graph show continuous data where there is a relationship between two interdependent variables?

Scatter **graphs** are used to display continuous **data** where there is a relationship between two interdependent **variables**.
- The data must be continuous for both variables.
- There is no **independent** (manipulated) **variable**, but the variables are often correlated, i.e. they vary together in some predictable way.
- Scatter graphs are useful for determining the relationship between two variables.
- The points on the graph should not be connected, but a line of best fit is often drawn through the points to show the relationship between the variables.

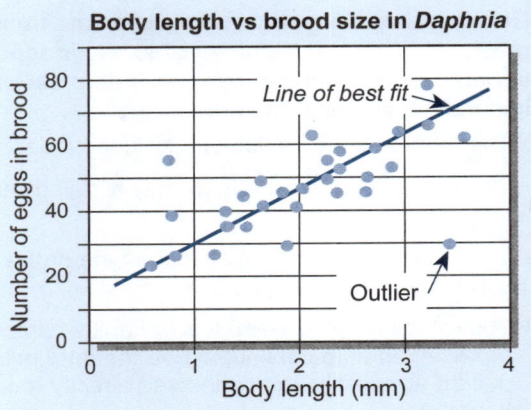

1. In the example below, metabolic measurements were taken from seven Antarctic fish *Pagothenia borchgrevinski*. The fish are affected by a gill disease, which increases the thickness of the gas exchange surfaces and affects oxygen uptake. The results of oxygen consumption of fish with varying amounts of affected gill (at rest and swimming) are tabulated below.

   (a) Plot the data on the grid (bottom right) to show the relationship between oxygen consumption and the amount of gill affected by disease. Use different symbols or colors for each set of data (at rest and swimming), and use only one scale for oxygen consumption.

   (b) Draw a line of best fit through each set of points.

2. Describe the relationship between the amount of gill affected and oxygen consumption in the fish:

   (a) For the "at rest" data set:

   _____
   _____
   _____
   _____
   _____

   (b) For the swimming data set:

   _____
   _____
   _____
   _____
   _____

Oxygen consumption of fish with affected gills

Fish number	Percentage of gill affected	Oxygen consumption (cm^3/g/hour)	
		At rest	Swimming
1	0	0.05	0.29
2	95	0.04	0.11
3	60	0.04	0.14
4	30	0.05	0.22
5	90	0.05	0.08
6	65	0.04	0.18
7	45	0.04	0.20

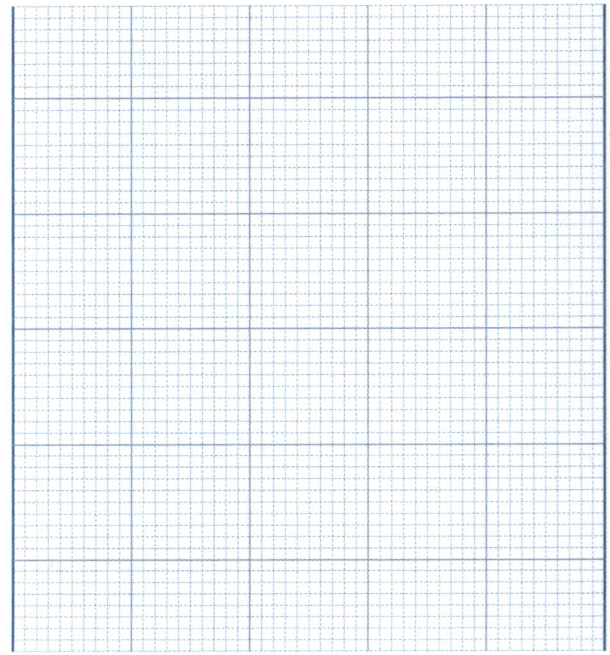

263 Correlation or Causation?

Key Question: What does correlation mean, and why can you not assume a correlation is the result of causation?

Correlation does not imply causation

- You may come across the phrase "correlation does not necessarily imply causation". This means that even when there is a strong correlation between variables (they vary together in a predictable way), you cannot assume that change in one variable caused change in the other.

- **Example**: When data from the organic food association and the office of special education programmes is plotted (below), there is a strong correlation between the increase in organic food and rates of diagnosed autism. However, it is unlikely that eating organic food causes autism, so we can not assume a causative effect here.

Drawing the line of best fit

Some simple guidelines need to be followed when drawing a line of best fit on your scatter plot.

- Your line should follow the trend of the **data** points.
- Roughly half of your data points should be above the line of best fit, and half below.
- The line of best fit does not necessarily pass through any particular point.
- A line of best fit should pivot around the point representing the mean of the x and y variables.

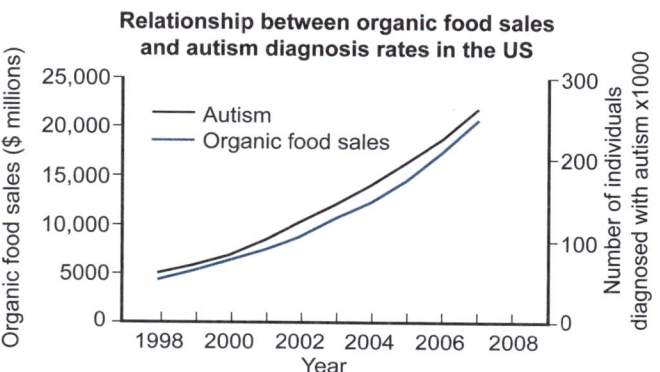

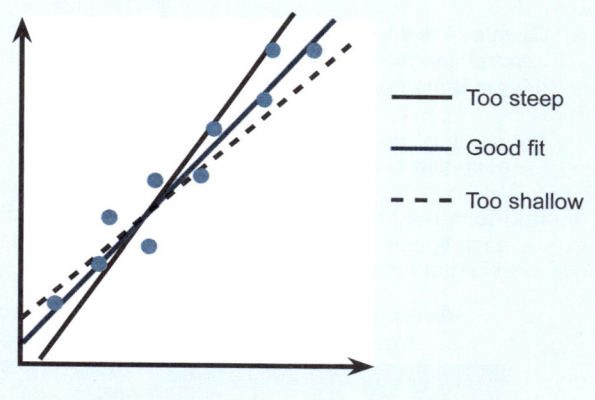

1. What does the phrase "correlation does not imply causation" mean? _____

_____

_____

2. A student measured the hand span and foot length measurements of 21 adults and plotted the data as a scatter graph (right).

    (a) Draw a line of best fit through the data:

    (b) Describe the results: _____

    _____

    _____

    _____

    _____

    (c) Using your line of best fit as a guide, comment on the correlation between hand span and foot length:

    _____

    _____

    _____

    _____

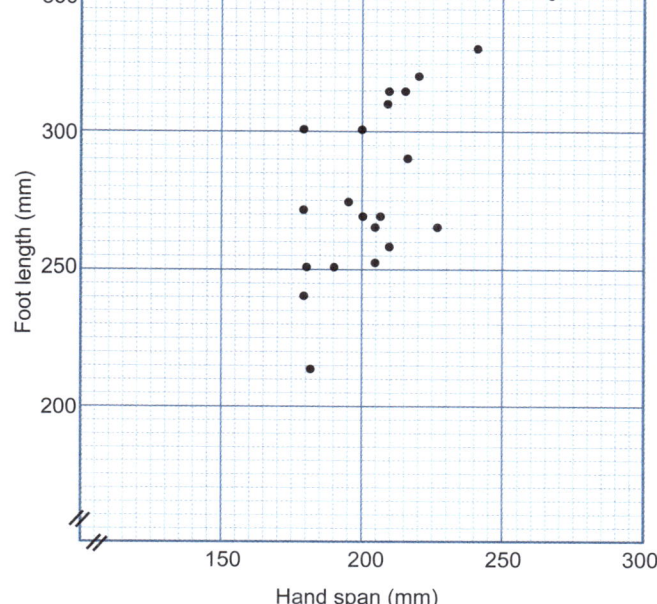

©2024 **BIOZONE** International
ISBN: 978-1-99-101405-4
Photocopying Prohibited

264 Drawing Bar Graphs

Key Question: What kind of data is shown on bar graphs?

Bar **graphs** are appropriate for **data** that is non-numerical and discrete for at least one **variable**.
- There are no **dependent** or **independent variables**.
- Data is collected for discontinuous, non-numerical categories, e.g. place, color, and species, so the bars do not touch.
- Multiple sets of data can be displayed side by side for direct comparison.
- Axes may be reversed, i.e. the bars can be vertical or horizontal. When they are vertical, these graphs are called column graphs.

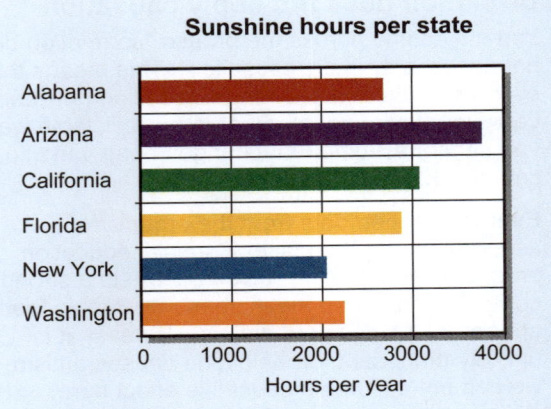

1. Counts of eight mollusk species were made from a series of quadrat samples at two sites on a rocky shore. The summary data are presented on the right.

   (a) Tabulate the **mean** (average) numbers per square meter at each site in the table (below).

   (b) Plot a bar graph of the tabulated data on the grid below. For each species, plot the data from both sites, side by side, using different colors to distinguish the two sites.

Average abundance of 8 mollusk species from two sites along a rocky shore

Species	Mean (no./m^2)	
	Site 1	Site 2

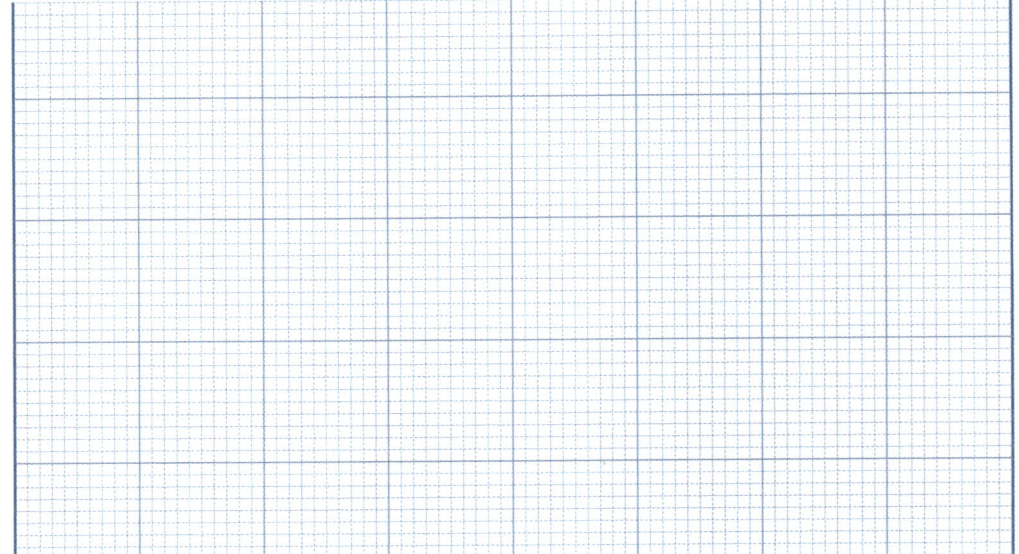

Field data notebook
Total counts at site 1 (11 quadrats) and site 2 (10 quadrats). Quadrats = 1 m^2

Species	Site 1 Total	Number/m^2 Mean	Site 2 Total	Number/m^2 Mean
Ornate limpet	232	21	299	30
Radiate limpet	68	6	344	34
Limpet sp. A	420	38	0	0
Cats-eye	68	6	16	2
Top shell	16	2	43	4
Limpet sp. B	628	57	389	39
Limpet sp. C	0	0	22	2
Chiton	12	1	30	3

265 Drawing Histograms

Key Question: What kind of data is shown on a histogram?

Histograms are plots of continuous **data** and are often used to represent frequency distributions, where the y-axis shows the number of times a particular measurement or value was obtained. For this reason, they are often called frequency histograms. Important features of histograms include:
- The data are numerical and continuous, e.g. height or weight, so the bars touch.
- The x-axis usually records the class interval. The y-axis usually records the number of individuals in each class interval (frequency).

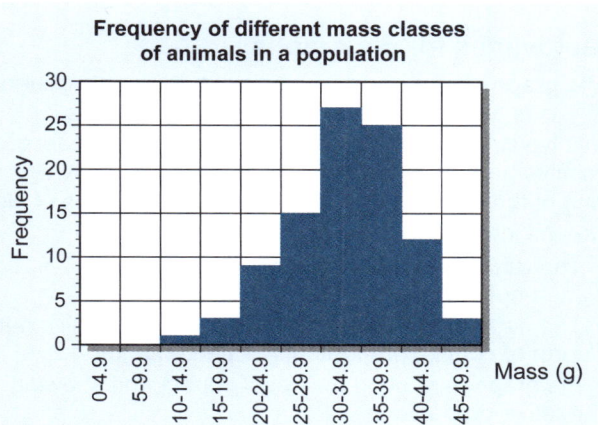

Frequency of different mass classes of animals in a population

1. The weight data provided below were recorded from 95 individuals (male and female), older than 17 years.
   (a) Create a tally chart (frequency table) in the table provided (right). An example of the tally for the weight grouping 55-59.9 kg has been completed for you. Note that the **raw data** values, once they are recorded as counts on the tally chart, are crossed off the data set in the notebook. It is important to do this in order to prevent data entry errors.

   (b) Plot a frequency histogram of the tallied data on the grid below.

Weight (kg)	Tally	Total
45-49.9		
50-54.9		
55-59.9	⊪⊪ II	7
60-64.9		
65-69.9		
70-74.9		
75-79.9		
80-84.9		
85-89.9		
90-94.9		
95-99.9		
100-104.9		
105-109.9		

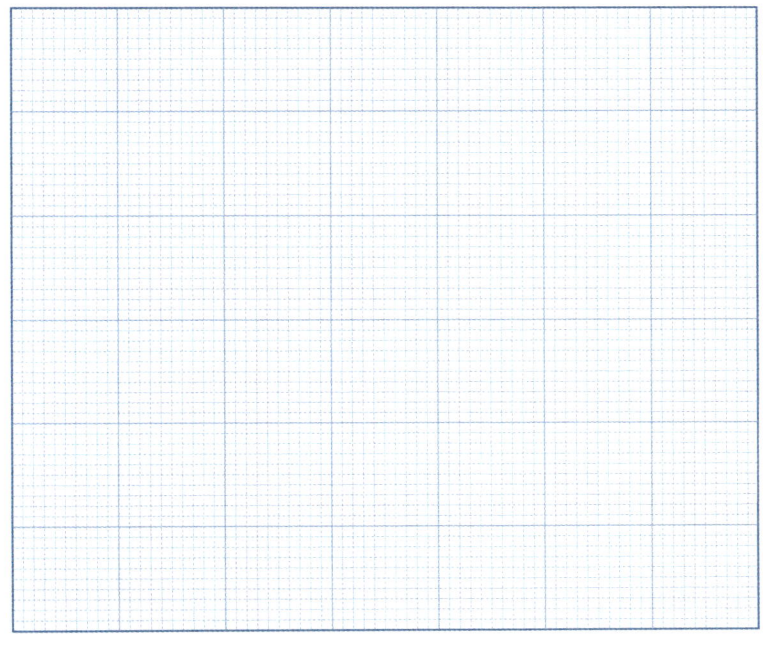

Lab notebook
Weight (in kg) of 95 individuals

63.4, 81.2, 65, 56.5, 83.3, 75.6, 84, 95, 76.8, 81.5, 105.5, 67.8, 73.4, 82, 68.3, 56, 73.5, 63.5, 60.4, 75.2, 58, 83.5, 63, 58.5, 82, 70.4, 50, 61, 82.2, 92, 55.2, 87.8, 91.5, 48, 86.5, 88.3, 53.5, 85.5, 81, 63.8, 87, 72, 69, 98, 66.5, 82.8, 71, 61.5, 68.5, 76, 66, 67.2, 72.5, 65.5, 82.5, 61, 67.4, 83, 60.5, 73, 78.4, 67, 67, 76.5, 86, 71, 83.4, 85, 70.5, 77.5, 93.5, 65.5, 77, 62, 68, 87, 62.5, 90, 89, 63, 83.5, 93.4, 60, 73, 83, 71.5, 66, 80, 73.8, 57.5, 76, 77.5, 76, 56, 74

©2024 **BIOZONE** International
ISBN: 978-1-99-101405-4
Photocopying Prohibited

266 Drawing Pie Graphs

Key Question: What kind of data is shown on pie graphs?

Guidelines for pie graphs

Pie **graphs** can be used instead of bar graphs, generally in cases where there are six or fewer categories involved. A pie graph provides strong visual impact of the relative proportions in each category, particularly where one of the categories is very dominant. Features of pie graphs include:

- The **data** for one **variable** are discontinuous (non-numerical or categories).
- The data for the dependent variable are usually in the form of counts, proportions, or percentages.
- Pie graphs are good for visual impact and showing relative proportions.
- They are not suitable for data sets with a large number of categories.

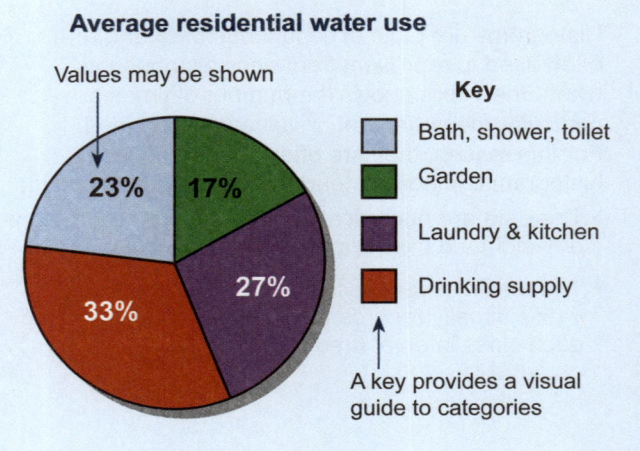

Average residential water use

1. The table provided below shows the total area of cereals, in hectares, planted in the United States in 1959 and 2021.
   (a) For each cereal, calculate the angle for each percentage, given that each percentage point is equal to 3.6° (the first example is provided: 23.6 x 3.6 = 85).
   (b) Plot a pie graph for each cereal in the circles provided. The circles have been marked at 5° intervals to enable you to do this exercise without a protractor. For the purposes of this exercise, begin your pie graphs at the 0° mark (in blue) and work in a clockwise direction from the largest to the smallest percentage. Use one key for both pie graphs.

Total area planted in United States (hectares x1000)

Cereal	1959	% of total area	Angle (°)	2021	% of total area	Angle (°)
Corn (maize)	34,612	38.5	138.5	37,780		
Wheat	23,489			18,900		
Sorghum	8047			2,956		
Barley	6875			1076		
Oats	14,625			1032		
Other	2266			2181		
Total	89,914			63,925		

Data: USDA

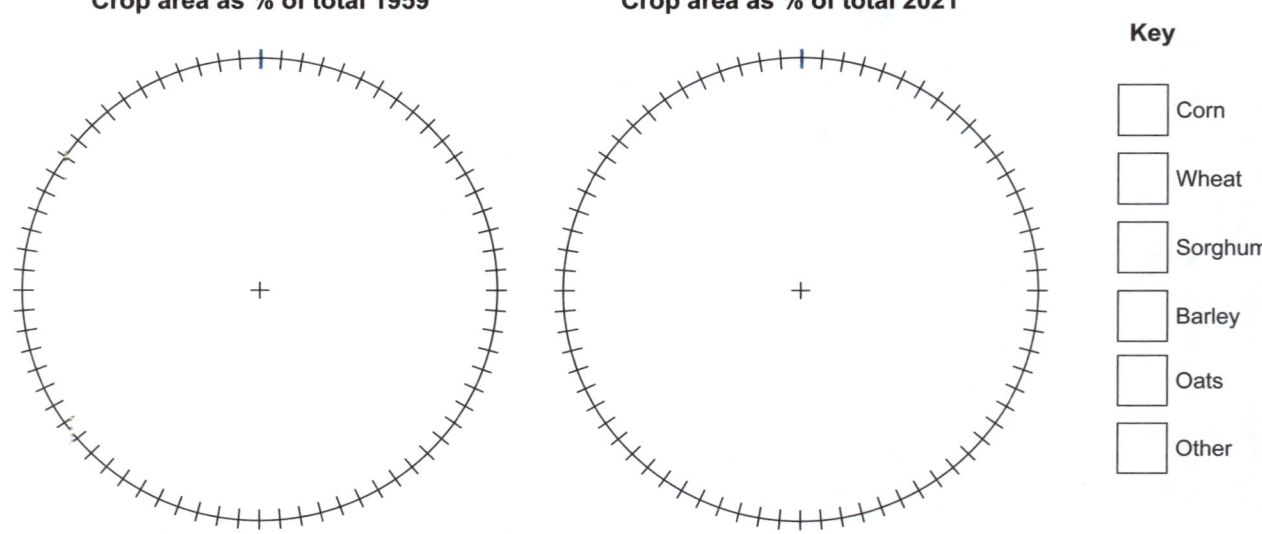

Crop area as % of total 1959   **Crop area as % of total 2021**

Key: Corn, Wheat, Sorghum, Barley, Oats, Other

267 Mean, Median, and Mode

Key Question: What are descriptive statistics and how are they used to summarize a data set and describe its basic features?

Descriptive statistics

▸ When we describe a set of **data**, it is usual to give a measure of central tendency. This is a single value identifying the central position within that set of data.

▸ **Descriptive statistics**, such as **mean**, **median**, and **mode**, are all valid measures of central tendency depending on the type of data and its distribution. They help to summarize features of the data, so are often called summary statistics.

▸ The appropriate statistic for different types of data **variables** and their distributions is described below.

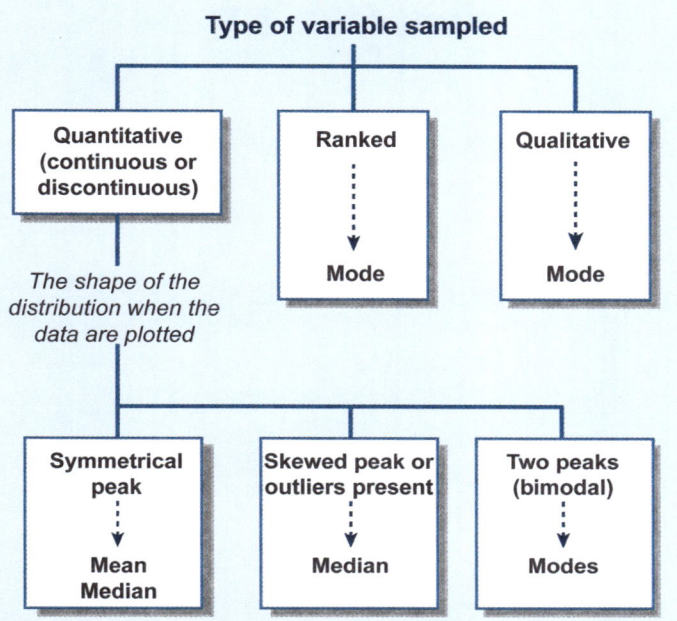

Distribution of data

Variability in continuous data is often displayed as a frequency distribution. There are several types of distribution.

- Normal distribution (A): Data has a symmetrical spread about the mean. It has a classical bell shape when plotted.
- Skewed data (B): Data is not centered around the middle but has a "tail" to the left or right.
- Bimodal data (C): Data which has two peaks.

The shape of the distribution will determine which statistic (mean, median, or mode) should be used to describe the central tendency of the sample data.

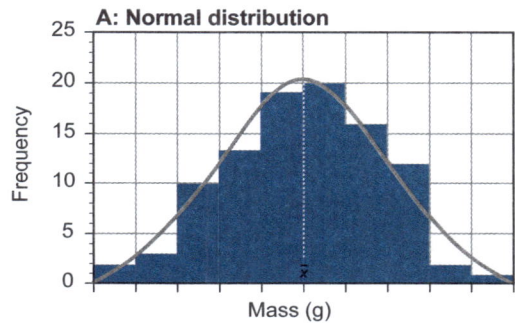

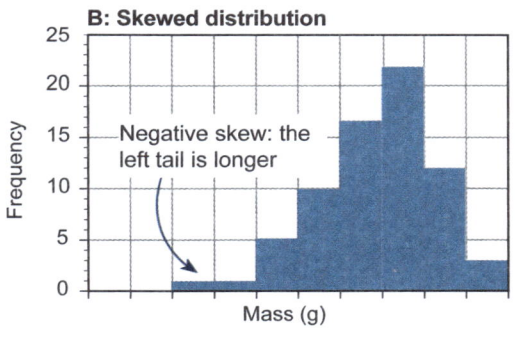

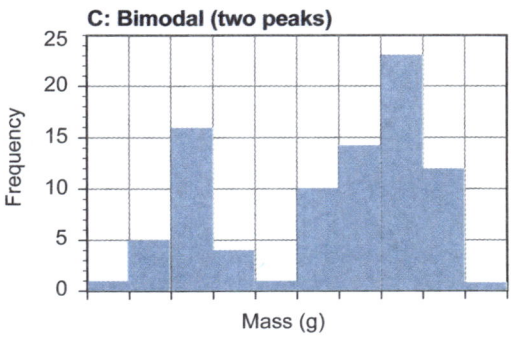

Statistic	Definition and when to use it	How to calculate it
Mean	• The average of all data entries. • Measure of central tendency for normally distributed data.	• Add up all the data entries. • Divide by the total number of data entries.
Median	• The middle value when data entries are placed in rank order. • A good measure of central tendency for skewed distributions.	• Arrange the data in increasing rank order. • Identify the middle value. • For an even number of entries, find the mid point of the two middle values.
Mode	• The most common data value. • Suitable for bimodal distributions and qualitative (categorical) data.	• Identify the category with the highest number of data entries using a tally chart or a bar graph.

1. The birth weights of 60 newborn babies are provided (right). Create a tally chart (frequency table) of the weights in the table provided below. Choose an appropriate grouping of weights.

Weight (kg)	Tally	Total

Birth weights (kg)

3.740	2.660	4.170	3.970	3.570
3.830	3.375	4.400	3.840	3.620
3.530	3.840	3.770	4.710	3.260
3.095	3.630	3.400	4.050	3.315
3.630	3.810	3.825	4.560	3.230
1.560	2.640	3.130	3.350	3.790
3.910	3.955	3.400	3.380	2.620
4.180	2.980	3.260	3.690	3.030
3.570	3.350	4.100	1.495	
2.660	3.780	3.220	3.260	
3.150	3.260	3.135	3.430	
3.400	4.510	3.090	3.510	
3.380	3.800	3.830	3.230	

NEED HELP? See Activity 260 & 265

2. (a) On the graph paper (right) draw a frequency histogram for the birth weight data.

   (b) What type of distribution does the data have?

   _____
   _____
   _____

   (c) Predict whether mean, median, or mode would be the best measure of central tendency for the data:

   _____

   (d) Explain your reason for your answer in (c):

   _____
   _____
   _____

   (e) Calculate the mean, median, and mode for the birth weight data:

   Mean: _____

   Median: _____

   Mode: _____

   (f) What do you notice about the results in (e)? _____
   _____
   _____

   (g) Explain the reason for this: _____
   _____
   _____

268 What is Standard Deviation?

Key Question: What does standard deviation measure, what is its purpose, and how is it calculated?

▸ While it is important to know the **mean** of a **data** set, it is also important to know how well the mean represents the data set as a whole. This is evaluated using a simple measure of the spread in the data called standard deviation.

▸ In general, if the standard deviation is small, the mean will more accurately represent the data than if it is large.

Standard deviation

- Standard deviation is usually presented as $\bar{x} \pm s$. In normally distributed data, 68% of all data values will lie within one standard deviation (s) of the mean ($\bar{x}$) and 95% of all data values will lie within two standard deviations of the mean (right).
- Different sets of data can have the same mean and range, yet a different data distribution. In both the data sets below, 68% of the values lie within the range $\bar{x} \pm 1s$ and 95% of the values lie within $\bar{x} \pm 2s$. However, in B, the data values are more tightly clustered around the mean.
- Standard deviation is easily calculated using a spreadsheet. Data should be entered as columns. In a free cell, type the formula for standard deviation (this varies depending on the program) and select the cells containing the data values, enclosing them in parentheses.

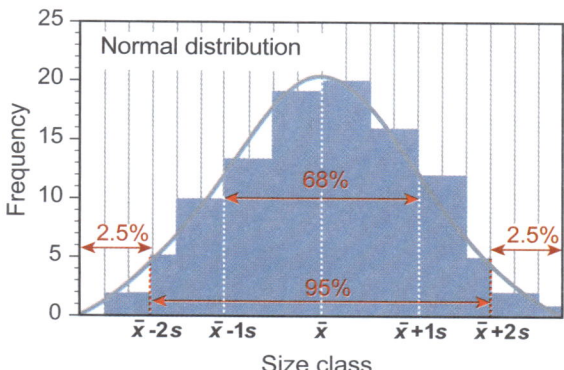

Histogram A has a larger standard deviation; the values are spread widely around the mean.

Both plots show a normal distribution with a symmetrical spread of values about the mean.

Histogram B has a smaller standard deviation; the values are clustered more tightly around the mean.

Calculating s

$$s = \sqrt{\frac{\sum(x - \bar{x})^2}{n - 1}}$$

$\sum(x - \bar{x})^2$ = sum of squared deviations from the mean

n = sample size. $n - 1$ provides a unbiased s for small sample sizes (large samples can use n).

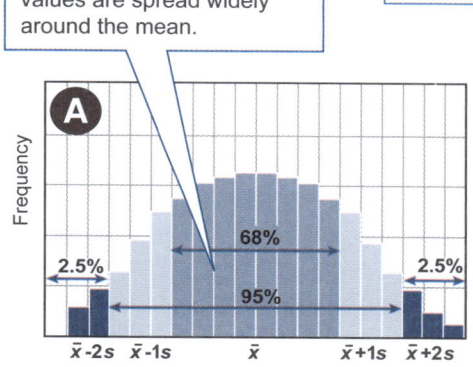

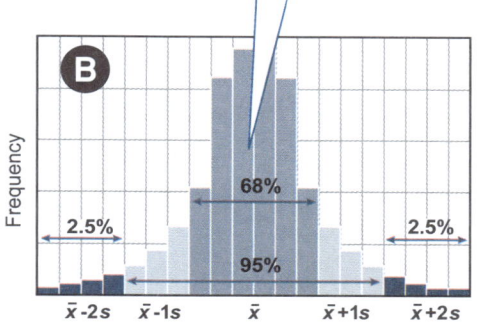

1. Two data sets have the same mean. The first data set has a much larger standard deviation than the second data set. What does this tell you about the spread of data around the mean in each case? Which data set is most reliable?

   _____

   _____

   _____

2. The data on the right shows the heights for 29 male swimmers.

   (a) Calculate the mean for the data: _____

   (b) Use manual calculation, a calculator, or a spreadsheet to calculate the standard deviation (s) for the data:

   _____

   (c) State the mean ± 1s: _____

   (d) What percentage of values are within 1s of the mean? _____

   (e) What does this tell you about the spread of the data? _____

   _____

Raw data: Height (cm)

178	177	188	176	186	175
180	181	178	178	176	175
180	185	185	175	189	174
178	186	176	185	177	176
176	188	180	186	177	

©2024 **BIOZONE** International
ISBN: 978-1-99-101405-4
Photocopying Prohibited

269 Detecting Bias in Samples

Key Question: What is sampling bias, and how can it be detected and eliminated?

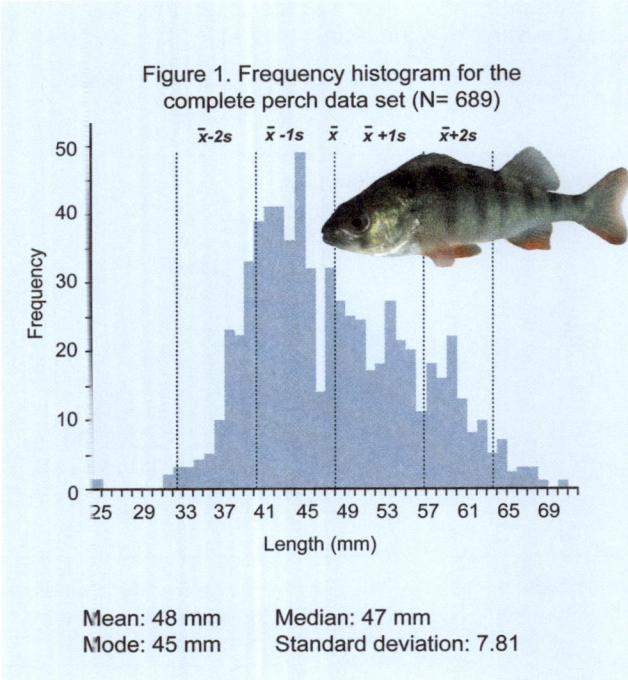

Figure 1. Frequency histogram for the complete perch data set (N= 689)

Mean: 48 mm   Median: 47 mm
Mode: 45 mm   Standard deviation: 7.81

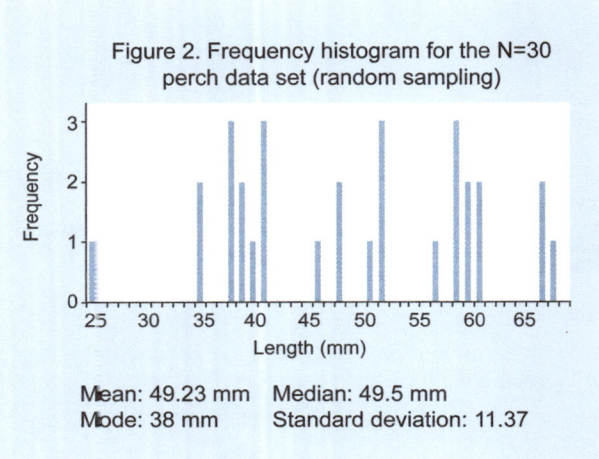

Figure 2. Frequency histogram for the N=30 perch data set (random sampling)

Mean: 49.23 mm   Median: 49.5 mm
Mode: 38 mm   Standard deviation: 11.37

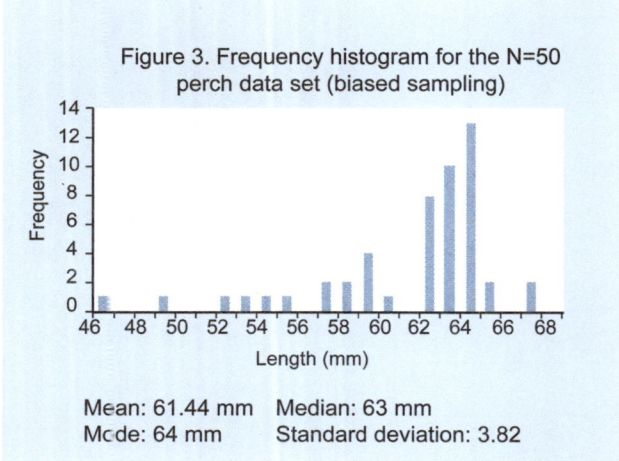

Figure 3. Frequency histogram for the N=50 perch data set (biased sampling)

Mean: 61.44 mm   Median: 63 mm
Mode: 64 mm   Standard deviation: 3.82

Bias is the selection for or against one particular group and can influence the findings of an investigation. Bias can occur when sampling is not random and certain members of a population are under- or over-represented. Small sample sizes can also bias results. Bias can be reduced by random sampling (sampling in which all members of the population have an equal chance of being selected). Using appropriate collection methods will also reduce bias.

- This exercise illustrates how random sampling, large sample size, and sampling bias affect our statistical assessment of variation in a population. In this exercise, perch were collected and their body lengths (mm) were measured. Data are presented as a frequency histogram and with **descriptive statistics** (**mean, median, mode** and standard deviation).
- Figure 1 shows the results for the complete data set. The sample set was large (N= 689) and the perch were randomly sampled. The data are close to having a normal distribution.
- Figures 2 and 3 show results for two smaller sample sets drawn from the same population. The data collected in Figure 2 were obtained by random sampling but the sample was relatively small (N = 30). The person gathering the data displayed in Figure 3 used a net with a large mesh size to collect the perch.

1. (a) Compare the results for the two small data sets (Figures 2 and 3). How close are the mean and median to each other in each sample set?

_____
_____
_____
_____

(b) Compare the standard deviation for each sample set:

_____
_____
_____

(c) Describe how each of the smaller sample sets compares to the large sample set (Figure 1):

_____
_____
_____
_____

(d) Why do you think the two smaller sample sets look so different from each other?

_____
_____
_____
_____

270 Biological Drawings

Key Question: What is the purpose of a good biological drawing when studying a specimen?

- Drawing is a very important skill to have in biology. Drawings record what a specimen looks like and give you an opportunity to record its important features. Often, drawing something will help you remember its features at a later date, e.g. in a test.
- **Biological drawings** require you to pay attention to detail. It is very important that you draw what you actually see, and not what you think you should see.
- Biological drawings should include as much detail as you need to distinguish different structures and types of tissue, but avoid unnecessary detail which can make your drawing confusing.
- Attention should be given to the symmetry and proportions of your specimen. Accurate labeling, a statement of magnification or scale, the view (section type), and type of stain used (if applicable) should all be noted on your drawing.
- Some key points for making good biological drawing are described on the example below. The drawing of *Drosophila* (right) is well executed but lacks the information required to make it a good biological drawing.

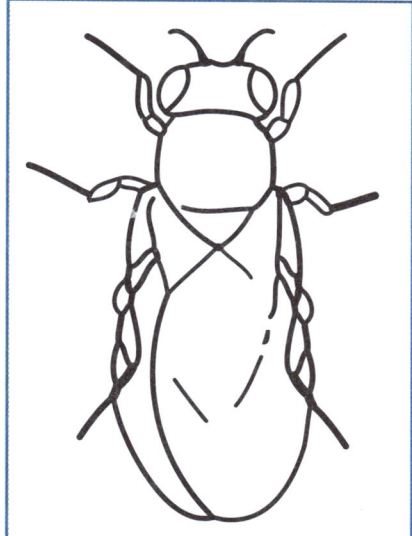

This drawing of *Drosophila* is a fair representation of the animal, but has no labels, title, or scale.

All drawings must include a title. Underline the title if it is a scientific name. → **Copepod**

Place your drawing on the left of the page. This will leave room to place all the labels to the right of the drawing.

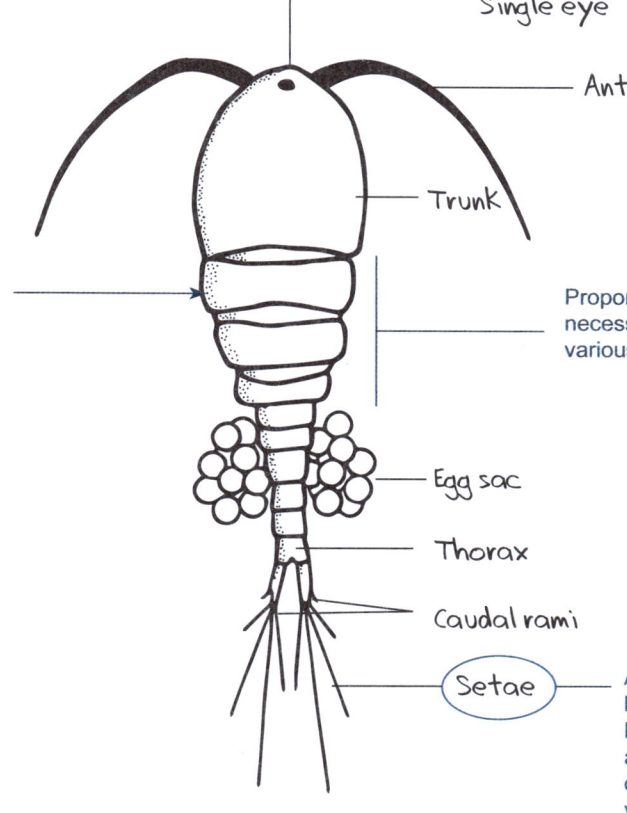

If you need to represent depth, use stippling (dotting). Do not use shading as this can smudge and obscure detail.

Proportions should be accurate. If necessary, measure the lengths of various parts with a ruler.

Use simple, narrow lines to make your drawings.

Use a sharp pencil for drawing. Make your drawing on plain white paper.

All parts of your drawing must be labeled accurately. Labeling lines should be drawn with a ruler and should not cross over other label lines. Try to use only vertical or horizontal lines, although this is not always possible.

Your drawing must include a scale or magnification to indicate the size of your subject. → Scale 0.2 mm

Annotated diagrams

An annotated diagram is a diagram that includes a series of explanatory notes. These provide important or useful information about your subject.

Transverse section through collenchyma of Helianthus stem. Magnification (× 450)

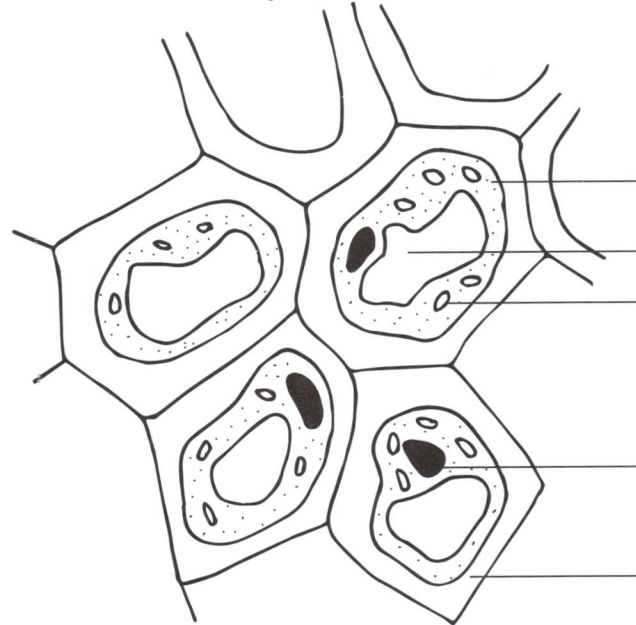

- Cytoplasm - Solution of dissolved substances, enzymes, and organelles.
- Vacuole containing cell sap.
- Chloroplast - Organelles containing chlorophyll where photosynthesis occurs.
- Nucleus - A large organelle containing most of the cell's DNA.
- Primary wall with secondary thickening.

Plan diagrams

Plan diagrams are drawings made of samples viewed under a microscope at low or medium power. They are used to show the distribution of the different tissue types in a sample without any cellular detail. The tissues are identified, but no detail about the cells within them is included. The example here shows a plan diagram produced after viewing a light micrograph of a transverse section through a dicot stem.

Light micrograph of a transverse section through a dicot stem.

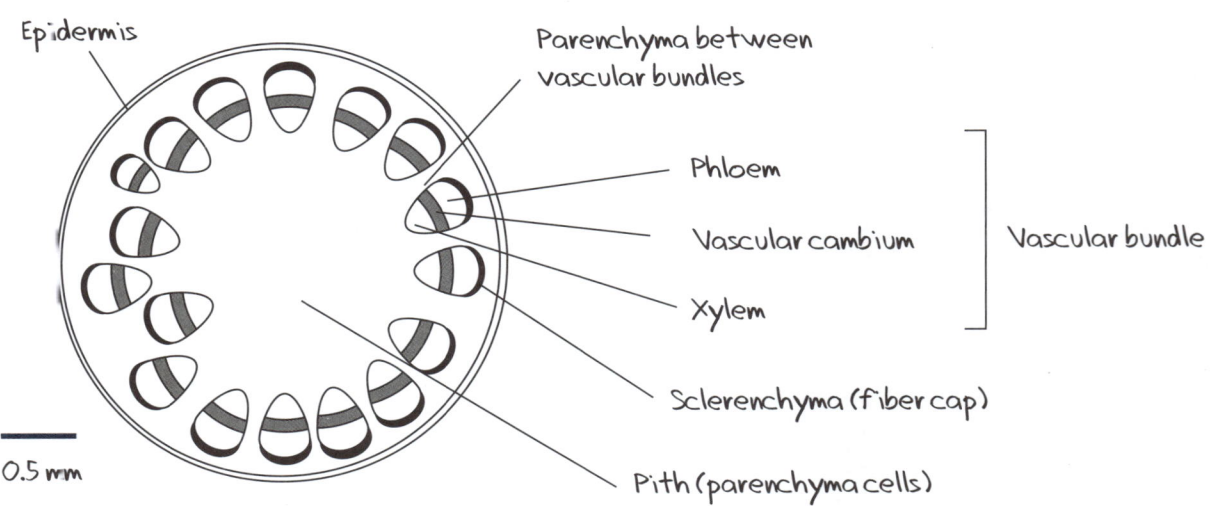

- Epidermis
- Parenchyma between vascular bundles
- Phloem
- Vascular cambium
- Xylem
- Sclerenchyma (fiber cap)
- Pith (parenchyma cells)
- Vascular bundle

0.5 mm

271 Practicing Biological Drawings

Key Question: What kind of detail is needed when making accurate and useful biological drawings?

Above: Use relaxed viewing when drawing at the microscope. Use one eye (the left for right handers) to view and the right eye to look at your drawing.

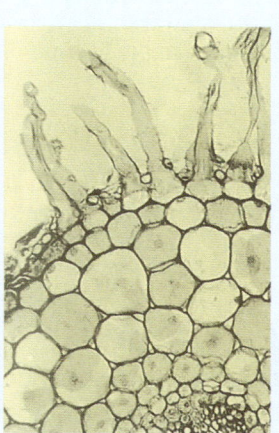

Above: Light micrograph Transverse section (TS) through a Ranunculus root. Right: A biological drawing of the same section.

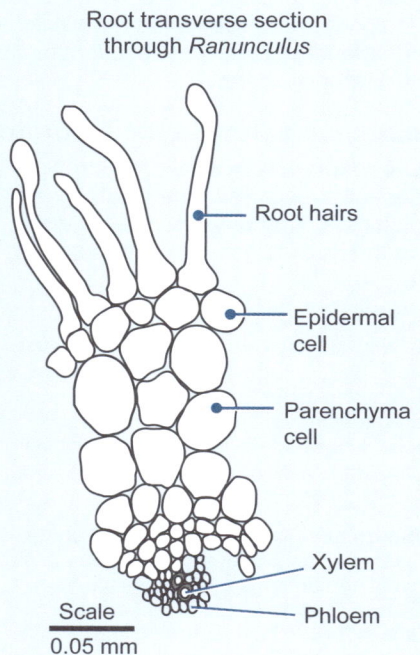

Root transverse section through *Ranunculus*
- Root hairs
- Epidermal cell
- Parenchyma cell
- Xylem
- Phloem

Scale 0.05 mm

1. The image below is a labeled photomicrograph (x50) showing a partial section through a dicot root. Use this image to construct a plan diagram:

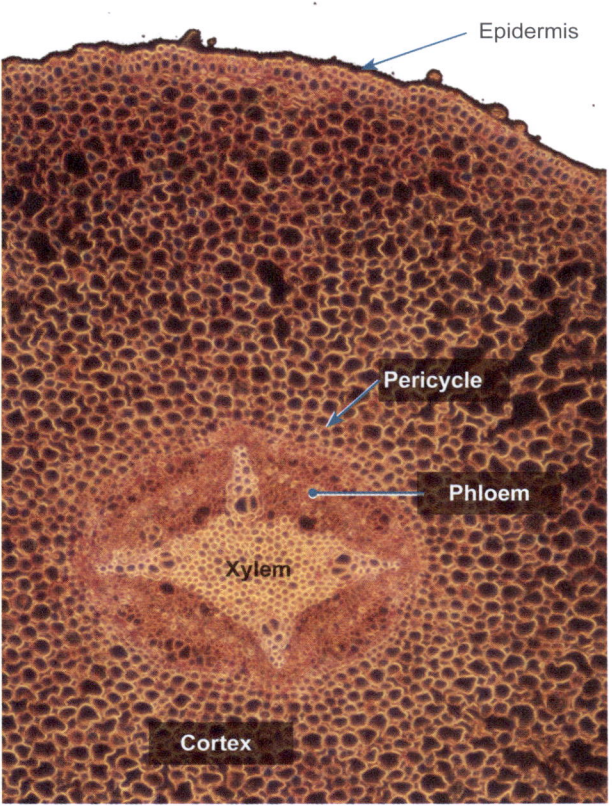

272 Safety and Ethics in Investigations

Key Question: What safety and ethical issues need to be addressed during investigations?

▸ Scientific research, no matter at what level, should be carried out in accordance with ethical and safety guidelines. These guidelines apply to health and safety in the laboratory and field, risk assessment, and correct use of equipment, as well as the ethical issues associated with animal welfare, privacy and personal information, and environmental impact. Ethical considerations also apply to reporting of **data** and honest use and acknowledgement of reference material.

Health and safety in the laboratory

▸ Laboratory hazards fall into three general categories: chemical, biological, and physical. Depending on the hazard, they have potential to cause harm to people, other organisms, or the environment.

▸ **Chemical**: Chemicals could be ingested, absorbed through the skin, or inhaled. Examples include cleaning agents, disinfectants, and reagents (powdered and liquid). Some chemicals can cause fires or explosions if not handled correctly.

▸ **Biological**: All biological material should be treated as potentially hazardous to avoid contamination and possible harm. Examples include microbial samples, animal tissue, fluid samples, and plant samples.

▸ **Physical**: There are numerous potential physical hazards ranging from the laboratory environment itself to the equipment you are using. Common hazards include injury caused by not using the equipment correctly (electrical, thermal, or sound hazards), cluttered working spaces, and tripping or slip hazards, e.g. wet floor.

Assessing and reducing risk in the lab

▸ Identify potential hazards before you start and use risk assessments informed by safety data sheets (SDS) held by your school.
▸ Wear appropriate personal protection equipment (PPE) such as lab coat, gloves, safety glasses, ear protection, and a mask as necessary.
▸ Ensure all chemicals and solutions are clearly labeled. Respect warnings and hazard notices.
▸ Know how to correctly use all equipment and machinery before you begin.
▸ Maintain clean work spaces and floors to reduce the risk of slips and spills. Keep access ways to emergency equipment clear.

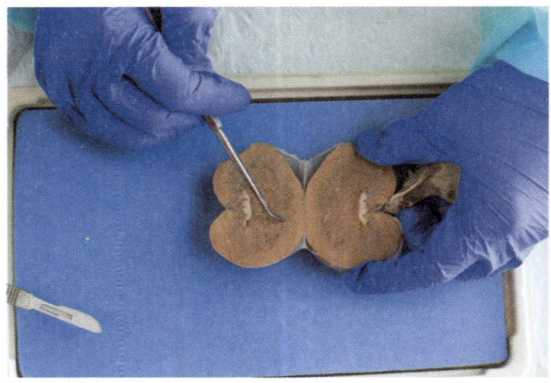

1. (a) Identify potential health and safety issues associated with the dissection of the pig kidney being carried out in the photo (left):
   _____
   _____
   _____

   (b) What has been done to reduce potential risks?
   _____
   _____

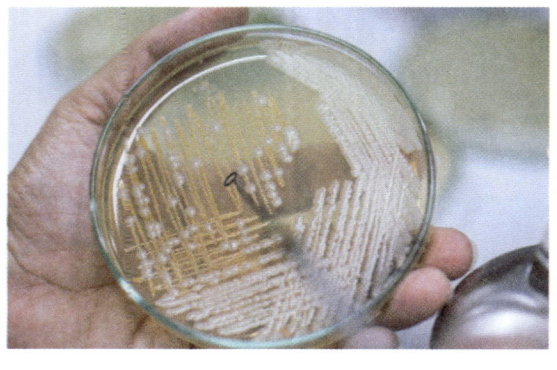

2. (a) Identify two potential safety or health hazards associated with the inspection of bacterial colonies in the photo (bottom left):
   _____
   _____

   (b) What could be done to reduce these risks?
   _____
   _____
   _____

Take care!

▶ The science laboratory is an exciting place to be. You will have a chance to carry out or observe many experiments during your course. Some are more spectacular than others! Regardless of the experiment or investigation, you need to follow safe laboratory practices to keep yourself and others safe. The cartoon below shows students working in a laboratory.

3. Identify at least 8 safety hazards in the cartoon above and provide solutions for how you could reduce risk:

___

Health and safety in fieldwork

Field studies present their own their own set of ethical and safety considerations. The outdoor environment can be harsh, and contain wildlife, plants, and geographic features that can be hazardous.

▸ Assess the potential hazards of the area before beginning any field studies. Field studies may also require some follow-up laboratory work, especially if samples found in the field need to be identified or processed. In these cases, follow lab health and safety guidelines.

▸ Identify potential hazards before you start and become knowledgeable about their risks. In the field, this includes the weather as well as your surroundings. Be aware of hidden hazards such as wasp nests, stinging plants or venomous animals.

Honesty and ethical issues

▸ If you are sampling or collecting live organisms, you must consider the environmental impact of any sampling procedures, return live organisms to the same place if possible, respect the natural environment, and handle animals in a way that minimizes stress or damage to them. Plan your study to minimize your impact on the natural environment.

▸ Report your true data and findings, even if they are not the results you were expecting. Changing results to fit your hypothesis is misleading and unethical.

▸ Acknowledge the intellectual property of others, e.g. photographs and data, and do not to copy directly from sources. Representing the work of others as your own is plagiarism.

Be meticulous in maintaining an accurate logbook, acknowledge all your sources, and reference cited works accurately. Act ethically and responsibly in all aspects of your research, including in the disposal of biological material.

4. Identify potential health and safety issues associated with the rocky shore study being carried out in the photo (left):

_____
_____
_____
_____
_____

5. Identify potential health and safety issues associated with carrying out a study in the area in the photo (left):

_____
_____
_____
_____
_____
_____

6. Describe the potential ethical issues associated with the following investigative scenario: a vegetation survey in a sensitive ecological area:

_____
_____
_____

273 A Qualitative Practical Task

Key Question: How can qualitative data be recorded and analyzed?

- Not all the experimental work you carry out in biology will yield **quantitative data**. For example, when recording color changes in simple biochemical tests for common components of foods.
- Two common tests for carbohydrates are the iodine/potassium iodide test for starch, and the Benedict's test for reducing sugars such as glucose. These tests indicate the presence of a substance with a color change.
- When a starchy fruit ripens, the starch is converted to simple reducing sugars.

The aim
To investigate the effect of ripening on the relative content of starch and simple sugars in bananas.

The tests

Iodine-potassium iodide test for starch

The sample is covered with the iodine in potassium iodide solution. The sample turns blue-black if starch is present.

Benedict's test for reducing sugars

The sample is heated with the reagent in a boiling water bath. After 2 minutes, the sample is removed and stirred, and the color recorded immediately after stirring. A change from a blue to a brick red color indicates a reducing sugar.

Summary of the method
Two 1 cm thick slices of banana from each of seven stages of ripeness were cut and crushed to a paste. One slice from each stage was tested using the I/KI test for starch, and the other was tested using the Benedict's test.

The color changes were recorded in a table. Signs (+/–) were used to indicate the intensity of the reaction relative to those in bananas that were either less or more ripe.

Green unripe and hard → bright yellow ripening but firm with green tip → mottled yellow/brown ripe and soft

Stage of ripeness	Starch-iodine test		Benedict's test	
1	blue-black	+++++	blue clear	–
2	blue-black	++++	blue clear	–
3	blue-black	+++	green	+
4	blue-black	++	yellow cloudy	++
5	slight darkening	+	orange thick	+++
6	no change	–	orangey-red thick	++++
7	no change	–	brick-red thick	+++++

1. Explain why each of the following protocols was important:

   (a) All samples of banana in the Benedict's reagent were heated for 2 minutes: _____

   _____

   _____

   (b) The contents of the banana sample and Benedict's reagent were stirred after heating: _____

   _____

   _____

2. Explain what is happening to the relative levels of starch and glucose as bananas ripen: _____

   _____

   _____

   _____

274 Analyzing Experimental Data

Key Question: What is the effect of fertilizer on the growth of radishes?

The aim
To investigate the effect of a nitrogen fertilizer on the growth of radish plants.

Radishes

Background
Inorganic fertilizers were introduced to crop farming during the late 19th century.

Fertilizer addition increased crop yields. An estimated 50% of crop yield is attributable to the use of fertilizer.

Nitrogen is a very important element for plant growth. Several types of nitrogen fertilizers are manufactured, e.g. urea.

Hypothesis
If plants need nitrogen to grow, radish growth will increase with increasing nitrogen concentration.

Experimental method

- Radish seeds were planted in separate identical pots (5 cm x 5 cm wide x 10 cm deep) and kept together in standard lab conditions. The seeds were planted into a commercial soil mixture and divided randomly into six groups, each with five sample plants (a total of 30 plants in six treatments).

- The radishes were watered every day at 10 am and 3 pm with 500 mL per treatment per watering. Water soluble nitrogen fertilizer was added to the 10 am watering on the 1st, 11th, and 21st days. The fertilizer concentrations used were: 0.00, 0.06, 0.12, 0.18, 0.24, and 0.30 g, and each treatment received a different concentration.

- The plants were grown for 30 days before being removed from the pots, washed, and the radish root weighed. The results are presented below.

Fertilizer concentration (g/L)	Sample number				
	1	2	3	4	5
0	80.1	83.2	82.0	79.1	84.1
0.06	109.2	110.3	108.2	107.9	110.7
0.12	117.9	118.9	118.3	119.1	117.2
0.18	128.3	127.3	127.7	126.8	DNG*
0.24	23.6	140.3	139.6	137.9	141.1
0.30	122.3	121.1	122.6	121.3	123.1

*DNG = did not germinate

† Based on data from M S Jilani, et al Journal Agricultural Research

1. Identify the independent variable for the experiment and its range: _____

2. Identify the dependent variable for the experiment: _____

3. What is the sample size for each concentration of fertilizer? _____

4. (a) One of the radishes recorded in the table on the previous page did not grow as expected and produced an extreme value. Record the outlying value here:

_____

(b) Why should this value not be included in future calculations? _____

_____

_____

5. Use Table 1 below to record the raw data from the experiment. You will need to include column and row headings and a title, and complete some simple calculations. Some headings have been entered for you.

Table 1: _____

	Mass of radish root (g)					Total mass	Mean mass

6. The students decided to collect more data by counting the number of leaves on each radish plant at day 30. The data are presented in Table 2.
Use the space below to calculate the mean, median and mode for the leaf data. Add these data to table 2.

Table 2: Number of leaves on radish plants under six different fertilizer concentrations.

| Fertilizer concentration (g/L) | Number of leaves at day 30 | | | | | Mean | Median | Mode |
| | Sample (n) | | | | | | | |
	1	2	3	4	5			
0	9	9	10	8	7			
0.06	15	16	15	16	16			
0.12	16	17	17	17	16			
0.18	18	18	19	18	DNG*			
0.24	6	19	19	18	18			
0.30	18	17	18	19	19			

* DNG: Did not germinate

7. Use the grid below to draw a line graph of the experimental results. Plot your calculated mean mass data from Table 1, and remember to include a title and correctly labelled axes.

8. Which fertilizer concentration appeared to produce the best growth of radish mass? _____

9. Which fertilizer concentration appeared to produce the best growth of leaves? _____

10. Write a short conclusion for the entire experiment: _____

11. (a) What assumptions were made in the design of this experiment? _____

(b) Can you suggest ways in which the design could be improved? How could you evaluate whether the mass differences between the fertilizer concentrations were significant?

Glossary: English/Spanish

abiotic factor: Non-living, physical features in an ecosystem, including temperature, humidity, and rainfall.
factor abiótico: Características físicas no vivas en un ecosistema, incluida la temperatura, la humedad y la lluvia.

accuracy: The correctness of a measurement; how close a measured value is to the true value.
exactitud: La exactitud de una medición; qué tan cerca está un valor medido del valor verdadero.

active site: Region of an enzyme where the substrate binds and undergoes a chemical reaction.
sitio activo: Región de una enzima donde el sustrato se une y sufre una reacción química.

active transport: The movement of molecules or ions across a cell membrane against a concentration gradient, requiring an expenditure of energy.
transporte activo: Movimiento de moléculas o iones a través de una membrana celular contra un gradiente de concentración, que requiere un gasto de energía.

adenosine triphosphate - ATP: An organic compound that serves as an energy source for metabolic processes.
ATP/trifosfato de adenosina: Un compuesto orgánico que sirve como fuente de energía para los procesos metabólicos.

aerobic: A biological process that requires oxygen.
aerobio: Un proceso biológico que requiere oxígeno.

aerobic respiration: type of respiration that requires oxygen.
Respiración aeróbica: tipo de respiración que requiere oxígeno.

allele: Any of the alternative versions of a gene that may produce distinguishable phenotypes.
alelo: Cualquiera de las versiones alternativas de un gen que puede producir fenotipos distinguibles.

amino acid: Any organic compound containing both an amino group and a carboxylic acid group.
aminoácido: Cualquier orgánico compuesto que contenga tanto un grupo amino como un grupo ácido carboxílico. Los bloques de construcción de las proteínas.

anabolic reaction / anabolism: A chemical reaction that constructs large, complex molecules from simpler molecules.
reacción anabólica: Una reacción química que construye moléculas grandes y complejas a partir de moléculas más simples.

anaerobic respiration: Type of respiration that does not require oxygen.
respiración anaerobica: Tipo de respiración que no requiere oxígeno.

antibody: A protein produced by the body in response to a specific antigen and aimed at targeting and destroying it.
anticuerpo: Una proteína producido por el cuerpo en respuesta a un antígeno específico y destinado a atacar y destruirlo.

anticodon: A sequence of three adjacent nucleotides in tRNA that binds to a corresponding codon in mRNA during protein synthesis.
anticodón: Secuencia de tres nucleótidos adyacentes en el ARNt que se une a un codón correspondiente en el ARNm durante la síntesis de proteínas.

antigen: A foreign molecule that stimulates an immune response in the body.
antígeno: Una molécula extraña que estimula una respuesta inmune en el cuerpo.

assumption: A statement that is assumed to be true but is not (or cannot be) tested.
presunción: Una afirmación que se supone que es verdadera pero que no se prueba (o no se puede probar).

ATP/adenosine triphosphate: An organic compound that serves as an energy source for metabolic processes.
ATP/trifosfato de adenosina: Un compuesto orgánico que sirve como fuente de energía para los procesos metabólicos.

auxin: Any of several plant hormones that regulate the growth and development of plants.
auxina: Cualquiera de varias hormonas vegetales que regulan el crecimiento y desarrollo de las plantas.

biodiversity: The amount of biological variation present in a region (includes genetic, species, and habitat diversity).
biodiversidad: La cantidad de variación biológica presente en una región (incluye genética, especies y diversidad de hábitat).

bioinformatics: The use of computer science, mathematics, and information theory to organize and analyze complex biological data.
bioinformática: El uso de las ciencias computacionales, las matemáticas y la teoría de la información para organizar y analizar datos biológicos complejos.

biological drawing: An illustration that visually communicates the structure of a subject being studied, showing specific details.
dibujo biológico: Una ilustración que comunica visualmente la estructura de un tema que se está estudiando, mostrando detalles específicos.

biotic factor: Relating to the living factors in an ecosystem, including distribution and abundance.
factor biótico: Relacionado con los factores vivos en un ecosistema, incluida la distribución y la abundancia.

cancer: The malignant growth of cells due to uncontrolled cell division.
cáncer: El crecimiento maligno de las células debido a la división celular incontrolada.

carbohydrate: Any of a class of molecules that contain carbon, hydrogen and oxygen, with a general formula $C_n(H_2O)_n$.
carbohidrato: Cualquiera de una clase de moléculas que contienen carbono, hidrógeno y oxígeno, con una fórmula general $C_n(H_2O)_n$.

carbon cycle: The process by which carbon is exchanged between living organisms, the earth and its atmosphere.
ciclo del carbono: El proceso por el cual el carbono se intercambia entre los organismos vivos, la tierra y su atmósfera.

catabolic reaction / catabolism: The breakdown of large, complex molecules into smaller, simpler molecules.
reacción catabólica: La descomposición de moléculas grandes y complejas en moléculas más pequeñas y simples.

catalyst: A substance that modifies and increases the rate of a chemical reaction without being consumed in the process.
catalizador: Sustancia que modifica y aumenta la velocidad de una reacción química sin ser consumida en el proceso.

cell: The smallest biological unit that can survive on its own. It is the base unit of all living organisms.
celda: La unidad biológica más pequeña que puede sobrevivir por sí sola. Es la unidad base de todos los organismos vivos.

cell cycle: The cycle of stages that occur in a cell as it grows and divides to produce new daughter cells.
célula: La unidad biológica más pequeña que puede sobrevivir por sí sola. Es la unidad base de todos los organismos vivos.

cell differentiation: The process by which a cell changes from one cell type to another, usually to a more specialized type of cell.
diferenciación celular: El proceso por el cual una célula cambia de un tipo de célula a otro, generalmente a un tipo de célula más especializada.

cell division: The cycle of stages that occur in a cell as it grows and divides to produce new daughter cells.
división celular: El ciclo de etapas que ocurren en una célula a medida que crece y se divide para producir nuevas células hijas.

cell specialization: the process by which a generic cell becomes a cell with a defined task.
especialzación celular: el proceso por el cual una célula genérica se convierte en una célula con una tarea definida.

cell wall: The rigid outermost cell layer found in plants and certain algae, bacteria, and fungi but absent from animal cells.
pared celular: Capa celular externa rígida que se encuentra en las plantas y ciertas algas, bacterias y hongos, pero ausente de las células animales.

cellular differentiation: The process by which a cell changes from one cell type to another, usually to a more specialized type of cell.
diferenciación celular: El proceso por el cual una célula cambia de un tipo de célula a otro, generalmente a un tipo de célula más especializada.

cellular respiration: The series of metabolic reactions that oxidize organic molecules to produce ATP.
respiración celular: La serie de reacciones metabólicas que oxidan las moléculas orgánicas para producir ATP.

chlorophyll: A green photosynthetic pigment found primarily in the chloroplasts of algae and plants, essential to photosynthesis.
clorofila: Un pigmento fotosintético verde que se encuentra principalmente en los cloroplastos de algas y plantas, esencial para la fotosíntesis.

chloroplast: An organelle within the cells of plants and green algae that contains chlorophyll and is the site of photosynthesis.
cloroplasto: Un orgánulo dentro de las células de las plantas y algas verdes que contiene clorofila y es el sitio de la fotosíntesis.

chromatid: One half of a replicated chromosome, held to its other half at the centromere.
cromátida: La mitad de un cromosoma replicado, retenido en su otra mitad en el centrómero.

chromatin: A complex of DNA and proteins, making up the chromosomes.
cromatina: Un complejo de ADN y proteínas, que componen los cromosomas.

chromosome: A cellular structure consisting of one DNA molecule and associated protein molecules.
cromosoma: Estructura celular que consiste en una molécula de ADN y moléculas de proteínas asociadas.

climate change: A change in the patterns of climate such as temperature and rainfall that is attributed to human activity, including the use of fossil fuels.
cambio climático: Un cambio en los patrones del clima, como la temperatura y las precipitaciones, que se atribuye a la actividad humana, incluido el uso de combustibles fósiles.

codon: A sequence of three adjacent nucleotides in a DNA or mRNA sequence that is part of the genetic code.
codón: Una secuencia de tres nucleótidos adyacentes en una secuencia de ADN o ARNm que es parte del código genético.

commensalism: A relationship between two organisms in which one benefits, and one is unaffected.
comensalismo: Una relación entre dos organismos en la que uno se beneficia y el otro no es afectado.

common ancestor: An ancestor that two or more descendant species have in common.
ancestro común: Un ancestro que dos o más especies descendientes tienen en común.

competition: Interaction within or between species in which individuals attempt to access the same limited resource.
competición: Interacción dentro o entre especies en la que los individuos intentan acceder al mismo recurso limitado.

condensation reaction: A chemical reaction in which two molecules are joined covalently with the removal of an -OH ion from one and a +H ion from the other to form water.
reacción de condensación: Reacción química en la que dos moléculas se unen covalentemente con la eliminación de un ion -OH de una y un ion +H de la otra para formar agua.

consumer/heterotroph: An organism that feeds on producers, other consumers, or non-living organic material.
consumidor: Un organismo que se alimenta de productores, otros consumidores o material orgánico no vivo.

continuous data: Quantitative data representing a scale of measurement that can consist of numbers other than whole numbers.
Datos continuos: datos cuantitativos que representan una escala de medida que puede consistir en números distintos de los números enteros.

continuous variation: Variation in phenotypic traits in which in which there is a complete spread of forms (phenotypes) across a range.
variación continua: Variación en los rasgos fenotípicos en los que hay una extensión completa de formas (fenotipos) a través de un rango.

control: A component of an experiment that is isolated from the effects of the independent variable. It demonstrates that any change to the dependent variable must come from the independent variable.
control: Componente de un experimento aislado de los efectos de la variable independiente. Demuestra que cualquier cambio en la variable dependiente debe provenir de la variable independiente.

controlled variable: Another word for independent variable. The variable being controlled or changed by the experimenter.
variable controlada: Otra palabra para variable independiente. El variable es controlado o cambiado por el experimentador.

crossing over: The reciprocal exchange of genetic material between non-sister chromatids during prophase 1 of meiosis.
cruzando: El intercambio recíproco de material genético entre cromátidas no hermanas durante la profase 1 de la meiosis.

cytokinesis: The part of the cell division process when the cytoplasm of a single eukaryotic cell divides to produce two daughter cells.
citocinesis: La parte del proceso de división celular cuando el citoplasma de una sola célula eucariota se divide para producir dos células hijas.

data: A set of values of qualitative or quantitative variables, collected through observation.
datos: Conjunto de valores de variables cualitativas o cuantitativas, recogidos a través de la observación.

deforestation: The removal of forests by cutting, burning or other large scale activity by humans, usually to make way for crops or monoculture plantations.
deforestación: La eliminación de bosques mediante la tala, quema u otra actividad a gran escala por parte de los seres humanos, generalmente para dar paso a cultivos o plantaciones de monocultivos.

denaturation: The alteration of a protein shape resulting in a loss of function.
desnaturalización: La alteración de la forma tridimensional de una proteína que resulta en la pérdida de su función.

dependent variable: The variable being tested and measured in an experiment, whose value depends on that of the independent variable.
variable dependiente: La variable que se está probando y medido en un experimento, cuyo valor depende del de la variable independiente.

descriptive statistics: Also called summary statistics, these are brief descriptors that help summarise features of data.
estadística descriptiva: También llamadas estadísticas resumidas, estas son descriptores breves que ayudan a resumir las características de los datos.

diabetes mellitus: A disease in which the body is unable to produce or respond to the hormone, insulin, to maintain optimum levels of glucose in the blood.
diabetes mellitus: Una enfermedad en la que el cuerpo es incapaz de producir o responder a la hormona, la insulina, para mantener niveles óptimos de glucosa en la sangre.

diffusion: The net movement of molecules from a region of high concentration to one of lower concentration.
difusión: Movimiento neto de moléculas de una región de alta concentración a una de menor concentración.

dihybrid cross: An organism that is heterozygous, with two different alleles at a genetic location.
dihíbrido: Un organismo que es heterogótico, con dos alelos diferentes en una ubicación genética.

discontinuous variation: Variation in phenotypes of individuals that fall into discrete categories.
variación discontinua: Variación en los fenotipos de los individuos que caen en categorías discretas.

DNA - deoxyribonucleic acid: A large molecule composed of two polynucleotide chains that carries the genetic code and enables cells to function.
ADN: La duplicación de una molécula de ADN, produciendo dos copias idénticas de una molécula de ADN original.

DNA replication: The duplication of a DNA molecule, producing two identical copies from one original DNA molecule.
replicación del ADN: La duplicación de una molécula de ADN, produciendo dos copias idénticas de una molécula de ADN original.

dominant allele/trait: An allele that is expressed (as a trait) even if the individual only has one copy of the allele.
alelo dominante: Un alelo que se expresa (como un rasgo) incluso si el individuo sólo tiene una copia del alelo.

ecological pyramid: A graphical representation of the relationship between different organisms in an ecosystem, and their trophic levels.
pirámide ecológica: Representación gráfica de la relación entre los diferentes organismos de un ecosistema y sus niveles tróficos.

ecosystem: All the organisms in a given area as well as the abiotic factors with which they interact.
ecosistema: Todos los organismos en un área determinada, así como los factores abióticos con los que interactúan.

electron transport chain: A series of protein complexes that transfer electrons from donors to acceptors across a membrane via redox reactions.
Cadena de transporte de electrones: Una serie de complejos de proteínas que transfieren electrones de donantes a aceptores a través de una membrana a través de reacciones redox.

endocytosis: A process of cellular ingestion by which the plasma membrane folds inward to bring substances into the cell.
endocitosis: Proceso de ingestión celular por el cual la membrana plasmática se pliega hacia adentro para introducir sustancias en la célula.

enzyme: Globular proteins that act as biological catalysts for specific reactions.
enzima: Proteínas globulares que actúan como catalizadores biológicos para reacciones específicas.

epigenetics: Heritable phenotypic changes that do not involve changes to the DNA sequence.
epigenética: Cambios fenotípicos hereditarios que no implican cambios en la secuencia de ADN.

eukaryotic cell: An animal cell in which the genetic material is contained within a distinct, membrane-bound nucleus.
célula eucariota: Célula animal en la que el material genético está contenido dentro de un núcleo distinto unido a la membrana.

evolution: The change in the heritable characteristics of populations over successive generations.
evolución: El cambio en las características hereditarias de las poblaciones a lo largo de generaciones sucesivas.

extocytosis: Secretion of intracellular molecules, contained within membrane-bound vesicles, to the outside of the cell by fusion of vesicles with the plasma membrane.
exocitosis: Secreción de moléculas intracelulares, contenidas dentro de vesículas unidas a la membrana, al exterior de la célula por fusión de vesículas con la membrana plasmática.

fermentation: An anaerobic metabolic process by which a carbohydrate, such as starch or a sugar, is converted into an alcohol or an acid.
fermentación: Un proceso metabólico anaeróbico por el cual un carbohidrato, como el almidón o un azúcar, se convierte en un alcohol o un ácido.

fitness: An organism's ability to survive to reproductive age and produce offspring. A mathematical measure of an organism's genetic contribution to the next generation.
aptitud: La capacidad de un organismo para sobrevivir hasta la edad reproductiva y producir descendencia. Una medida matemática de la contribución genética de un organismo a la próxima generación.

food chain: A model that is used to demonstrate the feeding relationships between organisms in an ecosystem.
cadena alimentaria: Un modelo que se utiliza para demostrar las relaciones de alimentación entre los organismos en un ecosistema.

food web: The combination of all the food chains in a particular ecosystem.
red trófica: La combinación de todas las cadenas alimentarias en un ecosistema en particular.

fossil: The preserved remains, found in the earth's crust, of animal or plants that lived a long time ago.
fósil: Los restos conservados, encontrados en la corteza de la tierra, de animales o plantas que vivieron hace mucho tiempo.

fossil record: The history of life as documented by fossils.
registro fósil: La historia de la vida documentada por los fósiles.

gel electrophoresis: The separation and analysis of protein molecules of varying sizes by moving them through a block of gel using an electric field.
electroforesis en gel: Separación y análisis de moléculas de proteínas de diferentes tamaños moviéndolas a través de un bloque de gel usando un campo eléctrico.

gene: A unit of hereditary information consisting of a specific nucleotide sequence in DNA.
genotipo: Unidad de información hereditaria que consiste en una secuencia específica de nucleótidos en el ADN.

gene expression: The transcription and translation of a gene.
expresión génica: La transcripción y traducción de un gen.

gene flow: Movement of individuals and their genes from one population to another.
flujo de genes: Movimiento de individuos y sus genes de una población a otra.

gene pool: The collective genetic information within a population of interbreeding organisms.
reserva genética: La información genética colectiva dentro de una población de organismos que se cruzan.

genetic diversity: The differences in the genetic makeup of individuals in a population.
diversidad genética: Las diferencias en la composición genética de los individuos en una población.

genetic drift: The random changes in allele frequency in a population over generations.
Deriva genética: Los cambios aleatorios en la frecuencia de alelos en una población a lo largo de generaciones.

genotype: The genetic makeup of an organism
genotipo: La composición genética de un organismo.

geologic time scale: A system for organizing the history of Earth into units of time, based on major geologic events.
escala de tiempo geológica: Un sistema para organizar la historia de la Tierra en unidades de tiempo, basado en los principales eventos geológicos.

glucose: A simple sugar that functions as the main source of metabolic energy in living things.
glucosa: Un azúcar simple que funciona como la principal fuente de energía metabólica en los seres vivos.

glycolysis: The metabolic pathway that converts glucose into pyruvate.
glucólisis: La vía metabólica que convierte la glucosa en piruvato.

graph: A diagram that is used to present scientific data, usually with two variables on an x and y axis. It allows relationships between variables to be clearly visualized.
gráfico: Un diagrama que se utiliza para presentar datos científicos, generalmente con dos variables en un eje x e y. Permite visualizar claramente las relaciones entre variables.

habitat: The natural environment in which an organism lives, including all of the biotic and abiotic factors.
hábitat: El medioambiente natural en el que vive un organismo, incluidos todos los factores bióticos y abióticos.

heterozygous: Having two different alleles for any hereditary characteristic.
heterocigótico: Tener dos alelos diferentes para cualquier característica hereditaria.

histone proteins: A small basic protein found in the nucleus of eukaryotic cells that organizes DNA strands to form chromatin.
histona: Una pequeña proteína básica que se encuentra en el núcleo de las células eucariotas que organiza las hebras de ADN para formar cromatina.

homeostasis: The steady-state physiological condition of the body.
homeostasis: La condición fisiológica de estado estacionario del cuerpo.

homologous structure: Structures found in different organisms that result from a common ancestor, e.g. forelimb of a seal and wing of a bird.
estructura homóloga: Estructuras que se encuentran en diferentes organismos que resultan de un ancestro común, por ejemplo, la extremidad anterior de una foca y el ala de un ave.

homology: Similarity between two different species of organisms due to shared ancestry.
homología: Similitud entre dos especies diferentes de organismos debido a la ascendencia compartida.

homozygous: Where both chromosomes possess identical alleles for a gene, at a specific locus.
homocigótico: Donde ambos cromosomas poseen alelos idénticos para un gen, en un locus específico.

hormone: Chemical messengers secreted directly into the blood, where they circulate to exert specific effects on target tissues and organs.
hormona: Mensajeros químicos secretados directamente en la sangre, donde circulan para ejercer efectos específicos sobre los tejidos y órganos diana.

hydrolysis: A reaction in which a molecule is split into smaller molecules by reacting with water.
hidrólisis: Reacción en la que una molécula se divide en moléculas más pequeñas reaccionando con agua.

hypothesis: A tentative explanation, proposition, or set of propositions capable of being tested by scientific experimentation.
hipótesis: Una explicación tentativa, proposición o conjunto de proposiciones capaces de ser probadas por experimentación científica.

independent assortment: With reference to inheritance, describing how alleles for separate traits are passed to the gametes independently of one another.
surtido independiente: Con referencia a la herencia, describiendo cómo los alelos para rasgos separados se pasan a los gametos independientemente unos de otros.

independent variable: The variable being set by the experimenter which is assumed to have a direct effect upon the dependent variable.
variable independiente: La variable establecida por el experimentador que se supone que tiene un efecto directo sobre la variable dependiente.

interphase: The period in the cell cycle when the cell is not dividing, which accounts for about 90% of the cell cycle. During interphase, cellular metabolic activity is high and cell size may increase.
interfase: El período en el ciclo celular cuando la célula no se está dividiendo, que representa aproximadamente el 90% del ciclo celular. Durante la interfase, la actividad metabólica celular es alta y el tamaño celular puede aumentar.

interspecific competition: Competition for resources between different species.
competencia interespecífica: Competencia por los recursos entre diferentes especies.

intraspecific competition: Competition for resources between member of the same species.
competencia intraespecífica: Competencia por los recursos entre miembros de la misma especie.

invasive species: A species that is not native to that ecosystem.
especies invasoras: Una especie que no es nativa de ese eccosistema.

keystone species: A species that occupies an essential role in an ecosystem and on which most or all of the other species in an ecosystem depend, directly or indirectly.
especies clave: Una especie que ocupa un papel esencial en un ecosistema y de la que dependen, directa o indirectamente, la mayoría o la totalidad de las demás especies de un ecosistema.

krebs cycle: A cycle of aerobic catalyzed reactions in respiration occurring within mitochondria. Generates ATP and reducing power.
ciclo de Krebs: Ciclo de reacciones aeróbicas catalizadas en la respiración que ocurren dentro de las mitocondrias. Genera ATP y potencia reductora.

light dependent reaction/phase: The phase of photosynthesis during which light energy is converted into chemical energy through chemical reactions.
fase dependiente de la luz: La fase de la fotosíntesis durante la cual la energía de la luz se convierte en energía química a través de reacciones químicas.

light independent reaction/phase: phase of photosynthesis during which chemical reactions convert carbon dioxide into sugars.
fase independiente de la luz: Fase de fotosíntesis independiente de la luz durante la cual las reacciones químicasconvierten el dióxido de carbono en azúcares.

link reaction: The stage in respiration that converts pyruvate into acetyl CoA, linking glycolysis to the Krebs cycle.
reacción de enlace: La etapa de la respiración que convierte el piruvato en acetil CoA, vinculando la glucólisis al ciclo de Krebs.

lipid: Any organic substance that does not dissolve in water, but dissolves well in nonpolar organic solvents. Lipids include fatty acids, fats, oils, waxes and steroids.
lípido: Cualquier sustancia orgánica que no se disuelve en agua, pero se disuelve bien en disolventes orgánicos no polares. Los lípidos incluyen ácidos grasos, grasas, aceites, ceras y esteroides.

mean: The sum of the data divided by the number of data entries; a measure of central tendency in a normal distribution.
promedio: La suma de los datos dividida por el número de entradas de datos; una medida de tendencia central en una distribución normal.

median: The middle number in an ordered sequence of numbers. For an odd number of values, it is the average of the two middle numbers.
mediana: El número medio en una secuencia ordenada de números. Para un número impar de valores, es el promedio de los dos números medios.

meiosis: The process of double nuclear division in sexually reproducing organisms which results in cells with half the original number of chromosomes (haploid).
meiosis: El proceso de doble división nuclear en organismos que se reproducen sexualmente que da como resultado células con el medio número eoriginal de cromosomas (haploide).

mitochondria: Organelles in eukaryotic cells that serves as the site of cellular respiration.
mitocondria: Orgánulos en células eucariotas que sirven como sitio de respiración celular.

mitosis: The phase of the cell cycle resulting in nuclear division.
mitosis: La fase del ciclo celular que resulta en la división nuclear.

mode: The value that occurs most often in a data set.
modo: El numero que se produce con mayor frecuencia en un conjunto de datos.

model: A conceptual, mathematical or physical representation of a real-world phenomenon.
modelo: Una representación conceptual, matemática o física de un fenómeno del mundo real.

monohybrid cross: An organism that is homozygous, with two alleles that are the same, at a genetic location.
monohíbrido: Un organismo que es homocigoto, con dos alelos que son iguales, en una ubicación genética.

mutation: A change in the nucleotide sequence of an organism's DNA (or RNA).
mutación: Un cambio en la secuencia de nucleótidos del ADN (o ARN) de un organismo.

mutualism: Biological interaction between (usually two) species that benefits both parties.
mutualismo: Interacción biológica entre (generalmente dos) especies que beneficia a ambas partes.

natural selection: The differential survival and reproduction of favourable phenotypes.
selección natural: La supervivencia diferencial y la reproducción de fenotipos favorables.

negative feedback: The differential survival and reproduction of favourable phenotypes.
selección natural: La supervivencia diferencial y la reproducción de fenotipos favorables.

niche: The position occupied by an organism within its specific environmental conditions.
nicho: La posición que ocupa un organismo dentro de sus condiciones ambientales específicas.

nitrogen cycle: The processes by which nitrogen, in different forms, is cycled between living and non-living things in marine, terrestrial and atmospheric ecosystems.
ciclo del nitrógeno: Los procesos por los cuales el nitrógeno, en diferentes formas, se recicla entre seres vivos y no vivos en ecosistemas marinos, terrestres y atmosféricos.

nuclear membrane: The double-layered membrane enclosing the nucleus of a cell.
membrana nuclear: La membrana de doble capa que encierra el núcleo de una célula.

nucleic acid: A polymer consisting of many nucleotide monomers. Serves as a blueprint for proteins and therefore for all cellular activities. The two types are DNA and RNA.
ácido nucleico: Un polímero que consiste en muchos monómeros de nucleótidos. Sirve como modelo para las proteínas y, por lo tanto, para todas las actividades celulares. Los dos tipos son ADN y ARN.

nucleotide: An organic molecule that is the building block of DNA and RNA. Consist of a sugar molecule (ribose in RNA or deoxyribose in DNA) attached to a phosphate group and a nitrogen-containing base.
nucleótido: Una molécula orgánica que es el bloque de construcción del ADN y el ARN. Consiste en una molécula de azúcar (ribosa en ARN o desoxirribosa en ADN) unida a un grupo fosfato y una base que contiene nitrógeno.

nucleus: The organelle of a eukaryotic cell that contains the genetic material in the form of chromosomes, made up of chromatin.
núcleo: El orgánulo de una célula eucariota que contiene el material genético en forma de cromosomas, formado por cromatina.

observation: The activity of watching or recording what is happening in a given, often experimental, setting.
observación: La actividad de observar o registrar lo que está sucediendo en un entorno dado, a menudo experimental.

optimum (enzyme): The conditions under which a particular enzyme is most active.
óptimo (enzima): Las condiciones bajo las cuales una enzima particular es más activa.

organ: Structures comprising two or more tissues with related functions.
organo: Estructuras que comprenden dos o más tejidos con funciones relacionadas.

organ system: A group of organs in the body that work together to perform a particular function, e.g. the circulatory system involves the heart and the different blood vessels.
sistema de órganos: Un grupo de órganos en el cuerpo que trabajan juntos para realizar una función particular, por ejemplo, el sistema circulatorio involucra el corazón y los diferentes vasos sanguíneos.

organelle: A subcellular structure with one or more specific jobs to perform in the cell.
orgánulo: Estructura subcelular con uno o más trabajos específicos para realizar en la célula.

osmosis: The diffusion of free water across a selectively permeable membrane, from a region of low solute concentration to a regions of high solution concentration.
ósmosis: La difusión de agua libre a través de una membrana selectivamente permeable.

overfishing: The practice of removing fish from an area at a rate higher than that which allows the population can replace itself.
sobrepesca: La práctica de retirar peces de un área a un ritmo superior al que permite a la población puede reemplazarse a sí misma.

parasitism: Biological interaction in which one organism, the parasite, benefits at the expense of the other, the host.
parasitismo: Interacción biológica en la que un organismo, el parásito, se beneficia a expensas del otro, el huésped.

pentadactyl limb: A limb with a specific arrangement of bones, containing five digits. It's used to provide evidence that all organisms possessing it derived from a common ancestor.
extremidad pentadactyl: Una extremidad con una disposición específica de huesos, que contiene cinco dígitos. Se utiliza para proporcionar evidencia de que todos los organismos que lo poseen derivan de un ancestro común.

phenotype: The observable physical and physiological traits of an organism, which are determined by its genetic makeup, environment and epigenetic factors.
fenotipo Los rasgos físicos y fisiológicos observables de un organismo, que están determinados por su composición genética, entorno y factores epigenéticos.

phospholipid: A polar lipid composed of glycerol joined to two fatty acids and a phosphate group. Phospholipids form bilayers that function as biological membranes.
fosfolípido: Un lípido polar compuesto de glicerol unido a dos ácidos grasos y un grupo fosfato. Los fosfolípidos forman bicapas que funcionan como membranas biológicas.

photosynthesis: A process used by green plants, algae, and some bacteria to convert light energy into chemical energy (carbohydrate).
fotosíntesis: Un proceso utilizado por plantas verdes, algas y algunas bacterias para convertir la energía de la luz en energía química (carbohidratos).

phylogenetic tree: A branching diagram showing evolutionary relationships among organisms. Can be based on cladistic analysis, in which case it is called a cladogram.
árbol filogenético: Un diagrama de ramificación que muestra las relaciones evolutivas entre los organismos. Puede basarse en el análisis cladístico, en cuyo caso se denomina cladograma.

plasma membrane: The membrane at the boundary of every cell that acts as a selective barrier, regulating the cell's chemical composition.
membrana plasmática: La membrana en el límite de cada célula que actúa como una barrera selectiva, regulando la composición química de la célula.

polymerase chain reaction (PCR): An in-vitro technique for rapid synthesis of a given DNA sequence.
reacción en cadena de la polimerasa (PCR): Técnica in vitro para la síntesis rápida de una secuencia de ADN dada.

population: A group of individuals of the same species living in a given area at a given time.
población: Un grupo de individuos de la misma especie que viven en un área determinada en un tiempo específico.

precision: How close repeated measurements are to each other, i.e. repeatability.
precisión: Qué tan cerca están las mediciones repetidas entre sí, es decir, la repetibilidad.

predation: Biological interaction in which one organism, the predator, kills and eats another, its prey.
predación: Interacción biológica en la que un organismo, el predador, mata y se come a otro, su presa.

prediction: What is expected to happen if the hypothesis of an experiment or scenario is true.
predicción: Lo que se espera que suceda si la hipótesis de un experimento o escenario es cierta.

producer: An organism that produces its own food using materials from inorganic sources (also known as an autotroph).
productor: Un organismo que produce su propio alimento utilizando materiales de fuentes inorgánicas (también conocido como autótrofo).

prokaryotic cell: Bacterial cells that lack any membrane-bound organelles or nucleus. And contain a simple chromosome of DNA.
célula procariota: Células bacterianas que cualquier orgánulo o núcleo unido a la membrana. Y contienen un simple cromosoma de ADN.

protein: A biologically functional molecule consisting of one or more polypeptides folded into a specific three-dimensional structure.
proteína: Una molécula biológicamente funcional que consiste en uno o más polipéptidos plegados en una estructura tridimensional específica.

qualitative data: Non-numerical data that describes qualities or characteristics.
datos cualitativos: Datos no numéricos que describen cualidades o características.

quantitative data: Numerical data expressing a certain quantity, amount, or range.
datos cuantitativos: Datos numéricos que expresan una determinada cantidad, monto o rango.

raw data: Experimental data that is collected in the field or laboratory and has not yet been processed.
datos brutos: Datos experimentales que se recogen en el archivo o laboratorio y que aún no han sido procesados.

recessive allele: An allele that is only expressed if the individual has two copies of the allele.
alelo recesivo: Un alelo que sólo se expresa si el individuo tiene dos copias del alelo.

recombination: The process by which genes are exchanged between different chromosomes to produce new combinations of alleles.
recombinación: El proceso por el cual los genes se intercambian entre diferentes cromosomas para producir nuevas combinaciones de alelos.

resilience: The property of ecosystems or populations to recover from disturbances.
resiliencia: La propiedad de los ecosistemas o poblaciones para recuperarse de las perturbaciones.

scientific method: The processes applied to the way in which scientists discover how the universe works. It involves a specific way of asking questions, observing, measuring, and interpreting data to formulate hypotheses and theories.
método científico: Los procesos aplicados a la forma en que los científicos descubren cómo funciona el universo. Implica una forma específica de hacer preguntas, observar, medir e interpretar datos para formular hipótesis y teorías.

selection pressure: Any factor that reduces or increases fitness in a portion of a population.
Presión de selección: Cualquier factor que reduce o aumenta la aptitud física en una porción de una población.

semi-conservative replication: The normal mechanism of DNA replication, where each strand acts as a template for a new double helix.
replicación semiconservadora: El mecanismo normal de replicación del ADN, donde cada hebra actúa como una plantilla para una nueva doble hélice.

specialized cell: A cell that has developed the characteristics needed to perform particular functions.
célula especializada: Una célula que ha desarrollado las características necesarias para realizar funciones particulares.

species: A group of organisms that can successfully interbreed and produce fertile offspring, and is reproductively isolated from other such groups.
especie: Un grupo de organismos que pueden cruzarse con éxito y producir descendencia fértil, y se aísla reproductivamente de otros grupos similares.

stem cell: A cell that has the ability to develop into many other different cell types.
célula madre: Una célula que tiene la capacidad de desarrollarse en muchos otros tipos de células diferentes.

table: A way of presenting data in a structured format that allows relationships and trends to be easily recognized.
tabla: Una forma de presentar los datos en un formato estructurado que permite reconocer fácilmente las relaciones y tendencias.

tissue: An integrated group of cells with a common structure, function, or both.
tejido: Un grupo integrado de células con una estructura, función o ambas comunes.

trait: A specific variant of a phenotype, controlled by one or more genes, e.g. flower color in plants or tongue rolling ability in humans.
rasgo: Una variante específica de un fenotipo, controlada por uno o más genes, por ejemplo, el color de las flores en las plantas o la capacidad de rodar la lengua en los seres humanos.

transcribing/transcription: The process of copying a segment of DNA into a strand of mRNA.
transcripción: El proceso de copiar un segmento de ADN en una hebra de ARNm.

transitional fossil: A fossil that exhibits traits common to both an ancestral group and its derived descendants.
fósil de transición: Un fósil que exhibe rasgos comunes tanto a un grupo ancestral como a sus descendientes derivados.

translation: The process of decoding a strand of mRNA to produce a sequence of amino acids.
traducción: El proceso de decodificación de una hebra de ARNm para producir una secuencia de aminoácidos.

transpiration: The evaporative loss of water from a plant.
transpiración: La pérdida evaporativa de agua de una planta.

trophic level: A level or position in a food chain, food web or ecological pyramid. An organism's trophic level is determined by its feeding behavior.
nivel trófico: Un nivel o posición en una cadena alimentaria, red alimentaria o pirámide ecológica. El nivel trófico de un organismo está determinado por su comportamiento de alimentación.

tropism (plant): Directional growth in response an external stimulus, such as light or gravity.
tropismo (plantas): crecimiento direccional en respuesta a un estímulo externo, como la luz o la gravedad.

variable: A measurable property that changes over time or can take on different values.
variable: Una propiedad medible que cambia con el tiempo o puede tomar diferentes valores.

variation: The diversity of phenotypes and genotypes within a population or species.
variación: La diversidad de fenotipos y genotipos dentro de una población o especie.

virus: An infectious agent that can only multiply within the living cells of its host.
virus: Un agente infeccioso que solo puede multiplicarse dentro de las células vivas de su huésped.

zygote: A fertilised egg.
cigoto: Un óvulo fertilizado.

Equipment List

1: Cells and Viruses

INVESTIGATION 1.1
Modeling protein structure

Per student/pair/group
Pipe cleaners (2 white, 2 pink, 2 purple, 4 blue)
Sticky tape
2 x binder clips or paper clips

INVESTIGATION 1.2
Preparing an onion slide

Per student/pair
Light microscope
Onion/onion leaf
Glass microscope slides
Coverslips
Scalpel or razor
Iodine stain
Filter paper/tissue paper

INVESTIGATION 1.3
Simple diffusion across a membrane

Per student/pair
200 mL beaker
1 mL pipette
Glucose dipsticks
Lugol's indicator
4 x test tubes
Dialysis tubing
Thread or nylon line
Distilled water
1% starch solution
10% glucose solution
Timer or watch

INVESTIGATION 1.4
Modeling disease outbreak and spread

Per student/pair
Computer
Spreadsheet application e.g. Excel

2: Cell Cycle

INVESTIGATION 2.1
Modeling mitosis

Per student/pair
4 x pipe-cleaners (2 colors) cut in half
Yarn or string
A3 sheet of paper
Marker

3: Photosynthesis and cellular respiration

INVESTIGATION 3.1
Measuring bubble production in Cabomba

Per pair/group
1.0 g *Cabomba aquatica*
Balance
Scissors
Water
1 x large beaker (large enough to hold the glass funnel)
1 x glass funnel
0.2 mol/L sodium hydrogen carbonate solution (enough to cover the plant)
1 x test tube
1 x lamp with a 60W bulb
Lux meter
Timer
1 x ruler or tape measure

INVESTIGATION 3.2
Measuring respiration in germinating seeds

Per group
3 x boiling tubes
Marker pen
6 x cotton balls
15% KOH solution
2 x eye dropper or plastic pipette
3 x gauze pieces
Germinated bean seeds (enough to fill one quarter of the boiling tube)
Ungerminated bean seeds (enough to fill one quarter of the boiling tube)
Glass beads (enough to fill one quarter of the boiling tube)
3 x 2-hole tube stoppers
3 x bent glass tubes or pipettes
3 x tubes (must be able to be clamped shut)
3 x screw clips
A few drops of colored liquid
3 x syringes (must fit tube with screw clamp attached)
3 x clamp stands or rack
Water bath (25°C)
Ruler, Timer

INVESTIGATION 3.3
Modeling photosynthesis and cellular respiration

Per individual, pair, or group
Scissors

INVESTIGATION 3.4
Effect of temperature on enzyme activity

Per group/temperature
1 x spotting plate/reaction plate
1 x test tube
1 x plastic pipette
Water bath
Timer
0.1 M iodine solution (I_2KI)
2 mL 1% amylase solution
1 mL buffer solution (pH 7.0)
1 mL 1% starch solution

4: Animal and Plant Structure and Function

INVESTIGATION 4.1
Investigating effect of exercise on heart rate.

1 x stopwatch per group

INVESTIGATION 4.2
Investigating effect of exercise on breathing rate.

Equipment depends on group method

INVESTIGATION 4.3
Investigating vascular tissue

Per student/pair
Light microscope
Dicot plants (e.g. buttercup sunflowers)
Monocot plant (e.g. maize or corn)
Glass microscope slides
Coverslips
Scalpel or razor
Access to a computer or device with internet connection

INVESTIGATION 4.4
Investigating plant transpiration

Per pair/group
250 mL conical flask with rubber bung
Petroleum jelly
1 cm³ pipette
Clamp stand
Leafy plant shoot
Water
Cooking oil (for optional set up)
Timer or watch
Lamp, or plastic bag and water spray bottle, or fan
A4 or graph paper

INVESTIGATION 4.5
Plant propagation

Per student/pair
9 × plant/seed containers or trays
3 × planting mediums (e.g. sand, bark, potting mix)
Rooting hormone
9 × ice block sticks
Secateurs or scissors
Measuring flask or container for water

INVESTIGATION 4.6
Germination investigation

4 × plant/seed trays
4 × sets of 100 tomato seeds or similar (e.g. mustard seeds)
Sterilized growing medium
Measuring flask or container for water

5: DNA and Gene Expression

INVESTIGATION 5.1
Creating a model of a DNA molecule

Per pair
Scissors
Tape or paste

INVESTIGATION 5.2
Modeling gene expression

LEGO® or similar model blocks
Camera

6: Patterns of Inheritance

INVESTIGATION 6.1
Phenotypic variation in your class

No equipment required

INVESTIGATION 6.2
Modeling meiosis using iceblock sticks

Per pair
8 × iceblock sticks
8 × sticky dots
Colored pencils or markers
Marker pen

8: Evolution and Natural Selection

INVESTIGATION 8.1
Modeling selection with M&M's®

Per group
100 M&Ms®
1 × lidded container
1 × plate

INVESTIGATION 8.2
Investigating gene pool changes

Per student/pair
Computer
Spreadsheet application e.g. Excel

9: Ecological Interactions

INVESTIGATION 9.1
Investigating predator-prey populations

Per student
Teaspoon
Cup
Per class
2 trays of shallow bowls
100 dried beans, peas, or beads etc.

INVESTIGATION 9.2
Mutualistic relationships in my community

Pens, notebook, camera for recording observations.

INVESTIGATION 9.3
Exploring biomass pyramids

Per student/pair
Computer and online access to Biointeractive interactive module: bit.ly/3LHITkW

INVESTIGATION 9.4
A model of the carbon cycle

Per group
1 × 2 L clear soda bottle with lid
1 scoop aquarium gravel
Several dead leaves
1 × aquatic plant (e.g. Cabomba)
3-4 small pond snails
2 L filtered pond water

INVESTIGATION 9.5
Investigating how dry ice affects pH

Per group
Dry ice (solid CO_2)
100 mL flasks
Universal indicator

INVESTIGATION 9.6
Investigating biodiversity and human impacts

Per student group
40 × red plastic tokens or bricks
30 × blue plastic tokens or bricks
20 × plastic tokens or bricks
10 × plastic tokens or bricks
1 × plastic tub
1 × plastic tray

INVESTIGATION 9.7
A model of human impacts on fish stocks

Per group
Small cup
Teaspoon
Per group:
2 bowls
120 red beads, 80 yellow beads, 80 blue beads
Chopsticks
Spatula
Stopwatch

Credits and Terms

We acknowledge the generosity of those who have provided photographs for this edition: • Louisa Howard and Katherine Connolly, Dartmouth College Electronic Microscope Facility • PASCO for photographs of probeware • Louisa Howard and Chuck Daghlian, Dartmouth College • Berkshire Community College Bioscience Image Library • Watson and Crick DNA photo: A. Barrington-Brown, © Gonville and Caius College, Cambridge / Colored by Science Photo Library, Marc King • PASCO for photographs of probeware • Martin D, Thompson A, Stewart

We also acknowledge the photographers that have made their images available through Wikimedia Commons under Creative Commons Licences 1.0, 2.0, 2.5, 3.0, or 4.0: Kristian Peters • Mnolf • Rufino Uribe • J. Miquel, D. Vilavella, Z.widerski, V. V. Shimalov and J. Torres • Jakob Suckale • SubtleGuest • Pengo • PloS • CDC: Dr Lucille K. Georg • Openstax College • Dartmouth College • Dr Graham Beards • Manuel Almagro Rivas • Jpatokal • CDC: Alissa Eckert & Dan Higgins • Pita, Erwin,Turon & Lopez-Legentil • Matthias Zepper • Bill Rhodes • Jpbarrass • Brocken Inaglory • dsworth Center: New York State Department of Health • Emmanuelm • Suseno • Sabina Bajracharya • Ernie • Neutr0nics • Luteus • Stan Shebs • Lazaregagnidze • Orikrin 1998 • Tangopaso • ESA • Vossman • Stoolhog • it:Utente:Cits • 25kartika • Andreas Eichler • Diacritica • Fred Wierum • H. Zell • DanielCD • Masahiro miyasaka • Michael Ryge • Cayambe • Junchang Lü and Stephen L. Brusatte • José Luis Barthelo • Ghedoghedo • MeegsC • Putneymark • J Podos • Olaf Leillinger • Allan and Elaine Wilson • UtahCamera • thinboyfatter • Brian Gratwicke • Wendy Kaveney • Marco vinci • Bramadi Arya • Mike Baird • Biglaci • Matt Reinbold • Radiogaga • US Fish and Wildlife Service • viamoi • The High Fin Sperm Whale • Althepal • 350z33 • Neil Wedge • Famartin • Derek Quinn • Bruno de Giusti • Nicholls H • Martin D • Roger Griffith • Ky1958 • Danio • Scott McD1 • BS Thurner Hof

Contributors identified by coded credits are: BCC: Berkshire Community College Image Library, BF: Brian Finerran (University of Canterbury), NOAA: National Oceanic and Atmospheric Administration, WMU: Waikato Microscope Unit, EII: Education Interactive Imaging, KP: Kent Pryor, CDC: Centers for Disease Control and Prevention, NCI: National Cancer Institute, RCN: Ralph Cocklin, USDA: United States Department of Agriculture, ESA: European Space Agency, NIH: National Institutes of Health, MPI: Max Planck Institute, NASA: National Aeronautics and Space Administration, USFWS: United States Fish and Wildlife Service, USGS: United States Geological Survey, NPS: National Park Service, FAOUN: Food and Agriculture Organization United Nations, RA: Richard Allan, VMRCVM Virginia-Maryland College of Veterinary Medicine

Royalty free images, purchased by BIOZONE International Ltd, are used throughout this workbook and have been obtained from the following sources: Corel Corporation from their Professional Photos CD-ROM collection; IMSI (Intl Microcomputer Software Inc.) images from IMSI's MasterClips® and MasterPhotos™ Collection, 1895 Francisco Blvd. East, San Rafael, CA 94901-5506, USA; ©1996 Digital Stock, Medicine and Health Care collection; © 2005 Jupiter Images Corporation www.clipart.com; ©Hemera Technologies Inc, 1997-2001; ©Click Art, ©T/Maker Company; ©1994., ©Digital Vision; Gazelle Technologies Inc.; PhotoDisc®, Inc. USA, www.photodisc.com. • TechPool Studios, for their clipart collection of human anatomy: Copyright ©1994, TechPool Studios Corp. USA (some of these images were modified by Biozone) • Totem Graphics, for their clipart collection • Corel Corporation, for use of their clipart from the Corel MEGAGALLERY collection • 3D images created using Poser and Pymol • iStock images • Art Today • Adobestock

Questioning terms in biology

The following terms are often used when asking questions in examinations and assessments.

Analyze:	Interpret data to reach stated conclusions.
Annotate:	Add brief notes to a diagram, drawing or graph.
Apply:	Use an idea, equation, principle, theory, or law in a new situation.
Calculate:	Find an answer using mathematical methods. Show the working unless instructed not to.
Compare:	Show similarities between two or more items, referring to both (or all) of them throughout.
Construct:	Represent or develop in graphical form.
Contrast:	Show differences. Set in opposition.
Define:	Give the precise meaning of a word or phrase as concisely as possible.
Derive:	Manipulate a mathematical equation to give a new equation or result.
Describe:	Define, name, draw annotated diagrams, give characteristics of, or an account of.
Design:	Produce a plan, object, simulation or model.
Determine:	Find the only possible answer.
Discuss:	Show understanding by linking ideas. Where necessary, justify, relate, evaluate, compare and contrast, or analyze.
Distinguish:	Give the difference(s) between two or more items.
Draw:	Represent by means of pencil lines. Add labels unless told not to do so.
Estimate:	Find an approximate value for an unknown quantity, based on the information provided and application of scientific knowledge.
Evaluate:	Assess the implications and limitations.
Explain:	Provide a reason as to how or why something occurs.
Identify:	Find an answer from a number of possibilities.
Illustrate:	Give concrete examples. Explain clearly by using comparisons or examples.
Interpret:	Comment upon, give examples, describe relationships. Describe, then evaluate.
List:	Give a sequence of answers with no elaboration.
Outline:	Give a brief account or summary. Include essential information only.
Predict:	Give an expected result.
Solve:	Obtain an answer using numerical methods.
State:	Give a specific name, value, or other answer. No supporting argument or calculation is necessary.
Suggest:	Propose a hypothesis or other possible explanation.
Summarize:	Give a brief, condensed account. Include conclusions and avoid unnecessary details.

©2024 **BIOZONE** International
Photocopying prohibited

Index

10% law 380

A
Abiotic factors 349
Abscisic acid, plant hormone 182
Absorption, of nutrients 145-147
Accuracy, of measurements 428
Activation energy 115
Active site, of enzyme 114
Active transport 40-43
Adaptation 317-318
Adaptive immune system 154
Adaptive radiation 331
Adenosine triphosphate 101
- structure 94-95
Aerobic respiration 103-105
Agriculture, genetic engineering 237-239
Allele, defined 248
Amino acid structure 9
Amylase activity 117
Anabolic reaction 113
Anaerobic respiration 103
Animal cell 24
Animal cell
- features of 27-28
- organelles 4
Antibodies, defined 155
Antigen 155
Apoptosis, and cancer 84
Asexual reproduction, plants 172
Assumptions 427
Atmospheric carbon dioxide 390-391
ATP 101
- structure 94-95
Autotrophs 374-375
Auxin, plant hormone 182-184

B
B-cells 155
Bacterial cell
- features of 26, 28
- structure 45
Bar graph 442
Base pairing rule 71, 202
Beneficial mutations 225
Bias, sampling 448
Biodiversity
- human impact on 404-418
- measures 400-401
- role of keystone species 402-403
Biogeochemical cycles 378, 386-391, 394-397
- disruptions to 379, 385, 396-397
Biogeography 294-295
Biological drawing, rules 449-451
Biological species 328
Biomolecules 4-15
Biotic factors 349
Birth, hormonal regulation 153
Blood cells, white 155
Blood clotting, role in defense 156
Blood glucose
- regulation of 136, 148-149
Blood, regulation of 138-139
Body systems, overview 129-131

C
Cancer 66, 82-84
- and estrogen 80-81
Cancer cell, features of 83
Carbohydrate 4, 6-7
Carbon 5
Carbon cycle 378, 388-391
Carbon dioxide
- atmospheric 390-391
- ocean 392-393
Catabolic reaction 113
Catalase activity 119
Catalyst, enzyme 114-115
Cell cycle
- disruption 82-84

- phases 65
- regulation of 66, 83
Cell differentiation 75-78
Cell membrane
- model 35
- structure of 33-35
Cell sizes, comparison 29
Cell specialization 75-78
Cell types, overview 24
Cell wall, plant 4
Cellular environments 25
Cellular functions 25
Cellular respiration 101-105
- measuring 106
- modeling 109-111
Cellulose 7
Central tendency, measures 445-446
Centriole 4
Chargaff's rules 202
Chi-Squared test 276-277
Chlorophyll 96
Chloroplast 4, 98
- structure of 93
Chromatid 70
Chromatin 199
Chromosomal mutations 224
Chromosome 4, 70-71, 199
Circulatory system 138, 140-141, 145-147, 152
Codominance 272-273
Codon, definition 215
Commensalism 356, 370
Common ancestry, evidence 286-305
Competition 81356, 363-366
Concentration gradient, defined 37
Consumers 374, 376
Continuous data 434
Continuous variation 250
Control, experimental 435
Converting numbers 429
Correlation, vs causation 441
Covid-19 55-56
- PCR testing 236
Crick, Francis 208
CRISPR, gene editing 233, 237
Crossing over, meiosis 254
Cytokinesis 65, 68
Cytosis 42-43

D
Damming, and biodiversity 416-417
Darwin, Charles 305
Darwin's finches 317-318
Data distribution, measurements 445
Data logger 436
Data, types 434
Defense, body 155
Deforestation
- and biodiversity 412-415
- effects 399
Denaturation, of enzymes 115
Deoxyribonucleic acid 8, 199-210
- discovery 206-208
- editing 227
- origin 209-210
- structure 201-205
Dependent variable 435
Developmental homology 298
Diabetes mellitus 148
Differentiation, of cells 75-78
Diffusion, in cells 37-38
Digestive system 145-147
Dihybrid cross 266-268
Directional selection 319-321
Discontinuous data 434
Discontinuous variation 250
Disease transmission 48-52
Disease
- cancer 66, 80-84
- Covid-19 236
- viral 48-49, 52
Disruptive selection 319-320

DNA 8, 199-210
- discovery 206-208
- editing 227
- mutations 221-226
- origin 209-210
- structure 201-205
DNA homology 296
DNA polymerase 228
DNA replication 70-71
- semi-conservative model 70, 72-73
Dominant allele 259
Double helix, DNA 208

E
E. coli structure 26
Ecological niche 351
Ecological pyramids 382-384
Ecosystem change 398-399
Ecosystem resilience 354-355
Ecosystem services 418
Ecosystem stability 353, 398-399
Ecosystem
- components 349
- dynamics in 352-353
Effector 132, 135
Electron transport chain 103-104
Elements 5
Endocytosis 43
Endosymbiosis theory 31-32
Energy flow, ecosystems 374, 376, 380-381, 385
Energy
- in cells 94-95
- transfer between systems 101
Environment, and phenotype 79-81
Environmental factors, and variation 249, 316
Enzyme 114-118
- active site 114
- amylase activity 117
- catalase activity 119
- denaturation 115
- DNA polymerase 228
- effect of temperature 117
- optimal conditions 116
Epidemic, defined 50
Epigenetics 79-81
Equipment, selecting 433
Estimates 429
Estrogen 150-151
- effect on phenotype 80-81
Eukaryotic cell, features of 27-28
Eukaryotic cells 24
Evolution
- defined 313
- evidence for 286-305
- patterns 331
- theory of 304-305
Exercise, and homeostasis 141-144
Exocytosis 42
Exponential data 432

F
Fermentation, types 103
Fertilization, plants 177
Fever, role of 158
Fibrous protein 9
Flower structure 174-176
Fluid mosaic model 33
Food chain 377
Food web 373, 377
Fossil record 288-289
- changes in 299-303
Fossils 286-291
Founder effect 337-338
Fractions, using 431
Franklin, Rosalind 208
Fructose 6
Fungal cell 24

G

Galápagos finches 317-318
Gametic mutation 222
Gel electrophoresis 229-230
Gene editing 233
Gene expression 212-219
Gene flow 329, 333-335
Gene mutations 223
Gene pool
- changes in 322
- factors affecting 332
Genes
- cancer 83-84
- linked 258
Genetic bottlenecks 339
Genetic code 215
Genetic drift, gene pools 332, 336
Genetic engineering
- agriculture 237-239
- insulin 234-235
Genetic recombination
- in gene pools 341
Genetic variation, types 250-251
Genetically modified organisms 227
Genotype, defined 249
Geographic isolation 330
Germination, seeds 178-179
Gibberellin, plant hormone 182, 187
Globular protein 9
Glucagon 136, 148-149
Glucose 6
Glycogen 7
Glycolysis 103-104
GMO 227
Graphs, types 438-444
Gravitropism 188-189
Greenhouse gas 391

H
Habitat 350
Harmful mutations 225-226
Hemoglobin
- homology 297
- structure 10
Heterotrophs 374, 376
Heterozygous, defined 248
Hierarchy of life 128
Histogram 443
Homeostasis 131-134
- and exercise 141-144
Homologous chromosomes 248
Homology
- anatomical 292-293
- anatomical 292-293
- developmental 298
- molecular 296-297
Homozygous, defined 248
Hormonal regulation 136-137, 147-153
Hormones
- birth 153
- digestion 147
- plants 182-187
- pregnancy 152
Hydrogen 5
Hypotheses 426

I
Immune system 154
Incomplete dominance 270-271
Independent assortment, meiosis 95
Independent variable 435
Induced fit model, enzyme activity 114
Inflammatory response 157-158
Innate immune system 154
Insulin 136, 148-149
- recombinant 234-235
Interacting systems
- humans 131, 137-153
- plants 160-161
Interactions, species 356-373

©2024 **BIOZONE** International
Photocopying prohibited

Interphase 65
Interspecific competition 363, 366-367
Intraspecific competition 363-365
Ion pump 41
Isolating mechanisms 330
Isomer, of glucose 6

K
Keystone species 402-403
Koch's postulates 22
Krebs cycle 103-104

L
Laws, scientific 426
Light dependent phase 97-99
Light independent phase 97-99
Line graph 439
Line of best fit 440-441
Link reaction 103
Linked genes 258
Lipid 4
 - role of 16
Log book, data collection 436
Log transformations 432
Lymphocyte 66
Lysogenic cycle 47
Lytic cycle 47

M
Magnification, in microscopy 17-18
Mean, calculating 445
Median, calculating 445
Meiosis 253-257
Mendel, experiments 259-261
Mendelian genetics 259-268
Menstrual cycle, hormonal regulation 150-151
Messenger RNA 200, 212, 214
Metabolism, defined 113
Microscope
 - early 17
 - slide preparation 20
 - stains 20
 - structure 17
Migration, gene pools 332-335
Mitochondria, structure of 92
Mitosis 65-66
 - functions of 64
 - modeling 69
 - phases 67
Mode, calculating 445
Models, scientific 425
Molecular homology 296-297
Molecular techniques 227-239
Monohybrid cross 262-263, 265
Monosaccharide 6
mRNA 200, 212, 214
Mt St Helens, disruption 398-399
Multicellularity, benefits of 30
Multiple alleles 273
Multipotent stem cell 76
Mutagens 222
Mutations 221-226, 249
 - and gene pools 332, 340
 - variation 316
Mutualism 356, 371-372

N
Nastic responses, plants 190-191
Natural selection
 - finches 317-318
 - mechanism of 313-314
 - modeling 315
 - rock pocket mice 323-327
Negative feedback 133-134, 148-151
Nervous system, regulation 135, 137
Niche 351
Nitrogen 5
Nitrogen cycle 378, 394-397
 - disruptions to 396-397
Non-Mendelian genetics 269-275
Normal distribution, of data 445
Nucleic acid 4, 8
Nucleotide bases 200
Numbers, working with 429
Nutrient absorption 145-147
Nutrient cycles 378, 386-391, 394-397

O
Observations 427
Ocean acidification 392-393
Optimal conditions, of enzymes 116
Organ system, plants 159
Organelles 4, 92-93, 98
 - animal cell 27
 - prokaryotic cell 26
Organs 128-129
Osmosis 39
 - in roots 170
Osmotic potential 39
Overfishing 408-411
Oxygen 5
Oxygen cycle 378
Oxytocin 153

P
p53 cancer gene 82
Pandemic, defined 50
Parasitism 356, 368-369
Passive transport 37-39
Pasteur, experiments 22
PCR 228
 - and Covid-19 testing 236
Pentadactyl limb 293
Peppered moth, selection 321
Peptide bond 9
Percentage errors, calculating 433
Percentages, calculating 430
Phenotype
 - and selection 316
 - defined 249
 - effect of environment 79-81, 247
Phloem 162, 171
Phospholipid 16
Phospholipid bilayer 33
Phosphorus 5
Photoautotrophs 374-375
Photosynthesis 96-101, 375
 - effect of light intensity 100
 - modeling 109-111
Phototropism 186
Phyletic gradualism 302
Phylogenetic species 328
Pie graph 444
Placenta 152
Plant cell 24
Plant cell wall 4
Plant hormones 182-187
Plant organelles 4
Plant propagation 173
Plant responses 181
Plant organ systems 159
Plasma membrane
 - model 35
 - structure of 33-35
Pluripotent stem cell 76
Pollination 177
 - by insects 174, 176
 - by wind 175-176
Polymerase chain reaction 228
 - and Covid-19 testing 236
Polypeptide chain 9
Polysaccharide 7
Populations, variation in 316
Positive feedback, and fever 158
Potometer 166
Precision, of measurements 428
Predation 356, 358-362
Predator-prey cycle 358-359
Pregnancy, interacting systems 152-153
Primary structure, protein 10
Probability, and inheritance 264
Progesterone 150-151, 153
Prokaryotic cell 24
 - features of 26, 28
 - structure 45
Propagation, plants 173
Prostaglandin 153
Protein 4, 9-15
 - functions of 12-15
 - hemoglobin 10
 - homology 297
 - RuBisCo 98
 - structure 9-11
Protist 17
Protist cell 24
Proto-oncogenes 82
Punctuated equilibrium 302
Pyramids, ecological 382-384

Q
Qualitative data 434, 455
Qualitative traits 250
Quantitative data 434, 456-458
Quantitative traits 250
Quaternary structure, protein 10

R
R group, of amino acid 9
Ranked data 434
Rates, calculating 430
Ratios, calculating 431
Receptor 132, 135
Recessive allele 259
Recombinant DNA 231-232
Recombination, meiosis 254
Regulation
 - blood 138-139
 - hormonal 136
 - menstrual cycle 150-151
 - nervous system 135
 - respiratory gases 140-141
Reproduction, asexual 172
Reproductive isolating mechanisms 330
Reproductive system 152
Resilience, in ecosystems 354-355
Resolution, microscopy 17
Resources, competition for 353-366
Respiratory gases, regulation of 140-141
Respiratory system 140-141
Restriction enzymes 231
Results, recording 436
Ribonucleic acid (RNA) 8, 200
Ribosomal RNA 200
Ribosomes
 - 70s 26, 28
 - 80s 27-28
 - role in translation 215
RNA (Ribonucleic acid) 8, 200
Root structure 163
Root system, plants 159
Roots, water uptake 169-170
RuBisCo, and photosynthesis 98

S
Safety practices 452-454
Sample bias 448
Scatter graph 440
Scientific drawing, rules 449-451
Scientific process 424
Secondary structure, protein 10
Seed
 - dispersal 180
 - germination 178-179
 - structure 178
Selection pressure 316
 - types 319-320
Semiconservative model, of DNA replication 70, 72-73
Sex linkage 274-275
Sexual reproduction 249
 - and variation 252-257, 316
Shoot system, plants 159
Simpson's Index, of biodiversity 401
Slides, microscopy 20
Somatic mutation 222
Specialization, of cells 75-78
Species interactions 356-373
Species
 - defined 328
 - formation 329-330
Stability, ecosystems 353
Stabilizing selection 319-320
Stains, microscopy 21
Standard deviation, calculating 447
Starch 7
Stasis, fossil record 301
Stem cells 75-76
Stem structure 163
Stomata, role of 164
Structure, of protein 9-11
Symbiosis 371-372

T
T-cells 55
Tables, data 437
Tally chart 430, 443
Temperature
 - effect on enzymes 117
 - effect on phenotype 79
Ten percent law 380
Tertiary structure, protein 10
Test cross 263
Theories, scientific 426
Thermoregulation 134
Tissue, vascular 160-162
Tolerance range 350
Totipotent stem cell 76
Traits 220
 - defined 247
 - types 250-251
Transcription 212, 213
Transfer RNA 200
Transitional fossils 290-291
Translation 212, 216
Translocation 171
Transpiration 160, 164-168
Triglyceride, structure 16
Trophic level 377-381
Tropism, plants 181-182, 186, 188-189
Tumor 82-83
Tumor suppression gene 82
Type 1 diabetes 148

U V
Urinary system 138
Valency shells 5
Van Leeuwenhoek 17
Variables, types 435
Variation
 - and natural selection 314
 - in populations 316
 - sexual reproduction 252-257
 - sources of 249
 - types 250
Vascular tissue, plants 160-162
Viral reproduction 46-47
Viral spread, modeling 53-54
Viral structural 44-45
Virus 24
 - reproduction of 46-47
 - structure of 44-45
Viruses
 - and cancer 84
 - and disease 48-49, 52, 55-56
 - transmission of 48-49, 51-52, 55

W X Y Z
Wallace, Alfred 305
Water 4
 - uptake in plants 169-170
Watson, James 208
White blood cells 155

Xylem 162, 165

Zoonotic disease 52